Library of Congress Cataloging-in-Publication Data

ASCHER, U.M. (URI M.)   (date)
   Numerical solution of boundary value problems for ordinary differential equations.

   Bibliography: p.
   Includes index.
   1. Boundary value problems — Numerical solutions.
I. Mattheij, Robert M. M. II. Russell,
R. D. (Robert D.)   III. Title.
QA379.A83 1988   515.3'52   87-29249
ISBN 0-13-627266-5

Editorial/production supervision
  and interior design: Gloria Jordan
Cover Design: Edsel Enterprises
Manufacturing buyer: Lorraine Fumoso

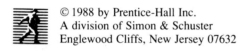

© 1988 by Prentice-Hall Inc.
A division of Simon & Schuster
Englewood Cliffs, New Jersey 07632

All rights reserved. No part of this book may be
reproduced, in any form or by any means,
without permission in writing from the publisher.

Printed in the United of America
10 9 8 7 6 5 4 3 2 1

ISBN 0-13-627266-5

Prentice-Hall International (UK) Limited, *London*
Prentice-Hall of Australia Pty. Limited, *Sydney*
Prentice-Hall Canada Inc., *Toronto*
Prentice-Hall Hispanoamericana, S.A., *Mexico*
Prentice-Hall of India Private Limited, *New Delhi*
Prentice-Hall of Japan, Inc., *Tokyo*
Simon & Schuster Asia Pte. Ltd., *Singapore*
Editora Prentice-Hall do Brasil, Ltda., *Rio de Janeiro*

Numerical Solution of Boundary Value Problems for Ordinary Differential Equations

URI M. ASCHER

University of British Columbia

ROBERT M. M. MATTHEIJ

Technische Universiteit
Eindhoven, Netherlands

ROBERT D. RUSSELL

Simon Fraser University
British Columbia

PRENTICE HALL, ENGLEWOOD CLIFFS, NEW JERSEY 07632

To Nurit, Marie-Anne, and Robin

Contents

LIST OF EXAMPLES   XV

PREFACE   XVII

1 INTRODUCTION   1

    1.1 Boundary Value Problems for Ordinary Differential Equations   1

       1.1.1 Model problems   *1*
       1.1.2 General forms for the differential equations   *3*
       1.1.3 General forms for the boundary conditions   *4*

    1.2 Boundary Value Problems in Applications   7

2 REVIEW OF NUMERICAL ANALYSIS AND MATHEMATICAL BACKGROUND   28

    2.1 Errors in Computation   28

2.2 Numerical Linear Algebra — 32

- 2.2.1 Eigenvalues, transformations, factorizations, projections *32*
- 2.2.2 Norms, angles, and condition numbers *36*
- 2.2.3 Gaussian elimination, LU-decomposition *39*
- 2.2.4 Householder transformations: QU-decomposition *43*
- 2.2.5 Least squares solution of overdetermined systems *45*
- 2.2.6 The QR algorithm for eigenvalues *45*
- 2.2.7 Error analysis *46*

2.3 Nonlinear Equations — 48

- 2.3.1 Newton's method *48*
- 2.3.2 Fixed-point iteration and two basic theorems *49*
- 2.3.3 Quasi-Newton methods *51*
- 2.3.4 Quasilinearization *52*

2.4 Polynomial Interpolation — 55

- 2.4.1 Forms for interpolation polynomials *55*
- 2.4.2 Osculatory interpolation *58*

2.5 Piecewise Polynomials, or Splines — 59

- 2.5.1 Local representations *59*
- 2.5.2 B-splines *61*
- 2.5.3 Spline interpolation *62*

2.6 Numerical Quadrature — 63

- 2.6.1 Basic rules *63*
- 2.6.2 Composite formulas *66*

2.7 Initial Value Ordinary Differential Equations — 67

- 2.7.1 Numerical methods *68*
- 2.7.2 Consistency, stability and convergence *71*
- 2.7.3 Stiff problems *73*
- 2.7.4 Error control and step selection *74*

2.8 Differential Operators and Their Discretizations — 77

  2.8.1 Norms of functions, spaces, and operators *77*
  2.8.2 Roundoff errors and truncation errors *80*

3 THEORY OF ORDINARY DIFFERENTIAL EQUATIONS 84

3.1 Existence and Uniqueness Results 85

  3.1.1 Results for boundary value problems *85*
  3.1.2 Results for initial value problems *87*

3.2 Green's Functions 94

  3.2.1 A first-order system *94*
  3.2.2 A higher-order ODE *96*

3.3 Stability of Initial Value Problems 99

  3.3.1 Stability and fundamental solutions *100*
  3.3.2 The constant coefficient case *102*
  3.3.3 The general linear case *104*
  3.3.4 The nonlinear case *109*

3.4 Conditioning of Boundary Value Problems 110

  3.4.1 Linear problems and conditioning constants *111*
  3.4.2 Dichotomy *115*
  3.4.3 Well-conditioning and dichotomy *121*
  Exercises *127*

4 INITIAL VALUE METHODS 132

4.1 Introduction: Shooting 132

  4.1.1 Shooting for a linear second-order problem *133*
  4.1.2 Shooting for a nonlinear second-order problem *134*

4.1.3 Shooting — general application, limitations, extensions *134*

4.2 Superposition and Reduced Superposition — 135

4.2.1 Superposition *135*
4.2.2 Numerical accuracy *137*
4.2.3 Numerical stability *140*
4.2.4 Reduced superposition *143*

4.3 Multiple Shooting for Linear Problems — 145

4.3.1 The method *146*
4.3.2 Stability *149*
4.3.3 Practical considerations *151*
4.3.4 Compactification *153*

4.4 Marching Techniques for Multiple Shooting — 155

4.4.1 Reorthogonalization *156*
4.4.2 Decoupling *157*
4.4.3 Stabilized march *161*

4.5 The Riccati Method — 164

4.5.1 Invariant imbedding *164*
4.5.2 Riccati transformations for general linear BVPs *166*
4.5.3 Method properties *168*

4.6 Nonlinear Problems — 170

4.6.1 Shooting for nonlinear problems *170*
4.6.2 Difficulties with single shooting *174*
4.6.3 Multiple shooting for nonlinear problems *175*
Exercises *180*

5 FINITE DIFFERENCE METHODS — 185

5.1 Introduction — 185

5.1.1 A simple scheme for a second-order problem *187*

- 5.1.2 Simple one-step schemes for linear systems *190*
- 5.1.3 Simple schemes for nonlinear problems *194*

5.2 Consistency, Stability, and Convergence — 198

- 5.2.1 Linear problems *198*
- 5.2.2 Nonlinear problems *203*

5.3 Higher-Order One-Step Schemes — 208

- 5.3.1 Implicit Runge-Kutta schemes *210*
- 5.3.2 A subclass of Runge-Kutta schemes *213*
- 5.3.3 On the implementation of Runge-Kutta schemes *217*

5.4 Collocation Theory — 218

- 5.4.1 Linear problems *219*
- 5.4.2 Nonlinear problems *222*

5.5 Acceleration Techniques — 226

- 5.5.1 Error expansion *228*
- 5.5.2 Extrapolation *230*
- 5.5.3 Deferred corrections *234*
- 5.5.4 More deferred corrections *238*

5.6 Higher-Order ODEs — 244

- 5.6.1 More on a simple second-order scheme *245*
- 5.6.2 Collocation *247*
- 5.6.3 Collocation implementation using spline bases *259*
- 5.6.4 Conditioning of collocation matrices *262*

5.7 Finite Element Methods — 266

- 5.7.1 The Ritz method *267*
- 5.7.2 Other finite element methods *271*
- Exercises *272*

6 DECOUPLING — 275

6.1 Decomposition of Vectors — 277

6.2 Decoupling of the ODE — 279

    6.2.1 Consistent fundamental solution   *260*
    6.2.2 The basic continuous decoupling algorithm   *284*

6.3 Decoupling of One-Step Recursions — 288

    6.3.1 Consistent fundamental solutions   *290*
    6.3.2 The basic discrete decoupling algorithm and additional considerations   *291*

6.4 Practical Aspects of Consistency — 293

    6.4.1 Consistency for separated BC   *293*
    6.4.2 Consistency for partially separated BC   *295*
    6.4.3 Consistency for general BC   *296*

6.5 Closure and its Implications — 297
Exercises   *299*

7 SOLVING LINEAR EQUATIONS — 303

7.1 General Staircase Matrices and Condensation — 306

7.2 Algorithms for the Separated BC Case — 308

    7.2.1 Gaussian elimination with partial pivoting   *308*
    7.2.2 Alternate row and column elimination   *310*
    7.2.3 Block tridiagonal elimination   *313*
    7.2.4 Stable compactification   *316*

7.3 Stability for Block Methods — 318

7.4 Decomposition in the Nonseparated BC Case — 319

    7.4.1 The LU-decomposition   *320*
    7.4.2 A general decoupling algorithm   *321*

7.5 Solution in More General Cases 322

    7.5.1 BVPs with parameters *322*
    7.5.2 Multipoint BC *323*
    Exercises *324*

8 SOLVING NONLINEAR EQUATIONS 327

8.1 Improving the Local Convergence of Newton's Method 329

    8.1.1 Damped Newton *329*
    8.1.2 Altering the Newton direction *338*

8.2 Reducing the Cost of the Newton Iteration 341

    8.2.1 Modified Newton *341*
    8.2.2 Rank-1 updates *343*

8.3 Finding a Good Initial Guess 343

    8.3.1 Continuation *344*
    8.3.2 Imbedding in a time-dependent problem *350*

8.4 Further Remarks on Discrete Nonlinear BVPS 353
    Exercises *355*

9 MESH SELECTION 358

9.1 Introduction 359

    9.1.1 Error equidistribution and monitoring *362*

9.2 Direct Methods 364

    9.2.1 Equidistributing local truncation error *365*

9.3 A Mesh Strategy for Collocation 367

    9.3.1 A practical mesh selection algorithm *367*
    9.3.2 Numerical examples *371*

9.4 Transformation Methods — 373

9.4.1 Explicit method *373*
9.4.2 Implicit method *375*

9.5 General Considerations — 380

9.5.1 Some practical considerations *380*
9.5.2 Coordinating mesh selection and nonlinear iteration *381*
9.5.3 Other approaches *383*
Exercises *384*

10 SINGULAR PERTURBATIONS — 386

10.1 Analytical Approaches — 389

10.1.1 A linear second-order ODE *390*
10.1.2 A nonlinear second-order ODE *395*
10.1.3 Linear first-order systems *397*
10.1.4 Nonlinear first-order systems *411*

10.2 Numerical Approaches — 414

10.2.1 Theoretical multiple shooting *415*
10.2.2 Stability and global error analysis, I *419*
10.2.3 Stability and global error analysis, II *422*
10.2.4 The discretization mesh as a stretching transformation *428*

10.3 Difference Methods — 432

10.3.1 One-sided schemes *432*
10.3.2 Symmetric schemes *440*
10.3.3 Exponential fitting *454*

10.4 Initial Value Methods — 457

10.4.1 Sequential shooting *457*
10.4.2 The Riccati method *460*
Exercises *467*

11 SPECIAL TOPICS — 469

11.1 Reformulation of Problems in "Standard" Form — 469

- 11.1.1 Higher-order equations, parameters, nonseparated BC, multipoint BC — 470
- 11.1.2 Conditions at special points — 471
- 11.1.3 Integral relations — 473

11.2 Generalized ODEs and Differential Algebraic Equations — 474

11.3 Eigenvalue Problems — 478

- 11.3.1 Sturm-Liouville problems — 478
- 11.3.2 Eigenvalues of first-order systems — 482

11.4 BVPs with Singularities; Infinite Intervals — 483

- 11.4.1 Singularities of the first kind — 484
- 11.4.2 Infinite interval problems — 486

11.5 Path Following, Singular Points and Bifurcation — 490

- 11.5.1 Branching and stability — 493
- 11.5.2 Numerical techniques — 496

11.6 Highly Oscillatory Solutions — 500

11.7 Functional Differential Equations — 505

11.8 Method of Lines for PDEs — 508

11.9 Multipoint Problems — 512

11.10 On Code Design and Comparison — 515

APPENDIX A: A MULTIPLE SHOOTING CODE — 517

APPENDIX B: A COLLOCATION CODE — 526

REFERENCES **534**

BIBLIOGRAPHY **566**

INDEX **587**

List of Examples

Example	Page no.	Example	Page no.
1.1	2	4.2	136
1.2	2	4.3	140
1.3	3, 4	4.4	143
1.4	7	4.5	143
1.5	8	4.6	145
1.6	9	4.7	147
1.7	9	4.8	151
1.8	11	4.9	154
1.9	12	4.10	159
1.10	13	4.11	160
1.11	14	4.12	165
1.12	16	4.13	169
1.13	16	4.14	171
1.14	17	4.15	172
1.15	19	4.16	174
1.16	21	4.17	179
1.17	21	4.18	179
1.18	22	5.1	192
1.19	22	5.2	193
1.20	23	5.3	196
1.21	23	5.4	209
1.22	24	5.5	211
1.23	25	5.6	214
1.24	26	5.7	216
2.1	31	5.8	223
2.2	55	5.9	232
2.3	57	5.10	236
2.4	76	5.11	241
2.5	78	5.12	251
2.6	78	5.13	251
2.7	80	5.14	257
2.8	80	5.15	263
3.1	87	5.16	268
3.2	89	6.1	280
3.3	92	6.2	280
3.4	95	6.3	295
3.5	98	6.4	296
3.6	102	7.1	306
3.7	106	7.2	315
3.8	110	8.1	336
3.9	112	8.2	338
3.10	113	8.3	342
3.11	116	8.4	345
3.12	117	8.5	345
3.13	121	8.6	346
3.14	123	8.7	352
4.1	134	8.8	354

Example	Page no.
9.1	360
9.2	371
9.3	372
9.4	372
9.5	374
9.6	376
9.7	378
9.8	382
10.1	387
10.2	389
10.3	390
10.4	394
10.5	395
10.6	396
10.7	412
10.8	413
10.9	420
10.10	424
10.11	435, 456
10.12	442
10.13	445
10.14	446
10.15	448
10.16	456
10.17	466
10.18	466
11.1	471
11.2	477
11.3	483
11.4	487
11.5	491
11.6	492
11.7	494
11.8	497
11.9	500
11.10	502
11.11	505
11.12	506
11.13	510
11.14	512
11.15	513

Preface

Our knowledge and understanding of methods for the numerical solution of boundary value problems (BVPs) for ordinary differential equations has increased significantly in the past few years. Although important theoretical and practical developments have taken place on a number of fronts, they have not previously been comprehensively described in any text. This book was written with the intention of bringing together basic theory and practice, and more advanced recent developments concerning these methods.

This field has been rather late in maturing as compared with the numerical solution of initial value problems (IVPs). In fact, years ago many considered BVPs as some sort of a subclass of IVPs, wherein one fiddles with the initial conditions in order to get things right at the other end. It gradually became clear that IVPs are actually a special and in some sense relatively simple subclass of BVPs. The fundamental difference is that for IVPs one has complete information about the solution at one point (the initial point), so one may consider using a marching algorithm which is always *local* in nature. For BVPs, on the other hand, no complete information is available at any point, so the end points have to be connected by the solution algorithm in a *global* way. Only after stepping through the entire domain can the solution at any point be determined.

This book was written for a broad range of readers, including the senior undergraduate or graduate student, the scientist or engineer with a practical orientation and probably a specific application in mind, the purer analyst whose interest lies in the mathematical properties of the various numerical methods, and the researcher working at the frontiers in this subject. In order to accommodate such diverse interests, we maintain several levels of presentation simultaneously. To do this, we try to provide ample motivation for the methods and to demonstrate and analyze them, often in significant detail. We make use of examples, some as practical case studies, and we give exercises. In the appendices we describe two general-purpose codes which have been used for most of our numerical calculations. Typically (but not always), the first one or two sections of each chapter are at a more elementary level and are more concerned with the description of methods than with their analysis. Historical notes and an annotated list of references for each chapter are given in the Notes and References, which have been collected together after the appendices. A more extensive bibliography of papers dealing with the general topics in this book has also been included.

OVERVIEW

The first three chapters of the book are preliminary to the actual introduction of the numerical methods. The preliminaries often involve some rather nontrivial mathematics. While a complete understanding of this material is not prerequisite to much of what follows, a general understanding of Sections 1.1 and 3.1 and much of Chapter 2 would be needed fairly early on.

Chapter 1 consists of two parts. In the first we introduce some basic forms for BVPs which are treated throughout the book. We observe some connections and conversions between different forms, from which we conclude that the standard form consisting of a first-order system of differential equations with (often separated) two-point boundary conditions is rather general.

In Section 1.2 we give a representative collection of BVPs which arise in applications. The early appearance of this collection serves two purposes. One is to emphasize the importance of applications as a guide in the development and analysis of the numerical methods, and the other is to enable us to use them for examples and exercises in subsequent chapters. Some of the applications are well documented in the literature, while others were brought to our attention more directly by scientists or engineers who had developed a model and wanted help with its numerical solution. The physical soundness of these problems is not the issue here.

The study of numerical methods for BVPs relies on a substantial amount of background knowledge. This includes a background in ordinary differential equations (ODEs), linear algebra, and basic numerical analysis. Since not all readers are presumed to have this knowledge, we gather relevant material (except for the basic theory for ODEs) in Chapter 2. This is not intended to replace a text in numerical analysis, but rather to briefly expose the material required later on. We recommend an early reading of Sections 2.7 and 2.8. The other sections may be referred to as needed.

Chapter 3 is devoted to theory of ODEs. Again the idea is not to replace a text on this topic, but rather to introduce the material necessary for the development and understanding of numerical methods presented later on. The reader should be familiar with the material of Section 3.1, which is elementary, and with some of Section 3.2 (especially Section 3.2.1), which is very useful for understanding basic conditioning and stability issues. The other two sections, especially Section 3.4, are more advanced. Indeed, the material of Section 3.4 does not appear in any other texts of which we are aware, but it is very important if one is to acquire a deeper theoretical knowledge of this subject.

In Chapter 4 we describe the first class of numerical methods for the solution of BVPs. The methods considered here are all based on solving related initial value problems (IVPs) in order to obtain a BVP solution. The simplest of these methods is single shooting. It can be easily applied in a very general setting, but it also suffers from potential numerical instabilities, as we show. Other methods, which moderate this instability with various degrees of success, are then proposed and analyzed. Some of them are rather powerful, but they are no longer so simple. In Section 4.3 we consider the method of multiple shooting, both as a method in its own right and as a framework for examining other methods in this chapter and the next. Marching techniques and the Riccati method are discussed in the next two sections, and finally, nonlinear problems are considered.

Chapter 5 describes the other class of basic numerical methods used in the book. Unlike Chapter 4, which assumes that one already knows how to integrate IVPs, Chapter 5 essentially develops the methods from scratch (using Taylor expansions). This is one reason for its length. After introducing some simple methods and concepts in Section 5.1, we give a general convergence and stability analysis in Section 5.2, followed by more methods and their analysis in Sections 5.3 and 5.4. Emphasis is placed on one-step finite difference schemes (including collocation) for first-order ODE systems. In Section 5.5 we consider methods for accelerating the convergence (that is, for increasing the order of convergence) of the schemes, and in Section 5.6, some methods for higher-order ODEs, concentrating mainly on collocation methods. A few words about finite element methods conclude this chapter. Our belief is that while finite element methods are essential in the numerical solution of some partial differential equations, they are not so important in the numerical solution of ODEs.

Chapter 6 is devoted to the fundamental principle of decoupling. It generally treats more advanced theoretical considerations than the previous two chapters, and it is not requisite for obtaining a good working knowledge and understanding of BVP methods. Nevertheless, this material is essential, indeed pivotal, for obtaining a deeper understanding of many numerical methods in this book. We discuss decoupling both on a continuous level (that is, "without discretization") and on a discrete level. The advanced methods in Chapter 4, in particular, are well understood only in light of the considerations here.

Chapter 7 is devoted to numerical methods for solving the linear systems of equations resulting from the discretizations (of linear or linearized BVPs) described in Chapters 4 and 5. A typical linear system here is large and sparse, with its nonzero elements concentrated around the main diagonal in a certain block form. Only direct (noniterative) methods need be considered, but the sparseness structure must be taken into account. Methods are described in Section 7.2, analysis for some of them (relying on the analysis in Chapter 6) is in Section 7.3. Nonseparated boundary conditions (which yield linear equations where not all elements are near the main diagonal) are considered in Section 7.4, and additional parameters are briefly discussed in Section 7.5.

The topic of Chapter 8 is the numerical solution of the nonlinear equations resulting from discretizing nonlinear BVPs. The basic method considered in this book is Newton's method. It has some significant advantages, but it has some drawbacks as well, and in this chapter we discuss various methods to overcome its shortcomings. In Section 8.1 we consider mainly the damped Newton's method, which is designed to increase the domain of attraction and robustness of the basic method. Then various techniques for reducing the cost of the iteration are presented. The more powerful, but potentially more expensive, tool of using the method of continuation for solving the nonlinear problems is discussed in Section 8.3. In Section 8.4 we make some further comments regarding the relation between the nonlinear algebraic equations and the underlying nonlinear BVPs from which they arise.

The finite difference methods of Chapter 5 are defined on a discretization mesh (or grid), and the choice of this mesh can crucially affect the efficiency of the method. Mesh selection is the topic of Chapter 9. In this chapter, and to a lesser extent in the previous one, our presentation is more practically oriented and less theoretical. One reason for this is that the available theory is more fragmented and less capable of explaining the observed phenomena. The examples in these two chapters are also more sophisticated than in previous chapters, dealing mostly with more difficult (and more practical) BVPs.

The current research on numerical methods for IVPs focuses on stiff problems, since these are considered to be the ones not yet handled satisfactorily in practice. The corresponding research in BVPs has claimed front stage much more recently, but with much vigor, and this is evidenced by the size of Chapter 10—devoted to this subject. A large volume of analytical work on the related singular perturbation problem has been done, and is summarized in Section 10.1. Before presenting specific methods, we perform some general analysis of numerical methods for this type of problem in Section 10.2. Finite difference methods, both one-sided and symmetric, are considered in Section 10.3, and special initial value approaches are discussed in Section 10.4.

This completes the description of material we attempted to cover in depth. However, a number of related topics, several being important enough to warrant a book in their own right, are briefly described in Chapter 11.

TEACHING A COURSE

This book contains a great deal of material, so there are a number of possibilities for teaching a course using it, depending on the level and orientation of the students and on the time allotted. Essential material is in Sections 1.1, 3.1, 3.2.1, 4.1, 4.2.1, 4.2.3, 4.3, 4.6, 5.1–5.3.2, 5.5.1–5.5.3 and 7.1–7.2.1. This can be covered adequately in half a semester, for example, in conjunction with IVPs in a course on numerical ODEs at the senior undergraduate level. In a full semester devoted to BVPs, one can do much more. A theoretically flavored approach could include parts of Sections 3.3 and 3.4, more advanced sections from either Chapter 4 or Chapter 5, Sections 6.2 and 6.3, some of Chapters 7 and 8, and as much of Chapter 10 as possible. A more practically oriented direction could include additional material from either Chapter 4 or Chapter 5, Sections 8.1–8.3 and 9.1–9.3, and as much as possible of Chapter 10 (with less emphasis on Section 10.1).

ACKNOWLEDGMENTS

In addition to our students who suffered through numerous early versions of this material, we wish to express our gratitude to those who have provided comments on parts of the manuscript: Georg Bader, Peter Deuflhard, Luca Dieci, Ian Gladwell, Frank de Hoog, Eugene Isaacson, Simon Jacobs, Doug Lawson, Paul van Loon, Mike Osborne, Steve Pruess, Larry Shampine, and Gustaf Söderlind.

Uri M. Ascher

Robert M. Mattheij

Robert D. Russell

Numerical Solution of Boundary Value Problems for Ordinary Differential Equations

1

Introduction

Our purpose in this opening chapter is to acquaint the reader with the class of mathematical problems discussed in this book. There are a number of general forms into which such problems are naturally cast. We consider these forms and discuss their relationships in Section 1.1. Then, in Section 1.2, a number of examples from various application areas are presented, which lead to instances of the general forms of Section 1.1. To a varying extent, we shall apply the general theory of Chapter 3 and the numerical methods of following chapters to these examples.

1.1 BOUNDARY VALUE PROBLEMS FOR ORDINARY DIFFERENTIAL EQUATIONS

We shall be concerned with *boundary value problems* (BVPs) for *ordinary differential equations* (ODEs). In this section several model problems and the most common general forms of BVPs are presented. Frequently, the special case of *initial value problems* (IVPs) will be discussed as well.

The treatment throughout deals almost exclusively with ODEs, so when we refer to IVPs or BVPs, "for ODEs" is to be implied.

1.1.1 Model problems

A boundary value problem consists of a differential equation (or equations) on a given interval and an explicit condition (or conditions) that a solution must satisfy at one or several points. The simplest instance of such explicit conditions is when they are all specified at one initial point. We refer to this as an initial value problem. Thus, a simple IVP would have the form

$$y' = f(x, y) \qquad x > a \qquad (1.1a)$$

$$y(a) = \alpha \qquad (1.1b)$$

where a is the initial point and α is a constant. Here, and throughout the book, we use the notation $y' \equiv \dfrac{dy}{dx}$. The IVP is called linear or nonlinear depending upon whether $f(x,y)$ is linear or nonlinear in y.

Frequently the variable x corresponds to time, and (1.1b) corresponds to the known initial position of the solution $y(x)$.

Example 1.1

If $y(x)$ represents the amount of the radioactive compound lead 210 present in a sample of ore at time x, then

$$\frac{dy}{dx} = \lambda y + r \qquad x > a$$

$$y(a) = \alpha$$

where α is the original amount present at the initial time $x = a$, the constant $\lambda \approx 22$ years is the half-life of lead 210, and $r(x)$ is the number of disintegrations of radium 226 (which produces lead 210) at time x. □

For a boundary value problem, information about a solution to the differential equation(s) may be generally specified at more than one point. Often there are two points, which correspond physically to the boundaries of some region, so that it is a *two-point boundary value problem*. A simple and common form for a two-point boundary value problem is

$$u'' = f(x, u, u') \qquad a < x < b \qquad (1.2a)$$

$$u(a) = \beta_1, \qquad u(b) = \beta_2 \qquad (1.2b)$$

where β_1 and β_2 are known constants and the known endpoints a and b may be finite or infinite. For the linear case of this BVP, (1.2a) takes the simpler form

$$u''(x) - c_1(x) u'(x) - c_0(x) u(x) = q(x) \qquad a < x < b \qquad (1.2c)$$

Example 1.2

Consider a tightly stretched string with ends represented by the points $(0,0)$ and $(b,0)$ in the (x,u) plane. If it is hanging at rest under its own weight, the static displacement $u(x)$ satisfies

$$au'' - q = 0 \qquad 0 \le x \le b$$

$$u(0) = u(b) = 0$$

where a and q are constants dependent upon the material properties. □

1.1.2 General forms for the differential equations

Usually one assumes that general ordinary differential equations can be written as a first-order system

$$\mathbf{y}' = \mathbf{f}(x, \mathbf{y}) \qquad a < x < b \qquad (1.3a)$$

where $\mathbf{y}(x) = (y_1(x), y_2(x), \ldots, y_n(x))^T$ is the unknown function and $\mathbf{f}(x, \mathbf{y}) = (f_1(x, \mathbf{y}), f_2(x, \mathbf{y}), \ldots, f_n(x, \mathbf{y}))^T$ is the (generally nonlinear) right-hand side. The interval ends a and b are finite or infinite constants. For linear problems, the ODE simplifies to

$$\mathbf{y}' = A(x)\mathbf{y} + \mathbf{q}(x) \qquad a < x < b \qquad (1.4a)$$

where the matrix A and the vector $\mathbf{q}$ are functions of x, $A(x) \in \mathbf{R}^{n \times n}$ and $\mathbf{q}(x) \in \mathbf{R}^n$. The linear system (1.4a) is called *homogeneous* if $\mathbf{q} \equiv 0$, and it is *inhomogeneous* otherwise.

High-order ODEs can normally be converted to the first order form (1.3a). Given any scalar differential equation

$$u^{(m)} = f(x, u, u', \ldots, u^{(m-1)}), \qquad a < x < b \qquad (1.5)$$

let $\mathbf{y}(x) = (y_1(x), y_2(x), \ldots, y_m(x))^T$ be defined by

$$\begin{aligned} y_1(x) &= u(x) \\ y_2(x) &= u'(x) \\ &\vdots \\ y_m(x) &= u^{(m-1)}(x) \end{aligned} \qquad (1.6)$$

Then the ODE can be converted to the equivalent first-order form

$$\begin{aligned} y_1' &= y_2 \\ y_2' &= y_3 \\ &\vdots \\ y_{m-1}' &= y_m \\ y_m' &= f(x, y_1, y_2, \ldots, y_m) \end{aligned}$$

This is in the form (1.3a).

Example 1.3

The equations for a three-body problem, such as a satellite moving under the influence of the earth and the moon in a coordinate system (u, v) and rotating so as to keep the positions of the earth and moon fixed, are

$$u'' = 2v + u - \frac{c_1(u + c_2)}{((u + c_2)^2 + v^2)^{1/2}} - \frac{c_2(u - c_1)}{((u - c_1)^2 + v^2)^{1/2}}$$

$$v'' = -2u + v - \frac{c_1 v}{((u + c_2)^2 + v^2)^{1/2}} - \frac{c_2 v}{((u - c_1)^2 + v^2)^{1/2}}$$

where c_1 is a given constant and $c_2 = 1 - c_1$. Letting $y_1 = u$, $y_2 = u'$, $y_3 = v$, and $y_4 = v'$, one obtains the first-order system

$$y_1' = y_2$$
$$y_2' = 2y_3 + y_1 - \frac{c_1(y_1+c_2)}{((y_1+c_2)^2+y_3^2)^{1/2}} - \frac{c_2(y_1-c_1)}{((y_1-c_1)^2+y_3^2)^{1/2}}$$
$$y_3' = y_4$$
$$y_4' = 2y_1 + y_3 - \frac{c_1 y_3}{((y_1+c_2)^2+y_3^2)^{1/2}} - \frac{c_2 y_3}{((y_1-c_1)^2+y_3^2)^{1/2}}$$
□

In the linear case, the higher-order ODE (1.5) simplifies to

$$u^{(m)} = \sum_{j=0}^{m-1} c_j(x) u^{(j)} + q(x), \qquad a < x < b \tag{1.7}$$

The transformation (1.6) to a first-order system (1.4a) remains the same.

The most general form of a boundary value problem which we shall consider involves a system of differential equations which are of different orders. This is called a *mixed-order system*. It has the form

$$y_i^{(m_i)} = f_i(x, y_1, \ldots, y_1^{(m_1-1)}, y_2, \ldots, y_d^{(m_d-1)}) \tag{1.8a}$$
$$= f_i(x, \mathbf{z}(\mathbf{y})) \qquad 1 \le i \le d \qquad a < x < b$$

where $\mathbf{y}(x) = (y_1(x), \ldots, y_d(x))^T$ and

$$\mathbf{z}(\mathbf{y}(x)) := (y_1(x), y_1'(x), \ldots, y_1^{(m_1-1)}(x), y_2(x), \ldots, y_2^{(m_2-1)}(x), \ldots, y_d^{(m_d-1)}(x))^T$$

The conversion of this system to a first-order form can proceed directly, as for one higher-order ODE. Letting $n := \sum_{i=1}^{d} m_i$, note that $\mathbf{z}(\mathbf{y}(x)) \in \mathbf{R}^n$ would be the vector of unknown functions of x in the first-order form.

1.1.3 General forms for the boundary conditions

A first-order system of ODEs like (1.3a) is normally supplemented by n boundary conditions (BC)

$$\mathbf{g}(\mathbf{y}(a), \mathbf{y}(b)) = \mathbf{0} \tag{1.3b}$$

where $\mathbf{g} = (g_1, \ldots, g_n)^T$ is a (generally nonlinear) vector function and $\mathbf{0}$ is a vector of n zeros. The simplest instance of $\mathbf{g}$ is the case for an IVP. Then the solution is given at the initial point; that is,

$$\mathbf{y}(a) = \boldsymbol{\alpha} \tag{1.3c}$$

where $\boldsymbol{\alpha} = (\alpha_1, \ldots, \alpha_n)^T \in \mathbf{R}^n$ is a known vector of initial conditions which uniquely determines $\mathbf{y}(x)$ near a.

Example 1.3 (continued)

For the three-body problem, typical boundary conditions would specify initial position and velocity of the body. If at time $x = 0$ the body is at $(u, v) = (1, 0)$ with velocity -1 in the v direction then the initial conditions are

$$y_1(0) = 1$$
$$y_2(0) = 0$$
$$y_3(0) = 0$$
$$y_4(0) = -1$$

Note that these conditions have the form (1.3c). □

The relative simplicity of IVPs lies in the fact that the entire solution is known at a certain point. For the general BC (1.3b) this is not necessarily the case, as we have already seen in Example 1.2. Both analytic theory and numerical methods are considerably more involved in the general case. Correspondingly, there are a number of special cases of (1.3b) which will be considered.

The general form of linear two-point BC for a first-order system (or for a higher-order ODE) is

$$B_a \mathbf{y}(a) + B_b \mathbf{y}(b) = \boldsymbol{\beta} \tag{1.4b}$$

Here B_a and $B_b \in \mathbf{R}^{n \times n}$ and $\boldsymbol{\beta} \in \mathbf{R}^n$. In Chapter 3 we shall see that for the linear BVP (1.4a,b) to have a unique solution, it is necessary but not sufficient that these BC be linearly independent; that is, that the matrix (B_a, B_b) have n linearly independent columns, or simply *rank* $(B_a, B_b) = n$. BC of the general form (1.4b) are called *nonseparated BC*, since each involves information about $\mathbf{y}(x)$ at both endpoints.

It is worthwhile to distinguish cases when some BC information is given at only one point. If rank $(B_a) < n$ or rank $(B_b) < n$, then the BC are called *partially separated*. In the case rank $(B_b) = q < n$, the BVP can be transformed to one where the BC have the form

$$B_{a1} \mathbf{y}(a) = \boldsymbol{\beta}_1$$
$$B_{b1} \mathbf{y}(a) + B_{b2} \mathbf{y}(b) = \boldsymbol{\beta}_2 \tag{1.4c}$$

where $B_{a1} \in \mathbf{R}^{p \times n}$ ($p := n - q$), B_{b1} and $B_{b2} \in \mathbf{R}^{q \times n}$, $\boldsymbol{\beta}_1 \in \mathbf{R}^p$, and $\boldsymbol{\beta}_2 \in \mathbf{R}^q$. The case rank $(B_a) < n$ is of course similar. Details of the transformation from (1.4b) to (1.4c) are given in Chapter 4.

The BC are called *separated* if they simplify further to

$$B_{a1} \mathbf{y}(a) = \boldsymbol{\beta}_1$$
$$B_{b2} \mathbf{y}(b) = \boldsymbol{\beta}_2 \tag{1.4d}$$

The nonlinear BC (1.3b) can also occur in partially separated or separated form. Thus, the boundary conditions are separated if they are of the form

$$\mathbf{g}_1(\mathbf{y}(a)) = \mathbf{0}_1$$
$$\mathbf{g}_2(\mathbf{y}(b)) = \mathbf{0}_2 \tag{1.3d}$$

where $\mathbf{g}_1, \mathbf{0}_1 \in \mathbf{R}^p$ and $\mathbf{g}_2, \mathbf{0}_2 \in \mathbf{R}^q$ with $n = p + q$. This latter case turns out to be particularly pleasant, both theoretically and practically. In fact, a significant portion of the currently available software for BVPs assumes that the BC are separated. Fortunately, a BVP with nonseparated or partially separated BC can be converted to one with separated BC (but with more ODEs), as we show next.

Consider the BVP with partially separated BC (1.4a,c) (which contains the nonseparated case (1.4b) as a special case with $p = 0$). Adding the $q = n - p$ trivial ODEs

$$\mathbf{z}' = \mathbf{0}$$

(implying $\mathbf{z}(a) = \mathbf{z}(b)$, *not* through the BC), we have an augmented system of order $n + q$ with separated BC which can be written as

$$\begin{bmatrix} \mathbf{y} \\ \mathbf{z} \end{bmatrix}' = \begin{bmatrix} A & 0 \\ 0 & 0 \end{bmatrix} \begin{bmatrix} \mathbf{y} \\ \mathbf{z} \end{bmatrix} + \begin{bmatrix} \mathbf{q} \\ \mathbf{0} \end{bmatrix} \quad (1.9a)$$

$$\begin{bmatrix} B_{a1} & 0 \\ B_{a2} & -I \end{bmatrix} \begin{bmatrix} \mathbf{y} \\ \mathbf{z} \end{bmatrix}(a) = \begin{bmatrix} \boldsymbol{\beta}_1 \\ \mathbf{0} \end{bmatrix}, \quad (B_{b2} \; I) \begin{bmatrix} \mathbf{y} \\ \mathbf{z} \end{bmatrix}(b) = \boldsymbol{\beta}_2 \quad (1.9b)$$

A demonstration of this trick is provided later on, in Example 1.10. Its validity is further discussed in Chapter 6. Note, however, that the enlarged size of the augmented system (1.9) is a disadvantage.

A general linear *multipoint* BVP consists of the ODE (1.4a) and multipoint BC

$$\sum_{j=1}^{J} B_j \, \mathbf{y}(\zeta_j) = \boldsymbol{\beta} \quad (1.4e)$$

where $B_1, \ldots, B_J \in \mathbf{R}^{n \times n}$, $\boldsymbol{\beta} \in \mathbf{R}^n$, and $a = \zeta_1 < \zeta_2 < \cdots < \zeta_J = b$. A multipoint BVP can be converted to a two-point problem by transforming each of the subintervals $[\zeta_j, \zeta_{j+1}]$ onto the interval $[0,1]$, say, and writing the ODEs (1.4a) for the independent variable

$$t = \frac{x - \zeta_j}{\zeta_{j+1} - \zeta_j}$$

for each j, $1 \leq j \leq J-1$. The obtained $(J-1)n$ ODEs are then subject to the n BC of (1.4e) which are now specified at the interval ends, plus $n(J-2)$ additional BC resulting from the requirement that the solution $\mathbf{y}(x)$ should be continuous at the interior break points $\zeta_j, j = 2, \ldots, J-1$. Obviously, this transformation is not without a cost, but it helps to justify an analysis for only two-point BVPs. This we do throughout most of the book.

The most general form of a boundary value problem which we shall consider involves a mixed-order system (1.8a) subject to separated, multipoint boundary conditions

$$g_j(\mathbf{z}(\mathbf{y}(\zeta_j))) = 0 \quad 1 \leq j \leq n \quad (1.8b)$$

$a = \zeta_1 \leq \zeta_2 \leq \ldots \leq \zeta_n = b$. Again there is a transformation which brings this BVP into one with separated two-point BC, at the cost of increased system size.

Typically, the theoretical and numerical treatment of initial value problems is done by assuming they are in the first-order form (1.3a,c) or (1.4a), (1.3c). It is also common to treat BVPs in the form of first-order systems (1.3a,b) or (1.4a,b). Occasionally we will also consider BVPs expressed in terms of high-order equations (1.5) or (1.7) because (a) problems usually arise in this form, (b) numerical methods can be easier to motivate and describe for these equations, and (c) these methods can be more efficient if applied directly to the high-order equations instead of the corresponding first-order systems. Nevertheless, the basic properties and definitions for BVPs will be given just for first-order systems, as they can then be related to a general BVP by applying them to its equivalent first-order formulation.

1.2 BOUNDARY VALUE PROBLEMS IN APPLICATIONS

In this section we have collected 22 instances of BVPs which arise in a variety of application areas. The reader can therefore have an appreciation of the types of problems and difficulties encountered in practice. These examples can be used in order to test proposed methods and codes for the numerical solution of BVPs. When presenting numerical techniques in later chapters, we will use some of the examples listed here to illustrate the discussion, and the reader will be invited to try and solve other examples. Reading this section is, however, *not a prerequisite* to any other part of the book.

We do not insist on a uniform notation here; in some cases the notation is natural to the application. In each case, the formulation of the BVP is brought into one of the usual general forms discussed in the previous section, which we shall refer to as a "standard form."

For the background, and sometimes the detailed derivation of the application, we rely on the literature cited in the notes and references at the end of the book, and our intention here is not to go into too much detail about the physical origins of each problem. (Many of the formulations originate as PDEs which by various techniques are reduced to ODEs.) Additional papers which discuss applications not included here are mentioned in the bibliography of the chapter. The list is certainly not complete, but the collection forms a significant test-bed for any code.

Example 1.4 Flow in a Channel

Consider the problem of fluid injection through one side of a long vertical channel. The Navier-Stokes and the heat transfer equations can be reduced to the following system:

$$f''' - R\,[(f')^2 - ff''] + RA = 0 \qquad (1.10a)$$

$$h'' + R\,fh' + 1 = 0 \qquad (1.11a)$$

$$\theta'' + Pf\,\theta' = 0 \qquad (1.12a)$$

$$f(0) = f'(0) = 0, \qquad f(1) = 1, \qquad f'(1) = 0 \qquad (1.10b)$$

$$h(0) = h(1) = 0 \qquad (1.11b)$$

$$\theta(0) = 0, \qquad \theta(1) = 1 \qquad (1.12b)$$

Here f and h are two potential functions, θ is a temperature distribution function, and A is an undetermined constant. There are two parameters with known values, $R =$ Reynolds number and $P =$ Peclet number (e. g., take $P = 0.7R$).

At first, note that the subproblem (1.10a, b) is separated from the rest and thus can be solved separately. Suppose that this is done. Then (1.11a, b) and (1.12a, b) are two separated, *linear* second-order problems in standard form. The original problem is thus effectively broken into three subproblems.

Now consider (1.10a, b). We have a nonlinear third-order ODE for f, with the constant A determined by the requirement that the *four* BC (1.10b) be satisfied. One way to bring this into standard form is to differentiate (1.10a), obtaining

$$f'''' = R\,[f'f'' - ff'''] \tag{1.10c}$$

The problem (1.10c, b) is now in standard form (and no longer explicitly involves A). Another, more general, trick is to treat the constant A as another dependent variable, adding the ODE

$$A' = 0 \tag{1.10d}$$

The problem (1.10a, b, d) is again in standard form.

The difficulty in solving the nonlinear problem (1.10) numerically depends, in a typical way, on the Reynolds number R. For moderate values of R, say $R = 10$, the problem is easy, but it gets tougher as R increases and for $R = 10{,}000$ there is a fast change in some solution values near $x = 0$. This is called a *boundary layer*. □

Example 1.5 Particle Diffusion and Reaction

The ODEs governing the reaction are

$$T'' + \frac{2}{x}T' = -\phi^2 \beta C e^{\gamma(1-T^{-1})} \tag{1.13a}$$

$$0 < x < 1$$

$$C'' + \frac{2}{x}C' = \phi^2 C e^{\gamma(1-T^{-1})} \tag{1.13b}$$

where x is time, C is the concentration, and T is the temperature. The constants ϕ, γ and β are known (they are the Thiele modulus, thermicity, and activation energy parameter, respectively). Representative values are $\phi = 14.44$, $\gamma = 20$, $\beta = 0.02$.

The BC at $x = 0$ are

$$T'(0) = C'(0) = 0 \tag{1.13c}$$

Note that the coefficient $\frac{2}{x}$ in (1.13a, b) is unbounded as $x \to 0$. This singularity typically comes from a reduction, due to cylindrical or spherical symmetry, of a partial differential equation to an ODE and is an artificial singularity: The solution is smooth near $x = 0$. The BC (1.13c) imply that

$$\frac{2}{x}T' \to 2T'', \qquad \frac{2}{x}C' \to 2C'' \text{ as } x \to 0 \tag{1.14}$$

In a numerical implementation, an expression giving $\frac{0}{0}$ should not be evaluated, of course, so (1.13a, b) should be modified by (1.14) if we intend to evaluate the ODE at the boundary.

At $x = 1$ we may have two types of BC:

(i) Dirichlet type

$$T(1) = C(1) = 1 \tag{1.13d}$$

The resulting BVP (1.13a, b, c, d) is not very difficult, numerically.

(ii) Mixed type

$$-T'(1) = B(T(1) - 1), \qquad -C'(1) = B_m(C(1) - 1) \tag{1.13e}$$

with (for instance) $B = 5$, $B_m = 250$. Here we get a thin boundary layer near $x = 1$ and the BVP (1.13a, b, c, e) is significantly more difficult to solve numerically. □

Example 1.6 Soil Problem

The problem is to determine moisture (water) transport in dessicated soil. The numerical BVP is tough to solve for dry desert soil, and the difficulty is enhanced as the soil becomes drier. The original problem is a PDE in time t and one space variable ξ. However, using the similarity transformation for small times

$$x = \frac{\xi}{\sqrt{t}}$$

one obtains the BVP

$$(K_r P')' = \frac{1}{2} x (-\frac{dS}{dP}) P' \qquad 0 < x < \infty \tag{1.15a}$$

$$P(0) = \beta_0, \qquad P(\infty) = \beta_1 \qquad (\text{e.g., } \beta_0 = 0, \qquad \beta_1 = -1) \tag{1.15b}$$

where P is the water pressure, K_r is the relative permeability, and S is the saturation. The latter two are given in terms of P by

$$\frac{S - S_r}{1 - S_r} = \frac{1}{1 + (-PL/A)^\eta}, \qquad K_r = \frac{1}{1 + (-PL/B)^\lambda} \qquad 0 < x < \infty \tag{1.15c}$$

where, typically,

$$S_r = 0.32, \quad A = 231, \quad B = 146, \quad \eta = 3.65, \quad \lambda = 6.65, \quad L = 100 \tag{1.15d}$$

The problem gets tougher for larger L (up to $L = 1000$ may be desired).

Note that the BVP is defined on an infinite interval. Here, however, this does not cause practical difficulties. Simply replace ∞ by a large enough value b, e.g., $b = 10$, and solve the BVP with

$$P(b) = \beta_1 \tag{1.15e}$$

We will have much more to say about infinite intervals later on (see Example 8.1, Section 11.4, and Examples 1.8 and 1.12). □

Example 1.7 Seismic Ray Tracing

The problem of determining when and where a relatively minor earthquake has occurred can sometimes be dealt with through ray theory. Suppose that the origin of the earthquake (in cartesian coordinates) is at the hypocenter (x_0, y_0, z_0) somewhere underneath the earth's surface ($z = 0$) and that the time at which the explosion has occurred is T_0. The explosion generates waves that propagate in all directions from the hypocenter; the time that a seismograph, located at a point (x_i, y_i, z_i) (with $z_i = 0$ if the seismograph is not buried underground), registers the earthquake, is equal to T_0 plus the time t_i it takes the wave front to travel from (x_0, y_0, z_0) to (x_i, y_i, z_i). Let t_i^0 be the actually observed time of tremor at the station located at (x_i, y_i, z_i). Then theoretically

$$T_0 + t_i = t_i^0$$

In practice, of course, t_i cannot be found exactly, but it can be calculated approximately, depending on the unknown (x_0, y_0, z_0). Thus, if we have N seismographs, $N \geq 4$, located at (x_i, y_i, z_i) and observing times t_i^0, $i = 1, \ldots, N$, then we can solve the nonlinear least squares problem

$$\text{minimize } \sum_{i=1}^{N} F_i^2 \qquad (1.16a)$$

with

$$F_i = F_i(x_0, y_0, z_0, T_0) = T_0 + t_i - t_i^0 \qquad (1.16b)$$

The question is then, how to calculate t_i.

Now, it can be shown that the normal to the wave front at any point (x, y, z) behaves, as a function of time t, like an optical ray and satisfies the following differential equations

$$\frac{dx}{ds} = v\xi, \qquad \frac{dy}{ds} = v\eta, \qquad \frac{dz}{ds} = v\zeta \qquad (1.17a)$$

$$0 < s < S$$

$$\frac{d\xi}{ds} = u_x, \qquad \frac{d\eta}{ds} = u_y, \qquad \frac{d\zeta}{ds} = u_z$$

where s is the arclength along the path, S is the (unknown) total arclength, $(x(s), y(s), z(s))$ are the coordinates of the ray, $v = v(x, y, z)$ is the velocity of a sound wave at a point (x, y, z) of the earth and $u \equiv 1/v$ is the slowness. The notation u_x stands for $\frac{\partial u}{\partial x}$, etc. How to obtain the velocity structure $v(x, y, z)$ of the medium is a nontrivial practical question, but we assume here that it is given; we note in passing that this velocity structure, or an approximation to it, can be obtained from a set of velocity measurements using three-dimensional interpolation.

Now, the time t_i it takes the ray to reach the ith seismograph is

$$t_i = \int_0^S u(s)\, ds \qquad (1.17b)$$

and we can integrate this knowing the ray path, which we obtain by solving the differential equations (1.17a) subject to the boundary conditions

$$x(0) = x_0, \qquad y(0) = y_0, \qquad z(0) = z_0 \qquad (1.17c)$$

$$\frac{dx}{ds}(0)^2 + \frac{dy}{ds}(0)^2 + \frac{dz}{ds}(0)^2 = 1 \qquad (1.17d)$$

$$x(S) = x_i, \qquad y(S) = y_i, \qquad z(S) = z_i \qquad (1.17e)$$

Here, the boundary conditions (1.17c, e) are obvious and (1.17d) comes from ray theory. Since S is a free parameter, it makes sense to have seven boundary conditions for six differential equations.

We now reformulate the problem (1.17) in standard form. For this we have to convert the interval of integration from a free-ended to a specified one, which we do by scaling the independent variable

$$\tau = s/S$$

and having S as a dependent variable, specifying

$$\frac{dS}{d\tau} \equiv S' = 0$$

Then, to incorporate (1.17b) we write

$$t_i = T(1)$$

where $T' = Su$ and $T(0) = 0$. The converted system is then

$$\begin{aligned}
x' &= Sv\xi, & y' &= Sv\eta, & z' &= Sv\zeta \\
\xi' &= Su_x, & \eta' &= Su_y, & \zeta' &= Su_z & 0 < \tau < 1 \quad (1.18a)\\
T' &= Su, & S' &= 0
\end{aligned}$$

subject to the boundary conditions

$$\begin{aligned}
x(0) &= x_0, & y(0) &= y_0, & z(0) &= z_0 \\
T(0) &= 0, & \xi(0)^2 + \eta(0)^2 + \zeta(0)^2 &= u(x(0), y(0), z(0))^2 & & \quad (1.18b)\\
x(1) &= x_i, & y(1) &= y_i, & z(1) &= z_i
\end{aligned}$$

The problem (1.18a, b) is now a nonlinear BVP in standard form.

To summarize, the location and origin time of the earthquake are determined by solving the minimization problem (1.16a), where for each i, $1 \leq i \leq N$, and each trial location (x_0, y_0, z_0) and origin time T_0, the function F_i of (1.16b) is evaluated by solving the boundary value problem (1.18a, b) and using $t_i = T(1)$. Since there are many boundary value problems to be solved, the code used for their solution should be very efficient.

The difficulty in solving (1.18) depends on the smoothness of the velocity structure v. For a constant v (a uniform medium) the problem is trivial, the ray being a straight line. However, if there are abrupt changes in the medium, then the problem may be difficult to solve numerically. □

Example 1.8 Theoretical Seismograms

This is another seismological application, where one attempts to calculate the ground displacements caused by a point moment seismic source. Assuming that the material properties of the medium (the earth) are a function of depth z only, we apply a Fourier-Bessell transform to the governing PDE (which is linear, in four independent variables), obtaining two uncoupled ODE systems of the form

$$\frac{d\mathbf{y}}{dz} \equiv \mathbf{y}' = A(z; \omega, k)\mathbf{y} \qquad 0 < z < b \qquad (1.19)$$

Here the angular frequency ω and the horizontal wave number k are parameters, $-\infty < \omega < \infty, 0 \leq k < \infty$. For each depth z, the solution of (1.19), under appropriate BC to be discussed below, is a function of k and ω. A double integral on k and ω is then taken to obtain the solution in terms of space and time. Thus, there are a very large number of BVPs to be solved.

We next specify these BVPs. Consider a half-space $0 < z < \infty$ in which P-wave velocity $\alpha(z)$, S-wave velocity $\beta(z)$, and density $\rho(z)$ are given piecewise continuous functions, which are constant for $z \geq b$. The first ODE system, called the SH equations, is given by (1.19) with

$$A = A_H = \begin{bmatrix} 0 & 1/\mu \\ \mu k^2 - \rho\omega^2 & 0 \end{bmatrix} \qquad \mu = \rho\beta^2 \qquad (1.20a)$$

while the second, more complex ODE system, called the P-SV equations, is given by (1.19) with

$$A = A_P = \begin{bmatrix} 0 & \frac{1}{\rho\alpha^2} & k(1-2\beta^2/\alpha^2) & 0 \\ -\omega^2\rho & 0 & 0 & k \\ -k & 0 & 0 & 1/\mu \\ 0 & -k(1-2\beta^2/\alpha^2) & 4\mu k^2(1-\beta^2/\alpha^2)-\rho\omega^2 & 0 \end{bmatrix} \quad (1.21a)$$

The BC at $z = b$ are derived from a radiation condition, that is, the requirement that only downgoing waves exist for $z \geq b$. This gives for the SH problem

$$v_\beta y_1 + \mu^{-1} y_2 = 0 \qquad \text{at } z = b \qquad (1.20b)$$

and for the P-SV problem

$$(\rho\omega^2 - 2\mu k^2)y_1 + 2\mu k v_\alpha y_3 - v_\alpha y_2 + k y_4 = 0 \qquad (1.21b)$$

$$\text{at } z = b$$

$$2\mu k v_\beta y_1 + (\rho\omega^2 - 2\mu k^2)y_3 + k y_2 - v_\beta y_4 = 0 \qquad (1.21c)$$

(When $\omega = 0$, (1.21b) and (1.21c) become linearly dependent and (1.21c) is then replaced by

$$(\rho\alpha^2)^{-1} y_2 + 2k(1-\beta^2/\alpha^2) y_3 + \mu^{-1} y_4 = 0 \qquad \text{at } z = b) \qquad (1.21d)$$

Here, $\pm v_\beta$ and $\pm v_\beta$, $\pm v_\alpha$ are the eigenvalues of A_H and A_P, respectively:

$$v_\beta = (k^2 - \frac{\omega^2}{\beta^2})^{1/2}, \qquad v_\alpha = (k^2 - \frac{\omega^2}{\alpha^2})^{1/2}$$

with the sign choice $\text{Re}(v) \geq 0$; $\text{Im}(v) \leq 0$ when $\text{Re}(v) = 0$. When $\text{Re}(v) > 0$, we have a decaying solution in z (a surface wave), while if $\text{Re}(v) = 0$, $v \neq 0$, we have an oscillatory solution (body wave). The rate of decay or oscillation increases as ω increases and the problem then gets tougher.

To complete the specification of the BVPs, solution values are given at the earth surface, viz.

$$y_2(0) = \beta_1 \qquad (1.20c)$$

for the SH problem and

$$y_2(0) = \beta_1, \qquad y_4(0) = \beta_2 \qquad (1.21e)$$

for the P-SV problem. The problem actually appears in two flavors in the seismology literature. One is where $\beta_1 \neq 0$ and/or $\beta_2 \neq 0$, corresponding to including an inhomogeneous source term, and a unique solution to (1.20) and (1.21) is sought. In the other approach, $\beta_1 = \beta_2 = 0$, and one solves for the eigenvalues and eigenfunctions of the (linear) problem. The double integral over ω and k then becomes a sum of residues at poles.

The large number of BVPs to be solved and their drastically different character for different (large) values of ω and k make this problem challenging, despite its linearity. A saving grace is the high degree of parallelism possible in these computations. □

Example 1.9 Meniscus in a Cylinder

Consider the equilibrium-free surface of a liquid inside a vertical cylinder of circular cross section (e. g., a capillary). The surface $f(r)$ satisfies the BVP

$$\frac{1}{r}\left[\frac{rf'(r)}{(1+(f'(r))^2)^{1/2}}\right]' - Bf(r) - 2\lambda = 0 \qquad 0 < r < 1 \qquad (1.22a)$$

$$f(0) = 0, \qquad f'(0) = 0 \qquad (1.22b)$$

$$f'(1) = \cot \theta \qquad (1.22c)$$

The independent variable r runs from the middle of the cylinder ($r=0$) to its boundary $r=1$, where the angle θ of contact with the fluid is given. There are two other parameters: The Bond number B is given, while the mean curvature λ of the surface at $r=0$ is unknown, accounting for the three BC (1.22b, c). A straightforward reformulation of (1.22a) is then (cf. Example 1.4)

$$\lambda' = 0 \qquad (1.22d)$$

$$f'' = (1+(f')^2)^{3/2}(Bf+2\lambda) - \frac{1}{r}((f')^3 + f') \qquad (1.22e)$$

The problem (1.22d, e, b, c) is in standard form. The ranges of interest for the parameters are $B_{cr} < B \le 1000$, where B_{cr} is a critical parameter, $-10 < B_{cr} < 0$, and $0 \le \theta \le \pi/2$ (wetting fluid). Unless $\theta \approx 0$, this BVP is not particularly difficult numerically. When $\theta = 0$, however, the end value in (1.22c) blows up, and another formulation is needed. This is done by letting x, the angle between the surface and the horizontal line, be the independent variable. From the relation

$$\tan x = f'(r)$$

we get

$$\dot{\lambda} = 0 \qquad (1.23a)$$

$$\dot{f} = D \sin x \qquad 0 < x < \pi/2 - \theta \qquad (1.23b)$$

$$\dot{r} = D \cos x \qquad (1.23c)$$

$$f(0) = r(0) = 0, \qquad r(\pi/2 - \theta) = 1 \qquad (1.23d)$$

where $(\dot{\ }) \equiv \dfrac{d}{dx}$ and

$$D = D(x) = [Bf(x) - r^{-1} \sin x + 2\lambda]^{-1} \qquad (1.23e)$$

This latter formulation (1.23) is now good even when $\theta = 0$. □

Example 1.10 Measles

Consider the following epidemiology model. Assume that a given population of constant size N can be divided into four categories: Susceptibles, whose number at time t is $S(t)$, infectives $I(t)$, latents $L(t)$, and immunes $M(t)$. We have

$$S(t) + I(t) + L(t) + M(t) = N \qquad t \in [0,1]$$

Under certain assumptions on the disease, its dynamics can be expressed as

$$y_1' = \mu - \beta(t) y_1 y_3 \qquad (1.24a)$$

$$y_2' = \beta(t) y_1 y_3 - y_2/\lambda \qquad 0 < t < 1 \qquad (1.24b)$$

$$y_3' = y_2/\lambda - y_3/\eta \qquad (1.24c)$$

where $y_1 = S/N$, $y_2 = L/N$, $y_3 = I/N$, $\beta(t) = \beta_0(1+\cos 2\pi t)$ and representative values of the appearing constants are $\mu=0.02$, $\lambda=0.0279$, $\eta=0.01$, and $\beta_0 = 1575$.

The solution sought is periodic; that is, the BC are

$$\mathbf{y}(1) = \mathbf{y}(0) \qquad (1.24d)$$

The BC (1.24d) are not separated. We can separate them by the general trick introduced in the previous section. Thus, let $\mathbf{c} = (c_1, c_2, c_3)^T$ be a vector of constants. We augment (1.24a, b, c) and replace (1.24d) by

$$\mathbf{c}' = \mathbf{0} \qquad (1.24e)$$

$$\mathbf{y}(0) = \mathbf{c}(0), \qquad \mathbf{y}(1) = \mathbf{c}(1) \qquad (1.24f)$$

The BVP (1.24a, b, c, e, f) now has separated BC. However, the size of the problem has doubled — a significant expense. The problem is not very difficult numerically. □

Example 1.11 Kidney Model

This problem is not only larger, but also much tougher than the previous two. The model describes mass and energy balance of the renal counterflow system. With F_{iv} the axial volume flow in the ith tube, J_{iv} the outward transmural volume flux, $F_{iv}C_{ik}$ the axial flow of the kth solute in the ith tube and J_{ik} the outward transmural flux per unit length of the kth solute from the ith tube, the ODEs for the steady state problem are

$$\frac{dF_{iv}}{dx} + J_{iv} = 0, \qquad 1 \le i \le 6 \qquad (1.25a)$$

$$\frac{d}{dx}(F_{iv}C_{ik}) + J_{ik} = 0, \qquad 1 \le i \le 6, \qquad 1 \le k \le 2 \qquad (1.25b)$$

and $0 < x < 1$.

Boundary conditions are as follows:

$$\begin{aligned}
F_{1v}(0) &= 1 & F_{5v}(0) &= 5 \\
C_{11}(0) &= 1 & C_{51}(0) &= 1 \\
C_{12}(0) &= 0.05 & C_{52}(0) &= 0.05 \\
F_{2v}(1) &= -F_{1v}(1) & & \\
C_{2k}(1) &= C_{1k}(1) & \text{for } k &= 1,2 \\
F_{6v}(0) &= -F_{2v}(0) & & \\
C_{6k}(0) &= C_{2k}(0) & \text{for } k &= 1,2 \\
F_{3v}(0) &= F_{6v}(1) & & \\
C_{3k}(0) &= C_{6k}(1) & \text{for } k &= 1,2 \\
F_{4v}(1) &= -F_{5v}(1) & & \\
C_{4k}(1) &= C_{5k}(1) & \text{for } k &= 1,2
\end{aligned} \qquad (1.25c)$$

This gives 18 BC for the 18 ODEs (1.25a, b). But we still have to specify the functions J_{i1}, J_{i2}, and J_{iv}, $1 \le i \le 6$. Transmural volume fluxes are defined as follows:

$$J_{iv} = h_{iv}\sum_{k=1}^{2}(C_{4k} - C_{ik}), \quad i = 1,2,3,5$$

$$J_{4v} = -\sum_{i \neq 4,6} J_{iv}$$

$$J_{6v} = (1.0 - C_{61}) + (0.05 - C_{62})$$

where $h_{1v} = h_{3v} = 10$ and $h_{iv} = 0$ for $i=2,5$.

Transmural solute fluxes are

$$J_{i1} = 0, \quad i = 1,3$$
$$= 0.75\, C_{i1}/(1. + C_{i1}), \quad i = 6$$
$$= 1000(C_{i1} - C_{41}), \quad i = 5$$

$$J_{21}(x) = \begin{cases} 1.8, & 0. \leq x \leq 0.4 \\ 1.8 + [-18. + 100(C_{21}(x) - C_{41}(x))]\cdot(x - 0.4), & 0.4 < x < 0.5 \\ 10[C_{21}(x) - C_{41}(x)], & 0.5 \leq x \leq 1 \end{cases}$$

$$J_{i2} = 0, \quad i = 1,2,6$$
$$= 1000(C_{i2} - C_{42}), \quad i = 5$$

$$J_{32}(x) = \begin{cases} 0., & 0. \leq x \leq 0.4 \\ 0.1[C_{32}(x) - C_{42}(x)]\cdot(x - 0.4), & 0.4 < x < 0.5 \\ 0.01[C_{32}(x) - C_{42}(x)], & 0.5 \leq x \leq 1 \end{cases}$$

$$J_{4k}(x) = -\sum_{i \neq 4,6} J_{ik}(x), \quad 0. \leq x \leq 1, \quad 1 \leq k \leq 2$$

This completes the specification of the problem. However, some simplification is possible. The reader can verify that the following BVP of order 13 is equivalent to (1.25).

$$C'_{12} = 200(C_{12})^2[C_{41} + C_{42} - 21C_{12}] \tag{1.26a}$$

$$C_{11} = 20C_{12}, \quad F_{1v} = \frac{0.05}{C_{12}}, \quad F_{2v} = -\frac{0.05}{C_{22}}$$

$$C'_{21} = 20C_{22}J_{21}, \quad C'_{22} = 0 \tag{1.26b}$$

$$C'_{31} = \frac{10}{K_1}(C_{31})^2[C_{41} + C_{42} - C_{31} - C_{32}] \tag{1.26c}$$

$$C'_{32} = \frac{C_{31}}{K_1}[J_{3v}C_{32} - J_{32}] \tag{1.26d}$$

$$K'_1 = 0, \quad F'_{4v} = -J_{4v} \tag{1.26e}$$

$$K_1 = \frac{C_{31}(0)}{2C_{32}(0)}, \quad F_{3v} = \frac{K_1}{C_{31}}, \quad F_{5v} = 5, \quad F_{6v} = \frac{0.05}{C_{62}}$$

$$C'_{41} = \frac{1}{F_{4v}}[J_{4v}C_{41} - J_{41}] \tag{1.26f}$$

$$C'_{42} = \frac{1}{F_{4v}}[J_{4v}C_{42} - J_{42}] \tag{1.26g}$$

$$C'_{51} = -200(C_{51} - C_{41}), \qquad C'_{52} = -200(C_{52} - C_{42}) \tag{1.26h}$$

$$C'_{61} = 20C_{62}[J_{6v}C_{61} - J_{61}] \tag{1.26i}$$

$$C'_{62} = 20(C_{62})^2 J_{6v} \tag{1.26j}$$

$$C_{12}(0) = 0.05, \quad C_{51}(0) = 1, \quad C_{52}(0) = 0.05, \quad F_{4v}(1) = -5 \tag{1.26k}$$

$$C_{31}(0) - 20K_1(0)C_{32}(0) = 0, \quad C_{22}(0) = C_{62}(0), \quad C_{61}(0) = C_{21}(0) \tag{1.26l}$$

$$C_{12}(1) = C_{22}(1), \quad C_{21}(1) = 20C_{12}(1), \quad C_{41}(1) = C_{51}(1) \tag{1.26m}$$

$$C_{42}(1) = C_{52}(1), \quad C_{61}(1) - 20K_1(1)C_{62}(1) = 0$$

$$C_{31}(0) = C_{61}(1) \tag{1.26n}$$

This BVP has one nonseparated BC. An equivalent problem of order 14 can be formed which has only separated BC, using the trick introduced in the previous section. □

Example 1.12 Magnetic Monopoles

The standard laws of electromagnetism (Maxwell's equations) forbid the possibility of magnetic monopoles. But classical solutions having the properties of monopoles can be found in the more general Yang-Mills theory. The governing equations are nonlinear partial differential equations, but an ODE over an infinite interval can be obtained in some special cases involving symmetry.

After mapping the independent variable onto the interval $(0, 1)$, the obtained BVP reads

$$y''_1 = \frac{2y'_1}{1-x} + \frac{y_1}{x^2(1-x)^2}[y_1^2 - 1 + \frac{y_3^2 - y_2^2}{(1-x)^2}] \tag{1.27a}$$

$$y''_2 = \frac{2y_2 y_1^2}{x^2(1-x)^2} \qquad\qquad 0 < x < 1 \tag{1.27b}$$

$$y''_3 = \frac{y_3}{x^2(1-x)^2}[2y_1^2 + \frac{\beta(y_3^2 - x^2)}{(1-x)^2}] \tag{1.27c}$$

$$y_1(0) = 1, \qquad y_2(0) = y_3(0) = 0 \tag{1.27d}$$

$$y_1(1) = 0, \qquad y_2(1) = \eta, \qquad y_3(1) = 1 \tag{1.27e}$$

where β and η are given constants. The mass for a monopole can be expressed in terms of an integral of these quantities.

Typically, one may want solutions of this BVP for a number of parameter values in the range $0 \le \beta \le 20$, $0 < \eta < 1$. For an efficient numerical solution procedure it then makes sense to use information obtained when solving for one pair of parameter values to expedite solving a neighboring problem. Such a neighboring problem would be the same BVP (1.27) with a slightly different pair of values for β and η. This leads to ideas of *continuation*, applicable in a natural way to many of the examples presented here, and discussed in Section 8.3.

The BVP (1.27) is not very difficult numerically. □

Example 1.13 Solitary Wave

The Fitzhugh-Nagumo equations are a simple mathematical model for the propagation of action potentials down the giant axon of the squid, Loligo:

$$V_t = V_{\xi\xi} + V - \frac{1}{3}V^3 - R + S$$

$$(t,\xi) \in D \subset \mathbf{R}^2$$

$$R_t = \phi(V + a - bR)$$

where t is time, ξ is distance, V is membrane potential, R is recovery variable, S is prescribed stimulating current and ϕ, a, b are given constants. Subscripts denote partial derivatives with respect to ξ and t.

Looking for travelling wave solutions, we introduce a single variable $x = \xi + ct$, $c > 0$, and obtain the problem

$$v'' - cv' - (V_R^2 - 1)v - r - V_R v^2 - \frac{1}{3}v^3 = 0 \tag{1.28a}$$

$$r' - \frac{\phi}{c}(v - br) = 0 \tag{1.28b}$$

$$v(-\infty) = v(\infty) = r(\infty) = 0 \tag{1.28c}$$

$$r(-\infty) = 0 \tag{1.28d}$$

Here, V_R is a given constant (rest state) and c is an unknown constant. Representative values for the constants are $a = 0.7$, $b = 0.8$, $\phi = 0.08$, $V_R = 1.1994080352440$. The sought solution is a single pulse solitary wave, but note that equations (1.28a, b, c, d) pin it down only to within a translation in x. Thus, to get a unique solution we treat c as another dependent variable and add the equation

$$c' = 0 \tag{1.28e}$$

and the boundary condition

$$v(0) = v_0 \neq 0 \tag{1.28f}$$

where v_0 is some nonzero value in the range of v.

Now, however, we have too many BC. The BC (1.28c, d) need to be replaced by three independent ones. Analysis (see Section 11.4.2) yields that one possible way to proceed is to drop (1.28d) (this is a redundant BC). The remaining BVP (1.28a, b, c, e, f) can then be solved for $-L \leq x \leq L$, with, say, $L = 70$. Note that we have here a *3-point* BVP. The BC are given at three points -L, 0, and L. □

Example 1.14 Nonlinear Elastic Beams

The deformation of a beam under the action of axial and transverse loading which is also resting on a nonlinear foundation is governed by the equations

$$x' = (1+e)\cos\theta \tag{1.29a}$$

$$y' = (1+e)\sin\theta \tag{1.29b}$$

$$s' = 1 + e \tag{1.29c}$$

$$\theta' = (1+e)\kappa \qquad 0 < t < L \tag{1.29d}$$

$$Q' = (1+e)[(ky - P)\cos\theta - \kappa T] \tag{1.29e}$$

$$M' = (1+e)Q \tag{1.29f}$$

$$T' = (1+e)[(ky - P)\sin\theta + \kappa Q] \tag{1.29g}$$

with $e = T/EA$, $\kappa = M/EI$, $P(t)$ and $k(y)$ are given functions and E, I, A are constants. Possible boundary conditions are

(i) Simple supports

$$y(0) = y(L) = 0, \ x(0) = 0, \ M(0) = M(L) = 0, \ s(0) = 0, \ T(0) = x_0 \tag{1.30a}$$

(with x_0 given)

(ii) Clamped ends

$$y(0) = y(L) = 0, \ x(0) = 0, \ \theta(0) = \theta(L) = 0, \ s(0) = 0, \ T(0) = x_0 \tag{1.30b}$$

(iii) Elastic support at the left end

$$Q(0) = K_L y(0), \ x(0) = 0, \ M(0) = -K_T \theta(0), \ s(0) = 0, \ T(0) = x_0 \tag{1.30c}$$

[with BC at $x = L$ as in (ii)].

This defines one BVP with three types of BC. Now, assuming that the deformation is inextensional, i. e., $e \equiv 0$ but $T \neq 0$ (which makes sense only after introducing appropriate scaling and taking appropriate limits) and introducing dimensionless variables

$$t := t/L, \quad y := y/L, \quad M := ML/EI, \quad T := T/X_0, \quad k := k/k_0$$

$$P := PL/x_0, \quad \lambda := \sqrt{k_0 L^2 / x_0}, \quad \varepsilon := \sqrt{EI/x_0 L^2}, \quad Q := Q/\varepsilon x_0$$

we get (1.29) in the form

$$x' = \cos\theta \tag{1.31a}$$

$$y' = \sin\theta \tag{1.31b}$$

$$\theta' = M \qquad 0 < t < 1 \tag{1.31c}$$

$$\varepsilon M' = -Q \tag{1.31d}$$

$$\varepsilon Q' = (\lambda^2 ky - P)\cos\theta - MT \tag{1.31e}$$

$$T' = (\lambda^2 ky - P)\sin\theta + \varepsilon MQ \tag{1.31f}$$

The last equation (1.31f) can be replaced by

$$T = \sec\theta + \varepsilon Q \tan\theta \tag{1.31g}$$

Note that the first equation (1.31a) is not coupled with the rest and may be integrated after we solve a 4th order system (1.31b, c, d, e) for y, θ, M and Q, using (1.31g) to substitute for T.

The BC are extracted, in an obvious way, from (1.30a), (1.30b) or (1.30c). For instance, the simple support BC are

$$y(0) = y(1) = 0, \quad M(0) = M(1) = 0$$

Also, for simplicity one can take $\lambda^2 k = P = 1$ in (1.31e).

When extension dominates bending, $\varepsilon \ll 1$ and boundary layers at $t = 0$ and at $t = 1$ appear in the solution. A first approximation to the solution corresponding to the boundary conditions (1.30a) which does not contain the boundary layers [they are, incidentally, of width $O(\sqrt{sec\ \theta_0})$] is given by the system

$$x_0' = \cos\theta_0$$
$$y_0' = \sin\theta_0$$
$$\theta_0' = M_0$$
$$M_0 = (\lambda^2 k y_0 - P)\cos^2\theta_0 \tag{1.32}$$
$$Q_0 = 0$$
$$x_0(0) = 0, \qquad y_0(0) = y_0(1) = 0$$

The BVP (1.32) is easy to solve numerically, whereas the full BVP (1.31) is not. □

Example 1.15 Semiconductors

One popular mathematical model for a semiconductor device in steady state consists of three second order differential equations. These are Poisson's equation for the potential ψ, a continuity equation for the electron current J_n, and a continuity equation for the hole current J_p. In one dimension they can be written as

$$\psi'' = \frac{q}{\varepsilon}(n - p - C(x)) \tag{1.33a}$$

$$J_n' = q\hat{R}(n,p) \qquad -l < x < l \tag{1.33b}$$

$$J_p' = -q\hat{R}(n,p) \tag{1.33c}$$

where n and p are the unknown electron and hole densities (of negative and positive charges, respectively), q, ε and l are known constants, $C(x)$ is a known doping profile function and $\hat{R}(n,p)$ is a given generation-recombination rate. The continuity equations (1.33b, c) become second-order ODEs for $n(x)$ and $p(x)$ upon use of the electron and hole current relations

$$J_n = q(D_n n' - \mu_n n \psi') \tag{1.33d}$$

$$J_p = -q(D_p p' + \mu_p p \psi') \tag{1.33e}$$

where D_n, D_p, μ_n and μ_p are additional diffusion and mobility functions, which we assume for simplicity to be known constants, satisfying $D_n/\mu_n = D_p/\mu_p = U_T$, with U_T a thermal voltage.

These ODEs for ψ, n and p are subject to boundary conditions

$$\psi(-l) = U_T \ln\frac{n_i}{p(-1)} + U \tag{1.33f}$$

$$\psi(l) = U_T \ln\frac{n(1)}{n_i} \tag{1.33g}$$

$$n(\pm l)p(\pm l) = n_i^2 \tag{1.33h}$$

$$n(\pm l) - p(\pm l) = C(\pm l) \tag{1.33i}$$

where U is the applied bias and n_i is an intrinsic number.

The BVP (1.33) is not well-scaled, because the doping profile may have values in a rather wide range, say $[-10^{14}, 10^{20}]$. Use of the scaling

$$D(x) := C(x)/\bar{C}, \qquad \bar{C} := \max |C(x)|, \qquad x := x/l, \qquad U := U/U_T$$

$$\lambda^2 := \frac{U_T \varepsilon}{l^2 \overline{C} q}, \quad (\lambda > 0), \quad \gamma := \frac{n_i}{\overline{C}}$$

and an appropriate scaling of the dependent variables gives

$$\lambda^2 \psi'' = n - p - D(x) \tag{1.34a}$$

$$(n' - n\psi')' = R(n,p) \tag{1.34b}$$

$$(p' + p\psi')' = R(n,p) \tag{1.34c}$$

One choice for R is the Shockley-Read-Hall term, which yields

$$R(n,p) = \frac{1}{4} \frac{np - \gamma^2}{n + p + 2\gamma} \tag{1.34d}$$

Typical values for the constants appearing in (1.34a, d) are $\lambda^2 = 0.4 \cdot 10^{-6}$, $\gamma = 10^{-7}$, and they can get as low as 10^{-10}. The BC are now

$$\psi(-1) = \ln \frac{\gamma}{p(-1)} + U, \quad \psi(1) = \ln \frac{n(1)}{\gamma} \tag{1.34e}$$

$$n(\pm 1) p(\pm 1) = \gamma^2 \tag{1.34f}$$

$$n(\pm 1) - p(\pm 1) = D(\pm 1) \tag{1.34g}$$

In a typical situation, we may consider $D(x)$ to be piecewise smooth. Locations of discontinuities in the doping profile are called *pn-* or *np-* junctions. Since λ is small, we may expect that sharp layers develop in the solution near the junctions. However, it is important to note that these are essentially boundary-type layers, unlike those in Examples 1.17, 1.23, and 1.24 below. In particular, there are no turning points here, despite the appearance of internal layers (cf. Chapter 10). The location of these junction layers is known, and a simple transformation of the independent variable can be used to transform them to the boundary. Thus suppose, for simplicity, that there is one discontinuity in $D(x)$ at $x = 0$. Then we may transform $[-1, 0] \to [0, 1]$ by $x := -x$. This yields three second-order ODEs in addition to the original (1.34a, b, c), obtaining a BVP of order 12 on (0, 1), with a boundary layer at $x = 0$.

The BVP (1.34) may be cautiously treated as a singular perturbation problem (cf. Chapter 10), but note that the boundary values in (1.34e) slowly blow up when $\lambda \to 0$ (i.e., when $\overline{C} \to \infty$, which also implies $\gamma \to 0$).

Due to the special form of the continuity equations (1.34b, c), some special transformations can be applied, which have proven useful for both theoretical and practical purposes. One such transformation is

$$n = \gamma e^{\psi - \phi_n}, \quad p = \gamma e^{\phi_p - \psi} \tag{1.35a}$$

The unknowns ϕ_n and ϕ_p replacing n and p are called (scaled) quasi-Fermi levels. This transformation yields in place of (1.34b, c),

$$\gamma (e^{\psi - \phi_n} \phi_n')' = R \tag{1.35b}$$

$$\gamma (e^{\phi_p - \psi} \phi_p')' = R \tag{1.35c}$$

The BVP in the new dependent variables turns out to have nicer properties for numerical approximation. A slight disadvantage is, however, that ϕ_n and ϕ_p do not appear linearly in (1.35b, c). A related transformation which yields linear forms (useful for analysis) is

$$n = \gamma e^{\psi} u, \quad p = \gamma e^{-\psi} v \tag{1.36a}$$

This gives

$$(e^{\Psi}u')' = R \quad (1.36b)$$

$$(e^{-\Psi}v')' = R \quad (1.36c)$$

The latter transformation is not without fault either, because it turns out that u and v are not sufficiently well-scaled for numerical use, and overflow often occurs in (1.36a).

The BVP in any of the forms (1.34) or (1.35) is numerically difficult, but not extremely so. It becomes much more computationally challenging in several independent variables. □

Example 1.16   Electron-Irradiated Silicon

Here is another BVP from semiconductor theory,

$$\varepsilon n' = (n+\beta p)[\alpha n \tilde{f} - \sum_{i=1}^{N_A} \hat{f}_i - \sum_{j=1}^{N_D} \bar{f}_j] \quad (1.37a)$$

$$0 < x < 1$$

$$\varepsilon p' = (n+\beta p)[\alpha p \tilde{f} + \frac{1}{\beta}\sum_{i=1}^{N_A} \hat{f}_i + \frac{1}{\beta}\sum_{j=1}^{N_D} \bar{f}_j] \quad (1.37b)$$

$$n(0) = 1, \quad p(1) = 0 \quad (1.37c)$$

Here $n(x)$ and $p(x)$ are as in the previous example and ε is a normalized current density. Values of interest for ε range from 1 to 10^{-12}. The functions and constants appearing on the right-hand sides of (1.37a, b) are given by

$$\tilde{f} = 1-n+p-\sum_{i=1}^{N_A} a_i(x)\frac{n+\alpha_i u_i}{n+v_i+\alpha_i(u_i+p)} + \sum_{j=1}^{N_D} d_j(x)\frac{z_j+\delta_j p}{n+z_j+\delta_j(y_j+p)} \quad (1.37d)$$

$$\hat{f}_i = \alpha_i A_i a_i(x)\frac{np-v_i u_i}{n+v_i+\alpha_i(u_i+p)} \quad \bar{f}_j = \delta_j D_j d_j(x)\frac{np-y_j z_j}{n+z_j+\delta_j(y_j+p)} \quad (1.37e)$$

$$\beta = 1/3, \quad N_A = 2, \quad N_D = 1, \quad \alpha = 0.05162 \quad (1.37f)$$

$$\alpha_i = \delta_j = 1, \quad A_i = D_j = 2.222 \cdot 10^{-3} \quad \text{all } i,j$$

$$a_1(x) = 15, \quad a_2(x) = 10, \quad d_1(x) = 400$$

$$u_1 = 1.854 \cdot 10^{-4}, \quad u_2 = 0.1021, \quad v_1 = 21.47, \quad v_2 = 3.899 \cdot 10^{-2}$$

$$y_1 = 2.902 \cdot 10^3, \quad z_1 = 1.371 \cdot 10^{-6}$$

For ε small, this BVP has a boundary layer at $x=0$ and an interior (turning point) layer near $x=1$. □

Example 1.17   Shock Wave

Consider a shock wave in a one-dimensional nozzle flow. The steady state Navier-Stokes equations give

$$\varepsilon A(x)uu'' - [1+\frac{\gamma}{2}-\varepsilon A'(x)]uu' + u'/u + \frac{A'(x)}{A(x)}(1-\frac{\gamma-1}{2}u^2) = 0 \quad 0 < x < 1 \quad (1.38a)$$

where x is the normalized downstream distance from the throat, u is a normalized velocity, $A(x)$ is the area of the nozzle at x, e.g., $A(x) = 1+x^2$, $\gamma = 1.4$, and ε is essentially the inverse of Reynolds number, e.g., $\varepsilon = 4.792 \cdot 10^{-8}$. The BC are

$$u(0) = 0.9129 \quad \text{(supersonic flow in throat)} \quad (1.38b)$$

$$u(1) = 0.375 \quad (1.38c)$$

Given its simple appearance, the BVP (1.38a, b, c) turns out to be a surprisingly difficult nut to crack numerically. An $O(\sqrt{\varepsilon})$-wide shock develops, whose location depends on ε. Singular-perturbation-type problems usually require a *continuation method* to solve them; i. e., the problem is solved successively for a decreasing sequence of values of ε, thereby permitting a methodical refinement of the solution profile (and adjustment of certain parameters of the numerical method). For this BVP, however, many ε-steps need to be taken. (This, of course, depends also on the particular numerical method used.) □

Example 1.18   Swirling Flow I

Consider the steady flow of a viscous, incompressible axisymmetric fluid ("swirling" flow) above an infinite rotating disk. Using a cylindrical coordinate system (r, ϕ, z), the disk is rotating at $z = 0$ with angular velocity Ω, and the fluid has angular velocity $\gamma\Omega$ at $z = \infty$. Defining

$$x = \sqrt{\Omega/\nu}\, z$$

where ν is viscosity, we find that the Navier-Stokes equations yield by similarity transformation,

$$f''' + 2ff'' - (f')^2 + g^2 = \gamma^2 \quad (1.39a)$$

$$0 < x < \infty$$

$$g'' + 2fg' - 2f'g = 0 \quad (1.39b)$$

with the velocity field of the fluid given by $(\Omega r f'(x), \Omega r g(x), -2\sqrt{\nu\Omega}f(x))$. The BC are

$$f(0) = 0, \quad f'(0) = 0, \quad g(0) = 1 \quad (1.39c)$$

$$f'(\infty) = 0, \quad g(\infty) = \gamma \quad (1.39d)$$

The task at hand is to find solutions to (1.39) as γ, called the Rossby number, varies. The value $\gamma = 0$ is of particular interest.

It turns out that there are (possibly infinitely) many solutions to this problem for $\gamma = 0$. To find many solutions to a BVP, parameter continuation techniques are used (see Chapter 8). □

Example 1.19   Swirling Flow II

We consider again a swirling flow over an infinite disk, but now the azymuthal velocity behaves like r^{-n} (so $n = -1$ corresponds to a solid body rotation and $n = 1$ corresponds to a potential vortex) and a magnetic field is applied in the direction of the axis of rotation. The disk is stationary. The resulting BVP is

$$f''' + 1/2(3-n)ff'' + n(f')^2 + g^2 - sf' = \gamma^2 \quad (1.40a)$$

$$0 < x < \infty$$

$$g'' + 1/2(3-n)fg' + (n-1)gf' - s(g-1) = 0 \quad (1.40b)$$

$$f(0) = f'(0) = g(0) = 0, \quad f'(\infty) = 0, \quad g(\infty) = \gamma \quad (1.40c)$$

Note that the ODEs (1.39a, b) are a special case of (1.40a, b) with $n = -1$, $s = 0$. Let us take $\gamma = 1$. Whereas a numerical solution of (1.39) is not difficult to obtain (the difficulty there is of a different kind, namely, obtaining *many* solutions for $\gamma = 0$), the BVP (1.40) becomes tough as $n \uparrow 1$ and $s \downarrow 0$. In fact, it can be shown that when $s = 0$ no solution exists for $n=1$, and it is believed that no solution exists if $n > n_0 \approx 0.1217$.

Determining n_0, as well as solving (1.40) for values like $s = 0.05$, $n = 0.3$ lead to difficult numerical tasks. □

Example 1.20   Swirling Flow III

This time we consider the swirling flow between two rotating, coaxial disks, located at $x = 0$ and at $x = 1$. The BVP is

$$\varepsilon f'''' + f''' + g' = 0 \qquad (1.41a)$$
$$0 < x < 1$$
$$\varepsilon g'' + fg' - f'g = 0 \qquad (1.41b)$$
$$f(0) = f(1) = f'(0) = f'(1) = 0 \qquad (1.41c)$$
$$g(0) = \Omega_0, \qquad g(1) = \Omega_1 \qquad (1.41d)$$

where Ω_0, Ω_1 are the angular velocities of the infinite disks, $|\Omega_0| + |\Omega_1| \neq 0$, and ε is a viscosity parameter, $0 < \varepsilon \ll 1$. Thus we have an interesting singular perturbation problem, which becomes numerically difficult for ε small (say $\varepsilon \leq 10^{-4}$), i.e., for large Reynolds numbers. Multiple solutions are possible. Taking, e.g., $\Omega_1 = 1$, we obtain different cases for different values of Ω_0. If $\Omega_0 < 0$ (with a special symmetry when $\Omega_0 = -1$) then the disks are counter-rotating; if $\Omega_0 = 0$ then one disk is at rest, while if $\Omega_0 > 0$ then the disks are corotating. □

Example 1.21   Re-entry of a Space Vehicle

In this optimal control problem, a control $u(t)$ has to be chosen as a function of time t, to minimize the heating

$$\int_0^T 10 \, v^3 \sqrt{\rho} \, dt$$

which a space vehicle experiences during the flight through the earth's atmosphere on the way back from outer space. In this functional, T is an unspecified final time, v is velocity and $\rho = \rho_0 e^{-\beta R \xi}$ is atmospheric density, $\rho_0 = 2.704 \cdot 10^3$, $R = 209$, $\beta = 4.26$. The minimization of the functional is subject to the equations of state

$$\frac{dv}{dt} \equiv \dot{v} = -s\rho v^2 C_D(u) - \frac{g \sin \gamma}{(1+\xi)^2} \qquad (1.42a)$$

$$\dot{\gamma} = s\rho v C_L(u) + \frac{v \cos \gamma}{R(1+\xi)} - \frac{g \cos \gamma}{v(1+\xi)^2} \qquad (1.42b)$$

$$\dot{\xi} = \frac{v \sin \gamma}{R} \qquad (1.42c)$$

$$v(0) = 0.36, \qquad \gamma(0) = -8.1 \cdot \pi/180, \qquad \xi(0) = 4/R \qquad (1.42d)$$

$$v(T) = 0.27, \qquad \gamma(T) = 0, \qquad \xi(T) = 2.5/R \qquad (1.42e)$$

where γ is the flight-path angle, ξ is a normalized altitude, $s = 26{,}600$, $g = 3.2172 \cdot 10^{-4}$, $C_D(u) = 1.174 - 0.9 \cos u$, $C_L(u) = 0.6 \sin u$.

To solve the optimization problem we use three adjoint variables (Lagrange multipliers) λ_v, λ_γ and λ_ξ (which are functions of t) and form the Hamiltonian

$$H = 10v^3 \sqrt{\rho} + \lambda_v \dot{v} + \lambda_\gamma \dot{\gamma} + \lambda_\xi \dot{\xi}$$

where for $\dot{v}$, $\dot{\gamma}$ and $\dot{\xi}$ we use the right-hand sides of (1.42a, b, c). Then by calculus of variations we have the ODEs

$$\dot{\lambda}_v = -\frac{\partial H}{\partial v} \qquad (1.42\text{f})$$

$$\dot{\lambda}_\gamma = -\frac{\partial H}{\partial \gamma} \qquad (1.42\text{g})$$

$$\dot{\lambda}_\xi = -\frac{\partial H}{\partial \xi} \qquad (1.42\text{h})$$

and a terminal BC

$$H = 0 \quad \text{at } t = T \qquad (1.42\text{i})$$

In (1.42) we have a *free* BVP (free flight time T) with 6 ODEs and 7 BC. To obtain the problem in standard form we can transform the independent variable

$$x = t/T$$

and treat T as another dependent variable, adding the ODE

$$\frac{dT}{dx} \equiv T' = 0 \qquad (1.43\text{a})$$

Using the already defined right-hand sides we write the remaining equations of the BVP as

$$v' = \dot{v}T, \qquad \gamma' = \dot{\gamma}T, \qquad \xi' = \dot{\xi}T \qquad (1.43\text{b})$$

$$\lambda_v' = \dot{\lambda}_v T, \quad \lambda_\gamma' = \dot{\lambda}_\gamma T, \quad \lambda_\xi' = \dot{\lambda}_\xi T \qquad (1.43\text{c})$$

The BVP is then the ODEs (1.43) subject to the BC (1.42d, e, i) (with $x = 1$ replacing $t = T$).

The numerical difficulty in this problem is of a somewhat different character than that of the previous example. Here there is no strong singular perturbation feature, however the nonlinear problem is sensitive. Convergence of a numerical technique using some variant of Newton's method can be expected only if the initial iterate (i. e., an initial solution profile which a user has to guess in a — we hope — educated way) is fairly close to the solution. □

Example 1.22 Optimal Harvesting

This problem arises in the optimal harvesting of a randomly fluctuating resource. The objective is to choose a harvesting effort function $y = y(x)$, $y_- \leq y \leq y_+$, so as to maximize the present value, $v(x)$, of the resource. The maximum principle gives that

$$y(x) = \begin{cases} y_- & n(x) < v'(x) \\ y_+ & n(x) > v'(x) \end{cases} \qquad (1.44\text{a})$$

where, e. g., $n(x) = e^x$. The present value v satisfies

$$\varepsilon v'' + (f(x) - y(x))v' - \gamma v + n(x)y(x) = 0 \qquad -\infty < x < \infty \qquad (1.45\text{a})$$

where $f(x) = 1 - e^x$ and the discount rate γ is a parameter, $0 < \gamma < 1$. The BC for (1.45a) are that the solution v be bounded as $|x| \to \infty$.

The problem may at a first glance look linear, but in fact it is not, because even though the values of $y(x)$ are known, the *switching points s*, where

$$n(s) = v'(s) \qquad (1.44b)$$

are not. Assume further that there is only one switching point s and that $-\infty < y_- < y_+ < \infty$ (other cases may be similarly treated). Then the BC are

$$(f(-L) - y_-)v'(-L) - \gamma v(-L) + n(-L)y_- = 0 \qquad (1.45b)$$

$$(f(L) - y_+)v'(L) - \gamma v(L) + n(L)y_+ = 0 \qquad (1.45c)$$

with $L > 0$, sufficiently large (e.g., $L = 10$).

One may attempt to solve the BVP (1.45a, b, c) numerically, but this is not simple because of the jump in $y(x)$ at the unknown point s (where v and v' are continuous!). Thus, it is preferable to transform s to a known location, say 0,

$$x := x - s$$

This yields

$$\varepsilon v'' + (f(x+s) - y)v' - \gamma v + n(x+s)y = 0 \qquad (1.46a)$$

$$y = \begin{cases} y_- & x < 0 \\ y_+ & x > 0 \end{cases} \qquad (1.46b)$$

$$s' = 0 \qquad (1.46c)$$

$$v'(0) = n(s(0)) \qquad (1.46d)$$

and (1.45b, c). The obtained three-point BVP may now be solved, e.g. by a finite difference technique with a mesh point at $x = 0$ (cf. Chapter 5). It is not very difficult anymore. □

Example 1.23 Spherical Shells

Consider a homogeneous, isotropic, thin spherical shell of constant thickness, subject only to an axisymmetric normal distributed surface load. With ξ the angle between the meridional tangent at a point of the midsurface of the undeformed shell and the base plane, ϕ the meridional angle change of the deformed middle surface, $\beta = \xi - \phi$, and ψ a stress function, the following BVP governs the deformation elastostatics of the shell,

$$\mu[\psi'' + \cot\xi\,\psi' + (\nu - \cot^2\xi)\psi] - \frac{1}{\sin\xi}(\cos\beta - \cos\xi) \qquad (1.47a)$$

$$= \mu[\nu P' + (1+\nu)P\cot\xi - \frac{1}{\sin\xi}(\gamma\sin^2\xi)' - \nu\gamma\cos\xi] \qquad 0 \le \xi \le \pi/2$$

$$\varepsilon^4/\mu[\phi'' + \cot\xi\,\phi' + \frac{\cos\beta}{\sin^2\xi}(\sin\beta - \sin\xi) - \frac{\nu}{\sin\xi}(\cos\beta - \cos\xi)] + \frac{\sin\beta}{\sin\xi}\psi = \frac{\cos\beta}{\sin\xi}P \qquad (1.47b)$$

$$\phi(0) = \psi(0) = \phi(\pi/2) = \psi(\pi/2) = 0 \qquad (1.47c)$$

Here

$$P(\xi) = -\int_0^\xi (1 - \delta\sin\eta)\cos\beta\,\sin\eta\,d\eta, \qquad \gamma = -\sin\beta(1 - \delta\sin\xi)$$

and $\delta > 1$ a constant, say $\delta = 1.2$ (we have assumed a particular load distribution). Also $\nu = 0.3$ is a typical value.

To evaluate $P(\xi)$ we introduce the simple trick of incorporating it as another ODE and BC

$$P' = -(1-\delta\sin\xi)\cos\beta\sin\xi \tag{1.47d}$$

$$P(0) = 0 \tag{1.47e}$$

The BVP (1.47) is now in standard form.

The parameters ε and μ are positive and small (they relate to the thickness vs radius of the shell). The solution sought has an interior layer in ϕ (i.e., a narrow region in ξ, away from the boundaries, where ϕ varies fast), corresponding to a *dimpling* of the spherical shell.

Numerically, the problem gets tougher as ε and μ get smaller. Some representative (ε,μ)-values for which the problem is fairly difficult are (0.01, 0.0001), (0.001, 0.001), (0.0001, 0.01). □

Example 1.24   Shallow Cap Dimpling

This is another example from the theory of shells of revolution. The ODEs are

$$\varepsilon^2[\psi'' + \frac{1}{x}\psi' - \frac{1}{x^2}\psi] - \frac{1}{x}\phi(\phi_0 - \frac{1}{2}\phi) = 0 \tag{1.48a}$$

$$0 < x < 1$$

$$\varepsilon^2[\phi'' + \frac{1}{x}\phi' - \frac{1}{x^2}\phi] + \frac{1}{x}\psi(\phi_0 - \phi) = 2\kappa P(x) \tag{1.48b}$$

with ϕ and ψ essentially as in the previous example; $\phi_0(x)$ is ϕ of the undeformed shell (for a spherical shell $\phi_0(x) = x$, but consider also $\phi_0(x) = x^m$, $m = 2, 3$), and

$$P(x) = x(1 - \gamma + \frac{\gamma}{2}x^2)$$

$\gamma = 1.2$, $\nu = 0.3$, $\kappa = 1$. The BC are

$$\phi(0) = \psi(0) = 0, \qquad \phi(1) = \psi'(1) - \nu\psi(1) = 0 \tag{1.48c}$$

As in the previous example, the BVP gets tough as ε gets small, and an interior layer (corresponding to dimpling) forms in a solution for ϕ. There is an additional boundary layer at $x = 1$, and more than one solution exist. The value $\varepsilon = 10^{-4}$ (which gives a rather thin shell) yields a challenging numerical problem. □

Example 1.25   Burner-Stabilized Flame

A simple, two-stage, unimolecular, one-dimensional flame may be represented by the mechanism

$$Y_r \to Y_i \to Y_p$$

Here Y_r and Y_i are the mass fraction concentrations of the reactant and intermediate, respectively. The product concentration Y_p is determined from conservation of mass by

$$Y_p = 1.0 - Y_r - Y_i$$

After appropriate coordinate transformations and nondimensionalizations, the system can be described under steady-state conditions by the BVP

$$M_0 \frac{dY_r}{dx} = \frac{1}{Le_r} \frac{d^2Y_r}{dx^2} - k_{ri} Y_r \exp(-E_r/T) \qquad (1.49\text{a})$$

$$M_0 \frac{dY_i}{dx} = \frac{1}{Le_i} \frac{d^2Y_i}{dx^2} + k_{ri} Y_r \exp(-E_r/T) - k_{ip} Y_i \exp(-E_i/T), \qquad 0 < x < \infty \qquad (1.49\text{b})$$

$$M_0 \frac{dT}{dx} = \frac{d^2T}{dx^2} + k_{ri}(h_r - h_i) Y_r \exp(-E_r/T) + k_{ip} h_i Y_i \exp(-E_i/T) \qquad (1.49\text{c})$$

$$Y_r(0) - \frac{1}{M_0 Le_r} \frac{dY_r}{dx}(0) = \varepsilon_r \qquad (1.49\text{d})$$

$$Y_i(0) - \frac{1}{M_0 Le_i} \frac{dY_i}{dx}(0) = \varepsilon_i \qquad (1.49\text{e})$$

$$T(0) = T_0 \qquad (1.49\text{f})$$

$$\frac{dY_r}{dx}(\infty) = \frac{dY_i}{dx}(\infty) = \frac{dT}{dx}(\infty) = 0 \qquad (1.49\text{g})$$

where T denotes the temperature. Typical values for the problem constants are as follows: the preexponential constants $k_{ri} = 5 \times 10^8$ and $k_{ip} = 10^2$, the activation energies $E_r = 80$ and $E_i = 10$, the Lewis numbers $Le_r = 0.75$ and $Le_i = 1.25$, the specific enthalpy differences between the reactant and the product $h_r = 4.4$ and between the intermediate and product $h_i = 4.5$, the mass flux fractions $\varepsilon_r = 1$ and $\varepsilon_i = 0$, the burner temperature $T_0 = 1.25$, and the initial mixture flow rate $M_0 = 0.985$. Quantities such as the thermal conductivity and the specific heat capacity do not appear explicitly in the model, because they are contained in the other nondimensionalized variables. In practice the solution domain is truncated at a large value of $x = L$ such that the zero gradient boundary conditions are "satisfied" (cf. Section 11.4.2); for the given parameter values, $L = 10$ is sufficient.

Asymptotic analyses of this reaction-diffusion system can be performed and resulting analytical expressions for the flame velocity and species concentrations obtained for various values of activation energies, preexponential constants, and Lewis numbers. Although the three-species model is rather schematic and has limited ability to account for the behavior of real complex flames, it still contains terms that represent the main processes occurring in larger, many-species, reaction-diffusion systems. □

2
Review of Numerical Analysis and Mathematical Background

The numerical solution of BVPs relies on a significant amount of background knowledge. This involves a variety of areas in numerical analysis, as well as some mathematical background in addition to ODE theory. Because this background material is varied, we provide here a brief survey as a convenient review or for later referencing. Most of this material appears in general numerical analysis textbooks (see references at the end of the book) and our intention here is certainly not to replace such books. For a reader with anything more than a superficial knowledge of numerical analysis, we recommend that this chapter be skipped at first and the material in it later perused as needed.

2.1 ERRORS IN COMPUTATION

Numerical analysis, the study of methods for finding approximate quantitative solutions to mathematically posed problems, includes the study of the resulting errors and bounds on the errors. These errors are of three basic types. The first, *input errors,* are the errors in the input data for a problem, caused for instance by imperfections in physical measurements. The other two, *roundoff error* and *discretization* or *truncation error,* are types which arise in the actual numerical computation.

Certain errors result from the fact that the computation is being performed in finite precision on a digital computer. That is, the computation is performed with quantities whose representation is restricted to a finite number of digits. This is the source of *roundoff error,* which is defined as the amount by which the computed solution differs from the solution one would get if exact arithmetic were used. The computations are typically performed with floating point arithmetic in base 2 or a power of 2. The number of digits used to represent floating point numbers can vary considerably from one computer to another. Moreover, often both short word length ("single precision") and long word length ("double precision") are provided by the hardware.

A useful machine-dependent quantity for measuring the effects of roundoff errors is the *machine epsilon* (or *machine unit roundoff*) ε_M. It is the smallest floating point quantity which, when added to 1, gives a sum which is greater than 1 (in the floating point arithmetic). One can show that the error in representing any number y with a floating point number $\bar{y}$ is bounded by

$$|y - \bar{y}| \leq \varepsilon_M |y|$$

The errors present in any computation are of a particular concern when a solution process is sensitive to small perturbations in the computation. There can be two major causes for such sensitivity to small perturbations. The first is when the problem being solved is *ill conditioned* (equivalently, not *well conditioned*); i.e., small perturbations in the input cause large perturbations in the output. This is considered for linear systems of algebraic equations in Section 2.2.6. Such difficulty cannot be circumvented by using a different algorithm for obtaining the solution. The other major cause of numerical sensitivity is *instability* in the algorithm used. This means that the output obtained cannot be associated with the exact solution of a nearby, slightly perturbed problem. A better algorithm should then be sought.

The concepts of well-conditioning and stability are fundamental to this book, and we will return to them again and again. Here they are introduced on an intuitive level. Later on, in Section 2.8.2, we will try to quantify them and discuss them further in a more restricted context.

Typically, a numerical method is designed only to *approximate* the exact solution to a given problem, and this gives rise to *discretization* or *truncation errors*. Thus, this is the error that would arise in the absence of input or roundoff errors. For example consider an interpolation process (see Section 2.4). The interpolation points can correspond to known values of an unknown function; another function, usually an interpolating polynomial or piecewise polynomial, is constructed which passes through these data points. The discretization error is then a measure of the amount that these two functions differ between the data points.

As another example, consider the problem of approximating the solution of the IVP

$$y'(x) = f(x,y) \qquad x > a \tag{2.1a}$$

$$y(a) = \alpha \tag{2.1b}$$

This is considered in more detail in Section 2.7. Numerical methods typically define an approximate solution to the IVP in terms of some discrete (finite) set of parameters, e.g., the approximate solution values at the points

$$x_i := a + (i-1)h \qquad (i = 1, 2, \cdots)$$

for some $h > 0$. Call y_i the discrete approximation to $y(x_i)$. The simplest method for solving (2.1a, b) is *Euler's method*, which defines these approximations to the solution values by

$$y_1 = \alpha \qquad (2.2a)$$

$$y_{i+1} = y_i + h f(x_i, y_i) \qquad i = 1, 2, \ldots \qquad (2.2b)$$

This approximation comes about by approximating the derivative $y'(x_i)$ by the divided difference $\dfrac{y(x_{i+1}) - y(x_i)}{h}$. Specifically, assuming that $y(x_i) = y_i$, the derivative approximation used in (2.1a) at $x = x_i$ yields the approximation y_{i+1} to $y(x_{i+1})$ given in (2.2b). Even if $y_i = y(x_i)$ exactly, there is a truncation error introduced in calculating y_{i+1}.

In a numerical algorithm where a large number of elementary arithmetic operations must be performed but the representation uses only approximates the desired values, we need to be aware of not only these error sources, but their overall impact on the result as well. This brings up the distinction between *local errors* and *global errors*. A local error arises at the *source;* local errors are input errors, rounding errors or errors caused by discretizing some analytical expression [as in (2.2b)]. The global error is the error in the final numerical result. In (2.1) we may have, e.g., an input error or rounding error in α, say δ. Then instead of $y(a)$ we have

$$\bar{y}(a) = y(a) + \delta$$

The initial value $y(a) + \delta$ defines, in principle, a solution of (2.1), say $\bar{y}(x)$. The local error δ thus induces a global error $\bar{y}(x) - y(x)$. Also, if we had the exact solution y at x, then proceeding from x to $x + h$ by Euler's method would give the value $\bar{y}(x + h) = y(x) + hf(x, y(x))$. By a Taylor's series argument we must have for the exact value $y(x+h)$,

$$y(x + h) = y(x) + hf(x, y(x)) + \frac{h^2}{2} y''(\xi) \qquad \xi \in (x, x+h)$$

so

$$\frac{y(x+h) - y(x)}{h} - f(x, y(x)) = \frac{h}{2} y''(\xi)$$

and there is a local (discretization) error $\dfrac{h}{2} y''(\xi)$. In practice we generally have a perturbed value of $y(x)$, say $\bar{y}(x)$, which is used in (2.2b) (and therefore lies on a different integral curve). Hence in Euler's method we move to a (usually) different integral curve each step. As a consequence the computed value $\bar{y}(x + h) = y_{i+1}$ may deviate substantially from the exact $y(x + h)$. The *global discretization error* is the total effect of all the local discretization errors, "propagated" by the differential equation. For Euler's method it is $O(h)$.

When applying Euler's method (2.2) to integrate the ODE in (2.1) from $x = 0$ to $x = 1$ say, there is not only a discretization (truncation) error, but also a roundoff error accumulation. For h small enough, the relative error due to roundoff is proportional to ε_M / h. Thus, as the step size is refined ($h \to 0$), the global discretization error tends to 0, but the roundoff error blows up. This is typical, and is essentially due to the differentiation operation appearing in (2.1a). In practice one has to choose the machine word length to be long enough (thus making ε_M small enough) so that discretization error dominates roundoff error for the value of h used.

In connection with global errors, two other concepts are important and indeed fundamental for numerical analysis in general. First, if the global error is much larger than the sum of local errors for a well conditioned problem, then the algorithm is "unstable."

Example 2.1

Let $f(x,y) = -100y + 100$ in (2.1a) and $y(0) = 1$ in (2.1b). Then apparently $y(x) = 1$ for all x, and small perturbations in $y(0)$ or in f do not affect the solution much. But choosing $h = 0.1$ in Euler's method, we obtain the recursion [c.f. (2.2b)]

$$y_{i+1} = -9y_i + 10, \quad y_1 = 1 \tag{2.3}$$

Now supposing $\bar{y}_1 = 1 + \delta$, we obtain a perturbed solution $\{\bar{y}_i\}$, satisfying

$$\bar{y}_{i+1} = -9\bar{y}_i + 10, \quad \bar{y}_1 = 1 + \delta$$

From (2.3) $\bar{y}_{i+1} - y_{i+1} = -9(\bar{y}_i - y_i)$, with $\bar{y}_1 - y_1 = \delta$. Hence we deduce that $|\bar{y}_{i+1} - y_{i+1}| = |9^i \delta|$. For i large this global error $|9^i \delta|$ becomes large compared to $|\delta|$, from which we conclude that (2.3) is unstable. Other perturbations than for y_1 will exhibit the same phenomenon. □

The second important concept is related to the amount of work involved in using an algorithm. Usually one is given a certain prescribed accuracy, the *tolerance,* which has to be achieved by the approximation; i.e., the tolerance should exceed the global error (in an absolute sense, or a relative sense, or a combination of both). Naturally, one wishes to design an algorithm with a minimal *complexity,* that is one which requires the minimum number of elementary arithmetic operations, usually counted in terms of one multiplication followed by an addition, the *flop.* An algorithm with a complexity reasonably close to the minimum is generally satisfactory. For a problem like (2.1), the design of a fast algorithm involves choices of the scheme as well as the step size used for discretization. It is often an undesirable property of an algorithm if the accuracy is much better than required, as this may involve more work than needed for lower accuracy. Consider, for instance, Euler's method (2.2). The cost of applying one step of (2.2b) is rather low, but the step size h may be restricted by stability requirements. Indeed for the example above we saw that $h = 0.1$ is too large a step. In Section 2.7 we shall see that Euler's method for this problem is stable only if $100h \leq 2$. Bearing in mind this stability restriction, the optimal step size to approximate $y(10)$ at minimal cost is $h = \frac{1}{50}$; i.e., we need 500 steps. This simple example shows that stability and optimal complexity may be conflicting interests. Of course, one may consider the optimality of the algorithm complexity for a set of problems with respect to an entire class of methods. One should not expect, however, that this necessarily leads to the selection of *one* superior algorithm. On the contrary, as will be demonstrated in this book there are usually a number of characteristics that make an algorithm more powerful for certain problems than for others, and it is one of our purposes to shed some light on these.

2.2 NUMERICAL LINEAR ALGEBRA

Numerical linear algebra is a very fundamental and important area of numerical analysis. Linear systems arise in most computations (since even if the problem is non-linear, one usually "solves" it by linearization). Here we shall treat some basic material, with an emphasis on what is needed later on.

2.2.1 Eigenvalues, transformations, factorizations, projections

In this section, unless otherwise specified, A is an $n \times n$ real matrix. We first define some basic concepts. The scalars $\lambda_1, \ldots, \lambda_n$ are the *eigenvalues* of A if $\psi(\lambda_j, A) := \det(A - \lambda_j I) = 0$, $j = 1, \ldots, n$. $\psi(\lambda, A)$ is called the *characteristic polynomial* of A and $\max |\lambda_j| =: \rho(A)$, the *spectral radius*. With each eigenvalue λ_j we associate an *eigenvector* $\mathbf{v}_j$, i.e.,

$$A \mathbf{v}_j = \lambda_j \mathbf{v}_j \qquad j = 1, \ldots, n \tag{2.4}$$

The eigenvalues and eigenvectors are in general complex-valued, but if A is symmetric then they are guaranteed to be real.

Given any polynomial $\phi(\lambda) = \lambda^n + \sum_{i=0}^{n-1} b_{n-i} \lambda^i$, we can find a matrix C such that $\det(C - \lambda I) = \phi(\lambda)$, viz.

$$C = \begin{bmatrix} 0 & 1 & 0 & \cdots & 0 \\ 0 & 0 & 1 & \cdots & 0 \\ \vdots & \vdots & \vdots & & \vdots \\ 0 & 0 & 0 & \cdots & 1 \\ -b_n & -b_{n-1} & -b_{n-2} & \cdots & -b_1 \end{bmatrix} \tag{2.5}$$

Such a matrix is called a *companion matrix*. It also arises naturally when reducing a scalar higher order ODE to a first-order system (see Section 1.1.2).

Two matrices A and B are called *similar* if there is a nonsingular matrix T such that

$$B = T^{-1} A T \tag{2.6}$$

The matrix T is called a *similarity transformation*. Let the n eigenvectors, $\mathbf{v}_1, \ldots, \mathbf{v}_n$, satisfy (2.4). Defining a matrix T

$$T := (\mathbf{v}_1 | \cdots | \mathbf{v}_n), \tag{2.7a}$$

whose columns are the eigenvectors, and a diagonal matrix Λ

$$\Lambda := \begin{pmatrix} \lambda_1 & & 0 \\ & \lambda_2 & \\ & & \ddots \\ 0 & & \lambda_n \end{pmatrix} \equiv \text{diag}\{\lambda_1, \ldots, \lambda_n\} \tag{2.7b}$$

we see that

$$AT = T\Lambda \qquad (2.7c)$$

If T is nonsingular, i. e., the eigenvectors of A are linearly independent, then by (2.7c) A is similar to the diagonal matrix Λ.

A more general situation arises when A is *defective;* that is, A does not have n linearly independent eigenvectors. This situation may arise when there are fewer than n distinct eigenvalues. In general we can still transform A to a *Jordan canonical form,* where for some nonsingular T,

$$AT = T \begin{bmatrix} \Lambda_1 & & \\ & \ddots & \\ & & \Lambda_k \end{bmatrix} \qquad k \leq n \qquad (2.8a)$$

with Λ_j an $n_j \times n_j$ Jordan block ($\sum_{j=1}^{k} n_j = n$)

$$\Lambda_j = \begin{bmatrix} \lambda_j & 1 & & 0 \\ & \ddots & \ddots & \\ & & \lambda_j & 1 \\ 0 & & & \lambda_j \end{bmatrix} \qquad (2.8b)$$

The number of times that a particular eigenvalue λ appears in the sequence $\lambda_1, \ldots, \lambda_n$ is called its *algebraic multiplicity,* whereas the largest number of linearly independent eigenvectors associated with λ is called its *geometric multiplicity.* Thus it follows that if λ_j is distinct from all other eigenvalues outside its Jordan block Λ_j then it has algebraic multiplicity n_j and geometric multiplicity 1 (because it has a one-dimensional *eigenspace*). A vector $\mathbf{w}$ is a *principal vector of grade* $p \geq 0$ belonging to the eigenvalue λ_j if $(\lambda_j I - A)^p \mathbf{w} = \mathbf{0}$ and there is no nonnegative integer $q < p$ for which $(\lambda_j I - A)^q \mathbf{w} = \mathbf{0}$. Thus, eigenvectors are of grade one, and each column of T in (2.8a) is a principal vector (why?).

If the elements of A are strongly weighted along the main diagonal, the following theorem is useful for estimating the location of the eigenvalues:

Theorem 2.9 (Gershgorin) Every eigenvalue λ of A satisfies at least one of the inequalities

$$|\lambda - a_{ii}| \leq \sum_{\substack{j=1 \\ j \neq i}}^{n} |a_{ij}| \qquad i=1, \ldots, n \qquad (2.9)$$

□

From a computational point of view and for geometrical insight, the form (2.8) is often unsatisfactory. In practice we usually restrict the class of similarity transformations considered to orthogonal or unitary matrices. A square, nonsingular matrix Q is *orthogonal* if $Q^T = Q^{-1}$, i.e., $Q^T Q = I$. Two popular 2×2 orthogonal matrices correspond to reflection and rotation of a vector by an angle θ. The *reflection* transformation is

$$Q = \begin{bmatrix} \cos\theta & \sin\theta \\ \sin\theta & -\cos\theta \end{bmatrix} \quad (2.10a)$$

and the *rotation* transformation is

$$Q = \begin{bmatrix} \cos\theta & \sin\theta \\ -\sin\theta & \cos\theta \end{bmatrix} \quad (2.10b)$$

The requirement $Q^T Q = I$ is useful when Q is real. If Q has complex elements, then the desired property is that $Q^H Q = I$, where Q^H is the *conjugate* transpose, i. e., that Q be *unitary*. In the sequel, whenever we refer to an orthogonal matrix, it is to be understood that the matrix is unitary if it is not real. Next we consider a very important matrix factorization method. Here and elsewhere it is helpful to look at a matrix as an n-tuple of column vectors [see also (2.7a)].

Theorem 2.11 (Gram-Schmidt) A general $m \times n$ matrix A can be factored as $A = QU$, where Q is orthogonal and U is upper triangular, (i. e., $u_{ij} = 0$ for $i > j$). If A is real, then Q and U are real. If A is nonsingular and $A = Q_1 U_1 = Q_2 U_2$ (Q_1, Q_2 orthogonal and U_1, U_2 upper triangular), then there is a diagonal matrix D whose diagonal elements are ± 1 such that $Q_1 = Q_2 D$ and $U_2 = DU_1$.

Proof: It is instructive to show part of the proof. We shall construct Q and U for the case when A is real, square, and nonsingular. Let $\mathbf{a}_1$ be the first column of A, and normalize it by $\mathbf{q}_1 = \dfrac{\mathbf{a}_1}{(\mathbf{a}_1^T \mathbf{a}_1)^{1/2}}$, $u_{11} = (\mathbf{a}_1^T \mathbf{a}_1)^{1/2}$. By projecting $\mathbf{a}_2$ (the second column of A) onto span($\mathbf{q}_1$) and subtracting this from $\mathbf{a}_2$, we obtain a direction orthogonal to $\mathbf{a}_1$, i. e., $\mathbf{b}_1 := \mathbf{a}_2 - (\mathbf{a}_2^T \mathbf{q}_1)\mathbf{q}_1$ and we normalize it by $\mathbf{q}_2 = \dfrac{\mathbf{b}_1}{(\mathbf{b}_1^T \mathbf{b}_1)^{1/2}}$ (see Fig. 2.1). Now define $u_{12} = \mathbf{a}_2^T \mathbf{q}_1$ and $u_{22} = (\mathbf{b}_1^T \mathbf{b}_1)^{1/2}$. If $n = 2$ we have indeed found $A = QU$. The general case proceeds by induction. Further, if $Q_1 U_1 = Q_2 U_2$, we obtain $Q_2^T Q_1 = U_2 U_1^{-1}$. Because $Q_2^T Q_1$ is orthogonal and $U_2 U_1^{-1}$ is upper triangular, both must equal an upper triangular orthogonal matrix, i. e., a diagonal matrix D whose elements are ± 1. □

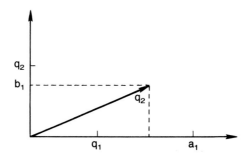

Figure 2.1 Orthogonalization

We do not recommend using this orthogonalization process numerically. In Section 2.2.4 we shall see a much more stable algorithm to obtain the so-called QU-*decomposition* (of Theorem 2.11).

We can now derive a computationally more attractive canonical form than the Jordan form, namely one where an upper triangular matrix is made similar to A via an orthogonal similarity transformation.

Theorem 2.12 (Schur) For each real, square matrix A there exists an orthogonal matrix Q such that $Q^T A Q = V$, where V is upper triangular.

Proof: Let $\Lambda = T^{-1}AT$ [cf. (2.8a)]. Then according to Theorem 2.11 we have a QU-decomposition $T = QU$. Hence $\Lambda = U^{-1}Q^T A Q U = U^{-1} V U$, and since U^{-1} and Λ are upper triangular, $V = U \Lambda U^{-1}$ is, too. □

Corollary If A is symmetric, then V and Q are real and V is diagonal. □

A special class of matrices is *symmetric positive definite* matrices. A symmetric A is positive definite if

$$\mathbf{x}^T A \mathbf{x} > 0, \quad \text{all } \mathbf{x} \neq \mathbf{0} \tag{2.13}$$

A is *positive semidefinite* if the inequality in (2.13) is relaxed to ≥ 0. If now λ and $\mathbf{v}$ are an eigenpair satisfying $A\mathbf{v} = \lambda \mathbf{v}$, then a multiplication by $\mathbf{v}^T$ yields the *Rayleigh quotient*

$$\lambda = \frac{\mathbf{v}^T A \mathbf{v}}{\mathbf{v}^T \mathbf{v}}$$

Hence, if A is positive definite then its eigenvalues are all positive, while if it is positive semidefinite then they are nonnegative.

Given a (possibly rectangular) matrix A, the matrix $A^T A$ is clearly symmetric positive semidefinite. In fact, $A^T A$ is positive definite if A has a full column rank. The eigenvalues of $A^T A$ have a special importance: Let $\sigma_1^2, \ldots, \sigma_n^2$ be the eigenvalues of $A^T A$ ordered so that $\sigma_1^2 \geq \sigma_2^2 \geq \cdots \geq \sigma_n^2$. With the choice $\sigma_j \geq 0$, $1 \leq j \leq n$, the values $\sigma_1, \ldots, \sigma_n$ are called the *singular values of A*. This leads to

Theorem 2.14 (Singular value decomposition) For each matrix A there exist orthogonal matrices Q_1 and Q_2 such that $Q_1^T A Q_2 = \text{diag}\{\sigma_1, \ldots, \sigma_n\}$.

Proof: Since $A^T A$ is symmetric, from the corollary to Theorem 2.12 there exists an orthogonal matrix Q_2 such that $Q_2^T A^T A Q_2 = \text{diag}\{\sigma_1^2, \ldots, \sigma_n^2\}$. By ordering the columns of Q_2 appropriately, $\sigma_1 \geq \cdots \geq \sigma_n$. Thus AQ_2 has orthogonal columns and, if we assume $\sigma_j \neq 0$ for all j, $AQ_2 \text{diag}\{\sigma_1^{-1}, \ldots, \sigma_n^{-1}\}$ is an orthogonal matrix, say Q_1. If the last few σ_j are zero, then we restrict ourselves to the nonzero σ_j, obtaining a matrix with orthogonal columns. This can be completed to a full orthogonal matrix. □

Note that Theorem 2.14 applies equally well to *rectangular* matrices. If A is an $m \times n$ real matrix with rank $(A) = r \leq \min(n,m)$, then in Theorem 2.14 Q_1 is $m \times m$, Q_2 is $n \times n$, and

$$Q_1^T A Q_2 = \begin{bmatrix} \Sigma & 0 \\ 0 & 0 \end{bmatrix}$$

with $\Sigma = \text{diag}\{\sigma_1, \ldots, \sigma_r\}$. We define the *Moore-Penrose pseudo-inverse* of A as the $n \times m$ matrix

$$A^+ := Q_2 \begin{bmatrix} \Sigma^{-1} & 0 \\ 0 & 0 \end{bmatrix} Q_1^T \qquad (2.15)$$

In future chapters we shall be concerned with projecting solution vectors into solution subspaces. Here we describe projections in vector spaces. Given any k vectors $\mathbf{v}_1, \ldots, \mathbf{v}_k$ in $\mathbf{R}^n$, let V be the $n \times k$ matrix $V := [\mathbf{v}_1 | \cdots | \mathbf{v}_k]$. Then the *subspace* of $\mathbf{R}^n$ *spanned* by $\mathbf{v}_1, \ldots, \mathbf{v}_k$ and denoted range $(V) \equiv \text{span}(\mathbf{v}_1, \ldots, \mathbf{v}_k)$ consists of all vectors which are linear combinations of $\mathbf{v}_1, \ldots, \mathbf{v}_k$. Now let S be a subspace of $\mathbf{R}^n$. The square $n \times n$ matrix P is an *orthogonal projection* (of $\mathbf{R}^n$) onto S if

$$\text{range}(P) = S \qquad (2.16a)$$

$$P^2 = P \qquad (2.16b)$$

$$P^T = P \qquad (2.16c)$$

It turns out that for a given subspace S, the projection P is unique. If V is an $n \times k$ matrix whose columns form an orthonormal basis for S, i.e., $V^T V = I$ and range $(V) = S$, then

$$P = VV^T \qquad (2.17)$$

For example, given an $m \times n$ real matrix A with rank r, the orthogonal projection onto range (A) is given by (2.17) where V is composed of the first r columns of the matrix Q_1 arising in the singular value decomposition of A. The same orthogonal projection is also given by

$$P = AA^+$$

[see (2.15)].

2.2.2 Norms, angles, and condition numbers

In order to measure the size of vectors and matrices (and in particular of errors involving them) we need to provide our linear space with a norm.

Definition 2.18 A *vector norm* $|\cdot| : \mathbf{R}^n \to \mathbf{R}_+ = \{x \in \mathbf{R}: x \geq 0\}$ is a nonnegative function for which

(i) $|\mathbf{x}| = 0$ iff $\mathbf{x} = \mathbf{0}$
(ii) $|\gamma \mathbf{x}| = |\gamma| |\mathbf{x}|$, for all $\gamma \in \mathbf{R}$
(iii) $|\mathbf{x} + \mathbf{y}| \leq |\mathbf{x}| + |\mathbf{y}|$ (triangular inequality)
    We mainly deal with so-called *Holder norms*,

$$|\mathbf{x}|_p = \left[\sum_{i=1}^{n} |x_i|^p\right]^{1/p}, \quad p \geq 1 \tag{2.19a}$$

and especially

$$|\mathbf{x}|_1 = \sum_{i=1}^{n} |x_i|, \text{ the } 1-\text{norm} \tag{2.19b}$$

$$|\mathbf{x}|_2 = \left[\sum_{i=1}^{n} |x_i|^2\right]^{1/2}, \text{ the Euclidean or 2- norm} \tag{2.19c}$$

$$|\mathbf{x}|_\infty = \max_{1 \leq i \leq n} |x_i|, \text{ the max (or sup) norm} \tag{2.19d}$$

All norms on a finite dimensional space define the same topology, since they are equivalent (i.e., given two norms $|\cdot|_\alpha$ and $|\cdot|_\beta$, there exist constants c_α and c_β such that $c_\alpha |\mathbf{x}|_\alpha \leq |\mathbf{x}|_\beta \leq c_\beta |\mathbf{x}|_\alpha$ for all $\mathbf{x} \in \mathbf{R}^n$). Nevertheless, for some applications one particular norm is preferred over another. Normally we use the max norm, and denote it simply by $|\mathbf{x}|$.

For matrices we can also define a norm.

Definition 2.200 A *matrix norm* $\|\cdot\| : \mathbf{R}^{m \times n} \to \mathbf{R}_+$ is a nonnegative function for which

(i) $\|A\| = 0$ iff $A = 0$
(ii) $\|\gamma A\| = |\gamma| \|A\|$ for all $\gamma \in \mathbf{R}$
(iii) $\|A + B\| \leq \|A\| + \|B\|$

All of the norms which we consider also satisfy

(iv) $\|AB\| \leq \|A\| \|B\|$

We shall mainly deal with matrix norms derived from *associated least upper bounds* for vector norms,

$$\|A\|_p = \text{lub}_p(A) = \max_{\mathbf{x} \neq 0} \frac{|A\mathbf{x}|_p}{|\mathbf{x}|_p} = \max_{|\mathbf{x}|_p = 1} |A\mathbf{x}|_p \tag{2.21}$$

Such norms are sometimes called *induced matrix norms*. Note that it trivially holds that $|A\mathbf{x}|_p \leq \|A\|_p |\mathbf{x}|_p$. In analogy to (2.21) we also define a *greatest lower bound*

$$\text{glb}_p(A) = \min_{\mathbf{x} \neq 0} \frac{|A\mathbf{x}|_p}{|\mathbf{x}|_p} = \min_{|\mathbf{x}|_p = 1} |A\mathbf{x}|_p \tag{2.22}$$

Note that glb is *not a norm*. In connection to (2.21) and (2.22) we have

Definition 2.23 $\mathbf{x}$ is called a *maximizing* vector of A with respect to $\|\cdot\|_p$ if $|A\mathbf{x}|_p = \|A\|_p |\mathbf{x}|_p$ and a *minimizing* vector if $|A\mathbf{x}|_p = \text{glb}_p(A) |\mathbf{x}|_p$.

Theorem 2.24 For $A = (a_{ij}) \in \mathbf{R}^{m \times n}$, $\|A\|_1 = \max_j \sum_i |a_{ij}|$, $\|A\|_\infty = \max_i \sum_j |a_{ij}|$, and $\|A\|_2 = (\rho(A^T A))^{1/2}$. □

Theorem 2.25 If Q_1 and Q_2 are arbitrary orthogonal matrices then $\|Q_1 A Q_2\|_2 = \|A\|_2$. □

Theorem 2.25 indicates why orthogonal matrices are so important in numerical analysis: In many applications we have to multiply a certain matrix by many transformation matrices. If the transformations are orthogonal, and we have a method to compute them reasonably well, the resulting product will have roughly the same 2-norm; i.e., the multiplication may be expected to be *numerically stable!*

We shall also need to formalize the notion of angles between vectors and matrices.

Definition 2.26 The angle between *two vectors* $\mathbf{u}, \mathbf{v} \in \mathbf{R}^n$ is defined by
$$\theta = \arccos \frac{|\mathbf{u}^T \mathbf{v}|}{|\mathbf{u}|_2 |\mathbf{v}|_2}$$

This generalizes to a notion of *angles between matrices or subspaces* as follows: Let U and V be in $\mathbf{R}^{n \times k}$ and $\mathbf{R}^{n \times l}$, respectively. Then the angle between range (U) and range (V) (or simply the angle between U and V) is defined by
$$\theta = \arccos \max_{\mathbf{u} \in U} \max_{\mathbf{v} \in V} \frac{|\mathbf{u}^T \mathbf{v}|}{|\mathbf{u}|_2 |\mathbf{v}|_2}$$
□

Note that $0 \leq \theta \leq \pi/2$. If range $(U) \cap$ range $(V) = \{\mathbf{0}\}$, then $\theta > 0$.

Theorem 2.27 Let $\{\mathbf{x}_1, \ldots, \mathbf{x}_k\}$ be an orthonormal basis for range (U) and denote $X = (\mathbf{x}_1 | \cdots | \mathbf{x}_k) \in \mathbf{R}^{n \times k}$. Similarly, let the columns of Y be an orthonormal basis for range (V). Then
$$\|X^T Y\|_2 = \cos \theta \tag{2.27a}$$
If Z is a matrix with orthonormal columns that are orthogonal to range $(X) =$ range (U) and such that range $(U | V) =$ range $(X | Z)$, then
$$\mathrm{glb}_2(Z^T Y) = \sin \theta \tag{2.27b}$$

Proof: Note that $Y = XX^T Y + ZZ^T Y$. Hence, for any $\mathbf{y} \in \mathbf{R}^l$ with $|\mathbf{y}|_2 = 1$, we obtain from Pythagoras' Theorem
$$|X^T Y \mathbf{y}|_2^2 + |Z^T Y \mathbf{y}|_2^2 = 1$$
so
$$\min_{\mathbf{y} \in \mathbf{R}^l} |Z^T Y \mathbf{y}|_2 \geq \sqrt{1 - \cos^2 \theta} = \sin \theta$$
Equality holds for the normalized minimizing vector $\mathbf{y}$ of $Z^T Y$. □

In analyzing numerical algorithms one frequently encounters so-called condition numbers.

Definition 2.28 Let A be square and nonsingular, and $\|\cdot\|$ a norm. Then cond$(A) := \|A\| \|A^{-1}\|$ is called the *condition number* of the matrix A. □

We note that cond$(A) \geq 1$ and cond$(\gamma A) = $ cond(A) for any $\gamma \in \mathbf{R}$. The value of cond(A) depends, of course, on the matrix norm used in its definition. But we often avoid being specific about the norm used for this concept, emphasizing its qualitative nature. For the norm $\|\cdot\|_p$ of (2.21) we sometimes use the notation cond$_p(A)$. For $p = 2$, the condition number can be expressed in terms of singular values as

$$\text{cond}_2(A) = \frac{\sigma_1}{\sigma_n} \qquad (2.29)$$

Note that if σ_n gets closer to zero (that is the matrix becomes closer to being numerically singular), the condition number gets larger. If cond(A) is large, then the matrix A is called *ill-conditioned*. Since cond$_2(A) = 1$ if A is orthogonal, we conclude that orthogonal matrices are optimally well-conditioned with respect to the 2-norm.

2.2.3 Gaussian elimination, LU-decomposition

Consider the linear system of n equations

$$\begin{aligned}
a_{11}x_1 + a_{12}x_2 + \cdots + a_{1n}x_n &= b_1 \\
a_{21}x_1 + a_{22}x_2 + \cdots + a_{2n}x_n &= b_2 \\
&\vdots \\
a_{n1}x_1 + a_{n2}x_2 + \cdots + a_{nn}x_n &= b_n
\end{aligned} \qquad (2.30)$$

which we write in more convenient form

$$A\mathbf{x} = \mathbf{b} \qquad (2.31)$$

where $A = (a_{ij})_{i,j=1}^n$, $\mathbf{x} = (x_1, \ldots, x_n)^T$, $\mathbf{b} = (b_1, \ldots, b_n)^T$. Assume that A is nonsingular. The most common method for solving for $\mathbf{x}$ is called *Gaussian elimination*, which we describe briefly: Let $(A\,|\,\mathbf{b})$ be the *augmented* matrix formed by adding $\mathbf{b}$ as an $(n+1)^{\text{st}}$ column of A. Suppose the *pivot* $a_{11} \neq 0$, then multiply the first row of $(A\,|\,\mathbf{b})$ by $-\dfrac{a_{i1}}{a_{11}}$ and add this to the ith row of $(A\,|\,\mathbf{b})$, for $i = 2, \ldots, n$. This yields the system

$$A^{(1)}\mathbf{x} = \mathbf{b}^{(1)} \qquad (2.32)$$

where $A^{(1)} = (a_{ij}^{(1)})$, $\mathbf{b}^{(1)} = (b_1^{(1)}, \ldots, b_n^{(1)})^T$ with

$$a_{ij}^{(1)} = a_{ij} - a_{1j}\frac{a_{i1}}{a_{11}} \qquad (2.33a)$$

$$b_i^{(1)} = b_i - b_1\frac{a_{i1}}{a_{11}} \qquad (2.33b)$$

so in particular

$$A^{(1)} = \begin{bmatrix} a_{11} & a_{12} & \cdots & a_{1n} \\ 0 & a_{22}^{(1)} & \cdots & a_{2n}^{(1)} \\ \vdots & \vdots & & \vdots \\ 0 & a_{n2}^{(1)} & \cdots & a_{nn}^{(1)} \end{bmatrix} \quad (2.33c)$$

Now assuming $a_{22}^{(1)} \neq 0$ ($a_{22}^{(1)}$ is again called the *pivot*), we can apply the same procedure to the lower right $(n-1) \times (n-1)$ block of $A^{(1)}$, i.e., $(a_{ij}^{(1)})_{i,j \geq 2}$, from which we obtain a matrix $A^{(2)}$ and a right-hand side $\mathbf{b}^{(2)}$. In matrix notation this continuing process can be described as follows: Define at the jth step

$$l_{ij} = a_{ij}^{(j-1)}/a_{jj}^{(j-1)} \quad (2.34a)$$

$$L_j = \begin{bmatrix} 1 & & & & & \\ & \ddots & & & & \\ & & 1 & & & \\ & & -l_{j+1,j} & \ddots & & \\ & & \vdots & & \ddots & \\ & & -l_{n,j} & & & 1 \end{bmatrix} \quad (2.34b)$$

(L_j is an *elementary lower triangular transformation*). Then

$$L_j A^{(j-1)} = A^{(j)} \quad (2.35a)$$

with

$$A^{(j-1)} = \begin{bmatrix} a_{11} & a_{12} & & & & a_{1n} \\ 0 & a_{22}^{(1)} & \cdots & & & a_{2n}^{(1)} \\ \vdots & & \ddots & & & \\ & & & 0 & a_{jj}^{(j-1)} & \cdots & a_{jn}^{(j-1)} \\ \vdots & & & & \vdots & & \vdots \\ 0 & & & 0 & a_{nj}^{(j-1)} & \cdots & a_{nn}^{(j-1)} \end{bmatrix} \quad (2.35b)$$

Assuming that all pivots are nonzero, we finally obtain $L_{n-1}A^{(n-2)} = A^{(n-1)}$, which is an upper triangular matrix. Hence

$$L_{n-1} \cdots L_1 A = A^{(n-1)} =: U \quad (2.36a)$$

or equivalently

$$A = L_1^{-1} \cdots L_{n-1}^{-1} U =: LU \quad (2.36b)$$

where L is a lower triangular matrix. Therefore, the factorization in (2.36) is also called LU-decomposition. It is easily seen that

$$L = \begin{bmatrix} 1 & & & & \\ l_{21} & 1 & & & \\ \vdots & \vdots & \ddots & & \\ l_{n1} & l_{n2} & l_{n,n-1} & 1 \end{bmatrix} \qquad (2.36c)$$

Generally, the assumption that the pivots encountered are nonzero is too restrictive. In fact, to obtain a numerically stable algorithm, these pivots cannot be allowed to be too small compared to the eliminated elements. But because A is nonsingular we know that at least one element in the first column of A must be nonzero. To prevent numerical instability, we look for that element which has the largest magnitude. Then we interchange the row of this element with the first row. Such a permutation can be described by a so-called permutation matrix; e.g., for the ith row, the 1st and i^{th} rows of the identity matrix are interchanged, so

$$P = \begin{bmatrix} 0 & 0 & \cdots & 1 & & \\ 0 & 1 & & & & \\ \vdots & \vdots & & \vdots & & \\ 1 & 0 & \cdots & 0 & & \\ & & & & 1 & \\ & & & & & 1 \end{bmatrix} \qquad (2.37a)$$

and

$$PA = \begin{bmatrix} a_{i1} & \cdots & a_{in} \\ \vdots & & \vdots \\ a_{i-1,1} & \cdots & a_{i-1,n} \\ a_{11} & \cdots & a_{1n} \\ a_{i+1,1} & \cdots & a_{i+1,n} \\ \vdots & & \vdots \\ a_{n,1} & \cdots & a_{n,n} \end{bmatrix} \qquad (2.37b)$$

Performing such a so-called *row partial pivoting* strategy for each elimination step, we have, instead of (2.36a),

$$L_{n-1}P_{n-1} \cdots L_2 P_2 L_1 P_1 A = U \qquad (2.38a)$$

where, moreover, $|l_{ij}| \leq 1$, all i, j. It is not difficult to see that there exist matrices $\hat{L}_j$ of the same form as L_j, such that

$$\hat{L}_{n-1} \cdots \hat{L}_1 P_{n-1} \cdots P_1 A = U \qquad (2.38b)$$

Hence partial pivoting can formally be described by

$$PA = LU \qquad (2.38c)$$

where P is a permutation matrix. In an implementation, we retain the effect of P in a vector **p** and overwrite A with its factors L and U as follows:

Algorithm (2.38d)   LU-decomposition with partial pivoting
FOR $k = 1, \ldots, n-1$ DO
    Determine index r, $k \leq r \leq n$, such that $|a_{rk}| = \max\limits_{k \leq i \leq n} |a_{ik}|$.
    $p_k := r$
    Swap a_{kj} and a_{rj}, $j = k, \ldots, n$.
    FOR $i = k+1, \ldots, n$ DO
        $m := a_{ik}/a_{kk}$
        $a_{ik} = m$
        FOR $j = k+1, \ldots, n$ DO
            $a_{ij} := a_{ij} - m a_{kj}$ □

Once we have factored A, the actual solution of (2.31) goes in two sweeps. First we find **y** from

$$L\mathbf{y} = P\mathbf{b} \tag{2.39a}$$

which is called the *forward substitution* step, and then find **x** from

$$U\mathbf{x} = \mathbf{y} \tag{2.39b}$$

which is called the *back substitution* step. The triangular form makes it easy to compute $y_1, y_2, \ldots, y_n$ and then $x_n, \ldots, x_1$. If we indicate the computation cost for one multiplication (or division) followed by an addition (or subtraction) by a *flop*, then the LU-decomposition has a *complexity* of $\sim \frac{n^3}{3}$ flops, whereas the forward and back substitutions account for $\sim n^2$ flops.

The decomposition algorithm (2.38) is row-oriented; i. e., the innermost loop manipulates rows of A. If the algorithm is implemented in a language like Fortran, which stores matrices by columns, then a column-oriented version of the algorithm is preferable because the array elements are then successively accessed. Such a version of (2.38d), obtained by switching the innermost loops, can be surprisingly faster on some machines, depending on the architecture.

There are some special matrices for which it is known in advance that LU-decomposition without pivoting can be safely carried out. One such class is the symmetric positive definite matrices, mentioned earlier. Another class is *strictly diagonally dominant* matrices. A is diagonally dominant if for all rows i

$$|a_{ii}| \geq \sum_{j \neq i} |a_{ij}|$$

and strictly diagonally dominant if these inequalities are all strict.

Finally, if the equations in (2.31) are poorly scaled, i. e., some row in $(A|b)$ has a much larger norm than another, one normally should *equilibrate* them, e. g., by forming

$$d_i = \max_j |a_{ij}| \tag{2.40a}$$

$$\hat{a}_{ij} = a_{ij}/d_i, \quad \hat{b}_i = b_i/d_i, \quad i,j=1, \ldots, n \tag{2.40b}$$

Then the matrix $\hat{A} = (\hat{a}_{ij})$ is "row equilibrated with respect to the max norm."

In solving BVPs one encounters so-called *sparse* matrices A, i.e., ones with many zero elements (at least 90 per cent, say, and normally appearing in a systematic way). If the nonzero elements are located only on the main diagonal and some adjacent diagonals, the matrix is said to have a *band structure:* If $a_{ij} = 0$, whenever $i-j > k_1$ or $j-i > k_2$ then k_1 (k_2) is called the *lower (upper) bandwidth* and $k_1 + k_2 + 1$ the *total bandwidth*. In such a case, LU-decomposition with partial pivoting results in L and U with bandwidths $k_1 + 1$ and $k_1 + k_2 + 1$ respectively. (Without pivoting the latter has bandwidth $k_2 + 1$.)

The decomposition algorithm (2.38d) is modified in an obvious way, by shortening the ranges of the $i-$ and $j-$loops. If k_1 and k_2 are fixed, then as a function of the matrix size n, the complexity drops to $O(n)$ (from $O(n^3)$). A special case of a banded matrix is a *tridiagonal* matrix, where $k_1 = k_2 = 1$. If no pivoting is needed, then the decomposition yields bidiagonal L and U.

2.2.4 Householder transformations: QU-decomposition

As we have seen in the previous subsection, we can reduce a matrix to upper triangular form by means of elementary lower triangular transformations and a permutation matrix (cf. (2.38)). Another way of solving (2.31) is to use a decomposition of A into a product of an orthogonal matrix and an upper triangular matrix, $A = QU$ (cf. Theorem 2.11). We then solve (2.31) via

$$\mathbf{y} := Q^T \mathbf{b} \qquad (2.41a)$$

(hence $Q\mathbf{y} = \mathbf{b}$),

$$U\mathbf{x} = \mathbf{y} \qquad (2.41b)$$

Note that (2.41a) is simple to carry out, while (2.41b) is again a back substitution.

Rather than implementing the Gram-Schmidt process to carry out the triangulation procedure, Householder's method employs plane reflections: Consider the matrix

$$H_{\mathbf{x}} = I - 2\frac{\mathbf{x}\mathbf{x}^T}{\mathbf{x}^T\mathbf{x}} \qquad (2.42)$$

Theorem 2.43 The following properties hold:

(i) $H_{\alpha\mathbf{x}} = H_{\mathbf{x}}$ for any real $\alpha \neq 0$
(ii) $H_{\mathbf{x}}$ is symmetric and orthogonal
(iii) $H_{\mathbf{x}}\mathbf{y} = \mathbf{y}$ iff $\mathbf{y}^T\mathbf{x} = 0$
(iv) $H_{\mathbf{x}}\mathbf{x} = -\mathbf{x}$ □

Because of these orthogonality and symmetry properties, $H_{\mathbf{x}}$ is often called an *elementary reflector, or an elementary Hermitian matrix* (Hermitian being the analogue of symmetric in the complex case). In Fig. 2.2 we see that $H_{\mathbf{x}}$ reflects any vector $\mathbf{z} = \mathbf{x} + \mathbf{y}$ with respect to the plane through the origin and orthogonal to $\mathbf{x}$ into the vector $\hat{\mathbf{z}} = -\mathbf{x} + \mathbf{y}$. For $n = 2$ we obtain (2.10a). This is used as follows: Given any pair of vectors $\mathbf{u}$, $\mathbf{v}$, with $\mathbf{u} \neq \mathbf{v}$, let λ be such that $\|\mathbf{u}\|_2 = \|\lambda\mathbf{v}\|_2$. Choosing $\mathbf{x} = \mathbf{u} - \lambda\mathbf{v}$ gives $H_{\mathbf{x}}\mathbf{u} = \lambda\mathbf{v}$, or choosing $\mathbf{x} = \mathbf{u} + \lambda\mathbf{v}$ gives $H_{\mathbf{x}}\mathbf{u} = -\lambda\mathbf{v}$. Now let $\mathbf{a}_1$ be the first

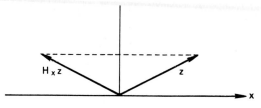

Figure 2.2 Reflection

column of A and let $\lambda = (\mathbf{a}_1^T\mathbf{a}_1)^{1/2}$. If we take

$$\mathbf{x} = \mathbf{a}_1 \pm \lambda \mathbf{e}_1 \tag{2.44a}$$

where

$$\mathbf{e}_1 = (1, 0, \ldots, 0)^T \tag{2.44b}$$

then $H^{(1)} := H_\mathbf{x}$ gives

$$H^{(1)}A = \begin{bmatrix} \pm\lambda & \times & \cdots & \times \\ 0 & \times & & \times \\ \vdots & \vdots & & \vdots \\ 0 & \times & & \times \end{bmatrix} =: A^{(1)} \tag{2.44c}$$

where $\times$ denotes a generally nonzero element. As in the previous subsection we can proceed with the lower right $(n-1)\times(n-1)$ block of $A^{(1)}$ and find an elementary reflector $H^{(2)}$,

$$H^{(2)} = \begin{bmatrix} 1 & 0 & \cdots & 0 \\ 0 & & & \\ \vdots & & H_{\mathbf{x}^1} & \\ 0 & & & \end{bmatrix} \tag{2.45a}$$

where $H_{\mathbf{x}^1}$ is an $(n-1)$st order elementary reflector such that

$$H^{(2)}A^{(1)} = \begin{bmatrix} \times & \times & \times & \cdots & \times \\ 0 & \times & \times & \cdots & \times \\ 0 & 0 & \times & \cdots & \times \\ \vdots & \vdots & \vdots & & \vdots \\ 0 & 0 & \times & \cdots & \times \end{bmatrix} \tag{2.45b}$$

Thus, after $(n-1)$ steps

$$H^{(n-1)} \cdots H^{(1)}A = A^{(n-1)} =: U \tag{2.45c}$$

Remarks

The sign in (2.44a) is chosen so that cancellation is avoided. Hence if $a_{11} \geq 0$, then $\mathbf{x} = \mathbf{a}_1 + \lambda\mathbf{e}_1$ (so $a_{11}^{(1)} > 0$) and if $a_{11} \leq 0$, then $\mathbf{x} = \mathbf{a}_1 - \lambda\mathbf{e}_1$ (so $a_{11}^{(1)} < 0$).

The complexity of Householder's method is $\sim \frac{2}{3}n^3$ flops, but it also includes $\sim n$ square roots.

It is easy to see that although the Gram-Schmidt and Householder methods are quite different, geometrically they lead to a similar decomposition: Suppose A is non-singular and let $A = Q_1 U_1 = Q_2 U_2$ (Q_j orthogonal, U_j upper triangular). Then $Q_1^T Q_2 = U_1 U_2^{-1}$, hence $Q_1^T Q_2$ must be a diagonal matrix consisting of ± 1 terms on the diagonal. This amounts to the choice $\pm \mathbf{q}_j$ as the only degree of freedom in the Gram-Schmidt algorithm. □

2.2.5 Least squares solution of overdetermined systems

If A is $m \times n$ with rank$(A) = r < m$, then the system of equations (2.31) is generally overdetermined. One may then attempt to solve it in the least squares sense, i.e., determine $\mathbf{x}$ to

$$\text{minimize } |A\mathbf{x} - \mathbf{b}|_2 \qquad (2.46)$$

If $r = n$ then $\mathbf{x}$, which solves this minimization problem, is uniquely determined by the *normal equations*

$$A^T A \mathbf{x} = A^T \mathbf{b} \qquad (2.47)$$

(Note that the symmetric $n \times n$ matrix $A^T A$ is then positive definite.) However, the algorithm resulting from forming and solving (2.47) is not as numerically stable as the (twice as expensive) alternative of using the QU-decomposition.

If $r < n$ then $A^T A$ is singular and (2.47) cannot be directly solved. In this case the solution of (2.46) is not unique either. A solution of (2.46), good also for this rank deficient case, is given by

$$\mathbf{x} = A^+ \mathbf{b} \qquad (2.48)$$

with the pseudo-inverse defined in (2.15). The distinguishing feature of (2.48) is that $\mathbf{x}$ is the *unique solution of (2.46) with the smallest 2-norm*.

2.2.6 The QR algorithm for eigenvalues

An important application of the QU-decomposition is as a building block in a method for computing eigenvalues of a matrix. The idea is to compute the Schur factorization of Theorem 2.12

$$A = QVQ^T$$

where Q is orthogonal and V is upper triangular. Then, since V is upper triangular and the similarity transformation preserves eigenvalues, the main diagonal of V contains the *eigenvalues of A*.

This method can be described briefly as follows: Let $A_0 := A$. At the ith step, $i = 0, 1, 2, \ldots,$ let σ_i be an appropriately chosen number, called the *shift*. The step consists of a QU-decomposition

$$A_i - \sigma_i I =: Q_i R_i \qquad (2.49a)$$

followed by a computation of the next iterate

$$A_{i+1} := R_i Q_i + \sigma_i I \qquad (2.49b)$$

It can be shown that under appropriate conditions $R_i \to V$ and $Q_i Q_{i-1} \cdots Q_0 \to Q$. We shall not analyze this so-called *QR algorithm* here, but restrict ourselves to two observations: (i) In order to improve efficiency (lower the complexity), the matrix A is first transformed by one orthogonal similarity transformation to an *upper Hessenberg matrix*, which is a matrix where only the upper triangular part and one additional codiagonal are not necessarily zero. The iterations (2.49) then only have to zero out that additional codiagonal. (ii) By cleverly choosing the shifts σ_i it is possible to get quadratic convergence to the eigenvalues of A.

2.2.7 Error analysis

Numerical linear algebra has probably been more extensively developed than any other branch of numerical analysis. A large collection of numerical algorithms as available (e.g., significant portions of the NAG and IMSL libraries are linear algebra routines), and the theory is fairly well understood. Here we shall briefly discuss some of the useful theoretical results.

One of the nice features of numerical linear algebra is that we can use so-called *backward error analysis.* That is, rather than trying to follow the effect of local errors and their contamination, one attempts to interpret the computed results as the exact solution of a slightly perturbed system and thus guarantee stability of the algorithm. Combining a stable algorithm with a well-conditioned problem then yields bounds for the global error. Suppose that $\mathbf{x} + \delta\mathbf{x}$ is an *exact* solution of the *perturbed* problem

$$(A + \delta A)(\mathbf{x} + \delta\mathbf{x}) = \mathbf{b} + \delta\mathbf{b} \qquad (2.50)$$

We relate $\mathbf{x} + \delta\mathbf{x}$ to $\mathbf{x}$ below. Then we give some backward error analysis results which show that when using the decomposition algorithms discussed before, one actually solves (2.50) when the deviations $\|\delta A\|$ and $|\delta\mathbf{b}|$ are small.

Lemma 2.51 Let A be nonsingular and δA be such that $\|A^{-1}\delta A\| < 1$. Then

$$\frac{\|A^{-1}\|}{1 + \|A^{-1}\delta A\|} \leq \|(A + \delta A)^{-1}\| \leq \frac{\|A^{-1}\|}{1 - \|A^{-1}\delta A\|} \qquad (2.51)$$

□

Writing out (2.50), we have $A\mathbf{x} + \delta A\mathbf{x} + (A + \delta A)\delta\mathbf{x} = \mathbf{b} + \delta\mathbf{b}$, so $\delta\mathbf{x} = (A + \delta A)^{-1}(\delta\mathbf{b} - \delta A\mathbf{x})$. This yields two types of bounds on $\delta\mathbf{x}$. The first is *absolute*,

$$|\delta\mathbf{x}| \leq \|(A + \delta A)^{-1}\| \, |\delta\mathbf{b} - \delta A\mathbf{x}|$$

Thus, for small perturbations in the data the perturbation in the solution $\mathbf{x}$ is essentially bounded by $\|A^{-1}\|$ times these perturbations. But when assessing roundoff error propagation, the *relative* error is often more meaningful. This essentially replaces $\|A^{-1}\|$ by cond(A). We have

$$|\delta\mathbf{x}|/|\mathbf{x}| \leq \frac{\|A^{-1}\|}{[1 - \|A^{-1}\delta A\|]} \left[|\delta\mathbf{b}|/|\mathbf{x}| + \|\delta A\| \right]$$

and since $|\mathbf{x}| \geq |\mathbf{b}|/\|A\|$, we obtain

Theorem 2.52 The relative perturbation in **x** is bounded by

$$|\delta x|/|x| \leq \frac{\text{cond}(A)}{[1 - \text{cond}(A)\|\delta A\|/\|A\|]} \left[|\delta b|/|b| + \|\delta A\|/\|A\|\right] \qquad \square$$

If the factor $[1 - \text{cond}(A)\frac{\|\delta A\|}{\|A\|}]^{-1}$ above is close to 1, we see that the *relative perturbation* in **x** is roughly bounded by the sum of the *relative perturbations* in A and **b**, multiplied by the condition number. This then explains the important role of the condition number $\text{cond}(A)$. These perturbations may come from a variety of sources, e.g., rounding errors, errors resulting from discretizations, or inherent errors in the problem data.

We give several perturbation results for the LU-decomposition and QU-decomposition:

Theorem 2.53 Let ε_M be the machine unit roundoff.

(i) If we perform LU-decomposition, then the numerical solution $\mathbf{x} + \delta\mathbf{x}$ satisfies

$$(A + \delta A)(\mathbf{x} + \delta\mathbf{x}) = \mathbf{b}$$

with

$$\|\delta A\|_\infty \leq n[3\|A\|_\infty + 5\|L\|_\infty \|U\|_\infty]\varepsilon_M$$

for the computed L and U.

(ii) For Gaussian elimination with partial pivoting, we have further that

$$\|\delta A\|_\infty \leq \frac{3}{2} n^2 g \, \varepsilon_M$$

where g is the maximal element encountered in the elimination process and is bounded by 2^{n-1} times the largest element of A.

(iii) For Householder's method, we have

$$\|\delta A\|_2 \leq \kappa_1 \varepsilon_M (1 + \kappa_2 \varepsilon_M)^{n-2} \left(\sum |a_{ij}|^2\right)^{1/2}$$

where κ_1 and κ_2 are of moderate size. $\qquad \square$

In practice g is rarely larger than $\kappa\|A\|_\infty$, where κ is a constant of moderate size.

We therefore conclude from (ii) and (iii) that both methods in Theorem 2.53 are *stable* (with a much nicer theoretical bound holding for Householder's method) in the sense that $\|\delta A\| \sim \|A\| \varepsilon_M$. More generally, we have from (i) that an LU-decomposition algorithm is stable if $\|L\| \|U\| \sim \|A\|$. For a well-conditioned problem we see from Theorems 2.52 and 2.53 that $\|\delta A\|$ should be $\sim \|A\| \, TOL$ at most, if the error $|\delta x|/|x|$ is not to exceed a required tolerance TOL.

2.3 NONLINEAR EQUATIONS

The solution of nonlinear BVPs by (discrete) numerical methods reduces at some point to the solution of a system of nonlinear equations. Proper treatment, especially for sensitive problems, is critical for a successful computation. In this section we consider only some straightforward methods. In Chapter 8 a number of important modifications will be discussed in more detail.

2.3.1 Newton's method

The best-known and most popular method for solving nonlinear equations is Newton's method. Let $f(s)$ be a real-valued scalar function and let s^* be a root, i.e.,

$$f(s^*) = 0 \qquad (2.54)$$

Having chosen an initial approximation s_0, we compute

$$s_{m+1} := s_m - f(s_m)/f'(s_m) \qquad m = 0, 1, 2, \ldots \qquad (2.55)$$

Newton's method arises from linearizing $f(s)$, i.e., replacing $f(s)$ by a linear approximation $l(s)$ consisting of the first two terms in its Taylor expansion about s_m, and solving $l(s_{m+1}) = 0$. Thus,

$$f(s) \approx l(s) := f(s_m) + f'(s_m)(s - s_m)$$

and s_{m+1} in (2.55) is the root of $l(s)$.

One can show that if $|s_0 - s^*|$ is small enough and f satisfies some (weak) conditions, this process gives second-order convergence; i.e., $|s_{m+1} - s^*| = O(|s_m - s^*|^2)$.

The same idea can be used to solve a *system* of equations

$$\mathbf{f}(\mathbf{s}^*) = \mathbf{0} \qquad (2.56)$$

where $\mathbf{f}(\mathbf{s}) = (f_1(\mathbf{s}), \ldots, f_n(\mathbf{s}))^T$, $\mathbf{s} = (s_1, \ldots, s_n)^T$. Expanding $\mathbf{f}(\mathbf{s})$ around an approximation $\mathbf{s}^m$ of $\mathbf{s}^*$ gives the linear expression $\mathbf{l}(\mathbf{s}) := \mathbf{f}(\mathbf{s}^m) + J(\mathbf{s}^m)(\mathbf{s} - \mathbf{s}^m)$, where

$$J(\mathbf{s}) = \left[\frac{\partial f_m}{\partial s_j} \right] = \begin{bmatrix} \frac{\partial f_1}{\partial s_1} & \cdots & \frac{\partial f_1}{\partial s_n} \\ \vdots & & \vdots \\ \frac{\partial f_n}{\partial s_1} & \cdots & \frac{\partial f_n}{\partial s_n} \end{bmatrix}$$

is the *Jacobian matrix* of $\mathbf{f}(\mathbf{s})$ (also denoted by $\partial \mathbf{f}/\partial \mathbf{s}$ or $\nabla \mathbf{f}$). Thus, given an initial approximation $\mathbf{s}^0$, the generalization of (2.55) is

$$J(\mathbf{s}^m)(\mathbf{s}^{m+1} - \mathbf{s}^m) = -\mathbf{f}(\mathbf{s}^m) \qquad (2.57)$$

An important factor in assessing an iterative method is how quickly it converges. The *rate of convergence* of a convergent iterative method is the (largest) constant $p \geq 1$ such that for some constant $c > 0$

$$|\mathbf{s}^{m+1} - \mathbf{s}^*| \leq c |\mathbf{s}^m - \mathbf{s}^*|^p \qquad (2.58a)$$

If $p = 1$ then the convergence is said to be *linear* (and we must require $c < 1$ for convergence) and if $p = 2$, it is *quadratic*. If $p > 1$ then we can fairly safely use the approximation

$$|\mathbf{s}^* - \mathbf{s}^m| \approx |\mathbf{s}^{m+1} - \mathbf{s}^m| =: e_{m+1} \qquad (2.58b)$$

and a practical estimate of the rate is given by

$$p \approx \frac{\ln(e_{m+1}/e_m)}{\ln(e_m/e_{m-1})} \qquad (2.58c)$$

For Newton's method, as in the scalar case, one can show that if $|\mathbf{s}^* - \mathbf{s}^0|$ is small enough and $\mathbf{f}$ satisfies suitable conditions, there is quadratic convergence (see Theorem 2.64 on page 52). Quadratic convergence, when present, makes for a very rapid process in an actual computation, because the number of accurate digits roughly doubles per iteration step.

Thus, Newton's method converges very rapidly when $\mathbf{s}^0$ is sufficiently close to $\mathbf{s}^*$. It should be realized, though, that this is a local property. Newton's method may behave poorly when $|\mathbf{s}^0 - \mathbf{s}^*|$ is not small or c in (2.58a) is large. In fact, it may not converge at all. Fortunately, there are modifications of the method that have better global convergence properties; they will be dealt with in Chapter 8.

2.3.2 Fixed-point iteration and two basic theorems

The iteration scheme (2.55) can be seen as a special case of the general iteration method

$$s_{m+1} = g(s_m) \qquad (2.59)$$

where $g(s)$ has a so-called *fixed point* s^* satisfying $s^* = g(s^*)$. So instead of (2.56) we are led to investigate some equivalent form

$$\mathbf{g}(\mathbf{s}) = \mathbf{s} \qquad (2.60)$$

Note that for a given $\mathbf{f}(\mathbf{s})$ there are many ways to construct $\mathbf{g}(\mathbf{s})$ such that (2.60) is equivalent to (2.56). Given such $\mathbf{g}(\mathbf{s})$ and an initial approximation $\mathbf{s}^0$ to the zero $\mathbf{s}^*$ of $\mathbf{s} - \mathbf{g}(\mathbf{s})$, we consider the *fixed point iteration*

$$\mathbf{s}^{m+1} = \mathbf{g}(\mathbf{s}^m) \qquad m = 0, 1, 2, \ldots \qquad (2.61)$$

The convergence of the iteration (2.61) depends on the choice of $\mathbf{g}(\mathbf{s})$. If $\mathbf{g}$ is a *contraction mapping*, i.e., there is a constant $\gamma < 1$ such that in some norm

$$|\mathbf{g}(\mathbf{x}) - \mathbf{g}(\mathbf{y})| \le \gamma |\mathbf{x} - \mathbf{y}|, \quad \text{all } \mathbf{x}, \mathbf{y} \in D \qquad (2.62a)$$

with D a suitable domain to be specified below, then

$$|\mathbf{s}^{m+1} - \mathbf{s}^m| = |\mathbf{g}(\mathbf{s}^m) - \mathbf{g}(\mathbf{s}^{m-1})| \le \gamma |\mathbf{s}^m - \mathbf{s}^{m-1}| \le \cdots \le \gamma^m |\mathbf{s}^1 - \mathbf{s}^0|$$

So, if we begin the iteration with, say,

$$|\mathbf{s}^0 - \mathbf{g}(\mathbf{s}^0)| \le \alpha \qquad (2.62b)$$

then as $m \to \infty$

$$|\mathbf{s}^{m+1} - \mathbf{s}^m| \le \gamma^m \alpha \to 0$$

Thus we have a Cauchy sequence, which converges if the domain D on which $\mathbf{g}$ is a contraction mapping is closed. This shows *existence* of a fixed point of (2.60) [i.e., a zero of (2.56)], as well as the convergence of the fixed point iteration (2.61). The above is made precise in the following.

Theorem 2.62 (Contraction Mapping) Suppose that $\mathbf{g} : D \to D$, D a closed subset of $\mathbf{R}^n$, and (2.62a) is satisfied. Then

(i) There exists precisely one fixed point $\mathbf{s}^*$ in D, where $\mathbf{g}(\mathbf{s}^*) = \mathbf{s}^*$.
(ii) Starting with any $\mathbf{s}^0$ in D, the sequence $\{\mathbf{s}^m\}$ generated by (2.61) stays in D and converges linearly to $\mathbf{s}^*$.
(iii) If (2.62b) holds, then

$$|\mathbf{s}^m - \mathbf{s}^*| \le \frac{\alpha \gamma^m}{1-\gamma}, \qquad m = 0, 1, \ldots$$

□

Fixed point iteration is useful as a general approach for showing convergence of iteration methods for solving systems of nonlinear equations. It is also useful as an analytic iteration process (Picard iteration) for proving existence and uniqueness results for solutions of ODEs.

Newton's method (2.57) may be considered as a special case of the fixed point iteration (2.61) with

$$\mathbf{g}(\mathbf{s}) = \mathbf{s} - J(\mathbf{s})^{-1}\mathbf{f}(\mathbf{s})$$

Let us define a *sphere* in $\mathbf{R}^n$ (for a given norm) with center $\mathbf{x}$ and radius r as

$$S(\mathbf{x}, r) := \{\mathbf{y} \in \mathbf{R}^n; \; |\mathbf{y} - \mathbf{x}| < r\} \tag{2.63a}$$

This is an open set with closure

$$\overline{S}(\mathbf{x}, r) = \{\mathbf{y} \in \mathbf{R}^n; \; |\mathbf{y} - \mathbf{x}| \le r\} \tag{2.63b}$$

It is not difficult to see that if $\mathbf{s}^*$ of (2.56) exists for $\mathbf{f}(\mathbf{s})$ continuously differentiable, and if we take D in Theorem 2.62 to be the closure of a sphere centered at $\mathbf{s}^*$ with a sufficiently small radius, then the contraction mapping can be applied. However, it does not yield quadratic convergence, nor does it yield the existence of $\mathbf{s}^*$. For these stronger results we need a more specialized approach.

Specifically, let α, β and γ be constants such that the following bounds are satisfied:

$$|J(\mathbf{s}^0)^{-1}\mathbf{f}(\mathbf{s}^0)| \le \alpha \tag{2.64a}$$

$$\|J(\mathbf{s}^0)^{-1}\| \le \beta \tag{2.64b}$$

$$\|J(\mathbf{x}) - J(\mathbf{y})\| \le \gamma |\mathbf{x} - \mathbf{y}|, \qquad \mathbf{x}, \mathbf{y} \in D \tag{2.64c}$$

Consider applying Newton's method to the scalar, quadratic function

$$\phi(t) = \frac{\gamma}{2}t^2 - \frac{t}{\beta} + \frac{\alpha}{\beta}$$

starting at $t_0 = 0$. The zeros of $\phi(t)$ are, of course,

$$r_{\pm} = \frac{1}{\beta\gamma}\{1 \pm \sqrt{1-2\sigma}\}, \qquad \sigma := \alpha\beta\gamma \qquad (2.64\text{d})$$

If $\sigma \leq \frac{1}{2}$, then one can easily show that the sequence of Newton iterates $t_0, t_1, \ldots, t_m, \ldots$ converges to r_-. Now, it turns out that the sequence $\{t_m\}$ *majorizes* the sequence $\{s^m\}$ generated by (2.57), i.e.,

$$|s^{m+1} - s^m| \leq |t_{m+1} - t_m| \qquad m = 0, 1, \ldots$$

This implies both the existence of a solution s^* for the nonlinear system of equations (2.56) and the quadratic convergence of the Newton sequence to it. The complete theorem is stated below, with the full proof omitted.

Theorem 2.64 (Newton-Kantorovich) Assume that $\mathbf{f}(\mathbf{s})$ is continuously differentiable on a sphere $D = S(\mathbf{s}^0, r)$ and that (2.64a, b, c) hold. If $\sigma \leq \frac{1}{2}$ and $r \geq r_-$, then

(i) There exists a unique root $\mathbf{s}^*$ in $\overline{S}(\mathbf{s}^0, r_-)$.
(ii) The Newton sequence of iterates starting from $\mathbf{s}^0$ converges to $\mathbf{s}^*$.

If $\sigma < \frac{1}{2}$ then

(iii) $\mathbf{s}^*$ is the unique root of $\mathbf{f}(\mathbf{s})$ in $\overline{S}(\mathbf{s}^0, \min(r, r_+))$
(iv)
$$|\mathbf{s}^m - \mathbf{s}^*| \leq (2\sigma)^{2^m}\frac{\alpha}{\sigma} \qquad m = 0, 1, \ldots$$
□

Further generalizations of these results are possible. For example, if the hypotheses of the theorem are satisfied except that errors are made in solving (2.57) (e.g., because a numerical method is used to approximate the solution, or simply because of roundoff errors), and if these errors are sufficiently small, then the conclusions of the theorem still hold, with the bounds suitably perturbed.

2.3.3 Quasi-Newton methods

There are two main drawbacks to Newton's method. The first is the fact that the method is guaranteed to converge only if the initial guess is already "sufficiently close" to the solution (i.e., a *local* convergence result). This is dealt with in Chapter 8. The second drawback is the expense of the method: Each iteration (2.57) requires the evaluation of the Jacobian $J(\mathbf{s}^m)$ and the right hand side $\mathbf{f}(\mathbf{s}^m)$ and the factorization of $J(\mathbf{s}^m)$. In many instances the first partial derivatives needed for $J(\mathbf{s}^m)$ may be unavailable. Even if they are, when $J(\mathbf{s}^m)$ is a dense matrix, each iteration requires roughly $n^2 + n$ scalar function evaluations and $O(n^3)$ arithmetic operations to solve the linear system (2.57). This provides motivation to find cheaper ways to perform the iteration.

Quasi-Newton methods approximate $J(\mathbf{s}^{m+1})$ by modifying $J(\mathbf{s}^m)$ in a simple way that allows iterates to be computed cheaply. Probably the simplest way to do this is to hold the Jacobian fixed, e. g., to use $J(\mathbf{s}^m)$ as the approximation to $J(\mathbf{s}^{m+1})$ when calculating $\mathbf{s}^{m+2}$, and repeat this process for several iterations. Another way is to use so-called *Broyden updates*, whereby an approximation $\hat{J}_{m+1}$ to $J(\mathbf{s}^{m+1})$ and a factorization of $\hat{J}_{m+1}$ are obtained from the previous approximate Jacobian $\hat{J}_m$ by performing only $O(n^2)$ arithmetic operations and evaluating only $\mathbf{f}(\mathbf{s}^m)$ and $\mathbf{f}(\mathbf{s}^{m+1})$. Specifically, if $\mathbf{w}^m = \mathbf{f}(\mathbf{s}^{m+1}) - \mathbf{f}(\mathbf{s}^m)$ then with $\Delta\mathbf{s}^m := \mathbf{s}^{m+1} - \mathbf{s}^m$

$$\hat{J}_{m+1} = \hat{J}_m + \frac{1}{|\Delta\mathbf{s}^m|_2^2} \Delta\mathbf{s}^m (\mathbf{w}^m - \hat{J}_m \Delta\mathbf{s}^m)^T \qquad (2.65)$$

Given, for example, a QU-factorization of $\hat{J}_m$, it is possible to find the corresponding factorization of $\hat{J}_{m+1}$ in $O(n^2)$ arithmetic operations before finding $\mathbf{s}^{m+2}$. Alternatively, if $H_m = \hat{J}_m^{-1}$ and $H_{m+1} = \hat{J}_{m+1}^{-1}$, then this inverse can be directly calculated from

$$H_{m+1} = H_m + \frac{(H_m \Delta\mathbf{s}^m)(\Delta\mathbf{s}^m - H_m \mathbf{w}^m)^T}{(\Delta\mathbf{s}^m)^T H_m \mathbf{w}^m} \qquad (2.66)$$

With this approach, Broyden's method can thus be implemented in the form

$$\mathbf{s}^{m+1} = \mathbf{s}^m - H_m \mathbf{f}(\mathbf{s}^m) \qquad (2.67)$$

This requires only n scalar function evaluations and $O(n^2)$ arithmetic operations per iteration although (2.65) has preferable stability properties (which in turn are inferior to those of Newton's method).

The Broyden update is sometimes called *rank-1 update*. Indeed, the matrix $\hat{J}_{m+1} - \hat{J}_m$ is of rank 1.

A price paid for using quasi-Newton methods such as (2.65) is that the convergence rate drops from quadratic to *superlinear; i.e.*,

$$|\mathbf{s}^{m+1} - \mathbf{s}^*| \le \gamma_m |\mathbf{s}^m - \mathbf{s}^*| \qquad (2.68)$$

where $\lim_{m \to \infty} \gamma_m = 0$. A property of superlinearly convergent iterates is that

$$\lim_{m \to \infty} \frac{|\mathbf{s}^{m+1} - \mathbf{s}^m|}{|\mathbf{s}^m - \mathbf{s}^*|} = 1$$

so it is reasonable to use distance between successive iterates as a measure of error, as in (2.58b)

Many quasi-Newton methods such as (2.67) suffer from the more serious disadvantage that the updates do not preserve matrix sparsity, making them impractical in many applications (including for most methods for solving BVPs).

2.3.4 Quasilinearization

While Newton's method has so far been considered only for solving a system of nonlinear equations (2.56) (with a solution $\mathbf{s}^* \in \mathbf{R}^n$), it is also possible to apply it in a direct way to a nonlinear *operator* equation, which in our case is a nonlinear BVP. Historically, this has been called *quasilinearization*, a name which of course refers to the fact that a nonlinear equation is in some way linearized. We also use this name, since it allows us to conveniently distinguish it from the case where Newton's method is applied to a discrete system of nonlinear equations.

To see how quasilinearization is done, it is desirable to express BVPs in operator notation. We define the nonlinear differential operator

$$\mathbf{N}\mathbf{y}(x) \equiv \mathbf{y}'(x) - \mathbf{f}(x, \mathbf{y}(x)) \qquad a < x < b \qquad (2.69a)$$

and boundary conditions

$$\mathbf{G}\mathbf{y} := \mathbf{g}(\mathbf{y}(a), \mathbf{y}(b)) \qquad (2.69b)$$

and consider the nonlinear BVP

$$\mathbf{N}\mathbf{y} = \mathbf{0} \qquad a < x < b \qquad (2.70a)$$

$$\mathbf{G}\mathbf{y} = \mathbf{0} \qquad (2.70b)$$

The linearization about a known function $\mathbf{y}^m(x)$ now has the form

$$\mathbf{N}\mathbf{y}(x) \approx \mathbf{N}\mathbf{y}^m(x) + N'(\mathbf{y}^m)(\mathbf{y}(x) - \mathbf{y}^m(x)) \qquad a < x < b$$

$$\mathbf{G}\mathbf{y} \approx \mathbf{G}\mathbf{y}^m + G'(\mathbf{y}^m)(\mathbf{y} - \mathbf{y}^m)$$

Setting the right-hand sides to $\mathbf{0}$ and solving for $\mathbf{y}^{m+1}$, we obtain

$$N'(\mathbf{y}^m)(\mathbf{y}^{m+1}(x) - \mathbf{y}^m(x)) = -\mathbf{N}\mathbf{y}^m(x) \qquad a < x < b \qquad (2.71a)$$

$$G'(\mathbf{y}^m)(\mathbf{y}^{m+1} - \mathbf{y}^m) = -\mathbf{G}\mathbf{y}^m \qquad (2.71b)$$

The operators $N'(\mathbf{y})$ and $G'(\mathbf{y})$ are called *Frechet derivatives* of $\mathbf{N}\mathbf{y}$ and $\mathbf{G}\mathbf{y}$ respectively, but it is not our intention to discuss the functional analysis explaining what properties they have. Instead, we just explain how one does the quasilinearization in practice: If D denotes differentiation, i.e., $D\mathbf{y} := \mathbf{y}'(x)$, then formally differentiating (2.69a) with respect to $\mathbf{y}$ gives a differential operator $N'(\mathbf{y})$ which depends on the function $\mathbf{y}$ where the linearization takes place and operates on any function $\mathbf{w} \in C^{(1)}[a,b]$,

$$N'(\mathbf{y})\mathbf{w}(x) \equiv [D - \frac{\partial \mathbf{f}}{\partial \mathbf{y}}(x, \mathbf{y}(x))]\mathbf{w}(x) \qquad (2.72a)$$

Also, differentiating (2.70a) with respect to $\mathbf{y}(a)$ and $\mathbf{y}(b)$, we have

$$G'(\mathbf{y})\mathbf{w} = \frac{\partial \mathbf{g}(\mathbf{y}(a), \mathbf{y}(b))}{\partial \mathbf{y}(a)} \mathbf{w}(a) + \frac{\partial \mathbf{g}(\mathbf{y}(a), \mathbf{y}(b))}{\partial \mathbf{y}(b)} \mathbf{w}(b) \qquad (2.72b)$$

Substituting (2.72) into (2.71), we get the *linear* BVP for $\mathbf{w}(x)$,

$$\mathbf{w}'(x) - A(x)\mathbf{w}(x) = -(D\mathbf{y}^m(x) - \mathbf{f}(x, \mathbf{y}^m(x))) \qquad a < x < b \qquad (2.73a)$$

$$B_a \mathbf{w}(a) + B_b \mathbf{w}(b) = -\mathbf{g}(\mathbf{y}^m(a), \mathbf{y}^m(b)) \qquad (2.73b)$$

where

$$A(x) = \frac{\partial \mathbf{f}}{\partial \mathbf{y}}(x, \mathbf{y}^m(x)) \tag{2.74}$$

$$B_a = \frac{\partial \mathbf{g}}{\partial \mathbf{y}(a)}(\mathbf{y}^m(a), \mathbf{y}^m(b)), \qquad B_b = \frac{\partial \mathbf{g}}{\partial \mathbf{y}(b)}(\mathbf{y}^m(a), \mathbf{y}^m(b))$$

Thus, (2.73) and

$$\mathbf{y}^{m+1}(x) = \mathbf{y}^m(x) + \mathbf{w}(x) \tag{2.75}$$

comprise one step of Newton's method applied to the nonlinear BVP (2.70). We see that this quasilinearization amounts to solving a nonlinear BVP by an iterative method where a sequence of linear BVPs (2.73), (2.75) is solved, $m = 0, 1, \ldots,$ with $\mathbf{y}^0(x)$ a given initial solution guess. In fact, we can rewrite (2.73) as

$$\mathbf{L}[\mathbf{y}^m]\mathbf{w} = -\mathbf{N}\mathbf{y}^m \tag{2.76a}$$

$$\mathbf{B}[\mathbf{y}^m]\mathbf{w} = -\mathbf{G}\mathbf{y}^m \tag{2.76b}$$

where we define the linear operators

$$\mathbf{L}[\mathbf{y}^m](\cdot) := [D - A(x)](\cdot) \tag{2.77a}$$

$$\mathbf{B}[\mathbf{y}^m](\cdot) := B_a(\cdot)_{x=a} + B_b(\cdot)_{x=b} \tag{2.77b}$$

using (2.74).

The Newton-Kantorovich Theorem can be used to guarantee convergence of the quasilinearization method under suitable assumptions. In this new context (2.64b) holds for $\mathbf{y}^0$ near $\mathbf{y}$ if the operator $[D - A(x)]$ is smooth and invertible at $\mathbf{y}(x)$. This translates into invertibility of the variational problem discussed in Section 3.1,

$$\mathbf{L}[\mathbf{y}]\mathbf{w} = \mathbf{0}, \qquad \mathbf{B}[\mathbf{y}]\mathbf{w} = \mathbf{0} \tag{2.78}$$

i.e., one requires that $\mathbf{y}(x)$ be isolated or that the unique solution to (2.78) is $\mathbf{w} \equiv \mathbf{0}$. The bound in (2.64b) reflects the effect that perturbations in the right-hand-sides of the two equations (2.78) have upon the solution. It turns out that the sensitivity of the nonlinear BVP near an isolated solution and the convergence properties of quasilinearization *are directly related to the conditioning of the variational problem for the desired solution.* Chapter 3 discusses this further.

An application where quasilinearization is carried out in detail is given in Example 4.12.

We conclude this section with a discussion of quasilinearization for the case of high-order scalar equations. Given the nonlinear BVP

$$Nu(x) \equiv u^{(n)}(x) - f(x, u, u', \ldots, u^{(n-1)}) = 0 \qquad a < x < b \tag{2.79a}$$

$$g(u(a), \ldots, u^{(n-1)}(a), u(b), \ldots, u^{(n-1)}(b)) = \mathbf{0} \tag{2.79b}$$

we can write (2.79b) as (2.69b), (2.70b) with

$$\mathbf{y}(x) = (y_1(x), \ldots, y_n(x)) := (u(x), \ldots, u^{(n-1)}(x))$$

Corresponding to the steps in (2.72) through (2.77), one can show that quasilinearization takes the form

$$L[u^m]w(x) = -Nu^m(x) \qquad a < x < b \qquad (2.80a)$$

$$B_a \mathbf{z}(a) + B_b \mathbf{z}(b) = -\mathbf{g}(\mathbf{y}^m(a), \mathbf{y}^m(b)) \qquad (2.80b)$$

$$u^{m+1}(x) := u^m(x) + w(x) \qquad a < x < b \qquad (2.80c)$$

where

$$L[u]w(x) \equiv w^{(n)}(x) - \sum_{l=1}^{n} \frac{\partial f(x, \mathbf{y})}{\partial y_l} w^{(l-1)}(x) \qquad (2.81a)$$

$$B_a := \frac{\partial \mathbf{g}(\mathbf{u}, \mathbf{v})}{\partial \mathbf{u}}, \quad B_b := \frac{\partial \mathbf{g}(\mathbf{u}, \mathbf{v})}{\partial \mathbf{v}} \quad \text{at} \quad \mathbf{u} = \mathbf{y}^m(a), \quad \mathbf{v} = \mathbf{y}^m(b) \quad (2.81b)$$

and

$$\mathbf{z}(x) := (w(x), \ldots, w^{(n-1)}(x))$$

Example 2.2

Consider the scalar BVP

$$u''' + \frac{3}{2} u u'' + (u')^2 - u' - 1 = 0 \qquad 0 < x < \infty$$

$$u(0) = u'(0) = u'(\infty) = 0$$

Quasilinearization gives

$$w''' + (\frac{3}{2} u) w'' + (\frac{3}{2} u'') w + 2u'w' - w' = -(u''' + \frac{3}{2} u u'' + (u')^2 - u' - 1)$$

$$w(0) = w'(0) = w'(\infty) = 0$$

for (2.80a, b). This is a linear BVP for $w(x)$. □

2.4 POLYNOMIAL INTERPOLATION

Given a set of n points $x_1, x_2, \ldots, x_n$ and function values $f(x_1), f(x_2), \ldots, f(x_n)$, the *polynomial interpolation* problem is that of finding a polynomial $p(x)$ which satisfies

$$p(x_i) = f(x_i) \qquad i = 1, 2, \ldots, n \qquad (2.82)$$

The points $\{x_i\}_{i=1}^n$ are called *interpolation points* and $p(x)$ is said to *interpolate* $f(x)$ at these points. Interpolation polynomials are frequently used in data fitting; moreover, they play a central role in the development and analysis of numerical methods for solving differential equations.

2.4.1 Forms for interpolation polynomials

In this section the existence of an interpolation polynomial $p(x)$ is shown, and useful ways of expressing $p(x)$ are developed.

Assume for now that the interpolation points $\{x_i\}_{i=1}^n$ are distinct. Then there is a unique polynomial $p(x)$ of order n (degree $< n$) satisfying (2.82). A simple constructive proof that $p(x)$ exists involves writing $p(x)$ in its *Lagrange form*,

$$p(x) = \sum_{i=1}^n f(x_i) L_i(x) \qquad (2.83a)$$

where

$$L_i(x) = \prod_{\substack{j=1 \\ j \neq i}}^n \frac{(x - x_j)}{(x_i - x_j)} \qquad (1 \leq i \leq n) \qquad (2.83b)$$

Since each $L_i(x)$ is a polynomial of degree $n-1$ which satisfies $L_i(x_j) = \delta_{ij}$, the function $p(x)$ is a polynomial of order n which satisfies (2.82). Uniqueness of the interpolation polynomial readily follows, since if $q(x)$ were another polynomial of order n (denoted $q \in P_n$) satisfying (2.82), $r(x) = p(x) - q(x)$ would be a polynomial of order n with zeros at $x_1, x_2, \ldots, x_n$. This implies $r(x) \equiv 0$.

Another obvious way to compute $p(x)$ is to write $p(x) = \sum_{j=1}^n a_j x^{j-1}$ and to determine the unknown coefficients by satisfying (2.82). This results in the system of equations

$$V\mathbf{a} = \mathbf{f} \qquad (2.84a)$$

where

$$V = (x_i^{j-1})_{i,j=1}^n, \qquad \mathbf{a} = (a_1, \ldots, a_n)^T, \qquad \mathbf{f} = (f(x_1), \ldots, f(x_n))^T \qquad (2.84b)$$

The matrix V is called the *Vandermonde matrix* and the ith column of its inverse can be shown to consist of the coefficients of $l_i(x)$ in (2.83a).

The Lagrange form (2.83) is convenient from a theoretical standpoint, but in practice it is expensive to evaluate and cumbersome to use when interpolation at additional points is desired. The Vandermonde system (2.84) can be ill-conditioned. This motivates developing a computationally more convenient form for $p(x)$.

Assume that the n distinct points $x_1, x_2, \ldots, x_n$ lie in an interval $[a,b]$ and that the values of $f(x)$ at these n points are given. Furthermore, assume that $f(x)$ is sufficiently smooth, so whenever $f^{(l)}(x)$ is used we are assuming that $f(x) \in C^{(l)}[a,b]$.

We define *divided differences* of $f(x)$ at the points $x_1, x_2, \ldots, x_n$ as follows: The 0th divided difference $f[x_l]$ at x_l is

$$f[x_l] := f(x_l) \qquad (1 \leq l \leq n)$$

The 1st divided difference $f[x_l, x_{l+1}]$ at x_l, x_{l+1} is

$$f[x_l, x_{l+1}] := \frac{f(x_{l+1}) - f(x_l)}{x_{l+1} - x_l} \qquad (1 \leq l \leq n-1)$$

Continuing this process, the kth divided difference $f[x_l, x_{l+1}, \ldots, x_{l+k}]$ at $x_l, \ldots, x_{l+k}$ is recursively defined as

$$f[x_l, \ldots, x_{l+k}] := \frac{f[x_{l+1}, \ldots, x_{l+k}] - f[x_l, \ldots, x_{l+k-1}]}{x_{l+k} - x_l} \quad (1 \le l \le n-k) \quad (2.85)$$

It can be shown that if $x_l', \ldots, x_{l+k}'$ is any rearrangement of $x_l, \ldots, x_{l+k}$ then

$$f[x_l', \ldots, x_{l+k}'] = f[x_l, \ldots, x_{l+k}] \quad (2.86)$$

A further useful property of divided differences is that

$$f[x_l, x_{l+1}, \ldots, x_{l+k}] = \frac{f^{(k)}(\xi)}{k!} \quad (2.87)$$

for some point ξ lying in $[a, b]$.

After constructing a table of divided differences for $f(x)$ at the interpolation points, we express the interpolating polynomial directly in its *Newton form*

$$p(x) = \sum_{i=1}^{n} f[x_1, \ldots, x_i] \prod_{j=1}^{i-1} (x - x_j) \quad (2.88)$$

$$= f(x_1) + f[x_1 x_2](x - x_1) + , \ldots + f[x_1, \ldots, x_n](x - x_1) \ldots (x - x_{n-1})$$

The Newton form for $p(x)$ overcomes the two disadvantages of the Lagrange form. To evaluate $p(x)$ at a point $x = z$, nested multiplication (Horner's algorithm) can be used. In particular if we rewrite (2.88) in the form

$$p(x) = f(x_1) + (x - x_1)(f[x_1, x_2] + \cdots + (x - x_{n-1})(f[x_1, \ldots, x_n]) \cdots)$$

the following algorithm produces $a_1 = p(z)$.

Algorithm 2.89 Evaluating polynomial in Newton's form

$a_{n+1} := 0$
FOR $j = n, \ldots, 1$ DO
$\quad a_j := f[x_1, \ldots, x_j] + (z - x_j) a_{j+1}$ □

If the polynomial of order $n+1$ interpolating $f(x)$ at $x_1, \ldots, x_n, x_{n+1}$ is needed, we only have to extend the divided difference table to give $f[x_1, \ldots, x_{n+1}]$, since

$$p(x) + f[x_1, \ldots, x_{n+1}] \prod_{j=1}^{n} (x - x_j) = \sum_{j=1}^{n+1} f[x_1, \ldots, x_j] \prod_{j=1}^{j-1} (x - x_j) \quad (2.90)$$

is the desired polynomial.

Example 2.3

The divided difference table for $f(x) = \cos x$ at the points $x_j = (j-1)\frac{\pi}{4}$, $j = 1, 2, 3, 4, 5$, constructed in five-decimal digit arithmetic, is given below.

l	x_l	$f[x_l]$	$f[x_l, x_{l+1}]$	$f[x_l, x_{l+1}, x_{l+2}]$	$f[,,,]$	$f[,,,,]$
1	0.0	1.0	-.37292	-.33575	-.14250	.09072
2	.78540	.70711	-.90032	0.0	.14250	
3	1.5708	0.0	-.90032	.33575		
4	2.3562	-.70711	-.37292			
5	3.1416	-1.0				

To within roundoff error, the quadratic polynomial interpolating $\cos x$ at $0, \frac{\pi}{4}$, and $\frac{\pi}{2}$ is

$$p(x) = 1.0 - .37292x - .33575x(x - .78540)$$

The quartic polynomial interpolating $\cos x$ at $0, \frac{\pi}{4}, \frac{\pi}{2}, \frac{3\pi}{4}$, and π is

$$p(x) = .14250x(x-.78540)(x-1.5708) + .09072x(x-.78540)(x-1.5708)(x-2.3562)$$

$\square$

The form of the error arising from interpolation is given in the following theorem.

Theorem 2.91 The unique polynomial $p(x)$ of order n interpolating $f(x)$ at $x_1, x_2, \ldots, x_n$ has, for any $\bar{x} \in [a,b]$, the error term

$$f(\bar{x}) - p(\bar{x}) = \frac{f^{(n)}(\xi_{\bar{x}})}{n!} \prod_{j=1}^{n} (\bar{x} - x_j) \qquad (2.91)$$

for some $\xi_{\bar{x}} \in [a,b]$. $\square$

If $f^{(n)}(x)$ is bounded by M_n on $[a,b]$, then from (2.91) the interpolation error is bounded by $M_n C_n h^n$, where $h = b - a$ and C_n is a constant which depends on n. The rate at which C_n actually grows with n depends upon how the interpolation points are selected. For example, for equally spaced points C_n grows like $e^{n/2}$, while for the so-called Chebyshev points C_n only grows linearly with n.

2.4.2 Osculatory interpolation

Frequently we want to interpolate derivative values of $f(x)$ in addition to function values. This is called *osculatory interpolation*. It can be done simply when the Newton form of the interpolating polynomial is used. The definition of the kth divided difference of $f(x)$ is extended to the case $x_l = x_{l+1} = \ldots = x_{l+k}$ by defining

$$f[x_l, \ldots, x_{l+k}] := \frac{f^{(k)}(x_l)}{k!} \qquad (2.92)$$

The basic properties of divided differences can be shown to still hold. Note that if $x_1 = \cdots = x_n$, the Newton form (2.88) is just the first n terms of the Taylor expansion about x_1,

$$p(x) = \sum_{i=0}^{n-1} \frac{f^{(i)}(x_1)}{i!} (x - x_1)^i$$

which is, of course, a polynomial matching the first n derivatives of $f(x)$ at x_1. This provides motivation for the next theorem, which we give without proof.

Theorem 2.93 Given points $x_1, \ldots, x_n$ in $[a,b]$, the polynomial $p(x) = \sum_{i=1}^{n} f[x_1, \ldots, x_i] \prod_{j=1}^{i-1} (x - x_j)$ interpolates $f(x)$ in the following sense: If the point z occurs l times among the points $\{x_i\}_{j=1}^{n}$, then

$$p^{(j)}(z) = f^{(j)}(z) \qquad j = 0, \ldots, l-1 \qquad (2.93)$$

In addition, the error formula (2.91) is still valid. □

2.5 PIECEWISE POLYNOMIALS, OR SPLINES

For various kinds of approximation problems, including interpolation, it is frequently advantageous to use piecewise polynomials instead of polynomials. This is because using low-degree polynomials locally is usually more accurate and/or more efficient than using a high-degree polynomial globally. We define a *spline* as any piecewise polynomial function.

2.5.1 Local representations

In this section we introduce convenient methods for locally representing splines.

Definition 2.94 Given a partition π of some interval $[a, b]$,

$$\pi : a = x_1 < x_2 < \cdots < x_{N+1} = b \qquad (2.94a)$$

$\mathbf{P}_{k,\pi}$ is defined as the family of functions which, on each subinterval of π, are polynomials of order k. Thus, $s(x) \in \mathbf{P}_{k,\pi}$ if

$$s(x) = P_i(x) := a_{i0} + a_{i1}(x-x_i) + \cdots + a_{i,k-1}(x-x_i)^{k-1} \qquad (2.94b)$$

on (x_i, x_{i+1}) for some constants $\{a_{ij}\}_{j=0}^{k-1}$, $1 \le i \le N$. At a mesh point x_i, when $s(x)$ is not continuous, we shall arbitrarily define $s(x)$ to be right continuous by setting $s(x_i) = P_i(x_i)$, i.e., $P_i(x)$ is defined on $[x_i, x_{i+1})$ $(1 \le i \le N)$, and $s(b) = P_N(b)$. The representation (2.94b) is called the *(local) monomial representation* for $s(x)$. □

Definition 2.95 The subfamily of $\mathbf{P}_{k,\pi}$ consisting of splines $s(x) \in C^{(m-1)}[a, b]$ is denoted by $\mathbf{P}_{k,\pi,m}$. Also, $\mathbf{P}_{k,\pi,0} \equiv \mathbf{P}_{k,\pi}$. □

In some literature, a spline of order k means a function in $\mathbf{P}_{k,\pi,k-1}$, but such a restriction is not useful for our purposes. We shall call $\mathbf{P}_{k,\pi,k-1}$ the space of *smooth splines* of order k on π. When k is even, $\mathbf{P}_{k,\pi,k/2}$ is called the space of *Hermite splines* of order k.

For $m \le \dfrac{k}{2}$, it is sometimes convenient to use an alternative to (2.94b) as a local representation, which can be derived as follows: Given the points

$$(\eta_0 <) \; 0 \le \eta_1 \le \eta_2 \le \cdots \le \eta_k = 1 \qquad (2.96)$$

let $\phi_j(x)$ (for $j = 1, \ldots, k$) be the polynomial of order k defined by

$$D^\mu \phi_j(\eta_l) = \delta_{jl} \qquad (2.97)$$

for $1 \le l, j \le k$, where μ is defined such that $\eta_{l-\mu-1} < \eta_{l-\mu} = \eta_{l-\mu+1} = \cdots = \eta_l$. The construction and evaluation of each $\phi_j(t)$ for $t \in (0, 1)$ can be efficiently done by forming a divided difference table for $\phi_j(t)$ corresponding to $\{\eta_j\}_1^k$ and (2.97), and then using the algorithm (2.89) to evaluate $\phi_j(t)$. On (x_i, x_{i+1})

$$s(x) = \sum_{j=1}^{k} \alpha_{(i-1)k+j} \phi_j \left(\frac{x-x_i}{h_i} \right) \qquad (2.98)$$

where $h_i = x_{i+1} - x_i$. By requiring $\{\eta_j\}_{j=1}^{k}$ to satisfy

$$0 = \eta_1 = \cdots = \eta_m \leq \cdots \leq \eta_{k-m} < \eta_{k-m+1} \qquad (2.99)$$

$$= \cdots = \eta_k = 1$$

we obtain that $s(x) \in \mathbf{P}_{k,\pi,m}$ if $s^{(j)}(x)$ is continuous at x_i, i.e., if $\dfrac{\alpha_{(i-1)k+j}}{h_i^{j-1}}$
$= \dfrac{\alpha_{(i-1)k+j-m}}{h_{i-1}^{j-1}}$, for $1 \leq j \leq m$, $2 \leq i \leq N$. In this case, (2.98) is called the *Hermite-type* representation for $s(x)$. When $m < \dfrac{k}{2}$, $\{\eta_j\}_{j=m+1}^{k-m}$ can be chosen arbitrarily, subject to (2.99). The functions $\{\psi_r(x)\}_{r=1}^{(k-m)N+m}$ defined by

$$\psi_j(x) = \begin{cases} \phi_j\left(\dfrac{x-x_1}{h_1}\right), & \text{if } x \in (x_1, x_2) \\ & \qquad 1 \leq j \leq m \\ 0, & \text{otherwise} \end{cases}$$

$$\psi_{(i-1)(k-m)+j}(x) = \begin{cases} \left(\dfrac{h_{i-1}}{h_i}\right)^{j-1} \phi_{j+(k-m)}\left(\dfrac{x-x_{i-1}}{h_{i-1}}\right), & \text{if } x \in (x_{i-1}, x_i) \quad 1 \leq j \leq m \\ \phi_j\left(\dfrac{x-x_i}{h_i}\right), & \text{if } x \in (x_i, x_{i+1}) \\ 0, & \text{otherwise} \qquad (2 \leq i \leq N) \end{cases} \qquad (2.100)$$

$$\psi_{(i-1)(k-m)+j}(x) = \begin{cases} \phi_j\left(\dfrac{x-x_i}{h_i}\right), & \text{if } x \in (x_i, x_{i+1}) \quad m+1 \leq j \leq k-m \\ 0, & \text{otherwise} \qquad (1 \leq i \leq N) \end{cases}$$

$$\psi_{N(k-m)+j}(x) = \begin{cases} \phi_{j+m}\left(\dfrac{x-x_N}{h_N}\right), & \text{if } x \in (x_N, x_{N+1}) \quad 1 \leq j \leq m \\ 0, & \text{otherwise} \end{cases}$$

constitute a *Hermite-type basis* for the spline space $\mathbf{P}_{k,\pi,m}$. Note that each $\psi_r(x) \in C^{(m-1)}[a,b]$, and from (2.99) and (2.100)

$$D^{r-1}\psi_{(i-1)k+j}(x_l) = \delta_{il}\delta_{rj}h_l^{1-r} \qquad 1 \leq j, r \leq \frac{k}{2}, \; 1 \leq i, l \leq N+1 \qquad (2.101)$$

2.5.2 B-splines

It may be desirable to construct a basis for $\mathbf{P}_{k,\pi,m}$, i.e., to construct a convenient set of splines which span $\mathbf{P}_{k,\pi,m}$. For the special case $m \leq \frac{k}{2}$, Hermite-type bases are one possibility. In this section we show briefly how to construct a more general alternative basis, the *B-spline basis*, which has historically been the most used general basis for $\mathbf{P}_{k,\pi,m}$.

The *truncated power function* $(t)_+^r$ $(r>0)$ is the function [in $C^{(r-1)}(-\infty,\infty)$] defined by

$$(t)_+^r := (\max\{t,0\})^r = \begin{cases} t^r & \text{if } t > 0 \\ 0 & \text{if } t \leq 0 \end{cases} \tag{2.102}$$

Let $\{t_j\}_{j=1}^{J+k}$, $J := (N-1)(k-m) + k$, be a sequence of nondecreasing points such that $t_1 \leq t_2 \leq \cdots \leq t_k = x_1$; $x_2, x_3, \ldots, x_N$ are each repeated $(k-m)$ times; and $x_{N+1} = t_{J+1} \leq t_{J+2} \leq \cdots \leq t_{J+k}$. For $1 \leq j \leq J$, define $M_j(x)$ by

$$M_j(x) := g_x[t_j, \ldots, t_{j+k}] \tag{2.103}$$

where

$$g_x(t) := (x-t)_+^{k-1} \tag{2.104}$$

The *B-spline of order k on $t_j, \ldots, t_{j+k}$* is

$$B_{j,k}(x) := (t_{j+k} - t_j)(-1)^k M_j(x) \quad 1 \leq j \leq J \tag{2.105}$$

The set $\{B_{j,k}(x)\}_{j=1}^{J}$ forms a basis for $\mathbf{P}_{k,\pi,m}$. This basis has a number of attractive features. For example, $B_{j,k}(x)$ is positive on (t_j, t_{j+k}) and zero elsewhere; i.e., $B_{j,k}(x)$ has *local support* (t_j, t_{j+k}), so computations using a B-spline basis generally lead to linear systems of equations with banded matrices. Evaluation of kth-order B-splines and their derivatives can be done by generating lower order ones first, utilizing a relation expressing pth-order B-splines in terms of $(p-1)$st-order ones. Specifically, the following algorithm can be used to evaluate the k nonzero kth-order B-splines $B_{j,k}(x)$ at $x \in (t_j, t_{j+1})$:

Algorithm 2.106 B-spline evaluation

$B_{j,1}(x) = 1$

FOR $l = 1, \ldots, k-1$ DO
  $B_{j-l,l+1}(x) = 0$
    FOR $i = 1, \ldots, l$ DO
      $M_{j+i-l,l}(x) = B_{j+i-l,l}/(t_{j+i} - t_{j+i-l})$

      $B_{j+i-l-1,l+1}(x) = B_{j+i-l-1,l+1}(x) + (t_{j+i} - x)M_{j+i-l,l}(x)$

      $B_{j+i-l,l+1}(x) = (x - t_{j+i-l})M_{j+i-l,l}(x)$ □

If $s(x) = \sum_{r=1}^{J} a_r B_{r,k}(x)$, then its derivatives can be found for $x \in (t_j, t_{j+k})$ from

$$s^{(i)}(x) = \sum_{l=j-k+i-1}^{j} a_{l,i+1} B_{l,k-i}(x) \qquad (2.107)$$

where

$$a_{l,i+1} := \begin{cases} a_l & \text{if } i = 0 \\ (k-i)\dfrac{a_{l,i} - a_{l-1,i}}{t_{l+k-i} - t_l} & \text{if } i > 0 \end{cases} \qquad (2.108)$$

To evaluate $B_{j,k}^{(i)}(x)$, (2.107) and (2.108) are used with $a_r = \delta_{rj}$, $1 \le r \le J$.

2.5.3 Spline interpolation

Consider the problem of interpolating a function $f(x)$ at $J = (N-1)(k-m) + k$ points $z_1 < z_2 < \cdots < z_J$ in $[a,b]$ with a spline function $s(x) \in \mathbf{P}_{k,\pi,m}$. Given a B-spline basis for $\mathbf{P}_{k,\pi,m}$, this problem is equivalent to solving the linear system of equations

$$s(z_j) = \sum_{i=1}^{J} B_{i,k}(z_j) a_i = f(z_j) \qquad 1 \le j \le J \qquad (2.109)$$

for the unknown coefficients $\{a_i\}_{i=1}^{J}$. It can be shown that the matrix $(B_{i,k}(z_j))_{i,j=1}^{J}$ is nonsingular iff $B_{i,k}(z_i) \ne 0$ for $1 \le i \le J$. From the local support property of B-splines, the matrix has bandwidth at most $2k-1$ whenever the interpolation problem has a unique solution. In fact, the matrix is *totally positive* (i.e., all minors have strictly positive determinants), which implies that Gaussian elimination without pivoting can be used to solve (2.109). Note that using a local representation like (2.94b) instead to find $s(x)$ would require forming a system of equations involving the J interpolation conditions *and* the $(N-1)m$ continuity conditions at $\{x_i\}_{i=2}^{N}$.

The form of the error for spline approximation is what one might expect from the comments following Theorem 2.91: if $f(x) \in C^{(k)}[a,b]$ then there is a spline $\hat{s}(x) \in \mathbf{P}_{k,\pi,m}$ satisfying

$$\max_{a \le x \le b} |f(x) - \hat{s}(x)| =: \|f - \hat{s}\| \le C_k h^k \|f^{(k)}\| \qquad (2.110)$$

for some constant C_k independent of $f(x)$ and π, and $h := \max_{1 \le i \le N}(x_{i+1} - x_i)$. The interpolating spline $s(x)$ in place of $\hat{s}(x)$ also satisfies (2.110). Under certain conditions, a spline approximation is determined locally (e.g., each B-spline coefficient a_i only depends upon $f(x)$ on $[t_{i+1-k}, t_{i+k}]$) and (2.110) can be strengthened to

$$\max_{t_i \le x \le t_{i+k}} |f(x) - \hat{s}(x)| =: \|f - \hat{s}\|_{[t_i, t_{i+k}]} \le \hat{C}_k \hat{h}_i^k \|f^{(k)}\|_{[t_{i+1-k}, t_{i+k}]} \qquad (2.111)$$

where $\hat{h}_i := t_{i+k+1} - t_{i+1-k}$ and $\hat{C}_k$ is a constant independent of π and $f(x)$. This form for the error motivates selection of a mesh which *equidistributes*, or "distributes equally" the local errors $\hat{h}_i^k \|f^{(k)}\|_{[t_{i+1-k}, t_{i+k}]}$. In fact, if $\pi^* : a = \tau_1 < \tau_2 < \cdots < \tau_{N+1} = b$ is chosen such that

$$\int_{\tau_j}^{\tau_{j+1}} |f^{(k)}(x)|^{1/k} dx \approx \frac{1}{N} \int_a^b |f^{(k)}(x)|^{1/k} dx \qquad 1 \le i \le N \qquad (2.112)$$

then there exists a function $s^*(x) \in \mathbf{P}_{k,\pi^*,m}$ satisfying

$$\|f - s^*\| \le \frac{C_k}{N^k} [\int_a^b |f^{(k)}(x)|^{1/k} dx]^k \qquad (2.113)$$

The idea of adapting an equidistributing mesh for a particular $f(x)$ to give an improved error bound like (2.113) is basic to much of Chapter 9.

A simple example of local spline approximation is osculatory interpolation of the first $\frac{k}{2}$ derivatives of $f(x)$ at $\{x_i\}_{i=1}^{N+1}$ by a Hermite spline $s(x) \in \mathbf{P}_{k,\pi,\frac{k}{2}}$. In this case, from (2.101) $s(x)$ is given explicitly by

$$s(x) = \sum_{i=1}^{N+1} \sum_{j=1}^{k/2} f^{(j-1)}(x_i) h_i^{1-j} \psi_{(i-1)k+j}(x) \qquad (2.114)$$

On (x_i, x_{i+1}) $s(x)$ is the kth order polynomial interpolating $f^{(j-1)}(x_i)$ and $f^{(j-1)}(x_{i+1})$ ($1 \le j \le \frac{k}{2}$), and so from (2.91)

$$\|f - s\|_{[x_i, x_{i+1}]} \le \frac{1}{k!} \left[\frac{h_i}{2}\right]^k \|f^{(k)}\|_{[x_i, x_{i+1}]} \qquad (2.115)$$

which is of the same form as (2.111).

2.6 NUMERICAL QUADRATURE

A *numerical quadrature method* is a method for numerically approximating a definite integral. For such methods a *basic rule* is usually applied within a *composite formula*, as we briefly describe below. Our interest in them here stems from the fact that finite difference schemes for ODEs are often based on numerical quadrature rules.

2.6.1 Basic rules

The basic strategy for approximating the integral $I(f) = \int_a^b f(x)w(x)dx$, where $w(x)$ is a given weight function, is to use analytic substitution. That is, $f(x)$ is replaced by a simple function such as a polynomial $p(x)$, and $I(p)$ is integrated exactly. Let $p(x)$ be the Lagrange interpolating polynomial (2.83) for $f(x)$ at $a \le \rho_1 < \rho_2 < \cdots < \rho_s \le b$. Then

$$I(f) \approx I(p) = \int_a^b p(x)w(x)dx = \int_a^b \sum_{i=1}^s f(\rho_i) L_i(x) w(x) dx \qquad (2.116)$$

$$= \sum_{i=1}^s w_i f(\rho_i) =: S(f)$$

where the *weights* are defined by

$$w_i := \int_a^b L_i(x)w(x)\,dx \tag{2.117}$$

Note that, from uniqueness of an interpolating polynomial, $I(q) = S(q)$ for any polynomial $q(x) \in \mathbf{P}_s$. Sometimes s is called the *precision* of the quadrature rule. From Theorem 2.91, the error is

$$I(f) - I(p) = \int_a^b f[\rho_1, \ldots, \rho_s, x]w(x)\prod_{j=1}^s (x-\rho_j)\,dx \tag{2.118}$$

$$= \int_a^b \frac{f^{(s)}(\xi_x)}{s!} w(x)\prod_{j=1}^s (x-\rho_j)\,dx$$

When $w(x) \equiv 1$ and equal spaced points are used, the resulting rules are are called *Newton Cotes Formulas*. Important examples are the *midpoint rule* ($s = 1$):

$$I(f) \approx (b-a)f\left(\frac{a+b}{2}\right) =: M(f) \tag{2.119a}$$

with error

$$I(f) - M(f) = \frac{f''(\eta)}{24}(b-a)^3, \quad \eta \in (a,b) \tag{2.119b}$$

the *trapezoidal rule* ($s = 2$):

$$I(f) \approx \frac{b-a}{2}[f(a)+f(b)] =: T(f) \tag{2.120a}$$

with error

$$I(f) - T(f) = -\frac{f''(\eta)}{12}(b-a)^3, \quad \eta \in (a,b) \tag{2.120b}$$

and *Simpson's rule* ($s = 3$):

$$I(f) \approx \frac{b-a}{6}\left[f(a)+4f\left(\frac{a+b}{2}\right)+f(b)\right] =: S(f) \tag{2.121a}$$

with error

$$I(f) - S(f) = \frac{f''''(\eta)}{2880}(b-a)^5, \quad \eta \in (a,b) \tag{2.121b}$$

In general, the odd order Newton Cotes formulas have a precision one higher than to be expected from the interpolation error.

For the *Gauss formulas*, the points $\{\rho_i\}_{i=1}^s$ are chosen so as to make the quadrature formula exact for as high order polynomials as possible. [Note that there are $2s$ degrees of freedom in $S(f)$ in (2.116).]

A sequence of polynomials $\{P_i(x)\}$ is a set of *orthonormal polynomials* on $[a, b]$ if each $P_i(x)$ is of degree i and

$$\int_a^b P_i(x)P_j(x)w(x)\,dx = \delta_{ij} \qquad i,j = 0,1,2,\cdots \tag{2.122}$$

The basic relation between general Gauss quadrature formulas and orthonormal polynomials is given in the next theorem.

Theorem 2.123 Given a weight function $w(x) \geq 0$ on $[a,b]$ and corresponding orthonormal polynomials $\{P_i(x)\}_{i=0}^{\infty}$, if $P_s(x) = c_s x^s + \cdots$ and $P_{s+1}(x) = c_{s+1} x^{s+1} + \cdots$ then the zeros $\{\rho_s\}$ of $P_s(x)$ are real and satisfy $a < \rho_1 < \rho_2 < \cdots < \rho_s < b$. If $w_j = \dfrac{c_{s+1}}{c_s} \dfrac{1}{P_{s+1}(\rho_j) P_s'(\rho_j)}$ $(1 \leq j \leq s)$, then the Gauss formula

$$\int_a^b f(x) w(x)\, dx \approx \sum_{j=1}^{s} w_j f(\rho_j) =: S(f) \tag{2.123a}$$

is exact for all polynomials of order $2s$. If $f(x) \in C^{(2s)}[a,b]$ then

$$I(f) - s(f) = \frac{f^{(2s)}(\eta)}{(2s)! c_s^2} (b-a)^{2s+1}, \; \eta \in (a,b) \tag{2.123b}$$

□

A number of choices of $[a,b]$ and $w(x)$ are useful in practice. In the case $w(x) \equiv 1$, $a = -1$, and $b = 1$, $\{P_i(x)\}$ are called the *Legendre polynomials,* the quadrature points

$$-1 < \rho_1 < \rho_2 < \cdots < \rho_s < 1 \tag{2.124}$$

are called the *Gauss-Legendre points,* or simply the *Gauss points,* and

$$I(f) - S(f) = \frac{2^{2s+1}(s!)^4}{(2s+1)[(2s)!]^3} f^{(2s)}(\eta), \; \eta \in (-1, 1) \tag{2.125}$$

Some Gauss-type formulas use certain preassigned quadrature points. In particular

$$\int_a^b f(x) w(x)\, dx \approx \sum_{j=1}^{s_1} v_j f(\alpha_j) + \sum_{j=1}^{s_2} w_j f(\beta_j) \tag{2.126}$$

where $\{\alpha_j\}$ are fixed in advance and the $s_1 + 2s_2$ constants $\{v_j, w_j, \beta_j\}$ are chosen such that (2.126) has the highest precision possible, which is $s_1 + 2s_2$. We also have

Theorem 2.127 The quadrature formula (2.126) is exact for all polynomials of order $s_1 + 2s_2$ iff

(i) It is exact for all polynomials of order $s_1 + s_2$ and

(ii) $\int_a^b p(x) \prod_{j=1}^{s_1} (x - \alpha_j) \prod_{j=1}^{s_2} (x - \beta_j) w(x)\, dx = 0$ for all $p(x) \in \mathbf{P}_{s_2}$ □

For $w(x) \equiv 1$, $a = -1$, and $b = 1$, an important special case is the *Radau points*

$$-1 < \rho_1 < \rho_2 < \cdots < \rho_s = 1 ,$$

where in the notation of (2.124) $w_s = \dfrac{2}{s^2}$, $w_j = \dfrac{1}{s^2} \dfrac{1-\rho_j}{(P_{s-1}(\rho_j))^2}$, and ρ_j is the jth zero of $\dfrac{P_{s-1}(x) + P_s(x)}{x+1}$ ($1 \leq j \leq s-1$) with the $P_l(x)$s the unnormalized Gauss Legendre polynomials defined by $P_0(x) = 1$

$$P_l(x) = \frac{1}{2^l l!} \frac{d^l}{dx^l}(x^2-1)^l \qquad (1 \leq l \leq s) \qquad (2.128)$$

The error for the Radau formula when $f(x) \in C^{(2s-1)}[-1, 1]$ is

$$I(f) - S(f) = \frac{2^{2s-1}}{[(2s-1)!]^3}[(s-1)!]^4 f^{(2s-1)}(\eta), \quad \eta \in (-1, 1) \qquad (2.129)$$

Another important case is the *Lobatto points*

$$-1 = \rho_1 < \rho_2 < \cdots < \rho_{s-1} < \rho_s = 1$$

where $w_1 = w_s = \dfrac{2}{s(s-1)}$, $w_j = \dfrac{2}{s(s-1)[P_{s-1}(\rho_j)]^2}$ and ρ_j is the $(j-1)$st zero of $P_{s-1}'(x)$, $2 \leq j \leq s-1$, with $P_s(x)$ defined again as in (128). For $f(x) \in C^{(2s-2)}[-1, 1]$, the error of the Lobatto formula is

$$I(f) - S(f) = \frac{s(s-1)^3 2^{2s-1}[(s-2)!]^4}{(2s-1)[(2s-2)!]^3} f^{(2s-2)}(\eta), \quad \eta \in (-1, 1) \qquad (2.130)$$

If follows from Theorem 2.127 that

$$\int_a^b p(x) \prod_{j=1}^s (x-\rho_j) dx = 0 \qquad (2.131)$$

for all $p(x) \in \mathbf{P}_\mu$, where

$$\mu = \begin{cases} s & \text{for *Gauss points*} \\ s-1 & \text{for *Radau points*} \\ s-2 & \text{for *Lobatto points*} \end{cases} \qquad (2.132)$$

2.6.2 Composite formulas

The basic quadrature rules in the last section are usually not sufficiently accurate, especially if $[a, b]$ is a large interval. *Composite formulas* are formed by dividing $[a, b]$ into N subintervals and applying a basic rule to each of the subintervals. This involves a transformation of the basic rule, so first we consider such a change of interval.

Given a quadrature rule

$$S(f) = \sum_{i=1}^s w_i f(\rho_i)$$

to approximate the integral

$$I(f) = \int_a^b f(x) dx$$

[cf. (2.116)], it is easy to derive a similar formula for approximating the integral

$$\tilde{I}(f) = \int_c^d f(y)\,dy \qquad (2.133\text{a})$$

Using the linear transformation
$$y = \phi(x) = (b-a)^{-1}[(d-c)x - (ad-bc)]$$
we obtain
$$\tilde{S}(f) = \int_a^b f(\phi(x))\frac{dy}{dx}dx = \frac{d-c}{b-a}[\sum_{i=1}^s w_i f(\phi(\rho_i))] \qquad (2.133\text{b})$$

which is an approximation for $\tilde{I}(f)$ of the same precision as $S(f)$ is to $I(f)$. Indeed, if the remainder term satisfies
$$I(f) - S(f) = K(b-a)^{m+1}f^{(m)}(\theta)$$
for some $\theta \in (a,b)$ and some K and m (which is true for all quadrature rules considered) then
$$\tilde{I}(f) - \tilde{S}(f) = K(d-c)^{m+1}f^{(m)}(\eta), \qquad \eta \in (c,d) \qquad (2.133\text{c})$$

For composite formulas, a basic rule is often used on $[0,h]$ instead of $[-1,1]$, so the corresponding error term has the factor $(h/2)^{m+1}$.

For example the composite trapezoidal rule is
$$I(f) = \int_a^b f(x)\,dx = \sum_{i=1}^N \int_{x_i}^{x_{i+1}} f(x)\,dx \qquad (2.134\text{a})$$
$$\approx \sum_{i=1}^N \frac{x_{i+1}-x_i}{2}[f(x_i)+f(x_{i+1})] = T_N(f)$$

For a uniform mesh, from (2.120b) and the Mean Value Theorem for sums, the error is
$$I(f) - T_N(f) = -\sum_{i=1}^N \frac{f''(\eta_i)}{12}h^3 = -\frac{f''(\hat{\eta})}{12}(b-a)h^2 \qquad (2.134\text{b})$$

where $\eta_i \in (x_i, x_{i+1})$ $(1 \le i \le N)$, and $\hat{\eta} \in (a,b)$.

Similar construction of composite formulas can be done based on other rules, e. g., on Gaussian quadrature. The derivation of corresponding error terms is equally simple.

2.7 INITIAL VALUE ORDINARY DIFFERENTIAL EQUATIONS

As we saw in the previous chapter, both the mathematical theory and the numerical methods for solving boundary value problems are intimately related to corresponding techniques for initial value problems in ordinary differential equations. The situation, though, is in many respects much simpler for IVPs: A unique solution is guaranteed to exist under very mild assumptions, and the fact that everything is known about the solution at the starting point enables us to numerically construct a local algorithm in the independent variable. This contrasts with the case of BVPs, where much more complex global algorithms are needed.

There are a number of general-purpose codes which solve, efficiently and reliably, most IVPs arising in practice (with the possible exception of stiff problems — see Section 2.7.3). These codes are based on one-step methods, multistep methods, or extrapolation methods, and we briefly describe some of the underlying schemes in Section 2.7.1. However, our purpose here is not to replace a thorough exposition of numerical techniques for IVPs. Rather, we prepare the groundwork needed in subsequent chapters. In particular, the description is motivated by (a) the techniques of Chapter 4, where the initial value methods are used for solving BVPs, and by (b) difference methods and stability analysis needed in subsequent chapters, which rely on the corresponding initial value techniques.

2.7.1 Numerical methods

The problem considered is

$$y' = f(x,y) \qquad a < x < b \qquad (2.135a)$$

$$y(a) = \alpha \qquad (2.135b)$$

where f is a (possibly) nonlinear function, assumed to be Lipschitz continuous in y, and α is a given initial value. We consider a single equation (2.135a) because simpler notation can then be used. Everything in this section can be generalized to systems of differential equations in a straightforward manner.

A numerical approximation $\{y_i : i = 1, 2, \ldots, N+1\}$ is to be calculated on a mesh

$$\pi : a = x_1 < x_2 < \cdots < x_{N+1} = b$$

such that $y_i \approx y(x_i)$. For simplicity we shall use a constant step size $h = h_i := x_{i+1} - x_i$, i.e., $x_i = a + (i-1)h$ for all i.

We have already seen a simple method (2.2) for solving (2.135a), namely Euler's method, which is based on the simple difference scheme

$$y_{i+1} = y_i + h f_i \qquad i = 1, \ldots, N$$

where $f_i := f(x_i, y_i)$. Starting with $y_1 = \alpha$, the formula (2.2b) provides an algorithm for obtaining $\{y_2, \ldots, y_{N+1}\}$ by explicitly stepping forward in i. More generally, all the schemes treated here have the form

$$y_{i+k} = \alpha_{k-1} y_{i+k-1} + \cdots + \alpha_0 y_i + h \Phi(x_{i+k}, \ldots, x_i; y_{i+k}, \ldots, y_i; h)$$
$$i = 1, 2, \ldots \qquad (2.136)$$

where $\alpha_0, \alpha_1, \ldots, \alpha_{k-1}$ are known constants, Φ is a known function depending on f, and $y_1, \ldots, y_k$ are given. (If $k > 1$, the additional initial values have to be obtained in some other way.) Clearly, for Euler's method, $k = 1$, $\alpha_0 = 1$, and $\Phi(x_{i+1}, x_i; y_{i+1}, y_i; h) = f(x_i, y_i)$.

Two classes of schemes of the general form (2.136) are considered. In the first, we take $k = 1$ to obtain *one-step schemes,* i.e., schemes which use only information about the approximate solution at x_i to obtain the approximate solution at x_{i+1}. A popular family of schemes of this type are the *Runge-Kutta* schemes.

The general form of a *q-stage Runge-Kutta method* is

$$y_1 = y(a) \tag{2.137a}$$

$$y_{i+1} = y_i + h \sum_{j=1}^{q} b_j g_j, \quad i = 1, 2, 3, \ldots \tag{2.137b}$$

where

$$g_j = f\left(x_i + h\rho_j, y_i + h \sum_{l=1}^{q} a_{jl} g_l\right), \quad (1 \le j \le q) \tag{2.137c}$$

Here $\{\rho_j, b_j, a_{jl}\}$ are known constants such that $0 \le \rho_1 \le \rho_2 \le \cdots \le 1$. We will often use the so-called *Butcher array* notation of Fig. 2.3 below to represent Runge-Kutta schemes.

$$\begin{array}{c|ccc} \rho_1 & a_{11} & \cdots & a_{1k} \\ \vdots & \vdots & & \vdots \\ \rho_k & a_{k1} & \cdots & a_{kk} \\ \hline & b_1 & \cdots & b_k \end{array}$$

Figure 2.3 A Runge-Kutta scheme

The difference scheme in (2.137) is called *explicit* if $a_{jl} = 0$, $l \ge j$; it is called *implicit* otherwise. Explicit schemes have the advantage that the computation of the new approximation y_{i+1} is straightforward. For implicit schemes we need to solve a nonlinear equation (if f is nonlinear). If $|h\frac{\partial f}{\partial y}| < 1$ (cf. Theorem 2.62), then this can be done by a simple fixed-point iteration, and otherwise by the more expensive Newton method. Note that an implicit scheme always needs another scheme predicting the starting value for the iteration. This other scheme is called a *predictor,* and the implicit scheme is called a *corrector*.

Euler's scheme (2.2) may be looked upon as a one-stage Runge-Kutta scheme with $\rho_1 = 0$, $b_1 = 1$, and $a_{11} = 0$. Two popular two-stage formulas are the explicit *(modified Euler)* scheme

$$y_{i+1} = y_i + \frac{h}{2}\{f_i + f(x_{i+1}, y_i + hf_i)\} \tag{2.138}$$

and the implicit *trapezoidal scheme*

$$y_{i+1} = y_i + \frac{h}{2}\{f_i + f_{i+1}\} \tag{2.139}$$

A popular pair of explicit Runge-Kutta formulas are the so-called *fourth- and fifth-order Fehlberg* schemes. Table 2.1 shows their coefficients. The a_{jl} are the same for both methods, but b_j applies to the "fourth order" one and b_j^* to the "fifth order" one. This pair of formulas is useful because of the inexpensive way it can be used to provide both an approximate solution and an *error estimate* (see Section 2.7.4). The most robust Runge-Kutta schemes for nonstiff IVPs are explicit. However, when attempting to use the same difference scheme for a boundary value problem, as we do in Chapter 5, any scheme becomes effectively implicit. Thus, the distinction between explicit and implicit initial value schemes becomes less important in the BVP context. Indeed, some implicit

TABLE 2.1

0						
$\frac{1}{4}$	$\frac{1}{4}$					
$\frac{3}{8}$	$\frac{3}{32}$	$\frac{9}{32}$				
$\frac{12}{13}$	$\frac{1932}{2197}$	$-\frac{7200}{2197}$	$\frac{7296}{2197}$			
1	$\frac{439}{216}$	-8	$\frac{3680}{513}$	$-\frac{845}{4104}$		
$\frac{1}{2}$	$-\frac{8}{27}$	2	$-\frac{3544}{2565}$	$\frac{1859}{4104}$	$-\frac{11}{40}$	
b_j	$\frac{25}{216}$	0	$\frac{1408}{2565}$	$\frac{2197}{4104}$	$-\frac{1}{5}$	
b_j^*	$\frac{16}{135}$	0	$\frac{6696}{12{,}825}$	$\frac{28{,}561}{56{,}430}$	$-\frac{9}{50}$	$\frac{2}{55}$

Runge-Kutta schemes offer more under these circumstances. For instance, by choosing $\rho_1, \ldots, \rho_q$ as the Gaussian points, i.e., the zeros of Legendre polynomial of degree q on $[0,1]$, we can obtain a q-stage Runge-Kutta method of order $2q$. The simplest of these schemes is the *midpoint scheme,* obtained for $q = 1$,

$$y_{i+1} = y_i + hf\left(x_i + \frac{1}{2}h_i \,,\, \frac{1}{2}(y_i + y_{i+1})\right) \tag{2.140}$$

With Radau points (cf. Section 2.6.1), order $2q-1$ is achievable, while with Lobatto points we can get schemes of order $2q-2$. These schemes are treated in more detail in Sections 5.3 and 5.4.

The second successful class of schemes which is covered by (2.136) is that of linear multistep schemes. These have the general form

$$y_{i+k} = \sum_{j=0}^{k-1} \alpha_j y_{i+j} + h \sum_{j=0}^{k} \beta_j f_{i+j} \tag{2.141}$$

with $\beta_0, \beta_1, \ldots, \beta_k$ known constants. If $\beta_k = 0$ then the scheme is explicit; otherwise, it is implicit. For instance, the specification $k = 1$, $\alpha_0 = 1$, $\beta_0 = 1$, $\beta_1 = 0$ defines once again Euler's scheme, while $k = 1$, $\alpha_0 = 1$, $\beta_0 = \beta_1 = 0.5$ defines the trapezoidal method (2.139). A very popular class of schemes are the *Adams schemes,* obtained by formally integrating the differential equation,

$$y_{i+k} - y_{i+k-1} \approx \int_{x_{i+k-1}}^{x_{i+k}} y'(x)\,dx = \int_{x_{i+k-1}}^{x_{i+k}} f(x, y(x))\,dx \tag{2.142}$$

and replacing $f(x, y(x))$ by an interpolating polynomial. For an explicit formula, the polynomial $P_{k-1} \in \mathbf{P}_k$ satisfying

$$P_{k-1}(x_{i+j}) = f(x_{i+j}, y_{i+j}) = f_{i+j} \qquad j = 0, \ldots, k-1$$

is used, and for an implicit formula, the polynomial $P_k \in \mathbf{P}_{k+1}$ satisfying

$$P_k(x_{i+j}) = f_{i+j} \qquad j = 0, \ldots, k$$

is used. For instance, the third-order (3-step) explicit formula of this family is

$$y_{i+3} = y_{i+2} + \frac{h}{12}[23f_{i+2} - 16f_{i+1} + 5f_i] \tag{2.143a}$$

while the fourth-order (3-step) implicit formula is

$$y_{i+3} = y_{i+2} + \frac{h}{24}[9f_{i+3} + 19f_{i+2} - 5f_{i+1} + f_i] \tag{2.143b}$$

A pair of such formulas can be used in a predictor-corrector combination, where y_{i+k} is "predicted" by the explicit formula and then "corrected" by applying one or more fixed-point iterations using the implicit formula. This also gives an estimate of the local truncation error. The resulting method generally requires fewer function evaluations, but more overhead, than a corresponding Runge-Kutta method.

2.7.2 Consistency, stability, and convergence

Consider a general scheme of the form (2.136). The *local truncation error* δ_i at x_i is defined as the residual when an exact solution $y(x)$ is substituted into that expression and divided by h, i.e.,

$$\delta_i := h^{-1}[y(x_{i+k}) - \sum_{j=0}^{k-1} \alpha_j y(x_{i+j}) - \Phi(x_{i+k}, \ldots, x_i; y(x_{i+k}), \ldots, y(x_i); h)] \tag{2.144}$$

The scheme is said to be *consistent of order* p if $\delta_i = O(h^p)$ for any differential equations (2.135a) with $y \in C^{(p+1)}[a,b]$ and f smooth. It can be shown that Euler's scheme is of order 1, the two stage schemes (2.139) and (2.140) are of order 2 and the Runge-Kutta Fehlberg schemes (see Table 2.1) are of order 4 and 5, respectively.

As we said in the introduction to this chapter, in a numerical algorithm one is interested in the *global error;* that is

$$e_i(h) := y_i(h) - y(x_i) \tag{2.145}$$

where h denotes the dependence of the numerical solution on the step size h. In particular we want the step size h to control the error. A complication arises when we have a multistep method and need to find the $(k-1)$ additional initial values. This may be done using the given initial value and a one-step method.

Definition 2.146

(i) A one-step method based on (2.136) is called *convergent* of order p if $e_i(h) = O(h^p)$, x_i fixed.

(ii) A multistep method based on (2.136) is called *convergent of order* p if $e_i(h) = O(h^p)$, x_i fixed, when $e_1(h), e_2(h), \ldots, e_k(h)$ are $O(h^p)$. □

Consistency is necessary for convergence. It is also sufficient for one-step methods, but for multistep methods we need an additional requirement (which-one step methods trivially satisfy): Define the characteristic polynomial

$$r(\xi) = \xi^k - \alpha_{k-1}\xi^{k-1} - \cdots - \alpha_0 \qquad (2.147)$$

Note that consistency of order 1 (for instance) means that

$$r(1) = 0 \qquad (2.148a)$$

$$\Phi(x, \ldots, x; y(x), \ldots, y(x); 0) = r'(1)f(x, y(x)) \qquad (2.148b)$$

for all $x \in [a, b]$. If we moreover require that $(k-1)$ zeros ξ_j of r satisfy $|\xi_j| < 1$, and one (essential) root $\xi_0 = 1$ [cf (2.148a)], the multistep method is called *strongly stable*. Consistency and strong stability are sufficient conditions for convergence. All Adams methods are easily seen to satisfy this so-called root condition and (2.148b) as well, which guarantees their convergence for h small enough.

Thus far we have assumed a constant step size. Often this is not a good strategy: For most difference schemes one can express the local truncation error δ_i of (2.144) in an expression like

$$\delta_i(h) = c y^{(p+1)}(x_i) h^p + O(h^{p+1}) \qquad (2.149)$$

c some constant, which shows that the *local error* depends on the *local smoothness* of the solution y. In other words, if $y^{(p+1)}$ is fairly small in magnitude, then we should take larger steps, and if $y^{(p+1)}$ is fairly large in magnitude then we should take smaller steps. Much effort has been devoted to implementing this idea in adaptive algorithms, and some of it is described in Section 2.7.4.

We now proceed with our discussion, assuming that all methods considered satisfy (2.148a, b). This means that convergence is guaranteed for $h > 0$ small enough, but it does not mean that there are no stability restrictions on how large h may be. In particular, consider the test equation

$$y' = \lambda y, \qquad x > 0 \qquad (2.150)$$

The scalar λ is allowed to be complex (thus representing an eigenvalue for a system of differential equations), and we assume that $\operatorname{Re}(\lambda) < 0$. The exact solution $y(x) = y_1 e^{\lambda x}$ decays as x increases, and so the approximate solution should not grow in magnitude. For Euler's scheme, for instance,

$$y_{i+1} = y_i + h\lambda y_i = (1+\lambda h)y_i$$

so to keep the approximate solution from growing requires

$$|1 + \lambda h| \leq 1$$

This imposes an absolute stability restriction on h. In particular, if λ is real (and negative), then we get the restriction $h \leq -2/\lambda$. This is not an accuracy restriction, but rather a stability one, and it may be severe if $\operatorname{Re}(\lambda) \ll -1$. This is the case for stiff problems, discussed next.

2.7.3 Stiff problems

Many applications involve initial value problems with fast and slow decay rates. For instance, a solution may look like

$$y(x) = e^{-x} + e^{-1000x}, \qquad x \geq 0$$

with the second component corresponding to a much faster time scale than the first. So, for x positive away from 0, the solution behaves essentially like e^{-x} and large step sizes may be taken for good accuracy. Nevertheless, the numerical method may be restricted to using very small steps, in the case that its absolute stability region is limited, because of the presence of a fast time scale in the differential equation. This is the problem of *stiffness*. An ODE system of the form

$$\mathbf{y}' = \mathbf{f}(x, \mathbf{y}) \qquad (2.151)$$

which is defined on the interval $[a, b]$, is said to be *stiff* in a neighborhood of a solution $\mathbf{y}$ if there exists a component of $\mathbf{y}$ whose variation is large compared to $[b-a]^{-1}$. Often stiffness can be directly related to the eigenvalues of the local Jacobian matrix $\frac{\partial \mathbf{f}}{\partial \mathbf{y}}(x, \mathbf{y}(x))$, viz. to the situation where

$$\min (Re\ (\lambda_i))\ (b-a) \ll -1 \qquad (2.152)$$

To examine the behavior of numerical methods, consider the test equation (2.150) when $Re(\lambda) \ll -1$. Then a restriction like $|1 + \lambda h| \leq 1$ requires taking extremely small steps h in order to approximate a smooth curve. Such severe stability restrictions apply not only to Euler's method, but to each and every one of the more efficient and useful methods for the non-stiff case. Thus, other methods are sought which do not have severe absolute stability limitations on the size of h.

A method is called *A-stable* if its region of absolute stability contains the entire left half plane of λh, i. e., for any choice of $h > 0$, the numerical solutions do not grow for the test equation (1.150) if $Re(\lambda) < 0$. For instance, the trapezoidal scheme (2.139) and the midpoint scheme (2.140) are A-stable, while the similar explicit scheme (2.138) is not. It turns out that all the implicit Runge-Kutta schemes based on Gauss, Radau, or Lobatto points, which have been mentioned earlier and are derived in Section 5.3.2, are A-stable. Unfortunately, as implied above, explicit schemes tend to have rather limited absolute stability regions, so implicit schemes for nonlinear problems have to be dealt with. Moreover, the resulting nonlinear equations to be solved at each step cannot be solved for a reasonably large h by a fixed-point iteration as in the predictor-corrector technique, and a modified Newton method has to be used. It is little wonder, then, that the numerical solution of stiff IVPs is still a very active research area.

The A-stability property relates to a rather simple test equation (2.150). It has been generalized in many ways (too many, it seems at times), a few of which we mention. If we let λ in (2.150) depend on x and still require $|y_{i+1}| \leq |y_i|$ for any $h > 0$, $Re(\lambda(x)) < 0$ (this is called *AN-stability*), then for Runge-Kutta schemes (2.137) with distinct ρ_j this is equivalent to algebraic stability: a scheme (2.137) is called *algebraically stable* if the matrix $M = (m_{jl})$ defined by

$$m_{jl} = b_j a_{jl} + b_l a_{lj} - b_j b_l \qquad 1 \leq j, l \leq q$$

is symmetric nonnegative definite.

The schemes based on Gauss points and Radau points are algebraically stable, while those based on Lobatto points are not.

When $\text{Re}(h\lambda) \to -\infty$, the requirement that the numerical solution not grow in magnitude may not be sufficient and one wants to also require that it should decay. To make this precise, consider the test problem

$$y' = \lambda(y - g(x)), \quad x > a \tag{2.153}$$

with any initial condition and with $g(x)$ smooth but otherwise arbitrary. This can be rewritten as $\varepsilon y' = -y + g(x)$, with $\varepsilon = \dfrac{-1}{\lambda}$, which is in singular perturbation form with ε as the perturbation parameter (see Chapter 10). If $\varepsilon = 0$, the *reduced* solution is $y = g(x)$, while for $\varepsilon > 0$ the integral curves move very rapidly to this reduced solution. A difference scheme is said to exhibit *stiff decay* if it has such a behavior as well, i.e., for ih fixed ($ih > a$),

$$|y_i - g(x_i)| \to 0, \quad \text{as } \text{Re}(h\lambda) \to -\infty \tag{2.154}$$

If (2.154) holds for $g \equiv 0$ then the scheme is *L-stable*. The implicit Runge-Kutta schemes of Section 5.3.2, which are based on Radau points, exhibit stiff decay, whereas those based on Gauss and Lobatto points do not (and are not L-stable either).

Currently, the most popular schemes for handling stiff initial value problems are *backward differentiation formulas* (BDF), which are linear multistep schemes with $\beta_0 = \beta_1 = \cdots = \beta_{k-1} = 0$ in (2.141) and β_k, α_0, ..., α_{k-1} determined such that the scheme is of order k ($1 \leq k \leq 6$). For $k = 1$ we get the so-called *backward Euler scheme*

$$y_{i+1} = y_i + h f_{i+1} \tag{2.155}$$

For $k > 2$, these schemes are not quite A-stable (only so-called $A(\alpha)$ *stable* sectors with angle $\dfrac{\pi}{2} - \alpha$ along the positive and negative imaginary axis have to be excluded from their absolute stability domain), but they do exhibit stiff decay. The basic advantage of these methods is that at each step only n unknowns are solved for, where n is the number of differential equations in (2.151). In contrast, a q-stage Radau Runge-Kutta method would involve qn unknowns at each step.

Note that for BVPs, stability notions like A-stability or stiff decay are only half of the story. We also need a counterpart to indicate what is happening in the positive half-plane of λh. To appreciate this, consider a BDF method like backward Euler. For $h\lambda > 2$ ($h > 0$) this method gives stiff decay as well, meaning that the unstable exponential is approximated by a "stable" (i.e., decaying) mode $\sim (1 - h\lambda)^{-i}$ ($x = ih$). In the next chapter we shall see that this is often highly undesirable, because the behavior of the growing solutions should be approximated relatively accurately as well.

2.7.4 Error control and step selection

The pride of a general-purpose code for automatically integrating IVPs (an "IVP solver") lies in its error control and step size selection. By specifying an appropriate tolerance *TOL* a user can require a more accurate (and more expensive) approximate solution or a less accurate (and cheaper) one. For a specified tolerance, the code can by its step selection calculate the solution efficiently by attempting to equalize the local error throughout the interval of integration. Here we briefly explain some of the details involved.

There are two basic premises to the usual IVP error control. One is that it is much simpler to estimate a local error than a global one, and this is what is done in practice. The other premise is that the discretization error dominates the roundoff error, so the latter is ignored. In order for this premise to hold, stable methods must be used, and a relative error tolerance must be a few orders of magnitudes larger than the machine constant ε_M. Fortuitously, when roundoff error is not negligible because an absolute stability restriction for the scheme is violated (as explained in the last paragraph of Section 2.7.2), then the generated "noise" results in a large local error estimate, so the step size is decreased. This in turn helps to prevent the violation of the absolute stability restriction.

Thus assume that step selection depends exclusively on local errors caused by the local truncation error $\delta_i(h)$, given by (2.149). Rather than estimating $\delta_i(h)$ from (2.149), we estimate its effect directly. Consider for simplicity a one-step scheme, and let y_i be the computed solution value at a point x_i. We want next to determine $h = h_i$ and find y_{i+1} at $x_{i+1} = x_i + h_i$. Let $\tilde{y}(x)$ be the exact solution of the given ODE for $x > x_i$ satisfying $\tilde{y}(x_i) = y_i$. The local error divided by h_i is then

$$\varepsilon_i := h_i^{-1}(\tilde{y}(x_{i+1}) - y_{i+1}) = \delta_i(1 + O(h_i)) \qquad (2.156)$$

Now suppose we have two schemes, one of order p and the other of order $p+1$. If the first one gives a numerical result y_{i+1} and the second $\bar{y}_{i+1}$ (both starting from y_i at x_i), then we have

$$\varepsilon_i = h_i^{-1}(\tilde{y}(x_{i+1}) - \bar{y}_{i+1} + \bar{y}_{i+1} - y_{i+1}) = O(h_i^{p+1}) + h_i^{-1}(\bar{y}_{i+1} - y_{i+1})$$

so the computable expression $h_i^{-1}(\bar{y}_{i+1} - y_{i+1})$ satisfies

$$h_i^{-1}(\bar{y}_{i+1} - y_{i+1}) = \varepsilon_i(1 + O(h_i)) \qquad (2.157)$$

This sets the stage for a local error estimation. Calculating the left-hand side of (2.157) for a given step size h_i (and assuming for the moment that *TOL* is an absolute error tolerance) we can accept the step (and the obtained approximate solution) if

$$|\bar{y}_{i+1} - y_{i+1}| \leq h_i TOL \qquad (2.158)$$

If (2.158) does not hold and we can reduce the step size to $\tilde{h}_i$, we predict from (2.149) that the local error will be reduced by the factor $(\tilde{h}_i/h_i)^p$, so we choose $\tilde{h}_i$ such that

$$(\tilde{h}_i/h_i)^p |\bar{y}_{i+1} - y_{i+1}| \approx h_i TOL \qquad (2.159)$$

and repeat the process. This is called *error per unit step control*.

We have assumed that *TOL* is an absolute tolerance. If, more generally, a mixture of an absolute tolerance *ATOL* and a relative tolerance *RTOL* is specified then

$$TOL = ATOL + RTOL \ |y_i| \qquad (2.160)$$

is inserted in (2.158) and (2.159).

For nonstiff IVPs a particularly popular pair of one-step schemes is the Runge-Kutta-Fehlberg 4-5 pair given in Table 1. Here $p = 4$. This is attractive from a practical point of view because the two schemes share the same coefficients a_{jl}, hence the g_j in (2.137) used for the fifth-order scheme are also usable for the fourth-order scheme. Similar advantages are shared by the multistep Adams predictor-corrector pairs [e.g., (2.143)].

We now wish to consider the effect of stiffness on the choice of step size. Moreover, in a BVP context we often have to deal with the additional complication that the IVPs can have exponentially *growing* solutions, a problem which is important for Chapter 4 while being unusual for IVPs in general.

Consider for simplicity the model problem
$$y' = \lambda y, \qquad y(0) = 1$$
with λ real. Then from (2.149), (2.156),
$$\varepsilon_i \approx c h_i^p |\lambda|^{p+1} e^{\lambda x_i}$$
and using (2.157), (2.158) with equality, and (2.160), we obtain
$$h_i \approx (c\,|\lambda|^{p+1})^{-1/p}\,[(ATOL + e^{\lambda x_i} RTOL)e^{-\lambda x_i}]^{-1/p} \qquad (2.161)$$
We assume that *ATOL* and *RTOL* are positive and comparable in size.

Corresponding to a stiff IVP, assume that $\lambda < 0$. Then, after a while the absolute tolerance dominates and increasingly large step sizes are taken.

But now assume that $\lambda > 0$. Then the relative tolerance dominates and the step size is essentially constant. If we use only an absolute tolerance (*RTOL* = 0) then the step size decreases exponentially. Indeed, using the mixed tolerance boils down to using an absolute error control for small and moderate values of $|y_i|$ and a relative error control for large values.

Note that we do not get a global error estimate from the procedure outlined above, only a local one. Usually, it is implicitly assumed that the accumulation of local errors is linear relative to the solution. This usually holds in practice, but not always, as the following example shows.

Example 2.4

Consider the problem
$$y' = \lambda(y - g(x)) + g'(x), \qquad x > 0$$
$$y(0) = g(0)$$
with $\lambda > 0$ and g a smooth function bounded by 1. Most IVP codes would yield large global errors for $\lambda = 50$, say, even for small input error tolerances, because while $y(x) = g(x)$, the error accumulates like $e^{\lambda x}$. □

The reason that local error control performs poorly for this example is that the IVP is ill-conditioned, as discussed in Section 2.8 and in Section 3.3, and the local control fails to detect this because of the bounded exact solution. Fortunately, local error control normally *does* work fine in practice. This, combined with the fact that global error control is very difficult, makes this type of local error control the strategy of many IVP solvers.

2.8 DIFFERENTIAL OPERATORS AND THEIR DISCRETIZATIONS

It is our declared intention in this book to use concepts from functional analysis as little as possible. However, some basic notions of function norms and function spaces have already crept into Section 2.5.3, and more will be needed since, after all, our discretizations are for differential operators. Here we introduce some basics and define norms of differential operators at an elementary level. This allows us to discuss again the basic sources of errors made when solving differential equations, highlighting the difference between truncation and roundoff errors. Attempts to quantify instabilities and control these errors are then further considered.

2.8.1 Norms of functions, spaces, and operators

Function norms may be defined in a similar way to the vector norms of (2.19). Thus, for a function

$$f : [a, b] \to \mathbf{R}$$

we let

$$\|f\|_{p,[a,b]} := [\int_a^b |f(x)|^p dx]^{1/p}, \qquad 1 \leq p < \infty \qquad (2.162a)$$

$$\|f\|_{\infty,[a,b]} := \sup_{a \leq x \leq b} |f(x)| \qquad (2.162b)$$

In a context where there is no confusion, we will omit writing p and/or $[a,b]$ in the norm notation. In (2.110) and (2.111) $p = \infty$ is omitted, and in (2.110) $[a,b]$ is also omitted. The extension of (2.162) for vector functions is straightforward. Also, for a matrix function $A(x)$, $\|A(x)\|_p$ denotes the matrix norm of $A(x)$, while $\|A\|_p$ or $\|A(\cdot)\|_p$ is the function norm, with the matrix norm appearing under the integral sign in the corresponding expression to (2.162a) This convention allows room for confusion when A is a constant matrix, but it is to be hoped that the choice of norm in such cases will be clear from the context.

The linear space consisting of all functions $f(x)$ such that $\|f\|_{p,[a,b]} < \infty$ is denoted by $L_p[a,b]$. For each p, $1 \leq p \leq \infty$, this is a Banach space (i.e., it is closed). The *Cauchy Schwartz inequality* is

$$\int_a^b f(x)g(x) dx \leq \|f\|_p \|g\|_q, \qquad \frac{1}{p} + \frac{1}{q} = 1 \qquad (2.163)$$

(where $p = 1$ if $q = \infty$), and $L_{p+1}[a,b]$ is contained in $L_p[a,b]$ for each p because

$$\|f\|_p \leq \|f\|_{p+1}$$

The space $C[a,b]$ of all continuous functions is contained in each $L_p[a,b]$, so any of the norms in (2.162) can be used on $C[a,b]$.

On rare occasions we will need more complicated function norms. For any function $u(x)$ which has m derivatives on $[a,b]$ we define the *Sobolev norms*

$$\|u\|_{m,p,[a,b]} := \sum_{j=0}^{m} \|u^{(j)}\|_{p,[a,b]} \qquad (2.164)$$

Next consider a linear operator from one linear function space to another,

$$L : U \to V \qquad (2.165a)$$

where the norms on U and V are $\|\cdot\|_U$ and $\|\cdot\|_V$, respectively. The norm of the operator L is then defined, just like the induced matrix norm (2.21), as

$$\|L\|_{U \to V} := \sup \{ \|Lu\|_V : \|u\|_U = 1 \} \qquad (2.165b)$$

If L is one-to-one and onto, the inverse operator is defined by

$$L^{-1}v := u \qquad \text{where} \quad Lu = v$$

For a given problem

$$Lu = q \qquad (2.166a)$$

we may consider the *conditioning constant*

$$\kappa := \|L^{-1}\|_{V \to U} \qquad (2.166b)$$

and the *condition number*

$$\chi := \|L\|_{U \to V} \|L^{-1}\|_{V \to U} \qquad (2.166c)$$

Both the conditioning constant and the condition number depend, sometimes crucially, on the *choice* of norms $\|\cdot\|_U$ and $\|\cdot\|_V$ for the linear function spaces U and V.

Example 2.5

Let $L \equiv \frac{d}{dx}$, i.e., $Lu = \frac{du}{dx}$, for any differentiable function $u(x)$ whose derivative lies in $L_\infty[a,b]$. If U is equipped with the Sobolev norm $\|\cdot\|_{1,\infty,[a,b]}$ and $V = L_\infty[a,b]$ then

$$\|L\|_{U \to V} = \sup \{ \|v\|_\infty : \|u\|_\infty + \|v\|_\infty = 1 \} = 1$$

But if U and V are both equipped with the sup norm, then $\|L\|_{U \to V}$ is unbounded because, for instance, $u(x) = \sin \omega x$ yields $\|u\| = 1$ and $\|Lu\| = \omega$ for any $\omega > 0$. □

Example 2.6

Let $Ly = q$ stand for the ODE

$$y' + y = q(x), \qquad a < x < b$$

plus the initial condition

$$y(a) = 0$$

Then L^{-1} is the operator which for a given $q(x)$ assigns a function $y(x)$ which solves this IVP.

If U consists of all the functions u satisfying $u(a) = 0$ and $\|u'\|_{\infty,[a,b]} < \infty$, with the Sobolev norm $\|\cdot\|_{1,\infty,[a,b]}$ and $V = L_\infty[a,b]$, then as in Example 2.5, $\|L\|_{U \to V} = 1$. Also, since for a given $q(x)$ the solution is

$$y(x) = \int_a^x e^{t-x} q(t)\,dt$$

we have

$$\|y\|_\infty \le \|q\|_\infty (1 - e^{a-b}) \le \|q\|_\infty$$

$$\|y'\|_\infty \le \|y\|_\infty + \|q\|_\infty \le 2\|q\|_\infty$$

so

$$\|L^{-1}\|_{V \to U} \le 3$$

and

$$\kappa = \chi \le 3$$

On the other hand, consider the choice of the sup norm for both U and V. Then the bound on $y(x)$ in terms of $q(x)$ yields $\kappa \le 1$, but χ is *infinite*, as in Example 2.5. □

It is straightforward to see that the essence of Example 2.6 remains unchanged when we generalize the problem to a linear BVP with properly scaled homogeneous BC (cf. Chapter 3),

$$\mathbf{y}' - A(x)\mathbf{y} = \mathbf{q}(x), \qquad a < x < b \tag{2.167a}$$

$$B_a \mathbf{y}(a) + B_b \mathbf{y}(b) = \mathbf{0} \tag{2.167b}$$

Again we write (2.167) in operator form as

$$\mathbf{L}\mathbf{y} = \mathbf{q}$$

and consider $V = L_\infty[a, b]$ and U with two possible norms.

The Sobolev norm $\|\cdot\|_{1,\infty,[a,b]}$ incorporates the additional smoothness of $\mathbf{y}(x)$ over $\mathbf{q}(x)$. We obtain

$$\|\mathbf{L}\|_{U \to V} \le \max(1, \|A\|_\infty) \tag{2.168a}$$

The conditioning constant κ of (2.166b), which we denote by $\hat{\kappa}$ for this particular choice of norm on U, provides an *absolute* bound on the solution $\mathbf{y}(x)$ in terms of the data $\mathbf{q}(x)$, namely

$$\|\mathbf{y}\|_{1,\infty,[a,b]} \le \hat{\kappa} \|\mathbf{q}\|_{\infty,[a,b]} \tag{2.168b}$$

The second choice of norm on U is the sup norm. Here we denote by κ the conditioning constant of (2.166b) and obtain

$$\|\mathbf{y}\|_\infty \le \kappa \|\mathbf{q}\|_\infty \tag{2.169}$$

In Chapter 3 and throughout the rest of the book we call BVPs for which κ of (2.169) is a constant of moderate-size *well-conditioned BVPs*. They are discussed extensively in Section 3.4. Given that $\|A\|_\infty$ is also bounded by a constant of moderate size, $\hat{\kappa}$ in (2.168b) and $\|\mathbf{L}\|$ in (2.168a) and hence the condition number χ of (2.166c) are all bounded by constants of moderate size as well. That is to say, *for a well conditioned BVP (2.167) with $\|A\|$ of moderate size, if U is equipped with the appropriate Sobolev norm then the conditioning constant as well as the condition number are constants of* *moderate size. At the same time,* $\|L\|$ *and hence the condition number of (2.166c) are unbounded when U is equipped with the sup norm.*

The importance of this distinction becomes clear in Section 2.8.2. First, we give a simple example of where A in (2.167) cannot be considered bounded.

Example 2.7

Consider the IVP

$$y' = \lambda y + g(x), \qquad a < x < b$$

$$y(a) = 0$$

where $\lambda \ll -1$. This form is not well-scaled, so denote [as for (2.153)] $\varepsilon = -\lambda^{-1}$, $q(x) = \varepsilon g(x)$, and let $Ly = q$ represent the same IVP in the form

$$\varepsilon y' + y = q(x) \qquad a < x < b$$

$$y(a) = 0$$

If we again use the sup norm for V, it is now natural to use a weighted Sobolev norm for U

$$\|y\|_U := \|y\|_\infty + \|\varepsilon y'\|_\infty$$

Then we have

$$\|L\|_{U \to V} \leq 1$$

and the condition number χ of (2.166c) can be verified to satisfy

$$\chi \leq 3$$

□

2.8.2 Roundoff errors and truncation errors

In Section 2.1 we have already mentioned the very different nature of roundoff errors as compared to truncation errors when one is solving differential equations. Essentially, as the discretization is refined the truncation error decreases, but the roundoff error grows. Let us demonstrate this on a simple example.

Example 2.8

Consider the numerical differentiation of $y''(x)$ for a function $y \in C^{(2)}[a,b]$. We have by Taylor's expansion

$$y''(x) = h^{-2}[y(x-h) - 2y(x) + y(x+h)] + O(h^2) \qquad (2.170)$$

For a given BVP

$$y'' = q(x)$$

$$y(a) = y(b) = 0$$

which we denote $Ly = q$, we may associate a discretization $L^h y^h = q^h$, given by

$$h^{-2}(y_{i-1} - 2y_i + y_{i+1}) = q(x_i), \qquad i = 2, \ldots, N \qquad (2.171a)$$

$$y_1 = y_{N+1} = 0 \qquad (2.171\text{b})$$

where $h := (b-a)/N$ and $x_i = a + (i-1)h$, $i = 1, \ldots, N+1$. The local truncation error is $O(h^2)$, which tends to 0 as $h \to 0$. But writing (2.171) as a tridiagonal system of linear equations

$$\mathbf{A}\mathbf{y}^h = \mathbf{q}^h \qquad (2.172)$$

for the vector of unknowns $\mathbf{y}^h = (y_1, \ldots, y_{N+1})^T$, we have that $\text{cond}_2(\mathbf{A}) = O(h^{-2})$. [While a straightforward row scaling in (2.171) is called for, it turns out not to improve the matrix condition number in this case.] Thus, even if a stable algorithm is used to solve (2.172), the roundoff error may be magnified by a factor which is $O(h^{-2})$, and hence unbounded as $h \to 0$. Indeed we see already from (2.170) that if $y(x-h)$, $y(x)$, and $y(x+h)$ are each perturbed by an arbitrary quantity of size ε_M, say, then the approximation to $y''(x)$ may be expect to have a roundoff error of size $\sim h^{-2}\varepsilon_M$. □

The basic point that Example 2.8 brings forward is that while the truncation error is usually "smooth," the roundoff error is not. (It may be said that the roundoff error is "random," while the truncation error is not.) Thus, to analyze the discretization operator L^h we may wish to consider two fundamentally different norms, analogous to the ones considered for the differential operator L in Section 2.8.1.

The truncation error is "smoother" in the solution y than in the inhomogeneity q; hence a *discrete Sobolev norm*, defined similarly to (2.164), may be used, yielding a bounded $\|L^h\|$ from consistency of the difference scheme (cf. Section 2.7.2). But the roundoff error is arbitrary, and may not be considered "smooth," so to analyze its effect the same norm (the sup norm, say) has to be used for the discrete solution space as for the discrete data space, and L^h then approximates an *unbounded* differential operator L.

Given a discretization for the differential operator, we define the *stability constant* κ^h of the numerical algorithm in an analogous way to the conditioning constant of (2.166b) with $\|\cdot\|_U = \|\cdot\|_V$. Thus we may expect a discrete analogue of (2.169) to hold. The *condition number* χ^h of the algorithm is defined analogously to (2.166c) with the same norms. Thus we may expect $\chi^h \to \infty$ as the discretization of an ODE is refined to the limit. The stability constant is an absolute quantity corresponding to $\|(L^h)^{-1}\|$, while the condition number is a relative quantity corresponding to $\|L^h\|\|(L^h)^{-1}\|$. In Example 2.8, applying the stable Gaussian elimination method to (2.172), we obtain essentially $\kappa^h = \|\mathbf{A}^{-1}\|$ and $\chi^h = \text{cond}(\mathbf{A})$.

Throughout this book we shall encounter numerical methods where some discretization of a BVP is applied, yielding a system of linear algebraic equations. The numerical algorithm is completely specified only when both the discretization scheme and the method for solving the linear equations are specified. There are many stable methods for solving the linear equations, and we assume here that such a stable method is used. For a "well-conditioned" BVP we concentrate on the conditioning of the resulting linear system (cf. Section 2.2.7). We say that a discretization algorithm is *stable* if its stability constant is of moderate size. (The "strong stability" requirement of a multistep method in Section 2.7.2 is in fact a condition which guarantees a moderate stability constant for a multistep scheme when the IVP being discretized has a moderate conditioning constant.) For a sufficiently fine discretization, the stability constant of a decent algorithm can be expected to be close to the conditioning constant of the BVP (cf. Section 5.2). To within a moderate constant of proportionality, it controls the amount by which *local, smooth* errors (i.e., local truncation errors) get magnified for the *global* error of the resulting approximate solution. Example 2.4 shows an IVP where the conditioning constant is large.

The stability constant of an algorithm appears also in the amplification factor χ^h of the roundoff error, but in view of the unbounded growth of $\|L^h\|$, it cannot also be said to characterize this error. In fact, given a BVP with a conditioning constant κ and an algorithm with a condition number χ^h for a mesh with N elements, a (relative) input error of ε_M can be expected to yield an output error $\chi^h \varepsilon_M$ satisfying at least

$$\chi^h \varepsilon_M \geq \kappa N \varepsilon_M \qquad (2.173)$$

Of course, the reason that we need to make N large (h small) is to make the local truncation error small (see Figure 2.4). Fortunately, in practice ε_M is sufficiently small that $\kappa N \varepsilon_M$ is normally well below the truncation error level. In fact, all we need is for $\varepsilon_M \chi^h / \kappa^h$ to be below the magnitude of the local truncation error.

We can now relate this back to the IVP error control in Section 2.7.4, as well as to finite difference schemes for BVPs in Chapters 5 and 10. With a stable difference scheme it is usually safe to neglect the effect of roundoff error and just deal with the truncation error. If the problem is well-conditioned then the global truncation error can be expected to be only slightly worse than the local truncation error, justifying the local error control in Section 2.7.4.

The situation is more complicated for the methods of Chapter 4. For reasons explained there, we attempt to integrate possibly ill-conditioned IVPs and *control* the amount of instability introduced in the process. In fact, some methods in Chapter 4

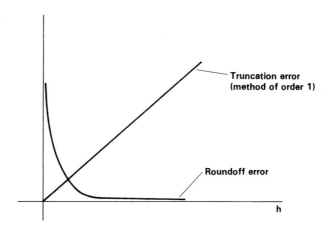

Figure 2.4 Truncation and roundoff errors

differ only in their roundoff error propagation (and not in their truncation error characteristics), so a roundoff error control becomes mandatory. When addressing the stability of an algorithm with respect to roundoff error propagation, we will speak of a "stable algorithm with condition number χ^h." The qualitative measure of such stability (or instability) is how much larger χ^h is than κN. The quantitative measure yields that an algorithm is *acceptable* (in this respect) in a particular situation as long as its condition number times ε_M stays well below the prescribed error tolerance.

3

Theory of Ordinary Differential Equations

The theory of numerical methods for ordinary differential equations is closely associated with certain aspects of the analytic theory of these equations. These aspects are highlighted in this chapter, which is the last preparatory chapter of the book.

In the first section, we dispense with some of the more basic theoretical considerations concerning solutions of initial and boundary value problems. Some fundamental existence and uniqueness results are given. Of course, our purpose here is not to duplicate the efforts of a basic text in ODEs. Thus we merely review these results.

Let us recall that the success of a numerical process depends on a combination of two factors: well-conditioned problem and a stable algorithm to solve it. The construction of stable algorithms is a major topic of this book and is considered in succeeding chapters. In the present chapter we address the question: What is a well-conditioned BVP and how can such a concept be quantified?

For linear BVP, we shall deal almost exclusively with the case where we have unique solutions (although this uniqueness does not hold for some classes of problems, like eigenvalue problems, cf. Section 11.3), but for well-posedness more is needed: The solution has to depend continuously on the data. To bring this dependence out, we consider Green's functions in Section 3.2. This enables us to define a conditioning constant for the BVP, which bounds perturbations in the solution in terms of perturbations in the data.

In Sections 3.3 and 3.4 we consider in some detail the concepts of well-posedness and conditioning for initial and boundary value problems, respectively. This we do because these concepts are very close to the central concept of stability in numerical methods and because they are not commonly exposed in ODE texts. The important notion of dichotomy is discussed in Section 3.4. This is a key concept in investigating conditions under which a given BVP is guaranteed to be well-posed.

We will consider, unless indicated otherwise, the general first-order form

$$\mathbf{y}' = \mathbf{f}(x,\mathbf{y}) \qquad a < x < b \qquad (3.1)$$

or for linear differential equations

$$\mathbf{y}' = A(x)\mathbf{y} + \mathbf{q}(x) \qquad a < x < b \qquad (3.2)$$

together with either initial conditions or two-point conditions.

3.1 EXISTENCE AND UNIQUENESS RESULTS

In this section we have collected together some basic theoretical results concerning the existence and uniqueness of solutions to BVPs. The special case where the entire solution is known at some point is particularly simple, so we begin with it.

3.1.1 Results for initial value problems

Here we consider the system of ODEs (3.1) or (3.2), subject to the initial conditions

$$\mathbf{y}(a) = \boldsymbol{\alpha} \qquad (3.3)$$

The existence and uniqueness of solutions for IVPs can be guaranteed under fairly general conditions. In particular, for the nonlinear IVP (3.1), (3.3) we have the following theorem, which is stated without proof.

Theorem 3.4 Suppose that $\mathbf{f}(x,\mathbf{y})$ is continuous on $D = \{(x,\mathbf{y}): a \leq x \leq b, |\mathbf{y}-\boldsymbol{\alpha}| \leq \rho\}$ for some $\rho > 0$, and suppose that $\mathbf{f}(x,\mathbf{y})$ is Lipschitz continuous with respect to $\mathbf{y}$; i.e., there exists a constant L such that for any $(x,\mathbf{y})$ and $(x,\mathbf{z})$ in D,

$$|\mathbf{f}(x,\mathbf{y}) - \mathbf{f}(x,\mathbf{z})| \leq L|\mathbf{y} - \mathbf{z}| \qquad (3.4a)$$

If $|\mathbf{f}(x,\mathbf{y})| \leq M$ on D and $c = \min\{b - a, \rho/M\}$ then the IVP (3.1), (3.3) has a unique solution for $a \leq x \leq a + c$. □

It follows that if the so-called Lipschitz condition (3.4a) holds uniformly for all $\mathbf{y}$ and $\mathbf{z}$, then the IVP has a unique solution for all $x \geq a$. By the Mean Value Theorem, if $\mathbf{y} = (y_1, \ldots, y_n)^T$ and $\mathbf{z} = (z_1, \ldots, z_n)^T$, then

$$f_i(x,\mathbf{y}) - f_i(x,\mathbf{z}) = \sum_{j=1}^{n} \frac{\partial f_i}{\partial y_j}(x,\mathbf{w})(y_j - z_j) \qquad (1 \leq i \leq n) \qquad (3.5)$$

where $\mathbf{w} = (w_1, \ldots, w_n)^T$ is such that w_j is between y_j and z_j ($1 \leq j \leq n$). Thus, a sufficient condition that $\mathbf{f}(x,\mathbf{y})$ be Lipschitz continuous is that it have bounded partial derivatives $\frac{\partial f_i}{\partial y_j}$ ($1 \leq i, j \leq n$), since (3.5) then implies (3.4a). Another direct consequence of Theorem 3.4 is that the linear IVP (3.2), (3.3) has a unique solution if $A(x)$ is continuous on $[a, b]$.

To characterize the solution for a linear IVP, consider the homogeneous first-order system corresponding to (3.2),

$$\mathbf{y}' = A(x)\mathbf{y} \tag{3.6}$$

where $A(x) \in \mathbf{R}^{n \times n}$ is a continuous function on the interval $[a, b]$. For any $t \in [a, b]$ a *fundamental solution (matrix)* $Y(x; t) \in \mathbf{R}^{n \times n}$ of (3.6) is defined as the function satisfying

$$Y'(x; t) = A(x)Y(x; t), \qquad a < x < b \tag{3.7a}$$

$$Y(t; t) = I \tag{3.7b}$$

In (3.7a) the differentiation is with respect to x, and t is a parameter. Thus, the ith column $\mathbf{Y}_i(x; t)$ of $Y(x; t)$ is the unique solution of (3.6) satisfying the initial condition $\mathbf{Y}_i(t; t) = \mathbf{e}_i := (\delta_{ij})_{j=1}^n$. The inverse $Y^{-1}(x)$ of $Y(x) \equiv Y(x; a)$ exists for all x (see Exercise 3.3), and direct substitution shows that

$$\mathbf{y}(x) = Y(x)[\boldsymbol{\alpha} + \int_a^x Y^{-1}(t)\mathbf{q}(t)\, dt] \tag{3.8a}$$

is the unique solution of the linear IVP (3.2), (3.3).[1]

It will sometimes be convenient to also call a fundamental solution any $Y(x) = Y(x; t) \in \mathbf{R}^{n \times n}$ satisfying the homogeneous ODE (3.7a) [but not necessarily the initial conditions (3.7b)] and having linearly independent columns. If $\Psi(x)$ and $\Phi(x)$ are two fundamental solutions, then there is a constant nonsingular matrix R such that

$$\Phi(x) = \Psi(x)R, \qquad x \geq a$$

If $\Psi(a) = I$ then simply

$$R = \Phi(a)$$

So, (3.8a) can be written more generally for *any* fundamental solution $Y(x)$ as

$$\mathbf{y}(x) = Y(x)[Y^{-1}(a)\boldsymbol{\alpha} + \int_a^x Y^{-1}(t)\mathbf{q}(t)\, dt\,] \tag{3.8b}$$

In anticipation of the corresponding treatment for BVPs, let us write (3.8) in a form which emphasizes the dependence of the solution $\mathbf{y}(x)$ on the data $\boldsymbol{\alpha}$ and $\mathbf{q}(t)$:

$$\mathbf{y}(x) = Y(x)Y^{-1}(a)\boldsymbol{\alpha} + \int_a^x Y(x)Y^{-1}(t)\mathbf{q}(t)\, dt$$

---

[1] Recall that $\dfrac{d}{dx}[\int_a^x \mathbf{r}(t)\, dt] = \mathbf{r}(x)$.

or

$$\mathbf{y}(x) = Y(x)Y^{-1}(a)\boldsymbol{\alpha} + \int_a^b G(x,t)\mathbf{q}(t)\,dt \qquad (3.9a)$$

where the matrix function $G(x,t)$ is defined by

$$G(x,t) := \begin{cases} Y(x)Y^{-1}(t) & t \le x \\ 0 & t > x \end{cases} \qquad (3.9b)$$

In view of the general results available for the exact solution of IVPs, one may wonder whether a numerical approximation is needed at all. However, although the theory of existence and uniqueness for IVPs is fairly complete, solution of such problems in practice usually requires numerical treatment. One reason is that even for simple ODEs such as $y' = x^2 + y^2$, analytic solutions are unavailable. Even if they exist in closed form, their calculation may be prohibitive. For instance, if the solution of the trivial IVP $y' = y$, $y(0) = 1$, is needed at many points, it could be more practical to use one of the numerical algorithms reviewed in Chapter 2 than to continually evaluate the exponential function.

3.1.2 Results for boundary value problems

Guaranteeing solution existence and uniqueness for boundary value problems is considerably more difficult than it is for initial value problems. This can be easily demonstrated even for a simple scalar, second-order linear BVP.

Example 3.1

Consider the problem

$$u'' + u = 0 \qquad 0 < x < b \qquad (3.10a)$$

$$u(0) = 0, \qquad u(b) = \beta_2 \qquad (3.10b)$$

The general solution to (3.10a) which vanishes at $x = 0$ is $u(x) = c\,\sin x$, where c is an arbitrary constant. Thus, if $b = n\pi$, then (3.10a, b) has no solution when $\beta_2 \ne 0$ and an infinite number of solutions when $\beta_2 = 0$ (one for each value of c). Note that if, on the other hand, we now replace the condition on $u(b)$ by the initial condition

$$u'(0) = s \qquad (3.10c)$$

then Theorem 3.4 guarantees that for any given scalar s, there exists a unique solution for all $x \ge 0$. That solution is, of course,

$$u(x) = s\,\sin x \qquad \square$$

The failure in this example to have existence or uniqueness of a solution for certain distinct values of b is typical of linear boundary value problems. For nonlinear problems, the situation can be much more complex. For example, it is possible that no solutions exist for all b sufficiently large (see Exercise 1).

There are various restrictive results about existence and uniqueness in special cases. Some relate to the important model problem (1.2a, b), namely

$$u'' = f(x, u, u') \qquad a < x < b$$
$$u(a) = \beta_1, \qquad u(b) = \beta_2$$

For example, we have

Theorem 3.11 Suppose that $f(x, y, z)$ is continuous on $D = \{(x, y, z): a \le x \le b, -\infty < y < \infty, -\infty < z < \infty\}$ and satisfies a Lipschitz condition on D, so that there exist constants L and M such that for any (x, y, z) and $(x, \hat{y}, \hat{z})$ in D,

$$|f(x, y, z) - f(x, \hat{y}, \hat{z})| \le L |y - \hat{y}| + M |z - \hat{z}|$$

If

$$b - a < 4 \begin{cases} \dfrac{1}{(4L - M^2)^{1/2}} \cos^{-1} \dfrac{M}{2\sqrt{L}} & \text{if } 4L - M^2 > 0 \quad (3.11a) \\ \dfrac{1}{(M^2 - 4L)^{1/2}} \cosh^{-1} \dfrac{M}{2\sqrt{L}} & \text{if } 4L - M^2 < 0 \qquad L, M > 0 \\ \dfrac{1}{M} & \text{if } 4L - M^2 = 0 \qquad M > 0 \\ \infty & \text{otherwise} \end{cases}$$

then the nonlinear BVP (1.2a, b) has a unique solution. The result is sharp in the sense that strict inequality in (3.11) is necessary. □

In more general circumstances, it is extremely difficult to provide such explicit existence and uniqueness results for the BVP solution. In many cases of practical interest, there are in fact a number of solutions (cf. Examples 1.18–1.20, 1.23, 1.24, and Example 3.2 below). It is possible, nonetheless, to derive some useful theoretical results by expressing the BVP in terms of an associated IVP. Consider the differential equations (3.1) subject to the BC

$$\mathbf{g}(\mathbf{y}(a), \mathbf{y}(b)) = \mathbf{0} \tag{3.12}$$

Associated with this BVP is the IVP

$$\mathbf{w}' = \mathbf{f}(x, \mathbf{w}) \qquad x > a \tag{3.13a}$$

$$\mathbf{w}(a) = \mathbf{s} \tag{3.13b}$$

where $\mathbf{s}$ is a parameter vector. Under the mild conditions of Theorem 3.4 (which we shall assume), for each $\mathbf{s} \in \mathbf{R}^n$ there is a unique solution of the IVP (13a, b), denoted $\mathbf{w}(x; \mathbf{s})$. Now, $\mathbf{w}(x; \mathbf{s})$ satisfies the ODE (3.1), so $\mathbf{y}(x) = \mathbf{w}(x; \mathbf{s})$ solves the BVP (3.1), (3.12) if $\mathbf{s}$ can be chosen to satisfy the BC, i.e.,

$$\phi(\mathbf{s}) \equiv \mathbf{g}(\mathbf{s}, \mathbf{w}(b, \mathbf{s})) = \mathbf{0} \tag{3.14}$$

The equations (3.14) are a set of n nonlinear algebraic equations for the n unknowns $\mathbf{s}$. It is well-known that in general such a system may have many solutions, one, or none at all.

Theorem 3.15 Suppose that $\mathbf{f}(x, \mathbf{y})$ is continuous on $D = \{(x,\mathbf{y}): a \leq x \leq b, |\mathbf{y}| < \infty\}$ and satisfies there a uniform Lipschitz condition in $\mathbf{y}$. Then the BVP (3.1), (3.12) has as many solutions as there are distinct roots $\mathbf{s}^*$ of (3.14). For each $\mathbf{s}^*$ satisfying $\phi(\mathbf{s}^*) = 0$, a solution of the BVP is given by

$$\mathbf{y}(\cdot) = \mathbf{w}(\cdot\,;\,\mathbf{s}^*)$$

□

The important solution property here is not global uniqueness, but rather *local uniqueness*. A solution $\mathbf{y}$ of the BVP (3.1), (3.12) is said to be locally unique if there is a "tube" around it where it is unique; i.e., there is a $\rho > 0$ such that in the class of functions

$$F = \{\mathbf{z}; \mathbf{z} \in C[a,b], \sup_{a \leq x \leq b} |\mathbf{z}(x) - \mathbf{y}(x)| \leq \rho\} \qquad (3.16)$$

$\mathbf{y}$ is the only solution of the BVP.

Example 3.2

Consider the problem

$$u'' + \lambda e^u = 0 \qquad 0 < x < 1$$

$$u(0) = u(1) = 0$$

where $\lambda > 0$ is a parameter. Substituting a function of the form

$$u(x) = -2 \ln\{\cosh[(x - \frac{1}{2})\frac{\theta}{2}]/\cosh(\frac{\theta}{4})\} \qquad (3.17)$$

which satisfies the BC, into the differential equation, we find that u is a solution if

$$\theta = \sqrt{2\lambda}\,\cosh(\frac{\theta}{4}) \qquad (3.18a)$$

The nonlinear algebraic equation (3.18a) for θ has two, one, or no solutions when $\lambda < \lambda_c$, $\lambda = \lambda_c$, or $\lambda > \lambda_c$, respectively. The critical value λ_c satisfies

$$1 = \frac{1}{4}\sqrt{2\lambda_c}\,\sinh(\frac{\theta}{4}) \qquad (3.18b)$$

because the two curves in Fig. 3.1 must then be tangential at their one intersection point. This gives $\lambda_c \approx 3.51383$. If (for instance) $\lambda = 1$, then we have two locally unique solutions, whose initial slopes are $s^* \approx 0.549$ and $s^* \approx 10.909$. If we convert the BVP into a first-order system (3.1), (3.12), then $\mathbf{s}^* = \begin{bmatrix} 0 \\ s^* \end{bmatrix}$ is a solution of the nonlinear equations (3.14) for each of the two values of s^*. □

The natural question to ask is, given a solution $\mathbf{y}(x)$, how can we know whether it is locally unique? This information is crucial for the feasibility of obtaining $\mathbf{y}$ numer-

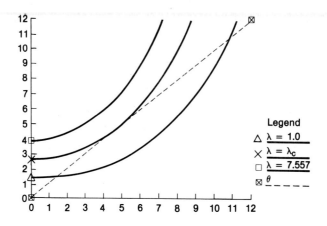

Figure 3.1 Roots of (3.18a)

ically by the methods described in the following chapters. For simplicity of notation we consider the nonlinear differential system (3.1) with *linear* BC

$$B_a \mathbf{y}(a) + B_b \mathbf{y}(b) = \boldsymbol{\beta} \tag{3.19}$$

For an answer it is necessary to look at the *variational problem* for the BVP (3.1), (3.19): Consider

$$\mathbf{z}' = A(x)\mathbf{z}, \quad a < x < b \tag{3.20a}$$

$$B_a \mathbf{z}(a) + B_b \mathbf{z}(b) = \mathbf{0} \tag{3.20b}$$

where

$$A(x) = A(x; \mathbf{y}(x)) = \frac{\partial \mathbf{f}(x, \mathbf{y}(x))}{\partial \mathbf{y}} = \left[\frac{\partial f_i(x, \mathbf{y}(x))}{\partial y_j}\right] \tag{3.21}$$

The matrix $A(x)$ is called the *Jacobian matrix* (cf. Section 2.3.4).

For instance, in Example 3.2 we get

$$\mathbf{z} = \begin{bmatrix} v \\ v' \end{bmatrix}, \quad A = \begin{bmatrix} 0 & 1 \\ -\lambda e^u & 0 \end{bmatrix}, \quad \mathbf{y} = \begin{bmatrix} u \\ u' \end{bmatrix} \tag{3.22}$$

with the exact solution u given by (3.17), (3.18a). The variational problem (3.20a, b) for $\mathbf{z}$ can be written in terms of v in this particular case, viz.

$$v'' + (\lambda e^u)v = 0 \quad 0 < x < 1 \tag{3.23a}$$

$$v(0) = v(1) = 0 \tag{3.23b}$$

Note that the BVP (3.23) for v is *linear*. This is also true in the general case (3.20).

The solution $\mathbf{y}(x)$ of the nonlinear problem (3.1), (3.19) is said to be *isolated* if the variational problem has the unique solution $\mathbf{z}(x) \equiv 0$. In terms of the associated IVP we then have $\mathbf{y}(x)$ given as in Theorem 3.15 with $\mathbf{s}^*$ a *simple* root of (3.14) [i.e., $\frac{\partial \phi}{\partial \mathbf{s}}(\mathbf{s}^*)$ is nonsingular]. It is possible to show

Theorem 3.24 Suppose that $\mathbf{f}(x, \mathbf{y})$ is sufficiently smooth that the variational problem is well-defined. If $\mathbf{y}(x)$ is an isolated solution of the BVP (3.1), (3.19), then it is locally unique. □

In practice, one often does not know how many solutions there are to a given nonlinear BVP. But the solution to the *linearized* problem (3.20), when the linearization is done at an isolated solution (or near it), is unique. Iterative methods for solving nonlinear equations usually utilize this fact and proceed locally.

If the boundary conditions associated with the nonlinear ODE (3.1) are also nonlinear, i. e., they are of the general form (3.12), then in deriving the variational problem they have to be linearized as well. We obtain the BC (3.20b) with

$$B_a = \frac{\partial \mathbf{g}(\mathbf{y}(a), \mathbf{y}(b))}{\partial \mathbf{y}(a)} \qquad B_b = \frac{\partial \mathbf{g}(\mathbf{y}(a), \mathbf{y}(b))}{\partial \mathbf{y}(b)} \qquad (3.25)$$

The definition of an isolated solution is directly extended. Theorem 3.24, with a slight modification of the smoothness conditions, holds here as well.

The discussion leading to Theorem 3.15 naturally suggests a numerical method: Solve numerically the nonlinear equations (3.14), where for each function evaluation of $\phi(\mathbf{s})$ for a given $\mathbf{s}$, the IVP (3.13a, b) is integrated numerically. (Fortuitously, there are robust packages commonly available for solving nonlinear algebraic equations and for integrating IVPs.) The method so obtained is called *single or simple shooting* and is analyzed in detail in Chapter 4. It is certainly a simple and intuitively appealing method for solving the general nonlinear BVP (3.1), (3.12). It actually works well for Example 3.2 with $\lambda = 1$. However, there are many other BVPs for which this method does not work. One basic reason is that the conditioning, or stability properties, of the IVP, linearized at the solution, may be quite different from the corresponding properties of the linearized BVP (3.20). The IVP may be extremely ill-conditioned while the BVP is agreeably well-conditioned (as described in detail in Sections 3.3 and 3.4). The single-shooting algorithm is unstable under such circumstances.

To understand in depth the performance of numerical methods on nonlinear problems, we must first consider linear problems and understand their conditioning. This is our main task in the remainder of this chapter.

The basic representation of the solution for the linear boundary value problem follows.

Theorem 3.26 Suppose $A(x)$ and $\mathbf{q}(x)$ in the linear differential equations (3.2) are continuous. The BVP (3.2),(3.19) has a unique solution $\mathbf{y}(x)$ iff [2] the matrix

$$Q = B_a Y(a) + B_b Y(b) \qquad (3.26a)$$

is nonsingular, in which case

---

[2] if and only if.

$$\mathbf{y}(x) = Y(x)Q^{-1}(\boldsymbol{\beta} - B_b Y(b) \int_a^b Y^{-1}(t)\mathbf{q}(t)\,dt\,) + Y(x) \int_a^x Y^{-1}(t)\mathbf{q}(t)\,dt \qquad (3.26b)$$

Here $Y(x)$ is any fundamental solution satisfying (3.7a), but not necessarily (3.7b).

Proof: From (3.9), any solution of (3.2) can be expressed in the form

$$\mathbf{y}(x) = Y(x)\mathbf{s} + \mathbf{y}_p(x)$$

i. e., as a sum of an arbitrary linear combination of (linearly independent) homogeneous solutions and the particular solution $\mathbf{y}_p(x)$. To satisfy the BC (3.19), $\mathbf{s}$ must be chosen to satisfy

$$B_a[Y(a)\mathbf{s} + \mathbf{y}_p(a)] + B_b[Y(b)\mathbf{s} + \mathbf{y}_p(b)] \equiv Q\mathbf{s} + B_b Y(b) \int_a^b Y^{-1}(t)\mathbf{q}(t)\,dt = \boldsymbol{\beta}$$

This is a system of linear algebraic equations for the parameters $\mathbf{s}$ of the general representation for $\mathbf{y}(x)$. Thus, there exists a unique solution if Q is nonsingular, and the expression (3.26b) follows directly by using (3.8). □

For the more general case of the multipoint boundary value problem (1.4a, e), Theorem 3.26 can be extended to read

Theorem 3.27 The BVP (1.4a, e) has a unique solution iff

$$Q := \sum_{j=1}^{J} B_j Y(\zeta_j) \qquad (3.27a)$$

is nonsingular, in which case

$$\mathbf{y}(x) = Y(x)Q^{-1}(\boldsymbol{\beta} - \sum_{j=2}^{J} B_j Y(\zeta_j) \int_a^{\zeta_j} Y^{-1}(t)\mathbf{q}(t)\,dt) + \mathbf{y}_p(x) \qquad (3.27b)$$

where the particular solution

$$\mathbf{y}_p(x) = Y(x) \int_a^x Y^{-1}(t)\mathbf{q}(t)\,dt \qquad (3.27c)$$

□

Example 3.3

Consider again the BVP (3.10) of Example 3.1, which can be rewritten as

$$\begin{bmatrix} y_1 \\ y_2 \end{bmatrix} = \begin{bmatrix} 0 & 1 \\ -1 & 0 \end{bmatrix} \begin{bmatrix} y_1 \\ y_2 \end{bmatrix} \qquad 0 < x < b$$

$$\begin{bmatrix} 1 & 0 \\ 0 & 0 \end{bmatrix} \begin{bmatrix} y_1(0) \\ y_2(0) \end{bmatrix} + \begin{bmatrix} 0 & 0 \\ 1 & 0 \end{bmatrix} \begin{bmatrix} y_1(b) \\ y_2(b) \end{bmatrix} = \begin{bmatrix} 0 \\ \beta_2 \end{bmatrix}$$

where $y_1(x) = u(x)$, $y_2(x) = u'(x)$. The fundamental solution for this system, which satisfies $Y(0) = I$ is

$$Y(x) = \begin{bmatrix} \cos x & \sin x \\ -\sin x & \cos x \end{bmatrix}$$

From (3.26a)

$$Q = \begin{bmatrix} 1 & 0 \\ 0 & 0 \end{bmatrix} I + \begin{bmatrix} 0 & 0 \\ 1 & 0 \end{bmatrix} \begin{bmatrix} \cos b & \sin b \\ -\sin b & \cos b \end{bmatrix} = \begin{bmatrix} 1 & 0 \\ \cos b & \sin b \end{bmatrix}$$

which is singular iff $b = n\pi$ for some integer n. □

Remarks

(i) From Theorem 3.26 it is easy to see that, if $A(x)$ and $\mathbf{q}(x)$ have r continuous derivatives, i. e., $A, \mathbf{q} \in C^{(r)}[a, b]$, then $\mathbf{y} \in C^{(r+1)}[a, b]$. Thus the solution has one more continuous derivative than the data A and $\mathbf{q}$.

(ii) If Q of Theorem 3.27, which contains Theorem 3.26 as a special case, is singular, then the problem (1.4a, e) has a solution iff the vector $\boldsymbol{\beta} - \sum_{j=2}^{J} B_j Y(\xi_j) \int_a^{\xi_j} Y^{-1}(t) \mathbf{q}(t)\, dt$ belongs to the range of Q, in which case it has infinitely many. For example, see (3.10) again, with $b = \pi$.

(iii) Although direct verification of the nonsingularity of Q can be virtually impossible in practice, there are theoretical results concerning how to approximate Q to determine its nonsingularity. In general, relating a boundary value problem to an initial value problem is a recurring theme, and in particular the form of Q will be exploited for this purpose later (e.g., see Chapter 4).

(iv) The fundamental solution $Y(x)$ in Theorem 3.26 is determined only up to $Y(a)$, its initial value. While the nonsingularity of Q does not depend on the choice of $Y(a)$, its condition number does — sometimes crucially! □

It is sometimes convenient to refer to the linear ODE (3.2) using a *differential operator* notation,

$$L\mathbf{y}(x) \equiv \mathbf{y}'(x) - A(x)\mathbf{y}(x), \qquad a < x < b \qquad (3.28a)$$

Then (3.2) reads

$$L\mathbf{y} = \mathbf{q} \qquad (3.29a)$$

Similarly, with a *boundary operator*

$$B\mathbf{y} \equiv B_a \mathbf{y}(a) + B_b \mathbf{y}(b) \qquad (3.28b)$$

the BC (3.19) reads

$$B\mathbf{y} = \boldsymbol{\beta} \qquad (3.29b)$$

A fundamental solution, scaled naturally for an IVP, satisfies

$$LY = 0, \qquad Y(a) = I \qquad (3.30a)$$

Correspondingly, a fundamental solution scaled naturally for the BVP satisfies

$$L\Phi = 0, \qquad B\Phi = I \qquad (3.30b)$$

The connection between $Y(x)$ of (3.30a) and $\Phi(x)$ of (3.30b) is

$$\Phi(x) = Y(x) Q^{-1} \qquad (3.30c)$$

(Why?).

3.2 GREEN'S FUNCTIONS

While the previous section discussed existence and uniqueness of solutions to BVPs, here the dependence of such solutions on the data is considered. Thus, suppose that the data $\mathbf{q}(x)$ and $\boldsymbol{\beta}$ of a linear BVP (3.29) are perturbed. We ask, what is the resulting perturbation of the solution $\mathbf{y}(x)$? A tight bound on the size of any such perturbation can be considered as a *conditioning constant* of the BVP. If the conditioning constant of a given problem is a constant of moderate size, then the problem is *well-conditioned*. (Precisely what "moderate size" means depends, to recall, on the accuracy required relative to the precision used in a calculation.)

The first-step towards obtaining a conditioning constant involves representing the solution of the BVP in terms of the data. This gives rise to a Green's function.

A Green's function is related to the inverse of a differential operator and can be used to express the solution to a linear BVP. First we describe it for a first-order system of ODEs, and then we consider a scalar higher-order differential equation.

3.2.1 A first-order system

From Theorem 3.26 we know that the BVP (3.2), (3.19) [or (3.29)] has a unique solution provided Q of (3.26a) is nonsingular [where $Y(x)$ is any fundamental solution]. In that case, the solution has the representation (3.26b). Let us now simplify (3.26b) by uniting the two integrals appearing there. Also, use of the fundamental solution $\Phi(x)$ of (3.30b) allows us to hide away Q [1]. We obtain [cf. (3.9)]

$$\mathbf{y}(x) = \Phi(x)\boldsymbol{\beta} + \int_a^b G(x,t)\mathbf{q}(t)\,dt \qquad (3.31)$$

where $G(x,t)$ is the $n \times n$ *Green's (matrix) function*, defined by

$$G(x,t) := \begin{cases} \Phi(x) B_a \Phi(a) \Phi^{-1}(t) & t \leq x \\ -\Phi(x) B_b \Phi(b) \Phi^{-1}(t) & t > x \end{cases} \qquad (3.32)$$

Note that $G(x,t)$ in (3.31) is independent of the choice of fundamental solution [see Exercise 10 and (3.36)].

The representation (3.31) is a desirable one, since it explicitly expresses the dependence of the solution on the data. Let us define the norm

$$\|G\|_\infty := \max_{a \leq x, t \leq b} \|G(x,t)\|_\infty$$

Then (3.31) yields

$$\max_{a \leq x \leq b} |\mathbf{y}(x)| =: \|\mathbf{y}\|_\infty \leq \kappa \left(|\boldsymbol{\beta}| + \int_a^b |\mathbf{q}(t)|\,dt \right) \qquad (3.33)$$

where
$$\kappa := \max \{ \|G\|_\infty, \|\Phi\|_\infty \} \tag{3.34}$$
is called the *conditioning constant*.

Remark The conditioning constant κ is also referred to in the literature as a *stability constant*, highlighting its absolute (rather than relative) nature. However, we have chosen in this book to use the term "stability" in a BVP context only when dealing with *numerical* methods. We delay a more precise treatment of conditioning constants (with numerical examples) to Sections 3.3 and 3.4. □

Example 3.4

For the simple BVP of Example 3.3 (and 1), set $b = \pi/2$. Thus we have
$$B_a = \begin{bmatrix} 1 & 0 \\ 0 & 0 \end{bmatrix} \quad B_b = \begin{bmatrix} 0 & 0 \\ 1 & 0 \end{bmatrix}$$
and, since $Q=I$ in Example 3.3,
$$\Phi(x) = \begin{bmatrix} \cos x & \sin x \\ -\sin x & \cos x \end{bmatrix}$$

Hence we obtain from (3.32)
$$G(x, t) := \begin{cases} F_1(x, t) & t \leq x \\ F_2(x, t) & t > x \end{cases}$$

where
$$F_1 = \begin{bmatrix} \cos x \cos t & -\cos x \sin t \\ -\sin x \cos t & \sin x \sin t \end{bmatrix} \quad F_2 = \begin{bmatrix} -\sin x \sin t & -\sin x \cos t \\ -\cos x \sin t & -\cos x \cos t \end{bmatrix}$$

Note that in this example, $\Phi(x)$ is a reflection matrix. It is not difficult to see that by (3.34), $\kappa = 1$. □

Let us extend the discussion to include the multipoint BVP (1.4a, e). If $Y(x)$ is a fundamental solution and Q in (3.27b) is nonsingular, then by Theorem 3.27 the unique solution of (1.4a, e) can be rewritten as
$$\mathbf{y}(x) = \int_a^b G(x, t) \mathbf{q}(t) \, dt + \mathbf{w}(x) \tag{3.35}$$
where $\mathbf{w}(x) = Y(x) Q^{-1} \boldsymbol{\beta}$ and the *Green's function* is defined by
$$G(x, t) = \begin{cases} Y(x) Q^{-1} \sum_{j=1}^{l} B_j Y(\zeta_j) Y^{-1}(t) & \text{if } t \leq x \\ -Y(x) Q^{-1} \sum_{j=l+1}^{J} B_j Y(\zeta_j) Y^{-1}(t) & \text{if } t > x \end{cases} \tag{3.36}$$

for $t \in [\zeta_l, \zeta_{l+1})$, $(1 \leq l \leq J-1)$. If $A(x) \in C^{(k)}[a, b]$ then $Y(x) \in C^{(k+1)}[a, b]$. Also $[Y^{-1}(t)] \in C^{(k+1)}[a, b]$ (e.g., see Exercise 3), so the next result follows.

Theorem 3.37 Suppose that $A(x), \mathbf{q}(x) \in C^{(k)}[a, b]$, $k \geq 0$, and that the multipoint BVP (1.4a, e) has a unique solution $\mathbf{y}(x)$. Then $\mathbf{y}(x) \in C^{(k+1)}[a, b]$ is given by (3.35), where $G(x, t)$ is defined in (3.36). Moreover, $G(x, t) \in C^{(k+1)}[a, t] \times C^{(k+1)}[t, b]$ as a function of x and $G(x, t) \in C^{(k+1)}[a, x] \times C^{(k+1)}[x, b]$ as a function of t. □

We note that the continuity requirements and statements in Theorem 3.37 can be readily relaxed to hold only in a piecewise manner on $[a, b]$, which is sufficiently general for our purposes.

Construction of the Green's function for a BVP involving a linear mixed-order system of ODEs (i. e., where each ODE can be of different order) with corresponding linear BC follows directly in theory. One considers the corresponding first-order system, as in Section 1.1, and applies Theorem 3.37. A conditioning constant κ for this BVP can be defined by considering $G(x, t)$ for the corresponding first-order system and using the definition (3.34). Instead, the Green's function for a higher-order ODE can be constructed directly, as seen in the next section.

3.2.2 A higher-order ODE

This subsection can be skipped without loss of continuity.

Consider the higher-order ODE

$$Lu := u^{(m)} - \sum_{j=0}^{m-1} c_j(x) u^{(j)} = q(x) \qquad a < x < b \qquad (3.38a)$$

subject to the homogeneous BC

$$B_a \mathbf{y}(a) + B_b \mathbf{y}(b) = \mathbf{0} \qquad (3.38b)$$

where $\mathbf{y}(x) := (u(x), \ldots, u^{(m-1)}(x))$. This is, of course, a special case of a mixed-order system for which we have already outlined how to obtain a Green's function as well as a conditioning constant, but we discuss it here separately, because we do not wish to consider the converted first-order system. Also, with this treatment we renew contact with standard ODE texts.

Under appropriate assumptions, there exists a unique Green's function $G(x, t)$—a scalar function of two variables such that the solution to (3.38a, b) can be expressed as

$$u(x) = \int_a^b G(x, t) q(t) \, dt \qquad (3.39)$$

The Green's function $G(x, t)$ is independent of the inhomogeneous term $q(x)$ and depends only upon the particular differential operator and BC operator. In principle, if $G(x, t)$ can be explicitly determined, then (3.39) is valid for an arbitrary piecewise continuous inhomogeneity $q(x)$, and it is thus possible (though not generally advisable) to solve BVPs by using numerical methods for integral equations.

Theorem 3.40 Assume that the homogeneous problem (3.38) with $q \equiv 0$ has only the trivial solution $u(x) \equiv 0$. Then there exists a *Green's function* $G(x, t)$ which is uniquely defined on $D = \{(x, t): a \leq x, t \leq b\}$ by the following conditions:

(i) $\dfrac{\partial^j G}{\partial x^j}$ and $\dfrac{\partial^j G}{\partial t^j}$ are continuous on D for $0 \leq j \leq m - 2$.

(ii) $\dfrac{\partial^{m-1}G(t^+,t)}{\partial x^{m-1}} - \dfrac{\partial^{m-1}G(t^-,t)}{\partial x^{m-1}} = \lim_{\varepsilon \to 0} \left[\dfrac{\partial^{m-1}G(t+\varepsilon,t)}{\partial x^{m-1}} - \dfrac{\partial^{m-1}G(t-\varepsilon,t)}{\partial x^{m-1}} \right] = 1$

for $a < t < b$. [3]

(iii) As a function of x, $G(x,t)$ satisfies (3.38a) with $q \equiv 0$ for $x \neq t$ and the BC (3.38b).

The following also hold:

(a) Any solution of (3.38a, b) with $q \not\equiv 0$ is equivalently determined by (3.39).

(b) If $c_j(x) \in C^{(k)}[a,b]$, $0 \le j \le m-1$ then $G(x,t) \in C^{(m+k)}[a,t] \times C^{(m+k)}[t,b]$ as a function of x, and similarly $G(x,t) \in C^{(m+k)}[a,x] \times C^{(m+k)}[x,b]$ as a function of t. □

Consider now the case where the BC in (3.38b) are nonhomogeneous; i. e., consider the BC (3.19), assuming that the m BC are linearly independent. Then there exists a unique polynomial $p_m(x)$ of degree $\le m$ satisfying the BC (3.19). (This polynomial is straightforward to calculate.) The function $w(x) = u(x) - p_m(x)$ satisfies

$Lw(x) = q(x) - Lp_m(x) =: \hat{q}(x)$ and (3.38b). Thus, $w(x) = \displaystyle\int_a^b G(x,t)\hat{q}(t)\,dt$, or

$$u(x) = p_m(x) + \int_a^b G(x,t)\hat{q}(t)\,dt = \hat{u}(x) + \int_a^b G(x,t)q(t)\,dt \qquad (3.41a)$$

with $\hat{u}$ given by

$$\hat{u}(x) := p_m(x) - \int_a^b G(x,t)Lp_m(t)\,dt \qquad (3.41b)$$

This is a generalization of the homogeneous boundary condition case.

The construction of a Green's function, though impractical for general BVPs, is possible when a complete set of solutions to the homogeneous problem is known (see Exercise 6). For simplicity, we consider only the case of the self-adjoint second-order BVP

$$Lu(x) := (c_2(x)u'(x))' - c_0(x)u(x) = q(x) \qquad a < x < b \qquad (3.42a)$$

$$a_1 u(a) + a_2 u'(a) = 0, \quad b_1 u(b) + b_2 u'(b) = 0 \qquad (3.42b)$$

Assume that (3.42a, b) has a unique solution.

If $u_1(x)$ and $u_2(x)$ are any two linearly independent solutions of

$$Lw(x) = 0 \qquad a < x < b \qquad (3.42c)$$

then any solution of this differential equation has the form

$$w(x) = v_1 u_1(x) + v_2 u_2(x), \qquad v_1, v_2 \in \mathbb{R}$$

We can choose $u_1(x)$ to satisfy

---

[3] The jump discontinuity is $\dfrac{1}{c_m(t)}$ if the ODE has a coefficient $c_m(x)$ for the term $u^{(m)}(x)$.

$$a_1 u_1(a) + a_2 u_1'(a) = 0 \qquad (3.43a)$$

and $u_2(x)$ to satisfy

$$b_1 u_2(b) + b_2 u_2'(b) = 0 \qquad (3.43b)$$

They are linearly independent, since otherwise $u_1(x) = c u_2(x)$ for some constant c, implying that $u_1(x)$ is a nontrivial solution to (3.42c,b) and thus contradicting the assumption that (3.42a, b) has a unique solution. From property (iii) of the Green's function in Theorem 3.40, $G(x, t)$ should satisfy (3.42c) for any fixed t, so using variation of parameters we seek continuous functions v_1, v_2, w_1, and w_2 such that

$$G(x,t) = \begin{cases} v_1(t)u_1(x) + v_2(t)u_2(x) & t > x \\ w_1(t)u_1(x) + w_2(t)u_2(x) & t < x \end{cases} \qquad (3.44)$$

Since $G(x, t)$ also satisfies (3.42b),

$$G(x,t) = \begin{cases} v_1(t)u_1(x) & t > x \\ w_2(t)u_2(x) & t < x \end{cases}$$

Properties (i) and (ii) imply

$$v_1(t)u_1(t) - w_2(t)u_2(t) = 0$$

and

$$v_1(t)u_1'(t) - w_2(t)u_2'(t) = \frac{1}{c_2(t)}$$

respectively. Solving for $v_1(t)$ and $w_2(t)$, we obtain

$$v_1(t) = \frac{u_2(t)}{c_2(t)W(t)}, \qquad w_2(t) = \frac{u_1(t)}{c_2(t)W(t)}$$

where the *Wronskian* $W(t) := \det [D(t)]$, $D(t) = \begin{bmatrix} u_1(t) & u_2(t) \\ u_1'(t) & u_2'(t) \end{bmatrix}$. It can be shown that $c_2(t)W(t) = k$, a constant independent of t, so

$$G(x,t) = \begin{cases} \dfrac{1}{k} u_1(t)u_2(x) & t \leq x \\ \dfrac{1}{k} u_1(x)u_2(t) & t > x \end{cases} \qquad (3.45)$$

Thus, by finding fundamental solutions to the homogeneous problem (3.42c, b), $G(x, t)$ can be determined, and the solution to the BVP (3.42a, b) is then determined by the integral relation (3.39) (see Exercise 7).

Example 3.5

Let us calculate the Green's function for

$$u''(x) + u(x) = [u'(x)]' + u(x) = 0 \qquad 0 < x < \frac{\pi}{2}$$

$$u(0) = 0 \qquad u(\frac{\pi}{2}) = 0$$

This is the same BVP as in Example 3.4, but with a higher-order scalar ODE instead of a first-order system. Here, $u_1(x) = \sin x$, $u_2(x) = \cos x$ [check (3.43)], so

$$W(t) = \det \begin{bmatrix} \sin x & \cos x \\ \cos x & -\sin x \end{bmatrix} = -1$$

Since $c_2(t) \equiv 1$ it follows from (3.45) that

$$G(x,t) = \begin{cases} -\sin x \cos t & t \geq x \\ -\sin t \cos x & t \leq x \end{cases}$$

This Green's function should be compared with the Green's matrix obtained for Example 3.4. Note that the scalar one appears in the matrix as the upper right element, and that its derivative with respect to x is at the lower right of the matrix Green's function.

The solution of the BVP

$$u'' + u(x) = 1 \qquad 0 < x < \frac{\pi}{2}$$

$$u(0) = 0, \qquad u(\frac{\pi}{2}) = 0$$

for instance, is

$$u(x) = \int_0^{\pi/2} G(x,t)\,dt = -\int_0^x \sin t \cos x\,dt - \int_x^{\pi/2} \sin x \cos t\,dt$$

$$= -\cos x [-\cos t]_0^x - \sin x [\sin t]_x^{\pi/x}$$

$$= \cos x [\cos x - 1] - \sin x [1 - \sin x]$$

$$= 1 - \cos x - \sin x \qquad \square$$

3.3 STABILITY OF INITIAL VALUE PROBLEMS

The effect that small changes in the data (e. g., initial or boundary values) have on the solution of a differential equation is of fundamental concern when one is approximating that solution numerically. Qualitatively speaking, a problem is called *stable* if "small" changes in the data produce only "small" changes in the solution, and it is *unstable* otherwise. When an unstable IVP is to be solved numerically, it is normally a source of grief regardless of the particular numerical method one uses. Even if the IVP is stable, "small" roundoff and truncation errors can still cause "large" errors in the computed solution if the *numerical method* is unstable. This has been discussed in Section 2.7. In this section we discuss the analytical problem of stability of IVPs. The notion of stability for IVPs is in some sense the parallel of well-conditioning for BVPs. The latter is discussed in Section 3.4.

3.3.1 Stability and fundamental solutions

Consider the general IVP (3.1),(3.3), rewritten here:

$$\mathbf{y}' = \mathbf{f}(x, \mathbf{y}) \qquad x > a \tag{3.46a}$$

$$\mathbf{y}(a) = \boldsymbol{\alpha} \tag{3.46b}$$

where $\mathbf{y}(x) = (y_1(x), \ldots, y_n(x))^T$. We shall need to distinguish between various kinds of stability and thus need the following:

Definitions 3.47 A solution $\mathbf{y}(x)$ is called *stable* [with respect to changes in the initial conditions $\mathbf{y}(a)$] if given any $\varepsilon > 0$ there is a $\delta > 0$ such that any other solution $\hat{\mathbf{y}}(x)$ of (3.46a) satisfying

$$|\mathbf{y}(a) - \hat{\mathbf{y}}(a)| \leq \delta \tag{3.47a}$$

also satisfies

$$|\mathbf{y}(x) - \hat{\mathbf{y}}(x)| \leq \varepsilon \quad \text{for all } x > a \tag{3.47b}$$

The solution $\mathbf{y}(x)$ is *asymptotically stable* if, in addition to (3.47b),

$$|\mathbf{y}(x) - \hat{\mathbf{y}}(x)| \to 0 \quad \text{as } x \to \infty \tag{3.47c}$$

Finally, $\mathbf{y}(x)$ is *relatively stable* if, instead of (3.47b),

$$|\mathbf{y}(x) - \hat{\mathbf{y}}(x)| \leq \varepsilon |\mathbf{y}(x)| \quad \text{for all } x > a. \tag{3.47d}$$

□

The stability definitions 3.47 relate to changes in the initial data, as (3.47a) indicates. To relate to perturbations in the right-hand side of (3.46a) as well, slightly stronger concepts are needed.

Definitions 3.48 A solution $\mathbf{y}(x)$ is called *uniformly stable* if, given any $\varepsilon > 0$, there is a $\delta > 0$ such that any other solution $\hat{\mathbf{y}}(x)$ of (3.46a) satisfying

$$|\mathbf{y}(c) - \hat{\mathbf{y}}(c)| \leq \delta \tag{3.48a}$$

at some point $c \geq a$, also satisfies

$$|\mathbf{y}(x) - \hat{\mathbf{y}}(x)| \leq \varepsilon \quad \text{for all } x > c \tag{3.48b}$$

The concepts of *asymptotic uniform stability* and *relative uniform stability* are analogously defined. □

Obviously, uniform stability implies stability. The converse does not hold in general.

For linear problems, the stability concepts (3.47) and (3.48) can be linked directly to properties of the fundamental solution. To see this, let $\mathbf{y}(x)$ and $\hat{\mathbf{y}}(x)$ both satisfy the linear ODE

$$\mathbf{y}' = A(x)\mathbf{y} + \mathbf{q}(x) \qquad x > a \tag{3.49}$$

First, note that only the homogeneous problem matters, because the difference $\mathbf{z}(x) = \mathbf{y}(x) - \hat{\mathbf{y}}(x)$ satisfies

$$\mathbf{z}' = A(x)\mathbf{z} \qquad x > a \qquad (3.50)$$

The following theorem indicates that, in the linear case, (uniform) stability or asymptotic (uniform) stability of one solution implies the same property for any solution.

Theorem 3.51 Let $Y(x) \equiv Y(x; a)$ be a fundamental solution of the ODE. Then the solution $\mathbf{y}(x)$ of (3.49), (3.46b) is stable iff $\sup_{x \geq a} \|Y(x)\|$ is bounded and is uniformly stable iff $\sup_{x \geq t \geq a} \|Y(x)Y^{-1}(t)\|$ is bounded. Moreover, $\mathbf{y}(x)$ is asymptotically stable iff $\|Y(x)\| \to 0$ as $x \to \infty$ and is asymptotically uniformly stable iff $\|Y(x)Y^{-1}(t)\| \to 0$ as $x \to \infty$ for all $t \geq a$.

Proof: The stability of $\mathbf{y}(x)$ follows directly from the definitions (3.47), because

$$|\mathbf{z}(x)| = |Y(x)(\mathbf{y}(a) - \hat{\mathbf{y}}(a))| \leq \|Y(x)\| \, |\boldsymbol{\alpha} - \hat{\mathbf{y}}(a)| \leq \|Y(x)\|\delta$$

Similarly, if (3.48a) holds then

$$|\mathbf{z}(x)| = |Y(x)Y^{-1}(c)(\mathbf{y}(c) - \hat{\mathbf{y}}(c))| \leq \|Y(x)Y^{-1}(c)\| \, \delta$$

Conversely if $\|Y(x)\|$ is unbounded, then one can always find a perturbation vector with a component in the unstable direction. □

For relative stability (which is sometimes more meaningful for a numerical approximation of the IVP) we have to establish

$$|Y(x)(\boldsymbol{\alpha} - \hat{\mathbf{y}}(a))| \leq \varepsilon |Y(x)\boldsymbol{\alpha}|$$

and this essentially holds *unless* $|Y(x)\boldsymbol{\alpha}| << \|Y(x)\| \, |\boldsymbol{\alpha}|$. In Exercise 8 this is made precise for the constant coefficient case.

The condition for uniform stability prompts the definition of a *stability constant*

$$\kappa := \sup_{a \leq t \leq x < \infty} \|Y(x)Y^{-1}(t)\| \qquad (3.52)$$

By (3.8) we can bound the solution in terms of the inhomogeneities using this stability constant, namely

$$|\mathbf{y}(x)| \leq \kappa \left(|\boldsymbol{\alpha}| + \int_a^x |\mathbf{q}(t)| \, dt \right) \qquad x \geq a \qquad (3.53)$$

[see also (3.9)]. Clearly this stability constant is a special case of the conditioning constant κ in (3.34) for a BVP with $b = \infty$, $B_a = I$ and $B_b = 0$.

Our next step is to characterize the premises of Theorem 3.51 more explicitly. Since $Y(x)$ satisfies the homogeneous ODE (3.50), this characterization must involve $A(x)$, and it is natural to try to give it in terms of the eigenvalues of A, which are computable. Thus let $T(x) \in \mathbf{C}^{n \times n}$ be defined for each $x \geq a$ as the similarity transformation which brings $A(x)$ into its Jordan canonical form [cf. (2.9)],

$$T^{-1}AT = \Lambda(x) = \begin{bmatrix} \Lambda_1(x) & & 0 \\ & \ddots & \\ 0 & & \Lambda_m(x) \end{bmatrix} \qquad (3.54)$$

where each Jordan block $\Lambda_i(x) \in \mathbf{C}^{r_i \times r_i}$ has the form

$$\Lambda_i = \begin{bmatrix} \lambda_i & 1 & & 0 \\ & \ddots & \ddots & \\ & & \ddots & 1 \\ 0 & & & \lambda_i \end{bmatrix} \qquad (3.55)$$

and $\lambda_i(x)$ is an eigenvalue of $A(x)$, $1 \le i \le m$ (N.B. $n = \sum_{i=1}^{m} r_i$). Define

$$\mathbf{w}(x) := T^{-1}(x)\mathbf{y}(x) \qquad x \ge a \qquad (3.56)$$

where $\mathbf{y}$ solves the homogeneous IVP

$$\mathbf{y}' = A(x)\mathbf{y}, \qquad x > a \qquad (3.57a)$$

$$\mathbf{y}(a) = \boldsymbol{\alpha} \qquad (3.57b)$$

Then, assuming that $T(x)$ is differentiable, we find that $\mathbf{w}(x)$ is the solution of

$$\mathbf{w}' = (\Lambda(x) - T^{-1}(x)T'(x))\mathbf{w} \qquad x > a \qquad (3.58a)$$

$$\mathbf{w}(a) = T^{-1}(a)\boldsymbol{\alpha} \qquad (3.58b)$$

Consider the relationship between the fundamental solutions $Y(x;t)$ for (3.57a) and $W(x;t)$ for (3.58a). Given the fundamental solution $W(x;t)$ [satisfying $W(t;t) = I$], the general form of a fundamental solution for (3.57a) is $T(x)W(x;t)R$, where $R \in \mathbf{C}^{n \times n}$ is a nonsingular constant matrix. The particular fundamental solution satisfying $Y(a) = I$ is

$$Y(x) := Y(x;a) = T(x)W(x;a)T^{-1}(a) \qquad (3.59)$$

Now, if $T' = 0$ then the differential equations (3.58a) for $\mathbf{w}(x)$ decouple. This occurs in the constant coefficient case, i. e., where A is independent of x, which is considered next.

3.3.2 The constant coefficient case

In this subsection, we show that the stability question can be described in terms of (the real parts of) the eigenvalues of A. Here also uniform stability is equivalent to stability (Why?). Specifically, we obtain

Theorem 3.60 The unique solution of the homogeneous IVP (3.57a, b) with A constant is (uniformly) stable iff all eigenvalues λ of A satisfy either $\text{Re}(\lambda) < 0$ or $\text{Re}(\lambda) = 0$ with λ simple (i. e., it belongs to a 1×1 Jordan block). Furthermore, the solution $\mathbf{y}(x)$ is asymptotically (uniformly) stable iff all eigenvalues of A satisfy $\text{Re}(\lambda) < 0$.

Proof: We verify the assumptions of Theorem 3.51. Consider first the fundamental solution $W(x) = W(x; a)$ for (3.58a) (with $T' = 0$). Since the fundamental solution for (3.57a) satisfies $Y(x) = Y(x; a) = TW(x)T^{-1}$, it is sufficient to prove boundedness for $W(x)$. We have

$$W(x) = e^{\Lambda(x-a)} = \begin{pmatrix} e^{\Lambda_1(x-a)} & & & 0 \\ & e^{\Lambda_2(x-a)} & & \\ & & \ddots & \\ 0 & & & e^{\Lambda_m(x-a)} \end{pmatrix}$$

Now, for any $B \in \mathbf{C}^{n \times n}$, $e^B := I + B + \frac{1}{2!}B^2 + \frac{1}{3!}B^3 + \cdots$, and since Λ_i^k is upper triangular with its $(j, j+l)$ element equal to $\binom{k}{l} \lambda_i^{k-l}$, we obtain

$$e^{\Lambda_i(x-a)} = e^{\lambda_i(x-a)} \begin{pmatrix} 1 & (x-a) & \frac{(x-a)^2}{2!} & \cdots & \frac{(x-a)^{r_i-1}}{(r_i-1)!} \\ & 1 & (x-a) & & \vdots \\ & & \ddots & & \vdots \\ & & & & (x-a) \\ 0 & & & & 1 \end{pmatrix}$$

Thus, $e^{\Lambda_i(x-a)} \to 0$ as $x \to \infty$ iff $\text{Re}(\lambda_i) < 0$ (where $\text{Re}(\lambda_i)$ is the real part of λ_i). Moreover, $e^{\Lambda_i x}$ remains bounded iff either $\text{Re}(\lambda_i) < 0$ or $\text{Re}(\lambda_i) = 0$ with λ_i a simple eigenvalue. The conclusions follow. $\square$

Example 3.6

Consider the IVP

$$\mathbf{y}' = \begin{bmatrix} 0 & 1 \\ 1 & 0 \end{bmatrix} \mathbf{y}, \quad x > 0 \tag{3.61}$$

$$\mathbf{y}(0) = \begin{bmatrix} \alpha_1 \\ \alpha_2 \end{bmatrix}$$

Since A has eigenvalues ± 1 and corresponding eigenvectors $\begin{bmatrix} 1 \\ \pm 1 \end{bmatrix}$, the solution is unstable regardless of the initial condition. The solution to this IVP is

$$\mathbf{y}(x) = \begin{bmatrix} \dfrac{\alpha_1+\alpha_2}{2} e^x + \dfrac{\alpha_1-\alpha_2}{2} e^{-x} \\ \dfrac{\alpha_1+\alpha_2}{2} e^x - \dfrac{\alpha_1-\alpha_2}{2} e^{-x} \end{bmatrix}$$

and the solution to (3.61) and $\mathbf{y}(0) = \begin{bmatrix} \alpha_1+\varepsilon_1 \\ \alpha_2+\varepsilon_2 \end{bmatrix}$ is

$$\hat{\mathbf{y}}(x) = \begin{bmatrix} \dfrac{\alpha_1+\varepsilon_1+\alpha_2+\varepsilon_2}{2} e^x + \dfrac{\alpha_1+\varepsilon_1-\alpha_2-\varepsilon_2}{2} e^{-x} \\ \dfrac{\alpha_1+\varepsilon_1+\alpha_2+\varepsilon_2}{2} e^x - \dfrac{\alpha_1+\varepsilon_1-\alpha_2-\varepsilon_2}{2} e^{-x} \end{bmatrix}$$

so

$$\mathbf{y}(x) - \hat{\mathbf{y}}(x) = - \begin{bmatrix} \dfrac{\varepsilon_1+\varepsilon_2}{2} e^x + \dfrac{\varepsilon_1-\varepsilon_2}{2} e^{-x} \\ \dfrac{\varepsilon_1+\varepsilon_2}{2} e^x - \dfrac{\varepsilon_1-\varepsilon_2}{2} e^{-x} \end{bmatrix}$$

Thus, while the absolute error grows exponentially large unless fortuitously $\dfrac{\varepsilon_1+\varepsilon_2}{2} = 0$, the relative error remains small unless $\alpha_1 + \alpha_2 = 0$ (assuming $|\varepsilon_1|$ and $|\varepsilon_2|$ are small compared to $|\alpha_1| + |\alpha_2|$). In other words, while no solution is stable, all solutions are relatively stable except for solutions with $\alpha_1 \approx -\alpha_2$. □

Relative stability of (3.47d) can be easily characterized in the constant coefficient case (see Exercise 8).

We conclude that, in the constant coefficient case, while $Y(x)$ gives an appropriate description of the propagation of the initial data $\boldsymbol{\alpha}$, the form $TW(x)$ [with $T(x) \equiv T$ in (3.59)] gives us more concrete information about the various types of solutions, or the *modes*. Specifically, the matrix T represents the *directions* of the various modes, while the matrix $W(x)$ represents their *growth*. Through $W(x)$, the eigenvalues of A with positive and negative real parts correspond to fundamental solution modes (vectors) which are exponentially increasing and decreasing, respectively. Nonzero eigenvalues with zero real part correspond to oscillatory fundamental solution modes, which are bounded as $x \to \infty$ only if the eigenvalues are simple.

For the variable coefficient case, which we consider next, the eigenvalues of $A(x)$ do not always give such a complete characterization of the stability of the BVP.

3.3.3 The general linear case

We have seen that the stability of the IVP (3.49), (3.46b) is analyzed by transforming its homogeneous counterpart (3.50) to the problem (3.58), where the unknowns are decoupled. The transformation matrix $T(x)$ in (3.54) which performs this decoupling consists of the principal vectors (sometimes called generalized eigenvectors) of A. When $A(x) \equiv A$, the matrix $\Lambda(x) - T(x)^{-1}T'(x) \equiv \Lambda$ is upper triangular, with diagonal elements being the eigenvalues of A.

In the variable coefficient case, when $T(x)$ consists of the principal vectors of $A(x)$, then $\Lambda(x) - T(x)^{-1}T'(x)$, even if it exists, does not necessarily have an upper triangular form. In particular, there is generally no reason to expect the unknown solution components corresponding to (3.58) to be decoupled. Thus, we need to consider more general transformations $T(x)$.

Suppose $T(x) \in \mathbf{R}^{n \times n}$ is a differentiable transformation with cond $(T;x,t) := \|T(x)\| \, \|T^{-1}(t)\|$ uniformly bounded for $x \geq t$. If $\mathbf{w}(x) := T^{-1}(x)\mathbf{y}(x)$, then we obtain as before that

$$\mathbf{w}' = V(x)\,\mathbf{w}, \qquad x > a \qquad (3.62)$$

where

$$V(x) := T^{-1}(x)(A(x)T(x) - T'(x)) \qquad (3.63)$$

It is easy to see that, since cond $(T;x,t)$ is bounded, the ODE systems (3.57a) and (3.62) have the same uniform stability properties. We make the following

Definitions 3.64 The systems (3.57a) and (3.62) are *kinematically similar*. When $V(x)$ is in upper triangular form, its diagonal elements are called the *kinematic eigenvalues* corresponding to $T(x)$. □

Note that, in the constant coefficient case, the eigenvalues of A coincide with the kinematic eigenvalues which correspond to the principal vector matrix $T(x)$. A simple generalization of the constant coefficient case occurs when $T'(x) \equiv 0$ but $\Lambda(x)$ of (3.54) varies with x. Asymptotic stability is then assured if, for some constant $c > 0$, $\mathrm{Re}\,(\int_t^x \lambda_i(s)\,ds) < 0$, $1 \leq i \leq m$, for all x, t with $x - t > c$ (why?); in particular we have asymptotic stability if for t fixed $\lim_{x \to \infty} \mathrm{Re}\,(\int_t^x (\lambda_i(s)\,ds) = -\infty$, $1 \leq i \leq m$. Note that this does allow $\mathrm{Re}\,(\lambda_i(s)) > 0$ in some intervals. This result can be generalized as follows:

Theorem 3.65 Suppose that the homogeneous ODEs (3.57a) and (3.62) are kinematically similar with $V(x)$ upper triangular and that $\|A(x)\|$ and $\|T'(x)\|$ are uniformly bounded in x. Let $Y(x)$ be any fundamental solution for (3.57a). Then we have uniform asymptotic stability, i. e.,

$$\|Y(x)\,Y^{-1}(t)\| \leq \gamma\, e^{-\lambda(x-t)} \quad \text{for } x \geq t \qquad (3.65a)$$

for some positive constants γ and λ, iff there are positive constants c and $\hat{\lambda}$ such that the kinematic eigenvalues $\{\lambda_i(x)\}_{i=1}^n$ satisfy

$$\mathrm{Re}\,(\int_t^x \lambda_i(s)\,ds) < -\hat{\lambda}(x-t) \quad \text{for } x - t > c \qquad (3.65b)$$

Proof: To show that (3.65a) implies (3.65b), let $W(x)$ be an upper triangular fundamental solution for (3.62), so $Y(x) = T(x)W(x)R$, R a constant, nonsingular matrix. From (3.65a),

$$\|W(x)W^{-1}(t)\| \leq \tilde{\gamma}\, e^{-\lambda(x-t)}, \qquad x \geq t$$

for some constant $\tilde{\gamma} > 1$. Since $\mathrm{diag}\,(W(x)W^{-1}(t)) = e^{\int_t^x \mathrm{diag}(V(s))ds}$, the diagonal elements $\lambda_i(s)$ of $V(s)$ satisfy

$$\left| e^{\int_t^x \lambda_i(s)\,ds} \right| \leq \tilde{\gamma}\, e^{-\lambda(x-t)} \quad \text{for } x \geq t$$

or $\operatorname{Re}\left(\int_t^x \lambda_i(s)\,ds \right) \leq \ln \tilde{\gamma} - \lambda(x-t)$. To obtain (3.65b), we choose c and $\hat{\lambda}$ such that $0 < \hat{\lambda} < \lambda$ and $c > \dfrac{\ln \tilde{\gamma}}{\lambda - \hat{\lambda}}$.

Conversely, assume (3.65b). From the boundedness assumptions of the theorem and (3.63), $\|V(x)\|$ is also uniformly bounded. Thus we may write

$$V(x) = \Lambda(x) + B(x)$$

where $\Lambda(x) = \operatorname{diag}\{V(x)\}$ and $B(x)$ is a bounded strictly upper triangular matrix. Now, since $W(x)$ is upper triangular, its first column, denoted say $\mathbf{w}_1(x)$, has only one nonzero element. Hence it follows from (3.65b) that

$$\frac{|\mathbf{w}_1(x)|}{|\mathbf{w}_1(t)|} \leq \left| e^{\int_t^x \lambda_1(s)\,ds} \right| \leq e^{-\hat{\lambda}(x-t)} \quad \text{if } x - t > c$$

When we assume only that $x > t$, then a translation $s := x + c$ yields $s - t > c$, and we have instead

$$\frac{|\mathbf{w}_1(x)|}{|\mathbf{w}_1(t)|} \leq \hat{\gamma}\, e^{-\hat{\lambda}(x-t)} \quad x > t$$

where $\hat{\gamma} := e^{\hat{\lambda} c}$. For the second column $\mathbf{w}_2(x)$, only the first two elements are nonzero. Since $\mathbf{w}_1(x)$ has already been treated and $\|B(x)\|$ is bounded, the $\mathbf{w}_1(x)$ contribution amounts to a bounded inhomogeneity, and the same treatment is repeated. We can continue this process for the rest of the columns of $W(x)$. A back transformation to $Y(x)$, given the boundedness of $\operatorname{cond}(T; x, t)$, completes the proof. □

Thus we see that the kinematic eigenvalues provide a criterion for uniform stability of the ODE (3.49), or (3.57a), in the same way that eigenvalues of A do in the constant coefficient case. Asymptotic uniform stability is obtained if $\lambda > 0$ in (3.65a). But before proceeding, it is natural to ask if one can deduce the stability properties of the linear IVP by looking at the eigenvalues of $A(x)$ alone. The answer is negative in general, as the following example shows.

Example 3.7

Consider the IVP

$$\mathbf{y}' = A(x)\mathbf{y} \quad x > a$$

with

$$A(x) = \begin{pmatrix} -\dfrac{1}{4} + \dfrac{3}{4}\cos 2x & 1 - \dfrac{3}{4}\sin 2x \\ -1 - \dfrac{3}{4}\sin 2x & -\dfrac{1}{4} - \dfrac{3}{4}\cos 2x \end{pmatrix}$$

and $\mathbf{y}(a)$ given. It can be verified (check!) that $A(x)$ has the eigenvalues $\dfrac{1}{4}(-1 \pm i\sqrt{7})$. Thus, both eigenvalues of $A(x)$ have negative real parts.

Nevertheless, the IVP is unstable. To see this, consider the transformation

$$\mathbf{w}(x) := T^{-1}(x)\mathbf{y}(x), \qquad T(x) = \begin{bmatrix} \cos x & \sin x \\ -\sin x & \cos x \end{bmatrix}$$

Then $\mathbf{w}$ satisfies the ODE

$$\mathbf{w}' = \begin{bmatrix} \frac{1}{2} & 0 \\ 0 & -1 \end{bmatrix} \mathbf{w} \equiv V\mathbf{w}, \qquad x > a$$

and the IVP is obviously unstable. □

Example 3.7 indicates that if $A(x)$ is "too far from being constant" then its eigenvalues are useless for determining stability. However, there are important special cases where stability can be established without explicitly determining kinematic eigenvalues. One such case is when A is *strictly diagonally dominant*, uniformly in x, i.e., there is a constant $\delta > 0$ independent of x such that

$$\sum_{\substack{j=1 \\ j \neq i}}^{n} |a_{ij}| \leq (1 - \delta)|a_{ii}| \qquad i = 1, \ldots, n \tag{3.66}$$

To see this, write

$$A = \Sigma(I + B) \tag{3.67a}$$

where

$$\Sigma = \text{diag}\{a_{11}, \ldots, a_{nn}\}, \qquad b_{ij} = \begin{cases} \dfrac{a_{ij}}{a_{ii}} & i \neq j \\ 0 & i = j \end{cases} \tag{3.67b}$$

Note that by (3.66), $\|B\|_\infty \leq 1 - \delta$. Assume that for all $x > a$

$$\text{Re}(a_{ii}) < 0 \qquad i = 1, \ldots, n \tag{3.68}$$

so that Theorem 3.65 indicates that the IVP for

$$\mathbf{y}' = \Sigma \mathbf{y} \qquad x > a \tag{3.69}$$

is uniformly stable. Thus, the fundamental solution $L(x) = e^{\int_a^x \Sigma(t)\,dt}$ for (3.69) is bounded. Writing (3.57a) as

$$\mathbf{y}' = A\mathbf{y} = \Sigma\mathbf{y} + \Sigma B\mathbf{y}$$

we treat $\Sigma B \mathbf{y}$ as an inhomogeneous term and write

$$\mathbf{y}(x) = L(x)[\mathbf{y}(a) + \int_a^x L^{-1}(t)\,\Sigma(t)B(t)\mathbf{y}(t)\,dt] \tag{3.70}$$

Then

$$|L(x)\int_a^x L^{-1}(t)\,\Sigma(t)B(t)\mathbf{y}(t)\,dt\,| \leq \|\int_a^x L(x)L^{-1}(t)\,\Sigma(t)\,dt\|\,\|B\|\,\|\mathbf{y}\|$$

where $\|\mathbf{y}\| = \|\mathbf{y}\|_\infty = \sup\limits_{x \geq a}|\,\mathbf{y}(x)\,|$, and

$$\|\int_a^x L(x)L^{-1}(t)\,\Sigma(t)\,dt\| = \|\int_a^x e^{\int_t^x \Sigma(\tau)d\tau}\,\Sigma(t)\,dt\| = \|-\int_a^x \frac{d}{dt}(e^{\int_t^x \Sigma(\tau)d\tau})\,dt\| \leq 1$$

In (3.70) we obtain

$$\|\mathbf{y}\| \leq \|L\|\,|\mathbf{y}(a)| + (1-\delta)\|\mathbf{y}\|$$

so

$$\|\mathbf{y}\| \leq \text{const}\,|\mathbf{y}(a)|$$

Noting that the initial point a may be replaced in the above argument by any point $c \geq a$, we have shown

Theorem 3.71 The unique solution of the IVP (3.57) is uniformly stable if (3.66) and (3.68) hold. □

Note that under the assumptions of Theorem 3.71, using Gershgorin's Theorem 2.9 for a real matrix, $A(x)$, we obtain that

$$\text{Re}\,(\lambda_i) < 0$$

for all eigenvalues of A. Hence, the eigenvalues do reflect the stability properties for this type of variable coefficient IVP. Still, we may conclude from the above discussion that it is not an easy task, in general, to establish stability for problems with variable coefficients.

Remarks

(a) One feature of the constant coefficient case that may be extended to the variable coefficient case is that of a decomposition of the fundamental solution $Y(x)$ of (3.57a) into a product of a "direction" matrix $T(x)$ and a "growth" matrix $W(x)$,

$$Y(x) = T(x)W(x) \tag{3.72}$$

We can proceed as follows: Take the 2-norms of the respective columns of $Y(x)$ and use these values as successive diagonal elements of a diagonal matrix $W(x)$. [So $W(x)$ is not the fundamental solution of a directly related problem any more and $T(x)$ is not the principal vector matrix of A.] The columns of $T(x)$ defined by (3.72) are then the normalized "directions" while the elements of $W(x)$ represent the growth. The question of how to find a fundamental solution of (3.57a) properly scaled so that $W(x)$ gives a good approximation of "pure" growth rates [as in (3.65)] is more or less equivalent to asking for a well-conditioned direction matrix $T(x)$. This is a difficult problem and is outside the scope of this presentation. Nonetheless, the directional dependence of solutions and its relation to their growth is often of primary importance for understanding the behavior of BVP solutions, and we will encounter this question again in the sequel.

(b) Given a decomposition (3.72) of a fundamental solution $Y(x)$ for the ODE (3.57a), we can obtain an interesting relation between $T(x)$ and $W(x)$ (the latter now not necessarily diagonal). First note that T is differentiable iff W is differentiable. Upon rewriting (3.63) we obtain

$$T' = AT - TV, \quad x > a \tag{3.73a}$$

where

$$V = W'W^{-1} \tag{3.73b}$$

Put another way, suppose that we are given $W(x)$ or $V(x)$. Then if the relation (3.72) is to hold, $T(x)$ must satisfy the matrix differential equation (3.73a). The ODE (3.73a) is called the *Lyapunov equation*. □

3.3.4 The nonlinear case

The previous discussion indicates that it is not an easy task to establish stability for problems with variable coefficients. This is even more the case for nonlinear IVPs. Here it is hard to say much in global terms, because the stability depends on the particular solution trajectory considered. Still, in the vicinity of an isolated solution $\mathbf{y}(x)$ of the IVP (3.46), it is possible to consider small perturbations and apply the linear theory just developed to the variational problem [cf. (3.20), (3.21)].

Thus, let $\hat{\mathbf{y}}(x)$ satisfy the nonlinear ODE (3.46a), i. e.,

$$\hat{\mathbf{y}}'(x) = \mathbf{f}(x, \hat{\mathbf{y}}(x)), \quad x > a$$

with $\hat{\mathbf{y}}(a)$ not too far from $\mathbf{y}(a) = \boldsymbol{\alpha}$, and consider the difference

$$\mathbf{z}(x) := \mathbf{y}(x) - \hat{\mathbf{y}}(x)$$

Expanding $\mathbf{f}(x, \hat{\mathbf{y}})$ in a Taylor series about $\mathbf{y}$, we get

$$\mathbf{f}(x, \hat{\mathbf{y}}) = \mathbf{f}(x, \mathbf{y}) + J(x, \mathbf{y})(\hat{\mathbf{y}} - \mathbf{y}) + \mathbf{r}(x, \mathbf{y}, \hat{\mathbf{y}}) \tag{3.74}$$

where the *Jacobian matrix* J is defined by

$$J(x, \mathbf{y}) := \frac{\partial \mathbf{f}}{\partial \mathbf{y}} \tag{3.75}$$

and $\mathbf{r}(x, \mathbf{y}, \hat{\mathbf{y}})$ is the *remainder term*, $\mathbf{r}(x, \mathbf{y}, \hat{\mathbf{y}}) = O(|\mathbf{z}(x)|^2)$. Ignoring $\mathbf{r}$, we find that (3.74) gives the so-called *variational equation* for (3.46a),

$$\mathbf{z}' = J(x)\mathbf{z} \tag{3.76}$$

where $J(x) \equiv J(x, \mathbf{y}(x))$. This describes the behaviour of the principal part of a small variation, and is of the form (3.50). A stability analysis requires investigating the fundamental solution $Y(x; a)$ satisfying

$$Y'(x; a) = J(x)Y(x; a) \quad x > a$$

$$Y(a; a) = I$$

The basic result is the following extension of Theorem 3.51.

Theorem 3.77 Suppose that $\frac{|\mathbf{r}(x, \mathbf{y}, \hat{\mathbf{y}})|}{|\mathbf{z}(x)|} \to 0$ as $|\mathbf{z}(x)| \to 0$. If $\|Y(x; a)\| < K$ ($x \geq a$) for some constant K then the solution $\mathbf{y}(x)$ is stable. If $\|Y(x; a)\| \to 0$ as $x \to \infty$ then $\mathbf{y}(x)$ is asymptotically stable. Obvious counterparts hold also for uniform stability. □

In short, we can extend the positive results of our linear stability analysis to the nonlinear case for small perturbations. But if $|\hat{\mathbf{y}}(a) - \boldsymbol{\alpha}|$ is not small, or if $|\mathbf{z}(x)|$ is not guaranteed to be small, then $\mathbf{r}(x, \mathbf{y}, \hat{\mathbf{y}})$ is not necessarily small in (3.74) and the linear model (3.76) may not be adequate.

Example 3.8

Occasionally, one is interested in instances where the solution $\mathbf{y}(x)$ of an IVP reaches a steady state; i.e., it becomes independent of x. We say that a solution $\mathbf{y}_e(x)$ is an *equilibrium solution* if $\mathbf{f}(x, \mathbf{y}_e(x)) = 0$ for all $x > a$. Note that an equilibrium solution is a constant vector.

It then becomes important to ask whether a given equilibrium solution is stable or not. For a concrete example, consider the scalar ODE

$$y' = y(1-y) \equiv f(y)$$

This is a special case of the Riccati equation, which will be encountered often in later chapters. Here we have the equilibrium solutions $y_e \equiv 0$ and $y_e \equiv 1$. Since $J(x, y) = 1 - 2y$, it is clear that the first is an unstable and the second is a stable equilibrium solution. Thus, even if we begin the integration of the ODE from an initial value close to 0, say $\alpha = 0.01$, the solution $y(x)$ will be repelled from the close unstable equilibrium 0 and attracted to steady state at the far but stable equilibrium 1. (If $\alpha < 0$, the solution never reaches any equilibrium and becomes unbounded.) □

3.4 CONDITIONING OF BOUNDARY VALUE PROBLEMS

In this section we investigate the concept of well-conditioning for BVPs in more depth. The ideas and results for IVPs from the previous section serve as guidelines here. In particular, the concept of dichotomy is introduced and is seen to play an analogous role to uniform stability for IVPs. While it may seem that a considerable amount of detail is included, a careful treatment here is well rewarded later when we investigate stability of the numerical methods for solving BVPs.

Intuitively we know what is meant by a "well-conditioned problem": A small change in $\mathbf{f}$ or $\mathbf{g}$ in (3.1), (3.12) should produce only a small change in the solution $\mathbf{y}$. (Thus, if a stable, accurate numerical algorithm is applied to solve a well-conditioned BVP, the results are accurate as well). The precise quantification of this concept proves to be fairly subtle, however, even under circumstances where much simpler types of problems are considered.

To study the well-conditioning of a nonlinear problem (3.1), (3.12), one would restrict attention to a neighborhood of an isolated solution (see Section 3.1.2). Under some smoothness assumptions on $\mathbf{f}$ and $\mathbf{g}$ this means that one would consider the sensitivity of the (linear) variational problem (3.20), whose coefficients are given by (3.21), (3.25). The treatment is not different in this respect than for IVPs, as done in Section 3.3.4. Thus, in this section we only consider *linear* problems.

3.4.1 Linear problems and conditioning constants

Consider the linear BVP (3.28b), (3.29b), rewritten here as

$$\mathbf{y}' = A(x)\mathbf{y} + \mathbf{q}(x) \qquad a < x < b \qquad (3.78a)$$

$$B_a \mathbf{y}(a) + B_b \mathbf{y}(b) = \boldsymbol{\beta} \qquad (3.78b)$$

Although we shall soon have to be more careful with the scaling in (3.78b), assume for now that

$$\max(\|B_a\|, \|B_b\|) = 1 \qquad (3.79)$$

Further, assume that the BVP has a unique solution $\mathbf{y}(x)$. Recall [cf. (3.31)] that this solution can be represented in terms of the data as

$$\mathbf{y}(x) = Y(x)Q^{-1}\boldsymbol{\beta} + \int_a^b G(x,t)\mathbf{q}(t)\,dt \qquad (3.80)$$

where $Y(x)$ is a fundamental solution, Q is defined in (3.26a), and the Green's function $G(x,t)$ is defined in (3.32). While the conditioning constant for the BVP can be defined as before in (3.34), now we use more care in the selection of stability (or conditioning) constants.

Recalling the definitions of Section 2.8, we take norms in (3.80) and find

$$\|\mathbf{y}\|_\infty \leq \kappa_1 |\boldsymbol{\beta}| + \kappa_2 \|\mathbf{q}\|_p \qquad (3.81a)$$

where

$$\kappa_1 := \|YQ^{-1}\|_\infty \qquad (3.82)$$

and

$$\kappa_2 := \sup_x \{[\int_a^b \|G(x,t)\|^q \, dt]^{1/q}\}, \qquad \frac{1}{p} + \frac{1}{q} = 1 \qquad (3.83a)$$

By $\|G(x,t)\|$ we mean the norm induced by the vector norm in $\mathbf{R}^n$. Thus, the conditioning constant

$$\kappa := \max\{\kappa_1, \kappa_2\} \qquad (3.84)$$

gives a bound on how perturbations in the right-hand sides of (3.78a, b) may be amplified, as we shall see below. One choice of p is $p = \infty$, so the same function norm is used on both sides of (3.81a). This gives

$$\|\mathbf{y}\|_\infty \leq \kappa_1 |\boldsymbol{\beta}| + \kappa_2 \|\mathbf{q}\|_\infty \qquad (3.81b)$$

with

$$\kappa_2 = \sup_x \{\int_a^b \|G(x,t)\| \, dt\} \qquad (3.83b)$$

[The constant κ in (3.34) is now seen to be a somewhat crude bound on the conditioning constant.] Perhaps a better choice is $p = 1$: Recall that generally $\mathbf{y}$ is smoother than $\mathbf{q}$ (cf. remark following Example 3.3), so it makes sense to demand a stricter norm for it. This gives

$$\| \mathbf{y} \|_\infty \leq \kappa_1 |\boldsymbol{\beta}| + \kappa_2 \| \mathbf{q} \|_1 \tag{3.81c}$$

with

$$\kappa_2 = \sup_{a \leq x, t \leq b} \| G(x,t) \| \tag{3.83c}$$

Comparing to the treatment of the IVP case in the previous section, note that κ_2 corresponds to uniform stability while κ_1 corresponds to stability. The best choice of p depends to a large extent on the particular BVP.

Example 3.9

Consider the simple problem

$$\varepsilon y' = -y + g(x) \qquad 0 < x < 1$$

$$y(0) = \alpha$$

with $g(x)$ an integrable function. Thus, the inhomogeneity corresponding to $\mathbf{q}(x)$ in (3.81a) is $\dfrac{g(x)}{\varepsilon}$. It is easy to see that $\kappa_1 = 1$. Also $\kappa_2 = \varepsilon(1 - e^{-1/\varepsilon})$ for $p = \infty$ and $\kappa_2 = 1$ for $p = 1$.

If $\varepsilon \approx 1$ then the choice $p = 1$ may be more appropriate, because a larger class of inhomogeneous functions g is permitted in (3.81c) than in (3.81b). On the other hand, if ε approaches 0, then choosing $p = 1$ yields an unbounded right-hand side in (3.81c), whereas $p = \infty$ yields a bounded quantity in (3.81b), which is a more realistic bound on $\| \mathbf{y} \|_\infty$ (when g is bounded). □

The form (3.81) itself is not always appropriate, because all components of $\mathbf{y}$ are weighted alike. For example, if one is given a BVP involving a solution u to a scalar m^{th} order ODE, then while conversion to an equivalent first-order system in $\mathbf{y} := (u, u', \ldots, u^{(m-1)})$ will lead to a conditioning constant κ given by (3.84), one may be interested only in perturbations of the solution $u(x)$ itself. In such a case, use of a "weighted norm" may be preferable (see Exercise 10).

To see that κ does indeed give a bound on the effect of perturbations in $\mathbf{q}(x)$ and $\boldsymbol{\beta}$ on $\mathbf{y}(x)$, consider the perturbed problem

$$\mathbf{w}' = A(x)\mathbf{w} + \mathbf{q}(x) + \delta\mathbf{q}(x) \qquad a < x < b \tag{3.85a}$$

$$B_a \mathbf{w}(a) + B_b \mathbf{w}(b) = \boldsymbol{\beta} + \delta\boldsymbol{\beta} \tag{3.85b}$$

The global error $\mathbf{e} := \mathbf{w} - \mathbf{y}$, i. e., the difference between the solutions to the perturbed and unperturbed problems, satisfies

$$\mathbf{e}' = A(x)\mathbf{e} + \delta\mathbf{q}(x) \qquad a < x < b \tag{3.86a}$$

$$B_a \mathbf{e}(a) + B_b \mathbf{e}(b) = \delta\boldsymbol{\beta} \tag{3.86b}$$

so the bound (3.81a) implies

$$\| \mathbf{e} \|_\infty \leq \kappa_1 |\delta\boldsymbol{\beta}| + \kappa_2 \| \delta\mathbf{q} \|_q \tag{3.87}$$

The perturbations as in (3.85) are, of course, not the most general ones. Perturbations in the BVP operator matrices $A(x)$, B_a and B_b must also be considered. A small change in $A(x)$, in fact, corresponds to a small change in the Green's function $G(x,t)$, hence in κ_2. However, under fairly general assumptions it can be shown that if we perturb $A(x)$ slightly, then $G(x, t)$ is perturbed slightly as well, so that the resulting change in κ_2 is also slight. The corresponding analysis of the effect of perturbations in the BC matrices B_a and B_b on κ_1 is easier, and is left as Exercise 11.

The conditioning constants κ_1 and κ_2 help us in considering the notion of well-conditioned BVPs. Intuitively, we would say that a given BVP is *well-conditioned* if $\kappa = \max \{\kappa_1, \kappa_2\}$ is a *constant of moderate size*. Just what is meant by a "constant of moderate size" is better understood qualitatively than quantitatively, because the actual size of allowed constants depends on a number of parameters in a given computation. This situation is not different, though, from that in less complicated contexts such as for linear systems of algebraic equations.

In a more rigorous but less intuitive attempt to define the notion of well-conditioning, it is often possible to be precise by considering a family of BVPs with $a \to -\infty$ and/or $b \to \infty$. For example, keeping a fixed and letting $b \to \infty$, one could call the problem well-conditioned if $\kappa_1(b) = O(1)$ and $\kappa_2(b) = O(1)$ as $b \to \infty$, and ill-conditioned if $\frac{1}{\kappa_1(b)} = o(1)$ or $\frac{1}{\kappa_2(b)} = o(1)$ as $b \to \infty$. Equivalently, each BVP in such a family can be transformed so that all BVPs are defined on the same finite interval, say $[0, 1]$. This yields a family of *singularly perturbed* problems with $\varepsilon = \frac{1}{b-a}$ a small parameter (see Exercise 12). This definition can be satisfactory for BVPs that are well-behaved for all sufficiently large values at b and is related to the problem of solving BVPs over infinite intervals. However, it does not always work:

Example 3.10

Consider the BVP

$$\begin{bmatrix} y_1 \\ y_2 \end{bmatrix}' = \begin{bmatrix} 0 & 1 \\ -1 & 0 \end{bmatrix} \begin{bmatrix} y_1 \\ y_2 \end{bmatrix} \quad 0 < x < b, \qquad y_1(0) = \beta_1, \qquad y_1(b) = \beta_2$$

A short calculation shows that for b away from any multiple of π, the solution is not sensitive to small changes in the boundary conditions, but when b gets close to a multiple of π, the problem is ill-conditioned because $\kappa_1(b) \approx \frac{\cos b}{\sin b}$. In fact, when b is a multiple of π, the homogeneous problem has infinitely many solutions. □

Remark Let us emphasize again that the notion of a well-conditioned BVP *is not a direct analogue* of a well-conditioned matrix, in that the conditioning constant κ is an absolute quantity, not a relative one (e. g., unlike a condition number of a matrix). To illustrate our point, consider first a system of linear equations

$$\mathbf{A}\mathbf{u} = \mathbf{v}$$

where the matrix $\mathbf{A}$ is square and nonsingular. The solution $\mathbf{u}$ is bounded, using consistent vector and matrix norms, by

$$|\mathbf{u}| \leq \|\mathbf{A}^{-1}\| \, |\mathbf{v}|$$

while the relative effect of rounding errors on the solution is bounded by the condition number

$$\text{cond}(\mathbf{A}) := \|\mathbf{A}\| \, \|\mathbf{A}^{-1}\|$$

We say (qualitatively) that the problem of solving the linear system of equations is *well-conditioned* if cond ($\mathbf{A}$) is not too large. Assuming that $\|\mathbf{A}^{-1}\|$ is of moderate size, we obtain that cond ($\mathbf{A}$) is not too large if $\|\mathbf{A}\|$ is also nicely bounded, say $\|\mathbf{A}\| \sim 1$.

Turning again to the BVP (3.29), let us assume for notational simplicity that the boundary operator $\mathbf{B}$ is included as part of the operator $\mathbf{L}$. Then the constant κ defined in (3.84) *is a bound on* $\mathbf{L}^{-1}$, and a well-conditioned BVP is one that, after an appropriate scaling of $\mathbf{L}$, has $\mathbf{L}^{-1}$ bounded by a constant of moderate size. However, no bound on $\mathbf{L}$ is mentioned in this chapter. In subsequent chapters, when considering the relative effect of rounding errors on the solution of a discretized version of (3.29), we will see that the bound for numerical methods is essentially a product of κ and a factor which grows unboundedly when the discretization error is made to approach 0. (Indeed, such a factor is not surprising, since we expect that rounding errors grow as the discretization becomes finer, and these errors must be manifested in the condition number of the discretization matrix.) □

While our intuitive notion of well-conditioning has been related to $\kappa = \max\{\kappa_1, \kappa_2\}$, it is natural to consider κ_1 and κ_2 individually, since they appear as possibly independent stability constants in (3.81). The question then is how these two constants are related. Since $G(x,t)$ in (3.32) depends upon the BC, one might expect for instance that large κ_1 would imply large κ_2, or likewise, that κ_2 of "moderate" size would imply κ_1 (and hence κ) of "moderate" size. Such an expression is also consistent with the IVP case, where uniform stability implies stability.

To show that this is the case under appropriate assumptions, it is first necessary to properly scale the BC. That (3.79) is not entirely satisfactory is not surprising if we realize that it does not preclude situations like

$$B_a = \begin{bmatrix} 1 & 0 \\ 0 & 10^{-6} \end{bmatrix}, \quad B_b = \begin{bmatrix} 1 & 0 \\ 0 & 10^{-5} \end{bmatrix}$$

Clearly,

$$B_a = \begin{bmatrix} 1 & 0 \\ 0 & 10^{-1} \end{bmatrix}, \quad B_b = \begin{bmatrix} 1 & 0 \\ 0 & 1 \end{bmatrix}$$

seems better scaled. The desirable scaling is provided by

Lemma 3.88 If rank $[B_a | B_b] = n$ then there exist a number k, $0 \leq k \leq n$, a nonsingular matrix S, orthogonal matrices Q_1, Q_2, and diagonal matrices D_1, D_2, all in $\mathbf{R}^{n \times n}$, such that

$$D_1 = \begin{pmatrix} D_{11} & 0 \\ 0 & I_k \end{pmatrix} \qquad D_2 = \begin{pmatrix} I_{n-k} & 0 \\ 0 & D_{22} \end{pmatrix}$$

where D_{11} and D_{22} have diagonal elements between 0 and 1, and

$$B_a = SD_1 Q_1^T, \qquad B_b = SD_2 Q_2^T \tag{3.88}$$

□

The proof of this is left as Exercise 13.

Thus, we may assume that the BC matrices in (3.78b) satisfy

$$B_a = D_1 Q_1^T, \; B_b = D_2 Q_2^T \tag{3.89}$$

The following allows us to bound κ_1 in terms of κ_2.

Theorem 3.90 If B_a and B_b are scaled as in (3.89), then

$$\max_{a \le x \le b} \| \Phi(x) \|_2 \le 2 \max_{a \le x, t \le b} \| G(x,t) \|_2 \tag{3.90}$$

where the fundamental solution $\Phi(x)$ is defined so that $B_a \Phi(a) + B_b \Phi(b) = I$ [cf. (3.30b)].

Proof: From Lemma 3.88 and (3.32),

$$\| \Phi(x) \|_2 \le \| \Phi(x)(D_1 + D_2) \|_2 \le \| \Phi(x) D_1 \|_2 + \| \Phi(x) D_2 \|_2$$

$$= \| \Phi(x) B_a \|_2 + \| \Phi(x) B_b \|_2 = \| G(x,a) \|_2 + \| G(x,b) \|_2 \qquad □$$

If κ_1 and κ_2 were defined by using 2-norms, then Theorem 3.90 implies (using $p = \infty$ as in (3.81b), (3.83b)) that $\kappa_1 \le 2\kappa_2$. For the maximum norm (which we use, for consistency), since $\| C \|_\infty \le \sqrt{n} \, \| C \|_2 \le n \| C \|_\infty$ for any $C \in \mathbf{R}^{n \times n}$, it implies that $\kappa_1 \le 2n \kappa_2$.

Hence, we see that the conditioning constant κ_2 is qualitatively as good as κ and can be taken as the conditioning constant of the BVP. The constant κ_1 does not play the same role of uniquely determining the conditioning of a BVP, since there are problems where κ_1 is of moderate size but κ_2 can be made arbitrarily large (see Exercise 14).

It remains to characterize conditions under which the conditioning constant κ can be bounded (by constants of moderate size). This is done after the concept of dichotomy is considered in the next subsection.

3.4.2 Dichotomy

In Section 3.3 we saw that stability of linear IVPs can be formulated in terms of the growth behavior of fundamental solutions and kinematic eigenvalues. Specifically, for uniform stability the fundamental solutions must not increase as x approaches ∞, so that κ of (3.52) is well-defined (bounded) [cf. (3.65a)]. If we keep the interval on which the IVP is defined fixed (and finite), then the requirement of stability is essentially that no rapidly increasing modes are allowed.

For well-conditioned BVPs, a generalization of the IVP uniform stability notion is called for, because there is no a priori preferred direction of integration any more, as the following simple example shows.

Example 3.11

The IVP

$$\varepsilon y' = -y \qquad 0 < x < 1$$
$$y(0) = \alpha$$

is clearly uniformly stable for any $\varepsilon > 0$ (and unstable for $\varepsilon < 0$). Thus, the BVP

$$\varepsilon y_1' = -y_1 \qquad 0 < x < 1$$
$$\varepsilon y_2' = y_2$$
$$y_1(0) = \beta_1, \qquad y_2(1) = \beta_2$$

is "uniformly stable" (i. e., well-conditioned) for any $\varepsilon > 0$. This is because the ODEs here are completely decoupled: For y_1 there is a uniformly stable IVP as above and for y_2 the change of variable $t := 1 - x$ yields a uniformly stable IVP

$$\varepsilon \frac{dy_2}{dt} = -y_2, \, y_2(0) = \beta_2$$

Note, though, that y_2 is rapidly increasing as a function of x and that if $y_2(0)$ had been specified instead of $y_2(1)$, the BVP would not have been well-conditioned. □

This example shows that in the generalization of the IVP condition for uniform stability to BVPs, we must accommodate *rapidly increasing as well as rapidly decreasing modes*. This splitting, or *dichotomy* notion, is made precise as follows:

Definitions 3.91 Suppose $Y(x)$ is a fundamental solution for the linear ODE

$$\mathbf{y}' = A(x)\mathbf{y} \tag{3.91}$$

where $A(x) \in C[a, b]$. The ODE has an *exponential dichotomy* if there exists a constant orthogonal projection matrix $P \in \mathbf{R}^{n \times n}$ (i. e., $P^2 = P$, cf. Section 2.2.1) of rank p, $0 \le p \le n$, and positive constants K, λ, μ with K of "moderate" size, such that

$$\| Y(x) P Y^{-1}(t) \| \le K e^{-\lambda(x-t)} \qquad \text{for } x \ge t$$
$$\| Y(x)(I-P) Y^{-1}(t) \| \le K e^{-\mu(t-x)} \qquad \text{for } x \le t \tag{3.92}$$

for $a \le x, t \le b$. It is said to have an *ordinary dichotomy* if (3.92) holds with $\lambda = 0$ and/or $\mu = 0$. In general, where (3.92) holds with λ and μ nonnegative but otherwise unspecified, we refer simply to a *dichotomy*. □

The dichotomy conditions (3.92) are an obvious generalization of (3.65a), which is a condition for uniform stability (asymptotic stability if $\lambda > 0$) in the IVP case. Note, though, that connecting dichotomy to well-conditioning of a BVP is not as immediate [because no BC appear in (3.92)]. This connection is made in Section 3.4.3.

In the case where $a = -\infty$ or $b = \infty$, dichotomy gives a qualitative behaviour of the solutions of the homogeneous ODE (3.91) as $x \to \pm\infty$. When a and b are finite, say $b - a = 1$, the definition is imprecise because it is possible to find P, K, λ, and μ which satisfy (3.92) for *any* given linear ODE. Still, the definition is useful if we remind ourselves that what is wanted is a qualitative description of solution behaviour, and K is required to be of moderate size, say the size of $\|A(x)\|(b-a)$ but not the size of $e^{\|A\|(b-a)}$.

Example 3.12

Consider the BVP of Example 3.11, written as (3.91) for $\mathbf{y} = \begin{bmatrix} y_1 \\ y_2 \end{bmatrix}$. A fundamental solution is given by

$$Y(x) = \begin{bmatrix} e^{-x/\varepsilon} & 0 \\ 0 & e^{x/\varepsilon} \end{bmatrix}$$

so the dichotomy bounds (3.92) are satisfied with the projection matrix

$$P = \begin{bmatrix} 1 & 0 \\ 0 & 0 \end{bmatrix}$$

and $K \approx 1$, $\lambda = \mu = \dfrac{1}{\varepsilon}$.

On the other hand, consider the ODE

$$\varepsilon y_1' = -2xy_1 \qquad -1 < x < 1$$

$$y_2' = y_2$$

$\varepsilon > 0$. A fundamental solution is given by

$$Y(x) = \begin{bmatrix} e^{\frac{-x^2}{\varepsilon}} & 0 \\ 0 & e^x \end{bmatrix}$$

It is easy to see that here there is no dichotomy because of the ODE for y_1 [i. e., (3.92) is not satisfied for any constants K, λ, and μ which remain of moderate size when $e^{1/\varepsilon}$ grows large]. However, on the interval $[0,1]$ there is a dichotomy for this system. □

Given a fundamental solution $Y(x)$ and an orthogonal projection matrix P of rank p such that (3.92) holds, we may construct another fundamental solution $Z(x)$ such that the dichotomy is explicit in Z; i. e., we can write

$$Z(x) = (Z^1(x) \quad Z^2(x)) \qquad (3.93)$$

where $Z^1(x) \in \mathbf{R}^{n \times (n-p)}$ contains the growing part of the fundamental solution and $Z^2(x) \in \mathbf{R}^{n \times p}$ the decaying part. This may be done as follows: The (orthogonal) projection P can be written as

$$P = H^2(H^2)^T \tag{3.94a}$$

where $H^2 \in \mathbf{R}^{n \times p}$ consists of the last p columns of an orthogonal matrix

$$H = (H^1 \mid H^2) \tag{3.94b}$$

It follows that

$$I - P = H^1(H^1)^T \tag{3.94c}$$

The matrix $Z(x)$ is now defined as

$$Z(x) := Y(x)H \tag{3.95a}$$

It is easy to see (Exercise 15) that

$$Z^1(x)(H^1)^T = Y(x)(I - P) \tag{3.95b}$$

$$Z^2(x)(H^2)^T = Y(x)P \tag{3.95c}$$

and

$$Y(x)PY^{-1}(t) = Z(x)\tilde{P}Z^{-1}(t) \tag{3.96a}$$

$$Y(x)(I - P)Y^{-1}(t) = Z(x)(I - \tilde{P})Z^{-1}(t) \tag{3.96b}$$

where

$$\tilde{P} = \begin{bmatrix} 0 & 0 \\ 0 & I_p \end{bmatrix} \tag{3.96c}$$

Thus, it is not restrictive to assume that, if the dichotomy conditions (3.92) are satisfied for a fundamental solution and corresponding projection, then P has the simple form of $\tilde{P}$ in (3.96).

In the constant coefficient case $A(x) \equiv A$, one can show that the ODE has an exponential dichotomy iff A has no purely imaginary eigenvalues and an ordinary dichotomy iff all purely imaginary eigenvalues are simple (see Exercise 16). This is analogous to Theorem 3.60 for IVPs, although here we have p eigenvalues with nonpositive real part and $n - p$ eigenvalues with nonnegative real part.

For problems with variable coefficients, one should not necessarily rely too much on the eigenvalues of A (recall Example 3.7). However, dichotomy does mean that the solution space can be split into two subspaces with the angle between them bounded away from 0. To see this, suppose that $Y(x)$ is a fundamental solution for our homogeneous ODE (3.91), and let P be the projection such that the bounds (3.92) hold. Denote the solution space $S := \{Y(x)\mathbf{c}; \mathbf{c} \in \mathbf{R}^n\}$, and let $S_2 := \{Y(x)P\mathbf{c}; \mathbf{c} \in \mathbf{R}^n\}$ and $S_1 := \{Y(x)(I - P)\mathbf{c}; \mathbf{c} \in \mathbf{R}^n\}$ be its subspaces. Then this solution space can be written as a direct sum, i. e., $S = S_1 + S_2$, so that any element of S is the sum of a unique element of S_1 and a unique element of S_2. Under the dichotomy assumptions we have

Theorem 3.97 Any solutions $\mathbf{u}(x) \in S_2$ and $\mathbf{w}(x) \in S_1$ satisfy

$$\frac{|\mathbf{u}(x)|}{|\mathbf{u}(t)|} \leq K e^{-\lambda(x-t)} \quad \text{if} \quad x \geq t \tag{3.97a}$$

$$\frac{|\mathbf{w}(x)|}{|\mathbf{w}(t)|} \leq K e^{-\mu(t-x)} \quad \text{if} \quad x \leq t \tag{3.97b}$$

The vector norm used here corresponds to the norm in which (3.92) is expressed.

Proof: We show only (3.97a): If $x \geq t$, then for some $\mathbf{c} \in \mathbf{R}^n$,

$$\frac{|\mathbf{u}(x)|}{|\mathbf{u}(t)|} = \frac{|Y(x)P\,\mathbf{c}|}{|Y(t)P\,\mathbf{c}|} = \frac{|Y(x)PY^{-1}(t)Y(t)P\,\mathbf{c}|}{|Y(t)P\,\mathbf{c}|} \leq \| Y(x)PY^{-1}(t) \| \qquad \square$$

Thus, S_1 and S_2 consist of the solution spaces of asymptotically increasing (nondecreasing) and decreasing (nonincreasing) solutions, respectively. Of course, if (3.92) holds with $\lambda = 0$ and/or $\mu = 0$, then the same is true in Theorem 3.97. Using the bounds (3.92) or (3.97) with the Euclidean norm, we can bound the *angle* between the two subspaces S_1 and S_2 away from 0.

Theorem 3.98 Let

$$\| Y(x)PY^{-1}(x) \|_2 \leq \bar{K}, \qquad a \leq x \leq b$$

and denote by θ the angle between the solution subspaces S_1 and S_2 [that is, θ is the minimal angle between $\mathbf{u}(x) \in S_2$ and $\mathbf{w}(x) \in S_1$, taken over all possible $\mathbf{u}(x)$ and $\mathbf{w}(x)$ and all x — see definition 2.24]. Then

$$\cot \theta \leq \bar{K} \tag{3.98}$$

Proof: Let the functions $\mathbf{u} \in S_2$ and $\mathbf{w} \in S_1$ be arbitrarily chosen. For each x, the angle $\eta(x)$ between the vectors $\mathbf{u}(x)$ and $\mathbf{w}(x)$ is defined by

$$\cos \eta(x) = \frac{\mathbf{u}^T(x)\mathbf{w}(x)}{|\mathbf{u}(x)|_2\,|\mathbf{w}(x)|_2}$$

We now consider a fixed x and show that the bound (3.98) holds for $\eta(x)$. If $\mathbf{u}(x)$ is orthogonal to $\mathbf{w}(x)$ then $\cot \eta(x) = 0$, so assume $\eta(x) < \pi/2$. Furthermore, since we are interested only in angles, we may assume that $|\mathbf{u}(x)|_2 = |\mathbf{w}(x)|_2 = 1$. Define $\hat{\mathbf{u}}(x) := \mathbf{u}(x)$ and $\hat{\mathbf{w}}(x) := -(\mathbf{u}(x)^T \mathbf{w}(x))^{-1}\mathbf{w}(x)$. It is easy to see that $\hat{\mathbf{u}}(x)$ is orthogonal to $\hat{\mathbf{u}}(x) + \hat{\mathbf{w}}(x)$, so

$$\cot \eta(x) = \frac{|\hat{\mathbf{u}}(x)|_2}{|\hat{\mathbf{u}}(x) + \hat{\mathbf{w}}(x)|_2}$$

Now let $\hat{\mathbf{u}}(x) + \hat{\mathbf{w}}(x) = Y(x)\mathbf{c}$ for some vector $\mathbf{c} \in \mathbf{R}^n$. Since $\hat{\mathbf{u}} \in S_2$ implies $\hat{\mathbf{u}}(x) = Y(x)P\,\mathbf{c}$,

$$\cot \eta(x) = \frac{|Y(x)P\,\mathbf{c}|_2}{|Y(x)\mathbf{c}|_2} = \frac{|Y(x)PY(x)^{-1}Y(x)\mathbf{c}|_2}{|Y(x)\mathbf{c}|_2} \leq \| Y(x)PY(x)^{-1} \|_2 \leq \bar{K} \qquad \square$$

Next, we would like to have a representation for a suitable fundamental solution which shows the directions and growths of both types of modes. We will be content to have a kinematically similar system in which the increasing solution components are decoupled from the decreasing ones [and P identified with $\tilde{P}$ of (3.96)]. Thus we look for a transformation

$$Y(x) = T(x)W(x) \qquad (3.99)$$

such that W satisfies (3.62) for $a < x < b$ and $V(x) \equiv T^{-1}(x)[A(x)T(x) - T'(x)]$ has the form

$$V(x) = \begin{bmatrix} V^{11}(x) & V^{12}(x) \\ 0 & V^{22}(x) \end{bmatrix} \qquad (3.100)$$

Here V^{11} and V^{22} are square matrices of order $n-p$ and p, respectively. (These are the dimensions of the subspaces of increasing and decreasing modes.) We have the following analogue to Theorem 3.65, which is given without proof.

Theorem 3.101 Suppose that the hypotheses of Theorem 3.65 are satisfied for the two ODEs (3.91) and (3.62). Then (3.91) has a dichotomy [i. e., (3.92) holds] iff there are positive constants c and λ such that for p kinematic eigenvalues

$$\text{Re} \left(\int_t^x \lambda_i(s)\, ds \right) < -\hat{\lambda}(x-t) \qquad \text{if } x - t > c \qquad (3.101\text{a})$$

and for the other $n-p$ kinematic eigenvalues

$$\text{Re} \left(\int_t^x \lambda_i(s)\, ds \right) > \hat{\lambda}(x-t) \qquad \text{if } x - t > c \qquad (3.101\text{b})$$

□

For most of our purposes, we shall fortunately not need to use kinematic eigenvalues; rather, it will be sufficient to decouple the increasing and decreasing fundamental solution modes. Theorem 3.101 gives us the tools to relate dichotomy and kinematic eigenvalue information to BVP conditioning, as kinematic eigenvalues were related to IVP uniform stability. Note that an additional complication here is that one has to keep track of the role played by the BC at each end; for IVPs, all BC are at the initial point.

The relationship between the kinematic eigenvalues of $A(x)$ [for a given transformation $T(x)$] and the eigenvalues of $A(x)$ is similar to that discussed for IVPs in Section 3.3. Thus, while the kinematic eigenvalues are more indicative of the actual dichotomy, the eigenvalues of $A(x)$ are easier to compute and do indicate the situation when A is a constant matrix. Moreover, Theorem 3.71 can be easily extended to the BVP case as well. Thus, if $A(x)$ is strictly diagonally dominant [i. e., it satisfies (3.66) uniformly in x], then the dichotomy structure can be read off from the main diagonal of $A(x)$:

Theorem 3.102 If $A(x)$ is strictly diagonally dominant uniformly in x, $a \leq x \leq b$, and there are $\lambda, \mu \geq 0$ and $0 \leq p \leq n$ such that

$$\text{Re}(a_{ii}) > \mu \qquad i = 1, \ldots, n-p \qquad (3.102\text{a})$$

$$\text{Re}(a_{ii}) < -\lambda \qquad i = n-p+1, \ldots, n \qquad (3.102\text{b})$$

then the dichotomy (3.92) holds for (3.91) with $P = \begin{bmatrix} 0 & 0 \\ 0 & I_p \end{bmatrix}$. □

As in the IVP case, Gerschgorin's Theorem can be used to make the connection between the diagonal elements of A and its eigenvalues.

The requirement that $A(x)$ be strictly diagonally dominant is quite restrictive. It can be relaxed as follows: If there is a matrix transformation $S(x)$ with a uniformly bounded condition number, then it is sufficient that diagonal dominance as above hold for $S^{-1}(x)A(x)S(x)$ in order to obtain a dichotomy (3.92) (with a transformed P) — see Exercise 20.

Example 3.13

We construct this example step by step, using two nonnegative parameters λ and ω. Beginning with the ODE

$$\mathbf{w}' = \begin{bmatrix} \lambda & 0 \\ 0 & -\lambda \end{bmatrix} \mathbf{w} \equiv V\mathbf{w}, \quad a < x < b$$

whose dichotomy is obvious, we *rotate* the two components of $\mathbf{w}$:

$$\mathbf{y}(x) := T(x)\mathbf{w}(x), \quad T(x) = \begin{bmatrix} \cos\omega x & \sin\omega x \\ -\sin\omega x & \cos\omega x \end{bmatrix}$$

Then $\mathbf{y}(x)$ satisfies the ODE

$$\mathbf{y}' = (TV + T')\mathbf{w} = T(V + T^{-1}T')T^{-1}\mathbf{y} \equiv A(x)\mathbf{y} \quad a < x < b$$

The matrix $A(x)$ is given for each x as an orthogonal similarity transformation of

$$V + T^{-1}T' = \begin{bmatrix} \lambda & 0 \\ 0 & -\lambda \end{bmatrix} + \omega \begin{bmatrix} 0 & 1 \\ -1 & 0 \end{bmatrix} = \begin{bmatrix} \lambda & \omega \\ -\omega & -\lambda \end{bmatrix}$$

viz.,

$$A(x) = \begin{bmatrix} \lambda\cos 2\omega x & -\lambda\sin 2\omega x + \omega \\ -\lambda\sin 2\omega x - \omega & -\lambda\cos 2\omega x \end{bmatrix}$$

The eigenvalues of $A(x)$ are those of $V + T^{-1}T'$, i. e., $\pm\sqrt{\lambda^2 - \omega^2}$, while the kinematic eigenvalues are $\pm\lambda$. So, we see that as the amount of rotation ω increases, the eigenvalues of $A(x)$ drift further away from the kinematic eigenvalues. When $\omega > \lambda$, the eigenvalues become imaginary and do not yield any information about the dichotomy (which does not change with ω, of course). This is reminiscent of Example 3.7. When $\omega < \lambda$, the eigenvalues of $A(x)$ still display the dichotomy of the associated ODE. It is interesting to note that in that case $V + T^{-1}T'$ is diagonally dominant. If, however, we restrict considerations to *constant* similarity transformations for $A(x)$, then there are values $\omega < \lambda$ for which there is no constant matrix S with $S^{-1}A(x)S$ diagonally dominant uniformly in $a \leq x \leq b$ (Exercise 21). □

3.4.3 Well-conditioning and dichotomy

Well-conditioning of BVPs is very closely related to the concept of dichotomy. In particular, if there is a dichotomy then it is easy to bound the conditioning constant κ_2 in terms of κ_1. (Recall that in Section 3.4.1 we have bounded κ_1 in terms of κ_2, showing that the bound κ_2 for the Green's function is the key quantity determining BVP conditioning.)

Theorem 3.103 If the homogeneous ODE (3.91) has a dichotomy (3.92), then the Green's function $G(x,t)$ for the BVP (3.78) satisfies

$$\|G(x,t)\|_2 \le \kappa_1 K(e^{-\lambda(b-t)} + e^{-\mu(t-a)}) + Ke^{-\mu(t-x)} \quad \text{if } x < t \quad (3.103a)$$

$$\|G(x,t)\|_2 \le \kappa_1 K(e^{-\lambda(b-t)} + e^{-\mu(t-a)}) + Ke^{-\lambda(x-t)} \quad \text{if } x > t \quad (3.103b)$$

(assuming the BC have been scaled).

Proof: We show only (3.103a); (3.103b) is similar. For $x < t$,

$$G(x,t) = -Y(x)Q^{-1}B_b Y(b)Y^{-1}(t)$$

$$= -Y(x)Q^{-1}B_b Y(b)PY^{-1}(t) - Y(x)Q^{-1}B_b Y(b)(I-P)Y^{-1}(t)$$

$$= -Y(x)Q^{-1}B_b Y(b)PY^{-1}(t) - Y(x)(I - Q^{-1}B_a Y(a))(I-P)Y^{-1}(t)$$

$$= -Y(x)Q^{-1}B_b Y(b)PY^{-1}(t) - Y(x)(I-P)Y^{-1}(t) + Y(x)Q^{-1}B_a Y(a)(I-P)Y^{-1}(t)$$

and (3.103a) follows by taking norms and using (3.82). □

The following corollary is obvious:

Corollary 3.104 Under the same conditions as in Theorem 3.103,

$$\kappa_2 \le K(2\kappa_1 + 1) \quad (3.104)$$

□

Remarks

(a) If there is an exponential dichotomy, then the norm of $G(x,t)$ also decays exponentially with $|x-t|$. This corresponds to the case where (3.65a) holds for the IVP, so the solution is asymptotically uniformly stable [cf. (3.47c)]. Thus, exponential dichotomy for the BVP corresponds to a particular form of asymptotic uniform stability for the IVP (cf. Section 3.3). Also, for the IVP case, κ_2 is bounded iff the ODE has a dichotomy.

(b) In order to practically estimate the conditioning constant $\kappa = \max\{\kappa_1, \kappa_2\}$ of a given BVP with the dichotomy bounds (3.92) holding, Corollary 3.104 implies that estimating κ_1 may be sufficient (when K is known). By (3.82),

$$\kappa_1 = \max_{a \le x \le b} \|Y(x)Q^{-1}\| \le \|Y\| \, \|Q^{-1}\|$$

for any fundamental solution $Y(x)$. For the choice of $Y(x)$, we would like one such that $\|Y\| \, \|Q^{-1}\|$ is as tight a bound as possible on κ_1. Consider three possible scalings:

(i) $Y(a) = I$. This is the IVP choice, but it should be generally avoided for BVPs because of the possible existence of rapidly increasing modes. In such a case, $\|Y\| \, \|Q^{-1}\|$ becomes large, even when κ_1 is of moderate size.

(ii) $Y(x) = \Phi(x)$, where $\Phi(x)$ is defined in (3.30). Thus $Q = I$ and
$$\kappa_1 = \|\Phi\|_\infty$$
This is a well-scaled choice which is used in the examples below.

(iii) $\|Y\|_\infty = 1$. This choice is also well-scaled. □

Example 3.14

Consider the constant coefficient ODE $u'' = \lambda^2 u$ with $\lambda > 0$. We convert this to a first-order system by setting $y_1 = u$, $y_2 = \lambda u'$, obtaining

$$\begin{bmatrix} y_1 \\ y_2 \end{bmatrix}' = \begin{bmatrix} 0 & \lambda \\ \lambda & 0 \end{bmatrix} \begin{bmatrix} y_1 \\ y_2 \end{bmatrix}$$

It is straightforward to show that

$$Y(x;0) = \begin{bmatrix} \cosh \lambda x & \sinh \lambda x \\ \sinh \lambda x & \cosh \lambda x \end{bmatrix}$$

This ODE has an exponential dichotomy. In fact, since A is constant and symmetric, the kinematic similarity transformation of Theorem 3.101 can be simply an orthogonal eigenvector matrix of A. We obtain

$$T = \sqrt{2} \begin{bmatrix} -\frac{1}{2} & \frac{1}{2} \\ \frac{1}{2} & \frac{1}{2} \end{bmatrix}, \quad V = T^{-1}AT = \begin{bmatrix} -\lambda & 0 \\ 0 & \lambda \end{bmatrix}$$

and the corresponding fundamental solution for (3.62) is

$$W(x) = \begin{bmatrix} e^{\lambda(x-b)} & 0 \\ 0 & e^{-\lambda x} \end{bmatrix}$$

Consider now the Dirichlet BC $u(0) = \beta_1$, $u(1) = \beta_2$; i. e., $b = 1$ and

$$B_a = \begin{bmatrix} 1 & 0 \\ 0 & 0 \end{bmatrix}, B_b = \begin{bmatrix} 0 & 0 \\ 1 & 0 \end{bmatrix}$$

It can be verified that $K = 1$ and

$$\Phi(x) = \begin{bmatrix} \cosh \lambda x - \dfrac{\sinh \lambda x \cosh \lambda}{\sinh \lambda} & \dfrac{\sinh \lambda x}{\sinh \lambda} \\ \sinh \lambda x - \dfrac{\cosh \lambda x \cosh \lambda}{\sinh \lambda} & \dfrac{\cosh \lambda x}{\sinh \lambda} \end{bmatrix}$$

So if $\lambda \gg 0$ then $\kappa_1 \approx 2$ and from (3.104) $\kappa_2 \leq 5$. □

Example 3.15

Consider Example 3.7 again for $0 < x < b$. Here, A is not constant and its eigenvalues do not tell us much. From the development in Example 3.7 we readily find that

$$Y(x) = \begin{bmatrix} \cos x & \sin x \\ -\sin x & \cos x \end{bmatrix} \begin{bmatrix} e^{x/2} & 0 \\ 0 & e^{-x} \end{bmatrix}$$

is a suitably scaled fundamental solution. We have an exponential dichotomy with $p = 1$, $\lambda = 1$, $\mu = \frac{1}{2}$ and $K = 1$ in (3.92).

Case 1. Consider the BC

$$\mathbf{y}(0) + \mathbf{y}(b) = \begin{bmatrix} 1 \\ 1 \end{bmatrix}, \, b = l\pi$$

l a large positive integer. Here, $B_a = B_b = I$, so

$$Q = \begin{bmatrix} 1 & 0 \\ 0 & 1 \end{bmatrix} \pm \begin{bmatrix} e^{l\pi/2} & 0 \\ 0 & e^{-l\pi} \end{bmatrix}$$

and

$$\Phi(x) \approx \begin{bmatrix} \cos x & \sin x \\ -\sin x & \cos x \end{bmatrix} \begin{bmatrix} e^{(x-l\pi)/2} & 0 \\ 0 & e^{-x} \end{bmatrix}$$

This yields $\kappa_1 \approx 1$, so $\kappa_2 \leq 2$. Again we may say that the problem is well-conditioned. It is not difficult to see that, if we now let the right endpoint b vary continuously, then no significant change occurs; in fact $\kappa_1 = 1 + O(e^{-b/2})$ (Exercise 22).

Case 2. Now consider another set of BC

$$B_a = \begin{bmatrix} 1 & 0 \\ 0 & 0 \end{bmatrix}, \, B_b = \begin{bmatrix} 1 & 0 \\ 0 & 1 \end{bmatrix}, \, \beta = \begin{bmatrix} 1 \\ 1 \end{bmatrix}$$

Here,

$$Q = \begin{bmatrix} 1 \pm e^{l\pi/2} & 0 \\ 0 & \pm e^{-l\pi} \end{bmatrix}$$

and $\|\Phi\|_\infty = \kappa_1 \approx e^{l\pi}$, which blows up exponentially as l grows. So this may be regarded as an ill-conditioned problem.

Case 3. If we further change B_a above to $B_a = 0$ then we obtain a terminal value problem. This is an IVP in the variable $x := b - x$, i.e. it involves integrating backwards from b to 0. The problem is ill-conditioned, that is to say unstable in the IVP terminology.

When comparing the various boundary conditions above, we observe that the BC in Cases 2 and 3, unlike that in Case 1, do not control the decaying mode. Indeed the matrix B_a in case 2 has row vectors which are orthogonal to the initial value vector $\begin{bmatrix} 0 \\ 1 \end{bmatrix}$ of the decaying mode. The terminal part of the BC cannot control this decaying mode effectively, because the actual magnitude of that mode at the terminal point $x = b$ is very small. $\square$

Our introduction of dichotomy gives the impression that it is a fundamental concept, being analogous to uniform stability in IVPs. Indeed, as it turns out, dichotomy is not only sufficient for well-conditioning (with κ_1 being bounded as well), it is also *necessary* for a BVP to be well-conditioned. This is particularly simple to show for the special but common case where the BC are separated, i. e., in (3.78)

$$B_a = \begin{bmatrix} B_{a1} \\ 0 \end{bmatrix} \quad B_b = \begin{bmatrix} 0 \\ B_{b2} \end{bmatrix} \tag{3.105}$$

where $B_{a1} \in \mathbf{R}^{m \times n}$, $B_{b2} \in \mathbf{R}^{(n-m) \times n}$.

Theorem 3.106 Consider the BVP (3.78) with separated BC. If there exist nonnegative constants M, ρ, and ν such that the Green's matrix satisfies

$$\|G(x,t)\| \le M e^{-\rho(t-x)} \quad \text{for } x < t$$

$$\|G(x,t)\| \le M e^{-\nu(x-t)} \quad \text{for } x > t$$

then (3.91) has a dichotomy, viz., there exist a fundamental solution $Y(x)$ and a projection matrix P such that

$$\|Y(x)PY^{-1}(t)\| \le M e^{-\rho(t-x)} \quad \text{for } x < t$$

$$\|Y(x)(I-P)Y^{-1}(t)\| \le M e^{-\nu(x-t)} \quad \text{for } x > t$$

The matrix P has the same rank as B_{a1}.

Proof: We have, for any fundamental solution $Z(x)$ of (3.91),

$$Q = B_a Z(a) + B_b Z(b) = \begin{bmatrix} B_{a1} Z(a) \\ B_{b2} Z(b) \end{bmatrix}$$

Hence the projection matrix P can be explicitly constructed as

$$B_a Z(a) Q^{-1} = \begin{bmatrix} I_p & 0 \\ 0 & 0 \end{bmatrix} =: P \tag{3.106a}$$

with $p := m$. Note that the projection P here is the complement of that in (3.102c). In particular, for the fundamental solution $Y(x) \equiv \Phi(x)$ which satisfies $B_a Y(a) + B_b Y(b) = I$, we have

$$P = B_a Y(a) \tag{3.106b}$$

Now for $t \le x$, $G(x,t) = Y(x) B_a Y(a) Y(t)^{-1} = Y(x) P Y(t)^{-1}$, and similarly for $t > x$, $G(x,t) = Y(x)(I-P)Y(t)^{-1}$. □

From this theorem we conclude that a well-conditioned BVP must have a corresponding dichotomy. The projection P in this theorem is a very natural one and is induced by the BC. This illustrates how closely conditioning and appropriate BC are intertwined. It is also possible to give a similar result for general BC, but this is more complicated and is left for the exercises.

The major purpose of this section may now be fairly simply summarized: We have been analyzing the relationship between the conditioning constants κ_1 and κ_2 and the fundamental solution dichotomy for the linear BVP. If the ODE has a dichotomy (3.92) with constant K, then an associated BVP has moderate size κ_2 if it has moderate size κ_1 and K. For the exponential dichotomy case, we have a further equivalence which involves the exponential decay of $\|G(x,t)\|$ with $|x-t|$.

Let us now further consider the separated BC (3.105). In this case we can make more definitive statements about the relationship between the BC and the various types of modes occurring. Specifically, if the BVP is well-conditioned, then the initial conditions

$$B_{a1}\mathbf{y}(a) = \boldsymbol{\beta}_1$$

must control the decreasing modes, while the terminal conditions

$$B_{b2}\mathbf{y}(b) = \boldsymbol{\beta}_2$$

must control the increasing modes. In particular, assuming for simplicity an exponential dichotomy in (3.92) with p decreasing modes and $(b-a)\min\{\lambda,\mu\}$ large, we must have $p = m$.

The above qualitative statements are made precise in the following:

Theorem 3.107 Let the BVP (3.78) with separated BC (3.105) be well-conditioned. For any fundamental solution $Y(x)$, denote by P the projection matrix of (3.106a). Then

(a) $m = p$.

(b) range $(B_{a1}^T)^\perp \cap$ range $(Y(a)P) = \{0\}$; i. e., no nonzero vector in range $(Y(a)P)$ is orthogonal to range (B_{a1}^T).

(c) range $(B_{b2}^T)^\perp \cap$ range $(Y(b)(I - P)) = \{0\}$.

(d) If the dichotomy (3.92) holds with $(b - a)\min\{\lambda, \mu\} \to \infty$, then the rank of the projection matrix must be m.

Proof: Consider for notational simplicity the fundamental solution $Y(x) \equiv \Phi(x)$ for which $Q = I$. Then by (3.106b), $P = B_a Y(a)$ with $p = m$, yielding (a). To obtain (d), note that Theorem 3.97 guarantees that, due to our assumption of *exponential dichotomy* (and the well-conditioning), there cannot be another projection P of a different rank for which the dichotomy bounds hold.

Now let the vector $\mathbf{z} \in$ range $(Y(a)P)$; i. e., $\mathbf{z} = Y(a)P \begin{bmatrix} \mathbf{c} \\ \mathbf{0} \end{bmatrix}$ for some $\mathbf{c} \in \mathbf{R}^m$. If also $\mathbf{z} \in$ range $(B_{a1}^T)^\perp$ then for any $\mathbf{x} \in \mathbf{R}^m$, $\mathbf{z}^T B_{a1}^T \mathbf{x} = 0$, so

$$0 = \mathbf{x}^T B_{a1} Y(a) P \begin{bmatrix} \mathbf{c} \\ \mathbf{0} \end{bmatrix} = \mathbf{x}^T \mathbf{c}$$

This implies $\mathbf{c} = \mathbf{0}$, hence proving (b). Part (c) follows similarly. $\square$

For large classes of problems the growth or decay rates of some fundamental solution modes are very rapid as compared to others. Such problems are frequently said to have greatly differing *time scales,* because these modes correspond to rates of decay or growth of physical quantities in a problem. (The independent variable in the BVP may be space instead of time, but the quantitative difficulty of the problem is the same.) Examples of problems of this type are chemical reactions where certain reactions occur at much more rapid rates than others, or in physical systems having mechanical and electronic components such that the time scale for a mechanical component is normally much slower than for an electronic component. If this difference in time scales is extreme, then numerical methods which are able to handle the pathological behaviour of the solution are necessary. This is the concern of Chapter 10. Many BVPs lie between the extreme case and the case with very slowly changing fundamental solution modes. For these, well-conditioning requires that exponentially decaying (increasing) modes be specified by the left (right) endpoint BC. But more importantly, the numerical method used to solve such BVPs must be sufficiently flexible to properly handle the exponential growth and decay. In the following chapters this underlying consideration will be very important.

EXERCISES

1. Consider the nonlinear BVP

$$u''(x) + |u(x)| = 0 \quad 0 < x < b$$

$$u(0) = 0, \quad u(b) = B$$

   (a) Show that if the solution satisfies $u(x_0) = 0$, $u'(x_0) < 0$, then it is never zero to the right of x_0.
   (b) Describe the solution behaviour for the three cases $b < \pi$, $b = \pi$, and $b > \pi$.
   Hint: For the most difficult case $b > \pi$, $B < 0$, consider

$$u(x) = \begin{cases} c_1 \sin x & \text{for } 0 \le x \le \pi \\ c_2(\sinh x - \tanh \pi \cosh x) & \text{for } \pi \le x \le b \end{cases}$$

2. Show Liouville's formula

$$\det Y(x) = \det Y(a) \exp\{\int_a^x \operatorname{tr}(A(s)) \, ds\}$$

   where $Y(x)$ is a fundamental solution for $y' = A(x)y$ with $A(x) = (a_{ij}(x))_{i,j=1}^n$, and $\operatorname{tr}(A(x)) = \sum_{i=1}^n a_{ii}(x)$.
   Hint: $\det Y'(x) = \sum_{j=1}^n \det Y_j(x)$, where $Y_j(x)$ is the $n \times n$ matrix obtained by replacing the jth row of $Y(x)$ by its derivative.

3. If $Y(x;t)$ is the fundamental solution for (3.6) satisfying $Y' = A(x)Y$, $Y(t;t) = I$, then show that
   (a) $Y(x;t) = Y(x;s)Y(s;t)$ for any x, s, and t.
   (b) $Y(x;t) = Y(t;x)^{-1}$.
   (c) $\dfrac{d}{dt}Y(x;t) + Y(x;t)A(t) = 0$.

4. Find the fundamental solution $Y(x;0)$ for the ODE (3.6) corresponding to $u''(x) - 100u(x) = 0$. If a two-point BVP were associated with this ODE over an interval $[0, b]$, the solution to that BVP could be calculated by using (3.26). Argue why the matrix Q defined there would be ill-conditioned if $b \gg 0$.

5. Construct the Green's functions for the following BVPs:

(a) $y'' = 0 \quad y(0) = y(1) = 0$.

(b) $y'' - n^2 y = 0 \quad y(0) = y(1) = 0$.

(c) $y'' + n^2 y = 0 \quad y(0) = y(1), \quad y'(0) = y'(1)$.
  Hint: Use (3.44) for (c).

6. Suppose that $u_1(x), \ldots, u_m(x)$ are linearly independent solutions of (3.38a) with $q \equiv 0$. Show that for any $t \in (a, b)$, the Green's function for (3.38a, b) is

$$G(x, t) = \begin{cases} \sum_{j=1}^{m} \alpha_j u_j(x) & \text{for } a \leq x < t \\ \sum_{j=1}^{m} \beta_j u_j(x) & \text{for } t < x \leq b \end{cases}$$

where $\alpha_j, \beta_j \, (1 \leq j \leq m)$ satisfy the following equations:

$$\sum_{j=1}^{m} (\beta_j - \alpha_j) u_j^{(p-1)}(t) = \delta_{pm} \quad 1 \leq p \leq m$$

$$\sum_{l=1}^{m} [a_{pl} (\sum_{j=1}^{m} \alpha_j u_j^{(l-1)}(a)) + b_{pl} (\sum_{j=1}^{m} \beta_j u_j^{(l-1)}(b))] = 0 \quad 1 \leq p \leq m$$

7. Show that if $W(t)$ is the Wronskian for (3.42a) then $c_2(t) W(t) \equiv k$, so that (3.45) is valid, and then show that (3.39) does indeed give the solution to (3.42a, b).

8. Consider the IVP (3.57a, b) for the case where A is a constant matrix, and let S be the set of eigenvalues of A with maximal real part. Show that the solution $\mathbf{y}(x)$ of (3.57a, b) is relatively stable iff $\boldsymbol{\alpha}$ has a component in the direction of a principal vector corresponding to some $\lambda_i \in S$, and that this vector has the maximal degree associated with S.
  Hint: First consider the case where Λ in (3.54) is diagonal.

9. Show that the conditioning constants κ_1 and κ_2 in (3.82) and (3.83) are independent of the particular choice of fundamental solution.

10. Suppose that the BVP (3.38a, b) with inhomogeneous BC is rewritten in the form (3.78a, b). Show that (3.81a) can be generalized to

$$\|W_1 \mathbf{y}\|_\infty \leq \kappa_1 |W_2 \boldsymbol{\beta}| + \kappa_2 \|W_3 \mathbf{q}\|_p$$

where $\kappa_1 := \|W_1 Y Q^{-1} W_2^{-1}\|_\infty$ and $\kappa_2 := \sup \|\{\int_a^b |W_1 G(\cdot, s) W_3^{-1}|^q \, ds\}^{1/q}\|_\infty$, with W_1, W_2 and W_3 arbitrary diagonal scaling matrices. If one were only interested in perturbations to the solution of (3.38a, b), what weighted norm might be appropriate?

11. Suppose that $\mathbf{y}(x)$ is the unique (smooth) solution of (3.78a, b) and what $\mathbf{w}(x)$ is the unique solution of (3.78a) with BC

$$(B_a + \delta B_a)\mathbf{y}(a) + (B_b + \delta B_b)\mathbf{y}(b) = \boldsymbol{\beta} + \delta\boldsymbol{\beta}$$

(so it is assumed that

$$\tilde{Q} = (Q + \delta Q) := (B_a + \delta B_a)Y(a) + (B_b + \delta B_b)Y(b)$$

is nonsingular). Show that if $\max\{\|\delta B_a\|, \|\delta B_b\|, |\delta\boldsymbol{\beta}|\} < \delta$ and $2\delta\kappa_1 < 1$, then the error $\mathbf{e}(x) := \mathbf{w}(x) - \mathbf{y}(x)$ satisfies

$$\|\mathbf{e}\|_\infty \le \frac{\kappa_1 |\delta\boldsymbol{\beta}^*|}{1-2\delta\kappa_1} \le (1 + |\mathbf{y}(a)| + |\mathbf{y}(b)|)\frac{\delta\kappa_1}{1-2\delta\kappa_1}$$

where $\delta\boldsymbol{\beta}^* := \delta\boldsymbol{\beta} - \delta B_a \mathbf{y}(a) - \delta B_b \mathbf{y}(b)$. (*Hint:* If $A = I + E$ and $\|E\| < 1$, then A^{-1} exists and $\|A^{-1}\| \le \frac{1}{1-\|E\|}$.)

12. Suppose that the conditioning constants $\kappa_1(b)$, $\kappa_2(b)$ for the family of BVPs (3.78a, b) with $b \to \infty$ satisfies $\kappa_1(b) \sim 1$ and $\kappa_2(b) \sim 1$. Transform these BVPs to an equivalent family of BVPs on $[0,1]$ involving a (suitably defined) small parameter ε. What can one say about this new family of BVPs, called singular perturbation problems? (See Chapter 10 for a long answer.)

13. (a) Assuming that rank $(B_a | B_b) = n$, show that for some nonsingular lower triangular matrix U, the matrix $[U^{-1}B_a | U^{-1}B_b]$ has orthonormal rows. Hence it is not restrictive to assume that $B_a B_a^T + B_b B_b^T = I$.
    (*Hint:* Perform a QU decomposition of $\begin{bmatrix} B_a^T \\ B_b^T \end{bmatrix}$.)

    (b) Show that if $(B_a | B_b)$ has orthonormal rows then there exist orthogonal matrices V, Q_1, and Q_2 and nonnegative diagonal matrices Σ_1, Σ_2 such that

    $$B_a^T = Q_1 \Sigma_1 V^T$$

    $$B_b^T = Q_2 \Sigma_2 V^T$$

    (*Hint:* Use the singular value decomposition and the orthogonality requirement.)

    (c) Prove Lemma 3.88.

    (d) Show that if $B_a B_a^T + B_b B_b^T = I$ and B_a, B_b satisfy (3.89), then $B_a^2 + B_b^2 = I$.

14. Consider the BVP

$$\mathbf{y}' = \begin{bmatrix} 2\lambda & 0 \\ 0 & \mu \end{bmatrix} \mathbf{y} \quad -1 \le x \le 1$$

$$y_1(-1) = y_2(1) = 1$$

where $\lambda > 0$ and $\mu < 0$. Show that for $\lambda \to \infty$, κ_1 is bounded but κ_2 becomes large.

15. Verify (3.95b, c) and (3.96a, b).

16. Consider the BVP (3.78a, b) for the constant coefficient case $A(x) \equiv A$. Show that there is an exponential dichotomy iff A has no purely imaginary eigenvalues and an ordinary dichotomy iff all purely imaginary eigenvalues are simple.

17. Given the BVP
$$\varepsilon u''(x) + \gamma u'(x) = 0 \qquad 0 < x < 1$$
$$u(0) = 1, \qquad u(1) = 0$$
with $0 < \varepsilon \ll 1$, convert to the form (3.78). Find the conditioning constants κ_1 and κ_2 calculating first $Y(x;0)$ and then $\Phi(x)$. Does the problem have an exponential dichotomy? Justify your answer.

18. Answer the questions of Exercise 17 for the BVP
$$\varepsilon^2 u''(x) - \gamma^2 u(x) = 0 \qquad 0 < x < 1$$
$$u(0) = 1, \qquad u(1) = 0$$

19. How are the sizes of κ_1 and κ_2 in Exercises 17 and 18 related to the width of the solutions' boundary layers for $0 < \varepsilon \ll 1$?

20. Suppose that there exists a matrix transformation $S(x)$, with $\|S(x)\|$ $\|S^{-1}(x)\|$ uniformly bounded for $a \le x \le b$, such that $\hat{A}(x) = S^{-1}(x) A(x) S(x)$ satisfies the hypotheses of Theorem 3.102. Show that (3.91) has an exponential dichotomy.

21. Show that, for the ODE in Example 3.13, if $\omega < \lambda$ then there is no constant matrix S with $S^{-1} A(x) S$ diagonally dominant uniformly for $a \le x \le b$.

22. For the BVP defined in Example 3.15, Case 1, show that $\kappa_1 = 1 + O(e^{-b/2})$.

23. Let $B_a B_a^T + B_b B_b^T = I$; i.e., let (B_a, B_b) have orthonormal columns (cf. Exercise 13).

   a) Show that $\|Y(t) Q^{-1}\|_2^2 = \|G(t,a) G(t,a)^T + G(t,b) G(t,b)^T\|_2$.

   b) Show that $\kappa_1 \le \sqrt{2} \kappa_2$.

24. Show that for any fundamental solution Y,

   (i) $Y(t) = G(t,s) Y(s) - G(t,u) Y(u)$, $a \le s < t < u \le b$ and

   (ii) $Y^{-1}(t) = Y^{-1}(u) G(u,t) - Y^{-1}(s) G(s,t)$, $\qquad a \le s < t < u \le b$.

25. Let $\tilde{y}$ satisfy the ODE (3.78a) and the BC
$$\tilde{B}_a \tilde{y}(a) + \tilde{B}_b \tilde{y}(b) = \tilde{\beta}$$

Then the ODE and this BC induce a Green's function, say $\tilde{G}$. Let G denote the Green's function associated with the BC (3.78b). Then

$$\tilde{G}(t,s) = G(t,s) - \tilde{\Phi}(t)[\tilde{B}_a G(a,s) + \tilde{B}_b G(b,s)]$$

where $\tilde{\Phi}$ satisfies the BC

$$\tilde{B}_a \tilde{\Phi}(a) + \tilde{B}_b \tilde{\Phi}(b) = I$$

(*Hint:* Use Exercise 24.)

26. Let the BVP (3.78) be well-conditioned and assume that B_a, B_b are scaled as in (3.89).

   a) Show that there exist separated BC $\tilde{B}_a$, $\tilde{B}_b$ such that $(\tilde{B}_a, \tilde{B}_b)$ has orthonormal rows, and $\|\tilde{G}(t,s)\|_2 \leq \kappa_2 + 2\sqrt{2}\kappa_2^2$.
   [*Hint:* Use Exercises 25 and 23(b).]

   b) Show that (3.78a) is dichotomic with $P = \begin{bmatrix} 0 & 0 \\ 0 & I_k \end{bmatrix}$ and $K = \kappa_2(1 + 2\sqrt{2}\kappa_2)$.

4

Initial Value Methods

In the previous chapter we saw that there is a close theoretical relationship between BVPs and IVPs. Indeed, the basic theorems of solution existence and representation for a BVP were given in terms of fundamental solutions, which are solutions of associated IVPs. At the same time, BVPs may have a more complex structure than IVPs (see Section 3.1). It makes sense, then, to construct a numerical method for a given BVP by relating the problem to corresponding "easier" IVPs, and solving the latter numerically. This chapter describes and analyzes such methods.

The intuitive appeal of this approach is strengthened by the advanced state of numerical analysis for IVPs: As described in Section 2.7, good numerical methods for such problems are well understood, and efficient, flexible, general-purpose software is readily available in any mathematical software library. Thus, one is able, at least in principle, to solve BVPs numerically with minimal problem analysis and preparation.

4.1 INTRODUCTION: SHOOTING

The simplest initial value method for BVPs is the *single* (or *simple*) *shooting* method, briefly described below for linear and nonlinear problems. The method is analyzed in more detail in further sections. It is a useful, easy-to-understand method. Unfortunately, it can also (and often does) have stability drawbacks. These drawbacks are alleviated by more complex initial value methods like *multiple shooting, the stabilized march,* and the *Riccati method.* Some of the resulting variants lead to powerful, general-purpose algorithms (unlike single shooting). At the same time, the charm of single shooting is its simplicity, and this is lost as well.

4.1.1 Shooting for a linear second-order problem

Consider the BVP (1.2c, b), rewritten here

$$u''(x) - c_1(x)u'(x) - c_0(x)u(x) = q(x) \qquad a < x < b \tag{4.1a}$$

$$u(a) = \beta_1 \tag{4.1b}$$

$$u(b) = \beta_2 \tag{4.1c}$$

Assume that (4.1a,b,c) has a unique solution $u(x)$. In order to use an initial value integrator for (4.1a) we need $u(a)$ and $u'(a)$. By (4.1b) we have $u(a)$, but $u'(a)$ is unknown. Let us therefore guess a "shooting angle" s_1,

$$u'(a) = s_1 \tag{4.1d}$$

The IVP (4.1a,b,d) can be solved, yielding a solution $v(x)$, say. But, in general, $v \neq u$ because $v(b) \neq \beta_2$. Let us then guess another value

$$u'(a) = s_2 \tag{4.1e}$$

and call $w(x)$ the solution of the IVP (4.1a,b,e). Again, in general $w(b) \neq \beta_2$, so $w \neq u$. Linearity of the problem implies that

$$u(x) = \theta w(x) + (1-\theta)v(x) \qquad a \leq x \leq b. \tag{4.2}$$

Assuming (reasonably) that $v(b) \neq w(b)$, we can adjust the shooting angle by defining

$$\theta := \frac{\beta_2 - v(b)}{w(b) - v(b)}, \qquad \text{i.e.,} \quad \beta_2 = \theta w(b) + (1-\theta)v(b) \tag{4.3}$$

A direct substitution into (4.1a,b,c) shows that $u(x)$ of (4.2) is indeed the sought solution of the BVP.

This is an instance of the single *shooting method*. The IVPs (4.1a,b,d) and (4.1a,b,e) are solved numerically to yield a numerical algorithm. If a stable scheme of order r is used (cf. Section 2.7) and the coefficients $c_1(x), c_0(x)$, and $q(x)$ of (4.1a) are sufficiently smooth, then it follows from (4.3) that the obtained approximation to $u(x)$ also has an error of $O(h^r)$, where h is the maximum step size used in the numerical integration of the IVPs.

A general exposition and analysis of the shooting method for linear systems of ODEs is given in Section 4.2. Here we note that instead of the IVP (4.1a,b,e) we could solve (4.1a) under the initial conditions

$$u(a) = 0, \qquad u'(a) = s_2 \tag{4.1f}$$

Calling the solution $\psi(x)$, say, we get

$$u(x) = v(x) + \eta\psi(x) \qquad a \leq x \leq b \tag{4.4}$$

for an appropriate constant η. We can identify η with θ of (4.2) and ψ(x) with $w(x) - v(x)$.

4.1.2 Shooting for a nonlinear second-order problem

In (4.3) we have explicitly used the linearity of the problem. Yet, the shooting principle extends to nonlinear problems as well, as the following example illustrates.

Example 4.1

Consider the following very simple model of a chemical reaction

$$u'' + e^u = 0, \qquad 0 < x < 1 \tag{4.5a}$$

$$u(0) = u(1) = 0 \tag{4.5b}$$

Recalling Example 3.2, we know that this nonlinear problem has two isolated solutions. As an initial value problem, however, with $u(0) = 0$ and $u'(0) = s$, the problem (4.5a) has a unique solution for each real s, denoted by $u(x;s)$. Now, if we find the correct "angle of shooting" s^* such that $u(1;s^*) = 0$ (there are two such values, in fact), then the solution of the IVP also solves the BVP (4.5a,b) (see Fig. 4.1). Thus, we get a nonlinear algebraic equation to solve: find $s = s^*$ which satisfies the equation

$$u(1;s) = 0 \tag{4.6}$$

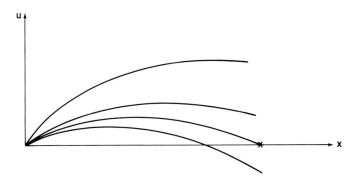

Figure 4.1 Shooting

This latter problem can be solved numerically by an iterative scheme (cf. Section 2.4). Note that each function evaluation in this iterative scheme involves the (numerical) solution of an IVP. □

4.1.3 Shooting — general application, limitations, extensions

The shooting approach remains rather straightforward even for a general system of first order nonlinear differential equations,

$$\mathbf{y}' = \mathbf{f}(x, \mathbf{y}) \qquad a < x < b \tag{4.7a}$$

subject to general two-point boundary conditions

$$\mathbf{g}(\mathbf{y}(a), \mathbf{y}(b)) = \mathbf{0} \qquad (4.7b)$$

Again, if we denote by $\mathbf{y}(x;\mathbf{s})$ the solution of (4.7a) subject to the initial conditions

$$\mathbf{y}(a;\mathbf{s}) = \mathbf{s} \qquad (4.7c)$$

the problem reduces to that of finding $\mathbf{s} = \mathbf{s}^*$ which solves the nonlinear algebraic equations

$$\mathbf{g}(\mathbf{s}, \mathbf{y}(b;\mathbf{s})) = \mathbf{0} \qquad (4.8)$$

Each function evaluation in (4.8) again involves the solution of an IVP (4.7a,c).

Now, a user can combine a standard library routine for solving nonlinear equations (e.g. using Broyden updates — see Section 2.3.3) and a standard initial value solver (cf. Section 2.7) to obtain a general code for solving BVPs, based on the method outlined above. This is so straightforward and general that, if the method did not have severe drawbacks, there would be hardly any justification for writing this book! However, as it turns out, the simple shooting method cannot always be applied, and its numerical results cannot always be trusted.

One major drawback already arises in the linear case. The IVPs integrated in the process could be unstable, even when the BVP is well-conditioned. The shooting method is unstable in that case, as discussed in Section 4.2. With nonlinear problems, there is another potential trouble: When shooting with the wrong initial values $\mathbf{s}$, the (exact) IVP solution $\mathbf{y}(x;\mathbf{s})$ might exist only on $[a,c]$, where $c < b$. The iteration process for (4.8) is obviously in trouble under such circumstances.

Initial value methods like multiple shooting and stabilized march attempt to alleviate these difficulties by restricting the size of the intervals over which IVPs are solved. The resulting schemes are described and analyzed in Sections 4.3, 4.4 and 4.6. The Riccati method, described in Section 4.5, involves solving different, usually more stable IVPs.

4.2 SUPERPOSITION AND REDUCED SUPERPOSITION

In this section we describe and analyze the shooting method for the general linear two-point BVP

$$\mathbf{y}' = A(x)\mathbf{y} + \mathbf{q}(x) \qquad a < x < b \qquad (4.9a)$$

$$B_a \mathbf{y}(a) + B_b \mathbf{y}(b) = \boldsymbol{\beta} \qquad (4.9b)$$

As in (4.3) and in (4.4), the solution is represented as a linear combination of solutions to associated IVPs, hence the method is referred to as the *superposition method*.

4.2.1 Superposition

We describe the extension of (4.4) for the general problem (4.9a,b), since it is mathematically more elegant than the extension of (4.3). Recall from Section 3.1 that the general solution of (4.9a) can be represented as

$$y(x) = Y(x)\mathbf{s} + \mathbf{v}(x) \qquad a \leq x \leq b \qquad (4.10)$$

where $Y(x)$ is a fundamental solution, $\mathbf{s}$ is a parameter vector ($\mathbf{s} \in \mathbf{R}^n$) and $\mathbf{v}(x)$ is a particular solution. For definiteness, let $Y(x) \equiv Y(x;a)$ be the fundamental solution satisfying

$$Y' = A(x)Y \qquad a < x < b \qquad (4.11\text{a})$$

$$Y(a) = I \qquad (4.11\text{b})$$

The particular solution $\mathbf{v}(x)$ may be determined as the solution of the IVP

$$\mathbf{v}' = A(x)\mathbf{v} + \mathbf{q}(x) \qquad (4.12\text{a})$$

$$\mathbf{v}(a) = \boldsymbol{\alpha} \qquad (4.12\text{b})$$

for some vector $\boldsymbol{\alpha}$ (e.g. $\boldsymbol{\alpha} = \mathbf{0}$). Thus, the n columns of $Y(x)$ and the vector $\mathbf{v}(x)$ can be computed as solutions of $n+1$ IVPs. The BVP solution $\mathbf{y}(x)$ is given as their superposition in (4.10). To determine the parameters $\mathbf{s}$ we substitute in (4.9b) to obtain

$$\boldsymbol{\beta} = B_a[Y(a)\mathbf{s} + \mathbf{v}(a)] + B_b[Y(b)\mathbf{s} + \mathbf{v}(b)] = [B_a + B_b Y(b)]\mathbf{s} + B_a \mathbf{v}(a) + B_b \mathbf{v}(b)$$

or

$$Q\mathbf{s} = \hat{\boldsymbol{\beta}} \qquad (4.13\text{a})$$

where

$$Q := B_a + B_b Y(b) \qquad (4.13\text{b})$$

$$\hat{\boldsymbol{\beta}} := \boldsymbol{\beta} - B_a \mathbf{v}(a) - B_b \mathbf{v}(b). \qquad (4.13\text{c})$$

Assuming that the BVP (4.9a,b) has a unique solution, we know that Q is nonsingular [cf. Theorem 3.30]; hence $\mathbf{s}$ is well-defined by (4.13). Then the solution for our BVP is given by (4.10).

Example 4.2

Consider the ODEs

$$\begin{bmatrix} y_1 \\ y_2 \end{bmatrix}' = \begin{bmatrix} 0 & 1 \\ \lambda^2 & 0 \end{bmatrix} \begin{bmatrix} y_1 \\ y_2 \end{bmatrix} + (1-\lambda^2) \begin{bmatrix} 0 \\ e^x \end{bmatrix} \qquad 0 < x < b \qquad (4.14\text{a})$$

$$\begin{bmatrix} 1 & 0 \\ 0 & 0 \end{bmatrix} \begin{bmatrix} y_1(0) \\ y_2(0) \end{bmatrix} + \begin{bmatrix} 0 & 0 \\ 1 & 0 \end{bmatrix} \begin{bmatrix} y_1(b) \\ y_2(b) \end{bmatrix} = \begin{bmatrix} 1 \\ e^b \end{bmatrix} \qquad (4.14\text{b})$$

for some $\lambda > 0$, $b > 0$. The reader should have no trouble in identifying here a special case of (4.1a, b, c). It is straightforward to show that

$$Y(x) = \begin{bmatrix} \cosh \lambda x & \lambda^{-1} \sinh \lambda x \\ \lambda \sinh \lambda x & \cosh \lambda x \end{bmatrix}$$

and

$$\mathbf{v}(x) = \begin{bmatrix} e^x - \cosh \lambda x - \lambda^{-1} \sinh \lambda x \\ e^x - \lambda \sinh \lambda x - \cosh \lambda x \end{bmatrix}$$

where in (4.12b) we choose $\boldsymbol{\alpha} = 0$. For (4.13a) we get

$$\begin{bmatrix} 1 & 0 \\ \cosh \lambda b & \lambda^{-1} \sinh \lambda b \end{bmatrix} \mathbf{s} = \begin{bmatrix} 1 \\ \cosh \lambda b + \lambda^{-1} \sinh \lambda b \end{bmatrix}$$

with the solution $\mathbf{s} = \begin{bmatrix} 1 \\ 1 \end{bmatrix}$. Hence the solution to (4.14a,b) is given by $\mathbf{y}(x) = \begin{bmatrix} e^x \\ e^x \end{bmatrix}$. □

The practical implementation of the superposition method involves, of course, numerical approximations to the quantities appearing in (4.10)-(4.13). Recall our convention to denote a numerical approximation of a quantity ψ by ψ^h. We obtain the following algorithm.

Algorithm: Single shooting

Input: A BVP (4.9a,b) and a sequence $a \leq x_1 < \cdots < x_J \leq b$ of evaluation points.
Output: The initial values $\mathbf{y}^h(a)$ and $\mathbf{y}^h(x_1)$, ..., $\mathbf{y}^h(x_J)$.

1. Integrate (4.11a,b), (4.12a,b) numerically, using an IVP code, obtaining $Y^h(b)$ and $\mathbf{v}^h(b)$.
2. Form Q^h and $\hat{\boldsymbol{\beta}}^h$ by (4.13b,c).
3. Solve $Q^h \mathbf{s}^h = \hat{\boldsymbol{\beta}}^h$ for $\mathbf{s}^h$.
4. $\mathbf{y}^h(a) := \mathbf{s}^h + \boldsymbol{\alpha}$
5. Integrate the IVP for (4.9a) numerically, obtaining $\mathbf{y}^h(x_1)$, ..., $\mathbf{y}^h(x_J)$. □

The last two steps of the algorithm deserve a comment. Step 4 follows directly from (4.10), but step 5, where the solution is evaluated (formed), solves an additional IVP, instead of using (4.10). There are two reasons for this. One is that if there are many evaluation points (e. g. we wish to plot or tabulate the solution), then storing $Y^h(x_j)$ and $\mathbf{v}^h(x_j)$ for each j during step 1 is impractical. The other reason is that when performing steps 1–4 we do not need to know what evaluation points would be desired. Note that theoretically, step 5 is equivalent to using (4.10), provided that the same mesh is used in the IVP integration.

The quality of the approximate solution obtained by this algorithm depends on the discretization error and on the roundoff error. In the next section we assume infinite precision, which enables us to deal with the discretization error alone. The following section then deals with roundoff errors.

4.2.2 Numerical accuracy

As indicated in the single shooting algorithm, the IVPs have to be integrated numerically, and this introduces a discretization error. We have already indicated, following

(4.3), that the discretization error (not the roundoff error!) of the computed BVP solution is expected to be of the order of the discretization errors in the calculated IVP solutions. We now make a more precise argument for the general case. This subsection may be skipped without loss of continuity.

For the actual numerical implementation, we use initial value integration codes, as described in Section 2.7. One of the attractive features of shooting is that this integration can be done automatically, as the step size is controlled by the integration routine for a given IVP. Usually, such a routine has a mixed absolute-relative error control. Roughly speaking, if the solution of the IVP grows significantly, then a relative control mechanism is preferred; otherwise, an absolute error control takes over (cf. Section 2.7.4).

Suppose that a tolerance TOL is given to the initial value integration routine. Then the routine would do its best to control the *local* error in $\mathbf{v}(x)$ and $Y(x)$ according to TOL. What can be then said about the error in $\mathbf{y}(x)$?

Assume, for simplicity, that the same mesh

$$a = x_1 < x_2 < \cdots < x_{\hat{N}} < x_{\hat{N}+1} = b$$

is used for all $n+1$ discretized IVP solutions, and suppose that the discrete solution $\{\mathbf{y}_i\}_{i=1}^{\hat{N}+1}$ satisfies

$$\mathbf{y}_{i+1} = \Gamma_i \mathbf{y}_i + \mathbf{g}_i \qquad i = 1, \ldots \hat{N} \qquad (4.15)$$

where the $n \times n$ matrices Γ_i and the inhomogeneity vectors $\mathbf{g}_i$ are determined by the discretization scheme used for the IVPs. Let $\{Y_i\}_{i=1}^{\hat{N}+1}$ denote the *discrete fundamental solution* of (4.15) defined by $Y_1 = I$, $Y_{i+1} = \Gamma_i Y_i$, $i = 1, \ldots, \hat{N}$; this implies $Y_i = Y^h(x_i)$. Define the *discrete Green's function* [cf. (3.32), (3.36)] by

$$G_{ij} = \begin{cases} Y_i(Q^h)^{-1} B_a Y_1 Y_j^{-1} & j \le i \\ -Y_i(Q^h)^{-1} B_b Y_{\hat{N}+1} Y_j^{-1} & j > i \end{cases} \qquad 1 \le i, \ j \le \hat{N} \qquad (4.16)$$

where Q^h is the discrete analogue of Q, viz.

$$Q^h = B_a + B_b Y_{\hat{N}+1} \qquad (4.17)$$

In a similar way to (3.80) for the continuous problem, we have

$$\mathbf{y}_i = Y_i(Q^h)^{-1} \boldsymbol{\beta} + \sum_{j=1}^{\hat{N}} G_{ij} \mathbf{g}_j \qquad (4.18)$$

Now, let the IVP integrator induce the (small) *local* truncation error $\boldsymbol{\tau}_i/(x_{i+1}-x_i)$ at x_i. Then

$$\mathbf{y}(x_{i+1}) - \mathbf{y}_{i+1} = \Gamma_i [\mathbf{y}(x_i) - \mathbf{y}_i] + \boldsymbol{\tau}_i \qquad (4.19)$$

Note that (4.19) has the form of (4.15) with the global error $\mathbf{y}(x_i) - \mathbf{y}_i$ replacing $\mathbf{y}_i$ and the local truncation error term $\boldsymbol{\tau}_i$ replacing inhomogeneity $\mathbf{g}_i$. Use of (4.18) then yields

$$\mathbf{y}(x_i) - \mathbf{y}_i = \sum_{j=1}^{\hat{N}} G_{ij} \boldsymbol{\tau}_j \qquad (4.20)$$

This expression relates the error in $\mathbf{y}_i$ to the local truncation errors in the IVP integrations. But we still have to consider G_{ij}. It often occurs, for instance, that a fundamental solution grows exponentially. Yet, it can be shown that under reasonable assumptions there is a constant C of moderate size such that

$$\|G_{ij}\| \leq C \|G(x_i, x_j)\| \qquad 1 \leq i, j \leq \hat{N}$$

Now, if we use a local error per step control, we can write

$$\tau_j = \mathbf{c}_j (x_{j+1} - x_j) TOL, \qquad |\mathbf{c}_j| \sim 1$$

so from (4.20),

$$\mathbf{y}(x_i) - \mathbf{y}_i = TOL \sum_{j=1}^{\hat{N}} G_{ij}(x_{j+1} - x_j) \mathbf{c}_j$$

Therefore

$$|\mathbf{y}(x_i) - \mathbf{y}_i| \leq K \, TOL \, \left\| \int_a^b G(x_i, t) dt \right\| \approx K \, TOL \, \kappa \qquad (4.21)$$

where κ is the conditioning constant of (3.34) and $K \sim 1$.

We conclude, therefore, that in general the shooting method will give appropriate global discretization errors if the problem (4.9a, b) is well-conditioned (i.e. κ is not large).

Before considering the propagation of roundoff errors, we make some practical observations for later use. First, it is not necessarily true that exponential behaviour of solution modes is discovered where it should be. For instance, if the solution contains a rapidly decreasing mode, this mode may be concealed both in $Y(x)$ and in $\mathbf{v}(x)$. Hence the high activity of the solution $\mathbf{y}(x)$ near $x = a$ would not be detected in this case, so the local errors may be large there. Second, in general it is very likely that the initial value choices both in (4.11b) and in (4.12b) induce a significant component of the most rapidly growing mode in all solution vectors. For instance, in Example 2 we have, for $\lambda x \geq 20$, say,

$$\cosh \lambda x \approx \sinh \lambda x \approx \frac{1}{2} e^{\lambda x}$$

and the mode $e^{-\lambda x}$ essentially disappears. As a consequence, the step size will mainly be determined by the dominating mode. It is sensible then to use the adaptive step selection of the integrator only once, thus constructing a mesh, and compute the other solution vectors on the same mesh. [Unexpectedly this supports the assumption made for convenience of the analysis before, when obtaining (4.15).] Third, as most standard integrators are designed for (usually stable, or at least not extremely unstable) IVPs, great care has to be taken with respect to the stability of the numerical computation of growing modes. In particular, methods for stiff IVPs are not necessarily good here. For instance, if one uses the *backward Euler* scheme (cf. Section 2.7) then $e^{\lambda x}$ is represented by a *decaying* discrete solution component if $h \lambda > 2$ (h the step size). This may have very undesirable consequences, as some terminal condition may no longer control this mode in the numerical approximation. An automatic stiff IVP integrator, even with a relative step size control, proceeds with a very small step size, keeping $h \lambda < 2$ for accuracy reasons. We return to this issue in Chapter 10.

Despite the negative tones in the above observations, (4.21) indicates that the superposition method is usually in a satisfactory state regarding the effect of discretization errors. The trouble is in roundoff error accumulation, which we treat next.

4.2.3 Numerical stability

Consider the linear BVP (4.9a,b) and assume that it is well-conditioned (i.e. κ of (3.34) is of moderate size). We shall see that accurate numerical solutions for such problems can be obtained, e.g., by the finite difference techniques of Chapter 5; so, if something goes wrong when one is using simple shooting, it is the method and not the problem which is to be blamed for the difficulty.

Stability difficulties arise when the differential operator of (4.9a) contains rapidly growing (and decaying) fundamental solution modes. Then the initial value solution is very sensitive to small changes in the initial values $\mathbf{s}$ of (4.10). Thus, even if a numerical approximation $\mathbf{s}^h$ is obtained for $|\mathbf{s}| \approx 1$ such that

$$|\mathbf{s}^h - \mathbf{s}| \approx \varepsilon_M$$

where ε_M is the machine precision (this is the best we may hope for), the roundoff error $e(x)$ when *forming* the approximation to $\mathbf{y}(x)$ by (4.10) can be expected to grow like

$$|e(x)| \sim \varepsilon_M \|Y(x)\|, \qquad (4.22)$$

which may be acceptable relative to $\|Y(x)\|$ but is *not acceptable relative to* $|\mathbf{y}(x)|$!

Example 4.3

We make a slight modification to Example 4.2 to yield a symmetric coefficient matrix

$$A(x) = \begin{bmatrix} 0 & \lambda \\ \lambda & 0 \end{bmatrix} \qquad 0 \le x \le 1$$

The fundamental solution modes for y_1 are $e^{\pm \lambda x}$, so the roundoff error when constructing the solution is expected to grow like

$$(\mathbf{s} - \mathbf{s}^h) e^{\lambda x}$$

The fundamental solution satisfying (4.11a, b) is

$$Y(x) = \begin{bmatrix} \cosh \lambda x & \sinh \lambda x \\ \sinh \lambda x & \cosh \lambda x \end{bmatrix}$$

Thus, as λx grows, $Y(x)$ rapidly becomes large in norm and nearly singular. This is just another manifestation of the presence of rapidly increasing and nonincreasing fundamental modes.

Taking the inhomogeneous term and the boundary conditions as

$$\mathbf{q}(x) = \begin{bmatrix} 0 \\ \lambda \cos^2 \pi x + \frac{2}{\lambda} \pi^2 \cos 2\pi x \end{bmatrix}, \qquad y_1(0) = y_1(1) = 0$$

we can show the BVP to be well-conditioned. It has the exact solution

$$\mathbf{y}(x) = \begin{bmatrix} \dfrac{e^{\lambda(x-1)} + e^{-\lambda x}}{1 + e^{-\lambda}} - \cos^2 \pi x \\ \dfrac{e^{\lambda(x-1)} - e^{-\lambda x}}{1 + e^{-\lambda}} + \dfrac{\pi}{\lambda} \sin 2\pi x \end{bmatrix}$$

so $s_1 = 0$ and $s_2 \approx -1$ if $\mathbf{v}(a) = \mathbf{0}$ and λ is large.

Table 4.1 lists numerical results for $\lambda = 20$ and 50. Absolute and relative tolerances for the initial value solver DE (see references in Chapter 2) were set to 10^{-10}, and this is roughly the error in s_2 (s_1 is obtained exactly). The computations were performed on an computer with a 14-hexadecimal-digit mantissa. Under "e" the absolute errors in y_1 are listed. The errors in y_2 are similar. Note that the error growth in the formed solution is exponential in λ, as predicted. These results are clearly unacceptable.

Note that the matrix

$$Q = \begin{bmatrix} 1 & 0 \\ \cosh \lambda & \sinh \lambda \end{bmatrix}$$

approximated by the numerical initial value solutions contains extremely large elements. Indeed, one thing to beware of when using the shooting method in general is the possibility of *floating-point overflow* in the course of initial value integration. This phenomenon certainly occurs in this example with $\lambda = 500$, say.

□

Having demonstrated, by means of a simple example, the potentially disastrous results when using the superposition method *carelessly,* we now turn to analyze the situation. Let $\mathbf{s}^h$ be the numerical approximation to $\mathbf{s}$ obtained for a given BVP (4.9a,b). Examining (4.10) in general, we see that we must study the behaviour of $Y(x)(\mathbf{s}^h - \mathbf{s})$ to get an idea of the error in $\mathbf{y}(x)$, given that $Y(x)$ has been well-approximated relatively, as discussed previously. For $\mathbf{v}(x)$ we obtain a large relative *and* absolute error. All this is modelled sufficiently well by assuming that there are matrices E and $\tilde{E}$ with a small norm (essentially $\approx \hat{N}\varepsilon_M$) and a vector $\mathbf{c}$ of norm 1, such that

$$[B_a + B_b Y(b)(I + E)]\mathbf{s}^h = \tilde{\boldsymbol{\beta}} := \hat{\boldsymbol{\beta}} - B_b Y(b) \tilde{E} \mathbf{c} \qquad (4.23a)$$

[see (4.13)]. Then, subtracting (4.13a) and manipulating a bit yields

$$Y(x)(\mathbf{s}^h - \mathbf{s}) = -Y(x)Q^{-1}B_b Y(b)[E\mathbf{s}^h - \tilde{E}\mathbf{c}] \qquad (4.23b)$$

Now, since we have assumed that (4.9a, b) is well-conditioned, $\|Y(x)Q^{-1}\|$ is of a

TABLE 4.1 Difficulties in simple shooting

λ	x	e	λ	x	e
20	0.1	.44-9	50	0.1	.19-7
	0.2	.30-8		0.2	.28-5
	0.3	.22-7		0.3	.41-3
	0.4	.26-6		0.4	.61-1
	0.5	.12-5		0.5	.90+1
	0.6	.87-5		0.6	.13+4
	0.7	.66-4		0.7	.20+6
	0.8	.48-3		0.8	.29+8
	0.9	.35-2		0.9	.44+10
	1.0	.26-1		1.0	.65+12

moderate size. However, since E, $\tilde{E}$ may be *any* (small norm) matrices, we cannot expect any special cancellation effect to occur in forming $B_b Y(b)[E\,\mathbf{s}^h - \tilde{E}\mathbf{c}]$. The large elements of $Y(b)$ magnify $[E\,\mathbf{s}^h - \tilde{E}\mathbf{c}]$ dramatically when exponentially increasing solution modes are present, as in Example 4.3.

It is important to realize that the large error seen in Table 4.1 is due to *roundoff, not discretization errors*. The large (absolute) discretization errors in the increasing fundamental modes cancel, just like the exact modes in (4.10) : This is what the bound in (4.21) is telling us. The roundoff errors, in contrast, generally do not cancel, as $E\,\mathbf{s}^h \neq \tilde{E}\mathbf{c}$.

One way to deal with the instability problem of the superposition method is to use multiple precision. Roughly speaking, given a machine precision ε_M, we have to require that $\varepsilon_M \|Y(b)\|$ is still less than the required tolerance TOL. For Example 4.2, the number of significant digits (size of the mantissa used) would grow linearly with λb. Due to the cost of such (software implemented) floating point operations, this approach does not seem very practical.

Let us now recall from its algorithm that the single shooting method can be viewed as a two-stage process: Obtaining $\mathbf{s}^h$, which approximates the solution of (4.13a), and then forming the approximate solution by (4.10) with $\mathbf{s}^h$ replacing $\mathbf{s}$. In Example 4.3, the obtained $\mathbf{s}^h$ was a very good approximation to the exact $\mathbf{s}$, and the trouble began only at the second stage, while the solution was formed. It is natural to ask whether it is generally true that $\mathbf{s}^h$ is a good approximation to $\mathbf{s}$. Since $\mathbf{s}^h$ is obtained from the perturbed (4.13a), our question relates to the conditioning of the matrix Q. Note that Q in Example 4.3 is badly scaled, but becomes well-conditioned after a diagonal rescaling. Thus, Gauss elimination with scaled partial pivoting works well here. Unfortunately, there is no guarantee that this is true in general.

To see what happens generally, consider a problem (4.9a, b) of order $n \geq 2$, with $n - v$ decaying and $v \geq 1$ increasing modes, with one mode increasing faster than any other, and with separated boundary conditions

$$B_a = \begin{bmatrix} B_{a1} \\ 0 \end{bmatrix}, \quad B_b = \begin{bmatrix} 0 \\ B_{b2} \end{bmatrix}$$

In order for such a problem to be well-conditioned, we must have

$$\text{rank}(B_{a1}) = n - v, \quad \text{rank}(B_{b2}) = v$$

Now, if there exists a mode that has grown much larger in magnitude than a complementary set of $(n-1)$ other modes at $x = b$, then $Y(b)$ is nearly a matrix of rank 1 [just like $Y(1)$ of Example 4.3]. In fact, in such a case the *effective* (or *numerical*) *rank* of $Y(b)$ is 1, due to rounding errors. Thus, the effective rank of $B_{b2}Y(b)$ is also 1 and that of

$$Q = \begin{bmatrix} B_{a1} \\ B_{b2}Y(b) \end{bmatrix}$$

is at most $n - v + 1$. So, if $v > 1$, the effective rank of Q is less than n, and we cannot expect $\mathbf{s}^h$ to be calculated accurately!

Example 4.4

Consider the problem

$$u''' = 7\lambda^2 u' - 6\lambda^3 u \qquad 0 < x < 1$$

which has modes that grow like $e^{-3\lambda x}$, $e^{\lambda x}$ and $e^{2\lambda x}$. Given one boundary condition at $x = 0$ and two (=v) conditions at $x = 1$, the simple shooting method using 15 decimal digit arithmetic fails to calculate $\mathbf{s}^h$ close to $\mathbf{s}$ for, say, $\lambda = 50$. □

If $A(x)$ varies with x, then even with one growing fundamental mode the shooting method may be unable to reproduce $\mathbf{s}$ accurately in some cases, as the following example shows.

Example 4.5

Consider the ODE

$$\mathbf{y}' = \pi \begin{bmatrix} 1 - \lambda \cos 2\pi x & 1 + \lambda \sin 2\pi x \\ -1 + \lambda \sin 2\pi x & 1 + \lambda \cos 2\pi x \end{bmatrix} \mathbf{y} \qquad 0 < x < 1$$

with λ a constant, $\lambda > 1$. Here,

$$Y(x) = \begin{bmatrix} e^{(1-\lambda)\pi x} \cos \pi x & e^{(1+\lambda)\pi x} \sin \pi x \\ -e^{(1-\lambda)\pi x} \sin \pi x & e^{(1+\lambda)\pi x} \cos \pi x \end{bmatrix}$$

Thus, the upper right element of $Y(x)$ is 0 at $x = 1$, but it is very large for $\lambda = 50$, say, when x is slightly perturbed from 1. This component of $Y^h(1)$, as well as its first column, would not approximate the corresponding elements of $Y(1)$ well, even relatively. Indeed, with boundary matrices $B_a = B_b = I$, which yield a well-conditioned problem, the simple shooting method fails to produce an $\mathbf{s}^h$ which accurately approximates $\mathbf{s}$ (see Example 9). □

Throughout the discussion of instability, we have been concerned with fast-growing fundamental modes. If we have only fast-growing, and no fast-decaying modes, then the shooting method can still be made to work well if we do the IVP integrations in the *reverse* direction, from b to a, and use a stiff integrator. That is so because if the BVP is well-conditioned, then a fast-growing mode from a to b is scaled down by the BC at b. Thus, looking from b back toward a, this same mode is a well-behaved, stable (stiff) mode. The trick of reversing the direction of integration would not work, however, if there are both fast-growing and fast-decaying modes present (Why?). This is the case in Examples 2.5.

4.2.4 Reduced superposition

For a BVP (4.9a,b) of order n, the superposition method calls for the integration of n+1 IVPs, namely, for n fundamental modes of (4.11) plus one particular solution of (4.12). When the BC are partially separated, e.g. if $n - v$ of the BC involve $\mathbf{y}(a)$ alone, for some $v < n$, then it is possible to integrate only $v + 1$ ODEs, as described below. The resulting procedure is called *reduced superposition*.

The reader may well ask at this point: Why waste time trying to improve upon the efficiency of a method which has just been shown to be unstable in many cases? The main answer is that similar considerations arise later on for methods which are designed to improve upon the stability problems of single shooting.

Suppose now that (4.9b) can be written as

$$\begin{bmatrix} B_{a1} \\ B_{a2} \end{bmatrix} y(a) + \begin{bmatrix} 0 \\ B_{b2} \end{bmatrix} y(b) = \begin{bmatrix} \beta_1 \\ \beta_2 \end{bmatrix} \qquad (4.24)$$

where $B_{a1} \in \mathbf{R}^{(n-\nu) \times n}$, $\text{rank}(B_{a1}) = n - \nu$, $B_{a2}, B_{b2} \in \mathbf{R}^{\nu \times n}$, $\beta_1 \in \mathbf{R}^{n-\nu}$ and $\beta_2 \in \mathbf{R}^{\nu}$, $0 < \nu < n$. Such partial separation is present in most applications in Section 1.2. Recall that the initial value α of the particular solution $v(x)$ was not specified in (4.12b) (subsequently, we chose $v(a) = 0$, for simplicity). Here we choose $v(a)$ to satisfy

$$B_{a1} v(a) = \beta_1 \qquad (4.25)$$

[How to implement this will be discussed following (4.28) below.] Then, instead of (4.10) we write the solution of (4.9a), (4.24) as

$$y(x) = \overline{Y}(x) \overline{s} + v(x) \qquad (4.26)$$

where $\overline{Y}(x)$ is an $n \times \nu$ matrix of linearly independent fundamental modes satisfying

$$\overline{Y}' = A(x) \overline{Y} \qquad a < x < b \qquad (4.27a)$$

$$B_{a1} \overline{Y}(a) = 0. \qquad (4.27b)$$

and $\overline{s} \in \mathbf{R}^{\nu}$.

The reason that we can write (4.26), in general, is that (4.25) (and the full rank of B_{a1}) imply that $v(x)$ should lie in a ν-dimensional subspace, and a complementary solution space is found by superposition of the ν columns of $\overline{Y}(x)$ which satisfy (4.27a,b).

Now, clearly $y(x)$ of (4.26) satisfies the first $n - \nu$ BC of (4.24). The parameters $\overline{s}$ are then determined so that the remaining ν BC are satisfied as well. Parallel to (4.13), we have the system of order ν

$$\overline{Q} \overline{s} = \overline{\beta} \qquad (4.28a)$$

$$\overline{Q} := B_{a2} \overline{Y}(a) + B_{b2} \overline{Y}(b) \qquad (4.28b)$$

$$\overline{\beta} := \beta_2 - B_{a2} v(a) - B_{b2} v(b). \qquad (4.28c)$$

To complete the description of the reduced superposition method, we now show how to get initial values out of (4.25) and (4.27b). We construct the QU decomposition of B_{a1}^T. Thus, let H be an orthogonal $n \times n$ matrix and R a lower triangular $(n-\nu) \times (n-\nu)$ matrix such that

$$H B_{a1}^T = \begin{bmatrix} R^T \\ 0 \end{bmatrix} \qquad (4.29)$$

(cf. Section 2.4). Now, let $\overline{Y}(a)$ be the last ν columns of H^T. Then, transposing (4.29) and considering the last ν columns, (4.27b) is seen to be satisfied. Writing

$$H^T =: [\hat{Y}(a) | \overline{Y}(a)] \qquad (4.30a)$$

and defining

$$\mathbf{v}(a) := \hat{Y}(a)R^{-1}\boldsymbol{\beta}_1 \qquad (4.30b)$$

we see that (4.25) is satisfied as well.

Example 4.6

Consider the Dirichlet problem (4.1a, b, c) again. When we describe the shooting method for this BVP in Section 4.1, only two IVPs were solved. Yet, when we convert to a first-order system (of order $n = 2$) and applying the superposition method, three IVPs ($=n+1$) need to be integrated! This discrepancy is now resolved: In Section 4.1 we merely showed a trivial case of reduced superposition. For $y_1 = u$, $y_2 = u'$ we have from (4.1b, c)

$$B_{a1} = B_{b2} = (1\ 0), \qquad B_{a2} = (0\ 0), \qquad \nu = 1$$

Hence $H = I$ and $R = 1$ in (4.29). From (4.30a, b) we then obtain

$$\overline{Y}(a) = \begin{bmatrix} 0 \\ 1 \end{bmatrix}, \qquad \mathbf{v}(a) = \begin{bmatrix} \beta_1 \\ 0 \end{bmatrix}$$

The reader can readily verify that the first components of $\overline{Y}(x)$ and $\mathbf{v}(x)$ here correspond to $\psi(x)$ and $v(x)$, respectively, in (4.4). □

It is not difficult to verify that the results obtained earlier for the full superposition method with respect to discretization and roundoff errors remain qualitatively the same for the reduced superposition method. The advantage of this method lies simply in having fewer IVPs to integrate and a smaller $\overline{Q}$ to invert. The minor disadvantage is in the added overhead of (4.29), (4.30).

4.3 MULTIPLE SHOOTING FOR LINEAR PROBLEMS

As we saw in the previous section, a major disadvantage of the single shooting method is the roundoff error accumulation, occurring when unstable IVPs have to be integrated. A rough (but often approximately achievable) bound on this error is of the order

$$\varepsilon_M e^{L(b-a)}$$

where $L = \max_x \|A(x)\|$ and ε_M is the machine precision. In an attempt to decrease this bound, it is natural to restrict the size of domains over which IVPs are integrated. Thus, the interval of integration [a, b] is subdivided by a mesh

$$a = x_1 < x_2 < \cdots < x_N < x_{N+1} = b \qquad (4.31)$$

and then, as in shooting, initial value integrations are performed on each subinterval $[x_i, x_{i+1}]$, $1 \le i \le N$ (see Fig. 4.2). Recall that $\hat{N}$ integration steps are performed on the one subinterval in Section 4.2.2. The resulting solution segments are patched up to form a continuous solution over the entire interval [a, b]. This leads to the method of *multiple shooting*.

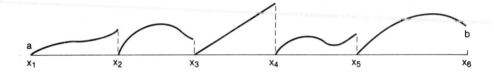

Figure 4.2 Multiple shooting

4.3.1 The method

We consider again the general linear BVP (4.9a, b) and generalize the superposition method (4.10)–(4.12) directly. On each subinterval $[x_i, x_{i+1}]$, $1 \leq i \leq N$, of the mesh (4.31), we can express the general solution of (9a) as

$$\mathbf{y}(x) = Y_i(x)\mathbf{s}_i + \mathbf{v}_i(x) \qquad x_i \leq x \leq x_{i+1} \qquad (4.32)$$

where $Y_i(x)$ is a fundamental solution, $\mathbf{s}_i$ is a parameter vector ($\mathbf{s}_i \in \mathbf{R}^n$) and $\mathbf{v}_i(x)$ is a particular solution. They are defined, for each i, by the IVPs

$$Y_i' = A(x)Y_i \qquad x_i < x < x_{i+1} \qquad (4.33a)$$

$$Y_i(x_i) = F_i \qquad (4.33b)$$

with F_i an $n \times n$ nonsingular matrix, and

$$\mathbf{v}_i' = A(x)\mathbf{v}_i + \mathbf{q}(x) \qquad x_i < x < x_{i+1} \qquad (4.34a)$$

$$\mathbf{v}_i(x_i) = \boldsymbol{\alpha}_i \qquad (4.34b)$$

Most common choices for F_i and $\boldsymbol{\alpha}_i$ are

$$F_i = I, \qquad \boldsymbol{\alpha}_i = \mathbf{0} \qquad 1 \leq i \leq N \qquad (4.35)$$

If the matrices F_i are chosen independently from each other, as in (4.35), then the problems (4.33a, b) can be integrated in parallel for $1 \leq i \leq N$. Indeed, the method is sometimes called *parallel shooting* in that case. Other variants involve *marching techniques*, discussed in the next section, where for each i, F_{i+1} is determined only after (33a, b) has been solved. If (4.35) holds then we call the method *standard multiple shooting*. Note that $Y_i(x) \equiv Y(x; x_i)$ when $F_i = I$. It should be mentioned that $Y_i(x)$ here *differs* from Y_i in (4.16).

The nN parameters

$$\mathbf{s}^T := (\mathbf{s}_1^T, \mathbf{s}_2^T, \ldots, \mathbf{s}_N^T) \qquad (4.36)$$

are determined so that the (numerical) solution is continuous over the entire interval [a,b] and so that the BC (4.9b) are satisfied. By using (4.32), the continuity (or matching) conditions $\mathbf{y}(x_{i+1}^-) = \mathbf{y}(x_{i+1}^+)$, $1 \leq i \leq N-1$, are

$$Y_i(x_{i+1})\mathbf{s}_i + \mathbf{v}_i(x_{i+1}) = Y_{i+1}(x_{i+1})\mathbf{s}_{i+1} + \mathbf{v}_{i+1}(x_{i+1}) \qquad 1 \leq i \leq N-1$$

Rearranging and substituting initial values, we get

$$F_{i+1}\mathbf{s}_{i+1} = Y_i(x_{i+1})\mathbf{s}_i + [\mathbf{v}_i(x_{i+1}) - \boldsymbol{\alpha}_{i+1}] \qquad 1 \le i \le N-1 \qquad (4.37a)$$

This recursion relation, together with the BC

$$B_a[Y_1(a)\mathbf{s}_1 + \mathbf{v}_1(a)] + B_b[Y_N(b)\mathbf{s}_N + \mathbf{v}_N(b)] = \boldsymbol{\beta}$$

give the linear system of nN equations

$$A\mathbf{s} = \hat{\boldsymbol{\beta}} \qquad (4.37b)$$

which in component form is

$$\begin{bmatrix} -Y_1(x_2) & F_2 & & & \\ & -Y_2(x_3) & F_3 & & \\ & & \ddots & \ddots & \\ & & & -Y_{N-1}(x_N) & F_N \\ B_a F_1 & & & & B_b Y_N(b) \end{bmatrix} \begin{bmatrix} \mathbf{s}_1 \\ \mathbf{s}_2 \\ \vdots \\ \vdots \\ \mathbf{s}_N \end{bmatrix} = \begin{bmatrix} \mathbf{v}_1(x_2) - \boldsymbol{\alpha}_2 \\ \mathbf{v}_2(x_3) - \boldsymbol{\alpha}_3 \\ \vdots \\ \mathbf{v}_{N-1}(x_N) - \boldsymbol{\alpha}_N \\ \boldsymbol{\beta} - B_a \boldsymbol{\alpha}_1 - B_b \mathbf{v}_N(b) \end{bmatrix} \qquad (4.37c)$$

Unlike single shooting, where after obtaining the initial values $\mathbf{s}$ the solution of the BVP still has to be formed (recall the last step of the single shooting algorithm and the discussion following), here the solution at x_i is given by

$$\mathbf{y}(x_i) = F_i \mathbf{s}_i + \boldsymbol{\alpha}_i$$

In particular, $\mathbf{y}(x_i) = \mathbf{s}_i$ for the standard multiple shooting method.

As before, let us denote by Y_i^h and $\mathbf{v}_i^h$ the approximations to Y_i and $\mathbf{v}_i$ obtained by using a numerical scheme. The corresponding solution to the linear system of algebraic equations (4.37b, c), denoted by $\mathbf{s}^h$, yields approximate solution values $\mathbf{y}^h(x_i)$, $1 \le i \le N$. Thus, if we assume that the evaluation points (i.e. the points where the solution is desired) are included in the mesh (4.31), the following algorithm is obtained.

Algorithm: Standard multiple shooting

Input: A BVP (4.9a, b) and a mesh (4.31).
Output: Solution values $\mathbf{y}^h(x_1), \ldots, \mathbf{y}^h(x_N)$ contained in $\mathbf{s}^h$.

1. FOR i = 1, ..., N DO
   Integrate (4.33a, b), (4.34a, b), (4.35) numerically, using an IVP code, obtaining $Y_i^h(x_{i+1})$ and $\mathbf{v}_i^h(x_{i+1})$.
2. Form A^h and $\hat{\boldsymbol{\beta}}^h$ of (4.37b, c).
3. Solve (4.37b, c) for $\mathbf{s}^h$. □

A discussion of the choice of shooting points (4.31) and further solution evaluation is delayed until the section on practical considerations.

Example 4.7

Consider Example 1.8 of Section 1.2 (theoretical seismograms). For simplicity, we concentrate on the SH case. To recall, the BVP is given by (1.20a, b, c). Thus, the differential equations are (with the notation of this Example 1.8)

$$y_1' = \mu^{-1} y_2$$
$$0 < x < b$$
$$y_2' = v^2 \mu y_1$$

$\mu = \rho \beta^2$, $v^2 = k^2 - \dfrac{\omega^2}{\beta^2}$. Here k and ω (horizontal wave number and angular frequency) are known constants and $\rho(x)$ and $\beta(x)$ (density and velocity) are given functions generated from measured data. The variable x is a space variable, pointing downward into earth. Let us assume that $v^2(b) > 0$. The BC are

$$y_2(0) = 1, \qquad v(b) y_1(b) + \mu^{-1}(b) y_2(b) = 0$$

(This corresponds to surface waves, in particular the so-called Love waves.)

Often a model is assumed where $\rho(x)$ and $\beta(x)$ are constant within each layer. Thus, the layer structure is defined by a mesh

$$0 = x_1 < x_2 < \cdots < x_N < x_{N+1} = b$$

and

$$\beta(x) = \beta_i, \qquad \rho(x) = \rho_i, \qquad x_i \le x < x_{i+1}, \qquad 1 \le i \le N$$

The given constants β_i and ρ_i vary with i, $1 \le i \le N$.

Since the functions $\beta(x)$ and $\rho(x)$ have jump discontinuities at the points x_i, $2 \le i \le N+1$, these points must be a part of the mesh of any difference scheme used to solve the problem numerically. For instance, if a simple shooting technique is used, these points have to be mesh points of the initial value integration. On the other hand, if we now choose the points x_i as multiple shooting points, we can obtain the fundamental solutions $Y_i(x)$ exactly.

In particular, defining positive constants v_i and μ_i by

$$v_i^2 := \left| k^2 - \dfrac{\omega^2}{\beta_i^2} \right|, \qquad \mu_i := \rho_i \beta_i^2, \qquad 1 \le i \le N,$$

then with the initial values (4.35) we get

$$Y_i(x) = \begin{bmatrix} \cosh(v_i(x-x_i)) & (\mu_i v_i)^{-1} \sinh(v_i(x-x_i)) \\ \mu_i v_i \sinh(v_i(x-x_i)) & \cosh(v_i(x-x_i)) \end{bmatrix}$$

if $k^2 - \dfrac{\omega^2}{\beta_i^2} \ge 0$, and

$$Y_i(x) = \begin{bmatrix} \cos(v_i(x-x_i)) & (\mu_i v_i)^{-1} \sin(v_i(x-x_i)) \\ -\mu_i v_i \sin(v_i(x-x_i)) & \cos(v_i(x-x_i)) \end{bmatrix}$$

otherwise. Also $v_i(x) \equiv 0$, $1 \le i \le N$, because the problem is homogeneous.

Thus, the exact solution for this particular model is obtained by solving the $2N \times 2N$ system (4.37b,c), where each vector s_i has two components, each F_i is a 2x2 identity matrix, the matrices $Y_i(x_{i+1})$ are given above,

$$B_a = \begin{bmatrix} 0 & 1 \\ 0 & 0 \end{bmatrix}, \qquad B_b = \begin{bmatrix} 0 & 0 \\ v(b) & \mu^{-1}(b) \end{bmatrix},$$

and the right-hand side vector is $\hat{\beta} = (0, \ldots, 0, 1, 0)^T$. The solution $\mathbf{y}(x)$ is given by (4.32). $\square$

4.3.2 Stability

A case like the simplified application in Example 4.7, where we know the exact solutions of (4.33a, b) and (4.34a, b), is of course rather rare in practice. Usually, these IVPs have to be integrated numerically as in the above algorithm. Thus we have, in fact, a *two-level discretization process:* First, by a coarse mesh (4.31), and then by a finer discretization in each subinterval for the numerical solution of the IVPs.

Consider now the resulting combined fine mesh for this two-level discretization process, and suppose we applied single shooting (a one-level process) based on the same fine mesh and using the same scheme for initial value integrations. It is not difficult to see that, in the absence of any roundoff error, the single shooting method would yield identical results to those of multiple shooting. Furthermore, since the main concern in Section 4.2 was the instability of the single shooting method, it remains to investigate how stability properties with respect to roundoff errors are improved with the current method. For notational convenience, let us then assume in the sequel that there is no discretization error, as in Example 4.7. Also assume that the initial value integrations each use $\hat{N}$ steps (as in Section 4.2.2), so the combined fine mesh of the two-level multiple shooting process has

$$\tilde{N} := N\hat{N} \qquad (4.38)$$

mesh points.

For simplicity, let us consider the standard multiple shooting method ($F_i = I$, $\alpha_i = 0$, for *all* i) and denote

$$\Gamma_i := Y(x_{i+1};x_i) \qquad 1 \leq i \leq N \qquad (4.39a)$$

Thus, (4.37a) yields

$$s_{i+1} = \Gamma_i s_i + \hat{\beta}_i \qquad 1 \leq i \leq N-1 \qquad (4.39b)$$

where $\hat{\beta}_i$ is the appropriate piece of $\hat{\beta}$ (see (4.37)). Again we denote by s^h the computed approximation of s, so the error in the solution $y(x_i)$ is $s_i^h - s_i$, $1 \leq i \leq N$. As for single shooting, we assume that for each i there are matrices E_i, $\tilde{E}_i$ with a small norm and a vector c_i, $|c_i| = 1$ such that

$$-\Gamma_i(I+E_i)s_i^h + s_{i+1}^h = \tilde{\beta}_i := \hat{\beta}_i + \Gamma_i \tilde{E}_i c_i \qquad 1 \leq i \leq N-1 \qquad (4.40a)$$

$$B_a s_1^h + B_b \Gamma_N(I+E_N)s_N^h = \tilde{\beta}_N := \hat{\beta}_N - B_b \Gamma_N \tilde{E}_N c_N \qquad (4.40b)$$

Thus, with the perturbation quantities

$$\mathbf{E} = \begin{bmatrix} -\Gamma_1 E_1 & & & & 0 \\ & -\Gamma_2 E_2 & & & \\ & & \cdots & & \\ 0 & & & -\Gamma_{N-1}E_{N-1} & \\ & & & & B_b \Gamma_N E_N \end{bmatrix}, \quad \mathbf{c}^h = \begin{bmatrix} c_1 \\ \vdots \\ c_N \end{bmatrix}$$

and $\tilde{\mathbf{E}}$ similarly constructed, we get [cf. (4.23)]

$$\mathbf{s}^h - \mathbf{s} = \mathbf{A}^{-1}[\mathbf{E}\mathbf{s}^h - \tilde{\mathbf{E}}\mathbf{c}^h], \qquad (4.41)$$

and we need to consider $\mathbf{A}^{-1}$.

Theorem 4.42 For the standard multiple shooting matrix **A**, the following hold:

(a) The inverse of **A** can be expressed in terms of Green's function (3.32), viz.

$$\mathbf{A}^{-1} = \begin{bmatrix} G(x_1, x_2) & \cdots & G(x_1, x_N) & Y_1(x_1)Q^{-1} \\ & \cdots & & \\ G(x_N, x_2) & \cdots & G(x_N, x_N) & Y_1(x_N)Q^{-1} \end{bmatrix} \quad (4.42a)$$

where Q is defined in (4.13b).

(b) Let κ be the conditioning constant of the BVP [cf. (3.34)]. Then

$$\|\mathbf{A}^{-1}\| \leq N \kappa \quad (4.42b)$$

(c) In the case of exponential dichotomy, the above bound can be extended to

$$\|\mathbf{A}^{-1}\| \leq \text{const } \underline{h} \, \kappa \quad (4.42c)$$

which holds even when $b - a \to \infty$, if $\min_{1 \leq i \leq N} x_{i+1} - x_i \geq \underline{h} > 0$. In (4.42b,c) const is a constant of moderate size, independent of N. □

The proof of this theorem is not difficult and is left as an exercise.

Theorem 4.42 indicates in a certain sense stability of the multiple shooting method. To understand how, recall the discussion of the algorithm condition number χ^h in Section 2.8.2. This number bounds roundoff error amplification, and the method performs well in this respect so long as this amplification is of a tolerable size (e.g., $\chi^h \varepsilon_M < TOL$). Consider then the relative error in (4.41). The other matrices appearing there are **E** and $\hat{\mathbf{E}}$. To consider their bounds, let

$$K := \max_{1 \leq i \leq N} \|\Gamma_i\| \quad (4.43)$$

Then we are in a position to estimate the error in $\mathbf{s}^h - \mathbf{s}$, assuming (reasonably) that $|\mathbf{s}| \sim 1$, $\|B_b\| = 1$ and $\max_{1 \leq i \leq N} \|E_i\| \approx \max_{1 \leq i \leq N} \|\hat{E}_i\| \approx \tilde{N}\varepsilon_M$. (The latter assumption on the size of perturbations makes sense when applying $\hat{N}$ steps of a stable IVP integrator between consecutive shooting points, in the absence of discretization errors.) Now, (4.41)–(4.43) yield (by also using $\|\mathbf{A}\| \approx K$)

$$|\mathbf{s}^h - \mathbf{s}| \leq \text{const } \tilde{N} \, \kappa \, K \, \varepsilon_M \approx \text{cond}(\mathbf{A})\hat{N}\varepsilon_M \quad (4.44)$$

where, to recall, $\text{cond}(\mathbf{A}) := \|\mathbf{A}\| \, \|\mathbf{A}^{-1}\|$.

The estimate (4.44) shows the following result.

Theorem 4.45 The standard multiple shooting method has a condition number $\chi^h \approx \tilde{N}\kappa K \approx \text{cond}(\mathbf{A})\hat{N}$. □

Thus, if the BVP is well-conditioned and if the shooting points are chosen so that K of (4.43) is of moderate size (with $\tilde{N}$ not extremely large), then the standard multiple shooting method is stable, in the sense that there is a tolerable roundoff error amplification. In contrast, with single shooting ($N = 1$) there are instances like Example 3 where K (and hence χ^h) becomes intolerably large.

4.3.3 Practical considerations

Theorem 4.45 states that it is feasible to control the roundoff error accumulation by using enough shooting points. Practically, the point is that given a well-conditioned problem and a tolerance TOL which is a few orders of magnitude larger than the machine precision ε_M, and choosing shooting points such that K is of moderate size, then normally either

$$\tilde{N} \kappa K \varepsilon_M < TOL$$

or $\tilde{N}$ is too large for the method to be cost-effective anyway. The effect of using more shooting points is demonstrated in the following example.

Example 4.8

Consider the BVP

$$\mathbf{y}'(x) = \begin{bmatrix} 1-19\cos 2x & 1+19\sin 2x \\ -1+19\sin 2x & 1+19\cos 2x \end{bmatrix} \mathbf{y}(x) + \mathbf{q}(x)$$

$$\mathbf{y}(0) + \mathbf{y}(\pi) = (1+e^\pi, 1+e^\pi)^T$$

where $\mathbf{q}(x)$ is chosen such that the solution is $\mathbf{y}(x) = e^x(1,1)^T$. A fundamental solution is given by

$$Y(x) = \begin{pmatrix} \cos x & \sin x \\ -\sin x & \cos x \end{pmatrix} \text{diag}(e^{-18x}, e^{20x})$$

and the problem is well-conditioned; in fact, $\kappa \approx 1$. Suppose we have N equally spaced shooting points. Then the stability of the multiple shooting system is quite well illustrated by Table 4.2. The reader should realize that $N=1$ actually gives single shooting, so the large $\|A\| \|A^{-1}\|$ is accounted for by the large $\|A\| = \|Q\|$. It is also seen that the algebraic condition number $\|A\| \|A^{-1}\|$ levels off already for quite moderate values of N. This seems to indicate that not very many shooting points are necessary for this problem, even if a very high accuracy is desired. □

TABLE 4.2 The effect of the number of shooting points on cond(A)

N	$\|A^{-1}\|$	$\|A\| \|A^{-1}\|$
1	1.0	1.9+27
3	1.0	1.2+9
6	1.0	3.5+4
9	1.0	1.1+3
12	1.0	1.9+2
15	1.0	6.9+1
18	1.0	3.5+1
21	1.1	2.2+1
24	1.1	1.6+1

The practical choice of shooting points is not an easy, or clearly decided matter. Some codes attempt to choose these points so that $\|\Gamma_i\| \approx K$, all i, where K depends on TOL / ε_M. That is, the IVP integration from a shooting point x_i proceeds until $\|Y_i^h(x)\|$ gets large ($\approx K$), whereupon the integration stops and a new shooting point x_{i+1} is set. However, taking K as large as possible in such a scheme (in an attempt to use as few shooting points as possible) is not always the most efficient way to go. If an IVP integration code with an *absolute* error control is used to solve a problem with increasing modes, then it is not difficult to see that the steps tend to decrease. Thus, longer shooting intervals may well be less desirable, as they require more integration steps, hence more function evaluations and an increased total amount of computation. On the other hand, one needs to restart the integration at each new shooting point, as well as to solve (4.37b,c), so some compromise is necessary.

Other factors are involved as well: Consider the question of evaluating the computed solution. If there are many output points, where the BVP solution value is desired (say, for graphing or tabulating purposes), then incorporating them all in the mesh (4.31), as in the standard multiple shooting algorithm, may be very wasteful. (There may even be more such evaluation points than points in the *fine* shooting mesh.) In that case, another solution evaluation procedure is necessary. One possibility is to integrate, for a given output point $\hat{x}$, $x_i < \hat{x} \leq x_{i+1}$, the IVP

$$\mathbf{y}' = A(x)\mathbf{y} + \mathbf{q}(x) \qquad x_i < x < \hat{x}$$

$$\mathbf{y}(x_i) = \mathbf{s}_i$$

which is similar to, but much more stable than, step 5 of the single shooting algorithm. Another possibility is to interpolate $\mathbf{y}^h(\hat{x})$ from the given values $\mathbf{y}^h(x_i) = \mathbf{s}_i$, $i = 1, \ldots, N$. This implies different, additional criteria for choosing the shooting points, viz., so that accurate interpolation can be made possible. None of these two alternatives is easily automated.

One important asset of the multiple shooting method is that the initial value integrations in different subintervals are independent of each other. Hence, they may be performed concurrently, if parallel computing is available. No such parallel computation can be done with the marching techniques of Section 4.4. But to utilize this parallelism feature, the shooting points have to be fixed in advance. An appropriate procedure in such circumstances is to first subdivide the interval $[a,b]$ coarsely, say uniformly. Then, on each subinterval, a marching technique can be used.

An application where multiple shooting is both natural and convenient is in problems where the solution has jumps at certain prescribed points. A typical example is an ODE

$$u'' = f(x, u, u')$$

with jump dicontinuities in $u'(x)$. Such is the case in some structural mechanics problems with layered material or cracks. Considering a conversion to a first-order system in (u, u'), if the jump in $u'(x)$ is known, then one should fix a shooting point there and apply the matching taking this jump into account. The only difference compared to the continuous case is a change in the right-hand side vector $\hat{\boldsymbol{\beta}}$.

As a final practical comment, let us note that *all* of the multiple shooting techniques of this chapter overcome the instability of the single shooting method only when this instability is not too extreme. To see this, consider Example 4.2 again and assume that $\lambda > 1, b \geq 1$. Here

$$\|\Gamma_i\| \approx \lambda e^{\lambda(x_{i+1} - x_i)}$$

and to keep K of (4.43) below some value M (say $M = 10^7$) we must choose $h := \max_{1 \leq i \leq N} (x_{i+1} - x_i)$ such that

$$h \leq \lambda^{-1} \ln(M/\lambda)$$

The smallest number of shooting points needed is obtained by a uniform distribution (because the BVP has constant coefficients). This number is

$$N = b/h \geq \lambda b / \ln(M/\lambda)$$

Now, if λb is "small," say $\lambda b \leq 10$, then single shooting is adequate; if λb is of "medium size," say $20 \leq \lambda b \leq 200$, then multiple shooting performs adequately where single shooting fails. But when $\lambda b = 20,000$, the lower bound on N means that, roughly, $N \geq 1000$. This leads to an extremely inefficient procedure, much like using a nonstiff integrator for a very stiff IVP. Methods for the latter case (which is a singular perturbation problem, if b is of a reasonably moderate size) are discussed in Chapter 10.

4.3.4 Compactification

In describing the multiple shooting method in Section 4.3.1, we have not discussed the important question of how to solve the linear system (4.37b, c). The corresponding step in the single shooting method requires no special attention, because Q is only $n \times n$ and is dense. Thus, the usual Gaussian elimination method (Section 2.3) is adequate. In (4.37b, c), on the other hand, only $2n^2N$ out of n^2N^2 elements of **A** are possibly nonzero. Thus, **A** is sparse when N is large, and a special scheme is needed to ensure its efficient storage and decomposition.

We defer the general treatment of this question to Chapter 7, because other BVP algorithms also lead to linear systems with a similar structure. An exception is made for two schemes, one discussed in Section 4.4 and the other, called the *compactification* or *condensing method*, discussed next.

The idea is dangerously appealing: Use the recursion relation (4.37a) to express (by repeated application for $i = N-1, N-2, \ldots, 1$) s_N in terms of s_1; substitute in the BC and obtain an $n \times n$ system of linear equations for s_1; solve for s_1; finally, use (4.37a) again (this time forward, $i = 1, \ldots, N-1$) to obtain **s**. This algorithm certainly avoids any need for sophisticated storage schemes for **A**! We express it, for the standard multiple shooting scheme, as a superposition method for the recursion (4.39b). Thus, we compute a particular solution $\{\xi_i\}_{i=1}^{N+1}$ and a fundamental solution $\{\Phi_i\}_{i=1}^{N+1}$ by

$$\xi_{i+1} = \Gamma_i \xi_i + \mathbf{v}_i(x_{i+1}) \qquad i = 1, \ldots, N \qquad (4.46a)$$

$$\xi_1 = 0 \tag{4.46b}$$

$$\Phi_{i+1} = \Gamma_i \Phi_i \quad i = 1, \ldots, N \tag{4.47a}$$

$$\Phi_1 = I \tag{4.47b}$$

and form

$$\mathbf{s}_i = \Phi_i \mathbf{s}_1 + \xi_i \quad i = 2, \ldots, N \tag{4.48a}$$

where $\mathbf{s}_1$ is obtained as the solution of the $n \times n$ linear system

$$[B_a + B_b \Phi_{N+1}]\mathbf{s}_1 = \beta - B_b \xi_{N+1} \tag{4.48b}$$

Alas, the compactification algorithm suffers from instability, much like the single shooting method. To see this, consider the case where there is no discretization error, as in Example 4.7. Then

$$\Phi_{N+1} = Y_N(b)Y_{N-1}(x_N)Y_{N-2}(x_{N-1}) \cdots Y_1(x_2) = Y(b;a)$$

and the matrix of (4.48b) is precisely Q of (4.13b). The solution formation by (4.48a) is equivalent to (4.10). When discretization errors are present as well, there is a similar error buildup in both $\{\Phi_i\}$ and $\{\xi_i\}$. But, just as for single shooting, we may not expect these errors to cancel when computing $\mathbf{s}$ in (4.48a, b).

Example 4.9

In Table 4.3 we list some results comparing the performance of simple shooting (SH), multiple shooting with Gaussian elimination and row-partial pivoting (MSH) and multiple shooting with compactification (MSC). The initial value solver, tolerances, and computer used are the same as in Example 4.3. The errors for the first solution component are listed for three problems:

(1) Example 4.3 with $\lambda = 50$.
(2) Example 4.5 with $\lambda = 50$, $A = B = I$, and $\beta = (1, 1)^T$, whose solution is

$$y_1(x) = \frac{e^{(1+\lambda)\pi(x-1)} \sin \pi x}{e^{-(1+\lambda)\pi} - 1} + \frac{e^{(1-\lambda)\pi x} \cos \pi x}{1 - e^{(1-\lambda)\pi}}$$

(3) Example 4.5 with $\lambda = 50$ and the boundary conditions

$$y_1(0) + y_2(0) = y_1(1) + y_2(1) = 1$$

whose solution is

$$y_1(x) = \frac{1}{e^{-2\pi\lambda} - 1} \{ (1+e^{(1-\lambda)\pi})e^{(1+\lambda)\pi(x-1)} \sin \pi x - (1+e^{-(1+\lambda)\pi})e^{(1-\lambda)\pi x} \cos \pi x \}$$

For MSH and MSC, 10 uniformly spaced shooting points were used. Note that $\mathbf{s}$ (or $\mathbf{s}_1$) is approximated well by all three methods for problems (1) and (3), whereas for problem (2) only MSH does a good job. In general, the full multiple shooting method performs well, whereas the other methods do not. The compactification algorithm fails, just like simple shooting, to approximate $\mathbf{s}_1$ well for problems (2). □

TABLE 4.3 Comparison of shooting, multiple shooting, and compactification.

	Problem (1)				Problem (2)				Problem (3)		
x	SH	MSH	MSC	x	SH	MSH	MSC	x	SH	MSH	MSC
0.1	.19−7	.25−10	.20−10	0.0	.10+1	.41−9	.10+1	0.0	.68−11	.68−11	.68−11
0.2	.28−5	.72−11	.10−10	0.1	.20−6	.13−12	.20−6	0.1	.33−9	.13−12	.13−12
0.3	.41−3	.16−10	.49−9	0.2	.34−13	.82−13	.34−13	0.2	.58−2	.82−13	.17−12
0.4	.61−1	.41−10	.71−7	0.3	.52−20	.11−17	.52−20	0.3	.72+5	.11−17	.11−5
0.5	.90+1	.73−11	.10−4	0.4	.28−21	.13−22	.56−27	0.4	.77+12	.13−22	.12+2
0.6	.13+4	.98−11	.16−2	0.5	.52−15	.41−38	.16−24	0.5	.74+19	.41−38	.11+9
0.7	.20+6	.25−10	.23	0.6	.50−9	.11−35	.14−17	0.6	.64+26	.12−35	.97+15
0.8	.29+8	.82−11	.34+2	0.7	.13−2	.12−29	.11−10	0.7	.49+33	.81−30	.75+22
0.9	.44+10	.31−10	.51+4	0.8	.84+4	.56−23	.72−4	0.8	.32+40	.27−23	.49+29
1.0	.65+12	.71−14	.75+6	0.9	.40+11	.13−16	.34+3	0.9	.16+47	.77−18	.24+36
				1.0	.48+9	.11−12	.41+1	1.0	.19+45	.11−12	.28+34

Despite this incriminating evidence, we should realize that MSC is not just SH. In particular, the IVP integrations are done in a safer way. This point becomes more important in nonlinear problems, as we shall see in Section 4.6.

The following two sections, Section 4.4 and Section 4.5, need a more involved notation and use somewhat more sophisticated mathematical tools. Those readers who are interested in an overview of initial value methods may want to read Section 4.6 first, but it is advisable to come back: The next section contains important practical material, including our most recommended numerical algorithm discussed in this chapter.

4.4 MARCHING TECHNIQUES FOR MULTIPLE SHOOTING

Let us consider again the solution of the BVP (4.9a, b) by multiple shooting. To recall, a marching technique for multiple shooting is one where the IVPs (4.33a, b)

$$Y_i' = A(x)Y_i \qquad x_i < x < x_{i+1}$$
$$Y_i(x_i) = F_i$$

are solved consecutively, with the initial matrix F_{i+1} for the $(i+1)^{st}$ subinterval being formed only *after* the IVP for the ith subinterval has been solved.

The motivation behind such a scheme is to construct not only F_{i+1}, but also the next shooting point x_{i+1}, using the following reasoning: In order to make a multiple shooting algorithm stable, it is necessary to monitor the growth of the fundamental solution modes to decide where the next shooting point should be. Since these basis functions tend to resemble more and more the most dominant one, they become more and more numerically linearly dependent as the integration interval gets larger (see Examples 2.5, for instance). Thus, when some tolerance of dependence is exceeded, a new shooting point x_{i+1} should be set and the columns of $Y_i(x_{i+1})$ should be reorthogonalized to restore their linear independence.

This intuitively appealing idea is fortified by the possibility of economizing the computation, precisely in the same way that reduced superposition economizes the superposition method (cf. Section 4.2.4). But there is more to it than that: The reorthogonalization turns out to produce a *decoupling* of growing and decreasing solution components, enabling us to construct a stable version of the compactification algorithm.

In the following sections we describe the general method, and then the decoupling of fundamental modes mentioned above, which yields the stability of the algorithm. Finally, we consider the most popular instance of these techniques, known as the *stabilized march*.

4.4.1 Reorthogonalization

We start integrating the IVPs (4.33a, b) and (4.34a, b) for $i = 1$, assuming that F_1 is an orthogonal matrix. For the general step i, $i = 1, \ldots, N$, we begin the IVP integrations with an orthogonal matrix F_i. Then, at the next shooting point x_{i+1}, we perform a QU-decomposition of $Y_i(x_{i+1})$, i.e.

$$Y_i(x_{i+1}) = F_{i+1}\Gamma_i \qquad 1 \le i \le N \tag{4.49}$$

with F_{i+1} orthogonal and Γ_i upper triangular (see Section 2.4). Regarding the particular solution $\mathbf{v}_i(x)$ of (4.34), we do not impose any special requirements on $\boldsymbol{\alpha}_i$ at this point (so, e.g., take $\boldsymbol{\alpha}_i = \mathbf{0}$).

Consider next the recursion relation (4.37a). Defining

$$\tilde{\boldsymbol{\beta}}_i := F_{i+1}^T(\mathbf{v}_i(x_{i+1}) - \boldsymbol{\alpha}_{i+1}) \qquad (\boldsymbol{\alpha}_{N+1} := \mathbf{0}) \qquad 1 \le i \le N \tag{4.50a}$$

$$\hat{\boldsymbol{\beta}}_N = \boldsymbol{\beta} - B_a\boldsymbol{\alpha}_1 - B_b\mathbf{v}_N(b), \qquad \tilde{\boldsymbol{\beta}}^T = (\tilde{\boldsymbol{\beta}}_1^T, \ldots, \tilde{\boldsymbol{\beta}}_{N-1}^T, \hat{\boldsymbol{\beta}}_N^T) \tag{4.50b}$$

we find

$$\mathbf{s}_{i+1} = \Gamma_i\mathbf{s}_i + \tilde{\boldsymbol{\beta}}_i \qquad 1 \le i \le N \tag{4.51a}$$

with Γ_i upper triangular. Incorporating the BC, we get

$$\mathbf{As} = \tilde{\boldsymbol{\beta}} \tag{4.51b}$$

$$\mathbf{A} = \begin{bmatrix} -\Gamma_1 & I & & & \\ & -\Gamma_2 & I & & \\ & & \ddots & & \\ & & & -\Gamma_{N-1} & I \\ B_aF_1 & & & & B_bF_{N+1}\Gamma_N \end{bmatrix} \tag{4.51c}$$

Note that $\mathbf{A}$ and Γ_i here correspond to, but are not the same as, $\mathbf{A}$ and Γ_i of Section 4.3. A skeptical reader may well ask at this point "so what?": $\mathbf{A}$ of (4.51c) does not seem to be very different from $\mathbf{A}$ of (4.37b, c). The gain in using (4.51) is explained next.

4.4.2 Decoupling

Let us examine the connection between Γ_i here and the fundamental solutions (4.38) of the standard multiple shooting method. To allow a simpler notation, let us again assume for the time being that no discretization errors occur. We have

$$Y(x_{i+1}; x_i) = Y_i(x_{i+1})[Y_i(x_i)]^{-1} = F_{i+1}\Gamma_i F_i^T \quad (F_i^{-1} = F_i^T)$$

so

$$Y(x_{i+1}; a) = Y(x_{i+1}; x_i)Y(x_i; x_{i-1}) \cdots Y(x_2; a) = F_{i+1}\Gamma_i \cdots \Gamma_1 F_1^T$$

$$Y_1(x_{i+1}) = F_{i+1}\Gamma_i \cdots \Gamma_1 \equiv F_{i+1}\hat{\Gamma}_i \quad (4.52)$$

Now, $\hat{\Gamma}_i = \Gamma_i \cdots \Gamma_1$ is upper triangular. Hence (4.52) gives the QU decomposition of $Y_1(x_{i+1})$ (see Section 2.4 for discussion of the uniqueness of this decomposition).

Consider the linear space spanned by the first k columns of F_{i+1}. Since $\hat{\Gamma}_i$ is upper triangular, the first k columns of $Y_1(x_{i+1})$ are, by (4.52), linear combinations of the first k columns of F_{i+1}, i.e. they are in the span of these columns of F_{i+1}. Thus, if F_1 is chosen such that its first k columns represent somehow initial values of the most dominant modes, then the span of the first k columns of F_{i+1} represents the directions of their integrated form up to $x = x_{i+1}$. Effectively, this means that the upper left $k \times k$ block of $\hat{\Gamma}_i$ has a norm of the same order of magnitude as the increment of these dominant modes.

There is a counterpart for the right lower $(n-k) \times (n-k)$ block of $\hat{\Gamma}_i$ in the case of a dichotomy with an $(n-k)$-dimensional decaying solution space. For this we need the following lemma.

Lemma 4.53 If an $n \times n$ matrix M can be written as

$$M = (M^1 | M^1 L + M^2)$$

where M^1 has k columns and L is some $k \times (n-k)$ matrix, then a QU decomposition of M, say $M = QR$, gives a right lower $(n-k) \times (n-k)$ block R^{22} of R such that $\|R^{22}\|_2 = \|M^2\|_2$. □

The reader will note that we are using here superscripts to denote partitions of matrices and vectors. This is because we will soon want to partition subscripted matrices. In order to get used to this, we give the proof of the lemma, despite its simplicity.

Proof: Writing Q^{-1} in the partitioned form

$$Q^{-1} = \begin{bmatrix} Q^{11} & Q^{12} \\ Q^{21} & Q^{22} \end{bmatrix}$$

and using $R = Q^{-1}M$, we have

$$(Q^{21} | Q^{22})M^1 = 0, \quad (Q^{21} | Q^{22})(M^1 L + M^2) = R^{22}$$

Hence, $(Q^{21} | Q^{22})M^2 = R^{22}$ and the proof is completed using the 2-norm preserving property of the orthogonal Q. □

Let us now utilize this lemma for (4.52) in case of a dichotomy. If the first k columns of $Y_1(x_{i+1})$ represent the dominant modes, then the remaining columns can be written as $M^1 L + M^2$ with M^2 representing values of decaying modes. Then the lemma implies that the right lower $(n-k) \times (n-k)$ block of $\hat{\Gamma}_i$ has a norm of the same order of magnitude as the decaying modes. [This is considered more precisely in Theorem 6.15.]

This means that in (4.51a) we have effectively *decoupled* the recursion (4.37a), where the decoupling corresponds to the splitting of growing and decaying modes. If we write

$$\Gamma_i = \begin{bmatrix} D_i & C_i \\ 0 & E_i \end{bmatrix} \qquad (4.54)$$

where D_i is $k \times k$, and partition vectors $\mathbf{g}_i \in \mathbf{R}^n$ as

$$\mathbf{g}_i = \begin{bmatrix} \mathbf{g}_i^1 \\ \mathbf{g}_i^2 \end{bmatrix} \qquad \mathbf{g}_i^1 \in \mathbf{R}^k \qquad (4.55)$$

then we can write (4.51a) as

$$\mathbf{s}_{i+1}^1 = D_i \mathbf{s}_i^1 + C_i \mathbf{s}_i^2 + \tilde{\boldsymbol{\beta}}_i^1 \qquad (4.56a)$$

$$1 \leq i \leq N-1$$

$$\mathbf{s}_{i+1}^2 = E_i \mathbf{s}_i^2 + \tilde{\boldsymbol{\beta}}_i^2 \qquad (4.56b)$$

Now, the recursion (4.56b) should be stable in a forward direction (i.e., i increasing from 1 to $N-1$), because $E_i \cdots E_1 = \hat{\Gamma}_i^{22}$, while (4.56a) is stable in a backward direction (i decreasing from $N-1$ to 1) because we expect $\|D_i^{-1} \cdots D_{N-1}^{-1}\| \sim 1$. It can be shown that this leads to a *stable* version of the compactification algorithm of Section 4.3.4, applied to the linear system (4.51c) (see Exercise 6.12).

Algorithm: Compactification for a decoupled system

Input: Upper triangular matrices Γ_i and inhomogeneities $\tilde{\boldsymbol{\beta}}_i$, $i = 1, \ldots, N$, matrices B_a, F_1, B_b, F_{N+1} and a partitioning parameter k.
Output: The solution $\mathbf{s}$ of (4.51b).

1. $\boldsymbol{\xi}_1^2 := 0$, $\boldsymbol{\xi}_{N+1}^1 := 0$, $\Phi_1^{22} := I_{n-k}$, $\Phi_{N+1}^{11} := I_k$, $\Phi_{N+1}^{12} := O$
2. FOR $i = 1, 2, \ldots, N$ DO
    $\boldsymbol{\xi}_{i+1}^2 := E_i \boldsymbol{\xi}_i^2 + \tilde{\boldsymbol{\beta}}_i^2$
    $\Phi_{i+1}^{22} := E_i \Phi_i^{22}$
3. FOR $i = N, N-1, \ldots, 1$ DO
    $\boldsymbol{\xi}_i^1 := D_i^{-1}(\boldsymbol{\xi}_{i+1}^1 - C_i \boldsymbol{\xi}_i^2 - \tilde{\boldsymbol{\beta}}_i^1)$
    $\Phi_i^{11} := D_i^{-1} \Phi_{i+1}^{11}$
    $\Phi_i^{12} := D_i^{-1}(\Phi_{i+1}^{12} - C_i \Phi_i^{22})$
4. $M := B_a F_1 \Phi_1 + B_b F_{N+1} \Phi_{N+1}$; $\boldsymbol{\mu} := \boldsymbol{\beta} - B_a(\boldsymbol{\alpha}_1 + F_1 \boldsymbol{\xi}_1) - B_b \left[F_{N+1} \boldsymbol{\xi}_{N+1} + \mathbf{V}_N(b) \right]$
5. Solve $M \boldsymbol{\eta} = \boldsymbol{\mu}$ (an $n \times n$ linear system).

6. FOR $i = 1, \ldots, N$ DO
$$y(x_i) := F_i \Phi_i \eta + F_i \xi_i + \alpha_i \qquad \square$$

This algorithm deserves some explanation. In principle, it is very similar to the compactification (4.46)–(4.48). The basic difference is that the recursions are evaluated only in stable directions. We have

$$\xi_i = \begin{bmatrix} \xi_i^1 \\ \xi_i^2 \end{bmatrix} \qquad \Phi_i = \begin{bmatrix} \Phi_i^{11} & \Phi_i^{12} \\ 0 & \Phi_i^{22} \end{bmatrix}$$

The lower rows' components of these are calculated in step 2 (after an initialization in step 1). The rest are calculated in step 3. Note that, since D_i is upper triangular, the multiplications by D_i^{-1} in this step amount to back substitution. Then steps 4 and 5 are the equivalent of (4.48b). Step 6 is a combination of the formulae

$$s_i = \Phi_i \eta + \xi_i \quad \text{and} \quad y(x_i) = F_i s_i + \alpha_i$$

Note that as soon as step 3 is over, we can work with

$$\Psi_i := F_i \Phi_i \quad \text{and} \quad z_i := F_i \xi_i \qquad (4.57)$$

alone. We now show that this algorithm is indeed much more stable than the previous compactification algorithm (4.46)–(4.48).

Theorem 4.58 If the BVP (4.9a, b) has a conditioning constant κ [cf. (3.34)], then the decoupling algorithm has a condition number $\chi^h \approx \kappa^2 K \tilde{N}$ [cf. (4.38), (4.43), (4.44)].

Proof: First, let us note that quantities involved in (4.56) are perturbed by $\approx \hat{N} K \varepsilon_M$. This and step 3 of the algorithm leave us to show that M^{-1} is bounded essentially by κ. The discussion preceding the algorithm shows that $\|\Phi_i\|$ and $|\xi_i|$ are bounded in terms of κ. Since $\|\Psi_i\|_2 = \|\Phi_i\|_2$ and $|z_i|_2 = |\xi_i|_2$, the quantities appearing in step 6 and in M are bounded by constants of moderate size as well (in Section 6.2 we give more precise bounds). So we are left to show that $\|M^{-1}\|$ is bounded, because this will imply that η can be calculated stably. Within numerical precision we may assume that

$$\|\Psi_j M^{-1}\|_2 \le \kappa \qquad j = 1, \ldots, N+1$$

Hence, using again the orthogonality of F_i,

$$\|(0 \mid I_{n-k}) M^{-1}\|_2 \le \|\Phi_1 M^{-1}\|_2 = \|\Psi_1 M^{-1}\|_2 \le \kappa$$

$$\|(I_k \mid 0) M^{-1}\|_2 \le \|\Phi_{N+1} M^{-1}\|_2 = \|\Psi_{N+1} M^{-1}\|_2 \le \kappa$$

Combining, we get $\|M^{-1}\|_2 \le 2\kappa$. $\qquad \square$

Example 4.10

Consider the problem from Example 4.8 again. Using the routine MUSL from Appendix A (where stable compactification is employed) with $TOL = 10^{-6}$ and $K = \max \|\Gamma_i\| \le 10^3$, we obtain the results listed in Table 4.4a, where e denotes the error in the max norm.

TABLE 4.4a Incremental matrices and errors for Examples 4.10, $K = 10^3$

i	x	Γ_{i-1}		e
0	.0			.19-8
1	.34	.11+4 .0	.25-2 .17-2	.12-8
2	.70	.70+3 .0	.85-3 .28-2	.18-8
⋮	⋮	⋮	⋮	⋮
9	2.83	.17+4 .0	.62-4 .12-2	.19-8
10	3.14	.16+2 .0	.38-6 .81-1	.19-8

If we increase K to 10^{30} then we obtain as output points only $x = 0$ and $x = \pi$, i.e., as in single shooting (but done in a stable way and for the discrete system only). The results are listed in Table 4.4b.

The algorithm described above in fact produces more than just the solution. The decoupling strategy implies $\|\Phi_i\| \sim 1$, and we know from (3.32) that the Green's function is essentially bounded by the "growth" of modes in appropriate directions. This in turn can be monitored quite well by checking the diagonal elements of Γ_i, to give an estimate of the conditioning constant of the BVP.

Example 4.11

Consider the BVP composed of the ODE

$$\mathbf{y}' = \begin{pmatrix} \psi(x) & 0 \\ 2\psi(x) & -\psi(x) \end{pmatrix} \mathbf{y} + \begin{pmatrix} (1-\psi(x))e^x \\ 2e^x \end{pmatrix}$$

where $\psi(x) = 20 \sin x + 20x \cos x$, and the BC

$$\mathbf{y}(0) + \mathbf{y}(T) = \begin{pmatrix} 1+e^T \\ 2(1+e^T) \end{pmatrix}, \quad T > 0$$

This BVP has the solution $\mathbf{y}(x) = \begin{pmatrix} 1 \\ 2 \end{pmatrix} e^x$, although the ODE has modes growing like $e^{20x \sin x}$ and $e^{-20x \sin x}$. Thus, we only have a monotonic behaviour of these modes (dichotomy with a dichotomy constant ≈ 1) on $[0, L]$, $L \approx 2.029$. For $x > L$ the increasing

TABLE 4.4b Incremental matrix and errors for Example 4.10, $K=10^{30}$

i	x	Γ_{i-1}		e
0	0			.19-8
1	3.14	.19+28 .0	.42+22 .28-24	.19-8

□

mode becomes decreasing and vice versa. Hence, we expect a moderate conditioning constant if $T \leq L$, but a large one if $T > L$. Table 4.5 shows output given by MUSL, where we asked for an absolute accuracy 10^{-6} and output at equally spaced points with distance 0.1 apart. For $T = 2$ we see the (common) slight overkill of a multiple shooting approach (i.e., the errors are smaller than the tolerance). For $T = 3$ we see a dramatic loss of accuracy, which is predicted by the estimate of the conditioning constant.

□

One of the questions left to be dealt with, if we want to use this algorithm, is to find an appropriate initial matrix F_1. Recall that the first k columns of F_1 have to span a subspace that generates the growing solutions (assuming that we have an exponential dichotomy). We will pursue this question for the case when the BC are separated, or partially separated, and leave more general cases to Section 6.4.3. We are now ready to generalize the reduced superposition method of Section 4.2.4.

4.4.3 Stabilized march

Let us consider again the ODE (4.9a) subject to the partially separated BC (4.24),

$$\mathbf{y}' = A(x)\mathbf{y} + \mathbf{q}(x)$$

$$\begin{bmatrix} B_{a1} \\ B_{a2} \end{bmatrix} \mathbf{y}(a) + \begin{bmatrix} 0 \\ B_{b2} \end{bmatrix} \mathbf{y}(b) = \begin{bmatrix} \beta_1 \\ \beta_2 \end{bmatrix} \updownarrow \nu$$

We proceed to generalize the reduced superposition method into a multiple shooting setting. Given a mesh (4.31) of shooting points, the solution $\mathbf{y}(x)$ is expressed in the form

$$\mathbf{y}(x) = \overline{Y}_i(x)\overline{\mathbf{s}}_i + \mathbf{v}_i(x) \qquad x_i \leq x \leq x_{i+1}, \qquad 1 \leq i \leq N \qquad (4.59)$$

where each $\overline{Y}_i(x)$ is now an $n \times \nu$, rather than $n \times n$, matrix of fundamental solutions, satisfying

$$\overline{Y}_i' = A\overline{Y}_i, \qquad x_i < x < x_{i+1}, \qquad \overline{Y}_i(x_i) = \overline{F}_i \qquad i = 1, \ldots, N \qquad (4.60)$$

and each $\overline{F}_i$ is an $n \times \nu$ matrix of rank ν. As in (4.25), (4.27b), we choose $\mathbf{v}_1(a)$ and $\overline{F}_1$ so that

$$B_{a1}\mathbf{v}_1(a) = \beta_1 \qquad (4.61a)$$

$$B_{a1}\overline{F}_1 = 0 \qquad (4.61b)$$

How this choice is implemented has already been discussed [see (4.29), (4.30)].

TABLE 4.5 Errors and conditioning constant estimates for Example 4.11

T	max|error|$_\infty$	min|error|$_\infty$	estimate for κ
2	.42-7	.40-9	.24+1
2.5	.24-4	.52-9	.40+6
3	.50+5	.18-1	.28+12

For the general step i, we integrate a particular solution $\mathbf{v}_i(x)$ and a (reduced) fundamental solution $\overline{Y}_i(x)$ until the next shooting point $x = x_{i+1}$. There, we have to make sure that the new (reduced) fundamental solution $\overline{Y}_{i+1}(x)$ is such that range $(\overline{Y}_{i+1}(x_{i+1}))$ = range $(\overline{Y}_i(x_{i+1}))$ and that the new particular solution $\mathbf{v}_{i+1}(x)$ is in range $(\overline{Y}_i(x_{i+1}))$ + $\mathbf{v}_i(x_{i+1})$. This is done as follows: First, we orthogonalize the columns of $\overline{Y}_i(x_{i+1})$ using, say, Householder transformations. This gives

$$\overline{Y}_i(x_{i+1}) = \overline{F}_{i+1}\overline{\Gamma}_i \tag{4.62}$$

with $\overline{\Gamma}_i$ an upper triangular $v \times v$ matrix and $\overline{F}_{i+1}$ an $n \times v$ matrix having orthonormal columns. Then, we choose the initial values for $\mathbf{v}_{i+1}(x)$ by

$$\boldsymbol{\alpha}_{i+1} := (I - \overline{F}_{i+1}\overline{F}_{i+1}^T)\mathbf{v}_i(x_{i+1}) \tag{4.63}$$

making this vector orthogonal to all v columns of $\overline{F}_{i+1}$.

Matching the solution segments of (4.59) to obtain a continuous $\mathbf{y}(x)$, we get

$$\overline{Y}_i(x_{i+1})\overline{\mathbf{s}}_i + \mathbf{v}_i(x_{i+1}) = \overline{F}_{i+1}\overline{\mathbf{s}}_{i+1} + \boldsymbol{\alpha}_{i+1}$$

Then, using (4.62), (4.63), and multiplying through by $\overline{F}_{i+1}^T$, yields

$$\overline{\mathbf{s}}_{i+1} = \overline{\Gamma}_i \mathbf{s}_i + \overline{F}_{i+1}^T \mathbf{v}_i(x_{i+1}) \qquad 1 \leq i \leq N-1 \tag{4.64a}$$

The BC now read

$$B_{a2}[\overline{F}_1\overline{\mathbf{s}}_1 + \boldsymbol{\alpha}_1] + B_{b2}[\overline{Y}_N(b)\overline{\mathbf{s}}_N + \mathbf{v}_N(b)] = \boldsymbol{\beta}_2 \tag{4.61c}$$

The relations (4.64a), (4.61c) give a linear system of order vN (not nN) for $\overline{\mathbf{s}}^T = (\overline{\mathbf{s}}_1^T, \ldots, \overline{\mathbf{s}}_N^T)$,

$$\overline{A}\,\overline{\mathbf{s}} = \overline{\boldsymbol{\beta}} \tag{4.64b}$$

where

$$\overline{A} = \begin{pmatrix} -\overline{\Gamma}_1 & I & & & \\ & -\overline{\Gamma}_2 & I & & \\ & & \ddots & & \\ & & & -\overline{\Gamma}_{N-1} & I \\ B_{a2}\overline{F}_1 & & & & B_{b2}\overline{F}_{N+1}\overline{\Gamma}_N \end{pmatrix} \tag{4.64c}$$

The construction of the right-hand-side vector $\overline{\boldsymbol{\beta}}$ is straightforward.

In (4.64) we recognize a system like (4.37) and (4.51), but of a *reduced order*. Hence, this implementation is more efficient with respect to the number of IVPs to be integrated and with respect to the algebraic system to be solved. The latter is even simpler if the BC are completely separated, i.e.,

$$B_{a2} = 0$$

Then $\overline{A}$ is a simple (block) upper triangular matrix, suggesting a computation of $\overline{\mathbf{s}}$ by (block) back-substitution. This, however, is just another application of the compactification algorithm, and would be judged unstable (in general), unless it happens that this is a special case of the decoupling algorithm of the previous section.

The latter, in fact, turns out to be the case. Indeed, the orthogonalization process here may be viewed as a special case of the process described in Section 4.4.1. Moreover, Theorem 3.107 implies that, if the BVP is well-conditioned, then $\overline{Y}_1(x_1)$ must generate the increasing solutions. Therefore, we may identify k and D_i of the previous section with $n - v$ and $\overline{\Gamma}_i$ here. The back substitution algorithm is further identified as the relevant part of the decoupling algorithm and the analysis of the previous section (and Section 6.2) shows that it is stable. We summarize these results in an algorithm and a theorem.

Algorithm: Stabilized march

Input: A BVP (4.9a, b) with separated BC and a mesh (4.31).
Output: Solution values $\mathbf{y}^h(x_1), \ldots, \mathbf{y}^h(x_N)$.

1. Determine $\overline{F}_1$ and $\boldsymbol{\alpha}_1$ by (4.29), (4.30).
2. FOR $i = 1, \ldots, N$ DO
    Integrate (4.60), (4.34) numerically, using an IVP code, obtaining $\overline{Y}_i^h(x_{i+1})$ and $\mathbf{v}_i^h(x_{i+1})$.
    Orthogonalize columns of $\overline{Y}_i^h(x_{i+1})$, as in (4.63), obtaining $\overline{F}_{i+1}$ and $\overline{\Gamma}_i$.
    Determine $\boldsymbol{\alpha}_{i+1}$ by (4.63).
3. Solve (4.64b) by (block) back substitution, obtaining $\overline{\mathbf{s}}^h$.
4. FOR $i = 1, \ldots, N$ DO
    $$\mathbf{y}^h(x_i) := \overline{F}_i \overline{\mathbf{s}}_i^h + \boldsymbol{\alpha}_i \ .$$
    □

Remark It is possible to show that Theorem 4.58 holds also for the stabilized march algorithm. [The proof is slightly more complicated than that shown for Theorem 4.58.] Thus, the method is stable in the sense that it yields, under very reasonable assumptions, agreeable roundoff error amplification. □

Note that in step 3 of the algorithm we need to solve N linear systems of order v; but only $B_{b2}\overline{F}_{N+1}\overline{\Gamma}_N$ is a possibly full matrix to be inverted. The other matrices encountered, $\overline{\Gamma}_i$, are all upper triangular, so this step is very fast. Most of the computational effort in this algorithm goes into the IVP integrations. The second largest expense is the orthogonalizations.

Note that in the above algorithm, we have assumed that the mesh (4.31) is fixed in advance. In practice, this algorithm should be modified so that the shooting points are chosen adaptively. One way to do that is to monitor the linear dependence of the columns of $\overline{Y}_i(x)$. (Instead, the growth of $\|\overline{Y}_i(x)\|$, i.e., the growth of $\|D_i\|$, could be monitored, essentially as discussed previously.) Another way is to use the IVP integrator only for a fixed (small) number of steps, letting the next shooting point be determined essentially by the (aggregated) adaptive step selection of the integrator. This introduces possibly more shooting points, but the integrator may perform better, using fewer steps overall.

The assumption of separated BC is crucial for Theorem 4.58 to hold for the method of stabilized march [because it relies on Theorem 3.107]. If $B_{a2} \neq 0$ then examples may be constructed where the compactification algorithm is unstable; for this, see Exercise 13 and a related discussion in Section 6.4.2.

Intermezzo. Let us reflect now on what we have done in this section. We have considered ways to implement multiple shooting by marching techniques. The desire to keep the fundamental modes from becoming numerically dependent has led us to consider reorthogonalization. Thus we have performed a QU decomposition of $Y_i(x_{i+1})$, using the obtained orthogonal matrix as starting values for the next shooting interval. The fact that the other matrix in the QU decomposition is upper triangular seemed to be merely a side effect.

But then, when stability is considered, it turns out that the *crucial* property of the decomposition is that the upper triangular matrix U induces decoupling. Moreover, the orthogonality of the Q-part of the decomposition has never been significantly used (except for the fact that this matrix is well conditioned), so *it* is the actual side effect! Other methods may be considered where an LU decomposition, for instance, is sought instead of QU decomposition.

The concept of decoupling is a rather fundamental one and we shall discuss and utilize it further in Chapter 6. The decoupling operation, which was done here in the framework of discretization, can be instead attempted at the ODE level, *before* discretization. One such method, using *Riccati transformations,* is discussed in the next section.

4.5 THE RICCATI METHOD

The last in our series of initial value methods for solving linear BVPs is related to what, for historical reasons, has been called *invariant imbedding*. A more adequate name for the method which we consider is the *Riccati method*. A common feature to all of the previous methods is that the discretization was first applied to the original BVP (4.9a, b), and only then some transformations were applied to resolve stability issues. In contrast, here the ODEs are transformed first and only then discretization is done. The hope is that the transformed system of ODEs is such that the IVPs are stable and hence can be stably and efficiently integrated.

In what follows, we first show how one can derive an invariant imbedding method. Then we give a general treatment of the Riccati method.

4.5.1 Invariant imbedding

Consider the system

$$y_1' = a_{11}y_1 + a_{12}y_2 \tag{4.65a}$$

$$0 < x < b$$

$$y_2' = a_{21}y_1 + a_{22}y_2 \tag{4.65b}$$

The system is subject to the (separated) BC

$$y_2(0) = c_1, \qquad y_1(b) = c_2 \tag{4.65c}$$

where c_1 and c_2 are given values. Since the problem is linear, the solution $\mathbf{y}(x)$ must depend linearly on $y_2(0)$ and $y_1(b)$. In particular, for $y_1(0)$ and $y_2(b)$ there is a relation of the form

$$y_1(0) = \alpha(b)y_1(b) + \beta(b)y_2(0) = c_2\alpha(b) + c_1\beta(b) \tag{4.66a}$$

$$y_2(b) = \gamma(b)y_1(b) + \delta(b)y_2(0) = c_2\gamma(b) + c_1\delta(b) \tag{4.66b}$$

In problems occurring in particle transport or wavelength computations in pipes, one is often interested in the terminal values of $\mathbf{y}(x)$ alone. For the BVP (4.65) this means the computation of $y_1(0)$ and $y_2(b)$. These are given by (4.66) if we know the coefficients α, β, γ and δ. The trick now is to use b as a *parameter*, thus *imbedding* the problem in a one-parameter family of BVPs. Under weak assumptions we may deduce that the coefficients in (4.66) are differentiable functions of b.

Differentiating (4.66a) with respect to b and using (4.65a) and (4.66b), we get

$$\begin{aligned} 0 &= \dot{\alpha} y_1 + \alpha \dot{y}_1 + \dot{\beta} y_2(0) = \dot{\alpha} y_1 + \alpha(a_{11}y_1 + a_{12}y_2) + \dot{\beta} y_2(0) \\ &= \dot{\alpha} y_1 + \alpha a_{11} y_1 + \alpha a_{12}(\gamma y_1 + \delta y_2(0)) + \dot{\beta} y_2(0) \\ &= (\dot{\alpha} + \alpha a_{11} + \alpha a_{12}\gamma) y_1 + (\alpha a_{12}\delta + \dot{\beta}) y_2(0) \end{aligned}$$

where the upper dot denotes differentiation with respect to b and all quantities are evaluated at b, unless otherwise specified. We now subject (4.66b) to a similar treatment and argue that the obtained coefficients of $y_2(0)$ and $y_1(x)$ at $x = b$ should equal zero. This yields the following ODEs for $b \geq 0$:

$$\dot{\alpha} = -(a_{11} + a_{12}\gamma)\alpha \tag{4.67a}$$

$$\dot{\beta} = -\alpha a_{12}\delta \tag{4.67b}$$

$$\dot{\gamma} = a_{21} + a_{22}\gamma - \gamma a_{11} - \gamma a_{12}\gamma \tag{4.67c}$$

$$\dot{\delta} = (-\gamma a_{21} + a_{22})\delta \tag{4.67d}$$

The way in which these ODEs are written is in anticipation for the more general case. The quantities appearing in (4.67) are all functions of our new independent variable, b.

The equations (4.66) also furnish us with initial values for (4.67a, b, c, d). We get

$$\alpha(0) = 1, \qquad \beta(0) = 0, \qquad \gamma(0) = 0, \qquad \delta(0) = 1 \tag{4.67e}$$

Therefore, by integrating the IVPs (4.67a, b, c, d), we can compute, in principle, the values $\alpha(b), \beta(b), \gamma(b)$, and $\delta(b)$. Then, a substitution in (4.66) completes the solution process.

We wish to stress that, unlike single shooting, for instance, the IVPs solved are not for the same ODEs (4.65a, b) as the BVP. In particular, a nonlinearity has been introduced in (4.67c) which, incidentally, is an ODE of *Riccati type*.

From the treatment above it is not clear why (4.67) constitutes a stable system. We defer a treatment of this question to the more general case treated in Section 4.5.2. Let us now demonstrate the method by an example.

Example 4.12

If we take

$$A(x) = \begin{bmatrix} 0 & 1 \\ 1 & 0 \end{bmatrix}$$

then (4.67a, b, c, d) read

$$\dot\alpha = -\gamma\alpha, \qquad \dot\beta = \alpha\delta, \qquad \dot\gamma = 1 - \gamma^2, \qquad \dot\delta = -\gamma\delta$$

The exact solution of these ODEs subject to (4.67e) is

$$\gamma(x) = \tanh(x), \qquad \alpha(x) = \delta(x) = \cosh^{-1}(x), \qquad \beta(x) = -\tanh(x)$$

so

$$y_1(0) = \frac{c_2}{\cosh(b)} - c_1 \tanh(b), \qquad y_2(b) = \frac{c_1}{\cosh(b)} + c_2 \tanh(b)$$

Note that the transformed IVP in this example is stable, even though the original ODE is not. Its kinematic eigenvalues are the eigenvalues of A, which are ±1. □

4.5.2 Riccati transformations for general linear BVPs

Let us consider now the ideas introduced above in a more general setting, which will enable us to attack a general BVP of the form (4.9a, b) and to obtain solution values also at intermediate points x.

For a given ODE

$$\mathbf{y}' = A(x)\mathbf{y} + \mathbf{q}(x)$$

let $T(x)$ be a linear transformation of the form

$$T(x) = \begin{bmatrix} I & 0 \\ R(x) & I \end{bmatrix} \qquad (4.68)$$

where $R(x)$ is an $(n-k) \times k$ matrix, to be defined later on. Define

$$\mathbf{w}(x) := T^{-1}(x)\mathbf{y}(x) \qquad (4.69)$$

Then $\mathbf{w}(x)$ satisfies

$$\mathbf{w}' = U(x)\mathbf{w} + \mathbf{g}(x) \qquad (4.70a)$$

where $U(x)$ (and $T(x)$) satisfy the Lyapunov equations [cf. (3.66)–(3.68)]

$$T' = AT - TU \qquad (4.70b)$$

and

$$\mathbf{g}(x) := T^{-1}(x)\mathbf{q}(x) \qquad (4.70c)$$

Now, we require that U be "block upper triangular" in the sense that its lower left block, corresponding in dimension to $R(x)$, is zero. Thus, we introduce decoupling. We write

$$U(x) = \begin{bmatrix} U^{11}(x) & U^{12}(x) \\ U^{21}(x) & U^{22}(x) \end{bmatrix} \updownarrow n-k \qquad \mathbf{g}(x) = \begin{bmatrix} \mathbf{g}^1(x) \\ \mathbf{g}^2(x) \end{bmatrix} \updownarrow n-k \qquad (4.71)$$

with U^{11} and U^{22} square submatrices and use a similar partition notation for other matrices and vectors in the sequel. Our requirement on $U(x)$ is that $U^{21}(x) \equiv 0$. This yields, upon substituting in (4.70b), the *Riccati differential equations* for $R(x)$,

$$R' = A^{21} + A^{22}R - RA^{11} - RA^{12}R \qquad (4.72)$$

and the block form of $U(x)$,

$$y_1(0) = \alpha(b)y_1(b) + \beta(b)y_2(0) = c_2\alpha(b) + c_1\beta(b) \quad (4.66a)$$

$$y_2(b) = \gamma(b)y_1(b) + \delta(b)y_2(0) = c_2\gamma(b) + c_1\delta(b) \quad (4.66b)$$

In problems occurring in particle transport or wavelength computations in pipes, one is often interested in the terminal values of $\mathbf{y}(x)$ alone. For the BVP (4.65) this means the computation of $y_1(0)$ and $y_2(b)$. These are given by (4.66) if we know the coefficients α, β, γ and δ. The trick now is to use b as a *parameter*, thus *imbedding* the problem in a one-parameter family of BVPs. Under weak assumptions we may deduce that the coefficients in (4.66) are differentiable functions of b.

Differentiating (4.66a) with respect to b and using (4.65a) and (4.66b), we get

$$0 = \dot{\alpha}y_1 + \alpha\dot{y}_1 + \dot{\beta}y_2(0) = \dot{\alpha}y_1 + \alpha(a_{11}y_1 + a_{12}y_2) + \dot{\beta}y_2(0)$$
$$= \dot{\alpha}y_1 + \alpha a_{11}y_1 + \alpha a_{12}(\gamma y_1 + \delta y_2(0)) + \dot{\beta}y_2(0)$$
$$= (\dot{\alpha} + \alpha a_{11} + \alpha a_{12}\gamma)y_1 + (\alpha a_{12}\delta + \dot{\beta})y_2(0)$$

where the upper dot denotes differentiation with respect to b and all quantities are evaluated at b, unless otherwise specified. We now subject (4.66b) to a similar treatment and argue that the obtained coefficients of $y_2(0)$ and $y_1(x)$ at $x = b$ should equal zero. This yields the following ODEs for $b \geq 0$:

$$\dot{\alpha} = -(a_{11} + a_{12}\gamma)\alpha \quad (4.67a)$$

$$\dot{\beta} = -\alpha a_{12}\delta \quad (4.67b)$$

$$\dot{\gamma} = a_{21} + a_{22}\gamma - \gamma a_{11} - \gamma a_{12}\gamma \quad (4.67c)$$

$$\dot{\delta} = (-\gamma a_{21} + a_{22})\delta \quad (4.67d)$$

The way in which these ODEs are written is in anticipation for the more general case. The quantities appearing in (4.67) are all functions of our new independent variable, b.

The equations (4.66) also furnish us with initial values for (4.67a, b, c, d). We get

$$\alpha(0) = 1, \quad \beta(0) = 0, \quad \gamma(0) = 0, \quad \delta(0) = 1 \quad (4.67e)$$

Therefore, by integrating the IVPs (4.67a, b, c, d), we can compute, in principle, the values $\alpha(b), \beta(b), \gamma(b)$, and $\delta(b)$. Then, a substitution in (4.66) completes the solution process.

We wish to stress that, unlike single shooting, for instance, the IVPs solved are not for the same ODEs (4.65a, b) as the BVP. In particular, a nonlinearity has been introduced in (4.67c) which, incidentally, is an ODE of *Riccati type*.

From the treatment above it is not clear why (4.67) constitutes a stable system. We defer a treatment of this question to the more general case treated in Section 4.5.2. Let us now demonstrate the method by an example.

Example 4.12

If we take

$$A(x) = \begin{bmatrix} 0 & 1 \\ 1 & 0 \end{bmatrix}$$

then (4.67a, b, c, d) read

$$\dot{\alpha} = -\gamma\alpha, \qquad \dot{\beta} = \alpha\delta, \qquad \dot{\gamma} = 1 - \gamma^2, \qquad \dot{\delta} = -\gamma\delta$$

The exact solution of these ODEs subject to (4.67e) is

$$\gamma(x) = \tanh(x), \qquad \alpha(x) = \delta(x) = \cosh^{-1}(x), \qquad \beta(x) = -\tanh(x)$$

so

$$y_1(0) = \frac{c_2}{\cosh(b)} - c_1 \tanh(b), \qquad y_2(b) = \frac{c_1}{\cosh(b)} + c_2 \tanh(b)$$

Note that the transformed IVP in this example is stable, even though the original ODE is not. Its kinematic eigenvalues are the eigenvalues of A, which are ± 1. □

4.5.2 Riccati transformations for general linear BVPs

Let us consider now the ideas introduced above in a more general setting, which will enable us to attack a general BVP of the form (4.9a, b) and to obtain solution values also at intermediate points x.

For a given ODE

$$\mathbf{y}' = A(x)\mathbf{y} + \mathbf{q}(x)$$

let $T(x)$ be a linear transformation of the form

$$T(x) = \begin{bmatrix} I & 0 \\ R(x) & I \end{bmatrix} \qquad (4.68)$$

where $R(x)$ is an $(n-k) \times k$ matrix, to be defined later on. Define

$$\mathbf{w}(x) := T^{-1}(x)\mathbf{y}(x) \qquad (4.69)$$

Then $\mathbf{w}(x)$ satisfies

$$\mathbf{w}' = U(x)\mathbf{w} + \mathbf{g}(x) \qquad (4.70a)$$

where $U(x)$ (and $T(x)$) satisfy the Lyapunov equations [cf. (3.66)–(3.68)]

$$T' = AT - TU \qquad (4.70b)$$

and

$$\mathbf{g}(x) := T^{-1}(x)\mathbf{q}(x) \qquad (4.70c)$$

Now, we require that U be "block upper triangular" in the sense that its lower left block, corresponding in dimension to $R(x)$, is zero. Thus, we introduce decoupling. We write

$$U(x) = \begin{bmatrix} U^{11}(x) & U^{12}(x) \\ U^{21}(x) & U^{22}(x) \end{bmatrix} \updownarrow n-k \qquad \mathbf{g}(x) = \begin{bmatrix} \mathbf{g}^1(x) \\ \mathbf{g}^2(x) \end{bmatrix} \updownarrow n-k \qquad (4.71)$$

with U^{11} and U^{22} square submatrices and use a similar partition notation for other matrices and vectors in the sequel. Our requirement on $U(x)$ is that $U^{21}(x) \equiv 0$. This yields, upon substituting in (4.70b), the *Riccati differential equations* for $R(x)$,

$$R' = A^{21} + A^{22}R - RA^{11} - RA^{12}R \qquad (4.72)$$

and the block form of $U(x)$,

$$U = \begin{bmatrix} A^{11}+A^{12}R & A^{12} \\ O & A^{22}-RA^{12} \end{bmatrix} \quad (4.73)$$

The transformed ODEs (4.70a) can be written in the decoupled form

$$\mathbf{w}^{1\prime} = U^{11}(x)\mathbf{w}^1 + U^{12}(x)\mathbf{w}^2 + \mathbf{g}^1(x) \quad (4.74a)$$

$$\mathbf{w}^{2\prime} = U^{22}(x)\mathbf{w}^2 + \mathbf{g}^2(x) \quad (4.74b)$$

and this is very much like the *continuous* analogue of (4.56a, b), as we discuss further in Chapter 6. The principle of the Riccati method is given below.

Outline: Riccati method

Input: A BVP (4.9a, b), initial value matrix $R(a)$
Output: Solution $\mathbf{y}(x)$

1. Find the transformation matrix $T(x)$ of (4.68), using (4.72).
2. Find solutions of (4.74b) by integrating in forward direction, from a to b.
3. Find solutions of (4.74a) by integrating in backward direction, from b to a.
4. Apply the inverse transformation (4.69) to find $\mathbf{y}(x)$.   □

Let us now describe how to choose the end values $R(a)$, $\mathbf{w}^2(a)$ and $\mathbf{w}^1(b)$ needed in the above outline, for the case of separated BC. Thus, let (4.9b) read

$$B_{a1}\mathbf{y}(a) = \boldsymbol{\beta}_1, \qquad B_{b2}\mathbf{y}(b) = \boldsymbol{\beta}_2 \quad (4.75a)$$

where

$$B_{a1} = (C \mid D) \quad (4.75b)$$

and D a nonsingular $k \times k$ matrix (see Exercise 4.16). Since by (4.69)

$$\mathbf{w}^2(a) = -R(a)\mathbf{y}^1(a) + \mathbf{y}^2(a)$$

the choice

$$R(a) := -D^{-1}C \quad (4.76a)$$

yields

$$\mathbf{w}^2(a) = D^{-1}\boldsymbol{\beta}_1 \quad (4.76b)$$

and so steps 1 and 2 of the method can be readily applied. After finding $R(b)$ and $\mathbf{w}^2(b)$, it is not difficult to find $\mathbf{w}^1(b)$ from (4.75a) and apply step 3.

Before proceeding to discuss the properties of this method, let us show how it corresponds to the invariant imbedding method described previously for the BVP (4.65a, b, c). This BVP is, of course, an instance of the BVP (4.9), (4.75) with $n = 2$, $k = 1$, $A_{ij} = a_{ij}$, $C = 0$, $D = 1$. Thus, the Riccati equations (4.67c) and (4.72) coincide. Moreover, (4.76a) yields $R(a) = 0$ here, so $R(x) \equiv \gamma(x)$. Further, we identify (4.67d) with (4.74b), which is integrated forward, i.e., $w_2(x) = c_1\delta(x)$. Finally, to integrate (4.74a) backward, consider the independent variable $t := b - x$, yielding

$$\frac{dw_1}{dt} = -U_{11}(t)w_1 - U_{12}(t)w_2(t) \tag{4.77}$$

Comparing this to (4.67a, b) we obtain [via (4.73)]

$$w_1(t) = c_2\alpha(t) + c_1\beta(t)$$

with $\alpha(t)$ a solution to the homogeneous part of (4.77) and $c_1\beta(t)$

$$= -\int_0^t \alpha(\tau)a_{12}(\tau)w_2(\tau)\,d\tau.$$

4.5.3 Method properties

We now discuss the properties of the Riccati method, outlined above, under the assumption that the BC are separated, i.e., (4.75) holds. Some disadvantages are apparent in (4.72): We have to solve a nonlinear *matrix* differential problem. This is potentially expensive or even infeasible, i.e., it is possible that the IVP (4.72), (4.76a) simply blows up for some c, $a < c < b$, as discussed further below. The question we address next is whether the method offers any advantages. The answer is positive: The instability problem which was most haunting with the single shooting method (and even with multiple shooting) is resolved here. Let us indicate why.

The method analysis is analogous to that of the stabilized march in Section 4.4.3. Using Theorem 3.107 we conclude that, provided the BVP is well-conditioned, range $\begin{bmatrix} I \\ R(a) \end{bmatrix}$ should induce initial values of increasing modes only. Hence the choice (4.76a) not only gives satisfactory initial values for $\mathbf{w}^2(a)$ in (4.76b) but also, in principle, a stable algorithm, in that the kinematic eigenvalues of $U^{11}(x)$ can be expected to have positive real parts.

The first k columns of the transformation matrix $T(x)$ of (4.68) should maintain the directions of the increasing modes. However, from a geometrical point of view, considering possible rotational activity of these solution directions, it is not clear whether or not this is true. To analyze this, let us consider a fundamental solution for (4.9a)

$$Y(x) = \begin{bmatrix} Y^{11}(x) & Y^{12}(x) \\ Y^{21}(x) & Y^{22}(x) \end{bmatrix}$$

such that

$$\text{range}\begin{bmatrix} Y^{11}(a) \\ Y^{21}(a) \end{bmatrix} = \text{range}\begin{bmatrix} I \\ R(a) \end{bmatrix}.$$

For stability of the Riccati method we need

$$\text{range}\begin{bmatrix} Y^{11}(x) \\ Y^{21}(x) \end{bmatrix} = \text{range}\begin{bmatrix} I \\ R(x) \end{bmatrix} \tag{4.78a}$$

But this implies that $Y^{11}(x)$ must be nonsingular, whence

$$R(x) = Y^{21}(x)[Y^{11}(x)]^{-1} \tag{4.78b}$$

Thus, we link the stability question of the method with the feasibility of integrating the nonlinear Riccati equation (4.72), starting with (4.76a). If difficulties arise, this would become apparent when integrating for $R(x)$.

Difficulties in integrating (4.72), (4.76a) may certainly occur, in general, and so the method may fail, unless a change in the algorithm is made. Such a change, sometimes referred to as a "change of the imbedding," would be to restart the integration of (4.72) at the trouble spot with a new (bounded) value of $R(x)$. This corresponds to calculating a new fundamental solution on the next subinterval, because the previous $[Y^{11}]^{-1}$ becomes unbounded. Of course, by this manipulation some merit of the algorithm, such as the straightforward calculation of $\mathbf{w}(x)$, is lost. Nevertheless, superposition may be used at each of the resulting subintervals, and a matching procedure similar to that for multiple shooting may be constructed. The resulting hybrid method bears some resemblance to the stabilized march procedure (Section 4.4.3), but there are some fundamental differences. Let us discuss them in the context of an example.

Example 4.13

Consider the ODE (cf. Example 3.12)
$$y' = \begin{bmatrix} -\lambda \cos 2\omega x & \omega + \lambda \sin 2\omega x \\ -\omega + \lambda \sin 2\omega x & \lambda \cos 2\omega x \end{bmatrix} y \quad 0 < x < \pi$$

A fundamental solution is given by
$$Y(x) = \begin{bmatrix} \cos \omega x & \sin \omega x \\ -\sin \omega x & \cos \omega x \end{bmatrix} \begin{bmatrix} e^{\lambda x} & 0 \\ 0 & e^{-\lambda x} \end{bmatrix}$$

from which we see that the (positive) parameter λ is "responsible" for growth and decay of modes, while ω controls the amount of rotation applied. Let us write the general solution as
$$\mathbf{y}(x) = Y(x) \begin{bmatrix} \alpha \\ \beta \end{bmatrix}$$

Now, considering (4.78b) (all quantities appearing there are now scalar), we realize that the solution of (4.72), (4.76a) is
$$R(x) = \frac{\alpha \cos \omega x \, e^{\lambda x} + \beta \sin \omega x \, e^{-\lambda x}}{\alpha \sin \omega x \, e^{\lambda x} - \beta \cos \omega x \, e^{-\lambda x}}$$

It is simple to check that $\alpha \neq 0$ is necessary for a well-conditioned problem. Hence $R(x)$ has poles at all points x for which
$$\frac{\beta}{\alpha} = e^{2\lambda x} \tan \omega x$$

From this expression we see that $O(\omega)$ restarts (or changes in imbedding) are needed for the solution of the Riccati equation. Away from the restart points, however, the integration should be fairly easy, as $R(x)$ is smooth here. More importantly, the integration of $R(x)$ is hardly affected by λ, regardless of its size. Thus, the Riccati method may overcome fast growth and decay of modes, but at the same time it could be plagued by rotational activity.

It is intriguing to note that the latter observation is contrary to the situation with multiple shooting techniques, even though the decoupling principle is so similar here and in the previous section. Let us consider two cases

Case (i): Set $\lambda = 1$ and let ω grow large.

Here the Riccati method performs poorly, as indicated above. On the other hand, the only effect on shooting techniques is that $O(\omega)$ steps are needed in the IVP integrations. The single shooting method may even be applied.

Case (ii): Set $\omega = 1$ and let λ grow large.
Here the Riccati method performs well, whereas multiple shooting techniques need $O(\lambda)$ shooting points. The reason is that to enable decoupling, a decent approximation of the fundamental modes is needed on the shooting subintervals, making their size $O(\lambda^{-1})$ as $\lambda \to \infty$. With the Riccati method, on the other hand, the decoupling is at the ODE level and so no such restriction arises.

□

The Riccati method is one of a larger class of methods which solve IVPs for transformed problems. Some of its desirable properties are its basic simplicity and the stability of (4.72), (4.74a, b). We defer a discussion of the latter to Section 10.4.1.

4.6 NONLINEAR PROBLEMS

In this section we return to the general nonlinear BVP (4.7a, b), briefly considered before in Section 4.1.3. Thus, the problem to be solved is

$$\mathbf{y}' = \mathbf{f}(x, \mathbf{y}) \qquad a < x < b \qquad (4.79a)$$

$$\mathbf{g}(\mathbf{y}(a), \mathbf{y}(b)) = \mathbf{0} \qquad (4.79b)$$

When applying single or multiple shooting discretizations to (4.79a, b), a nonlinear system of algebraic equations results. Below we consider using Newton's method for the solution of these nonlinear equations, first with single shooting and then with multiple shooting. At each Newton iteration we encounter the linear methods considered in previous sections. It turns out that in the nonlinear case, the single shooting method suffers from a new drawback, which is again alleviated by the multiple shooting method.

4.6.1 Shooting for nonlinear problems

Let us recall (4.7), (4.8). For any initial vector $\mathbf{s}$, let $\mathbf{y}(x; \mathbf{s})$ satisfy

$$\frac{d\mathbf{y}}{dx}(x; \mathbf{s}) = \mathbf{f}(x, \mathbf{y}(x; \mathbf{s})) \qquad a < x \qquad (4.80a)$$

$$\mathbf{y}(a; \mathbf{s}) = \mathbf{s} \qquad (4.80b)$$

The required $\mathbf{s} = \mathbf{s}^*$ is that for which the BC (4.79b) are satisfied, i.e.

$$\mathbf{g}(\mathbf{s}^*, \mathbf{y}(b; \mathbf{s}^*)) = \mathbf{0} \qquad (4.81)$$

From Theorem 3.16, the existence of isolated solutions for the BVP (4.79) corresponds to the existence of simple roots of (4.81).

In general $\mathbf{s}^*$ is not known, and therefore we may try to find it by an iterative scheme like Newton's method. We postpone a thorough treatment of methods for solving nonlinear equations to Chapter 8. Here we only briefly consider the basic Newton algorithm to obtain a solution of

$$\mathbf{F}(\mathbf{s}) = \mathbf{0} \tag{4.82a}$$

where

$$\mathbf{F}(\mathbf{s}) := \mathbf{g}(\mathbf{s}, \mathbf{y}(b, \mathbf{s})) \tag{4.82b}$$

Recall that, given an initial guess $\mathbf{s}^o$, Newton's method generates a sequence of iterates $\mathbf{s}^1, \mathbf{s}^2, \ldots, \mathbf{s}^m, \ldots$, hopefully converging to a solution $\mathbf{s}^*$ of (4.82a), by

$$\mathbf{s}^{m+1} := \mathbf{s}^m + \xi \tag{4.83a}$$

where ξ solves the linear system

$$F'(\mathbf{s}^m)\xi = -\mathbf{F}(\mathbf{s}^m) \tag{4.83b}$$

and F' is the $n \times n$ Jacobian matrix defined by

$$F'(\mathbf{s}) = \frac{\partial \mathbf{F}(\mathbf{s})}{\partial \mathbf{s}} \tag{4.84}$$

By (4.82) we have

$$F'(\mathbf{s}) = B_a + B_b Y(b) \tag{4.85}$$

where

$$B_a = \frac{\partial \mathbf{g}(\mathbf{u},\mathbf{v})}{\partial \mathbf{u}}, \quad B_b = \frac{\partial \mathbf{g}(\mathbf{u},\mathbf{v})}{\partial \mathbf{v}} \quad \text{at } \mathbf{u} = \mathbf{s}, \quad \mathbf{v} = \mathbf{y}(b; \mathbf{s}) \tag{4.86}$$

and $Y(x) \equiv \dfrac{\partial \mathbf{y}(x; \mathbf{s})}{\partial \mathbf{s}}$ is the $n \times n$ fundamental solution defined by

$$Y' = A(x)Y \quad a < x < b \tag{4.87a}$$

$$Y(a) = I \tag{4.87b}$$

where

$$A(x) = A(x, \mathbf{s}) \equiv \frac{\partial \mathbf{f}}{\partial \mathbf{y}}(x, \mathbf{y}(x; \mathbf{s})) \tag{4.87c}$$

In general, B_a, B_b, and $Y(x)$ all depend on $\mathbf{s}$. Also, note that in actual computation $\mathbf{y}(x; \mathbf{s})$ and $Y(x)$ are replaced by their numerical approximations obtained by solving the initial value problems.

Examining (4.83)–(4.85), we see that it is now apparent that the nonlinear problem solution proceeds by solving a sequence of linear problems by the method of superposition, as discussed in Section 4.2.

Example 4.14

Recall the seismic ray tracing problem, presented as Example 1.7 in Section 1.2. In the notation of (4.79), but with τ as the independent variable and x, y as dependent ones, we have

$$\mathbf{y} = (x, y, z, \xi, \eta, \zeta, T, S)^T$$

$$\mathbf{f} = (Sv\xi, Sv\eta, Sv\zeta, Su_x, Su_y, Su_z, Su, 0)^T$$

$$\mathbf{g} = (x(0) - x_0, y(0) - y_0, z(0) - z_0, T(0), \xi(0)^2 + \eta(0)^2 + \zeta(0)^2$$
$$-u(x(0), y(0), z(0))^2, x(1) - x_i, y(1) - y_i, z(1) - z_i)^T$$

with $a = 0$, $b = 1$. Thus for a given $\mathbf{s}$ we solve

$$\mathbf{y}' = \mathbf{f}, \qquad \mathbf{y}(0) = \mathbf{s}$$

for $\mathbf{y} = \mathbf{y}(\tau; \mathbf{s})$ and define

$$B_a = \begin{bmatrix} 1 & 0 & 0 & 0 & 0 & 0 & 0 & 0 \\ 0 & 1 & 0 & 0 & 0 & 0 & 0 & 0 \\ 0 & 0 & 1 & 0 & 0 & 0 & 0 & 0 \\ 0 & 0 & 0 & 0 & 0 & 0 & 1 & 0 \\ -2uu_x & -2uu_y & -2uu_z & 2\xi(0) & 2\eta(0) & 2\zeta(0) & 0 & 0 \\ 0 & 0 & 0 & 0 & 0 & 0 & 0 & 0 \\ 0 & 0 & 0 & 0 & 0 & 0 & 0 & 0 \\ 0 & 0 & 0 & 0 & 0 & 0 & 0 & 0 \end{bmatrix}$$

where all functions are evaluated at $\tau = 0$,

$$B_b = \begin{bmatrix} 0 & 0 & 0 & 0 & 0 & 0 & 0 & 0 \\ 0 & 0 & 0 & 0 & 0 & 0 & 0 & 0 \\ 0 & 0 & 0 & 0 & 0 & 0 & 0 & 0 \\ 0 & 0 & 0 & 0 & 0 & 0 & 0 & 0 \\ 0 & 0 & 0 & 0 & 0 & 0 & 0 & 0 \\ 1 & 0 & 0 & 0 & 0 & 0 & 0 & 0 \\ 0 & 1 & 0 & 0 & 0 & 0 & 0 & 0 \\ 0 & 0 & 1 & 0 & 0 & 0 & 0 & 0 \end{bmatrix}$$

and

$$A = \frac{\partial \mathbf{f}}{\partial \mathbf{y}} = \begin{bmatrix} Sv_x\xi & Sv_y\xi & Sv_z\xi & Sv & 0 & 0 & 0 & v\xi \\ Sv_x\eta & Sv_y\eta & Sv_y\eta & 0 & Sv & 0 & 0 & v\eta \\ Sv_x\zeta & Sv_y\zeta & Sv_y\zeta & 0 & 0 & Sv & 0 & v\zeta \\ Su_{xx} & Su_{xy} & Su_{xz} & 0 & 0 & 0 & 0 & u_x \\ Su_{yx} & Su_{yy} & Su_{yz} & 0 & 0 & 0 & 0 & u_y \\ Su_{zx} & Su_{zy} & Su_{zz} & 0 & 0 & 0 & 0 & u_z \\ Su_x & Su_y & Su_z & 0 & 0 & 0 & 0 & u \\ 0 & 0 & 0 & 0 & 0 & 0 & 0 & 0 \end{bmatrix}$$

Then, we solve the eight linear initial value problems (4.87), form $F'(\mathbf{s})$ of (4.85), solve the linear system (4.83b) for $\bm{\xi}$, add $\bm{\xi}$ to $\mathbf{s}$, and continue iterating until $|\bm{\xi}|$ is sufficiently small.

Note that if an analogue of reduced superposition (cf. Section 4.2.4) is used, then only three, not eight, additional initial value problems have to be solved at each iteration step. However, extra caution has to be exercised with the latter approach, as the stability properties of the integrated IVPs for $\mathbf{s}$ "far away" from $\mathbf{s}^*$ may vary significantly. Thus, no guarantee exists that the computed particular solution remains on the stable manifold. □

Example 4.15

Consider once again the BVP (4.5) in Example 4.1. Converting to a first-order system, we get

$$\begin{bmatrix} u \\ v \end{bmatrix}' = \begin{bmatrix} v \\ -e^u \end{bmatrix}$$

$$\begin{bmatrix} 1 & 0 \\ 0 & 0 \end{bmatrix} \begin{bmatrix} u \\ v \end{bmatrix}(0) + \begin{bmatrix} 0 & 0 \\ 1 & 0 \end{bmatrix} \begin{bmatrix} u \\ v \end{bmatrix}(1) = \begin{bmatrix} 0 \\ 0 \end{bmatrix}$$

so

$$F(s) = \begin{bmatrix} u(0;s) \\ u(1;s) \end{bmatrix}, \quad A(x;s) = \begin{bmatrix} 0 & 1 \\ -e^{u(x;s)} & 0 \end{bmatrix}$$

The first component of $F(s)$ is equal to 0 if the first component of s is 0, so we concentrate on the second component of s (call it σ) and the second component of $F(s)$.

We illustrate the shooting method for this easy problem by simply using the classical Runge-Kutta method of order 4 with a fixed step size h to integrate the initial value problems for $y(x;s^m)$ and $\dfrac{\partial y(x;s^m)}{\partial s}$ and applying Newton's method. The results are compared to the exact solution, which is

$$u(x) = -2 \ln \{ \cosh [(x-\tfrac{1}{2})\theta/2]/\cosh(\theta/4) \}$$

where θ is a solution of the nonlinear algebraic equation

$$\theta = \sqrt{2} \cosh(\theta/4)$$

This equation has, in fact, two solutions, corresponding to two solutions of the differential problem (4.5) (see Example 3.2). For the physically stable one we have $\theta \approx 1.5171646$, $\sigma \approx 0.54935$.

Table 4.6 displays some runs for different values of h and σ^o, where "itn" lists the number of iterations needed for convergence, and $e := u^h(.5) - u(.5)$.

Observe the expected fourth-order reductions in the error in the last column. Also, for this simple problem there are no convergence difficulties and the number of iterations is essentially independent of h. If the iteration is started with $\sigma^o = 10$, convergence to the second solution with $\sigma \approx 10.8469$ is obtained. □

The reader should realize from Example 4.14 that Newton's method may sometimes be quite cumbersome to use. In other practical examples, the matrices B_a, B_b, and especially A, may be much more complicated than in Example 4.14. In any case, the need to solve $n+1$ initial value problems at each iteration is not appealing. (Note,

TABLE 4.6 Single shooting computations

σ^o	h	itn	e	σ^o	h	itn	e
0	.2	6	$-.426-5$	5	.2	8	$-.426-5$
0	.1	5	$-.279-6$	5	.1	7	$-.279-6$
0	.05	5	$-.174-7$	5	.05	7	$-.174-7$
0	.025	5	$-.109-8$	5	.025	7	$-.109-8$

though, that n of these are linear, so an accurate implicit method may be used conveniently). One way to get around this is to use a secant method to solve (4.82) (cf. Section 2.3.3). A sometimes useful alternative is to approximate $A(\mathbf{s})$ by finite differences, e.g., the jth column $\partial \mathbf{F}/\partial s_j$ is approximated by

$$\varepsilon_j^{-1}(\mathbf{F}(\mathbf{s}+\varepsilon_j \mathbf{e}_j) - \mathbf{F}(\mathbf{s})) \tag{4.88}$$

where $\mathbf{e}_j$ is the jth unit vector and ε_j is small, but not too small. Frequently, one can choose $\varepsilon_j \approx s_j \sqrt{\varepsilon_M}$ with ε_M the machine precision. This still necessitates evaluating $\mathbf{F}$, i.e., solving an initial value problem, $n+1$ times per iteration (in fact, now the $n+1$ IVPs are all nonlinear). But the requirement to evaluate the Jacobian matrix $F'(\mathbf{s})$ explicitly is avoided; hence the need to evaluate $A(x)$ in (4.87c) is avoided as well.

4.6.2 Difficulties with single shooting

The stability difficulties encountered when one is using shooting for linear problems were analyzed in Section 4.2.3. Here, using the same method for nonlinear problems, we find that the same difficulties appear again. There is an additional concern, namely that when one is solving (4.80) with the "wrong" initial values $\mathbf{s}^m$, the solution $\mathbf{y}(x; \mathbf{s}^m)$ may not exist for all $a \leq x \leq b$ (i.e., it becomes unbounded before x reaches the right end b). In such a case it is not known how to correct $\mathbf{s}^m$, because the nonlinear iteration cannot be completed. Newton's method (or any of its variants) then fails. The following example is famous for its simple and yet extreme nature.

Example 4.16

Consider

$$u'' = \lambda \sinh \lambda u \qquad 0 < x < 1$$
$$u(0) = 0, \qquad u(1) = 1$$

where λ is a positive constant such that the problem's difficulty rapidly increases as λ is increased. The solution remains almost constant for $x \geq 0$ until x gets very close to 1, where it rises sharply to $u(1) = 1$. The missing initial value $s^* = u'(0)$ satisfies $s^* > 0$ and $|s^*|$ is very small. Now, if we "overshoot," i.e., solve the initial value problem

$$u'' = \lambda \sinh \lambda u$$
$$u(0) = 0, \qquad u'(0) = s$$

calling the solution $u(x;s)$, with $s > s^*$, then the solution is likely to blow up before $x=1$ is reached. This is because, as it turns out, the initial value solution $u(x;s)$ has a singular point at

$$x_s = \frac{1}{\lambda} \int_0^\infty \frac{d\xi}{\sqrt{s^2 + 2 \cosh \xi - 2}} \approx \frac{1}{\lambda} \ln \frac{8}{|s|}$$

Thus, to ensure that $x_s > 1$ and hence that the initial value integration can be successfully completed, we need

$$|s| \leq 8e^{-\lambda}$$

— a very severe restriction indeed. □

As can be seen from the basic Theorem 3.4, the longer the interval [a, b] is, the smaller the chance that the initial value solutions will reach $x = b$. A sufficient condition for solution existence over [a, b], which intimately involves the size of $b - a$, is expressed in the following theorem.

Theorem 4.89   Let $\mathbf{y}(x)$ be an isolated solution of (4.79) and assume that $\mathbf{f} \in C^{(2)}(D)$, where D is a tube of radius $r > 0$ about $\mathbf{y}(x)$,

$$D = \{ (x, \mathbf{w}): a \leq x \leq b, |\mathbf{w} - \mathbf{y}(x)| \leq r \} \qquad (4.89a)$$

Assume that $\mathbf{f}$ satisfies a Lipschitz condition in D,

$$|\mathbf{f}(x, \mathbf{v}) - \mathbf{f}(x, \mathbf{w})| \leq L |\mathbf{v} - \mathbf{w}| \qquad (x, \mathbf{v}), (x, \mathbf{w}) \in D \qquad (4.89b)$$

Then the initial value problem (4.80) has a solution on $a \leq x \leq b$ if $\mathbf{s} \in S_\rho(\mathbf{y}(a))$, where $S_\rho(\mathbf{y}(a))$ is a sphere about $\mathbf{y}(a)$ of radius ρ,

$$S_\rho(\mathbf{y}(a)) = \{ \mathbf{s}: |\mathbf{y}(a) - \mathbf{s}| < \rho \} \qquad (4.89c)$$

and

$$\rho = re^{-L(b-a)} \qquad (4.89d)$$

Also, the fundamental solution $Y(x; \mathbf{s}) = \dfrac{\partial \mathbf{y}}{\partial \mathbf{s}}(x; \mathbf{s})$ exists then and satisfies (4.87). □

In short, to guarantee successful completion of a nonlinear iteration from $\mathbf{s}^m$, it is necessary that $\mathbf{s}^m$ be already sufficiently close to $\mathbf{s}^* = \mathbf{y}(a)$, as indicated by (4.89c), (4.89d). Note that if L is large and/or r is not large, then ρ may easily be excessively small!

An initial value solution of (4.80) satisfying (4.89b) may have exponentially growing solutions, bounded only by

$$|\mathbf{y}(x; \mathbf{s})| \leq |e^{L(x-a)} \mathbf{s}| \qquad (4.90)$$

An inspection of (4.89d) and (4.90) then suggests that one way to try to deal with the difficulties of the shooting method is to restrict the size of the interval of integration of initial value problems. As long as b is close enough to a, $e^{L(b-a)}$ can be kept reasonably small. This again leads to the idea of multiple shooting, discussed next.

4.6.3 Multiple shooting for nonlinear problems

We have seen that the two major drawbacks of the single shooting method for nonlinear problems lead to the same basic cure: The length of intervals over which IVPs are integrated should be shortened. Thus we are led to a direct extension of the standard multiple shooting method of Section 4.3.

Let us then subdivide the interval [a, b] by a mesh $a = x_1 < x_2 < \cdots < x_{N+1} = b$ [as in (4.31)] and consider the initial value problems for $1 \leq i \leq N$,

$$\mathbf{y}' = \mathbf{f}(x, \mathbf{y}) \qquad x_i < x < x_{i+1} \qquad (4.91a)$$

$$y(x_i) = s_i \tag{4.91b}$$

denoting the solution of (4.91a, b) by $y_i(x; s_i)$ (see Fig. 4.2).

The unknowns to be found are the nN parameters

$$s^T = (s_1^T, s_2^T, \ldots, s_N^T)$$

[as in (4.36)], and they are to be determined so that the solution is continuous over the entire interval $[a, b]$ and so that the BC (4.79b) are satisfied. Thus, the desired solution is defined by

$$y(x) := y_i(x; s_i) \qquad x_i \le x \le x_{i+1}, \qquad 1 \le i \le N \tag{4.92a}$$

where the requirements on s are

$$y_i(x_{i+1}; s_i) = s_{i+1} \qquad 1 \le i \le N - 1 \tag{4.92b}$$

$$g(s_1, y_N(b; s_N)) = 0 \tag{4.92c}$$

So, defining

$$F(s) := \begin{bmatrix} s_2 - y_1(x_2; s_1) \\ s_3 - y_2(x_3; s_2) \\ \vdots \\ s_N - y_{N-1}(x_N; s_{N-1}) \\ g(s_1, y_N(b; s_N)) \end{bmatrix} \tag{4.93a}$$

we are left to solve a set of nN nonlinear algebraic equations

$$F(s) = 0 \tag{4.93b}$$

Suppose that Newton's method is used to solve (4.93). The process is exactly the same as described for single shooting in (4.83) and (4.84), except that the matrix $F'(s)$ now has the form similar to A in (4.37b, c),

$$F'(s) = \begin{bmatrix} -Y_1(x_2) & I & & & \\ & -Y_2(x_3) & I & & \\ & & \ddots & \ddots & \\ & & & -Y_{N-1}(x_N) & I \\ B_a & & & & B_b Y_N(b) \end{bmatrix} \tag{4.94}$$

Here, $Y_i(x) = Y_i(x; x_i, s_i)$ is the $n \times n$ fundamental solution defined by

$$Y_i'(x) = A(x) Y_i \qquad x_i < x < x_{i+1} \tag{4.95a}$$

$$Y_i(x_i) = I \qquad 1 \le i \le N \tag{4.95b}$$

with $A(x) = A(x; s)$, $B_a = B_a(s)$ and $B_b = B_b(s)$ defined, as in simple shooting, by (4.87c) and (4.86).

Thus, as before, Newton's method reduces the nonlinear problem to a sequence of linear problems which are solved by multiple shooting as described in Section 4.3. In the above description we have chosen the matrices F_i in (4.33) to be the unit matrix I, i.e., we use the standard multiple shooting scheme. The extension for more general F_i, which includes marching techniques (Section 4.4), is straightforward.

An extensive treatment of methods for solving nonlinear equations is given in Chapter 8. Here, however, note that the matrix $F'(\mathbf{s})$ is large and sparse (if N is not very small). Thus, a secant method can be inefficient in solving (4.93) because it is generally hard to take advantage of sparseness with such methods. On the other hand, with Newton's method, or the finite difference variant (4.88) which was designed to avoid forming $A(x)$, sparseness can be used to advantage. This finite difference variant is the usual way to proceed, and is now described in more detail.

On each shooting interval (x_i, x_{i+1}) we first compute an approximate solution of (4.91)

$$\mathbf{y}_i(x, \hat{\mathbf{s}}_i)$$

with

$$\mathbf{y}_i(x_i, \hat{\mathbf{s}}_i) = \hat{\mathbf{s}}_i$$

($\hat{\mathbf{s}}_i$ being the previously obtained, or initially guessed approximate value of $\mathbf{y}(x_i)$). Concurrently we compute n more solutions of (4.91a), written in matrix form as

$$\tilde{Y}_i(x, \mathbf{s}_i) \tag{4.96a}$$

with

$$\tilde{Y}_i(x_i, \hat{\mathbf{s}}_i) = (\hat{\mathbf{s}}_i | \ldots | \hat{\mathbf{s}}_i) + \sqrt{\varepsilon_M}\, I$$

Then we can find an *approximate* $Y_i(x_{i+1})$ from

$$Y_i(x_{i+1}) \approx \varepsilon_M^{-1/2}[\tilde{Y}_i(x_{i+1}, \hat{\mathbf{s}}_i) - (\mathbf{y}_i(x_{i+1}, \hat{\mathbf{s}}_i)| \ldots |\mathbf{y}_i(x_{i+1}, \hat{\mathbf{s}}_i))] \tag{4.96b}$$

Consider now the two major potential concerns which arise when applying the single shooting method. The first one has to do with the propagation of roundoff errors by the initial value integrations. While it was only treated in the linear case, both for single and for multiple shooting, it is obviously relevant for nonlinear problems as well. The second concern, occurring only for nonlinear problems, is that the initial value problems encountered may only have a solution on a small interval. This is reflected by Theorem 4.89, in particular see (4.89c, d).

We show next that the multiple shooting method, by using initial value integrations over smaller subintervals, improves upon this latter difficulty as well. If we transform (for theoretical insight only!) each problem (4.91) onto [0, 1] by letting

$$\Delta_i := x_{i+1} - x_i, \qquad t = \frac{x - x_i}{\Delta_i}, \qquad 1 \leq i \leq N \tag{4.97}$$

we obtain a problem of the general form (4.79):

$$\mathbf{y}_i'(t) = \mathbf{f}_i(t, \mathbf{y}_i(t)) \qquad 0 < t < 1, \quad 1 = 1, \ldots, N \tag{4.98a}$$

$$\mathbf{y}_{i+1}(0) - \mathbf{y}_i(1) = \mathbf{0}, \qquad i = 1, \ldots, N-1 \tag{4.98b}$$

$$\mathbf{g}(\mathbf{y}_1(0), \mathbf{y}_N(1)) = \mathbf{0} \tag{4.98c}$$

where $\mathbf{y}_i(t) := \mathbf{y}(x_i + t\Delta_i)$ and $\mathbf{f}_i(t, \mathbf{y}_i(t)) := \Delta_i \mathbf{f}(x_i + t\Delta_i, \mathbf{y}_i(t))$, $1 \leq i \leq N$. Applying Theorem 4.89 to (4.97) we obtain that the radius ρ of (4.89d) is replaced in the multiple shooting technique by

$$\hat{\rho} = re^{-L\Delta} \tag{4.99}$$

where

$$\Delta := \max_{1 \leq i \leq N}(x_{i+1} - x_i) \tag{4.100}$$

Thus, an exponential (!) improvement in the radius of the sphere S_ρ has been obtained: The initial value solutions of (4.91) are guaranteed to exist if

$$|\mathbf{y}(x_i) - \mathbf{s}_i| < \hat{\rho} \qquad i = 1, \ldots, N \tag{4.101}$$

This corresponds naturally to an enlarged domain of attraction for Newton's method.

Note that, with regard to this improved convergence property, a compactification algorithm like MSC discussed in Section 4.3 is as good as the full multiple shooting method MSH there. In fact, using compactification to decompose $F'(\mathbf{s})$ of (4.94) has a further advantage when $F'(\mathbf{s})$ is nearly singular, as discussed in Chapter 8. In that case, essentially, a generalized inverse of $F'(\mathbf{s})$ has to be computed, and it is hard to maintain the sparseness structure of $F'(\mathbf{s})$. With compactification, however, the condensed matrix resembles the simple shooting matrix and is only $n \times n$, so no such problem arises. Thus, a multiple shooting implementation with the compactification of Section 4.3.4 may not be discarded easily, despite its potential instability. At the same time, it may well be that $F'(\mathbf{s})$ becomes nearly singular *because* of the compactification algorithm. A number of codes have been written which employ this compactification idea, and in practice they perform well for many problems. Yet the aforementioned instability problem can be avoided if the linear system is decoupled first (cf. Section 7.2.4). Furthermore, a resulting condensed matrix can then be used to compute a generalized inverse of $F'(\mathbf{s})$ in more general cases. This latter possibility is lost with straightforward Gaussian elimination, which some codes use (see references).

There are a few more comments to be made on the practical implementation of multiple shooting. First, we remark that in principle the number of shooting points as well as their locations should remain fixed throughout the iteration process. However, stability considerations or convergence considerations [cf. (4.99)–(4.101)] may necessitate the insertion of an extra shooting point, say in the middle of the shooting interval (x_i, x_{i+1}), in order to keep the solution increment below TOL/ε_M or $\hat{\rho}$ in (4.101) small enough. This can be done during the iteration process as follows: Integrate both forward from x_i on and backward from x_{i+1} on (note that we have some approximate value at the latter point) and match at $x = (x_i + x_{i+1})/2$.

This idea of *bidirectional shooting* is sometimes used right from the start. Note that if we choose the shooting points according to stability arguments only, we would need fewer of them than for one directional shooting. The price to be paid for this is the more complicated structure of $F'(\mathbf{s})$; for example, if $\mathbf{s}_3$ is such a bidirectional shooting point then the first and second terms of $\mathbf{F}(\mathbf{s})$ in (4.93a) are replaced by $\mathbf{y}_3(x_2, \mathbf{s}_3) - \mathbf{y}_1(x_2, \mathbf{s}_1)$. Then, as it turns out, matching at a bidirectional shooting point now involves two matrices which differ from the identity in general.

A second remark is that we may relax on the tolerance used during the iteration process. Indeed, the bulk of the efforts in obtaining a sufficiently accurate approximation to $\mathbf{y}(x)$ goes into computing the first few iterates. An intuitively attractive idea much related to what normally is done with methods discussed in Chapter 5 (see also Chapter 9) is to start off with a crude tolerance and try to get an approximation for this tolerance first. This in turn is an excellent initial value for a Newton process where a finer tolerance is used, etc.

As a last remark, note that the decoupling idea may also be employed more directly in the nonlinear case, and not just as a means to solve the linear system (4.94). This can be done as follows: Given a matrix $\tilde{Y}_{i-1}(x_i)$ as in (4.96b) (but $i-1$ replacing i), perform a QU-decomposition and use the orthogonal matrix for the perturbation directions at the next interval. In more detail, suppose

$$\tilde{Y}_{i-1}(x_i) = Q_i U_i$$

then instead of (4.96a) we use

$$\tilde{Y}_i(x_i, \hat{\mathbf{s}}_i) = (\hat{\mathbf{s}}_i | \ldots | \hat{\mathbf{s}}_i) + \sqrt{\varepsilon_M} Q_i$$

Of course, this choice has consequences for the variables in the Newton process. However, their transformation is closely related to what we say in Section 4.4. The advantages are obvious: we obtain a simple sparse matrix that can, e.g., be compactified in a stable way. All aspects mentioned above are implemented in the code MUSN, described in Appendix A.

Example 4.17

Consider Example 4.16 again. As an initial profile we choose

$$y = \frac{\lambda \sin h\lambda x}{\sin h\lambda}$$

If $\lambda = 5$ and we let the shooting points be 0, .2, .4, .6, .8, .85, .9, .95, 1, for an accuracy of 10^{-6} the code MUSN needs 2634 function calls when the tolerance is initially chosen equal to 10^{-2}, then 10^{-4} and only then 10^{-6} (where the result for 10^{-2} is used as an initial approximation for 10^{-4}, etc.). Using 10^{-6} directly we have 4505 function calls. If the tolerance is decreased to 10^{-10}, these numbers are 9679 function calls and 25,489 function calls, respectively (in the former case we choose $TOL = 10^{-2}$, 10^{-4}, 10^{-8}, 10^{-10}). This problem gets harder the larger λ is. As can be seen from linearization already around the initial approximation, there is a rapid increase of the increment of the fundamental solution over an interval. Moreover, the solution is more active for x values close to 1 than to 0. Hence it is not very easy to pick the appropriate shooting points in advance. In such a situation it might be useful to have a device that picks a new shooting point when this growth is beyond a certain threshold, K, say. For instance, if we choose $\lambda = 10$, we will find that we have to invest 50 percent of the points on $[0, .972]$ and 50 percent on $[.972, 1]$ alone! The sensitivity with respect to K is clearly demonstrated by the fact that MUSN takes 37 shooting points in this second interval when $K = 1000$ and 46 when $K = 100$. □

Example 4.18

A famous example for multiple shooting is the optimal reentry problem Example 1.21. As an initial guess for the Newton process employing MUSN, we use a sufficiently good approximation with $TOL = 10^{-1}$. How to obtain a "sufficiently good" approximation for Newton convergence is not a trivial question at all, but we concentrate here on other aspects. If we let the code determine the shooting points (increment $\leq TOL/\varepsilon_M$, $\varepsilon_M \approx 10^{-16}$) then we obtain the following table:

TABLE 4.7

Newton iterations	7	5	4	4
Shooting points	6	9	20	31
TOL	10^{-5}	10^{-6}	10^{-8}	10^{-9}

For a fixed tolerance $TOL = 10^{-5}$ and equispaced shooting points we obtain the Table 4.8.

TABLE 4.8

Newton iterations	6	5	4
Shooting points	12	23	40

Hence, we may cautiously conclude that increasing the number of shooting points improves the performance of the Newton process. On the other hand, choosing them in advance may not give a proper distribution as far as activity of the modes of the variational system is concerned. Hence we slightly favor the former type of choice. □

EXERCISES

1. Describe shooting schemes for the second-order ODE (4.1a) under the following BC:
   (a) $u(a) = \beta_1$, $\alpha_1 u(b) + \alpha_2 u'(b) = \beta_2$
   (b) $\alpha_1 u(a) + \alpha_2 u'(a) = \beta_1$, $u(b) = \beta_2$
   (c) $\alpha_1 u(a) + \alpha_2 u'(a) = \beta_1$, $\gamma_1 u(b) + \gamma_2 u'(b) = \beta_2$
   In all cases, assume that the BVP is well-conditioned and use only two IVPs.

2. Consider Example 1.4. Problem (1.10) for f can be solved by shooting *without* converting it to a standard form. Describe a scheme to do this.

3. The superposition method is described in the text for two-point boundary value problems (4.9a, b). Extend the method to multipoint boundary value problems (1.5a, e).

4. Consider the problem
$$u''' = 2u'' + u' - 2u \qquad 0 < x < b$$
$$u(0) = 1, \qquad u(b) - u'(b) = 0, \qquad u(b) = 1$$

   Let $b = 100$.

   (a) Convert the ODE into a first-order system, and find its fundamental solution satisfying $Y(0) = I$.

   (b) Using this exact fundamental solution, construct (approximately) the system $Q\hat{s} = \beta$ of (4.13). Show that the resulting Q is extremely ill-conditioned and that, therefore, s cannot be accurately obtained by the superposition method.

(c) Show that the BVP itself is well-conditioned, by constructing $\widetilde{Q} = B_a \Psi(a) + B_b \Psi(b)$ and showing that $\widetilde{Q}$ is well-conditioned. Here $\Psi(x)$ is the well-scaled fundamental solution

$$\Psi(x) = \begin{pmatrix} e^{-x} & e^{x-b} & e^{2(x-b)} \\ -e^{-x} & e^{x-b} & 2e^{2(x-b)} \\ e^{-x} & e^{x-b} & 4e^{2(x-b)} \end{pmatrix}$$

(d) Solve the problem numerically for $b = 1, 10$ and 100. What are your conclusions?

5. Solve the problem of the previous exercise again, this time performing the numerical integrations from b to 0. Explain the observed results.

6. Show that the reduced superposition matrix $\overline{Q}$ of (4.28) is nonsingular if the BVP (4.9a), (4.24) has a unique solution.

7. Consider a BVP (4.9a, b), where $\text{rank}(B_b) = \nu < n$. Show that the BC can be converted to the form (4.24) (and therefore, reduced superposition can be applied).

8. Consider the P-SV problem of application 1.8. The BVP is given in (1.21). Assuming, as in Example 4.7, a piecewise constant medium, construct the linear system (4.37b, c) for this case.

9. Prove Theorem 4.42. *Hints:* For part (a), consider (4.37) with unit vectors in place of $\hat{\boldsymbol{\beta}}$. Use part (a) for part (b). For part (c) show that

$$\sum_{j=2}^{N+1} e^{-\alpha|x - x_j|} \leq \hat{K} \cdot \underline{h}^{-1} \int_a^b e^{-\alpha|x-t|} dt$$

for some $\alpha > 0$, $a \leq x \leq b$, $\underline{h} := \min_{1 \leq i \leq N}(x_{i+1} - x_i)$ and $\hat{K} \approx 1$ an appropriate constant.

10. Show that the inverse of the multiple shooting matrix $\mathbf{A}$ of (4.37b, c) is equal to $\text{diag}(F_1^{-1}, \ldots, F_N^{-1})$ times the matrix given in (4.42a).

11. Theorem 4.42 is formulated for the exact $\mathbf{A}$, without discretization errors. Let us introduce discretization errors as well, bounded by an integration tolerance TOL, and consider $\mathbf{A}^h$ of the standard multiple shooting algorithm.

    (a) Let G_{ij} be the discrete Green's function, as in (4.16), (4.17). Show

    $$(\mathbf{A}^h)^{-1} = \begin{pmatrix} G_{12} & \cdots & G_{1N} & Y_1^h(x_1)(Q^h)^{-1} \\ \vdots & & \vdots & \vdots \\ G_{N2} & \cdots & G_{NN} & Y_1^h(x_N)(Q^h)^{-1} \end{pmatrix}$$

    (b) Let κ be the conditioning constant of the BVP and suppose $G_{ij} = G(x_i, t_j)(I + E_{ij})$, $Y_1^h(x_i)(Q^h)^{-1} = Y(x_i; a)Q^{-1}(I + E)$, where $\|E\|$, $\|E_{ij}\| \leq C$. Then show that

    $$\|(\mathbf{A}^h)^{-1}\| \leq \text{const }(1 + C)N\,\kappa$$

    Furthermore, show that in the case of exponential dichotomy the factor N in this bound can be replaced by $\underline{h}^{-1}$, even on an interval of infinite length.

12. Following (4.43) we have assumed $|s| \sim 1$, in order to obtain (4.44). Yet, it is s^h, not s, which appears on the right-hand side of (4.41). Show that $|s^h| \sim 1$ too.

13. Describe how the stabilized march algorithm with compactification would perform for the following BVP

Section 4.6   Exercises

$$\mathbf{y}' = \begin{bmatrix} -10 & 0 & 0 \\ 0 & -20 & 0 \\ 20 & 0 & 10 \end{bmatrix} \mathbf{y} + \begin{bmatrix} 10 \\ 20 \\ -30 \end{bmatrix} \qquad 0 < x < b$$

$$y_1(0) + y_3(b) = 2, \quad y_2(0) + y_2(b) = 2, \quad y_3(0) = 1$$

which has the solution $\mathbf{y} = (1, 1, 1)^T$. This BVP is well-conditioned.

*14. Let (4.9a) have a dichotomy constant K. Let there be p "nondecreasing" modes and $(n - p)$ "nonincreasing" modes, denoted by Φ^1 and Φ^2 respectively.

(a) Show that the dichotomy implies that there is a suitable BC operator $\hat{B}$, $\hat{B}\mathbf{y} := \hat{B}_a \mathbf{y}(a) + \hat{B}_b \mathbf{y}(b)$, rank $(\hat{B}_a) = (n - p)$, rank $(\hat{B}_b) = p$, such that

$$\hat{B}_a \Phi^1(a) = 0, \quad \hat{B}_b \Phi^1(b) = I_p$$
$$\hat{B}_a \Phi^2(a) = I_{j-p}, \quad \hat{B}_b \Phi^2(b) = 0$$

and that these BC induce a BVP for each of these modes with conditioning constant bounded by K.

(b) Let

$$\mathbf{y}_{i+1} = \mathbf{y}_i + h\Phi(x_i; \mathbf{y}_i; h)$$

be an explicit one-step approximation for (4.9a) (cf. Section 2.7). Define the following perturbation

$$\delta_{\mathbf{y}_i}(K) := \Phi(x_i; \mathbf{y}_i; h) - A(x)\mathbf{z}(x), \quad x_i \le x \le x_{i+1}$$

where $\mathbf{z}(x)$ is a $C^{(1)}$ function with

$$\mathbf{z}(x_i) = \mathbf{y}_i$$

Show that

$$\mathbf{z}' = A\mathbf{z} + \delta_{\mathbf{y}_i}, \quad x_i < x < x_{i+1}$$

and also that the local truncation error at x_{i+1} is given by $\delta_{\mathbf{y}_i}(x_{i+1})$.

(c) Let $\varsigma(x)$ be a basis mode as in (a). Denote

$$\|\delta_\varsigma\| = \max_{x \in [a,b]} |\delta_\varsigma(x)|$$

Then show that there exists a discrete approximation to ς such that

$$|\varsigma(x_i) - \varsigma_i| \le K \|\delta_\varsigma\|$$

(d) Let Φ_i denote the discrete approximation to Φ composed of modes as in (c). Then show that if the local error of each of the various modes is bounded by TOL we have

$$|\Phi(x_i) - \Phi_i| < K\ TOL$$

(e) If $|\Phi^{-1}(x_i)|\ K\ TOL$ is sufficiently smaller than 1, then for the discrete fundamental solution [cf (4.16)] we have $|G_{ij}| \le CG(x_i, x_j)$, C a moderate constant.

15. Using a multiple shooting code solve the SH and P-SV problems of Example 5 in Section 1.2 for the following earth model: $\alpha(z)$, $\beta(z)$, and $\rho(z)$ are piecewise linear functions whose values at break points are given in Table 4.9.

TABLE 4.9

z(km)	α (km/s)	β (km/s)	ρ (gm/cm^3)
0	2.50	1.44	1.39
4	5.00	2.89	2.28
4	6.00	3.46	2.58
≥ 4.924	6.00	3.46	2.58

(Thus we have a constant slope for $0 \leq z \leq 4$, a discontinuity in the coefficients at $z = 4$, and a uniform medium below.) The solution is to be evaluated at $z = 1.012$ and at $z = 4.924$. Try to obtain wave solutions for the phase velocities $w/k = 0.5, 1.5, 2.5, 3.5$ for each of the frequencies $w = \pi, 2\pi, 5\pi, 10\pi$. What do you observe?

16. Suppose that the BVP with separated BC (4.9a), (4.75a) has a unique solution. Show that the BVP can be written such that B_{a1} satisfies (4.75b) with D nonsingular.

17. Consider the nonlinear BVP

$$u'' = 15(u^2 - x^2)u' \qquad -1 < x < 0$$
$$u(-1) = 0.96 \qquad u(0) = 0.001$$

It turns out that this BVP has 3 isolated solutions and that the corresponding IVP with $u(-1) = 0.96$, $u'(-1) = \sigma$ is not very sensitive to the value of σ.

Using the single shooting method, find all three solutions and graph them.

18. In the multiple shooting method presented in Section 4.6.3 there are two basic operations: *discretization*, i.e., reduction by (4.91) of the problem to that of solving (4.93), and then *linearization*, i.e., Newton's iteration. Consider now *reversing* the order of these two operations. Thus, quasilinearization (Section 2.3.4) is applied to the BVP first, and the resulting linear BVP is solved by the standard multiple shooting technique.

(a) Show that precisely the same results are obtained, i.e., linearization and discretization commute here.

(b) Write down an algorithm for standard multiple shooting to solve (4.79), using the algorithm given in the text.

(c) Design an algorithm to solve nonlinear BVPs, using the stabilized march technique. (Note, however, that the convergence properties of the latter nonlinear algorithm are not necessarily the same as those of standard multiple shooting. Why?)

*19. Use an available multiple shooting code to solve the following problems:

(a) Example 1.4. Observe the need for more and more shooting points as the Reynolds number R increases. For which value of R does the code appear to be inefficient?

(b) Example 1.5 with the BC (1.13d). You will have to figure out what to do near $x = 0$ (cf. Section 11.4.1).

(c) Example 1.9. Attempt to solve for values of θ near and away from 0.

(d) Example 1.10 in the form (1.24a, b, c, d). Construct a BVP with the change of dependent variables $z_i = \log y_i$ and solve it numerically. Do you observe a difference in conditioning?

(e) Example 1.21. You will have to figure out a good initial guess for Newton's method to converge.

* These are more difficult exercises.

5

Finite Difference Methods

The methods considered in this chapter are conceptually different from those of the previous chapter. Here, no initial value problems are explicitly integrated. Rather, an approximate solution representation is sought over the entire interval of interest. Thus, these methods are sometimes referred to as *global methods*. We shall see, nonetheless, that there is a close theoretical relationship between these methods and initial value techniques. This is particularly highlighted in Chapters 6 and 7.

Compared to initial value methods, the class of finite difference methods presented in this chapter is more uniform and admits a more unified theory. Also, the development here must begin with more basic concepts, not assuming that any major part of the process is already taken care of.

5.1 INTRODUCTION

The basic steps of a finite difference method are outlined as follows.

Outline: Finite difference method

Input: A BVP
Output: An approximate solution $\mathbf{y}_\pi(x)$

1. Choose a mesh

$$\pi: a = x_1 < x_2 < \cdots < x_N < x_{N+1} = b \tag{5.1}$$

Approximate solution values are then sought at these mesh points x_j.

2. Form a set of algebraic equations for the approximate solution values by replacing derivatives with difference quotients in the differential equations and boundary conditions that the exact solution satisfies.

3. Solve the resulting system of equations for the approximate solution. This gives a set of discrete solution values $\mathbf{y}_i \equiv \mathbf{y}_\pi(x_i)$ and, optionally, interpolation can be used to construct $\mathbf{y}_\pi(x)$ for any $x \in [a,b]$. □

A simple illustration of the above process is given in Section 5.1.1. We give there a simple method for a second-order linear problem using a uniform mesh. This gives us the opportunity to introduce some concepts on a simple structure, as well as to discuss a model problem frequently occurring in the solution of partial differential equations (PDEs).

However, the method of Section 5.1.1 is not easy to generalize. To be able to come up with general methods, we consider in most of this chapter first-order systems of ODEs. Two simple difference schemes for linear ODEs are introduced in Section 5.1.2. These are the *midpoint* scheme and the *trapezoidal* scheme of (2.139), (2.140). Their application for nonlinear problems is described in Section 5.1.3.

Let us recall from Section 2.7 that the midpoint and trapezoidal schemes are *one-step* schemes. A unified theory of stability and convergence for one-step schemes is given in Section 5.2 for linear and nonlinear problems.

A disadvantage of the basic schemes introduced in this section is that they all have only second-order accuracy. This is not very efficient for many applications, where higher-order methods can be exploited to obtain accurate answers on relatively coarse meshes. Two types of extensions are possible. The first is to define families of higher-order methods, of which the basic schemes of Section 5.1.2 are the simplest members. This is done in Section 5.3 for first-order ODEs, where implicit Runge-Kutta schemes are introduced. Their relationship with certain collocation methods is explored in Section 5.4, and a convergence analysis is given.

A second approach towards obtaining higher-order methods is to take a basic scheme and to accelerate its convergence by applying extrapolation or deferred correction to it, repeatedly if necessary. These techniques are considered in Section 5.5.

The treatment in Sections 5.2–5.5 is for first-order systems of ODEs. In Section 5.6 we consider higher-order ODEs. Although many different methods have been proposed in the literature, we consider only an extension of the simple second-order scheme of Section 5.1.1, because of its usefulness as a special-purpose scheme and extendability to PDEs, and collocation, because of its generality and systematic presentation.

In the final section of this chapter, we briefly consider finite element methods. A number of books and *many* papers are available on this topic. However, we feel that the main power of finite element methods is in PDEs, rather than ODEs. Hence we give in this book a brief description of some main ideas only.

In this chapter we use, unless explicitly noted otherwise, max or sup norms. We also use c as a generic constant.

5.1.1 A simple scheme for a second-order problem.

Consider a linear scalar boundary value problem, as in (4.1)

$$u'' - p(x)u' - r(x)u = q(x) \qquad 0 < x < 1 \qquad (5.2a)$$

$$u(0) = \beta_1, \qquad u(1) = \beta_2 \qquad (5.2b)$$

where $p(x)$, $r(x)$, and $q(x)$ are continuous functions on $[0, 1]$. We illustrate a simple difference scheme on a uniform mesh for this problem: Thus, in (5.1) we take $x_i = (i-1)h$, $i = 1, 2, \ldots, N+1$, where $h = 1/N$ (so π is a function of h), and we seek a mesh function $u_\pi \equiv \{u_j\}_{j=1}^{N+1}$ such that $u_i \sim u(x_i)$, $i = 1, 2, \ldots, N+1$, where $u(x)$ is the exact solution of (5.2), assumed to exist.

By Taylor series,

$$u(x_i \pm h) = u(x_i) \pm hu'(x_i) + \frac{h^2}{2}u''(x_i) \pm \frac{h^3}{6}u'''(x_i) + O(h^4)$$

so we can express first and second derivatives as

$$u'(x_i) = \frac{u(x_{i+1}) - u(x_{i-1})}{2h} + O(h^2) \qquad (5.3a)$$

$$u''(x_i) = \frac{u(x_{i+1}) - 2u(x_i) + u(x_{i-1})}{h^2} + O(h^2) \qquad (5.3b)$$

Using (5.3), we can approximate the ODE (5.2a) at $x = x_i$ by

$$\frac{u_{i+1} - 2u_i + u_{i-1}}{h^2} - p(x_i)\frac{u_{i+1} - u_{i-1}}{2h} - r(x_i)u_i = q(x_i) \qquad 2 \le i \le N \qquad (5.4a)$$

and the BC (5.2b) give

$$u_1 = \beta_1, \qquad u_{N+1} = \beta_2 \qquad (5.4b)$$

This completes the second step of the outline above.

For the third step of the general method outline note that (5.4a, b) are $N+1$ linear equations for $N+1$ unknowns,

$$\mathbf{A}\mathbf{u}_\pi = \mathbf{b} \qquad (5.5a)$$

where $\mathbf{A}$ is a *tridiagonal* matrix given by

$$\mathbf{A} = \begin{bmatrix} a_1 & c_1 & & & & \\ b_2 & a_2 & c_2 & & & \\ & b_3 & a_3 & c_3 & & \\ & & & \cdot & & \\ & & & b_N & a_N & c_N \\ & & & & b_{N+1} & a_{N+1} \end{bmatrix} \qquad (5.5b)$$

$$a_i = -(\frac{2}{h^2} + r(x_i)), \quad b_i = \frac{1}{h^2} + \frac{p(x_i)}{2h}, \quad c_i = \frac{1}{h^2} - \frac{p(x_i)}{2h}, \quad 2 \leq i \leq N$$

$$a_1 = a_{N+1} = 1, \quad c_1 = b_{N+1} = 0$$

$$\mathbf{u}_\pi = (u_1, u_2, \ldots, u_{N+1})^T, \quad \mathbf{b} = (\beta_1, q(x_2), \ldots, q(x_N), \beta_2)^T \qquad (5.5c)$$

Before solving (5.5), we could perform row equilibration by multiplying the 2nd through the Nth equations by h^2. The $(i+1)^{\text{st}}$ row of the equilibrated $\mathbf{A}$ would then have

$$1 + \frac{h}{2} p(x_i) \qquad -2 - h^2 r(x_i) \qquad 1 - \frac{h}{2} p(x_i)$$

as its nonzero elements.

Finite difference schemes generally lead to systems of algebraic equations involving banded matrices, and in this particular case the matrix $\mathbf{A}$ is tridiagonal. Gaussian elimination strategies which utilize this special structure are very simple and efficient, cf. Section 2.2. No pivoting is needed if $\mathbf{A}$ is diagonally dominant, which occurs here if $h \leq \frac{2}{\|p\|}$ and $r \geq 0$ (why?).

Thus the approximate mesh function $\mathbf{u}_\pi$ exists, at least under the restrictions above. But does it really approximate the values $u(x_i)$, and how well? To answer this we turn to the equations (5.4), which are supposed to approximate the BVP (5.2). Since we cannot ask how well $\mathbf{u}_\pi$ satisfies the ODE (5.2a) (the mesh function is only defined at a discrete set of points), we ask how well the exact solution $u(x)$ satisfies the difference equations (5.4).

It is convenient here to use simple operator notation. Thus, the *differential operator* L for (5.2a) is

$$Lu(x) := u''(x) - p(x) u'(x) - r(x) u(x) \qquad (5.6a)$$

while the corresponding *difference operator* L_π for (5.4a) is

$$L_\pi u(x) := \frac{u(x+h) - 2u(x) + u(x-h)}{h^2} - p(x) \frac{u(x+h) - u(x-h)}{2h} - r(x) u(x) \qquad (5.6b)$$

So, for instance, we may write (5.4a) as

$$L_\pi u_i = q(x_i) \qquad 2 \leq i \leq N \qquad (5.6c)$$

For any smooth function $v(x)$, the *local truncation error* $\tau_i[v]$ is defined by

$$\tau_i[v] := L_\pi v(x_i) - Lv(x_i) \qquad 2 \leq i \leq N \qquad (5.6d)$$

By our construction [see (5.3)] it is clear that if $v \in C^{(4)}[0,1]$, then

$$\tau[v] := \max_{2 \leq i \leq N} |\tau_i[v]| \leq ch^2, \qquad c = \text{constant} \qquad (5.7)$$

Thus $\tau[v] \to 0$ as $h \to 0$. Furthermore, there is no error in the approximation of the BC in (5.4b). We may then say that the difference operator L_π approaches the differential operator L in the limit as $h \to 0$. A difference method satisfying this is called *con-*

sistent. From (5.7) we see that the particular method described here is not only consistent but is also of *order* 2; i. e., $\tau[v]$ approaches 0 as fast as h^2.

Consistency is necessary, but not sufficient, to get discrete *convergence*, i. e., that the approximate solution values at a point x_i approach $u(x_i)$ as $h \to 0$ with ih kept fixed. Since consistency ensures that the difference operator is "close" to the differential operator, all we need in addition is to be able to bound the solution values in terms of the values of the operator defining it. For the solution $u(x)$ of (5.2), this is assumed as part of the well-posedness of the problem to be solved. Indeed, from Section 3.2 we have

$$\|u\| \le \kappa \max\{|\beta_1|, |\beta_2|, \|Lu\|\} \tag{5.8}$$

where the BVP is well-conditioned if κ is of "moderate" size. We expect the finite difference scheme to preserve this conditioning (at least for h small) and say that it is *stable* if for all meshes π with h sufficiently small and all corresponding mesh functions $\mathbf{v}_\pi$,

$$|v_i| \le K \max\{|v_1|, |v_{N+1}|, \max_{2 \le j \le N} |L_\pi v_j|\} \quad 1 \le i \le N+1 \tag{5.9}$$

with K a constant (independent of π) which is of moderate size when κ is. By (5.5), (5.9) is obtained iff

$$\|A^{-1}\| \le K \tag{5.10}$$

Note that stability is a property inherited from the differential operator (plus the BC) and does not depend on the inhomogeneities $q(x)$ and β. Also, stability in this chapter is an *absolute* concept, analogous to well-conditioning of the differential problem.

For our scheme (5.4), it is not difficult to show stability when $h \le \dfrac{2}{\|p\|}$ and $r(x)$ is positive and bounded away from 0 (cf. Exercise 1). This is actually a weak result, because K is not related to κ in a natural way. But we shall reserve attempts at a more careful stability analysis to the more general schemes in Section 5.2. Here, let us see what to do with stability and consistency, once we have them.

For the global error

$$e_i := u(x_i) - u_i \quad 1 \le i \le N+1 \tag{5.11a}$$

we have by (5.6), (5.4), (5.2)

$$L_\pi e_i = L_\pi u(x_i) - q(x_i) = \tau_i[u] \tag{5.11b}$$

$$e_1 = e_{N+1} = 0 \tag{5.11c}$$

Thus, the error satisfies the difference equations with the local truncation error as an inhomogeneity! We now use the stability bound (5.9) for e_i to obtain, with (5.7),

$$|e_i| \le K\tau[u] \le Kch^2 \quad 2 \le i \le N \tag{5.11d}$$

if u has a bounded 4th derivative.

The finite difference scheme is said to be *convergent* if[1]

$$\max_{1 \le i \le N+1} |u(x_i) - u_i| \to 0 \quad \text{as } h \to 0 \tag{5.12}$$

Here we have obtained the basic result that

$$\text{consistency} + \text{stability} \Rightarrow \text{convergence}$$

Moreover, in (5.11d) we see how fast e_i approaches 0 as $h \to 0$. The *order* of the method is p if $e_i \to 0$ like $0(h^p)$.

5.1.2 Simple one-step schemes for linear systems

The attraction of the finite difference scheme of the previous section is in its simplicity. When it can be applied, the scheme is very efficient, and so it is useful as a basic discretization rule in PDEs as well. However, it is too restricted for most applications of ODEs, because it has been applied only to one second-order differential equation. Note also that the meshes are uniform. Most nontrivial applications arise as mixed-order systems of differential equations, and many require very dense (fine) meshes in some sections of the interval of definition. If such a mesh is required to be uniform then it becomes dense everywhere, making the computation very expensive. A generalization of the simple scheme (5.6) for nonuniform meshes is considered in Section 5.6.1. A few more involved issues arise there.

We consider now the linear first order system (3.2), (3.19) which we rewrite here,

$$\mathbf{y}' = A(x)\mathbf{y} + \mathbf{q}(x) \quad a < x < b \tag{5.13a}$$

$$B_a \mathbf{y}(a) + B_b \mathbf{y}(b) = \boldsymbol{\beta} \tag{5.13b}$$

where $A(x)$, B_a and $B_b \in \mathbf{R}^{n \times n}$, and we seek numerical methods which work equally well for nonuniform meshes. This naturally leads to one-step schemes (cf. Section 2.7), i.e., schemes which define the difference operator based only on values related to one subinterval $[x_i, x_{i+1}]$ of the mesh π of (5.1) at a time. The two simplest such finite difference schemes are the midpoint and the trapezoidal schemes, discussed next.

For a (generally nonuniform) mesh π of (5.1), a discrete numerical solution $\mathbf{y}_\pi = (\mathbf{y}_1, \mathbf{y}_2, \ldots, \mathbf{y}_{N+1})^T \in \mathbf{R}^{n(N+1)}$ is sought, where $\mathbf{y}_i$ is to approximate component-wise the exact solution $\mathbf{y}(x)$ at $x = x_i$. [A unique solution to (5.13) is assumed to exist.] The numerical solution (in all methods based on one-step schemes) is required to satisfy the boundary conditions:

$$B_a \mathbf{y}_1 + B_b \mathbf{y}_{N+1} = \boldsymbol{\beta} \tag{5.14}$$

For the interior mesh points, two difference schemes are presented [cf. (2.139), (2.140)]. For each subinterval $[x_i, x_{i+1}]$ of π, the derivative in (5.13a) is replaced by $\frac{\mathbf{y}_{i+1} - \mathbf{y}_i}{h_i}$. This approximation is centered at $x_{i+1/2} := x_i + \frac{1}{2}h_i$, with $h_i := x_{i+1} - x_i$, i. e., at the middle of the subinterval. Then $A(x)\mathbf{y}(x) + \mathbf{q}(x)$ is approximated by a centered approximation, yielding second-order accuracy. The *trapezoidal* scheme is defined by

---

[1] In (5.12) we have preferred simplicity over rigor. We really consider the method on a family of meshes with $h \to 0$, and vary i with h so that $\hat{x} = (i-1)h$ remains fixed. The sequence of approximate solution values u_i thus obtained should approach $u(\hat{x})$.

$$\frac{\mathbf{y}_{i+1} - \mathbf{y}_i}{h_i} = \frac{1}{2}[A(x_{i+1})\mathbf{y}_{i+1} + A(x_i)\mathbf{y}_i] + \frac{1}{2}[\mathbf{q}(x_{i+1}) + \mathbf{q}(x_i)] \qquad 1 \le i \le N \qquad (5.15)$$

and the *midpoint* scheme is defined by

$$\frac{\mathbf{y}_{i+1} - \mathbf{y}_i}{h_i} = \frac{1}{2}A(x_{i+1/2})(\mathbf{y}_{i+1} + \mathbf{y}_i) + \mathbf{q}(x_{i+1/2}) \qquad 1 \le i \le N \qquad (5.16)$$

The latter scheme (5.16) is also known as the *box scheme*. In matrix form, both of these methods can be written as

$$\mathbf{A}\mathbf{y}_\pi = \hat{\boldsymbol{\beta}} \qquad (5.17a)$$

and, in detail,

$$\begin{pmatrix} S_1 & R_1 & & & \\ & S_2 & R_2 & & \\ & & \ddots & & \\ & & & S_N & R_N \\ B_a & & & & B_b \end{pmatrix} \begin{pmatrix} \mathbf{y}_1 \\ \mathbf{y}_2 \\ \vdots \\ \mathbf{y}_{N+1} \end{pmatrix} = \begin{pmatrix} \mathbf{q}_1 \\ \mathbf{q}_2 \\ \vdots \\ \mathbf{q}_N \\ \boldsymbol{\beta} \end{pmatrix} \qquad (5.17b)$$

where S_i, R_i are $n \times n$ matrices. For the trapezoidal scheme

$$S_i = -h_i^{-1}I - \frac{1}{2}A(x_i), \qquad R_i = h_i^{-1}I - \frac{1}{2}A(x_{i+1})$$

$$\qquad 1 \le i \le N \qquad (5.18)$$

$$\mathbf{q}_i = \frac{1}{2}[\mathbf{q}(x_{i+1}) + \mathbf{q}(x_i)]$$

while for the midpoint scheme

$$S_i = -h_i^{-1}I - \frac{1}{2}A(x_{i+1/2}), \qquad R_i = h_i^{-1}I - \frac{1}{2}A(x_{i+1/2})$$

$$\qquad 1 \le i \le N \qquad (5.19)$$

$$\mathbf{q}_i = \mathbf{q}(x_{i+1/2})$$

Note that **A** has the same block structure as the matrices **A** for multiple shooting and stabilized march discussed in Chapter 4 [see (4.37)]. A similar block structure also arises from other one-step schemes, as will be discussed in Sections 5.3 and 5.6.

In Fig. 5.1a we depict the structure of **A** for the case $n = 2$, $N = 3$. Here, the symbol × stands for a possibly nonzero element in the matrix. If the BC are separated, as is often the case in applications, then it is natural to move the BC at the left endpoint a to the top of the matrix (and split $\boldsymbol{\beta}$ accordingly). Thus, if $n = 2$ and we have one boundary condition at each end, then we obtain the structure depicted in Fig. 5.1b.

The existence of an approximate solution depends on the invertibility of **A**. This in turn is a weaker requirement than stability, discussed in the next section. Once we show stability of the two schemes, the existence of the approximate solutions will follow. Here, we just give some numerical examples. Before doing that, we remark that the solution of the large sparse system of equations (5.17) is discussed in Chapter 7 rather than here (see also Exercise 3), since similar considerations hold for a variety of different discretization methods.

$$\begin{bmatrix} \times & \times & \times & \times & & & & & & \\ \times & \times & \times & \times & & & & & & \\ & & \times & \times & \times & \times & & & & \\ & & \times & \times & \times & \times & & & & \\ & & & & \times & \times & \times & \times & & \\ & & & & \times & \times & \times & \times & & \\ \times & \times & & & & & & & \times & \times \\ \times & \times & & & & & & & \times & \times \end{bmatrix}$$

Figure 5.1a. Structure of **A**.

$$\begin{bmatrix} \times & \times & & & & & & & \\ \times & \times & \times & \times & & & & & \\ \times & \times & \times & \times & & & & & \\ & & \times & \times & \times & \times & & & \\ & & \times & \times & \times & \times & & & \\ & & & & \times & \times & \times & \times & \\ & & & & \times & \times & \times & \times & \\ & & & & & & \times & \times & \end{bmatrix}$$

Figure 5.1b. **A** with separated BC

Example 5.1

When a partial differential equation is reduced to an ordinary one by (cylindrical) symmetry considerations, often a harmless singularity appears in the coefficient matrix $A(x)$. As a simple example, consider the problem

$$u'' = -\frac{1}{x}u' + \left(\frac{8}{8-x^2}\right)^2$$

$$u'(0) = u(1) = 0$$

The exact solution

$$u(x) = 2 \ln\left(\frac{7}{8-x^2}\right)$$

is an analytic function on [0, 1], but even without knowing it we can use l'Hôpital's rule to find that

$$\frac{u'(x)}{x} \to u''(0) \quad \text{as} \quad x \to 0$$

so at $x = 0$ the differential equation behaves regularly and reads

$$u''(0) = 1/2$$

This modification of the differential equation at $x = 0$ needs to be explicitly specified when the trapezoidal scheme (but not the midpoint scheme) is used, in order to avoid a division by 0 in the computation.

Converting the problem to a first-order system for $y_1 = u$ and $y_2 = u'$, we obtain

$$\begin{bmatrix} y_1 \\ y_2 \end{bmatrix}' = \begin{bmatrix} 0 & 1 \\ 0 & -1/x \end{bmatrix} \begin{bmatrix} y_1 \\ y_2 \end{bmatrix} + \left(\begin{bmatrix} 0 \\ \dfrac{8}{8-x^2} \end{bmatrix} \right)^2$$

In Table 5.1 we list results when using uniform meshes of $N + 1$ points, $N = 10, 20, 40, 80$. The maximum absolute errors in y_1 and y_2 are listed under the heading $e(y_1)$ and $e(y_2)$, respectively, for the midpoint and trapezoidal schemes.

TABLE 5.1: Midpoint and trapezoidal schemes for Example 5.1.

N	Midpoint		Trapezoidal	
	$e(y_1)$	$e(y_2)$	$e(y_1)$	$e(y_2)$
10	.10−4	.44−3	.31−3	.29−3
20	.26−5	.11−3	.76−4	.73−4
40	.65−6	.27−4	.19−4	.18−4
80	.16−6	.69−5	.47−5	.45−5

The second-order convergence rate in h (i. e., an error reduction factor of 4 each time h is halved) is clearly demonstrated here. Observe also that for high accuracy, higher-order methods may be desirable. The application of shooting techniques to this problem is not so straightforward. □

Example 5.2

Consider, once again, Example 4.3. For $\lambda = 50$ and uniform meshes as in the previous example, the results listed in Table 5.2 clearly leave something to be desired: The errors are comparatively large. Inspecting the exact solution, we see that the higher derivatives of $y(x)$, to which the local truncation error relates, are much larger near the ends than in the middle, so a nonuniform mesh with h_i smaller near the interval ends is suggested. We pick such a mesh of size $N = 10$ as

$$\pi(\lambda) = \{0, \frac{1}{\lambda}, \frac{3}{\lambda}, \frac{8}{\lambda}, .25, .5, .75, 1 - \frac{8}{\lambda}, 1 - \frac{3}{\lambda}, 1 - \frac{1}{\lambda}, 1\}$$

and then generate more meshes by repeatedly halving each subinterval of the current mesh. The results for $\lambda = 50$ and $\lambda = 5000$ are displayed in Table 5.2. Note the improvement over the uniform mesh results. An even better nonuniform mesh can be chosen to reduce the errors further, with the same number of mesh points. This topic of mesh selection is considered in Chapter 9.

Recall that, using initial value techniques for this example, only the full multiple shooting method with $N = 10$ performs satisfactorily for $\lambda = 50$. (Note that the adequate "mesh" of shooting points in Example 4.9 is uniform. It is constructed for different purposes than the finite difference mesh, despite the similar structure of **A**). With the mesh $\pi(\lambda)$ properly adjusted, the finite difference methods perform almost as well for $\lambda = 5000$ as for $\lambda = 50$ (see Section 10.3.2). In contrast, the performance of the standard multiple shooting method would rapidly deteriorate as λ increases.

TABLE 5.2: Midpoint and trapezoidal schemes for Example 5.2.

λ	mesh	N	Midpoint		Trapezoidal	
			$e(y_1)$	$e(y_2)$	$e(y_1)$	$e(y_2)$
50	uniform	10	.47	.45	.44	.44
		20	.20	.19	.19	.19
		40	.57 − 1	.55 − 1	.56 − 1	.56 − 1
		80	. − 1	.12 − 1	.12 − 1	.12 − 1
50	nonuniform	10	.16	.51 − 1	.55	.56
		20	.42 − 1	.14 − 1	.15 − 1	.15 − 1
		40	.96 − 2	.34 − 2	.36 − 2	.36 − 2
		80	.24 − 2	.85 − 3	.91 − 3	.91 − 3
5000	nonuniform	10	.95	.95	.45	.45
		20	.82 − 1	.15 − 1	.15 − 1	.15 − 1
		40	.19 − 1	.36 − 2	.36 − 2	.36 − 2
		80	.44 − 2	.91 − 3	.91 − 3	.91 − 3

□

5.1.3 Simple schemes for nonlinear problems

Here we consider the (generally) nonlinear first-order system of n ODEs as in (3.1), (3.12),

$$\mathbf{y}' = \mathbf{f}(x, \mathbf{y}) \qquad a < x < b \qquad (5.20a)$$

$$\mathbf{g}(\mathbf{y}(a), \mathbf{y}(b)) = \mathbf{0} \qquad (5.20b)$$

For numerical approximation we again consider a mesh π of (5.1) and denote the vector of approximate solution values at mesh points by $\mathbf{y}_\pi$. The two schemes presented in Section 5.1.2 are extended in a straightforward way: The trapezoidal scheme is given by

$$\frac{\mathbf{y}_{i+1} - \mathbf{y}_i}{h_i} = \frac{1}{2}(\mathbf{f}(x_{i+1}, \mathbf{y}_{i+1}) + \mathbf{f}(x_i, \mathbf{y}_i)) \qquad 1 \leq i \leq N \qquad (5.21)$$

$$\mathbf{g}(\mathbf{y}_1, \mathbf{y}_{N+1}) = \mathbf{0} \qquad (5.22)$$

while the midpoint scheme is given by

$$\frac{\mathbf{y}_{i+1} - \mathbf{y}_i}{h_i} = \mathbf{f}(x_{i+1/2}, \frac{1}{2}(\mathbf{y}_i + \mathbf{y}_{i+1})) \qquad 1 \leq i \leq N \qquad (5.23)$$

and (5.22).

Thus we obtain a system of $n(N+1)$ algebraic equations for the $n(N+1)$ unknowns $\mathbf{y}_\pi$. Unlike before, though, these equations are nonlinear. It is not uncommon to have, for instance, $n = 5$ and $N = 200$, yielding more than 1000 nonlinear equations! Fortunately, the Jacobian matrix of this system is rather sparse, as we shall see below.

The basic method for solving the nonlinear problems arising in this book is Newton's method (Section 2.3). It is also intimately related to basic questions of stability of the difference scheme and existence of its solution, as discussed in Section 5.2.2. For these reasons we next consider the application of Newton's method in the context of a one-step difference scheme for (5.20). More sophisticated practical considerations and modifications are deferred to Chapter 8.

Recall from Sections 2.3 and 4.6.1 that for the system of equations

$$\mathbf{F}(\mathbf{s}) = \mathbf{0} \tag{5.24a}$$

Newton's method is a fixed-point iteration

$$\mathbf{s}^{m+1} = \mathbf{G}(\mathbf{s}^m) \qquad m = 0, 1, 2, \ldots,$$

with the iteration function $\mathbf{G}(\mathbf{s})$ defined by

$$\mathbf{G}(\mathbf{s}) := \mathbf{s} - [\mathbf{F}'(\mathbf{s})]^{-1} \mathbf{F}(\mathbf{s})$$

where $\mathbf{F}'(\mathbf{s}) = \dfrac{\partial \mathbf{F}(\mathbf{s})}{\partial \mathbf{s}}$ is the Jacobian matrix. This can be written as [cf. (4.85)].

$$\mathbf{F}'(\mathbf{s}^m) \, \xi = -\mathbf{F}(\mathbf{s}^m) \tag{5.24b}$$

and

$$\mathbf{s}^{m+1} := \mathbf{s}^m + \xi \tag{5.24c}$$

Let the nonlinear algebraic equations (5.24a) be given, for instance, by the trapezoidal scheme (5.21), (5.22), with $\mathbf{s} \equiv \mathbf{y}_\pi = (\mathbf{y}_1, \ldots, \mathbf{y}_{N+1})^T$, an $n(N+1)$ vector. Using a difference operator notation for (5.21), we obtain

$$\mathbf{N}_\pi \mathbf{y}_i := \frac{\mathbf{y}_{i+1} - \mathbf{y}_i}{h_i} - \frac{1}{2}(\mathbf{f}(x_{i+1}, \mathbf{y}_{i+1}) + \mathbf{f}(x_i, \mathbf{y}_i)) \tag{5.25a}$$

and $\mathbf{F}(\mathbf{s}) := (\mathbf{N}_\pi \mathbf{y}_1, \ldots, \mathbf{N}_\pi \mathbf{y}_N, \mathbf{g}(\mathbf{y}_1, \mathbf{y}_{N+1}))^T$. Newton's iteration (5.24b) gives

$$\frac{\mathbf{w}_{i+1} - \mathbf{w}_i}{h_i} - \frac{1}{2}[A(x_{i+1})\mathbf{w}_{i+1} + A(x_i)\mathbf{w}_i] = -\mathbf{N}_\pi \mathbf{y}_i^m \qquad 1 \le i \le N \tag{5.25b}$$

$$B_a \mathbf{w}_1 + B_b \mathbf{w}_{N+1} = -\mathbf{g}(\mathbf{y}_1^m, \mathbf{y}_{N+1}^m) \tag{5.25c}$$

Here $\xi \equiv \mathbf{w}_\pi = (\mathbf{w}_1, \ldots, \mathbf{w}_{N+1})^T$, $\mathbf{y}_\pi^m$ are known values from a previous iteration ($\mathbf{y}_\pi^0$ is an initial guess) and

$$A(x_j) := \frac{\partial \mathbf{f}}{\partial \mathbf{y}}(x_j, \mathbf{y}_j^m) \tag{5.26a}$$

$$B_a = \frac{\partial \mathbf{g}(\mathbf{u}, \mathbf{v})}{\partial \mathbf{u}}, \quad B_b = \frac{\partial \mathbf{g}(\mathbf{u}, \mathbf{v})}{\partial \mathbf{v}} \quad \text{at } \mathbf{u} = \mathbf{y}_1^m, \quad \mathbf{v} = \mathbf{y}_{N+1}^m \tag{5.26b}$$

[The reader not yet used to the derivation of (5.26) should refer back to Section 3.1 at this point.] The next iterate is given, according to (5.24c), by $\mathbf{y}_i^{m+1} := \mathbf{y}_i^m + \mathbf{w}_i$, $i = 1, \ldots, N+1$.

The system (5.25) for the correction vector $\mathbf{w}_\pi$ is a linear system of equations which looks like a trapezoidal discretization of some linear problem. In fact, note that in each iteration we perform two operations in succession—discretization and linearization (i. e., Newton's method).

It now becomes apparent that the possibly large Jacobian matrix $\mathbf{F}'(\mathbf{s})$ is indeed sparse: It has the form of $\mathbf{A}$ in (5.17). Let us assume that we have an efficient linear system solver for (5.17). (This is discussed in Chapter 7.) Then we can present, as an example, the following crude algorithm.

Algorithm: Trapezoidal scheme with Newton's method

Input: A BVP (5.20), a mesh π, an initial guess of solution values $\mathbf{y}_\pi$ at mesh points, and a tolerance *TOL*.

Output: Solution values at mesh points, $\mathbf{y}_\pi$.

REPEAT
1. Generate B_a, B_b by (5.26b) and set $\hat{\boldsymbol{\beta}} := -\mathbf{g}(\mathbf{y}_1, \mathbf{y}_{N+1})$.
2. FOR $i = 1, \ldots, N$ DO
   Generate S_i, R_i and $\mathbf{q}_i$ of (5.18) using (5.26a) and
   $$\mathbf{q}_i := -\mathbf{N}_\pi \mathbf{y}_i$$
   {At this point, $\mathbf{A}$ and $\hat{\boldsymbol{\beta}}$ of (5.17) have been generated for the current iteration}.
3. Solve $\mathbf{A}\mathbf{w}_\pi = \hat{\boldsymbol{\beta}}$ for $\mathbf{w}_\pi$ {using the linear system solver the existence of which has been postulated above}.
4. FOR $i = 1, \ldots, N+1$ DO
   $$\mathbf{y}_i := \mathbf{y}_i + \mathbf{w}_i$$

UNTIL $|\mathbf{w}_\pi| \leq TOL$ (or iteration limit exceeded).
□

A similar algorithm can be constructed, of course, with the midpoint scheme replacing the trapezoidal scheme. Thus, in step 1.2.1 we would use (5.19) instead of (5.18).

Example 5.3

Let us consider again the simple problem (4.5) of Example 4.1, converted to a first-order system as in Example 4.15. Thus we have (5.20) with

$$n = 2, \quad \mathbf{y} = \begin{bmatrix} u \\ v \end{bmatrix}, \quad \mathbf{f} = \begin{bmatrix} v \\ -\exp(u) \end{bmatrix}, \quad \mathbf{g} = \begin{bmatrix} u(0) \\ u(1) \end{bmatrix}$$

and (5.25b, c) reads, with $\mathbf{y}_i^m = \begin{bmatrix} u_i^m \\ v_i^m \end{bmatrix}$ given $(1 \leq i \leq N + 1)$

$$\frac{\mathbf{w}_{i+1} - \mathbf{w}_i}{h_i} - \frac{1}{2}\left[\begin{pmatrix} 0 & 1 \\ -\exp(u_{i+1}^m) & 0 \end{pmatrix}\mathbf{w}_{i+1} + \begin{pmatrix} 0 & 1 \\ -\exp(u_i^m) & 0 \end{pmatrix}\mathbf{w}_i\right] = -\left[\frac{\mathbf{y}_{i+1}^m - \mathbf{y}_i^m}{h_i}\right.$$

$$\left. - \frac{1}{2}\begin{pmatrix} v_{i+1}^m \\ -\exp(u_{i+1}^m) \end{pmatrix} - \frac{1}{2}\begin{pmatrix} v_i^m \\ -\exp(u_i^m) \end{pmatrix}\right] \qquad 1 \leq i \leq N$$

$$\begin{pmatrix} 1 & 0 \\ 0 & 0 \end{pmatrix}\mathbf{w}_1 + \begin{pmatrix} 0 & 0 \\ 1 & 0 \end{pmatrix}\mathbf{w}_{N+1} = -\begin{pmatrix} u_1^m \\ u_{N+1}^m \end{pmatrix}$$

These equations are *linear* for the unknowns $\mathbf{w}_\pi$.

Now, recall that this BVP has two solutions. The initial guess $\mathbf{y}_\pi^0$ determines to which of them (if at all) our algorithm converges. It turns out that taking a uniform mesh with stepsize h and $\mathbf{y}_\pi^0 \equiv \mathbf{0}$, the algorithm converges to the physically stable solution (cf. Example 4.15). Corresponding to Table 4.4 we obtain the numerical data in Table 5.3 below.

The observed reductions in the error as h is decreased are expected, since the trapezoidal and the midpoint schemes are second-order accurate. This is also the reason for the observed errors, which are "worse," for a given h, than the ones in Table 4.4: The shooting method was used with a fourth-order integrator (cf. Example 5.8 below). However, the amount of work per step h is smaller here.

Note that the number of Newton iterations needed for this simple example is minimal. The initial guess here is different from the one used for the shooting scheme in Table 4.4. □

Recall now the quasilinearization method, discussed in Section 2.3.4: In the arguments leading to (5.25) we discretized first and then linearized, but now consider reversing the order of these operations. Letting $\mathbf{y}^m(x)$ be an appropriately smooth function satisfying

$$\mathbf{y}^m(x_i) = \mathbf{y}_i^m \qquad i = 1, \ldots, N+1$$

and linearizing the problem as in Section 2.3.4, (2.73), we obtain

$$\mathbf{w}' - A(x)\mathbf{w} = -[(\mathbf{y}^m)'(x) - \mathbf{f}(x; \mathbf{y}^m(x))] \qquad (5.27a)$$

TABLE 5.3: Nonlinear Example 3.2 with $\lambda = 1$

h	itn	e (trapezoidal)*
.2	2	.11 – 2
.1	2	.28 – 3
.05	2	.71 – 4
.025	2	.18 – 4
.05	2	.44 – 5

*Similar results are obtained with the midpoint scheme.

$$B_a \mathbf{w}(a) + B_b \mathbf{w}(b) = -\mathbf{g}(\mathbf{y}^m(a), \mathbf{y}^m(b)) \qquad (5.27b)$$

where $A(x)$, B_a and B_b are given in (5.26) with an obvious extension. Now use the trapezoidal scheme (5.15), (5.14) for the *linear* problem (5.27). The result is precisely (5.25)! Thus the two operations of linearization and discretization commute for the trapezoidal scheme. The next iterate $\mathbf{y}^{m+1}(x)$ for the quasilinearization process is again needed only at the mesh points π, and is obtained by $\mathbf{y}^{m+1}(x_i) := \mathbf{y}^m(x_i) + \mathbf{w}_i$

It is easy to verify that quasilinearization is equivalent to Newton's method applied to the discretized difference equations for most methods considered in this chapter. In practice, the quasilinearization approach is the recommended way of implementation because it leads to a modular program design. Once we have a program module to solve linear problems with a given method, it can be invoked repeatedly for each iteration of a nonlinear problem by first linearizing as in (5.27), (5.26).

5.2 CONSISTENCY, STABILITY, AND CONVERGENCE

We have already seen in Section 5.1.1 that when investigating the convergence properties of a finite difference scheme, two concepts naturally emerge. One is *consistency*, which says that the difference operator approaches the differential operator as the mesh width, or stepsize, $h \to 0$. More specifically, the *local truncation error*, which is the residual when substituting the exact solution of the BVP into the difference equations, is $O(h^p)$ for some $p \geq 1$. The other concept is that of *stability*, which requires that the inverse of the difference operator, including the BC, is suitably bounded.

Below we define these terms more precisely and discuss the conditions under which they can be guaranteed to hold. We do this for general one-step schemes for solving linear and nonlinear first-order systems of ODEs. It turns out that stability for such schemes essentially follows from consistency, under fairly weak assumptions. We shall show this for both linear and nonlinear BVPs and, in both cases, return to the fundamental theorem:

Theorem 5.28

$$\text{consistency} + \text{stability} \Rightarrow \text{convergence}$$

□

5.2.1 Linear problems

Consider the BVP (5.13). Here again it is convenient to speak in terms of the differential operator

$$\mathbf{L}\mathbf{y}(x) \equiv \mathbf{y}'(x) - A(x)\mathbf{y}(x) \qquad a < x < b \qquad (5.29a)$$

and a corresponding one-step difference operator

$$\mathbf{L}_\pi \mathbf{y}_i \equiv \frac{\mathbf{y}_{i+1} - \mathbf{y}_i}{h_i} - \Psi(\mathbf{y}_i, \mathbf{y}_{i+1}; x_i, h_i) \qquad 1 \leq i \leq N \qquad (5.29b)$$

For example, the midpoint scheme gives

$$\Psi(\mathbf{y}_i, \mathbf{y}_{i+1}; x_i, h_i) = \frac{1}{2}A(x_{i+1/2})(\mathbf{y}_i + \mathbf{y}_{i+1})$$

Using the notation of (5.17), we can write the general one-step scheme as

$$\mathbf{L}_\pi \mathbf{y}_i \equiv S_i \mathbf{y}_i + R_i \mathbf{y}_{i+1} = \mathbf{q}_i \qquad 1 \leq i \leq N \qquad (5.30\text{a})$$

$$B_a \mathbf{y}_1 + B_b \mathbf{y}_{N+1} = \boldsymbol{\beta} \qquad (5.30\text{b})$$

The two schemes which we have already seen are given by (5.30) with (5.18) or (5.19). Note that by the notation $\mathbf{L}_\pi \mathbf{y}_i$ we mean the ith segment of the nN vector resulting from applying $\mathbf{L}_\pi$ to the $n(N+1)$ vector $\mathbf{y}_\pi$. Similarly,

$$\mathbf{L}_\pi \mathbf{y}(x_i) \equiv S_i \mathbf{y}(x_i) + R_i \mathbf{y}(x_{i+1})$$

Definitions 5.31

The *local truncation error* $\boldsymbol{\tau}_i[\mathbf{y}]$ is defined as

$$\boldsymbol{\tau}_i[\mathbf{y}] := \mathbf{L}_\pi \mathbf{y}(x_i) - \mathbf{q}_i \qquad 1 \leq i \leq N \qquad (5.31\text{a})$$

The finite difference scheme (5.30) is said to be *consistent* of *order p* (p a positive integer) if for every (smooth) solution of (5.13a) there exist constants c and $h_0 > 0$ such that for all meshes π with $h := \max_{1 \leq i \leq N} h_i \leq h_0$,

$$\tau[\mathbf{y}] := \max_{1 \leq i \leq N} |\boldsymbol{\tau}_i[\mathbf{y}]| \leq ch^p \qquad (5.31\text{b})$$

[cf. (5.7)].

The scheme is said to be *stable* if there exist constants K and $h_0 > 0$ such that for all meshes π with $h \leq h_0$ and all mesh functions $\mathbf{v}_\pi$,

$$|\mathbf{v}_i| \leq K \max \{|B_a \mathbf{v}_1 + B_b \mathbf{v}_{N+1}|, \max_{1 \leq j \leq N}|\mathbf{L}_\pi \mathbf{v}_j|\} \qquad 1 \leq i \leq N+1 \qquad (5.31\text{c})$$

and K is of moderate size if the BVP conditioning constant κ is.

The solution of (5.30) is said to *converge* to the solution of the BVP (5.13) if

$$\max_{1 \leq i \leq N+1} |\mathbf{y}_i - \mathbf{y}(x_i)| \to 0 \qquad \text{as } h \to 0 \qquad (5.31\text{d})$$

$\square$

(Note again that the notation in (5.31d) is a bit sloppy, but, we hope, clear.)

Remarks

(a) From our definition of a consistent method of order p it follows that the local truncation error $\tau[\mathbf{y}]$ approaches 0 at least as fast as h does [see (5.31b)]. In fact, for Theorem 5.28 to hold it is sufficient to require only that

$$\tau[\mathbf{y}] \to 0 \qquad \text{as } h \to 0$$

for consistency. However, we have no interest in schemes that do not satisfy (5.31b), so we require that consistency imply (at least) $\boldsymbol{\tau}_i[\mathbf{y}] = O(h_i)$. Similarly, we have defined stability in a rather strict way, requiring K to be of moderate size when approximating a well-conditioned BVP, because we are interested only in schemes that satisfy this requirement. [In order for Theorem 5.28 to hold, any constant K would do.]

(b) Establishing consistency and order of a scheme is usually straightforward. In fact, schemes are often derived in such a way that consistency naturally follows. For the midpoint and trapezoidal schemes, second-order accuracy is easily shown by expanding $\mathbf{y}(x_i)$ and $\mathbf{y}(x_{i+1})$ in a Taylor series about the midpoint $x_{i+1/2}$ (see Exercise 5).

(c) In (5.29b) we have emphasized the homogeneous part of the difference equations, because this is what is essential for stability. In other circumstances, particularly in the nonlinear case, it is more natural to write the general one-step scheme as

$$\frac{\mathbf{y}_{i+1} - \mathbf{y}_i}{h_i} = \Phi(\mathbf{y}_i, \mathbf{y}_{i+1}; x_i, h_i) \qquad (5.32a)$$

$$1 \leq i \leq N$$

$$\Phi(\mathbf{y}_i, \mathbf{y}_{i+1}; x_i, h_i) = \Psi(\mathbf{y}_i, \mathbf{y}_{i+1}; x_i, h_i) + \mathbf{q}_i \qquad (5.32b)$$

The consistency requirement can then be written as

$$\Phi(\mathbf{y}(x), \mathbf{y}(x); x, 0) = A(x)\mathbf{y}(x) + \mathbf{q}(x) \qquad a \leq x \leq b \qquad (5.33)$$

(d) The definitions above can be easily extended to more general schemes for more general ODE forms. A thing to note, though, is that in general there may be a local truncation error $\tau_0[\mathbf{y}]$ from the BC as well [here it is zero - see (5.30b), (5.13b)]. Then in (5.31b) the max is taken for $0 \leq i \leq N$; see Exercise 6.

(e) Proving Theorem 5.28 here is straightforward, and can be done as in Section 5.1.1: Writing the difference equations for the global error

$$\mathbf{e}_i := \mathbf{y}(x_i) - \mathbf{y}_i \qquad 1 \leq i \leq N+1 \qquad (5.34a)$$

we have

$$\mathbf{L}_\pi \mathbf{e}_i = \boldsymbol{\tau}_i[\mathbf{y}] \qquad 1 \leq i \leq N \qquad (5.34b)$$

$$B_a \mathbf{e}_1 + B_b \mathbf{e}_{N+1} = \mathbf{0} \qquad (5.34c)$$

Now inserting in (5.31c), with $\mathbf{e}_i$ playing the role of $\mathbf{v}_i$, (5.31b) yields the result

$$|\mathbf{e}_i| \leq Kch^p \qquad 1 \leq i \leq N+1. \qquad (5.34d)$$

We therefore have obtained not only Theorem 5.28 with the order of convergence, but also that *for h small enough, the maximum error is essentially bounded by the maximum local truncation error times the conditioning constant of the BVP*. □

Let us consider now a general, consistent one-step scheme (5.32), where we assume that

$$|\Psi(\mathbf{u}, \mathbf{v}; x, h)| \leq c(|\mathbf{u}| + |\mathbf{v}|) \qquad h \leq h_0, \, a \leq x \leq b \qquad (5.35)$$

c a constant independent of h and x. Then

$$h_i R_i = I + O(h_i), \qquad h_i S_i = -I + O(h_i)$$

so (5.30a) can be written as

$$\mathbf{y}_{i+1} = \Gamma_i \mathbf{y}_i + \mathbf{r}_i \qquad (5.36a)$$

$$\Gamma_i = -R_i^{-1} S_i, \qquad \mathbf{r}_i = R_i^{-1} \mathbf{q}_i = O(h_i) \qquad (5.36b)$$

The form (5.36a) reminds us of standard multiple shooting (4.38), so it is natural to ask whether Γ_i here also approximates the fundamental solution $Y(x_{i+1}; x_i)$. To recall, the jth column of $Y(x; x_i)$, denoted $\mathbf{v}(x)$, satisfies

$$\mathbf{L}\mathbf{v} = \mathbf{0}, \qquad \mathbf{v}(x_i) = \mathbf{e}^j$$

where $\mathbf{e}^j = (0, \ldots, 0, 1, 0, \ldots, 0)^T$ is the jth unit vector, $1 \le j \le n$. For a one-step approximation of $\mathbf{v}$, consistency implies that

$$S_i \mathbf{e}^j + R_i \mathbf{v}(x_{i+1}) = O(h_i)$$

because any consistent scheme of order $p \ge 1$ is at least first-order accurate. So, by (5.36),

$$\mathbf{v}(x_{i+1}) = \Gamma_i \mathbf{e}^j + O(h_i^2) \qquad 1 \le j \le n$$

i. e.,

$$\Gamma_i = Y(x_{i+1}; x_i) + O(h_i^2) \qquad (5.37)$$

The estimate (5.37) shows, in fact, that a one-step difference scheme can be viewed as an instance of the standard multiple shooting method! The mesh points correspond to the shooting points in Section 4.3 and the IVP integrations over each subinterval $[x_i, x_{i+1}]$ are done here in one step. Of course, this particular form of IVP integration is rather special, and the entire approach towards implementation of the finite difference method, in particular mesh selection, is completely different here than in the previous chapter. Note, also, that we have assumed that h^{-1} is much larger than any other quantity in (5.13) or (5.30). Still, we can capitalize on this theoretical equivalence to multiple shooting and obtain stability:

Theorem 5.38 Suppose that the finite difference scheme (5.30) satisfies (5.35) and that the BVP (5.13) is well-posed with conditioning constant κ [cf. (3.34)]. Then the scheme is stable with a stability constant

$$K = \kappa + O(h) \qquad (5.38a)$$

Moreover, $\mathbf{A}$ of (5.17) can be equilibrated so that for the resulting matrix $\hat{\mathbf{A}}$,

$$\text{cond}(\hat{\mathbf{A}}) \le \text{const } \kappa N \qquad (5.38b)$$

Proof: We equilibrate $\mathbf{A}$ by multiplying its ith block of n rows by R_i^{-1}. Calling the equilibrated matrix $\hat{\mathbf{A}}$ we have, by (5.36), (5.37),

$$\hat{\mathbf{A}} = \mathbf{D}\mathbf{A} = \mathbf{M} + \mathbf{E} \qquad (5.39a)$$

where

$$\mathbf{D} = \begin{pmatrix} R_1^{-1} & & & \\ & \cdot & & \\ & & \cdot & \\ & & & R_N^{-1} \\ & & & & I \end{pmatrix}, \quad \mathbf{E} = \begin{pmatrix} E_1 & & & \\ & \cdot & & \\ & & \cdot & \\ & & & E_N \\ & & & & 0 \end{pmatrix} \quad (5.39b)$$

$$\mathbf{M} = \begin{pmatrix} -Y(x_2;x_1) & I & & & \\ & -Y(x_3;x_2) & I & & \\ & & \cdot & & \\ & & & -Y(x_{N+1};x_N) & I \\ B_a & & & & B_b \end{pmatrix}$$

with $\|R_i^{-1}\| = h_i(1 + O(h_i))$, $\|E_i\| = O(h_i^2)$, $1 \le i \le N$. The inverse of $\mathbf{M}$ is explicitly given [cf. Theorem 4.42] by

$$\mathbf{M}^{-1} = \begin{pmatrix} G(x_1, x_2) & \cdots & G(x_1, x_{N+1}) & Y(x_1)Q^{-1} \\ \vdots & & \vdots & \vdots \\ G(x_{N+1}, x_2) & \cdots & G(x_{N+1}, x_{N+1}) & Y(x_{N+1})Q^{-1} \end{pmatrix} \quad (5.39c)$$

Manipulating (5.39a) a bit, we write

$$\mathbf{A}^{-1} = \hat{\mathbf{A}}^{-1}\mathbf{D} = (\mathbf{M} + \mathbf{E})^{-1}\mathbf{D} = (\mathbf{I} + \mathbf{M}^{-1}\mathbf{E})^{-1}\mathbf{M}^{-1}\mathbf{D} \quad (5.40a)$$

Now, recalling (3.34), we have

$$\|\mathbf{M}^{-1}\mathbf{D}\| = \max_{1 \le i \le N+1} \{\sum_{j=1}^{N} h_j \|G(x_i, x_{j+1})\| + O(h) + \|Y(x_i)Q^{-1}\|\} = \kappa + O(h) \quad (5.40b)$$

Similarly,

$$\|\mathbf{M}^{-1}\mathbf{E}\| \le c\kappa h \quad (5.40c)$$

and therefore, if $c\kappa h \ll 1$,

$$\|(\mathbf{I} + \mathbf{M}^{-1}\mathbf{E})^{-1}\| = 1 + O(h) \quad (5.40d)$$

Substituting (5.40b) and (5.44d) in (5.44a) we have for some constant c

$$\|\mathbf{A}^{-1}\| \le \kappa + ch =: K \quad (5.40e)$$

This proves stability.

Now, the condition number of the nonequilibrated $\mathbf{A}$ is $\sim K\underline{h}^{-1}$ with $\underline{h} := \min_{1 \le i \le N} h_i$. But this can be improved by considering $\hat{\mathbf{A}}$. Here

$$\|\hat{\mathbf{A}}\| \le \max\{1 + O(h), \|B_a\|, \|B_b\|\}$$

and for $\hat{\mathbf{A}}^{-1}$ we utilize (5.40d) together with (4.43b) to obtain (5.38b). □

Thus we have stability for virtually all sensible one-step schemes. Note the contrast with Chapter 4, where some innocent-looking methods (namely, single shooting and multiple shooting with compactification) are not always stable. Note also that the absolute stability bound (5.31c), (5.38a) yields a rather satisfactory bound in (5.38b) on the *relative* growth of roundoff errors, namely a linear growth as a function of the mesh size N (with no other dependence on the mesh) with a modest constant of proportionality.

Remark Consider again the relationship between a one-step finite difference method and multiple shooting. Recall that the matrix **M** of (5.39b) could still have elements with large magnitude if some kinematic eigenvalue λ of $A(x)$ satisfies $\text{Re}(\lambda) \gg 0$, since $\|Y(x_{i+1}; x_i)\|$ reflects the rapid growth of certain fundamental solution modes over the subinterval $[x_i, x_{i+1}]$. Indeed, one would choose shooting points so that $e^{\lambda h_i}$ is of an acceptable size. Yet in Example 5.2 we are able to perform the computations with the particular one-step difference schemes for $\lambda = 5000$ and $h > .01$ (cf. Table 5.2). Many more mesh points would be needed for the standard multiple shooting technique to work properly for that case. We wish to emphasize, though, that this performance by the midpoint and trapezoidal schemes is not covered by Theorem 5.38, because h is not "small enough" and Γ_i of (5.36b) does not closely approximate $-Y(x_{i+1}; x_i)$. This case is treated in Chapter 10. □

5.2.2 Nonlinear problems

Here we consider the nonlinear BVP (5.20). Let us define the differential operator

$$\mathbf{Ny}(x) \equiv \mathbf{y}'(x) - \mathbf{f}(x, \mathbf{y}(x)) \qquad (5.41a)$$

and the corresponding one-step difference operator [cf. (5.32a)]

$$\mathbf{N}_\pi \mathbf{y}_i \equiv \frac{\mathbf{y}_{i+1} - \mathbf{y}_i}{h_i} - \Phi(\mathbf{y}_i, \mathbf{y}_{i+1}; x_i, h_i) \qquad 1 \leq i \leq N \qquad (5.41b)$$

We can look at Φ as follows: Integrating (5.20a), we have

$$\mathbf{y}(x_{i+1}) - \mathbf{y}(x_i) = \int_{x_i}^{x_{i+1}} \mathbf{f}(x, \mathbf{y}(x)) \, dx$$

so $h_i \Phi(\mathbf{y}(x_i), \mathbf{y}(x_{i+1}); x_i, h_i)$ is an approximation to $\int_{x_i}^{x_{i+1}} \mathbf{f}(x, \mathbf{y}(x)) \, dx$. Taking $h_i \to 0$, it is clear that a consistency requirement on Φ is

$$\Phi(\mathbf{y}(x), \mathbf{y}(x); x, 0) = \mathbf{f}(x, \mathbf{y}(x)) \qquad a \leq x \leq b \qquad (5.42)$$

Two examples for Φ are given in (5.21) and in (5.23).

Assumption. We will assume, for the rest of this chapter, the following *smoothness properties:* The functions $\mathbf{f}(x, \mathbf{u})$ and $\mathbf{g}(\mathbf{u}, \mathbf{v})$ appearing in (5.20) have continuous second partial derivatives, while $\Phi(\mathbf{u}, \mathbf{v}; x, h)$ of (5.41b) has continuous first partial derivatives with respect to **u** and **v**. □

As was noted in Chapter 3, with nonlinear problems certain properties hold only in a local sense: The differential problem (5.20) may have a number of solutions, and all we can hope for in general is that a solution $\mathbf{y}(x)$ is isolated (or, locally unique) and that if we begin our iteration sufficiently close to $\mathbf{y}(x)$, it will converge to an approximation of that solution. Recall from Section 3.1 that $\mathbf{y}(x)$ is an isolated solution of the BVP (5.20) if the linearized homogeneous problem at $\mathbf{y}$,

$$\mathbf{z}' = A(x)\mathbf{z} \qquad (5.43a)$$

$$B_a \mathbf{z}(a) + B_b \mathbf{z}(b) = \mathbf{0} \qquad (5.43b)$$

has only the trivial solution. In (5.43) we have used

$$A(x) := \frac{\partial \mathbf{f}}{\partial \mathbf{y}}(x, \mathbf{y}(x)) \qquad (5.44a)$$

$$B_a := \frac{\partial \mathbf{g}(\mathbf{u}, \mathbf{v})}{\partial \mathbf{u}}, \quad B_b := \frac{\partial \mathbf{g}(\mathbf{u}, \mathbf{v})}{\partial \mathbf{v}}, \quad \text{at } \mathbf{u} = \mathbf{y}(a), \quad \mathbf{v} = \mathbf{y}(b) \qquad (5.44b)$$

Note that in (5.44) the linearization is about the (unknown) exact solution, while in (5.26), (5.27), a similar linearization about $\mathbf{y}^m$ is used for the $(m+1)$st Newton iteration.

Let us denote the linear operator in (5.43a) by

$$L[\mathbf{y}]\mathbf{z}(x) \equiv \mathbf{z}'(x) - A(x)\mathbf{z}(x) \qquad (5.45)$$

while that of (5.27a) would be $L[\mathbf{y}^m]\mathbf{w}(x)$. Similarly, the difference operator of the left-hand side of (5.25b) is denoted by $L_\pi[\mathbf{y}^m]\mathbf{w}_i$. We can write, as in (5.30a),

$$L_\pi[\mathbf{y}^m]\mathbf{w}_i \equiv S_i \mathbf{w}_i + R_i \mathbf{w}_{i+1} \qquad 1 \leq i \leq N \qquad (5.46a)$$

where for a general nonlinear one-step scheme (5.41b)

$$S_i = -h_i^{-1}I - \frac{\partial \Phi}{\partial \mathbf{y}_i}(\mathbf{y}_i^m, \mathbf{y}_{i+1}^m; x_i, h_i), \quad R_i = h_i^{-1}I - \frac{\partial \Phi}{\partial \mathbf{y}_{i+1}}(\mathbf{y}_i^m, \mathbf{y}_{i+1}^m; x_i, h_i) \qquad (5.46b)$$

When using a difference operator $L_\pi[\mathbf{u}]$ linearized about some $\mathbf{u}$ with similarly linearized BC [see (5.27b), (5.43b)], we obtain a linear system as in (5.17). We denote the corresponding matrix A by $A[\mathbf{u}]$. This matrix contains the linearizations of both $\mathbf{f}(x, \mathbf{y})$ and $\mathbf{g}(\mathbf{y}(a), \mathbf{y}(b))$.

Definition 5.47 The *local truncation error* $\boldsymbol{\tau}_i[\mathbf{y}]$ is defined as

$$\boldsymbol{\tau}_i[\mathbf{y}] := N_\pi \mathbf{y}(x_i) \qquad 1 \leq i \leq N \qquad (5.47a)$$

The finite difference scheme

$$N_\pi \mathbf{y}_i = \mathbf{0} \qquad 1 \leq i \leq N$$

$$\mathbf{g}(\mathbf{y}_1, \mathbf{y}_{N+1}) = \mathbf{0}$$

is said to be *consistent of order p* if for every smooth solution of (5.20a) there exist positive constants c and h_0 such that for all meshes π with $h \leq h_0$

$$\tau[\mathbf{y}] := \max_{1 \leq i \leq N} |\tau_i[\mathbf{y}]| \leq ch^p \tag{5.47b}$$

Thus, consistency and order of accuracy are defined precisely as in the linear case. Stability, though, is defined only in the vicinity of a solution. Consider a "discrete tube" around $\mathbf{y}(x)$,

$$S_\rho^\pi(\mathbf{y}) := \{\mathbf{u}_\pi; \; |\mathbf{u}_i - \mathbf{y}(x_i)| \leq \rho, \; 1 \leq i \leq N+1 \} \tag{5.48}$$

for some radius $\rho > 0$.

Definition 5.49 The difference scheme is said to be *stable* around $\mathbf{y}(x)$ if there are positive constants K^*, ρ, and h_0 such that for all meshes π with $h \leq h_0$ and all mesh functions $\mathbf{u}_\pi, \mathbf{v}_\pi$ in $S_\rho^\pi(\mathbf{y})$,

$$|\mathbf{u}_i - \mathbf{v}_i| \leq K^* \max\{ \; |\mathbf{g}(\mathbf{u}_1, \mathbf{u}_{N+1}) - \mathbf{g}(\mathbf{v}_1, \mathbf{v}_{N+1})|, \tag{5.49}$$
$$\max_{1 \leq j \leq N} |\mathbf{N}_\pi \mathbf{u}_j - \mathbf{N}_\pi \mathbf{v}_j| \} \quad 1 \leq i \leq N+1$$

[cf. (5.31c)] and K^* is of moderate size if the constant κ of (3.34) associated with the variational BVP (5.43) is. □

The definition of *convergence* is the same as in the linear case [see (5.31d)]. Theorem 5.28 once again follows directly, provided $\mathbf{y}_\pi$ exists. To see this, consider the (nonlinear) difference equations which the error mesh function satisfies, and observe that the local truncation error is the corresponding residual vector [cf. (5.34)]. Substituting in the stability bound (5.49), we obtain the convergence bound

$$|\mathbf{y}_i - \mathbf{y}(x_i)| \leq K^* \tau[\mathbf{y}] \leq K^* ch^p \quad 1 \leq i \leq N+1 \tag{5.50}$$

To guarantee existence and stability we make the following assumption on the linearized scheme in the vicinity of the exact solution:

Assumption 5.51 For some $\rho > 0$ and all $h \leq h_0$, the linearized finite difference scheme

$$\mathbf{L}_\pi[\mathbf{y}]\mathbf{z}_i = \mathbf{0} \quad 1 \leq i \leq N$$

$$B_a \mathbf{z}_1 + B_b \mathbf{z}_{N+1} = \mathbf{0}$$

with B_a and B_b defined in (5.44b), is consistent and stable, with $\left\| \dfrac{\partial \Phi}{\partial \mathbf{y}_i}(\mathbf{u}_i, \mathbf{u}_{i+1}; x_i, h) \right\|$ and $\left\| \dfrac{\partial \Phi}{\partial \mathbf{y}_{i+1}}(\mathbf{u}_i, \mathbf{u}_{i+1}; x_i, h) \right\|$ bounded for all $\mathbf{u}_\pi$ in $S_\pi^\rho(\mathbf{y})$. Furthermore, there is a Lipschitz constant K_L such that for all such $\mathbf{u}_\pi$,

$$\| A[\mathbf{y}] - A[\mathbf{u}_\pi] \| \leq K_L \max_{1 \leq i \leq N+1} |\mathbf{y}(x_i) - \mathbf{u}_i| \tag{5.51}$$
□

This assumption is quite reasonable for one-step schemes if, for instance, $\mathbf{f}(x, \mathbf{u})$ and $\mathbf{g}(\mathbf{u}, \mathbf{v})$ have continuous second derivatives, as we have assumed. We use it next to show stability.

Theorem 5.52. Under assumption 5.51, the nonlinear scheme (5.41b), (5.22) is stable. Moreover, $\mathbf{A}[\mathbf{u}_\pi]$ has a uniformly bounded inverse for all $\mathbf{u}_\pi \in S_\rho^\pi(\mathbf{y})$ (i.e., the linearized schemes in a tube around $\mathbf{y}(x)$ are stable).

Proof: First we show that $\mathbf{A}[\mathbf{u}_\pi]$ has a uniformly bounded inverse. Assumption 5.51 and (5.40) yield

$$\| \mathbf{A}^{-1}[\mathbf{y}] \| \leq K, \qquad K = \kappa + O(h)$$

We now use (5.51) for a contraction argument. For arbitrary $\mathbf{r}$, writing $\mathbf{A}[\mathbf{u}_\pi]\mathbf{z} = \mathbf{r}$ as

$$\{\mathbf{A}[\mathbf{y}] + \mathbf{A}[\mathbf{u}_\pi] - \mathbf{A}[\mathbf{y}]\}\mathbf{z} = \mathbf{r}$$

we have

$$\mathbf{A}[\mathbf{y}]\mathbf{z} = \mathbf{r} - \{\mathbf{A}[\mathbf{u}_\pi] - \mathbf{A}[\mathbf{y}]\}\mathbf{z}$$

The bound on $\mathbf{A}^{-1}[\mathbf{y}]$ means that we can write

$$|\mathbf{z}| \leq K\{|\mathbf{r}| + \|\mathbf{A}[\mathbf{u}_\pi] - \mathbf{A}[\mathbf{y}]\| \, |\mathbf{z}|\}$$

and using (5.51), (5.48),

$$|\mathbf{z}| \leq K\{|\mathbf{r}| + K_L \rho |\mathbf{z}|\}$$

This proves the stability of the linearized schemes for ρ so small that $KK_L\rho < 1$, because then

$$|\mathbf{z}| \leq \frac{K}{1 - KK_L\rho} |\mathbf{r}|$$

which means

$$\|\mathbf{A}^{-1}[\mathbf{u}_\pi]\| \leq \frac{K}{1 - KK_L\rho} =: K^* \qquad \mathbf{u}_\pi \in S_\rho^\pi(\mathbf{y}) \qquad (5.53\text{a})$$

Now write the nonlinear difference scheme (5.41b), (5.22) as a system of nonlinear algebraic equations

$$\mathbf{F}(\mathbf{y}_\pi) := \begin{bmatrix} N_\pi y_1 \\ \vdots \\ N_\pi y_N \\ g(y_1, y_{N+1}) \end{bmatrix} = 0 \qquad (5.53\text{b})$$

Using our smoothness assumption on Φ and $\mathbf{g}$, the Mean Value Theorem (for vector-valued functions) can be invoked to yield

$$\mathbf{F}(\mathbf{u}_\pi) - \mathbf{F}(\mathbf{v}_\pi) = \mathbf{A}[\mathbf{u}_\pi, \mathbf{v}_\pi](\mathbf{u}_\pi - \mathbf{v}_\pi) \qquad (5.53\text{c})$$

where

$$\mathbf{A}[\mathbf{u}_\pi, \mathbf{v}_\pi] := \int_0^1 \mathbf{A}[t\mathbf{u}_\pi + (1-t)\mathbf{v}_\pi] \, dt$$

But, for $\mathbf{u}_\pi, \mathbf{v}_\pi \in S_\rho^\pi(\mathbf{y})$, (5.51) yields

$$\|A[\mathbf{u}_\pi, \mathbf{v}_\pi] - A[\mathbf{y}]\| \leq K_L \rho$$

so the bound

$$\|A^{-1}[\mathbf{u}_\pi, \mathbf{v}_\pi]\| \leq K^*$$

is obtained, precisely as in (5.53a). The stability bound (5.49) now follows from (5.53c). □

Since the hypotheses of Theorem 5.52 are quite realistic, we have been able to show stability in a neighborhood of the exact solution $\mathbf{y}(x)$ for virtually any sensible one-step difference scheme. However, this is worthwhile only if we are also able to show that there exists a solution $\mathbf{y}_\pi$ for the numerical scheme in a small neighborhood of the isolated solution. For this, let us recall Newton's method, already introduced in Section 5.1.3, which we rewrite for (5.53b) as

$$A[\mathbf{y}_\pi^m]\mathbf{w}_\pi = -\mathbf{F}(\mathbf{y}_\pi^m) \tag{5.54a}$$

$$m = 0, 1, \ldots$$

$$\mathbf{y}_\pi^{m+1} = \mathbf{y}_\pi^m + \mathbf{w}_\pi \tag{5.54b}$$

[cf. (5.24)]. The Newton-Kantorovich Theorem 2.64 can be used to show *existence* of $\mathbf{y}_\pi$, if we can show that, with $\mathbf{y}_\pi^0$ an initial guess, there are positive constants ρ_0, α, β, and γ such that

$$\alpha\beta\gamma \leq \frac{1}{2}, \qquad \rho_0 \alpha\gamma \leq 1 - \sqrt{1 - 2\alpha\beta\gamma} \tag{5.55a}$$

$$\|A^{-1}[\mathbf{y}_\pi^0]\| \leq \alpha \tag{5.55b}$$

$$|A^{-1}[\mathbf{y}_\pi^0]\mathbf{F}(\mathbf{y}_\pi^0)| \leq \beta \tag{5.55c}$$

$$\|A[\mathbf{u}_\pi] - A[\mathbf{v}_\pi]\| \leq \gamma |\mathbf{u}_\pi - \mathbf{v}_\pi| \qquad \text{for all } \mathbf{u}_\pi, \mathbf{v}_\pi \in S_{\rho_0}^\pi(\mathbf{y}_\pi^0) \tag{5.55d}$$

But this is straightforward, using the stability Theorem 5.52, if we choose $\mathbf{y}_\pi^0$ close enough to the vector of exact solution values at mesh points, $(\mathbf{y}(x_1), \ldots, \mathbf{y}(x_{N+1}))^T$. We have

Theorem 5.56 Let $\mathbf{y}(x)$ be an isolated solution of the BVP (5.20) and assume that (5.51) holds for a consistent scheme (5.53b). Then for some $\rho_0 \leq \rho$ and h_0 sufficiently small, (5.53b) has a unique solution $\mathbf{y}_\pi \in S_\rho^\pi(\mathbf{y})$ for all meshes π with $h \leq h_0$. This solution can be computed by Newton's method (5.54), which converges quadratically. □

The proof of this theorem is left as a useful exercise involving the concepts introduced above.

Remarks

(a) Theorem 5.56 relies on $\mathbf{y}_\pi^0$ being "sufficiently close" to $\mathbf{y}$. The big practical problem of how to choose $\mathbf{y}_\pi^0$ is ignored here. We will return to this in Chapter 8, but note that if $\mathbf{y}_\pi^0$ is not "close" to $\mathbf{y}$ and to $\mathbf{y}_\pi$, then the promised quadratic convergence may not be realized, and even if there is eventually a convergence of the nonlinear iteration, not until near the termination of the iteration process.

(b) The requirement that $\mathbf{y}_\pi^0$ is close enough to $\mathbf{y}(x)$ makes sense only when $\mathbf{y}_\pi$ is also close to $\mathbf{y}(x)$. This is hidden in the requirement that h_0 is "small enough." In practice, the closeness of $\mathbf{y}_\pi$ to $\mathbf{y}(x)$ depends on the mesh chosen. This is discussed in Chapter 9, but note here that if $\mathbf{y}_\pi$ is not close to $\mathbf{y}(x)$, even when $\mathbf{y}_\pi$ does exist, then we do not have the premises of Theorem 5.56.

(c) If ρ is small enough so that $KK_L\rho \ll 1$, then $K^* \approx K \approx \kappa$ [see (5.53a)]. This means that close to the solution, the condition of the difference method for the nonlinear BVP is essentially that of the linearized BVP (5.43), (5.44). This is reflected in the actual error bound (5.50). So, when h is small enough, under mild assumptions everything is under control. However, once again life is sometimes more difficult with certain challenging practical problems. If $KK_L\rho$ is not small (but $KK_L\rho < 1$) for a given $\mathbf{y}_\pi^0$, which is the best available guess in a particular instance, then the conditioning constant K^* may be quite different from K.

5.3 HIGHER-ORDER ONE-STEP SCHEMES

In the previous section, a detailed convergence and stability analysis was given for methods based on general one-step schemes. However, the only actual schemes we have seen so far are the trapezoidal and the midpoint schemes, introduced in Section 5.1. These are simple schemes, but their order is only 2; i. e., the error is $O(h^2)$. If high accuracy is desired, then higher-order schemes can be much more effective. As previously mentioned, there are two ways to obtain higher-order methods. One is to use convergence acceleration, which effectively gives a higher-order method. This is considered in Section 5.5. The other way, considered here, is to use higher-order schemes directly. We concentrate on one-step schemes, because nonuniform meshes are easily handled with them, and discretization near boundaries requires no special treatment.

Recall that a natural way to construct such methods is to integrate the ODE (5.20a)

$$\mathbf{y}' = \mathbf{f}(x, \mathbf{y})$$

over each subinterval (x_i, x_{i+1}) and replace the integral by a quadrature rule [cf. (5.41a,b)]. Thus

$$\mathbf{y}(x_{i+1}) - \mathbf{y}(x_i) = \int_{x_i}^{x_{i+1}} \mathbf{y}'(x)\,dx = \int_{x_i}^{x_{i+1}} \mathbf{f}(x, \mathbf{y}(x))\,dx$$

yields, when we replace the right-hand integral by a quadrature formula and divide by h_i, a scheme

$$h_i^{-1}(\mathbf{y}_{i+1} - \mathbf{y}_i) = \Phi(\mathbf{y}_i, \mathbf{y}_{i+1}; x_i, h_i) \qquad 1 \le i \le N$$

Note that this is precisely the same starting point as when we design one-step schemes for IVPs. The approximation of the BC is always by (5.22) and this avoids boundary truncation errors. The order of the scheme depends on the precision of the quadrature formula.

One choice is to replace $\mathbf{f}$ by its Hermite interpolant of order $p = 2k$ (cf. Section 2.5) and then integrate exactly. For this we need $\mathbf{f}$ and its derivatives up to order $k - 1$ at the subinterval ends, x_i and x_{i+1}. We get

$$\mathbf{f}(x, \mathbf{y}(x)) = \sum_{j=0}^{k-1} h_i^j \{ \mathbf{f}^{(j)}(x_i, \mathbf{y}(x_i)) \phi_j(\frac{x - x_i}{h_i})$$

$$+ (-1)^j \mathbf{f}^{(j)}(x_{i+1}, \mathbf{y}(x_{i+1})) \phi_j(\frac{x_{i+1} - x}{h_i}) \} + O(h_i^p)$$

where the functions $\phi_j(t)$ are polynomials of order $p = 2k$ determined on $[0, 1]$ by interpolation conditions as in Section 2.5. Thus, the numerical method defined by

$$\frac{\mathbf{y}_{i+1} - \mathbf{y}_i}{h_i} = \sum_{j=0}^{k-1} h_i^j \{ \alpha_j \mathbf{f}^{(j)}(x_i, \mathbf{y}_i) + \beta_j \mathbf{f}^{(j)}(x_{i+1}, \mathbf{y}_{i+1}) \} \qquad (5.57a)$$

with

$$\alpha_j := \int_0^1 \phi_j(t)\, dt, \qquad \beta_j = (-1)^j \int_0^1 \phi_j(1-t)\, dt \qquad (5.57b)$$

is of order $2k$.

Example 5.4

For $k = 1$ we get $\phi_0(t) = 1 - t$ and so $\alpha_0 = \beta_0 = \frac{1}{2}$ in (5.57b). The resulting method in (5.57a) is the trapezoidal scheme.

For $k = 2$ we get $\phi_0(t) = (2t+1)(1-t)^2$, $\phi_1(t) = t(1-t)^2$ and so $\alpha_0 = \beta_0 = \frac{1}{2}$, $\alpha_1 = -\beta_1 = \frac{1}{12}$. The resulting method of order 4 is

$$\frac{\mathbf{y}_{i+1} - \mathbf{y}_i}{h_i} = \frac{1}{2}(\mathbf{f}(x_i, \mathbf{y}_i) + \mathbf{f}(x_{i+1}, \mathbf{y}_{i+1})) + \frac{h_i}{12}(\mathbf{f}'(x_i, \mathbf{y}_i) - \mathbf{f}'(x_{i+1}, \mathbf{y}_{i+1})) \qquad (5.58)$$

$\square$

The obvious problem with (5.58) as well as with higher-order schemes of the form (5.57a) is how to obtain $\mathbf{f}^{(j)}(x_i, \mathbf{y}_i)$ for $j > 0$. This can be achieved by differentiating $\mathbf{f}$ given in the BVP formulation (5.20a), but note that total derivatives are needed, e.g.,

$$\mathbf{f}'(x_i, \mathbf{y}_i) = [\frac{\partial \mathbf{f}}{\partial x} + \frac{\partial \mathbf{f}}{\partial \mathbf{y}} \cdot \mathbf{f}]_{(x_i, \mathbf{y}_i)}$$

If $\mathbf{f}$ is easy to differentiate then schemes like (5.58) can be very efficient. However, this is hard to automate, and for general purposes this family of schemes is rather limited.

More hope lies with another family of higher-order methods, obtained by using values of $\mathbf{f}(x, \mathbf{y}(x))$ at intermediate points of the interval $[x_i, x_{i+1}]$. This leads to Runge-Kutta schemes, and we will concentrate on these for the remainder of this section.

5.3.1 Implicit Runge-Kutta schemes

As for IVPs (Section 2.7) we define a *k-stage Runge-Kutta scheme* for $\mathbf{y}' = \mathbf{f}(x, \mathbf{y})$ by

$$\mathbf{y}_{i+1} = \mathbf{y}_i + h_i \sum_{j=1}^{k} \beta_j \mathbf{f}_{ij} \qquad 1 \leq i \leq N \tag{5.59a}$$

where

$$\mathbf{f}_{ij} = \mathbf{f}(x_{ij}, \mathbf{y}_i + h_i \sum_{l=1}^{k} \alpha_{jl} \mathbf{f}_{il}) \qquad 1 \leq j \leq k \tag{5.59b}$$

The points x_{ij} are given by

$$x_{ij} = x_i + h_i \rho_j \qquad 1 \leq j \leq k, \qquad 1 \leq i \leq N \tag{5.59c}$$

with

$$0 \leq \rho_1 \leq \rho_2 \leq \cdots \leq \rho_k \leq 1 \tag{5.59d}$$

the "canonical points." Thus, the points x_{ij}, which are sometimes called *collocation points*, are N scaled translations of the canonical set of k points $\rho_1, \ldots, \rho_k$ into each subinterval of the mesh π.

We have used the subscript i as a mesh interval index, while j and l are stage counters within a particular subinterval. Thus, $1 \leq i \leq N$ and $1 \leq j, l \leq k$. We will continue to use the indices in this way throughout this section.

We will use the standard tableau of Fig. 5.2 below to represent Runge-Kutta schemes.

Let us recall from Section 2.7 that the scheme is called *explicit* if $\rho_1 = 0$ and $\alpha_{jl} = 0$, $j \leq l$, and *implicit* otherwise. So, for explicit schemes the matrix (α_{jl}) in Fig. 5.2 is strictly lower triangular. For initial value problems, explicit methods have obvious advantages over implicit ones, and so implicit methods are normally used only when there is no other choice (e. g., for very stiff ODEs; cf. Section 2.7). But for boundary value problems, implicitness is inherent in the problem in the sense that solution values on π are obtained simultaneously, so we might as well give up "explicitness" altogether and try to choose the $k(k+2)$ parameters ρ_j, β_l, and α_{jl}, $1 \leq j, l \leq k$ so as to get the most out of the method in terms of order of accuracy, efficiency, and stiff stability (which is elaborated upon in Section 10.3).

To interpret (5.59a, b) note that we can write

$$\mathbf{f}_{ij} = \mathbf{f}(x_{ij}, \mathbf{y}_{ij}) \tag{5.60a}$$

where

$$\begin{array}{c|ccc}
\rho_1 & \alpha_{11} & \cdots & \alpha_{1k} \\
\vdots & \vdots & & \vdots \\
\rho_k & \alpha_{k1} & \cdots & \alpha_{kk} \\
\hline
 & \beta_1 & \cdots & \beta_k
\end{array}$$

Figure 5.2 A Runge-Kutta scheme

$$y_{ij} := y_i + h_i \sum_{l=1}^{k} \alpha_{jl} f_{il} \qquad 1 \le j \le k \qquad (5.60b)$$

is an approximation of $y(x_{ij})$. Thus, the sum in (5.59a) is a quadrature rule for $\int_{x_i}^{x_{i+1}} f$ with quadrature weights $\beta_1, \ldots, \beta_k$, while the sum in (5.60b) is a quadrature rule for $\int_{x_i}^{x_{ij}} f$ with quadrature weights $\alpha_{j1}, \ldots, \alpha_{jk}$. Let us assume that the *precision* of the latter quadrature rule is the same for all j, $1 \le j \le k$, and denote it by s, $1 \le s \le k$. Thus

$$\int_0^{\rho_j} \phi(t)\, dt = \sum_{l=1}^{k} \alpha_{jl} \phi(\rho_l) \qquad \text{all } \phi \in \mathbf{P}_s, \qquad 1 \le j \le k \qquad (5.61a)$$

(recall that by $\mathbf{P}_s$ we denote all polynomials of order s, i. e., degree $<s$). Let us further assume that the precision of the quadrature rule in (5.59a) is p, $p \ge s$, i. e.,

$$\int_0^1 \phi(t)\, dt = \sum_{l=1}^{k} \beta_l \phi(\rho_l) \qquad \text{all } \phi \in \mathbf{P}_p \qquad (5.61b)$$

We are interested only in schemes which satisfy $p \ge s \ge 1$. Thus, in particular,

$$\sum_{l=1}^{k} \beta_l = 1 \quad \text{and} \quad \sum_{l=1}^{k} \alpha_{jl} = \rho_j, \qquad 1 \le j \le k \qquad (5.61c)$$

Example 5.5

Consider the choice $k = 1$, $\rho_1 = \frac{1}{2}$, $\beta_1 = 1$, $\alpha_{11} = \frac{1}{2}$. We want to find the values of s and p for which (5.61) holds. Now, any polynomial of order s, say, can be written as

$$\phi(t) = c_0 + c_1 t + c_2 t^2 + \cdots + c_{s-1} t^{s-1}$$

So, to verify a property like (5.61a) we need to consider only the monomials $1, t, t^2, \ldots, t^{s-1}$. Checking (5.61a) we see that

$$\int_0^{1/2} 1\, dt = \frac{1}{2} = \alpha_{11}, \qquad \int_0^{1/2} t\, dt = \frac{1}{8} \ne \alpha_{11} \cdot \frac{1}{2}$$

so $s = 1$. Checking (5.61b) we see that

$$\int_0^1 1\, dt = 1 = \beta_1, \qquad \int_0^1 t\, dt = \frac{1}{2} = \beta_1 \cdot \frac{1}{2}, \qquad \int_0^1 t^2\, dt = \frac{1}{3} \ne \beta_1 \cdot \frac{1}{4}$$

so $p = 2$.

Let us now take a closer look at this scheme. From (5.59) we have

$$y_{i+1} = y_i + h_i \mathbf{f}_{i1}$$

$$\mathbf{f}_{i1} = \mathbf{f}(x_{i+1/2}, \mathbf{y}_{i1}) = \mathbf{f}(x_{i+1/2}, \mathbf{y}_i + \frac{1}{2} h_i \mathbf{f}_{i1})$$

In this case we can eliminate $\mathbf{f}_{i1}$ quite simply, thus obtaining the previous form of a one-step scheme. From the first equation above,

$$\frac{1}{2} h_i \mathbf{f}_{i1} = \frac{1}{2}(\mathbf{y}_{i+1} - \mathbf{y}_i)$$

and substituting in the second equation, we have

$$\mathbf{f}_{i1} = \mathbf{f}(x_{i+1/2}, \frac{1}{2}(\mathbf{y}_i + \mathbf{y}_{i+1}))$$

so the scheme is

$$\mathbf{y}_{i+1} = \mathbf{y}_i + h_i \mathbf{f}(x_{i+1/2}, \frac{1}{2}(\mathbf{y}_i + \mathbf{y}_{i+1}))$$

It is important to note that the above elimination of $\mathbf{f}_{i1}$ can be done *locally*, without relating to any information outside the subinterval $[x_i, x_{i+1}]$.

The reader will identify the last expression as none other than the midpoint scheme (5.23). Thus, we see that the midpoint scheme is a particular instance of an implicit Runge-Kutta method. Another instance is the trapezoidal scheme (see Exercise 5.8). Note also that the precision $p = 2$ agrees with the known second-order accuracy of the midpoint scheme at mesh points. At the midpoint of each subinterval we have $s = 1$ only, which implies first-order accuracy; i.e.,

$$h_i^{-1}(\mathbf{y}(x_{i+1/2}) - \mathbf{y}(x_i)) = \frac{1}{2}\mathbf{f}(x_{i+1/2}, \mathbf{y}(x_{i+1/2})) + O(h_i)$$

It can be shown that this and stability yield

$$\mathbf{y}_{i1} = \mathbf{y}(x_{i+1/2}) + O(h)$$

But an improved, second-order accuracy is later obtained at the midpoints as well [see (5.79b)]. Note from (5.59a) and (5.60b) that

$$\mathbf{y}_{i1} - \frac{1}{2}(\mathbf{y}_i + \mathbf{y}_{i+1})$$

$\square$

Let us now treat the *linear case*, i.e., consider the BVP (5.13). In the Runge-Kutta formulation (5.59) there are two types of unknowns. One is the vector $\mathbf{y}_\pi^T = (\mathbf{y}_1^T, \ldots, \mathbf{y}_{N+1}^T)$ of mesh values. These variables are *global* in that their determination is in some sense done simultaneously, over the entire interval $[a, b]$. The other variables are *local* to $[x_i, x_{i+1}]$, for each i, $1 \leq i \leq N$. These are

$$\mathbf{f}_i^T := (\mathbf{f}_{i1}^T, \ldots, \mathbf{f}_{ik}^T) \qquad 1 \leq i \leq N \tag{5.62}$$

Alternatively, we can use (5.60) to express the scheme in terms of the local unknowns $\mathbf{y}_{i1}, \ldots, \mathbf{y}_{ik}$. We then obtain

$$h_i^{-1}(\mathbf{y}_{i+1} - \mathbf{y}_i) = \sum_{l=1}^{k} \beta_l \{A(x_{il})\mathbf{y}_{il} + \mathbf{q}(x_{il})\} \tag{5.63a}$$

$$h_i^{-1}(\mathbf{y}_{ij} - \mathbf{y}_i) = \sum_{l=1}^{k} \alpha_{jl} \{A(x_{il})\mathbf{y}_{il} + \mathbf{q}(x_{il})\} \qquad 1 \leq j \leq k \tag{5.63b}$$

For both practical and theoretical reasons, we now proceed to *eliminate the local unknowns*. This is sometimes referred to in the literature of finite element methods as *parameter condensation*. We have done it already for Example 5.5. To see how it can be done more generally, we can write (5.59b) for a linear BVP in vector form as

$$W\mathbf{f}_i = V\mathbf{y}_i + \mathbf{q}_i \tag{5.64a}$$

where $W \in \mathbf{R}^{nk \times nk}$, $V \in \mathbf{R}^{nk \times n}$ and $\mathbf{q}_i \in \mathbf{R}^{nk}$ are defined by

$$W = I - h_i \begin{bmatrix} \alpha_{11} A(x_{i1}) & \cdots & \alpha_{1k} A(x_{i1}) \\ & \vdots & \\ \alpha_{k1} A(x_{ik}) & \cdots & \alpha_{kk} A(x_{ik}) \end{bmatrix}, \quad V = \begin{bmatrix} A(x_{i1}) \\ \vdots \\ A(x_{ik}) \end{bmatrix}, \quad \mathbf{q}_i = \begin{bmatrix} \mathbf{q}(x_{i1}) \\ \vdots \\ \mathbf{q}(x_{ik}) \end{bmatrix} \quad (5.64b)$$

The obvious dependence of the matrices W and V on the index i is suppressed for notational simplicity. Clearly, for h_i small enough, W is invertible, with

$$W^{-1} = I + O(h_i) \tag{5.64c}$$

Again using linearity of the ODE, we find that substitution in (5.59b) yields, as in (5.36),

$$\mathbf{y}_{i+1} = \Gamma_i \mathbf{y}_i + \mathbf{r}_i \quad 1 \leq i \leq N \tag{5.65a}$$

with

$$\Gamma_i := I + h_i D W^{-1} V, \qquad \mathbf{r}_i := h_i D W^{-1} \mathbf{q}_i \tag{5.65b}$$

and $D \in \mathbf{R}^{n \times nk}$ is defined by

$$D := (\beta_1 I, \ldots, \beta_k I) \tag{5.65c}$$

Combining (5.65a) with the BC (5.14) yields, once again, a linear system of algebraic equations (5.17) with a multiple shooting type matrix **A**. Indeed, the process here resembles multiple shooting even more than before (cf. Section 5.2), because in multiple shooting, the IVP integrations from one shooting point to the next can also be considered as a local elimination process. Alternately, the process here can be considered as multiple shooting where IVP integrations are done by one step of a Runge-Kutta scheme.

Now, with (5.65) our schemes fall within the theory of Section 5.2.1. Thus, all we need for a convergence bound like (5.34d) is to show consistency of order p. This is where the precision requirements (5.61) come in. However, the general case is complicated. We will therefore restrict this aspect of our discussion to a subclass of the Runge-Kutta schemes, one which fortunately contains many schemes of interest.

5.3.2 A subclass of Runge-Kutta schemes

The class of Runge-Kutta schemes considered here is equivalent to the class of collocation schemes discussed in Section 5.4. These schemes are required to satisfy

Assumption 5.66 The points ρ_j are distinct; i. e.,

$$0 \leq \rho_1 < \rho_2 < \cdots < \rho_k \leq 1$$

and the precision s of (5.61a) equals k (so $p \geq k$). □

Under this assumption we will prove the claimed equivalence to collocation schemes in Section 5.4. Therefore, Theorem 5.79 will apply here. In particular, the error at mesh points is $O(h^p)$. We will delay the convergence proofs until the next section and concentrate here instead on the construction of such schemes. The distinction between them and corresponding collocation schemes will be intentionally blurred.

First, note that given the points $\rho_1, \ldots, \rho_k$, the quadrature weights β_j and α_{jl} are uniquely determined under assumption 5.66. For if (5.61) is to be satisfied for the k monomials $\phi(t) = 1, t, t^2, \ldots, t^{k-1}$ (cf. Example 5.5) then this gives a $k \times k$ linear system of constraints for each k coefficients $\alpha_{j1}, \ldots, \alpha_{jk}$ $(1 \leq j \leq k)$ or $\beta_1, \ldots, \beta_k$. The matrix of each such system of equations is a nonsingular Vandermonde matrix [see (2.84)], and therefore precisely one solution exists.

Let us now construct the weights β_j and α_{jl}. In the beginning of Section 5.3 we mention the idea of replacing $\mathbf{f}(x, \mathbf{y}(x))$ by a polynomial interpolant in order to obtain a quadrature rule for $\int_{x_i}^{x_{i+1}} \mathbf{f}$. We have already seen what a Hermite interpolant does. Here consider a Lagrange interpolant. Thus, given distinct points $\rho_1, \ldots, \rho_k$ of (5.59d) we define for each mesh subinterval $[x_i, x_{i+1}]$ the collocation points x_{ij} as in (5.59c). We then write the function $\mathbf{y}'(x)$ on $[x_i, x_{i+1}]$ as a sum of its Lagrange interpolant of order k plus a remainder term [recall Section 2.4, eqs. (2.83), (2.91)],

$$\mathbf{y}'(x) = \sum_{l=1}^{k} \mathbf{y}'(x_{il}) L_l\left(\frac{x - x_i}{h_i}\right) + \psi(x) \qquad x_i \leq x \leq x_{i+1} \qquad (5.67a)$$

where

$$L_l(t) := \frac{(t - \rho_1)\cdots(t - \rho_{l-1})(t - \rho_{l+1})\cdots(t - \rho_k)}{(\rho_l - \rho_1)\cdots(\rho_l - \rho_{l-1})(\rho_l - \rho_{l+1})\cdots(\rho_l - \rho_k)} \qquad 1 \leq l \leq k \qquad (5.67b)$$

and the remainder term $\psi(x)$ is expressed in a divided difference form

$$\psi(x) = \mathbf{y}'[x_{i1}, \ldots, x_{ik}, x] \prod_{l=1}^{k} (x - x_{il}) \qquad (5.67c)$$

Then, from

$$\mathbf{y}(x) - \mathbf{y}(x_i) = \int_{x_i}^{x} \mathbf{y}'(\xi)\, d\xi = \int_{x_i}^{x} \mathbf{f}(\xi, \mathbf{y}(\xi))\, d\xi$$

we obtain the implicit Runge-Kutta scheme (5.59), with

$$\beta_j = \int_0^1 L_j(t)\, dt \qquad \alpha_{jl} = \int_0^{\rho_j} L_l(t)\, dt \qquad 1 \leq j, \quad l \leq k \qquad (5.68)$$

With this construction, assumption 5.66 clearly holds. As noted before, there is no other Runge-Kutta scheme satisfying (5.66) for the same $\rho_1, \ldots, \rho_k$.

Example 5.6

It is easy to verify that for $k = 1$, $\rho_1 = \frac{1}{2}$, (5.68) gives (in the notation of Fig. 5.2) the midpoint scheme

1/2	1/2
	1

For the choice $k = 2$, $\rho_1 = 0$, $\rho_2 = 1$, (5.68) gives the trapezoidal scheme

0	0	0
1	1/2	1/2
	1/2	1/2

The scheme resembling Simpson's rule of integration has $k = 3$, $\rho_1 = 0$, $\rho_2 = 1/2$, $\rho_3 = 1$. We get

$$L_1(t) = \frac{(t - 1/2)(t - 1)}{1/2} = 2t^2 - 3t + 1$$

$$L_2(t) = \frac{(t - 0)(t - 1)}{-1/4} = -4t^2 + 4t$$

$$L_3(t) = \frac{(t - 0)(t - 1/2)}{1/2} = 2t^2 - t$$

and this used in (5.68) gives

0	0	0	0
1/2	5/24	1/3	-1/24
1	1/6	2/3	1/6
	1/6	2/3	1/6

We note that $p = 4$ for Simpson's scheme. (Check!) Also, when $\rho_k = 1$, $\alpha_{kl} = \beta_l$ (Why?). □

We now proceed with the construction (5.68). To obtain schemes with a higher-order of convergence, we need to use families of points which give high-precision quadrature. (See Section 2.6.1, and in particular Theorems 2.126, 2.127 and the following discussion.)

(a) **Gauss schemes:** The points $\rho_1, \ldots, \rho_k$ are chosen as the zeroes of a Legendre polynomial, (Section 2.6.1). Note that $\rho_1 > 0$, $\rho_k < 1$, so mesh points are not collocation points. For $k = 1$, we have $\rho_1 = \frac{1}{2}$, so the *midpoint scheme* (5.16) is obtained as the simplest instance of this family of schemes. For a k-stage scheme, the accuracy is $O(h^{2k})$.

(b) **Radau schemes:** The point $\rho_k = 1$ is fixed. Here $\rho_1 > 0$ and mesh points (except a) are particular collocation points. For $k = 1$ we obtain the *backward Euler* scheme

$$h_i^{-1}(\mathbf{y}_{i+1} - \mathbf{y}_i) = \mathbf{f}(x_{i+1}, \mathbf{y}_{i+1}) \tag{5.69}$$

which has some distinguished properties for stiff IVPs and accuracy $O(h)$ in general. For a k-stage scheme, the accuracy is $O(h^{2k-1})$.

(c) **Lobatto schemes:** The points $\rho_1 = 0$ and $\rho_k = 1$ are fixed. Here mesh points are particular collocation points, and internal ones are in some sense repeated twice. With careful implementation, a k-stage Lobatto scheme is roughly as efficient as a $(k-1)$-stage Gauss scheme. For $k = 2$ the *trapezoidal scheme* is obtained. For a k-stage scheme, the accuracy is $O(h^{2(k-1)})$.

Example 5.7

Let us return to the BVP of Example 5.1. In Table 5.1 we have seen the performance of the one-stage Gauss and the two-stage Lobatto schemes. Below we list numerical results for a two-stage Gauss scheme, given by

$$
\begin{array}{c|cc}
\frac{1}{2} - \frac{\sqrt{3}}{6} & \frac{1}{4} & \frac{1}{4} - \frac{\sqrt{3}}{6} \\
\frac{1}{2} + \frac{\sqrt{3}}{6} & \frac{1}{4} + \frac{\sqrt{3}}{6} & \frac{1}{4} \\
\hline
 & 1/2 & 1/2
\end{array}
$$

a three-stage Gauss scheme, given by

$$
\begin{array}{c|ccc}
\frac{1}{2} - \frac{\sqrt{15}}{10} & \frac{5}{36} & \frac{2}{9} - \frac{\sqrt{15}}{15} & \frac{5}{36} - \frac{\sqrt{15}}{30} \\
\frac{1}{2} & \frac{5}{36} + \frac{\sqrt{15}}{24} & \frac{2}{9} & \frac{5}{36} - \frac{\sqrt{15}}{24} \\
\frac{1}{2} + \frac{\sqrt{15}}{10} & \frac{5}{36} + \frac{\sqrt{15}}{30} & \frac{2}{9} + \frac{\sqrt{15}}{15} & \frac{5}{36} \\
\hline
 & \frac{5}{18} & \frac{4}{9} & \frac{5}{18}
\end{array}
$$

and a three-stage Lobatto scheme, given in Example 5.6.

The additional accuracy obtained by using higher-order methods is clearly demonstrated when we compare Tables 5.4 and 5.1. Note that the errors in the two-stage Gauss scheme and the three-stage Lobatto scheme are reduced by a factor of 16 when N is doubled (i. e., h halved), because these schemes are of order 4. The three-stage Gauss scheme is of order 6 and the reduction factor for the error is (roughly) $2^6 = 32$ when h is halved. This reduction occurs as long as the discretization error is much larger than the roundoff error. □

TABLE 5.4: Higher-order collocation schemes for Example 5.1

N	2-stage Gauss		3-stage Lobatto		3-stage Gauss	
	$e(y_1)$	$e(y_2)$	$e(y_1)$	$e(y_2)$	$e(y_1)$	$e(y_2)$
2	.20−3	.71−4	.17−4	.11−3	.14−6	.37−6
5	.64−5	.19−5	.57−6	.29−5	.70−9	.17−8
10	.46−6	.12−6	.37−7	.18−6	.13−10	.27−10
20	.33−7	.77−8	.23−8	.11−7	.27−12	.42−12
40	.23−8	.48−9	.15−9	.72−9	.60−14	.71−14
80	.16−9	.30−10	.91−11	.45−10	.13−14*	.94−15*

* These values are mainly roundoff errors.

5.3.3 On the implementation of Runge-Kutta schemes

For the nonlinear BVP (5.20), since $\mathbf{f}(x, \mathbf{y})$ is nonlinear, the Runge-Kutta scheme (5.59) yields a set of nonlinear algebraic equations, for which we consider Newton's method.

In the linear case, we have proceeded by eliminating the local unknowns $\mathbf{f}_{ij}$, defined by (5.59b). This is possible analytically, even in the nonlinear case, when there is only one point ρ_j satisfying $0 < \rho_j < 1$. We have seen one such instance with the midpoint scheme in Example 5.5. Other instances are Simpson's scheme (Lobatto with $k = 3$) and Radau scheme with $k = 2$; see Exercises 10 and 11. For these schemes, the theory of Section 5.2.2 can be applied directly.

Alas, the general picture is not so rosy. In (5.59b) there are nk nonlinear equations which need to be solved if local parameter elimination is to take place. This multiple-shooting-type approach would then yield a one-step scheme falling within the theory of Section 5.2.2.

When (5.59b) is solved for a stiff IVP, assuming that $\mathbf{y}_i$ is known, some variant of Newton's method is usually employed to obtain an accuracy which depends upon the prescribed overall tolerance. The resulting expense incurred is considerable and in fact is the main reason for the original disenchantment with fully implicit Runge-Kutta schemes for IVPs.

The situation is different for BVPs. Here $\mathbf{y}_i$, $\mathbf{y}_{i+1}$, and $\mathbf{f}_i$ are all unknown, the need to solve (5.59b) occurs within each Newton iteration for $\mathbf{y}_\pi$, and it makes little sense to try to solve (5.59b) with an accuracy higher than that produced by one Newton iteration. In other words, we perform Newton iterations on the entire set of unknowns

$$\mathbf{y}_1, \mathbf{f}_1, \mathbf{y}_2, \mathbf{f}_2, \ldots, \mathbf{y}_N, \mathbf{f}_N, \mathbf{y}_{N+1}$$

and carry out the local parameter elimination of the $\mathbf{f}_i$ only *after* the linearization. This process is also the one which the quasilinearization method yields. A quasilinearization algorithm with collocation is described in the next section.

There are a number of methods for solving nonlinear equations which are derived as cheaper approximate alternatives to Newton's method, or the quasilinearization method. Some such methods are discussed in Chapter 8. One way to make an iteration like (5.24) cheaper, which is particular to the Runge-Kutta schemes discussed here, is simply to take the Jacobian constant along each subinterval $[x_i, x_{i+1})$. Thus, the same Jacobian value, $\frac{\partial \mathbf{f}}{\partial \mathbf{y}}(x_i, \mathbf{y}_i)$ say, is used for all k collocation points in the ith subinterval of the mesh. Note that approximating the Jacobian $\mathbf{F}'(\mathbf{s})$ does not alter the value of the root of $\mathbf{F}(\mathbf{s})$ in (5.24). But the convergence properties of the resulting method may change. Still, if $\mathbf{F}'(\mathbf{s})$ does not vary a lot on one mesh subinterval, then this is a cheaper variant of Newton's method which closely approximates its properties.

Other, specially designed Runge-Kutta methods, which do not necessarily satisfy assumption 5.66, are also dicussed in the literature.

5.4 COLLOCATION THEORY

Recall that the Runge-Kutta scheme (5.59) gives a discrete solution with values $\mathbf{y}_i$ at mesh points x_i and values $\mathbf{y}_{ij}$ at collocation points x_{ij} satisfying

$$\mathbf{y}_{ij} = \mathbf{y}_i + h_i \sum_{l=1}^{k} \alpha_{jl} \mathbf{f}(x_{il}, \mathbf{y}_{il})$$

[see (5.60)]. Let $\mathbf{y}_\pi(x)$ be a polynomial of order $k+1$ defined on $[x_i, x_{i+1}]$ by the interpolation conditions

$$\mathbf{y}_\pi(x_i) = \mathbf{y}_i \qquad (5.70a)$$

$$\mathbf{y}_\pi'(x_{ij}) = \mathbf{f}(x_{ij}, \mathbf{y}_{ij}) \qquad 1 \leq j \leq k \qquad (5.70b)$$

It is simple to verify that such an interpolation process is well-defined. Indeed, if we write $\mathbf{y}_\pi(x)$ in terms of its first derivative, $\mathbf{y}_\pi(x) = \mathbf{y}_i + \int_{x_i}^{x} \mathbf{y}_\pi'(\xi)\, d\xi$, and replace $\mathbf{y}_\pi'(x)$ with its Lagrange interpolant form as in (5.67) [with no remainder because $\mathbf{y}_\pi'(x) \in \mathbf{P}_k$], then we obtain with (5.68)

$$\mathbf{y}_\pi(x_{ij}) = \mathbf{y}_i + \int_{x_i}^{x_{ij}} \sum_{l=1}^{k} \mathbf{f}(x_{il}, \mathbf{y}_{il}) L_l(\frac{\xi - x_i}{h_i})\, d\xi = \mathbf{y}_i + h_i \sum_{l=1}^{k} \alpha_{jl} \mathbf{f}(x_{il}, \mathbf{y}_{il}) = \mathbf{y}_{ij} \qquad (5.71a)$$

$$\mathbf{y}_\pi(x_{i+1}) = \mathbf{y}_i + h_i \sum_{l=1}^{k} \beta_l \mathbf{f}(x_{il}, \mathbf{y}_{il}) = \mathbf{y}_{i+1} \qquad (5.71b)$$

If we now extend the polynomial to the subinterval $[x_{i+1}, x_{i+2}]$ by using (5.70) with i replaced by $i+1$, the two polynomial pieces will be continuously matched at x_{i+1}, according to (5.71b) and (5.70a). Extending to all i, $1 \leq i \leq N$, we obtain a *continuous piecewise polynomial function* of order $k+1$ over $[a, b]$, which we still denote by $\mathbf{y}_\pi(x)$. [Note that $\mathbf{y}_\pi(x)$ is a continuous function, whereas $\mathbf{y}_\pi$ is the vector of mesh values $\mathbf{y}_i$. Confusion between the two is not disastrous, however, due to Theorem 5.73 below.]

Combining (5.70b) with (5.71a) we see that the piecewise polynomial function $\mathbf{y}_\pi(x)$ satisfies the ODE at the collocation points, i. e.,

$$\mathbf{y}_\pi'(x_{ij}) = \mathbf{f}(x_{ij}, \mathbf{y}_\pi(x_{ij})) \qquad 1 \leq i \leq N, \qquad 1 \leq j \leq k \qquad (5.72a)$$

The BC are satisfied as well,

$$\mathbf{g}(\mathbf{y}_\pi(a), \mathbf{y}_\pi(b)) = \mathbf{0} \qquad (5.72b)$$

Definition Given a mesh π and points $0 \leq \rho_1 < \cdots < \rho_k \leq 1$, a *collocation solution* for (5.20) is a continuous, piecewise polynomial function $\mathbf{y}_\pi(x)$ which reduces to a polynomial of degree at most k (order $k+1$) on each mesh subinterval and satisfies (5.72). □

What we have shown above is

Theorem 5.73 Given a mesh π, the unique Runge-Kutta method (5.59) satisfying assumption 5.66 is equivalent to the collocation method (5.72). Moreover,

$$\mathbf{y}_\pi(x_i) = \mathbf{y}_i, \qquad \mathbf{y}_\pi(x_{ij}) = \mathbf{y}_{ij} \qquad 1 \leq i \leq N, \qquad 1 \leq j \leq k \qquad \square$$

Note that up to this point we have not shown *existence* of the approximate solution in any of its two equivalent forms, the Runge-Kutta or the collocation. We do this next for linear problems; nonlinear problems are considered in Section 5.4.2. After showing existence of the approximate solution we will be interested in the approximation error. We will use the notation

$$\mathbf{e}_\pi(x) := \mathbf{y}_\pi(x) - \mathbf{y}(x) \tag{5.74a}$$

$$\mathbf{e}_{ij} := \mathbf{e}_\pi(x_{ij}), \qquad \mathbf{e}_i := \mathbf{e}_\pi(x_i) \tag{5.74b}$$

5.4.1 Linear problems

First we prove a basic existence and stability result. Then we show higher-order convergence.

Theorem 5.75 The k-stage collocation solution exists for the linear BVP (5.13) and is obtained in a stable way. Moreover, the error and its derivatives satisfy, for $1 \leq i \leq N$,

$$\max_{x_i \leq x \leq x_{i+1}} |\mathbf{e}_\pi(x)| = O(h^k) \tag{5.75a}$$

$$\max_{x_i \leq x \leq x_{i+1}} |\mathbf{e}_\pi^{(j)}(x)| = O(h^{k+1-j}) \theta_i^{j-1} \qquad 1 \leq j \leq k \tag{5.75b}$$

with

$$\theta_i := h/h_i \tag{5.75c}$$

Proof: We view the Runge-Kutta form of the collocation scheme as a one-step scheme. Thus, local parameters are eliminated as in (5.64), and (5.65) is considered. Using (5.61a) with $s = k$ and (5.67), we obtain

$$\mathbf{y}(x_{ij}) - \mathbf{y}(x_i) = \int_{x_i}^{x_{ij}} \mathbf{y}'(x)\,dx = h_i \sum_{l=1}^{k} \alpha_{jl} \mathbf{y}'(x_{il}) + \int_{x_i}^{x_{ij}} \psi(x)\,dx \qquad 1 \leq j \leq k \tag{5.76a}$$

Similarly,

$$\mathbf{y}(x_{i+1}) - \mathbf{y}(x_i) = h_i \sum_{l=1}^{k} \beta_l \mathbf{y}'(x_{il}) + \int_{x_i}^{x_{i+1}} \psi(x)\,dx \tag{5.76b}$$

In (5.76a, b) the rightmost terms are the residuals, i. e., the amount by which the exact solution fails to satisfy the difference equations, cf. (5.63). Clearly $|\psi(x)| = O(h_i^k)$. Repeating the elimination process which leads to (5.65a), one obtains (Exercise 9)

$$\mathbf{y}(x_{i+1}) = \Gamma_i \mathbf{y}(x_i) + \mathbf{r}_i + h_i \tau_i \tag{5.77a}$$

$$|\tau_i| = O(h_i^k) \tag{5.77b}$$

Since $k \geq 1$, the one-step method is consistent and thus stable by Theorem 5.38. Therefore, the discrete solution exists and satisfies

$$\max_{1 \leq i \leq N+1} |\mathbf{y}_i - \mathbf{y}(x_i)| = O(h^k)$$

Furthermore, using the ODE for $\mathbf{y}'(x_{il})$ in (5.76a) and subtracting (5.63b) yields the error equations

$$\mathbf{e}_{ij} := \mathbf{e}_i + h_i \sum_{l=1}^{k} \alpha_{jl} A(x_{il}) \mathbf{e}_{il} - \int_{x_i}^{x_{ij}} \psi(x)\, dx \tag{5.78a}$$

so

$$\mathbf{e}_{ij} = (1 + O(h_i))\mathbf{e}_i + O(h_i^{k+1}) \tag{5.78b}$$

and this gives a similar bound at collocation points

$$\max_{\substack{1 \leq i \leq N \\ 1 \leq j \leq k}} |\mathbf{y}_{ij} - \mathbf{y}(x_{ij})| = O(h^k)$$

Since $\mathbf{y}_\pi(x)$ and $\mathbf{y}(x)$ satisfy the same ODE at collocation points, we have

$$\mathbf{e}_\pi'(x_{ij}) = A(x_{ij}) \mathbf{e}_{ij} \tag{5.78c}$$

so

$$\mathbf{e}_\pi'(x_{ij}) = O(h^k), \qquad 1 \leq j \leq k$$

Next, we express the error derivative $\mathbf{e}_\pi'(x)$ on $[x_i, x_{i+1}]$ as in (5.67). Since $\mathbf{y}_\pi'(x)$ is in $\mathbf{P}_k$, it is equal to its interpolant, so the residual $\psi(x)$ is that of (5.67c),

$$\mathbf{e}_\pi'(x) = \sum_{l=1}^{k} \mathbf{e}_\pi'(x_{il}) L_l\left(\frac{x - x_i}{h_i}\right) + \psi(x) \tag{5.78d}$$

Finally, taking derivatives in (5.78d) and using the estimates of $\mathbf{e}_\pi'(x_{il})$ [and of $\mathbf{e}_i$ for (5.75a)], we obtain the desired results (5.75a, b). □

Theorem 5.75 provides the basic existence and stability result for the collocation solution. However, the orders of the convergence estimates in (5.75a, b) are not sharp. Higher-order estimates can be achieved when $p > k$, as promised in Section 5.3.2 and already demonstrated numerically in Example 5.7. We state the result in terms of the equivalent Runge-Kutta method, relating back to the theory developed in Section 5.2.

Theorem 5.79 If a Runge-Kutta scheme satisfies assumption 5.66 then, as a one-step scheme (5.65), it is accurate of order p for a BVP satisfying $A(x), q(x) \in C^{(p)}[a,b]$. Hence,

$$|\mathbf{y}_i - \mathbf{y}(x_i)| = O(h^p) \qquad 1 \leq i \leq N \tag{5.79a}$$

Also, at collocation points,

$$|\mathbf{y}_{ij} - \mathbf{y}(x_{ij})| = O(h_i^{k+1}) + O(h^p) \qquad 1 \leq j \leq k, \quad 1 \leq i \leq N \tag{5.79b}$$

Proof: Let $\mathbf{v}(x)$ satisfy

$$\mathbf{L}\,\mathbf{v}(x) = \mathbf{q}(x) \qquad x_i \le x$$

$$\mathbf{v}(x_i) = \mathbf{y}_i$$

Then, from (5.77a) and (5.65a),

$$\boldsymbol{\tau}_i = h_i^{-1}(\mathbf{v}(x_{i+1}) - \mathbf{y}_{i+1}) \tag{5.80}$$

Let $Y(x)$ be the fundamental solution satisfying

$$\mathbf{L}\,Y = 0, \qquad Y(x_i) = I$$

Then

$$\mathbf{v}(x_{i+1}) - \mathbf{y}_{i+1} = \int_{x_i}^{x_{i+1}} Y(x_{i+1}) Y^{-1}(t)(\mathbf{q}(t) - \mathbf{L}\,\mathbf{y}_\pi(t))\,dt$$

The integrand can be written, by (5.72a), as

$$Y(x_{i+1}) Y^{-1}(t)(\mathbf{q}(t) - \mathbf{L}\,\mathbf{y}_\pi(t)) = -Y(x_{i+1}) Y^{-1}(t)(\mathbf{L}\,\mathbf{e}_\pi(t)) =: \mathbf{w}(t) \prod_{l=1}^{k} (t - x_{il})$$

We claim that the function $\theta_i^{-k+1}\mathbf{w}(t)$ has $p - k$ bounded derivatives and can therefore be written, using for instance a local Taylor expansion, as

$$\theta_i^{-k+1}\mathbf{w}(t) = \boldsymbol{\phi}(t) + O(h_i^{p-k})$$

for some $\boldsymbol{\phi}(t) \in \mathbf{P}_{p-k}$, $x_i \le t \le x_{i+1}$. To see this, note that by assumption $A(x)$, $\mathbf{q}(x) \in C^{(p)}[a,b]$, which implies that $\mathbf{y}(x)$, $Y(x) \in C^{(p+1)}[a,b]$. This leaves us to consider $\mathbf{L}\,\mathbf{e}_\pi(t)$. Since $\mathbf{w}(t)$ contains the kth derivative of $\mathbf{L}\,\mathbf{e}_\pi(t)$, we must show that $\theta_i^{-k+1}\mathbf{L}\,\mathbf{e}_\pi(t)$ has p bounded derivatives. This will follow if we show that the derivatives of $\mathbf{e}_\pi(t)$ are appropriately bounded. The latter follows from the estimate (5.75b) of Theorem 5.75, the smoothness of $\mathbf{y}(t)$, and the fact that for any positive integer j

$$\mathbf{y}_\pi^{(k+j)}(t) \equiv \mathbf{0}$$

Now, (5.61b) implies that

$$\int_{x_i}^{x_{i+1}} \boldsymbol{\phi}(t) \prod_{l=1}^{k} (t - x_{il})\,dt = 0$$

so

$$|\mathbf{v}(x_{i+1}) - \mathbf{y}_{i+1}| = O(h_i^{p+1})\theta_i^{k-1} = O(h_i^{p+2-k} h^{k-1})$$

and (5.79a) is obtained from (5.80). Finally, (5.79b) follows directly from (5.78b). □

From (5.79a, b) we see that, for h small enough, the error at mesh points is particularly small compared to the error at collocation points when $p > k + 1$ and $h = O(h_i)$. The higher-order convergence at mesh points is called *superconvergence*. The proof of superconvergence given above is perhaps less elegant than possible, because we have tied it to the local truncation error and to the theory of Section 5.2.1. Instead, we can write

$$\mathbf{y}_\pi(x) - \mathbf{y}(x) = \int_a^b G(x,t) \mathbf{L}(\mathbf{y}_\pi(t) - \mathbf{y}(t))\, dt$$

or

$$\mathbf{y}_\pi(x) - \mathbf{y}(x) = \sum_{j=1}^{N} \int_{x_j}^{x_{j+1}} G(x,t)(\mathbf{L}\,\mathbf{y}_\pi(t) - \mathbf{q}(t))\, dt \tag{5.81a}$$

where $G(x,t)$ is the Green's function, given in (3.32). By the smoothness assumptions of Theorem 5.79, $G(x,t)$ is smooth as a function of t in $[a,x]$ and in $(x,b]$, but it has a jump discontinuity at $t = x$. Thus, if $x \in (x_i, x_{i+1})$ then for $j \neq i$ we obtain, precisely as in the proof of Theorem 5.79,

$$\int_{x_j}^{x_{j+1}} G(x,t)(\mathbf{L}\,\mathbf{y}_\pi(t) - \mathbf{q}(t))\, dt = O(h_j^{p+2-k} h^{k-1}) \qquad x \notin (x_j, x_{j+1}) \tag{5.81b}$$

while

$$\int_{x_i}^{x_{i+1}} G(x,t)(\mathbf{L}\,\mathbf{y}_\pi(t) - \mathbf{q}(t))\, dt = O(h^{k+1}) \qquad x \in (x_i, x_{i+1}) \tag{5.81c}$$

The beauty of this argument is in the way that superconvergence is obtained: When x is a mesh point, there is simply no i such that $x \in (x_i, x_{i+1})$! Hence the sum in (5.81a) involves only terms estimated by (5.81b), and the superconvergence result (5.79a) follows. The estimate (5.79b) is obtained as before from (5.78b). Note that using (5.78c), a similar $O(h_i^{k+1}) + O(h^p)$ estimate is obtained at collocation points for $\mathbf{e}_\pi'(x_{ij})$. This yields, in fact,

$$|\mathbf{y}_\pi(x) - \mathbf{y}(x)| = O(h_i^{k+1}) + O(h^p) \qquad x_i < x < x_{i+1}, \qquad 1 \leq i \leq N \tag{5.82}$$

5.4.2 Nonlinear problems

For the solution of nonlinear BVPs by collocation, we now describe quasilinearization in algorithmic form.

Algorithm: Collocation with quasilinearization

Input: (a) A procedure which returns values of $\mathbf{f} = \mathbf{f}(x, \mathbf{y})$ and $A = \dfrac{\partial \mathbf{f}}{\partial \mathbf{y}}(x, \mathbf{y})$ for given values of x and $\mathbf{y}$.

(b) A procedure which returns values of $\mathbf{g}(\mathbf{u},\mathbf{v})$, $B_1 = \dfrac{\partial \mathbf{g}}{\partial \mathbf{u}}(\mathbf{u},\mathbf{v})$ and $B_2 = \dfrac{\partial \mathbf{g}}{\partial \mathbf{v}}(\mathbf{u},\mathbf{v})$ for given values of $\mathbf{u}$ and $\mathbf{v}$.

(c) An initial solution profile $\mathbf{y}_\pi(x)$ [or, equivalently, the values $\mathbf{y}_1, \mathbf{f}_1, \mathbf{y}_2, \mathbf{f}_2, \ldots, \mathbf{y}_N, \mathbf{f}_N, \mathbf{y}_{N+1}$, where $\mathbf{f}_{ij} = \mathbf{y}_\pi'(x_{ij})$].

(d) A mesh π and a tolerance TOL.

Output: Collocation solution which satisfies the collocation equations (5.72) to the extent that the difference between two consecutive Newton iterates is at most TOL.

REPEAT
1. {BC equations}
    Set B_1 and B_2 at $\mathbf{u} = \mathbf{y}_1$, $\mathbf{v} = \mathbf{y}_{N+1}$;
    $\boldsymbol{\beta} := -\mathbf{g}(\mathbf{y}_1, \mathbf{y}_{N+1})$.
2. FOR $i = 1, \ldots, N$ DO
    {For each subinterval, assemble collocation equations}
    FOR $j = 1, \ldots, k$ DO
    {Get relevant quantities at a collocation point}

    $$x_{ij} := x_i + h_i \rho_j$$

    $$\mathbf{y}_{ij} := \mathbf{y}_i + h_i \sum_{l=1}^{k} \alpha_{jl} \mathbf{f}_{il}$$

    $$A(x_{ij}) := \frac{\partial \mathbf{f}}{\partial \mathbf{y}}(x_{ij}, \mathbf{y}_{ij})$$

    $$\mathbf{q}(x_{ij}) = \mathbf{f}(x_{ij}, \mathbf{y}_{ij}) - \mathbf{f}_{ij}$$

    END FOR {j}

    {Local elimination by (5.64)}

    $$U_i := W_i^{-1} V_i; \qquad \mathbf{p}_i := W_i^{-1} \mathbf{q}_i$$

3. Set up and solve the linear system
    {with Γ_i and $\mathbf{r}_i$ given by (5.65)}

$$\begin{bmatrix} -\Gamma_1 & I & & & \\ & -\Gamma_2 & I & & \\ & & \ddots & & \\ & & & -\Gamma_N & I \\ B_1 & & & & B_2 \end{bmatrix} \begin{bmatrix} \mathbf{w}_1 \\ \mathbf{w}_2 \\ \vdots \\ \vdots \\ \mathbf{w}_{N+1} \end{bmatrix} = \begin{bmatrix} \mathbf{r}_1 \\ \mathbf{r}_2 \\ \vdots \\ \mathbf{r}_N \\ \boldsymbol{\beta} \end{bmatrix}$$

4. {Update approximate solution for each subinterval}
    FOR $i = 1, \ldots, N$ DO

    $\mathbf{y}_i := \mathbf{y}_i + \mathbf{w}_i$;

    $\mathbf{f}_i := \mathbf{f}_i + U_i \mathbf{w}_i + \mathbf{p}_i$

5. {Finish update}

    $\mathbf{y}_{N+1} := \mathbf{y}_{N+1} + \mathbf{w}_{N+1}$;

UNTIL $\|\mathbf{w}_\pi\| \leq TOL$ (or iteration limit exceeded) □

Example 5.8

For the nonlinear Example 3.2, treated numerically also in Examples 5.3 and 4.13, we obtain the results shown in Table 5.5 (with the error e being the maximum error at mesh points).

TABLE 5.5: Higher-order collocation for a simple nonlinear problem

h	e (2-Gauss)	e (3-Lobatto)	e (3-Gauss)	e (4-Lobatto)
.2	.26 – 6	.16 – 5	.10 – 8	.10 – 9
.1	.18 – 7	.11 – 6	.16 – 10	.14 – 11
.05	.11 – 8	.67 – 8	.26 – 12	.21 – 13
.025	.69 – 10	.42 – 9	.45 – 14	.46 – 15
.0125	.43 – 11	.26 – 10	.87 – 15*	.79 – 15*

* These values are mainly roundoff errors.

In all cases listed here, we use the initial guess $\mathbf{y}_\pi(x) \equiv \mathbf{0}$ and need two Newton iterations to converge. Comparing these results to those in Table 5.3 or Table 4.4, the advantage of higher-order collocation schemes is clear. Note, though, that we do more work here per step h than for the midpoint scheme, say (cf. Section 5.3.3).

The fourth-order accuracy of the two-stage Gauss and the three-stage Lobatto schemes, and the sixth-order accuracy of the three-stage Gauss and four-stage Lobatto schemes are demonstrated in this nonlinear case. We also observe a difference, by roughly a constant factor, between the errors for schemes of the same order. This is due to some symmetry that holds for this particular example and is of no general importance. □

Let us now turn to the theory supporting the algorithm of collocation and quasilinearization. As mentioned before, the one-step theory of Section 5.2.2 does not quite cover everything here, because the local elimination of $\mathbf{f}_i$, which makes the remaining process a one-step scheme, is done only after linearization in each iteration.

Nevertheless, the modifications required to the one-step theory are not essential, and we discuss them in an informal manner. To simplify notation, we assume that the BC are linear; i. e., we consider the nonlinear problem

$$\mathbf{y}' = \mathbf{f}(x, \mathbf{y}) \qquad a < x < b \qquad (5.83a)$$

$$B_a \mathbf{y}(a) + B_b \mathbf{y}(b) = \boldsymbol{\beta} \qquad (5.83b)$$

We denote by $\mathbf{y}(x)$ an isolated solution of (5.83), on which we concentrate. Recall the linear operator

$$L[\mathbf{u}]\mathbf{z} \equiv \mathbf{z}' - A[\mathbf{u}]\mathbf{z} \qquad (5.84a)$$

where

$$A[\mathbf{u}](x) \equiv \frac{\partial \mathbf{f}}{\partial \mathbf{y}}(x, \mathbf{u}(x)) \qquad (5.84b)$$

[cf. (5.45), (5.44a)]. The corresponding collocation discretization (5.59) is denoted by $L_\pi[\mathbf{u}]$. Thus, $L_\pi[\mathbf{u}]$ is the collocation operator for the linear differential operator obtained by linearizing (5.83a) about $\mathbf{u}(x)$.

The algorithm of quasilinearization and collocation may be described as follows: Given an initial solution $\mathbf{y}_\pi^0(x)$, successive solution iterates $\mathbf{y}_\pi^{m+1}(x)$ are generated, where $\mathbf{y}_\pi^{m+1}(x)$ is the collocation solution of the linear BVP

$$\mathbf{L}\,[\mathbf{y}_\pi^m]\,\mathbf{z}(x) = \mathbf{f}(x,\mathbf{y}_\pi^m(x)) - A\,[\mathbf{y}_\pi^m](x)\,\mathbf{y}_\pi^m(x) \qquad a < x < b \qquad (5.85a)$$

$$B_a\,\mathbf{z}(a) + B_b\,\mathbf{z}(b) = \boldsymbol{\beta} \qquad (5.85b)$$

$m = 0, 1, \ldots$.

For the linearized BVP, the collocation method with local parameter elimination is a one-step method. It is not difficult to see that those parts of assumption 5.51 and Theorem 5.52 which relate to linearizations hold here. The stability results obtained at mesh points are then extended for the entire collocation solution everywhere, as in Theorem 5.75. Thus, corresponding to $A[\mathbf{y}]$ in (5.51)–(5.52), we have the collocation discretization for the linear problem

$$\mathbf{L}\,[\mathbf{y}]\mathbf{z}(x) = \mathbf{f}(x,\mathbf{y}(x)) - A\,[\mathbf{y}](x)\,\mathbf{y}(x) \qquad (5.86)$$

and (5.85b); i.e., the linearization is done about the exact solution $\mathbf{y}(x)$. Note that $\mathbf{y}(x)$ satisfies (5.86), (5.85b). Of course, $\mathbf{y}(x)$ is not known, but we may still consider this linear collocation process, denote its solution by $\mathbf{y}_\pi^0(x)$, and obtain from Theorems 5.75 and 5.79 that, with θ_i given in (5.75c),

$$\max_{x_i \le x \le x_{i+1}} \left| \frac{d^j}{dx^j}(\mathbf{y}_\pi^0(x) - \mathbf{y}(x)) \right| = O(h^{k+1-j})\theta_i^{j-1} \qquad 1 \le j \le k, \qquad 1 \le i \le N \qquad (5.87a)$$

$$|\mathbf{y}_\pi^0(x_i) - \mathbf{y}(x_i)| = O(h^p) \qquad 1 \le i \le N+1 \qquad (5.87b)$$

$$|\mathbf{y}_\pi^0(x) - \mathbf{y}(x)| = O(h_i^{k+1}) + O(h^p) \qquad x_i \le x \le x_{i+1}, \qquad 1 \le i \le N \qquad (5.87c)$$

Using this unknown-but-existing $\mathbf{y}_\pi^0(x)$ as a starting solution profile for the quasi-linearization (5.85), we can apply the Newton-Kantorovich Theorem 2.64, because the bounds in (2.64) [cf. (5.55)] are satisfied for h small enough, with α and γ constants and $\beta = O(h^k)$ according to (5.87a). This yields the existence of a solution $\mathbf{y}_\pi(x)$ to the *nonlinear* collocation equations for (5.83) and the quadratic convergence of the Newton iterates to it. We also obtain the convergence estimate (5.75).

Finally, we write (5.83a) as

$$\mathbf{z}' - A\,[\mathbf{y}]\mathbf{z} = \mathbf{f}(x,\mathbf{z}) - A\,[\mathbf{y}]\mathbf{z}$$

and use a Taylor expansion

$$\mathbf{f}(x,\mathbf{z}) = \mathbf{f}(x,\mathbf{y}) + A\,[\mathbf{y}](\mathbf{z}-\mathbf{y}) + O(|\mathbf{z}-\mathbf{y}|^2)$$

to obtain that $\mathbf{y}_\pi(x)$ is the collocation approximation to the solution of

$$\mathbf{L}\,[\mathbf{y}]\mathbf{z}(x) = \mathbf{f}(x,\mathbf{y}(x)) - A\,[\mathbf{y}](x)\,\mathbf{y}(x) + O(|\mathbf{z}-\mathbf{y}|^2)$$

and (5.85b). Comparing this to (5.86), which has the collocation solution $\mathbf{y}_\pi^0(x)$, and using the stability of $\mathbf{L}_\pi[\mathbf{y}]$, we have

$$\max_{a \le x \le b} |\mathbf{y}_\pi^0(x) - \mathbf{y}_\pi(x)| \le c \max_{a \le x \le b} |\mathbf{y}_\pi(x) - \mathbf{y}(x)|^2 = O(h^{2k})$$

So for each x

$$|\mathbf{y}(x) - \mathbf{y}_\pi(x)| \le |\mathbf{y}(x) - \mathbf{y}_\pi^0(x)| + |\mathbf{y}_\pi^0(x) - \mathbf{y}_\pi(x)|$$

$$= |\mathbf{y}(x) - \mathbf{y}_\pi^0(x)| + O(h^{2k})$$

and using (5.87) we retrieve all the results of Theorems 5.75 and 5.79 for the nonlinear problem as well. We summarize this as follows:

Theorem 5.88 Let $\mathbf{y}(x)$ be an isolated solution of the BVP (5.20) with smoothness assumptions as before. Then for each k-stage collocation scheme considered, there are positive constants ρ and h_0 such that the following hold for all meshes with $h \leq h_0$.

(a) There is a unique solution $\mathbf{y}_\pi(x)$ to the collocation equations (5.72) in a tube of radius ρ around $\mathbf{y}(x)$.

(b) This solution $\mathbf{y}_\pi(x)$ can be obtained by Newton's method, which converges quadratically provided the initial iterate $\mathbf{y}_\pi^0(x)$ is sufficiently close to $\mathbf{y}(x)$.

(c) The following error estimates hold:

$$|\mathbf{y}_i - \mathbf{y}(x_i)| = O(h^p) \qquad 1 \leq i \leq N$$

$$|\mathbf{y}_\pi(x) - \mathbf{y}(x)| = O(h_i^{k+1}) + O(h^p) \qquad x_i \leq x \leq x_{i+1}, \qquad 1 \leq i \leq N \qquad \square$$

5.5 ACCELERATION TECHNIQUES

Acceleration techniques are a very powerful tool for improving the accuracy of a basic, simple finite difference method, such as that based on the trapezoidal scheme or on the midpoint scheme. Recall that these basic schemes, presented in Section 5.1, were only second-order accurate. One way of increasing the order of accuracy, considered in the previous section, is to use higher-order (but per step more expensive) Runge-Kutta schemes. The alternative approach considered here is to use the basic scheme of choice, but in a more involved way. We describe two such ways, extrapolation and deferred corrections.

To understand the underlying principles, let us recall how convergence is shown in Section 5.2. For a one-step scheme (our basic scheme) we have a local truncation error $\tau_i[\mathbf{y}]$ and a stability bound [see, e.g., (5.31)]. We then observe that the error satisfies the difference equations with the local truncation error as the inhomogeneity [see, e. g., (5.34)]. Therefore, substituting into the stability bound, the error is seen to be of the same order in h as the local truncation error.

But suppose now that we know more about $\tau_i[\mathbf{y}]$ than just its order. If we have a way of *estimating* the local truncation error, then perhaps this knowledge can also be translated to a corresponding estimate of the error! Such an estimate for the error can, in turn, be used for an error control in an adaptive code (as discussed in Chapter 9), or it can be added to the approximate solution to improve its accuracy.

In order to be more specific, we consider throughout this section the trapezoidal scheme (5.15) or (5.21) as the *basic scheme*, i. e., the scheme applied in the *basic method*. Note, though, that other symmetric one-step schemes could be used as well. The local truncation error for the trapezoidal scheme is $O(h_i^2)$. This is all that is needed to show $O(h^2)$ error in the solution. But, in fact, there is much more structure in $\tau_i[\mathbf{y}]$. For the general system of ODEs

$$\mathbf{y}' = \mathbf{f}(x, \mathbf{y})$$

the local truncation error can be expanded in a series

$$\tau_i[\mathbf{y}] = \sum_{j=1}^{r} h_i^{2j} \mathbf{T}_j [\mathbf{y}(x_{i+1/2})] + O(h_i^{2r+2}) \qquad 1 \le i \le N \qquad (5.89)$$

if $\mathbf{f}$ possesses continuous partial derivatives up to order $2r + 2$ [i. e., the smoothness of $\mathbf{f}$ determines the positive integer r in (5.89)]. A simple exercise with Taylor expansions shows that for the trapezoidal scheme,

$$\mathbf{T}_j[\mathbf{z}(x)] = \frac{-j}{2^{2j-1}(2j+1)!} \mathbf{f}^{(2j)}(x, \mathbf{z}(x)) \qquad (5.90)$$

where

$$\mathbf{f}^{(2j)}(x, \mathbf{z}(x)) \equiv \frac{d^{2j}}{dx^{2j}} \mathbf{f}(x, \mathbf{z}(x))$$

Thus, it is natural to ask whether stability yields a series expansion corresponding to (5.89) for the global error, viz.,

$$\mathbf{y}(x_i) - \mathbf{y}_i = \sum_{j=1}^{r} h_i^{2j} \mathbf{d}_j(x_i) + O(h^{2r+2}), \qquad 1 \le i \le N+1 \qquad (5.91)$$

where $\mathbf{d}_j(x)$ are bounded functions on [a, b] which are independent of the mesh. In Section 5.5.1 we show that the answer is affirmative, provided that the family of meshes considered is restricted as follows:

Assumption 5.92 There is a constant $K > 0$ independent of h such that for all meshes considered,

$$h/h_i \le K \qquad \text{all } i \qquad (5.92)$$

A mesh satisfying (5.92) will be called *quasiuniform*. □

Note that in (5.91), the $O(h^{2r+2})$ remainder term is guaranteed to be smaller than the other terms in the expansion (as $h \to 0$) only under the quasiuniformity assumption. We thus assume throughout this section that (5.92) holds. Also, for convenience of presentation, we assume that any derivatives used are continuous.

Given that an error expansion (5.91) exists, we may attempt to find suitable approximations to the so-called *principal error functions* $\mathbf{d}_j(x)$ and improve the accuracy of the basic scheme. For instance, if we find an $O(h^2)$ approximation $\mathbf{v}_i$ to $\mathbf{d}_1(x_i)$, $1 \le i \le N+1$, then $\mathbf{y}_i + h_i^2 \mathbf{v}_i$ is an $O(h^4)$ approximation to $\mathbf{y}(x_i)$. Similarly, if $\mathbf{v}_i - \mathbf{d}_1(x_i) = O(h^4)$ and $\mathbf{w}_i - \mathbf{d}_2(x_i) = O(h^2)$, $1 \le i \le N+1$, then $\mathbf{y}_i + h_i^2 \mathbf{v}_i + h^4 \mathbf{w}_i$ is an $O(h^6)$ approximation to $\mathbf{y}(x_i)$ (assuming $r \ge 3$), because from (5.91)

$$\mathbf{y}(x_i) - (\mathbf{y}_i + h_i^2 \mathbf{v}_i + h_i^4 \mathbf{w}_i) = h_i^2 (\mathbf{d}_1(x_i) - \mathbf{v}_i) + h_i^4 (\mathbf{d}_2(x_i) - \mathbf{w}_i) + O(h^6)$$

$$= O(h^6) \qquad 1 \le i \le N+1$$

An alternative way of using $\mathbf{v}_i$ and $\mathbf{w}_i$ is to obtain a fourth-order method with an *error estimate*, because

$$\mathbf{y}(x_i) - (\mathbf{y}_i + h_i^2 \mathbf{v}_i) = h_i^4 \mathbf{w}_i + O(h^6) \qquad 1 \le i \le N+1$$

For this latter purpose it would suffice that $\mathbf{v}_i$ be an $O(h^3)$ approximation to $\mathbf{d}_1(x_i)$ and that $\mathbf{w}_i$ be an $O(h)$ approximation to $\mathbf{d}_2(x_i)$.

In Section 5.5.2 we consider one technique to approximate the principal error functions and thereby accelerate convergence. This is the method of *extrapolation*, also known as *Richardson's extrapolation*. It is well-known also in numerical quadrature and in the solution of IVPs. The basic advantage of this method is its simplicity. Another technique, considered in Section 5.5.3 and in Section 5.5.4, is that of *deferred corrections*. The basic advantage with it is that the same mesh is used for all corrections, while with extrapolation finer and finer meshes need to be used in order to obtain the desired increase in the order of convergence.

5.5.1 Error expansion

Before considering the specific techniques for convergence acceleration, we prove that the error in the numerical solution can be expanded in the form (5.91), given that a similar expansion (5.89) holds for the local truncation error. Since the proof is somewhat long, we present it for linear BVPs only and merely state the theorem for the nonlinear case. The more casual reader is asked to at least read the statements of these two theorems. As usual, we consider the linear case first.

Theorem 5.93 Suppose that the linear BVP

$$\mathbf{y}' = \mathbf{f}(x, \mathbf{y}) \equiv A(x)\mathbf{y} + \mathbf{q}(x) \qquad a < x < b$$

$$B_a \mathbf{y}(a) + B_b \mathbf{y}(b) = \boldsymbol{\beta}$$

has a unique solution $\mathbf{y}(x)$ and that $\mathbf{f} \in C^{(2r+2)}[a, b]$. Then the trapezoidal solution $\mathbf{y}_\pi$ on quasiuniform meshes with h small enough satisfies (5.91). The principal error functions are defined as the solutions of the BVPs

$$\mathbf{L}\mathbf{d}_j(x) \equiv \mathbf{d}_j'(x) - A(x)\mathbf{d}_j(x) = \mathbf{T}_j[\mathbf{y}(x)] - \sum_{l=1}^{j-1} \mathbf{T}_l[\mathbf{d}_{j-l}(x)] \qquad a < x < b \qquad (5.93a)$$

$$B_a \mathbf{d}_j(a) + B_b \mathbf{d}_j(b) = \mathbf{0} \qquad (5.93b)$$

for $1 \leq j \leq r$.

Proof: Note first that the functions $\mathbf{d}_j(x)$ are (recursively) well-defined in (5.93), independent of any mesh considerations. Given a trapezoidal approximation $\mathbf{y}_\pi$ on a mesh π, we define $\mathbf{v}_\pi = (\mathbf{v}_1, \ldots, \mathbf{v}_N + 1)^T$ by

$$\mathbf{v}_i := \mathbf{y}(x_i) - \mathbf{y}_i - \sum_{j=1}^{r} h_i^{2j} \mathbf{d}_j(x_i) \qquad 1 \leq i \leq N+1$$

Our task is then to show that $|\mathbf{v}_\pi| = O(h^{2r+2})$. It suffices to show that

$$\mathbf{L}_\pi \mathbf{v}_i = O(h_i^{2r+2}) \qquad 1 \leq i \leq N \qquad (5.94)$$

[using the operator notation of (5.29b)], because

$$B_a \mathbf{v}_1 + B_b \mathbf{v}_{N+1} = \mathbf{0}$$

and the basic scheme is stable. [Indeed, the local step h_i in (5.94) is replaced by the global step h in the remainder term of (5.91) *because* of the use of the stability bound (5.31c).]

To show (5.94) we write

$$\mathbf{L}_\pi \mathbf{v}_i = \mathbf{L}_\pi(\mathbf{y}(x_i) - \mathbf{y}_i) - \sum_{j=1}^{r} h_i^{2j} \mathbf{L}_\pi \mathbf{d}_j(x_i) = \tau_i[\mathbf{y}] - \sum_{j=1}^{r} h_i^{2j} \mathbf{L}_\pi \mathbf{d}_j(x_i)$$

and using the expansion (5.89) for $\tau_i[\mathbf{y}]$,

$$\mathbf{L}_\pi \mathbf{v}_i = \sum_{j=1}^{r} h_i^{2j} \{\mathbf{T}_j[\mathbf{y}(x_{i+1/2})] - \mathbf{L}_\pi \mathbf{d}_j(x_i)\} + O(h_i^{2r+2}) \tag{5.95}$$

For the functions $\mathbf{T}_j$ given by (5.90), $\mathbf{T}_j \in C^{(2r+2(1-j))}$ because $\mathbf{f} \in C^{(2r+2)}$. From (5.93a, b), a simple induction argument shows that $\mathbf{d}_j \in C^{(2(r-j)+3)}[a,b]$. Therefore, we can use the local error expansion (5.89) with $\mathbf{d}_j(x)$ replacing $\mathbf{y}(x)$ to obtain

$$\mathbf{L}_\pi \mathbf{d}_j(x_i) = \mathbf{L} \mathbf{d}_j(x_{i+1/2}) + \sum_{l=1}^{r-j} h_i^{2l} \mathbf{T}_l[\mathbf{d}_j(x_{i+1/2})] + O(h_i^{2(r-j+1)})$$

For $\mathbf{L} \mathbf{d}_j(x_{i+1/2})$ we use (5.93a), obtaining

$$\mathbf{L}_\pi \mathbf{d}_j(x_i) = \mathbf{T}_j[\mathbf{y}(x_{i+1/2})] - \sum_{l=1}^{j-1} \mathbf{T}_l[\mathbf{d}_{j-l}(x_{i+1/2})] + \sum_{l=1}^{r-j} h_i^{2l} \mathbf{T}_l[\mathbf{d}_j(x_{i+1/2})]$$

$$+ O(h_i^{2(r-j+1)})$$

and substituting into (5.95), we obtain

$$\mathbf{L}_\pi \mathbf{v}_i = \sum_{j=1}^{r} h_i^{2j} \{\sum_{l=1}^{j-1} \mathbf{T}_l[\mathbf{d}_{j-l}(x_{i+1/2})] - \sum_{l=1}^{r-j} h_i^{2l} \mathbf{T}_l[\mathbf{d}_j(x_{i+1/2})] + O(h_i^{2(r-j+1)})\}$$

Comparing equal powers of h_i in the last expression, we see that all but the highest powers vanish. The estimate (5.94) is therefore obtained. □

Consider now the nonlinear case.

Theorem 5.96 Let the nonlinear BVP

$$\mathbf{y}' = \mathbf{f}(x, \mathbf{y}) \qquad a < x < b$$

$$\mathbf{g}(\mathbf{y}(a), \mathbf{y}(b)) = \mathbf{0}$$

have an isolated solution $\mathbf{y}(x)$ and suppose that the assumptions of Theorem 5.93 hold. Then the error expansion (5.91) is valid here as well. The principal error functions are solutions of the linearized BVPs

$$\mathbf{L}[\mathbf{y}]\mathbf{d}_j \equiv \mathbf{d}_j' - A(x)\mathbf{d}_j = \mathbf{s}_j(x) \qquad a < x < b \tag{5.96a}$$

$$B_a \mathbf{d}_j(a) + B_b \mathbf{d}_j(b) = \mathbf{\gamma}_j \tag{5.96b}$$

for $1 \leq j \leq r$, where $A(x)$, B_a, and B_b are defined in (5.44); the inhomogeneous terms $\mathbf{s}_j(x)$ and $\mathbf{\gamma}_j$ depend on $\mathbf{T}_l$ $(1 \leq l \leq j)$, $\mathbf{d}_l$ $(1 \leq l < j)$, and their derivatives. □

We omit the proof of this theorem because of its technical nature and because the important aspects have already been covered in Theorem (5.93).

Remarks

(a) In the expansions (5.89) and (5.91), the remainder term $O(h^{2r+2})$ appears with an even power. Correspondingly, we have to assume in Theorems 5.93 and 5.96 that $\mathbf{f} \in C^{(5,2r+2)}[a,b]$. This can be relaxed to $\mathbf{f} \in C^{(2r+1)}[a,b]$, but then the remainder term in (5.91) is $O(h^{2r+1})$ in general.

(b) The values $\mathbf{y}_i$ of the approximate solution in (5.91) have been tacitly assumed to satisfy the difference equations *exactly*. In practical computation there are also roundoff errors and, in the case of a nonlinear BVP, a nonlinear convergence error. If we want to make use of (5.91) to obtain, say, an $O(h^{2k})$ approximation of $\mathbf{y}(x_i)$, then we must ensure that sufficiently many nonlinear iterations have been performed in obtaining $\mathbf{y}_i$ and that the machine precision constant is sufficiently small that the difference between $\mathbf{y}_i$ and its calculated value is below the $O(h^{2k})$ level.

(c) Let us add a few words regarding the quasiuniformity assumption 5.92. What it says is that the mesh cannot be much coarser in one part of the interval than in another. This is potentially annoying and contrary to the spirit in which we would like to conduct our adaptive mesh selection in Chapter 9 (namely, based on local accuracy needs alone). The practical question is how K of (5.92) appears in the actual error for a given computation. If $K \sim 1$ is necessary for a given method, then we would have to abandon adaptive mesh selection for that method. Note, though, that to obtain an $O(h^{2k})$ approximation ($k \leq r$) using approximations to the principal error functions $\mathbf{d}_1, \ldots, \mathbf{d}_{k-1}$ in (5.91), we can relax (5.92) somewhat to read

$$h/h_i \leq K/h^{r-k+1/k}$$

(see Exercise 16). Thus, if $\mathbf{f}$ is so smooth that r can be taken significantly larger than k, then the mesh restriction becomes much more tolerable. For instance, if we want a sixth-order approximation and $\mathbf{f}$ has 12 continuous derivatives, then we only need $h_i \geq Kh^2$ for all i.

5.5.2 Extrapolation

The essence of the extrapolation technique is the construction of successively higher-order approximate solutions from linear combinations of approximate solutions obtained by using the basic scheme on different meshes. Its success depends upon the existence of an error expansion like (5.91). The principal error functions $\mathbf{d}_j(x)$ are not known (or else we could use them directly to improve upon the accuracy of the approximate solution), but the knowledge of the powers of h_i appearing in the error expansion is sufficient to eliminate them.

Specifically, given any quasiuniform mesh π, we use the basic scheme to compute solutions on π and on refinements of π. These refinements are formed by subdividing each subinterval of π into some fixed number of subintervals (perhaps we should say sub-subintervals).

The simplest example of this is when each subinterval of π is halved. If we denote by $\{y_i^{(1)}\}_{i=1}^{N}+1$ the trapezoidal solution on $\pi_1 \equiv \pi$ and by $\{y_i^{(2)}\}_{i=1}^{2N}+1$ the trapezoidal solution on $\pi_2 = \{\hat{x}_j : \hat{x}_{2i-1} := x_i, i = 1, \ldots, N+1, \hat{x}_{2i} := x_{i+1/2}, i = 1, \ldots, N\}$, then (5.91) gives

$$y(x_i) - y_i^{(1)} = \sum_{j=1}^{r} h_i^{2j} d_j(x_i) + O(h^{2r+2})$$

$$y(x_i) - y_{2i-1}^{(2)} = \sum_{j=1}^{r} (\frac{h_i}{2})^{2j} d_j(x_i) + O(h^{2r+2})$$

Thus, the mesh function $y_\pi^{(1,2)} = (y_1^{(1,2)}, \ldots, y_{N+1}^{(1,2)})^T$ defined by

$$y_i^{(1,2)} := \frac{1}{3}(4y_{2i-1}^{(2)} - y_i^{(1)}) \tag{5.97}$$

is a fourth-order approximate solution, because d_1 has been eliminated and

$$y(x_i) - y_i^{(1,2)} = -\frac{h_i^4}{4}d_2(x_i) - \frac{5h_i^6}{16}d_3(x_i) + O(h^8) = O(h^4) \tag{5.98}$$

Similarly, an $O(h^{2k})$ approximate solution is obtained on the mesh π (provided that $\mathbf{f} \in C^{(2k)}[a,b]$) by computing k solutions $\{y_i^{(1)}\}, \ldots, \{y_i^{(k)}\}$ using the same basic scheme on k different meshes $\pi_1, \ldots, \pi_k$. The mesh π_j is obtained by subdividing each subinterval $[x_i, x_{i+1}]$ of π into p_j equal subintervals $(1 = p_1 < p_2 < \cdots < p_k)$. Two possible choices for p_j are

$$p_j = j \quad 1 \leq j \leq k \tag{5.99a}$$

and

$$p_j = 2^{j-1} \quad 1 \leq j \leq k \tag{5.99b}$$

As before, we have for each approximate solution $y^{(j)}$ and each mesh point x_i

$$y(x_i) - y_{(i-1)p_j+1}^{(j)} = \sum_{l=1}^{k-1} (\frac{h_i}{p_j})^{2l} d_l(x_i) + O(h^{2k}) \quad 1 \leq j \leq k \tag{5.100a}$$

In (5.100a) we have k linear equations for k unknowns, $\mathbf{d}_1(x_i), \ldots, \mathbf{d}_{k-1}(x_i)$ and $\mathbf{y}(x_i)$. We write this as

$$\begin{bmatrix} -p_1^{-2} & -p_1^{-4} & \cdots & -p_1^{-2(k-1)} & 1 \\ -p_2^{-2} & -p_2^{-4} & \cdots & -p_2^{-2(k-1)} & 1 \\ \vdots & \vdots & & \vdots & \vdots \\ -p_k^{-2} & -p_k^{-4} & \cdots & -p_k^{-2(k-1)} & 1 \end{bmatrix} \begin{bmatrix} h_i^2 \mathbf{d}_1(x_i) \\ h_i^4 \mathbf{d}_2(x_i) \\ \vdots \\ h_i^{2(k-1)} \mathbf{d}_{k-1}(x_i) \\ \mathbf{y}(x_i) \end{bmatrix} = \begin{bmatrix} \mathbf{y}_{(i-1)p_1+1}^{(1)} \\ \mathbf{y}_{(i-1)p_2+1}^{(2)} \\ \vdots \\ \\ \mathbf{y}_{(i-1)p_k+1}^{(k)} \end{bmatrix} + O(h^{2k}) \tag{5.100b}$$

Note that the matrix in (5.100b) is essentially a nonsingular Vandermonde matrix [cf. (2.84]. It is independent of the mesh π or the particular mesh point x_i and depends only on the p_js and on the powers of h_i in the expansion (5.91) (or (5.89)). Thus we can (analytically) apply Gauss elimination once and for all to transform this matrix into an upper triangular one. Then the last row of (5.100b) gives an $O(h^{2k})$ approximation to $\mathbf{y}(x_i)$ as a linear combination of the known values $\mathbf{y}_{(i-1)p_j+1}^{(j)}$.

An alternative use of (5.100) is to both obtain an $O(h^{2(k-1)})$ approximation on π and an *error estimate*. Writing (5.100) for the rearranged vector of unknowns $(h_i^2 \mathbf{d}_1(x_i), \ldots, h_i^{2(k-2)}\mathbf{d}_{k-2}(x_i), \mathbf{y}(x_i), h_i^{2(k-1)}\mathbf{d}_{k-1}(x_i))^T$ and using the Gauss elimination procedure to obtain an upper triangular matrix, we obtain from the kth row an estimate for the unknown $\mathbf{d}_{k-1}(x_i)$ expressed in terms of computed $\mathbf{y}^{(j)}$ values. Then an approximation for $\mathbf{y}(x_i)$ is obtained from the $k-1^{st}$ row in terms of $\mathbf{y}^{(j)}$ values and the computed estimate for $\mathbf{d}_{k-1}(x_i)$.

Example 5.9

Choosing $p_j = j$, $1 \le j \le 3$, $k = 3$, we get (if $\mathbf{f} \in C^{(6)}[a, b]$)

$$\mathbf{y}(x_i) - \mathbf{y}_i^{(1)} = h_i^2 \mathbf{d}_1(x_i) + h_i^4 \mathbf{d}_2(x_i) + O(h^6)$$

$$\mathbf{y}(x_i) - \mathbf{y}_{2i-1}^{(2)} = \frac{1}{4} h_i^2 \mathbf{d}_1(x_i) + \frac{1}{16} h_i^4 \mathbf{d}_2(x_i) + O(h^6)$$

$$\mathbf{y}(x_i) - \mathbf{y}_{3(i-1)+1}^{(3)} = \frac{1}{9} h_i^2 \mathbf{d}_1(x_i) + \frac{1}{81} h_i^4 \mathbf{d}_2(x_i) + O(h^6)$$

Thus, defining

$$\mathbf{y}_i^{(1,2)} := \frac{1}{3}(4\mathbf{y}_{2i-1}^{(2)} - \mathbf{y}_i^{(1)}), \qquad \mathbf{y}_i^{(1,3)} := \frac{1}{8}(9\mathbf{y}_{3(i-1)+1}^{(3)} - \mathbf{y}_i^{(1)})$$

[this corresponds to zeroing the subdiagonal elements in the first column of the matrix in (5.100b)], we get

$$\mathbf{y}(x_i) - \mathbf{y}_i^{(1,2)} = -\frac{1}{4} h_i^4 \mathbf{d}_2(x_i) + O(h^6)$$

$$\mathbf{y}(x_i) - \mathbf{y}_i^{(1,3)} = -\frac{1}{9} h_i^4 \mathbf{d}_2(x_i) + O(h^6)$$

Similarly, defining

$$\mathbf{y}_i^{(2,3)} := \frac{1}{5}(9\mathbf{y}_i^{(1,3)} - 4\mathbf{y}_i^{(1,2)})$$

[which corresponds to the completion of the Gauss elimination process for (5.100)], we get an approximation of order 6,

$$\mathbf{y}(x_i) - \mathbf{y}_i^{(2,3)} = O(h^6)$$

Alternatively, subtracting the error expressions for $\mathbf{y}_i^{(1,2)}$ and $\mathbf{y}_i^{(1,3)}$ from each other, we have

$$\mathbf{y}_i^{(1,3)} - \mathbf{y}_i^{(1,2)} = (-\frac{1}{4} + \frac{1}{9}) h_i^4 \mathbf{d}_2(x_i) + O(h^6)$$

and so the error in the fourth-order approximation $\mathbf{y}_i^{(1,3)}$ is estimated by the computable values

$$\mathbf{y}(x_i) - \mathbf{y}_i^{(1,3)} \approx -\frac{1}{9}(-\frac{1}{4} + \frac{1}{9})^{-1}(\mathbf{y}_i^{(1,3)} - \mathbf{y}_i^{(1,2)}) = \frac{4}{5}(\mathbf{y}_i^{(1,3)} - \mathbf{y}_i^{(1,2)})$$

Below we list some calculations made for the linear BVP of Examples 5.1 and 5.7. The mesh π is chosen to be uniform with $N = 10$. Then π_2 has 20 subintervals and π_3 has 30. We list only errors in $u = y_1$, using the exact solution (which is known for this simple example) to find the maximum errors at mesh points. Under *EST* we list the error estimate, computed as described above.

TABLE 5.6 Extrapolation for Example 5.1

j	N	$e(u^{(j)})$	$e(u^{(1,j)})$	$e(u^{(2,j)})$	$EST(u^{(1,j)})$
1	10	.31 – 3	–	–	–
2	20	.76 – 4	.34 – 6	–	–
3	30	.34 – 4	.16 – 6	.74 – 8	.15 – 6

□

In Example 5.9 we have used some notation which may be generalized. Thus, if $\mathbf{y}^{(1)}, \ldots, \mathbf{y}^{(k)}$ are the basic scheme solutions on the meshes $\pi_1, \ldots, \pi_k$, then we may define $\mathbf{y}^{(l,j)}$ recursively, as shown in Fig. 5.3.

$$\mathbf{y}^{(1)}$$

$$\mathbf{y}^{(2)} \longrightarrow \mathbf{y}^{(1,2)}$$

$$\mathbf{y}^{(3)} \longrightarrow \mathbf{y}^{(1,3)} \longrightarrow \mathbf{y}^{(2,3)}$$

$$\vdots$$

$$\mathbf{y}^{(k)} \longrightarrow \mathbf{y}^{(1,k)} \longrightarrow \cdots \longrightarrow \mathbf{y}^{(k-1,k)}$$

Figure 5.3 Extrapolation

Here, $\mathbf{y}^{(l,j)}$ is an $O(h^{2(l+1)})$ approximation to $\mathbf{y}(x)$, obtained from extrapolating $\mathbf{y}^{(l-1,j)}$ and $\mathbf{y}^{(l-1,j-1)}$ (with $\mathbf{y}^{(0,j)} \equiv \mathbf{y}^{(j)}$) and corresponding to intermediate steps of the Gauss elimination process for (5.100b).

Remarks

(a) We emphasize that the solution of (5.100b) is done as part of *defining* the algorithm, not as part of the solution process for a given BVP. This is apparent in Example 5.9 and Fig. 5.3.

(b) When applying extrapolation for quadrature, or for the solution of IVPs, the choice (5.99b) is usually preferred. This is because (5.99b) is more stable for IVPs and because each mesh π_j is then contained in the next one, π_{j+1}, allowing for savings in the number of evaluations of $\mathbf{f}$ and for a simple adaptive, recursive, local step refinement. For BVPs the situation is different and it is hard to benefit from the advantages of the choice (5.99b). The choice (5.99a) can be preferred then, because it leads to the fewest number of points in the meshes π_j for a given basic mesh π.

(c) The use of (5.100), whereby an $O(h^{2(k-1)})$ approximate solution with an error estimate is obtained, is particularly attractive. The error estimate can be used to determine whether a given error tolerance has been satisfied, or to refine the mesh in regions where the error is too large (cf. Chapter 9).

(d) From the above description of extrapolation techniques, it should be clear that the use of the trapezoidal scheme for the basic method is not essential. In particular, everything remains the same if we use the midpoint scheme instead. If a higher-order Runge-Kutta scheme is used for the basic method, then the error expansion (5.91) has different powers of h_i. This in turn is reflected in some of the details of the technique, but not in anything essential.

(e) One main criticism of extrapolation is that the computation becomes expensive if many acceleration steps are performed, because the meshes become increasingly fine. For nonlinear problems, this difficulty is somewhat offset because in the majority of applications most iterations are done on the mesh π alone. After $\mathbf{y}_\pi$ is obtained, an $O(h^2)$ approximation of $\mathbf{y}(x)$ is available as an initial solution profile on the next mesh π_2. Generally, after solving for $\{\mathbf{y}_i^{(j)}\}$, $1 \le j \le k$, we may extend it (or $\mathbf{y}_\pi$) by linear interpolation to π_{j+1} and obtain an initial solution profile which is only $O(h^2)$ away from the approximate solution $\{\mathbf{y}_i^{(j+1)}\}$. If this is close enough to give quadratic convergence with Newton's method, then only $\log_2 k + 1$ iterations are needed to find $\{\mathbf{y}_i^{(j+1)}\}$ to $O(h^{2k})$. In fact, a cheaper, modified Newton method may be used here as well (see Chapter 8). Note, however, that the solutions $\{\mathbf{y}_i^{(j)}\}$, $1 \le j \le k$, must all be calculated to accuracy $O(h^{2k})$ [see remark (b) at the end of Section 5.5.1]. Moreover, unlike the method of deferred corrections which is considered next, we cannot reduce the number of Newton iterations by gradually improving the approximation. That is, the use of the $O(h^{2(l+1)})$ solution $\mathbf{y}^{(l,j)}$ as an initial guess for the nonlinear iteration to find $\mathbf{y}^{(j+1)}$ (see Fig. 5.3) has no particular advantage over using $\mathbf{y}^{(j)}$ for that purpose. (Why?)

5.5.3 Deferred corrections

The deferred corrections method is a special case of the *defect correction* method. The basic idea of the latter is to approximate the local truncation error resulting from applying a basic scheme (the "defect" when one is substituting the exact solution into the difference scheme), and then solve for a corrected approximation using this *same* discretization.

Specifically, let $\mathbf{y}_\pi$ be an approximate solution on a mesh π for a nonlinear BVP (5.20) using some basic scheme, so that

$$\mathbf{N}_\pi \mathbf{y}_i = \mathbf{0}, \qquad 1 \le i \le N \qquad (5.101a)$$

$$\mathbf{g}(\mathbf{y}_1, \mathbf{y}_{N+1}) = \mathbf{0} \qquad (5.101b)$$

Recall that the exact solution satisfies

$$\mathbf{N}_\pi \mathbf{y}(x_i) = \boldsymbol{\tau}_i[\mathbf{y}], \quad 1 \le i \le N$$

If we know an approximation $\hat{\boldsymbol{\tau}}_i$ to the local truncation error $\boldsymbol{\tau}_i[\mathbf{y}]$ such that

$$\hat{\boldsymbol{\tau}}_i = \boldsymbol{\tau}_i[\mathbf{y}] + O(h^p), \quad 1 \le i \le N \tag{5.102}$$

then we can compute a corrected solution $\hat{\mathbf{y}}_\pi$ by solving the equations

$$\mathbf{N}_\pi \hat{\mathbf{y}}_i = \hat{\boldsymbol{\tau}}_i, \quad 1 \le i \le N \tag{5.103a}$$

$$\mathbf{g}(\hat{\mathbf{y}}_1, \hat{\mathbf{y}}_{N+1}) = \mathbf{0} \tag{5.103b}$$

Substitution into the stability bound (5.49) readily yields

$$|\hat{\mathbf{y}}_i - \mathbf{y}(x_i)| \le K^* \max_j |\hat{\boldsymbol{\tau}}_j - \boldsymbol{\tau}_j[\mathbf{y}]| = O(h^p) \tag{5.104}$$

so if p is larger than the order of the basic scheme then we obtain a higher-order method, using the same mesh π and the same difference operator $\mathbf{N}_\pi$.

This defect correction idea is quite general, and (5.89) and (5.91) do not a priori have to hold. But the question is how to obtain approximations $\hat{\boldsymbol{\tau}}_i$ satisfying (5.102) with large p. The method of *(iterated) deferred corrections* gives an adequate answer by using the series expansion (5.89) of the local truncation error and approximating the functions $\mathbf{T}_j$ which appear in that expansion. Here is one reason why we have chosen the trapezoidal scheme as the basic scheme for this section: The functions $\mathbf{T}_j$ given in (5.90) have a particularly simple form, and those of the midpoint scheme are more complicated.

To obtain an $O(h^{2k})$ approximation to $\boldsymbol{\tau}_i[\mathbf{y}]$ we need, by (5.89), an $O(h^{2(k-j)})$ approximation to each $\mathbf{T}_j[\mathbf{y}(x_{i+1/2})]$, $1 \le j \le k-1$. (We assume, of course, that $k \le r$). Suppose we knew the values of $\mathbf{f}(x, \mathbf{y}(x))$ at all mesh points. Then, by passing an interpolating polynomial of order $2k$ through these values at the points $x_{i-k+1}, \ldots, x_{i+k}$ (ignoring for the moment the treatment of near-boundary mesh points) and evaluating its derivatives at $x_{i+1/2}$, we could form the desired approximations to $\mathbf{T}_j$ according to (5.90).

Alas, we do not know $\mathbf{f}(x, \mathbf{y}(x))$, only values of $\mathbf{f}(x_i, \mathbf{y}_i)$, $1 \le i \le N+1$. Still, our goal can be achieved. Initially, $\mathbf{y}_i$ is $O(h^2)$ away from $\mathbf{y}(x_i)$, so we cannot approximate $\mathbf{T}_j[\mathbf{y}(x_{i+1/2})]$ to a higher-order. Fortunately, for $k = 2$ we only need an $O(h^2)$ approximation of $\mathbf{T}_1$ to obtain an $O(h^4)$ method. More generally, if $\mathbf{y}_i$ is $O(h^{2(k-1)})$ away from $\mathbf{y}(x_i)$ for all mesh points, then the interpolating polynomial using these values can yield $O(h^{2(k-j)})$ approximations for $\mathbf{T}_j$, $1 \le j \le k-1$. The approximate solution can therefore be improved, but gradually. Let us summarize this.

Outline: Deferred corrections

Input: A BVP (5.20), a mesh π, and a positive integer k.
Output: Solution values at mesh points, $\mathbf{y}_\pi^{(k)}$, accurate to $O(h^{2k})$.

1. {Obtain $\mathbf{y}_\pi^{(1)}$ using the basic scheme}
   Solve (5.101), denote solution by $\mathbf{y}_\pi^{(1)}$.
2. {Gradually improve accuracy}
   FOR $l = 2, \ldots, k$ DO
   {Find $O(h^{2l})$ approximation for $\boldsymbol{\tau}[\mathbf{y}]$}

2.1 FOR $i = 1, \ldots, N$ DO

Pass a polynomial of order $2l$ through $2l$ neighboring $(x_j, \mathbf{f}(x_j, \mathbf{y}_j^{(l-1)}))$ points; obtain approximations $\mathbf{T}_i^{(j,l)}$ to $\mathbf{T}_j[\mathbf{y}(x_{i+1/2})]$, $1 \leq j < l$, by differentiating the polynomial interpolant and evaluating at $x_{i+1/2}$, according to (5.90).

$$\hat{\boldsymbol{\tau}}_i := \sum_{j=1}^{l-1} h_i^{2j} \mathbf{T}_i^{(j,l)}$$

2.2 Solve (5.103), denote solution by $\mathbf{y}_\pi^{(l)}$.

□

Example 5.10

Consider the case where $k = 2$ and the mesh is uniform. Then from (5.90)

$$\mathbf{T}_1[\mathbf{y}(x_{i+1/2})] = -\frac{1}{12}\mathbf{f}''(x_{i+1/2}, \mathbf{y}(x_{i+1/2}))$$

Step 2.1 in the above outline calls for a cubic interpolation of $\mathbf{f}_j = \mathbf{f}(x_j, \mathbf{y}_j)$ where $\mathbf{y}_j = \mathbf{y}_j^{(1)}$ are solution values from the basic trapezoidal method. Thus we interpolate $\mathbf{f}_j$ values at x_{i-1}, x_i, x_{i+1} and x_{i+2} and evaluate the second derivative of the resulting cubic at $x_{i+1/2}$. This gives

$$\hat{\boldsymbol{\tau}}_i = h^2 \mathbf{T}_i^{(1,2)} = \frac{1}{24}(-\mathbf{f}_{i-1} + \mathbf{f}_i + \mathbf{f}_{i+1} - \mathbf{f}_{i+2}) \quad 2 \leq i \leq N-1$$

(A more complicated expression would result if the mesh were nonuniform.) For $i = 1$ we need the value of $\mathbf{f}_0$ for this centered formula, so we use instead the interpolant at x_1, x_2, x_3, x_4, evaluate it at $x_{3/2}$, and obtain a noncentered formula

$$\hat{\boldsymbol{\tau}}_1 = \frac{1}{24}(-3\mathbf{f}_1 + 7\mathbf{f}_2 - 5\mathbf{f}_3 + \mathbf{f}_4)$$

Similarly,

$$\hat{\boldsymbol{\tau}}_N = \frac{1}{24}(-3\mathbf{f}_{N+1} + 7\mathbf{f}_N - 5\mathbf{f}_{N-1} + \mathbf{f}_{N-2})$$

In Table 5.7 we list numerical results for this deferred correction method applied to the BVP of Example 5.1, to be compared to those of Examples 5.7 and 5.9. We have computed solutions on three meshes, as in Table 5.6; but unlike Table 5.6, here they are three independent calculations. The error $e(u^{(1)})$ is identical to that in the first column of Table 5.6. (Why?) After calculating $\mathbf{y}_\pi^{(1)}$, $\hat{\boldsymbol{\tau}}_i$ is found and added to the right-hand side of (5.17). The system is solved again, this time for the corrected solution $\mathbf{y}_\pi^{(2)}$. Only forward- and back-substitutions are needed for this, since $\mathbf{A}$ is already decomposed. The resulting maximum error is listed under $e(u^{(2)})$, and the maximum difference between $u^{(1)}$ and $u^{(2)}$ is listed under $EST(u^{(1)})$. Clearly, this is a reasonable estimate for the error in $u^{(1)}$.

TABLE 5.7 Deferred correction for Example 5.1

N	$e(u^{(1)})$	$e(u^{(2)})$	$EST(u^{(1)})$
10	.31 – 3	.19 – 5	.31 – 3
20	.76 – 4	.16 – 6	.76 – 4
30	.34 – 4	.34 – 7	.35 – 4

□

The reason that the above is an "outline," not an algorithm, is that we have avoided specifying a number of important details which we now discuss.

A good deal of the interpolation process in step 2.1 of the outline can be done ahead of time by first computing and storing interpolation (and differentiation) coefficients. (This is called *preprocessing*.) Still, the overhead may be significant. The cost of this step forms an important part of the total cost of the method and involves some nontrivial decisions, as described later on and in Section 5.5.4.

On the other hand, the method also has some beautiful properties. Notice that, as with extrapolation, a similar process can be used to yield, instead of $\mathbf{y}_\pi^{(k)}$, an approximate solution $\mathbf{y}_\pi^{(k-1)}$ with an error estimate (cf. Example 5.10). Moreover, unlike extrapolation, all discretizations are done on the same mesh and using the same basic scheme $\mathbf{N}_\pi$.

For linear BVPs, the linear algebra in steps 1 and 2.2 of the outline involves the same matrix $\mathbf{A}$ of (5.17), (5.18). This matrix is decomposed for the basic scheme (5.101) anyway, so each of the correction solutions (5.103) involves only forward- and back- substitutions. This is economical compared to what is needed to obtain higher-order approximations with extrapolation.

Note that we have not actually used the error expansion (5.91), only the local truncation error expansion (5.89). Still, one useful way to understand the method is to interpret the solution process with corrections as a solution process for the principal error functions $\mathbf{d}_j(x)$ via (5.93). This also helps to explain the recursive nature of the method: In order to define the right-hand side for $\mathbf{d}_j(x)$ in (5.93a), previous $\mathbf{d}_{j-l}(x)$ functions are needed. Thus, a *gradual* buildup of accuracy is at the heart of the process.

For nonlinear BVPs, the discretization $\mathbf{N}_\pi$ in (5.101) and (5.103) involves nonlinear equations that must be solved iteratively. However, we do not use a full Newton iteration. Rather, we relate to the right hand sides in (5.101) and (5.103) as if they were simple inhomogeneities [even though $\hat{\boldsymbol{\tau}}_i$ certainly depends on the unknown solution — see (5.102)]. The quasilinearization process is applied to $\mathbf{N}_\pi$ alone. This gives a fixed-point iteration scheme with Jacobian matrices which have the structure of those in a trapezoidal scheme and can be viewed as approximations for the full Jacobian matrices for the higher-order scheme. The linearized problem (5.25), (5.26) depends, of course, on the previous iterate, but we may still view the solutions of (5.101), (5.103) as one continuous process. Thus, the basic scheme is used as usual for $\mathbf{y}^{(1)}$ and then, in step 2.2 of the outline, additional fixed-point (approximate Newton) iterations are performed. An important question is, how many iterations are needed per value of l in step 2.2?

This is another point where the deferred correction method enjoys an advantage over extrapolation. Here $\mathbf{y}_\pi^{(l-1)}$ may be used as an initial guess for $\mathbf{y}_\pi^{(l)}$, with $|\mathbf{y}_\pi^{(l-1)} - \mathbf{y}_\pi^{(l)}| = O(h^{2(l-1)})$. Also, the error tolerance for the nonlinear convergence for $\mathbf{y}_\pi^{(l)}$ can be just $o(h^{2l})$, not $O(h^{2k})$. Thus we expect one full Newton iteration per correction 2.2 to be sufficient, and we may hope one or two approximate Newton iterations per correction to be sufficient as well! Just what is needed for this expectation to be fulfilled when the approximate Newton's iterations described above are used will be discussed in the next subsection.

While its implications are perhaps not obvious at first, one of the most important considerations for deferred corrections is deciding how to perform the polynomial interpolation (on a generally nonuniform mesh) in step 2.1, and in particular, which "neighboring points" to choose. Away from boundaries we would certainly use "centered" formulas, i. e., the points $x_{i-k+1}, \ldots, x_{i+k}$. But near boundaries (when $i-k+1 < 1$ or $i+k > N+1$) something special needs to be done. One possibility is to use noncentered formulas as we did in Example 5.10. Another possibility involves obtaining approximate solution values outside the interval $[a, b]$ by stepping outside the boundary with the basic discretization and then using only centered formulas. This requires an *a priori* choice of k. A third possibility is to use extrapolation for values outside $[a, b]$, which (as it turns out) leads to noncentered differences involving more than the minimum number of points. Additional discussion of this question is provided in the next section.

5.5.4 More deferred corrections

All of the alternatives mentioned at the end of Section 5.5.3 for the implementation of deferred correction are somewhat awkward. The main advantage of those which use only centered difference formulas is that it is possible to prove (under appropriate smoothness assumptions) that the method of deferred corrections as described in the outline does indeed give an $O(h^{2k})$ accurate solution $\mathbf{y}_\pi^{(k)}$. The proof utilizes the asymptotic expansion for the error. The main disadvantage in using only centered formulas is the need to find approximations outside the interval. This presents a potentially serious difficulty for BVPs like the one in Example 4.14, which has a singularity just outside $[a, b]$.

Largely for this reason, the major implementations of deferred corrections to date have used noncentered differences near boundaries. In this case the theory using asymptotic expansions is unsatisfactory because, as it turns out, it permits only one correction. We shall now look at deferred corrections from a more detailed and slightly different viewpoint which suggests a more general theoretical framework for proving convergence results.

We begin as before by constructing a high-order finite difference method

$$\mathbf{N}_{\pi,k}\mathbf{y}_i = \mathbf{0} \qquad 1 \leq i \leq N \qquad (5.105a)$$

$$\mathbf{g}(\mathbf{y}_1, \mathbf{y}_{N+1}) = \mathbf{0} \qquad (5.105b)$$

with local truncation error

$$\mathbf{N}_{\pi,k}\mathbf{y}(x_i) = \boldsymbol{\tau}_i^k[\mathbf{y}] = O(h^{2k}) \qquad 1 \leq i \leq N \qquad (5.106)$$

A way to achieve a scheme like this is to approximate $\mathbf{f}(x, \mathbf{y}(x))$ by a polynomial of order $2k$ which interpolates $\mathbf{f}$ at $2k$ points centered (to the extent possible) at $x_{i+1/2} := \dfrac{x_i + x_{i+1}}{2}$, and use this in

$$\frac{1}{h_i} \int_{x_i}^{x_{i+1}} (\mathbf{y}'(x) - \mathbf{f}(x, \mathbf{y}(x))) \, dx = 0$$

obtaining (see Exercise 5.23)

$$\mathbf{N}_{\pi,k}\mathbf{y}(x_i) = \frac{1}{h_i}(\mathbf{y}(x_{i+1}) - \mathbf{y}(x_i)) - \sum_{j=1}^{2k} \beta_{ij} \mathbf{f}(x_{p_i+j}, \mathbf{y}(x_{p_i+j})) \qquad (5.107)$$

$$= \mathbf{N}_\pi \mathbf{y}(x_i) - \sum_{j=1}^{2k} \alpha_{ij} \mathbf{f}(x_{p_i+j}, \mathbf{y}(x_{p_i+j})) \qquad 1 \leq i \leq N$$

Here

$$p_i = \begin{cases} 0 & 1 \leq i \leq k-1 \\ i - k & \text{if} \quad k \leq i \leq N-k+1 \\ N - 2k + 1 & N-k+2 \leq i \leq N \end{cases}$$

the coefficients $\{\alpha_{ij}\}_{j=1}^{2k}$ satisfy the Vandermonde system

$$\sum_{j=1}^{2k} \left(\frac{x_{p_i+j} - x_{i+1/2}}{h_i}\right)^p \alpha_{ij} = \begin{cases} \dfrac{p}{2^p(p+1)} & p = 0, 2, \ldots, 2k-2 \\ 0 & p = 1, 3, \ldots, 2k-1 \end{cases} \quad (5.108)$$

and from (5.89)[2]

$$\mathbf{C}_k(\mathbf{y}(x_i)) := \sum_{j=1}^{2k} \alpha_{ij} \mathbf{f}(x_{p_i+j}, \mathbf{y}(x_{p_i+j})) = \sum_{j=1}^{k-1} h_i^{2j} \mathbf{T}_j[\mathbf{y}(x_{i+1/2})] + O(h_i^{2k}) \quad (5.109)$$

The scheme (5.105) is consistent of order $2k$; however, it is computationally expensive, since the Jacobian matrix has bandwidth $2kn$.

Rather than try to solve (5.105) with the standard approach, we could solve as follows: Given an initial iterate $\mathbf{y}_\pi^{(1)}$, let

$$\mathbf{N}_\pi \mathbf{y}_i^{(l)} = \mathbf{C}_k(\mathbf{y}_i^{(l-1)}) \quad 1 \leq i \leq N$$

$$\mathbf{g}(\mathbf{y}_1^{(l)}, \mathbf{y}_{N+1}^{(l)}) = \mathbf{0}$$

for $l = 2, 3, \ldots, k$ or even

$$\mathbf{N}_\pi \mathbf{y}_i^{(l)} = \mathbf{C}_l(\mathbf{y}_i^{(l-1)}) \quad 1 \leq i \leq N \quad (5.110a)$$

$$\mathbf{g}(\mathbf{y}_1^{(l)}, \mathbf{y}_{N+1}^{(l)}) = \mathbf{0} \quad (5.110b)$$

for $l = 2, \ldots, k$ (where $\mathbf{C}_l$ is defined in the obvious way). But if $\mathbf{y}_\pi^{(1)}$ is the solution of (5.101) using the basic scheme, then (5.110) is just a particular way to carry out the deferred corrections algorithm outlined before, using (5.103). Indeed, under appropriate assumptions on $\mathbf{y}_\pi^{(l-1)}$ (as we discuss below),

$$\hat{\boldsymbol{\tau}}_i := \mathbf{C}_l(\mathbf{y}_i^{(l-1)}) = \mathbf{C}_l(\mathbf{y}(x_i)) + O(h^{2l}) = \mathbf{N}_\pi \mathbf{y}(x_i) + O(h^{2l})$$

$$= \boldsymbol{\tau}[\mathbf{y}(x_i)] + O(h^{2l}) \quad (5.111)$$

Thus, by constructing higher-order finite difference schemes of the form $\mathbf{N}_{\pi,l} \equiv \mathbf{N}_\pi - \mathbf{C}_l$, we are able to bootstrap our way up to successively higher-order approximations $\mathbf{y}_\pi^{(l)}$ and corresponding truncation error approximations. It *is* a deferred corrections process, since all solution steps use the same basic discretization $\mathbf{N}_\pi$. Note that the method is stable. The stability property is inherited from that of the basic scheme and does not depend on the stability of the higher-order scheme $\mathbf{N}_{\pi,k}$.

---

[2] Notice that our notation $\mathbf{C}_k(\mathbf{y}_i)$, or for that matter $\mathbf{N}_\pi \mathbf{y}_i$ or $\mathbf{N}_{\pi,k} \mathbf{y}_i$, does not relate the natural dependence upon the entire vector $\mathbf{y}_\pi = (\mathbf{y}_1, \ldots, \mathbf{y}_{N+1})$, but we use it for simplicity.

As previously mentioned, the deferred corrections algorithm needs modification, since the use of noncentered differences is not without consequence. The problem is that the iterates $\mathbf{y}_\pi^{(l-1)}$ in (5.110) are not "smooth" enough (in a sense explained later) for (5.111) to be valid. Instead, some inner iteration with $\mathbf{C}_l$ fixed is necessary. Let I_l be the number of such inner iterations. Then we have

Algorithm: Deferred corrections using noncentered differences

Input: A BVP (5.20), a mesh π, and a positive integer k.
Output: A solution $\mathbf{y}_\pi^{(k)}$ accurate to $O(h^{2k})$.

1. Solve (5.101); denote the solution by $\mathbf{y}_\pi^{(1)}$.
2. FOR $l = 2, \ldots, k$ DO
   2.1 $\mathbf{z}_\pi^{l,1} := \mathbf{y}_\pi^{(l-1)}$
   2.2 FOR $i = 1, \ldots, N$ DO
          Solve a Vandermonde system [like (5.108)] for $\mathbf{C}_l$.
   2.3 FOR $j=1, \ldots, I_l$ DO
          Solve
   $$\mathbf{N}_\pi \mathbf{z}_i^{l,j+1} = \mathbf{C}_l(\mathbf{z}_i^{l,j}) \quad 1 \le i \le N \tag{5.112}$$
   $$\mathbf{g}(\mathbf{z}_1^{l,j+1}, \mathbf{z}_{N+1}^{l,j+1}) = 0$$
   2.4 $\mathbf{y}_\pi^{(l)} := \mathbf{z}_\pi^{l,I_l}$

□

The following theorem describes the convergence of this strategy. The proof is fairly technical and is omitted.

Theorem 5.113 Suppose that the hypotheses of Theorem 5.96 are satisfied. Then the above deferred corrections algorithm generates $\{\mathbf{y}_\pi^l\}_{l=1}^k$ such that $\mathbf{y}_\pi^{(l)}$ has accuracy $O(h^{2l})$ for $1 \le l \le k$, if one of the following holds:

(i) $I_2 = 1$ and $I_3 = \cdots = I_k = 2$; or
(ii) $I_2 = I_3 = 1, I_4 = \cdots = I_k = 2$, and [in addition to (5.92)]
$$\sum_{j=1}^{N-1} \left| \frac{h_{j+1}}{h_j} - 1 \right| \le K \tag{5.113a}$$

or

(iii) $I_2 = I_3 = I_4 = 1, I_5 = \cdots = I_k = 2$ and [in addition to (5.92)]
$$\max_{1 \le j \le N-1} \left| \frac{h_{j+1}}{h_j} - 1 \right| \le Kh \tag{5.113b}$$

□

Remarks

(a) We see from the theorem that two inner iterations are always sufficient to achieve the desired orders of accuracy. Only one inner iteration is sufficient for the first few iterations if the additional mesh restriction (5.113a) or (5.113b) is imposed.

Of course, the constant K here is assumed to be independent of the mesh used. Note that (5.113a) is not too severe and holds, for instance (with K not large), for many piecewise uniform meshes. On the other hand, (5.113b) is a rather severe restriction, requiring the mesh to be almost uniform.

(b) In describing the method, we have been imprecise in assuming that the nonlinear equations in step 2.3 are solved exactly. In practice one would generally discretize (5.112) in the standard way to obtain, say, $\mathbf{z}_\pi^{l,j+1}$ from $\mathbf{z}_\pi^{l,j}$ (cf. Section 5.1.3). However, if $\mathbf{z}_\pi^{l,j}$ is sufficiently close to $\mathbf{y}_\pi^{(l)}$ [a reasonable assumption, since $\mathbf{z}^{l,1}$ is an $O(h^{2l-2})$ approximation to $\mathbf{y}_\pi^{(l)}$], then the rapid convergence of Newton's method assures us that one iteration in (5.112) gives essentially $\mathbf{z}_\pi^{l,j+1}$. Thus it is not surprising that, as it turns out, the theorem can be extended to assume that the nonlinear equations are solved sufficiently accurately (but not exactly), without modifying the conclusions.

(c) It is easy to obtain an *a posteriori* error estimate $\mathbf{e}$ for $\mathbf{y}_\pi^{(k)}$: Compute the coefficients of $\mathbf{C}_k(\mathbf{y}_i^{(k)})$, solve the linearized version of

$$\mathbf{N}_\pi \mathbf{z}_i = \mathbf{C}_k(\mathbf{y}_i^{(k)}) \qquad 1 \le i \le N$$

$$g(\mathbf{z}_1, \mathbf{z}_{N+1}) = \mathbf{0}$$

once, and let $\mathbf{e} := \mathbf{y}_\pi^{(k)} - \mathbf{z}$. Again, the linearized system inherits the simple matrix structure of the basic scheme.

(d) It can be shown that, if a deferred corrections method with noncentered differences is implemented by using (5.110) instead of the iteration in steps 2.3 and 2.4 of the algorithm, then for the three cases in Theorem 5.113, $\mathbf{y}_\pi^{(l)}$ has accuracy $O(h^{p_l})$, where the sequence $\{p_l\}_{l=1}^k$ is

(i) $\{2, 3, 4, 5, \ldots, k+1\}$, or
(ii) $\{2, 4, 5, 6, \ldots, k+2\}$, or
(iii) $\{2, 4, 6, 7, 8, \ldots, k+3\}$, respectively.

□

We demonstrate the convergence properties given in Theorem 5.113 and in this last remark with a numerical example.

Example 5.11

The simple nonlinear BVP

$$y_1' = y_2$$

$$0 < x < 1$$

$$y_2' = -y_2 - y_1^2 + e^{2x}$$

$$y_1(0) = 1, \qquad y_1(1) = e^{-1}$$

has the solution $y_1(x) = -y_2(x) = e^{-x}$. We solve for it on a uniform mesh π using the deferred corrections algorithm, first with all $I_l = 1$ and then with all $I_l = 2$. The resulting maximum errors in π, corresponding to case (iii) for remark (d) and for Theorem 5.113, respectively, are given in Table 5.8 below, along with the measured rates of convergence.

TABLE 5.8 Deferred corrections for Example 5.11

	One iteration					Two iterations					
k\N	12		24		48		12		24		48
0	.39−3		.96−4	2.00	.24−4	2.00	x		x	x	x
1	.82−6	3.64	.56−7	3.87	.36−8	3.95	x	x	x	x	x
2	.25−8	5.92	.42−10	5.92	.69−12	5.92	.22−8	5.89	.37−10	5.88	.61−12	5.93
3			.45−12	6.91	.37−14	6.92			.45−13	7.64	.21−15	7.74
4			.32−14	8.57	.11−16	8.22			.49−16	9.73	.58−19	9.73
5			.91−17	10.35	.21−19	8.74			.60−19	11.87	.19−22	11.62
6					.45−22	10.89					.61−26	13.61
7					.16−24	11.73					.20−29	15.66

□

The above observations show the method of deferred corrections to be a strong candidate for actual implementation. With the error estimation capabilities, it is straightforward to do adaptive mesh selection. Indeed, by comparing $y_i^{(l)} - y_i^{(l-1)}$ on successive meshes [as $y(x_i) - y_i^{(l-1)}$ was compared in Table 5.8 above], we can see if the theoretical orders of convergence are being realized and thus adapt the order k to our particular solution's smoothness.

Remark

It is useful at this point to make a superficial comparison of some aspects of the deferred corrections method and a high-order collocation method, two of our favorite global methods. But first, a word of caution: It is very difficult to "compare" methods broadly (i. e., it is difficult to make definitive statements), especially for solving BVPs, because of the large number of factors involved. For example, the class of BVPs selected for comparison, the type of linear system solver used and the way it is implemented, the storage requirements, the output (say, a continuous versus discrete solution and the number of points where solution values are desired), and the ease of implementation are but a few of the factors which are difficult to assess.

For deferred corrections, steps 2.2 and 2.3 are required at each intermediate step l, $2 \leq l \leq k$, while the fixed-order collocation method requires nonlinear iteration on the one large system. For BVPs with smooth solutions, Example 4.11 above shows how easily we can obtain extremely high accuracy with the acceleration method, so deferred correction would generally be preferable for nontrivial but simple problems. For less simple BVPs, the situation is not so straightforward. A potential difficulty of all acceleration methods is that the initial mesh π has to be sufficiently fine for the basic method to make sense on it. (Recall that our theory requires that h be "small enough".) For some applications this implies a dense mesh π (e.g., see Section 10.3.2) and the collocation solution could be obtained more cheaply than even the basic solution $y_\pi^{(1)}$. If mesh selection is used, collocation requires no mesh restrictions like (5.92) [and certainly not (5.113)] because of its local approximation nature. The deferred corrections scheme above uses $2k$ adjacent mesh points for its difference formulas, and so has difficulties in a transition region where the mesh changes from coarse to fine. Finally, we mention that in the nonlinear BVP case, a situation similar to that for deferred corrections occurs for a high-order collocation scheme with mesh selection. Namely, most Newton iterations are usually done on a coarse initial mesh, and for refined meshes a good initial guess for the nonlinear iteration is available from the solution on the previous mesh. One full Newton iteration plus a few fixed Jacobian corrections (cf. Section 8.2) are normally sufficient on all but the coarsest mesh when using collocation. □

The basic idea of the approach for showing convergence of the deferred corrections method is fairly simple: Suppose that $\mathbf{y}_\pi^{(l-1)}$ is accurate to $O(h^{2l-2})$ and that $\mathbf{y}_\pi^l$ is defined from (5.110). Stability of the basic scheme $\mathbf{N}_\pi$ implies

$$|\mathbf{y}_i^{(l)} - \mathbf{y}(x_i)| \leq K \max_{1 \leq i \leq N} |\mathbf{N}_\pi \mathbf{y}_i^{(l)} - \mathbf{N}_\pi \mathbf{y}(x_i)|$$

$$\leq K \max_{1 \leq i \leq N} \{|\mathbf{C}_l(\mathbf{y}_i^{(l-1)}) - \mathbf{C}_l(\mathbf{y}(x_i))| + |\mathbf{C}_l(\mathbf{y}(x_i)) - \mathbf{N}_\pi \mathbf{y}(x_i)|\}$$

$$\leq K \max_{1 \leq i \leq N} |\mathbf{C}_l(\mathbf{y}_i^{(l-1)}) - \mathbf{C}_l(\mathbf{y}(x_i))| + K \max_{1 \leq i \leq N} |\tau_i^l(\mathbf{y})| \quad (5.114)$$

from (5.106) and (5.107). Thus $\mathbf{y}_\pi^{(l)}$ is potentially accurate to $O(h^{2l})$, depending upon how well $\mathbf{C}_l(\mathbf{y}_i^{(l-1)})$ approximates the truncation error [see (5.111)] and how large $|\tau_i^l(\mathbf{y})|$ is. The actual analysis is complicated by the fact that the accuracy of $\mathbf{C}_l(\mathbf{y}_i^{(l-1)})$ depends upon the "smoothness" of $\mathbf{y}_\pi^{(l-1)}$. This smoothness is a discrete analogue of continuity of the function and some of its derivatives. Using divided differences of mesh values, it is adversely affected when the difference formula at a point near the boundary switches from centered to noncentered. A similar situation holds for $\tau_i^l(\mathbf{y})$. A careful description of these smoothness conditions becomes quite technical, and we give references at the end of the chapter. One attractive feature of the method of proof which uses (5.112) and (5.114) (instead of asymptotic expansions) is its "robustness"—the only restriction placed upon an initial iterate $\mathbf{z}^{l,1}$ at any stage is that it has to be sufficiently close to the solution. This is realistic, for example, if $\mathbf{z}^{l,1}$ is an interpolate of a solution from a different *basically unrelated* mesh, as happens with mesh selection. Assuming an asymptotic expansion (5.91) for the error, on the other hand, is less natural in such a case. Indeed, we are not really using a sequence of related meshes as we did for extrapolation.

Remarks

(a) While we have described deferred corrections for the particular scheme (5.107), it should be appreciated that the method and convergence approach (5.114) are applicable for virtually any higher-order scheme. In particular, one could use deferred corrections with one of the implicit Runge-Kutta schemes of Section 5.3 as a basic scheme. Efficient implementation and analysis are far from straightforward, especially for nonlinear BVPs, but nevertheless such an approach has a number of possible merits. In addition to the opportunity for utilizing the advantages of each, the corrections are now local. Thus, one would expect only one iteration (5.112) per step (i. e., $I_l = 1$ for all l).

(b) Deferred corrections (5.112) can be interpreted as a matrix fixed-point iteration (cf. also Section 2.3.2) under fairly general circumstances. For instance, for a linear problem (with appropriate boundary conditions) if the basic scheme looks like

$$\mathbf{L}_\pi \mathbf{y}_\pi^{(1)} = \mathbf{q}_\pi \qquad (5.115)$$

and the higher-order scheme is

$$(\mathbf{L}_\pi - \mathbf{L}_{\pi,l})\mathbf{y}_\pi^{(k)} = \mathbf{q}_\pi \qquad (5.116)$$

then the deferred corrections iteration can be written as

$$\mathbf{z}^{l,j+1} = \mathbf{z}^{l,j} - \mathbf{L}_\pi^{-1}[(\mathbf{L}_\pi - \mathbf{L}_{\pi,l})\mathbf{z}^{l,j} - \mathbf{q}_\pi] \qquad (5.117)$$

(check!). Letting $\mathbf{v}^{l,j} := [(\mathbf{L}_\pi - \mathbf{L}_{\pi,k})\mathbf{z}^{l,j} - \mathbf{q}_\pi]$, this is

$$\mathbf{v}^{l,j+1} = \mathbf{L}_{\pi,k}\mathbf{L}_\pi^{-1}\mathbf{v}^{l,j} \qquad (5.118)$$

Now, $\mathbf{L}_\pi^{-1}\mathbf{v}^{l,j}$ has the same magnitude as $\mathbf{v}^{l,j}$ and can be shown to be in some sense "smoother" (as $\mathbf{L}^{-1}\mathbf{v}$ is smoother than $\mathbf{v}$, cf. Section 2.8). It can be shown that applying $\mathbf{L}_{\pi,k}$ makes $\mathbf{v}^{l,j+1}$ smaller. □

5.6 HIGHER-ORDER ODEs

A brief glance at the examples in Chapter 1 suggests that in many applications ODEs appear in the form of mixed-order systems rather than as first-order systems. It is therefore natural to look for numerical methods to approximate BVPs for higher-order ODEs. A simple finite difference method for a second-order ODE has already been introduced in Section 5.1.1. The tridiagonal linear system of equations which results has a very attractive form, accounting for the popularity of the scheme and its extensions to PDEs. But it is not immediately apparent how to generalize the scheme for a nonuniform mesh, and this is discussed in Section 5.6.1.

Usually a BVP appears in a more involved form than one to which the scheme in Section 5.6.1 can be applied. At the same time, a transformation from a general mixed-order form to a first order form is rather straightforward (see Section 1.1.2), and for the latter some very powerful techniques have already been introduced in this chapter. Thus, the rest of the section focuses on methods for a higher-order ODE which are sufficiently flexible and robust to be of general-purpose use.

Under these restrictions, few methods remain. We choose to consider a collocation method which is a generalization of the method introduced in Section 5.4 for first-order ODEs. In Section 5.6.2 we introduce the method for a higher-order ODE and show that, indeed, some efficiency can be gained by applying it directly to the higher-order ODE, rather than to the transformed first-order system. In Section 5.6.3 we consider other good implementations for the same collocation method, and in Section 5.6.4 we discuss the conditioning of the various implementations introduced.

Let us introduce the general form of the BVP considered here. The differential equation of order m comes in two flavours: linear,

$$Lu \equiv u^{(m)} - \sum_{l=1}^{m} c_l(x) u^{(l-1)} = q(x), \qquad a < x < b \qquad (5.119a)$$

and nonlinear,

$$Nu \equiv u^{(m)} - f(x, u, u', \ldots, u^{(m-1)}) = 0, \qquad a < x < b \qquad (5.120a)$$

Denoting

$$\mathbf{y}(x) := (u(x), u'(x), \ldots, u^{(m-1)}(x))^T \qquad (5.121)$$

(this is the vector of unknowns for the corresponding first-order system—see Section 1.1.2) we can write (5.119a) as

$$Lu \equiv u^{(m)} - \mathbf{c}^T(x)\,\mathbf{y} = q(x)$$

and (5.120a) as

$$Nu \equiv u^{(m)} - f(x, \mathbf{y}) = 0$$

As usual, the linear form of the two-point BC is

$$B_a \mathbf{y}(a) + B_b \mathbf{y}(b) = \boldsymbol{\beta} \qquad (5.119b)$$

while the nonlinear form is

$$\mathbf{g}(\mathbf{y}(a), \mathbf{y}(b)) = \mathbf{0} \qquad (5.120b)$$

It is assumed that the coefficients $c_l(x)$ of the linear operator L are smooth and that there is a unique solution u for (5.119), $u(x)$ generally has m more continuous derivatives than the inhomogeneity $q(x)$. Thus, if $q(x)$ is a piecewise continuous function then $u \in C^{(m-1)}[a,b]$.

5.6.1 More on a simple second-order scheme

In Section 5.1.1 we present a simple scheme for a scalar, second-order Dirichlet BVP. The scheme has the advantage of simplicity, being second-order accurate and yielding a tridiagonal system of linear algebraic equations. But in addition to being defined only for a special BVP form, it is defined only on uniform meshes. Here we generalize it to an arbitrary mesh π of (5.1), retaining the second-order accuracy. The way to go about this is not obvious because the usual 3-point formula for u'' gives

$$\frac{2}{h_i + h_{i-1}} \left[\frac{u(x_{i+1}) - u(x_i)}{h_i} - \frac{u(x_i) - u(x_{i-1})}{h_{i-1}} \right] = u''(x_i) + O(h_i^l)$$

where $l = 2$ if $h_i = h_{i-1}$, but $l = 1$ if $h_i \neq h_{i-1}(1 + O(h_i))$. Indeed, if we use this approximation for $u''(x_i)$ in (5.4a), considering a case where $p \equiv 0$ and leaving everything else in (5.4) unchanged, then for an arbitrary mesh a first-order scheme results (i.e., the truncation error grows significantly as compared to a uniform mesh). Still, by a generalization the second-order accuracy can be salvaged, as we show below.

It is convenient to restrict the BVP (5.2) to the form

$$u'' = (p(x)u)' + r(x)u + q(x) \qquad a < x < b \qquad (5.122a)$$

$$u(a) = \beta_1, \qquad u(b) = \beta_2 \qquad (5.122b)$$

To derive a second-order formula we first integrate (5.122a) and write it as

$$u' = p(x)u + v \qquad (5.123a)$$

$$v' = r(x)u + q(x) \qquad (5.123b)$$

Then we introduce a *staggered mesh* (see Fig. 5.4) and define approximants $u_i \sim u(x_i)$ and $v_{i+1/2} \sim v(x_{i+1/2})$, $(x_{i+1/2} := x_i + \frac{1}{2}h_i)$ using the midpoint rule.

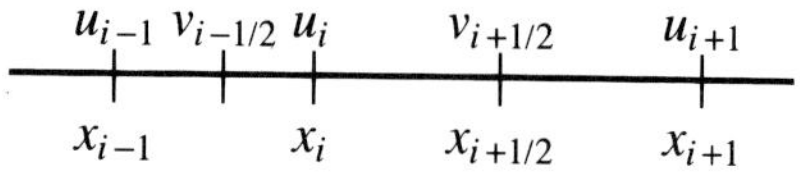

Figure 5.4 Staggered mesh and unknowns

For (5.123a) the midpoint rule on $[x_i, x_{i+1}]$ yields

$$h_i^{-1}[u_{i+1} - u_i] = \frac{1}{2}p_{i+1/2}(u_i + u_{i+1}) + v_{i+1/2} \qquad (5.124a)$$

where $p_{i+1/2} := p(x_{i+1/2})$. For (5.123b) the midpoint rule is less standard, being defined on $[x_{i-1/2}, x_{i+1/2}]$. The midpoint of this subinterval is

$$\tilde{x}_i = x_i + \frac{1}{4}(h_i - h_{i-1}) \qquad (5.125a)$$

If we approximate $u(\tilde{x}_i)$ in terms of $u(x_{i-1})$, $u(x_i)$ and $u(x_{i+1})$ using quadratic interpolation, this yields

$$\tilde{u}_i := \alpha_i u_{i-1} + \beta_i u_i + \gamma_i u_{i+1} \qquad (5.125b)$$

$$\beta_i = \frac{1}{16}\frac{(3h_i + h_{i-1})(h_i + 3h_{i-1})}{h_{i-1}h_i} \qquad (5.125c)$$

$$\alpha_i = \frac{1}{16}\frac{(3h_i + h_{i-1})(h_{i-1} - h_i)}{h_{i-1}(h_i + h_{i-1})}, \qquad \gamma_i = \frac{1}{16}\frac{(h_i + 3h_{i-1})(h_i - h_{i-1})}{h_i(h_i + h_{i-1})}$$

Then the midpoint rule approximation for (5.123b) becomes

$$\frac{v_{i+1/2} - v_{i-1/2}}{\frac{1}{2}(h_i + h_{i-1})} = r(\tilde{x}_i)\tilde{u}_i + q(\tilde{x}_i) \qquad (5.124b)$$

We may now use (5.124a) to define $v_{i+1/2}$, and substitution into (5.124b) gives

$$\frac{2}{h_{i-1}+h_i}[h_i^{-1}(u_{i+1} - u_i) - h_{i-1}^{-1}(u_i - u_{i-1}) - \frac{1}{2}p_{i+1/2}(u_{i+1} + u_i) + \frac{1}{2}p_{i-1/2}(u_{i-1} + u_i)]$$

$$= r(\tilde{x}_i)\tilde{u}_i + q(\tilde{x}_i) \qquad (5.126)$$

This is a 3-point formula involving only u-values and hence generalizing (5.4a).

Remark

This simple derivation actually brings up two curious points not encountered hitherto. One is the staggered mesh (grid): While a tool commonly used in deriving schemes in numerical fluid mechanics, it is rarely needed for the material covered in this book (but see Section 10.3.3).

The other point is that the global error seems at first sight to be of higher-order than the local truncation error! Substituting the exact solution into (5.124a), we obtain the local truncation error term

$$\frac{h_i^2}{24} u'''(x_{i+1/2}) - p_{i+1/2} \frac{h_i^2}{16} u''(x_{i+1/2}) + O(h_i^4)$$

Substituting this into (5.126) yields, corresponding to (5.6), an $O(h)$ local truncation error for an arbitrary mesh. Still, the error is expected to be $O(h^2)$, because the difference scheme (5.124a, b) as a discretization for (5.123a, b) is second-order and stable, and (5.126) (in exact arithmetic) is equivalent to (5.124). The second-order accuracy cannot be directly obtained with the approach in Section 5.1.1. This suggests that a less mysterious way to interpret (5.126) is as a clever implementation of (5.124). □

Substituting (5.125) into (5.126) we obtain the second-order 3-point formula

$$a_i u_{i-1} + b_i u_i + c_i u_{i+1} = g_i \qquad 2 \leq i \leq N \qquad (5.127a)$$

$$a_i = \frac{2}{h_{i-1}(h_i + h_{i-1})} + \frac{p_{i-1/2}}{h_i + h_{i-1}} - r(\tilde{x}_i)\alpha_i \qquad (5.127b)$$

$$c_i = \frac{2}{h_i(h_i + h_{i-1})} - \frac{p_{i+1/2}}{h_i + h_{i-1}} - r(\tilde{x}_i)\gamma_i \qquad (5.127c)$$

$$b_i = -\frac{2(h_i^{-1} + h_{i-1}^{-1})}{h_i + h_{i-1}} + \frac{p_{i-1/2} - p_{i+1/2}}{h_i + h_{i-1}} - r(\tilde{x}_i)\beta_i \qquad (5.127d)$$

$$g_i = q(\tilde{x}_i) \qquad (5.127e)$$

Note that if $h_i = h_{i-1}$ then

$$\tilde{x}_i = x_i, \, u_i = u_i, \, \alpha_i = \gamma_i = 0, \, \beta_i = 1, \quad \frac{h_i + h_{i-1}}{2} = h_i$$

and the scheme reduces to the usual one.

5.6.2 Collocation

The method we describe below for one ODE (5.119) or (5.120) can be directly extended to handle mixed-order systems of ODEs with multipoint BCs.

The basic idea of collocation is quite general: An approximate solution is sought in the form

$$u_\pi(x) = \sum_{j=1}^{M} \alpha_j \phi_j(x) \qquad a \le x \le b \tag{5.128}$$

where $\phi_j(x)$ are known linearly independent functions defined on $[a, b]$, and α_j are parameters. These parameters are determined by requiring $u_\pi(x)$ to satisfy M conditions—the m BC and the ODE at $M - m$ points (the *collocation points*) in $[a, b]$. It is sometimes convenient to say that $u_\pi(x)$ is an element in a *linear space* of dimension M which is *spanned* by *basis functions* $\phi_1(x), \ldots, \phi_M(x)$.

For the remainder of this chapter we choose this linear space to consist of piecewise polynomial functions. Thus, there is a partition π of $[a, b]$ as in (5.1) [justifying the notation u_π in (5.128)] such that any linear combination of the basis functions reduces to a polynomial on each subinterval $[x_i, x_{i+1}]$, $1 \le i \le N$. In this section we further restrict the functions $\phi_j(x)$, and therefore any of their linear combinations, to be in $C^{(m-1)}[a, b]$ (like the exact solution). Further, the order of the polynomial pieces is restricted to be $k + m$, for some $k \ge m$. Recall that $v(x)$ is in $\mathbf{P}_{k+m,\pi}$ if $v(x)$ is a piecewise polynomial of order $k + m$ on π.

The dimension M of the resulting approximation space is easy to calculate: Each polynomial piece has $k + m$ parameters and there are m matching constraints for $\mathbf{y}$ across each interior mesh point, so

$$M = N(k + m) - (N - 1)m = Nk + m \tag{5.129}$$

The M conditions imposed on $u_\pi(x)$ of (5.128) are:

(a) That u_π satisfy the m BC (5.119b) or (5.120b).
(b) That u_π satisfy the ODE (5.119a) or (5.120a) at k points in each of the N subintervals of the mesh π. These points are the collocation points x_{ij} of (5.59c, d) with the canonical points ρ_j being distinct.

Let us summarize this for the general BVP (5.120), denoting (similarly to (5.121))

$$\mathbf{y}_\pi(x) := (u_\pi(x), u_\pi'(x), \ldots, u_\pi^{(m-1)}(x))^T \tag{5.130}$$

Given a mesh π of (5.1) and k distinct points $\rho_1, \ldots, \rho_k$ of (5.59d), the collocation method determines an approximate solution $u_\pi(x)$ defined on $[a, b]$ such that

$$\mathbf{y}_\pi(x) \in C[a, b]; \qquad u_\pi(x) \in \mathbf{P}_{k+m,\pi} \tag{5.131a}$$

$$\mathbf{g}(\mathbf{y}_\pi(a), \mathbf{y}_\pi(b)) = \mathbf{0} \tag{5.131b}$$

$$Nu_\pi(x_{ij}) \equiv u_\pi^{(m)}(x_{ij}) - f(x_{ij}, \mathbf{y}_\pi(x_{ij})) = 0 \quad 1 \le j \le k, \quad 1 \le i \le N \tag{5.131c}$$

One important question is how to choose the basis functions $\phi_j(x)$ so as to obtain an efficient, stable method. Choices of Hermite-type bases and of B-splines (recall Section 2.5) are discussed in Section 5.6.3. With these we specify basis functions $\phi_j(x)$ with local support, and the continuity conditions on u_π are already imbedded in the basis functions, while the collocation equations are satisfied only later. But here we prefer not to explicitly specify any basis functions. Instead, we consider local representations of the polynomial pieces. It enables us to relate more directly to the collocation method for first-order systems already discussed, and to see that the method introduced here is just a fancy finite difference method. We first impose the collocation equations, followed by local parameter elimination, and only then connect to the action in adjacent subintervals. This is a multiple-shooting-type approach which allows us to capitalize on previous theoretical results and to avoid introducing heavier functional analysis machinery.

Let us focus on one subinterval $[x_i, x_{i+1}]$ and express the polynomial $u_\pi(x)$ in terms of its Taylor series about x_i,

$$u_\pi(x) = \sum_{j=1}^{k+m} \frac{(x-x_i)^{j-1}}{(j-1)!} u_\pi^{(j-1)}(x_i) \qquad x_i \leq x \leq x_{i+1}$$

This can be written as

$$u_\pi(x) = \sum_{j=1}^{m} \frac{(x-x_i)^{j-1}}{(j-1)!} y_{ij} + h_i^m \sum_{j=1}^{k} \psi_j\left(\frac{x-x_i}{h_i}\right) z_{ij} \qquad (5.132)$$

where, corresponding to (5.130),

$$\mathbf{y}_\pi(x_i) \equiv \mathbf{y}_i = (y_{i1}, \ldots, y_{im})^T$$

and

$$z_{ij} = h_i^{j-1} u_\pi^{(m+j-1)}(x_i), \qquad \mathbf{z}_i := (z_{i1}, \ldots, z_{ik})^T$$

The functions $\psi_j(t)$ are therefore defined as

$$\psi_j(t) = \frac{t^{m+j-1}}{(m+j-1)!} \qquad 0 \leq t \leq 1, \qquad 1 \leq j \leq k \qquad (5.133)$$

Note that $\psi_1(t), \ldots, \psi_k(t)$ are linearly independent polynomials of order $k+m$ on $[0, 1]$ satisfying

$$\psi_j^{(l-1)}(0) = 0 \qquad 1 \leq l \leq m, \qquad 1 \leq j \leq k \qquad (5.134)$$

These functions are independent of i and can be used for each subinterval as in (5.132).

Now, we can write down the constraints (5.131) which define the approximate solution $u_\pi(x)$ in terms of the parameters $\mathbf{y}_i$ and $\mathbf{z}_i$ of the representation (5.132). We do this for a linear problem first.

Linear problems For the linear problem (5.119) we write

$$Lu_\pi(x) = h_i^m \sum_{j=1}^{k} z_{ij} L\left[\psi_j\left(\frac{x-x_i}{h_i}\right)\right] - \sum_{l=1}^{m} c_l(x) \sum_{j=l}^{m} \frac{y_{ij}(x-x_i)^{j-l}}{(j-l)!}$$

so the collocation conditions (5.131c) give

$$V\mathbf{y}_i + W\mathbf{z}_i = \mathbf{q}_i \qquad 1 \leq i \leq N \qquad (5.135a)$$

where $\mathbf{q}_i = (q(x_{i1}), \ldots, q(x_{ik}))^T$, V is a $k \times m$ matrix with entries

$$V_{rj} = -\sum_{l=1}^{j} \frac{c_l(x_{ir})(h_i \rho_r)^{j-l}}{(j-l)!} \qquad 1 \leq r \leq k, \qquad 1 \leq j \leq m \qquad (5.136a)$$

and W is a $k \times k$ matrix with entries

$$W_{rj} = \psi_j^{(m)}(\rho_r) - \sum_{l=1}^{m} c_l(x_{ir})h_i^{m+1-l}\psi_j^{(l-1)}(\rho_r) \qquad 1 \leq r, j \leq k \qquad (5.136b)$$

The continuity conditions in (5.131a) are even easier to write down. We evaluate $u_\pi(x)$ and its first $m-1$ derivatives at $x = x_{i+1}$ by (5.132) and equate to $\mathbf{y}_{i+1}$, the corresponding values at the $(i+1)$st subinterval. This yields

$$\mathbf{y}_{i+1} = C \mathbf{y}_i + D \mathbf{z}_i \qquad 1 \leq i \leq N \qquad (5.135b)$$

where C is an $m \times m$ upper triangular matrix with entries

$$C_{rj} = \frac{h_i^{j-r}}{(j-r)!} \qquad j \geq r \qquad (5.136c)$$

and D is an $m \times k$ matrix with entries

$$D_{rj} = h_i^{m+1-r}\psi_j^{(r-1)}(1) \qquad 1 \leq r \leq m, \qquad 1 \leq j \leq k \qquad (5.136d)$$

Note that the obvious dependence of C, D, W, and V on i has been suppressed in the notation.

The specification of the collocation constraints (5.131) for the linear problem is completed by writing for (5.131b)

$$B_a \mathbf{y}_1 + B_b \mathbf{y}_{N+1} = \boldsymbol{\beta} \qquad (5.135c)$$

The reader should realize that, despite the unfortunate number of subscripts which we have to use here, there is no significant conceptual difference between this process and (5.64). Our next step is to eliminate the local unknowns, in a similar way to that used to obtain (5.65). We note that as $h \to 0$ in (5.136b),

$$W_{rj} \to \rho_r^{j-1}/(j-1)! =: W_{rj}(0)$$

The matrix $W(0)$ is a nonsingular Vandermonde matrix, so for h_i small enough W is nonsingular and

$$W^{-1} = W(0)^{-1} + O(h_i)$$

Eliminating $\mathbf{z}_i$ from (5.135a) and substituting in (5.135b), we arrive at the form

$$\mathbf{y}_{i+1} = \Gamma_i \mathbf{y}_i + \mathbf{r}_i \qquad 1 \leq i \leq N \qquad (5.137a)$$

with

$$\Gamma_i := C - DW^{-1}V, \qquad \mathbf{r}_i := DW^{-1}\mathbf{q}_i \qquad (5.137b)$$

The similarity between (5.137) and (5.65) is quite clear, noting that $C = I + O(h_i)$ and $D = O(h_i)$.

In (5.137), (5.135c) we have, once again, a linear system of equations of the form (5.17) for $u_\pi(x)$ and its first $m-1$ derivatives at mesh points. After obtaining the values of $\mathbf{y}_i$ we can easily obtain $\mathbf{z}_i$ from (5.135a), and hence $u_\pi(x)$ from (5.132), if we store the values of $W^{-1}V$ and $W^{-1}\mathbf{q}_i$ which are computed while assembling each Γ_i, $1 \leq i \leq N$.

We summarize the collocation procedure outlined above in an algorithm. For the efficient assembly of Γ_i and $\mathbf{r}_i$ we compute and store in advance all mesh independent quantities like $\psi_j^{(l)}(\rho_r)$, $\psi_j^{(l)}(1)$, ρ_r^j and $j!$. The distinct points $\rho_1, \ldots, \rho_k$ are therefore determined at preprocessing time. Then we obtain the following algorithm.

Algorithm: Collocation for a linear higher-order BVP

Input: A BVP (5.119) and a mesh π.

Output: Solution values $\mathbf{y}_1, \ldots, \mathbf{y}_{N+1}$ consisting of $u_\pi(x)$ and its first $m-1$ derivatives at mesh points.

1. {Assemble the difference equations}
   FOR $i = 1, \ldots, N$ DO
       Assemble V, W, C and D according to (5.130a–d).
       Find $U_i := W^{-1}V$ and $\mathbf{p}_i := W^{-1}\mathbf{q}_i$.
       Set $\Gamma_i := C - DU_i$; $\mathbf{r}_i := D\mathbf{p}_i$.

2. Set up and solve the linear system

$$\begin{pmatrix} -\Gamma_1 & I & & & \\ & -\Gamma_2 & I & & \\ & & \ddots & \ddots & \\ & & & -\Gamma_N & I \\ B_1 & & & & B_2 \end{pmatrix} \begin{pmatrix} \mathbf{y}_1 \\ \mathbf{y}_2 \\ \vdots \\ \mathbf{y}_N \\ \mathbf{y}_{N+1} \end{pmatrix} = \begin{pmatrix} \mathbf{r}_1 \\ \mathbf{r}_2 \\ \vdots \\ \mathbf{r}_N \\ \boldsymbol{\beta} \end{pmatrix}$$

3. Optionally, obtain $\mathbf{z}_i$ values from

$$\mathbf{z}_i := \mathbf{p}_i - U_i \mathbf{y}_i \qquad 1 \leq i \leq N$$

$\square$

Example 5.15

The BVP presented in Example 5.1 is a scalar second-order problem. In Example 5.7 we have computed approximate solutions, using various collocation schemes for the transformed first-order BVP, and listed the results in Table 5.4. We repeat these calculations here, using the corresponding collocation schemes for the original problem formulation. The result is that the obtained maximum errors at mesh points for $u_\pi (\equiv y_1)$ and $u_\pi' (\equiv y_2)$ are *identical* (to the number of digits shown) to those listed in Table 5.4. $\square$

Example 5.16

The following BVP describes, under suitable assumptions, a uniformly loaded beam of variable stiffness, simply supported at both ends:

$$(x^3 u'')'' = 1 \qquad 1 < x < 2$$

$$u(1) = u''(1) = u(2) = u''(2) = 0$$

The exact solution is

$$u(x) = \frac{1}{4}(10 \ln 2 - 3)(1-x) + \frac{1}{2}[\frac{1}{x} + (3+x)\ln x - x]$$

so this BVP too is well-behaved and has a rather smooth solution. Here we demonstrate the power of high-order methods for very smooth examples. Using Gauss points on uniform meshes, the maximum error in u at mesh points is $.24-13$ for $k = 4$, $N = 16$, and $.96-14$ for $k = 6$, $N = 4$. $\square$

Remark The representation for the polynomial $u_\pi(x)$ in (5.132) is not unique. In particular, other choices for $\psi_j(t)$ satisfying (5.134) can be considered. Two such choices are as follows: Requiring

$$\psi_j^{(m)}(\rho_r) = \delta_{jr} \quad 1 \leq j, r \leq k \tag{5.138a}$$

implies $z_{ij} = u_\pi^{(m)}(x_{ij})$. (Why?). It is easy to see that $\psi_j(t)$ is well-defined by (5.138a) and (5.134). Moreover, $W(0) = I$, and for $m = 1$ we have

$$\psi_j(\rho_r) = \alpha_{rj}, \qquad \psi_j(1) = \beta_j$$

with α_{rj}, β_j given by (5.68). We call this *a Runge-Kutta representation*. Another choice is

$$\psi_j^{(r-1)}(1) = \delta_{j-k+m,r} \quad 1 \leq j, r \leq k \tag{5.138b}$$

The importance of this is that D of (5.136d) becomes very simple. But for a mixed-order system of ODEs, the Runge-Kutta representation is preferred, because the same functions may be used for more than one order (see Exercise 21). □

The results in Example 5.15 indicate a rather strong connection between the collocation method for the higher-order BVP (5.119) and the corresponding collocation method using the same k points ρ_j ($k \geq m$) for the transformed first-order system

$$\mathbf{y}' = \begin{bmatrix} 0 & 1 & & \\ & & \ddots & \\ & & 0 & 1 \\ c_1(x) & c_2(x) & \cdots & c_m(x) \end{bmatrix} \mathbf{y} + \begin{bmatrix} 0 \\ \vdots \\ 0 \\ q(x) \end{bmatrix} \tag{5.139}$$

and (5.119b). Indeed, the linear equations (5.137a) form a *one-step difference scheme* for (5.139). Two questions immediately arise: (a) Have we gained anything in the treatment here; indeed, is anything different from the corresponding collocation methods of Section 5.3? and (b) Can we capitalize on the theory of Sections 5.2 and 5.3?

Clearly we should answer question (a) before spending time on (b). We first point out that while here $u_\pi(x) \in \mathbf{P}_{k+m,\pi} \cap C^{(m-1)}[a, b]$, the corresponding collocation approximation of y_1 in (5.139) is in $\mathbf{P}_{k+1,\pi} \cap C[a, b]$, so the approximations are generally not the same when $m > 1$. Next, note that in (5.132) we have $k + m$ parameters per mesh subinterval, whereas the corresponding method for a first-order system would have $m(k+1)$, i. e., $(m-1)k$ additional parameters. Of course the treatment of the parameters is slightly more cumbersome here; still, the matrix W of (5.64) for (5.139) is $mk \times mk$ and A_i is $mk \times m$, whereas in (5.136) W is merely $k \times k$ and V is $k \times m$. If $m = 4$ as in Example 5.16, for instance, then the saving in computation and storage is rather substantial. We can view (5.137) as *a sophisticated high-order one-step scheme for* (5.139), which takes advantage of the special structure of the coefficients in the transformed first-order system. We hasten to point out, however, that this interpretation is beneficial only for mesh point values.

The answer to question (b) is affirmative, as indicated by the proof to the following.

Theorem 5.140 Assume that there are integers $p \geq k \geq m$ such that
(a) the linear BVP of order m (5.119) is well-posed, in the sense that the equivalent first-order BVP (5.139), (5.119b) has a conditioning constant κ of moderate size; (5.119) has coefficients in $C^{(p)}[a,b]$ and has a unique solution $u(x)$ in $C^{(p+m)}[a,b]$; and
(b) the k canonical collocation points $\rho_1, \ldots, \rho_k$ satisfy the orthogonality conditions

$$\int_0^1 \phi(t) \prod_{l=1}^k (t-\rho_l)\, dt = 0 \qquad \phi \in \mathbf{P}_{p-k} \tag{5.140a}$$

(note that $p \leq 2k$).

Then for h small enough the collocation method given in the above algorithm is stable with stability constant $c\kappa N$, c a constant of moderate size, and it has a unique solution $u_\pi(x)$. Furthermore, the following error estimates hold: At mesh points

$$|u^{(j)}(x_i) - u_\pi^{(j)}(x_i)| = O(h^p) \qquad 0 \leq j \leq m-1, \qquad 1 \leq i \leq N+1 \tag{5.140b}$$

while elsewhere, for $1 \leq i \leq N$,

$$|u^{(j)}(x) - u_\pi^{(j)}(x)| = O(h_i^{k+m-j}) + O(h^p) \qquad x_i \leq x \leq x_{i+1}, \qquad 0 \leq j \leq k+m-1 \tag{5.140c}$$

Proof: First we show that the analogue of Theorem 5.75 holds. It is easy to see, by looking at a local Taylor expansion, that $u(x)$ has the representation (5.132) up to $O(h_i^{k+m})$ terms. From this we obtain (Exercise 18) that $\mathbf{y}(x)$ of (5.121), which is the exact solution of the first-order BVP (5.139), (5.119b), satisfies

$$\mathbf{y}(x_{i+1}) = \Gamma_i \mathbf{y}(x_i) + \mathbf{r}_i + O(h_i^{k+1})$$

This means that the finite difference scheme (5.137a), (5.135c) is a one-step scheme for (5.139), (5.119b) which is consistent of order k. The sound construction of Γ_i of (5.137) allows application of Theorem 5.38. Thus we obtain existence of a unique collocation solution and stability with the claimed stability bound. Moreover, the error at mesh points satisfies

$$|\mathbf{y}_i - \mathbf{y}(x_i)| = O(h^k) \qquad 1 \leq i \leq N+1 \tag{5.141a}$$

Let us use the notation $d_\pi(x) := u_\pi(x) - u(x)$ and write as in (5.132)

$$d_\pi(x) = \sum_{j=1}^m \frac{(x-x_i)^{j-1}}{(j-1)!} e_{ij} + h_i^m \sum_{j=1}^k \psi_j\left(\frac{x-x_i}{h_i}\right) g_{ij} - \mu(x) \tag{5.141b}$$

with

$$(e_{i1}, \ldots, e_{im})^T = \mathbf{e}_i = \mathbf{y}_i - \mathbf{y}(x_i)$$

and $\mu^{(j)}(x) = O(h_i^{k+m-j})$, $0 \leq j \leq k+m-1$. Equations (5.135a) for the error give the estimate

$$|g_{ij}| = O(h^k)$$

so taking derivatives in the expression (5.141b) for $d_\pi(x)$, we obtain

$$|u_\pi^{(j)}(x) - u^{(j)}(x)| = \begin{cases} O(h^k) & 0 \leq j \leq m \\ & x_i \leq x \leq x_{i+1} \\ O(h^{k+m-j})\theta_i^{j-m} & m \leq j \leq k+m-1 \end{cases} \quad (5.141c)$$

with θ_i defined in (5.75c).

We now proceed as in (5.81). Let $G(x,t)$ be Green's function for (5.119) and denote $G_j(x,t) \equiv \dfrac{\partial^j}{\partial x^j} G(x,t)$. Then from the assumption on the problem coefficients, $G_j(x,t)$ is smooth as a function of t in $[a,x)$ and in $(x,b]$; but at $t = x$, $G_j(x,t)$ has only $m-1-j$ derivatives. Writing for $0 \leq j \leq m-1$

$$u_\pi^{(j)}(x) - u^{(j)}(x) = \int_a^b G_j(x,t) L(u_\pi(t) - u(t)) \, dt = \sum_{i=1}^N E_i(x)$$

with

$$E_i(x) = \int_{x_i}^{x_{i+1}} G_j(x,t)(Lu_\pi(t) - q(t)) \, dt$$

we note that $Lu_\pi(t) - q(t) = 0$ at k collocation points $t = x_{il}$. Thus, if $x \notin (x_i, x_{i+1})$ then by (5.140a) and (5.140d)

$$F_i(x) = O(h_i^{p+2-k} h^{k-1}) \leq \text{const } h_i h^p \quad (5.141d)$$

while if $x \in (x_i, x_{i+1})$ then the limited smoothness of $G_j(x,t)$ allows us to conclude only that

$$E_i(x) = O(h_i^{k+m-j})\theta_i^{k-1} \quad (5.141e)$$

The conclusion (5.140b) follows, because for a mesh point each $E_i(x)$ satisfies (5.141d) and so $\sum_{i=1}^N E_i(x) = O(h^p)$. But to obtain (5.140c) without imposing any mesh restriction, we cannot just use (5.141e). Instead, knowing that (5.140b) holds, we again use (5.135a) for the error, obtaining

$$|g_{ij}| = O(h^p) + O(h_i^k)$$

and substitute into (5.141b). This yields (5.140c) and completes the proof. □

Remarks

(a) The result (5.140c) shows that, when x is not a mesh point, $u_\pi(x)$ is a *better approximation* to $u(x)$ than the corresponding collocation approximation for the equivalent first-order system.

(b) The approximation space to which $u_\pi(x)$ belongs is a linear space of piecewise polynomials of order $k+m$. A result from approximation theory states that, unless $u(x)$ itself is in the approximation space, we cannot get a global approximation order of more than $O(h^{k+m})$. In other words, for any piecewise polynomial $v_\pi(x)$ of order $k+m$, we cannot guarantee a better result than

$$\max_{a \le x \le b} |v_\pi(x) - u(x)| = O(h^{k+m})$$

If $p \ge k+m$ then the estimate (5.140c) shows that the collocation solution achieves the *optimal* global convergence order. Furthermore, if $p > k+m$ then at mesh points we obtain a *superconvergence* order, i. e., an order of convergence higher than the best possible global order. It is interesting to note that the superconvergence result, which (as the name suggests) is perhaps less natural from the point of view of approximation spaces, is most natural from the point of view of one-step difference schemes.

(c) A restriction on the mesh under which the convergence results of Theorem 5.140 hold is remarkably absent. We have not assumed that the mesh is quasiuniform (recall Section 5.5). Also, Theorem 5.38 which was used in the proof of (5.140) indicates that the condition number of the matrix $\mathbf{A}$ involved is $O(N)$. In many methods for higher-order BVPs one finds condition numbers which depend on $\frac{h}{\min h_i}$ (see Section 5.6.4, for instance) and/or on N^m (see Exercise 1).

(d) The proof of Theorem 5.140 yields an $O(N)$ conditioning for the approximate solution $\mathbf{y}_\pi(x)$ of the first-order system (5.139). To obtain from this a similar conditioning for the approximate solution $u_\pi(x)$ of a given BVP (5.119), note that it is easy to bound

$$\| u_\pi^{(j)} \| \le K \| u_\pi \|, \qquad 1 \le j \le m-1$$

where K is a constant which depends on κ and on $\frac{\|q\| + |\boldsymbol{\beta}|}{\|u\|}$, but not on the mesh (for $h > 0$ small enough).

(e) The requirements of Theorem 5.140 include fairly strong (but not unusual) smoothness assumptions on the BVP coefficients. These assumptions can be relaxed to allow a finite number of jump discontinuities, *provided* that mesh points are placed at each such discontinuity. Then $u_\pi(x)$ will be a polynomial in regions where $u(x)$ is highly smooth, but merely in $C^{(m-1)}[a,b]$ globally, just like $u(x)$. For instance, Example 4.7 can be solved in this way. The method cannot handle δ-functions in the coefficients, though. (Why?) $\square$

While the superconvergence in (5.140b) is of a high-order, the global convergence result in (5.140c) has a *local* nature (at least as long as $h_i^{k+m} \gg h^p$). This is important for mesh selection. In fact, if $p > k+m$ then we can refine (5.140c) by explicitly giving the leading term of the error:

Corollary 5.142 Let $p > k+m$. With the assumptions of Theorem 5.140, the collocation error satisfies

$$u^{(j)}(x) - u_\pi^{(j)}(x) = h_i^{k+m-j} u^{(k+m)}(x_i) P^{(j)}\left(\frac{x-x_i}{h_i}\right) + O(h_i^{k+m-j+1}) + O(h^p) \quad (5.142a)$$

$$x_i \le x \le x_{i+1}, \qquad 1 \le i \le N, \qquad 0 \le j \le k+m-1$$

where

$$P(\xi) = \frac{1}{k!(m-1)!} \int_0^\xi (t-\xi)^{m-1} \prod_{l=1}^k (t-\rho_l)\, dt \qquad (5.142b)$$

Proof: We will show (5.142a) for $j = 0$; the general case follows by taking derivatives in (5.141b). First, since both u_π and u satisfy the ODE at collocation points, it follows from (5.140c) that

$$u^{(m)}(x_{ir}) - u_\pi^{(m)}(x_{ir}) = O(h_i^{k+1}) + O(h^p) \qquad 1 \le r \le k \qquad (5.143)$$

It is therefore useful to consider here the Runge-Kutta choice (5.138a) for the functions $\psi_j(t)$, because the leading term of the error representation (5.141b) is then $\mu(x)$. The latter is the residual of $u(x)$ substituted in (5.132), so

$$\mu^{(m)}(x) = u^{(m)}[x_{i1}, \ldots, x_{ik}, x] \prod_{l=1}^k (x - x_{il})$$

$$\mu(x_i) = \cdots = \mu^{(m-1)}(x_i) = 0$$

Noting that $u^{(m)}[x_{i1}, \ldots, x_{ik}, x] = \frac{1}{k!} u^{(k+m)}(x_i) + O(h_i)$ and integrating m times for $\mu(x)$ yields the required result. □

The above corollary indicates that, if we are willing to "give up" the superconvergence order, then we get a *localized representation for the global error*. This is unlike the usual case, where the error can only be bounded; also, the leading error term depends on local quantities in (5.142). An estimate of $u^{(k+m)}(x_i)$ would produce an estimate of the error locally. This is utilized in Section 9.3 for mesh selection.

Another consequence of (5.142) is that there are additional "superconvergence" points, i.e., points other than mesh points where the order of convergence is higher than what is possible everywhere. In particular, any roots of $P^{(j)}(\xi)$ in (5.142b) correspond to such points for the jth derivative of the error in (x_i, x_{i+1}), $1 \le i \le N$. We have already mentioned the roots of $P^{(m)}$ in (5.143). Also, $P^{(k+m-1)}(\bar{t}) = 0$ at

$$\frac{\bar{x}_i - x_i}{h_i} \equiv \bar{t} := \frac{1}{k} \sum_{l=1}^k \rho_l \qquad (5.144a)$$

so the piecewise constant function $u_\pi^{(k+m-1)}$ satisfies

$$u_\pi^{(k+m-1)}(x) = u^{(k+m-1)}(\bar{x}_i) + O(h_i^2) \qquad x_i \le x \le x_{i+1}, \quad 1 \le i \le N \qquad (5.144b)$$

If the collocation points are symmetric then $\bar{x}_i = x_{i+1/2}$.

Nonlinear problems For the nonlinear BVP (5.120) we once again consider the method of quasilinearization. The principle is precisely the same as in Section 5.3.2, but the linearized differential system appears as a higher-order ODE. We denote, as in (5.43)–(5.45),

$$L[u]z(x) \equiv z^{(m)}(x) - \sum_{l=1}^m c_l(x) z^{(l-1)}(x) \qquad a < x < b \qquad (5.145a)$$

$$c_l(x) := \frac{\partial f(x, \mathbf{y}(x))}{\partial y_l} \qquad (5.145\text{b})$$

where u and $\mathbf{y}$ are related by (5.121). The linearization of the BC is as in Section 5.2.2.

Recall how the method is derived: Given a known solution profile $\hat{u}(x)$, we formally expand

$$u^{(m)}(x) = f(x, \mathbf{y}(x)) \approx f(x, \hat{\mathbf{y}}(x)) + \sum_{l=1}^{m} \frac{\partial f}{\partial y_l}(x, \hat{\mathbf{y}}(x))(u^{(l-1)}(x) - \hat{u}^{(l-1)}(x))$$

where the vector $\hat{\mathbf{y}}(x)$ has components $\hat{y}_l := \hat{u}^{(l-1)}$. This yields an approximate *linear* ODE for the correction $z(x) \equiv u(x) - \hat{u}(x)$. The method of collocation with quasilinearization then reads as follows: Given an initial approximate solution $u_\pi(x)$, *repeat* (i) solving by collocation the linearized problem

$$L[u_\pi]z(x) = -[u_\pi^{(m)}(x) - f(x, \mathbf{y}_\pi(x))] \qquad a < x < b \qquad (5.146\text{a})$$

$$B_a \mathbf{w}(a) + B_b \mathbf{w}(b) = -\mathbf{g}(\mathbf{y}_\pi(a), \mathbf{y}_\pi(b)) \qquad (5.146\text{b})$$

where $\mathbf{w}(x) := (z(x), z'(x), \ldots, z^{(m-1)}(x))$, and (ii) improving the approximate solution by

$$u_\pi(x) := u_\pi(x) + z_\pi(x) \qquad a \leq x \leq b \qquad (5.146\text{c})$$

until $\|z_\pi\|$ is below an error tolerance.

The quasilinearization method outlined above can be seen to be equivalent to Newton's method for solving (5.131).

The basic convergence results for the method of collocation and quasilinearization are a combination of Theorems 5.88 and 5.140: The nonlinear analysis is the same as for Theorem 5.88, while Theorem 5.140 applies for the linearized problem. We state the results but omit the proof.

Theorem 5.147 Let $u(x)$ be an isolated solution of the BVP (5.120), where f and $\mathbf{g}$ have continuous second partial derivatives, and assume that the hypotheses of Theorem 5.140 hold for u and $L[u]$. Consider a k-stage collocation method, satisfying (5.140a), for (5.131). Then there are positive constants ρ and h_0 such that for all meshes with $h \leq h_0$:

(a) There is a unique solution $u_\pi(x)$ to the collocation equations (5.131) in a tube $S_\rho(u)$ of radius ρ around $u(x)$.
(b) This solution $u_\pi(x)$ can be obtained by Newton's method, which converges quadratically provided that the initial guess for $u_\pi(x)$ is sufficiently close to $u(x)$.
(c) The error estimates (5.140b, c) and (5.142a) hold. □

Example 5.17

We have solved the nonlinear part of application 1.4 using a 5-stage collocation method with Gaussian points. The BVP (1.10c, b) is rewritten here,

$$u'''' = R[u'u'' - uu''']$$

$$u(0) = u'(0) = 0, \qquad u(1) = 1, \qquad u'(1) = 0$$

so $m = 4$, $k = 5$, $k + m = 9$, and $p = 10$.

This problem is slightly more difficult, when the Reynolds number R is large, than those in other examples in this chapter. The solution profile of $u'(x)$, shown in Fig. 5.5, exhibits a narrow region of fast change near $x = 0$, which gets narrower as R is increased.

Two observations come to mind. The first is that the mesh chosen should reflect this solution behaviour. As in Example 5.2, we make the mesh fine near $x = 0$ and coarse elsewhere. For $R = 10{,}000$, 40 mesh points were sufficient to obtain an accuracy of more than 6 significant digits in all components of $\mathbf{y}(x)$.

The second observation is that if we know nothing in advance about the solution profile, then our initial guess may be far from the actual solution, and Newton's method may not converge. This turns out not to be a serious problem here (at least for $R \leq 10{,}000$), but for other examples it is a rather acute difficulty. Figure 5.5 suggests one way to proceed: Solve a number of similar BVPs, gradually increasing R, and using the obtained solution and mesh for one R to start the Newton iteration for the next, larger R. This is a process of simple *continuation*, which we discuss further in Section 8.3. □

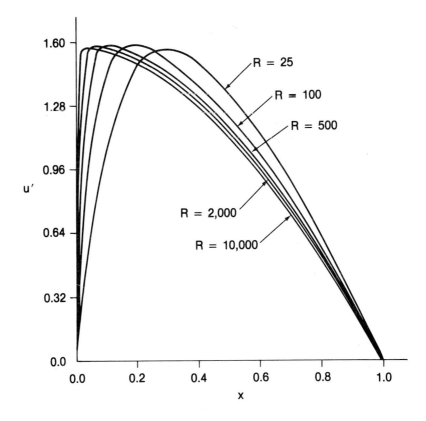

Figure 5.5 Example 5.17, tangential velocity u'

5.6.3 Collocation implementation using spline bases

The collocation algorithm of the previous subsection uses the local representation (5.132), but this is not the only possibility for implementing the method. Here we consider a different idea which arises naturally from an approximation point of view, when the numerical solution is viewed as an element from an approximation space. Thus we wish to construct good basis functions for the piecewise polynomial space in which the collocation solution lies.

The spline collocation method for solving (5.119) or (5.120) can be viewed as follows: We seek a function $u_\pi(x)$ in the space of piecewise polynomial functions with $m-1$ continuous derivatives $\mathbf{P}_{k+m,\pi} \cap C^{(m-1)}[a,b] =: \mathbf{P}_{k+m,\pi,m}$. After choosing a basis $\phi_1(x), \ldots, \phi_M(x)$ for this space, we write $u_\pi(x)$ as a linear combination of these functions as in (5.128) and determine the coefficients $\alpha_1, \ldots, \alpha_M$ by requiring $u_\pi(x)$ to satisfy (5.131b,c). Thus, collocation is a generalization of function interpolation, where we now "interpolate" (i. e., collocate) an ODE. Relative to the other global methods, collocation is in a sense midway between finite difference methods and finite element methods (see Section 5.7), since it may be considered a finite difference method as shown in Section 5.6.2, and at the same time involves finding a solution in a spline approximation space.

A natural approach is to try to compute the smooth solution $u_\pi(x)$ of (5.128) directly (and not only solution and derivative values $\mathbf{y}_i$ at mesh points). Thus, we must choose some basis functions $\phi_1(x), \ldots, \phi_M(x)$. With this approach we can naturally evaluate $u_\pi(x)$ for any $x \in [a,b]$, which is useful for nonlinear iterations or for just plotting the solution. Some popular implementations of collocation have used this approach. We shall see that while it offers some minor advantages over the local representation approach of the previous section, it also has drawbacks (which in our opinion usually outweigh the advantages).

When choosing the functions $\phi_j(x)$, we have to pay attention to considerations of efficiency as well as stability. A choice of spline basis functions which have *local support* (see Sections 2.5.1, 2.5.2) answers both of these concerns, particularly that of efficiency, because the resulting collocation matrix is banded. (Such a choice also enables us to again view the collocation method as a finite difference method.) The Hermite-type basis (well-defined if $k \geq m$) and the B-spline basis, both described in Section 2.5, turn out to have a minimal local support, and we consider them for the case $u_\pi(x) \in \mathbf{P}_{k+m,\pi,m}$, $k \geq m$. We consider only linear BVPs (5.119), since quasilinearization for (5.120) is done precisely as in Section 5.6.2.

We first consider using B-splines to represent $u_\pi(x)$, so let $\phi_j(x) = B_{j,k+m}(x)$, $1 \leq j \leq M := kN + m$. Algorithm 2.89 together with (2.90) describes how $\phi_j(x)$ and its derivatives are evaluated. The collocation matrix $\mathbf{A} \in \mathbf{R}^{M \times M}$ is formed by imposing the kN collocation conditions

$$Lu_\pi(x_{ij}) = q(x_{ij}) \quad 1 \leq j \leq k, \quad 1 \leq i \leq N \quad (5.148a)$$

and the BC

$$B_a \mathbf{y}_\pi(a) + B_b \mathbf{y}_\pi(b) = \boldsymbol{\beta} \quad (5.148b)$$

where $\mathbf{y}_\pi(x) := (u_\pi(x), u_\pi'(x), \ldots, u_\pi^{(m-1)}(x))^T$.

Since the multiplicity of an interior mesh point x_i in the knot sequence $\{t_j\}_{j=1}^{m+k+m}$ is k (see Section 2.5.2), m of the $k+m$ nonzero B-splines on (x_i, x_{i+1}) have support (x_{i-1}, x_{i+1}), $k-m$ have support (x_i, x_{i+1}), and m have support (x_i, x_{i+2}). The resulting collocation matrix $\mathbf{A}$ has the form

$$A = \begin{bmatrix} V_1 & & & \\ & V_2 & & \\ & & \cdot & \\ & & & V_N \\ W_1 & & & W_2 \end{bmatrix} \qquad (5.149)$$

where $V_1, \ldots, V_N \in \mathbf{R}^{k \times k+m}$ and $W_1, W_2 \in \mathbf{R}^{m \times (k+m)}$. Each V_i corresponds to the k collocation equations in the ith subinterval (x_i, x_{i+1}) and overlaps with V_{i-1} and V_{i+1} (for $V_1, \ldots, V_N$) in m columns. An example is depicted in Fig. 5.6. As before, the reader should picture the structure of $\mathbf{A}$ with the 3×5 blocks as indicated in the figure, but with N large.

$$A = \begin{bmatrix} \times & \times & \times & \times & \times & & & & & \\ \times & \times & \times & \times & \times & & & & & \\ \times & \times & \times & \times & \times & & & & & \\ & & & \times & \times & \times & \times & \times & & \\ & & & \times & \times & \times & \times & \times & & \\ & & & \times & \times & \times & \times & \times & & \\ \times & \times & & & & & & & \times & \times \\ \times & \times & & & & & & & \times & \times \end{bmatrix}$$

Figure 5.6 Structure of $\mathbf{A}$ using B-splines with $m = 2$, $k = 3$ and $N = 2$

If the BC are separated, the equations should be reordered so that those involving only information at $x = a$ are listed first, and then $\mathbf{A}$ is banded.

A second choice is to use a Hermite-type basis instead of a B-splines basis to represent $u_\pi(x)$. Then $\phi_j(x)$, $1 \leq j \leq M$, coincide with the functions described in Section 2.5.1. It is easy to see that these basis functions have the same support as the B-splines do, so that the resulting collocation matrix $\mathbf{A}$ again has the structure (5.149).

The numerical method used to solve the resulting linear system of collocation equations should utilize the sparseness structure of $\mathbf{A}$ as much as possible in order to provide economy both in terms of storage and in terms of computational effort. This issue is discussed in Chapter 7.

We have seen by now three implementations of the collocation method for a higher-order ODE. The first, presented in Section 5.6.2 and based on local solution representation, is markedly different from the other two, just introduced. A comparison of these implementations would have to take into account considerations of conditioning, computational effort, storage, and ease of implementation. As it turns out, the Hermite-type basis is simpler and cheaper to implement than the B-spline basis, and they both have essentially the same conditioning, as discussed in Section 5.6.4. We later comment on the (more intricate) comparison between the local solution representation and the Hermite-type basis. But it should be remarked that all three approaches produce reasonable, practical implementations.

Let us consider in some detail how the collocation matrix $\mathbf{A}$ is constructed in the case of a Hermite-type basis. From (2.83), the derivatives of the basis functions $\{\phi_j(x)\}_{j=1}^M$ involve subinterval-dependent factors times the terms $\{\psi_j^{(l-1)}(\rho_r): 1 \leq r \leq k - m, 1 \leq l \leq m+1, 1 \leq j \leq k\}$. These latter terms are mesh-independent and need only to be evaluated once (which is why the work required to construct the collocation matrix $\mathbf{A}$ is considerably less than for the B-splines). The approximate solution $u_\pi(x)$ of (5.128) and its derivatives are evaluated at collocation points by

$$u_\pi^{(l-1)}(x_{ir}) = \sum_{j=1}^{k+m} \alpha_{(i-1)k+j} \phi_{(i-1)k+j}^{(l-1)}(x_{ir})$$

$$1 \leq r \leq k, \quad 1 \leq l \leq m+1, \quad 1 \leq i \leq N$$

$$= \sum_{j=1}^{k+m} \alpha_{(i-1)k+j} w_{ij} \psi_j^{(l-1)}(\rho_r)/h_i^{l-1}$$

where from (2.83)

$$w_{ij} = \begin{cases} \left(\dfrac{h_{i-1}}{h_i}\right)^{j-1} & \text{if } 1 \leq j \leq m \text{ and } 2 \leq i \leq N \\ 1 & \text{otherwise} \end{cases} \quad (5.150)$$

The block $V_i = (V_{rj}^i)$ for $\mathbf{A}$ corresponding to the k collocation conditions in $[x_i, x_{i+1}]$ has the form

$$V_{rj}^i = w_{ij} \{h_i^{-m} \psi_j^{(m)}(\rho_r) - \sum_{l=1}^m c_l(x_{ir}) h_i^{1-l} \psi_j^{(l-1)}(\rho_r)\}$$

From (5.150)

$$w_{ij} \alpha_{(i-1)k+j} = h_i^{j-1} y_{ij} \quad 1 \leq j \leq m, \quad 1 \leq i \leq N$$

where y_{ij} are solution and derivative values at mesh points, as before. Thus, the (scaled) coefficients $\{\alpha_{(i-1)k+j}\}_{j=1;i=1}^{m;N+1}$ correspond to the unknowns $\{\mathbf{y}_i\}_{i=1}^{N+1}$ of the local representation (5.132). The other unknown coefficients correspond to basis functions which are nonzero over only one subinterval and can therefore be locally eliminated. Thus, $k - m$ Gaussian elimination steps (with row partial pivoting) per subinterval i can be carried out in order to locally eliminate some of the unknowns, leaving a system of equations of the form (5.17) for unknowns directly related to nodal solution values. This latter process is referred to as *parameter condensation*. Note that the Hermite-type basis implementation with condensation resembles (but is not the same as) the process in Section 5.6.2, where also some unknowns $\mathbf{z}_i$ are locally eliminated to produce the difference scheme (5.137a), yielding a linear system of equations with the same size and structure.

After the nodal solution values are obtained (when condensation is used), we may need to evaluate $u_\pi(x)$ in (x_l, x_{l+1}), particularly if the BVP is nonlinear. This can be done by (i) forming a local polynomial approximation by interpolating the nodal values, or (ii) recovering the eliminated unknowns by local back substitution for each subinterval and using (5.128). The first of these methods has the advantages that it is more efficient, both in terms of speed and in terms of storage (the upper triangular matrices resulting from the parameter condensation may be discarded), and that it uses only superconvergent nodal values [see Theorem 5.140]. This method applies equally well here and for the local solution representation (5.132). Its disadvantage is that it is possibly less robust in a general-purpose implementation. The second method above requires the same amount of effort as applying the usual Gaussian elimination to **A**; it simply does things in a different order.

Condensation of parameters could also be used in the B-spline case, whereby unknowns corresponding to B-splines having support over only one subinterval are eliminated. The nodal values $\mathbf{y}_i$ can be determined from the remaining unknowns (see Exercise 22). Still, condensation of parameters is not as natural in this case because these nodal values are not determined directly.

The use of condensation of parameters also allows us to examine relative merits of the Hermite-type basis and the local solution representation of Section 5.6.2. (The B-spline implementation is a somewhat distant third.) Whereas $k - m$ local elimination steps per subinterval are made here, k steps are needed in Section 5.6.2. Thus, if the first method above is used for solution evaluation, then the Hermite-type basis has a storage advantage. It is also cheaper in operation count, but only by a slight amount. The ease of carrying out the two implementations is comparable. But the local solution representation has a significant conditioning advantage, which in our opinion generally outweighs the slight disadvantages when compared to the Hermite-type basis. This is discussed next.

5.6.4 Conditioning of collocation matrices

We are considering the BVP (5.119) for an mth order ODE on a finite interval, assuming that as a first-order system it has a conditioning constant κ of moderate size. Using collocation as described above, it turns out that both the Hermite-type basis and the B-spline basis yield linear systems of equations with condition numbers which, as functions of the mesh π, are less than optimal. An indication of this is obtained if we consider $L\phi_j(x)$ for *any* (scaled) basis function of $\mathbf{P}_{k+m,\pi,m}$ with compact support on a uniform mesh π. When h is small enough, we obtain

$$L\phi_j(x) \sim \phi_j^{(m)}(x) \sim h^{-m}$$

and this factor h^{-m} appears in **A** when imposing the collocation equations (5.148a). It is not difficult to see that this yields

$$\text{cond}(\mathbf{A}) \sim h^{-m}$$

at best. (N.B. $N \sim h^{-1}$.) In general, when the mesh π is not necessarily uniform, the ith block of **A** corresponding to the collocation equations on the ith subinterval contains the factor h_i^{-m}. It makes sense to attempt to row-scale blocks of **A** differently, because they involve different scaling factors, and because the BC equations involve different conditioning, computational effort, storage, and ease of implementation. As it turns out, the Hermite-type basis is simpler and cheaper to implement than the B-spline basis, and they both have essentially the same conditioning, as discussed in Section 5.6.4. We later comment on the (more intricate) comparison between the local solution representation and the Hermite-type basis. But it should be remarked that all three approaches produce reasonable, practical implementations.

Let us consider in some detail how the collocation matrix $\mathbf{A}$ is constructed in the case of a Hermite-type basis. From (2.83), the derivatives of the basis functions $\{\phi_j(x)\}_{j=1}^{M}$ involve subinterval-dependent factors times the terms $\{\psi_j^{(l-1)}(\rho_r)\colon 1 \le r \le k-m, 1 \le l \le m+1, 1 \le j \le k\}$. These latter terms are mesh-independent and need only to be evaluated once (which is why the work required to construct the collocation matrix $\mathbf{A}$ is considerably less than for the B-splines). The approximate solution $u_\pi(x)$ of (5.128) and its derivatives are evaluated at collocation points by

$$u_\pi^{(l-1)}(x_{ir}) = \sum_{j=1}^{k+m} \alpha_{(i-1)k+j} \phi_{(i-1)k+j}^{(l-1)}(x_{ir})$$

$$1 \le r \le k, \quad 1 \le l \le m+1, \quad 1 \le i \le N$$

$$= \sum_{j=1}^{k+m} \alpha_{(i-1)k+j} w_{ij} \psi_j^{(l-1)}(\rho_r)/h_i^{l-1}$$

where from (2.83)

$$w_{ij} = \begin{cases} \left(\dfrac{h_{i-1}}{h_i}\right)^{j-1} & \text{if } 1 \le j \le m \quad \text{and} \quad 2 \le i \le N \\ 1 & \text{otherwise} \end{cases} \quad (5.150)$$

The block $V_i = (V_{rj}^i)$ for $\mathbf{A}$ corresponding to the k collocation conditions in $[x_i, x_{i+1}]$ has the form

$$V_{rj}^i = w_{ij}\{h_i^{-m}\psi_j^{(m)}(\rho_r) - \sum_{l=1}^{m} c_l(x_{ir})h_i^{1-l}\psi_j^{(l-1)}(\rho_r)\}$$

From (5.150)

$$w_{ij}\alpha_{(i-1)k+j} = h_i^{j-1} y_{ij} \quad 1 \le j \le m, \quad 1 \le i \le N$$

where y_{ij} are solution and derivative values at mesh points, as before. Thus, the (scaled) coefficients $\{\alpha_{(i-1)k+j}\}_{j=1;i=1}^{m;N+1}$ correspond to the unknowns $\{\mathbf{y}_i\}_{i=1}^{N+1}$ of the local representation (5.132). The other unknown coefficients correspond to basis functions which are nonzero over only one subinterval and can therefore be locally eliminated. Thus, $k-m$ Gaussian elimination steps (with row partial pivoting) per subinterval i can be carried out in order to locally eliminate some of the unknowns, leaving a system of equations of the form (5.17) for unknowns directly related to nodal solution values. This latter process is referred to as *parameter condensation*. Note that the Hermite-type basis implementation with condensation resembles (but is not the same as) the process in Section 5.6.2, where also some unknowns $\mathbf{z}_i$ are locally eliminated to produce the difference scheme (5.137a), yielding a linear system of equations with the same size and structure.

After the nodal solution values are obtained (when condensation is used), we may need to evaluate $u_\pi(x)$ in (x_i, x_{i+1}), particularly if the BVP is nonlinear. This can be done by (i) forming a local polynomial approximation by interpolating the nodal values, or (ii) recovering the eliminated unknowns by local back substitution for each subinterval and using (5.128). The first of these methods has the advantages that it is more efficient, both in terms of speed and in terms of storage (the upper triangular matrices resulting from the parameter condensation may be discarded), and that it uses only superconvergent nodal values [see Theorem 5.140]. This method applies equally well here and for the local solution representation (5.132). Its disadvantage is that it is possibly less robust in a general-purpose implementation. The second method above requires the same amount of effort as applying the usual Gaussian elimination to $\mathbf{A}$; it simply does things in a different order.

Condensation of parameters could also be used in the B-spline case, whereby unknowns corresponding to B-splines having support over only one subinterval are eliminated. The nodal values $\mathbf{y}_i$ can be determined from the remaining unknowns (see Exercise 22). Still, condensation of parameters is not as natural in this case because these nodal values are not determined directly.

The use of condensation of parameters also allows us to examine relative merits of the Hermite-type basis and the local solution representation of Section 5.6.2. (The B-spline implementation is a somewhat distant third.) Whereas $k-m$ local elimination steps per subinterval are made here, k steps are needed in Section 5.6.2. Thus, if the first method above is used for solution evaluation, then the Hermite-type basis has a storage advantage. It is also cheaper in operation count, but only by a slight amount. The ease of carrying out the two implementations is comparable. But the local solution representation has a significant conditioning advantage, which in our opinion generally outweighs the slight disadvantages when compared to the Hermite-type basis. This is discussed next.

5.6.4 Conditioning of collocation matrices

We are considering the BVP (5.119) for an mth order ODE on a finite interval, assuming that as a first-order system it has a conditioning constant κ of moderate size. Using collocation as described above, it turns out that both the Hermite-type basis and the B-spline basis yield linear systems of equations with condition numbers which, as functions of the mesh π, are less than optimal. An indication of this is obtained if we consider $L\phi_j(x)$ for *any* (scaled) basis function of $\mathbf{P}_{k+m,\pi,m}$ with compact support on a uniform mesh π. When h is small enough, we obtain

$$L\phi_j(x) \sim \phi_j^{(m)}(x) \sim h^{-m}$$

and this factor h^{-m} appears in $\mathbf{A}$ when imposing the collocation equations (5.148a). It is not difficult to see that this yields

$$\text{cond}(\mathbf{A}) \sim h^{-m}$$

at best. (N.B. $N \sim h^{-1}$.) In general, when the mesh π is not necessarily uniform, the ith block of $\mathbf{A}$ corresponding to the collocation equations on the ith subinterval contains the factor h_i^{-m}. It makes sense to attempt to row-scale blocks of $\mathbf{A}$ differently, because they involve different scaling factors, and because the BC equations involve different derivatives of $u_\pi(x)$ than the collocation equations. If we do the equilibration so that the largest element in each row has magnitude 1, and call the scaled matrix $\hat{\mathbf{A}}$, then it can be shown that the Hermite-type basis (and, under stronger assumptions, the B-spline basis) yields

$$\text{cond}(\hat{\mathbf{A}}) \sim \max_{1 \le i \le N+1} \sum_{j=1}^{N} \sum_{l=0}^{m-1} h_j^{-m} \hat{h}_i^l \int_{x_j}^{x_{j+1}} |G_l(x_i, t)| \, dt \tag{5.151}$$

where $\hat{h}_i := \max(h_i, h_{i+1})$ and as before $G(x, t)$ is the (scalar) Green's function of (5.119) and $G_l(x, t) \equiv \dfrac{\partial^l}{\partial x^l} G(x, t)$. It can also be shown that the parameter condensation procedure does not affect the conditioning of the algorithm significantly, so (5.151) is satisfied for this variant too. From (5.151) we can write

$$\text{cond}(\hat{\mathbf{A}}) \le \kappa N \underline{h}^{-m+1} \tag{5.152}$$

where

$$\underline{h} := \min_{1 \le i \le N} h_i$$

The approximate bound (5.152) is sharp when $\kappa \approx \|G\|$, and when there are $O(N)$ subintervals of size $h_i \sim \underline{h}$.

This is to be contrasted with the $\sim \kappa N$ condition number obtained with the local solution representation of Section 5.6.2 ! We emphasize that we are discussing different *implementations* of the *same* collocation method. Thus, the various implementations would all yield the same approximate solution $u_\pi(x)$ if exact arithmetic were used. However, as the condition numbers indicate, the roundoff error effects for the implementations considered in Section 5.6.3 differ considerably from those of Section 5.6.2. This is demonstrated in the following example.

Example 5.18

Consider the simple BVP

$$u'' - 4u = 16x + 12x^2 - 4x^4, \quad 0 < x < 1$$

$$u(0) = u'(1) = 0$$

which has the unique solution

$$u(x) = x^4 - 4x.$$

We solve this problem using collocation at Gaussian points in $\mathbf{P}_{6,\pi,2}$. Since the exact solution lies in the approximation space, there is no discretization error, and we can examine pure roundoff error effects. The spline solution is represented using (i) a B-spline basis, (ii) a Hermite-type basis, and (iii) a Runge-Kutta local representation. For uniform meshes, doubling N should quadruple the condition number for (i) and (ii), but only double it for (iii), according to the above estimates. To demonstrate the effect of highly nonuniform meshes in (5.151), (5.152), the solution is also computed on

$$\pi_1: 0 < 10^{-4} < .25 < .5 < .75 < 1$$

$$\pi_2: 0 < 10^{-6} < .25 < .5 < .75 < 1$$

$$\pi_3: 0 < .25 < .5 < .75 < 1-10^{-4} < 1$$

$$\pi_4: 0 < .25 < .5 < .75 < 1-10^{-6} < 1$$

Since the Green's function for this example is

$$G(x,t) = \begin{cases} \sinh 2x \cosh 2(1-t)/(2\cosh 2), & x \le t \\ \sinh 2t \cosh 2(1-x)/(2\cosh 2), & x > t \end{cases}$$

we have that for the first two meshes $G(x,t) \sim \underline{h}$ for $x \in [0, x_2]$, so the expression in (5.151) (with $m=2$) yields a condition number estimate of $\sim N$. Thus, the condition number should be roughly the same for π_1 and π_2, if all three implementations are used. On the other hand, $G(1,t)$ is not small, and since there is only one very small subinterval near $x=1$ in the last two meshes, the condition number should increase by a factor of ~ 100 when going from π_3 to π_4, using implementations (i) or (ii). No such deterioration is expected with (iii).

The calculated condition numbers and solution errors, when floating-point arithmetic with 56 binary digits in the mantissa is used, are listed in Table 5.9. Note that a fairly long wordlength has been used. If a short wordlength is used, the loss of precision from using the first two implementations (namely those of Section 5.6.3) may be significant. More severe effects for implementations (i) and (ii) [but not (iii)] can also occur if higher-order ODEs are encountered.

TABLE 5.9 Condition numbers and errors for Example 5.18

		Condition numbers			Errors in u		
	N	(i)	(ii)	(iii)	(i)	(ii)	(iii)
Uniform meshes	10	.42 + 3	.22 + 3	.20 + 2	.50 − 13	.21 − 13	.24 − 14
	20	.17 + 4	.87 + 3	.34 + 2	.26 − 12	.11 − 12	.33 − 14
	40	.69 + 4	.34 + 4	.64 + 2	.23 − 11	.72 − 13	.82 − 14
	80	.28 + 5	.14 + 5	.12 + 3	.68 − 11	.18 − 11	.13 − 13
Nonuniform meshes	π_1	.68 + 2	.43 + 2	.17 + 2	.10 − 13	.62 − 14	.67 − 15
	π_2	.68 + 2	.43 + 2	.12 + 2	.40 − 14	.58 − 14	.18 − 14
	π_3	.96 + 5	.54 + 5	.12 + 2	.41 − 10	.94 − 11	.18 − 14
	π_4	.96 + 7	.54 + 7	.12 + 2	.38 − 8	.15 − 8	.18 − 14

□

It is important to realize that truncation error usually dominates roundoff error when one is solving BVPs using collocation, regardless of the choice of implementation from amongst the ones introduced in this chapter. (Note also that it is extremely rare to find an ODE of order higher than 4 in practice.) Nevertheless, it is not unreasonable to expect the collocation method to perform well even if the mesh is extremely fine or highly nonuniform, and in such situations the local solution representation is considerably more secure. The issue becomes more important if use of a short wordlength is contemplated.

The different conditioning occurring with the various collocation implementations raises more global theoretical issues, particularly if we recall from Section 5.6.2 that the local representation implementation can be considered as a sophisticated discretization of the corresponding first-order ODE. Note that in Section 5.6.1 we also have essentially the same discretization for a second-order ODE and for a corresponding first-order system [viz. (5.126) and (5.124), respectively], and there too the condition numbers differ: that of (5.126) for a uniform mesh is $\sim N^2$, while that of (5.124) is only $\sim N$. Still, we would expect the *stability constant* of the algorithm not to change.

Indeed, recall the discussion in Section 2.8.2 on the fundamental difference between the "smooth" truncation error and the "rough" ("high frequency") roundoff error. The truncation error, controlled by the stability constant, is the same for the three collocation implementations [or for (5.124) and (5.126)]. It satisfies a discrete analogue of the smoothing property of the inverse of the differential operator (cf. Section 5.5.4), and it is bounded (in fact, vanishes) as $h \to 0$.

Not so for roundoff error. Consider a stable finite difference or collocation discretization of a BVP involving a first-order ODE system, and denote for simplicity the differential operator together with the boundary operator by **L** [cf. (2.167), (2.168)]. We have a system of linear algebraic equations

$$\mathbf{Au} = \mathbf{v} \qquad (5.153)$$

where **A** is a discretization of **L**, **u** relates to solution values of $y(x)$ at mesh points, say, and **v** relates to inhomogeneity values of $q(x)$ and **β**. The stability of the scheme yields that $\|\mathbf{A}^{-1}\| \leq K$, where the constant K is of moderate size because κ is. The qualitative size of the condition number of **A** is then essentially $\|\mathbf{A}\|$. When assessing the effect of *roundoff errors*, the vectors **u** and **v** in (5.153) must be considered as belonging to the same normed space, so the smoothing effect of $\mathbf{L}^{-1}$ is in this sense lost. Correspondingly, **A** is an approximation of *an unbounded differential operator* **L**. Indeed, reasonable one-step schemes yield matrices which can be equilibrated, as we have seen, to have condition numbers

$$\text{cond}(\mathbf{A}) \sim \kappa N$$

where $N \to \infty$ as $h \to 0$. In practice values of κN are quite tolerable for well-conditioned BVPs (i. e., with κ not large), but as we ask for a diminishing truncation error for a BVP, the roundoff error generally grows unboundedly (regardless of the numerical method and implementation used).

Now, consider the higher-order differential operator L. Its inverse L^{-1} is bounded, in an appropriate norm, by the conditioning constant κ of its equivalent first-order system form. (Let us assume for simplicity that $\kappa = 1$.) This says nothing about roundoff error accumulation resulting from (the discretization of) the unbounded operator L. If L is discretized directly, by evaluating $L \phi_j(x)$ at collocation points [as in (5.148), (5.128)], then a numerical differentiation of an mth order derivative is encountered, leading to the estimate (5.151). This condition number *cannot be improved* by any method of implementation which forms basis functions $\phi_j(x)$ with compact support and then differentiates them m times.

Now we can better understand also the conditioning of the local (or Runge-Kutta) representation of the collocation solution. Here the unfavourable condition number (5.152) (for $m > 1$) is avoided by never constructing basis functions $\phi_j(x)$ explicitly. Instead, we can view the implementation as starting from the mth *derivative* of the approximate solution (which is generally discontinuous at mesh points), representing it on each mesh subinterval i as a linear combination

$$u_\pi^{(m)}(x) = \sum_{j=1}^{k} z_{ij} \psi_j^{(m)}(\frac{x-x_i}{h_i}), \quad x_i \le x \le x_i+1 \tag{5.154}$$

where $\psi_j^{(m)}(t)$ are k linearly independent polynomials of order k (k being the number of collocation points per subinterval). Then $u_\pi(x)$ and its derivatives are formed by *integration* of (5.154), thus avoiding a high-order numerical differentiation. The k unknowns z_i are local in each subinterval i and can be locally eliminated upon imposing the k collocation equations, thereby eliminating $u_\pi^{(m)}$. The inevitable factor N in the condition number appears when we patch the solution segments together to form a global solution $u_\pi(x) \in C^{(m-1)}[0, 1]$. [When we attempt to simulate this argument for the Hermite-type basis of Section 5.6.3, the representation corresponding to (5.154) has more than k parameters, which are therefore not completely local to the ith subinterval alone, and the argument does not work.]

5.7 FINITE ELEMENT METHODS

Finite element methods are among the most extensively studied methods for solving differential equations. They have enjoyed widespread use in solving BVPs in partial differential equations (PDEs), originally steady state problems in structural engineering. While such problems can be described as BVPs, they also satisfy an equivalent variational formulation. Specifically, an integral relationship in the unknown function variable (a *functional*) is optimized over an appropriate class of functions by the solution of the BVP.

The variational formulation incorporates some of the BC information and involves lower derivatives than the BVP does. Although not so critical for ODEs, both features are important for PDEs because lower derivatives are much easier to handle and because satisfying the BC for irregularly shaped regions can pose difficulty. More importantly for the latter case, the variational formulation allows geometrically flexible discretizations of domains, e. g., using triangles (in two independent variables) which do not necessarily lie along rectangular grid lines. Again, this feature is dormant in the ODE case.

The basic idea of the classical *Ritz method* for elliptic problems is to take a convenient space of functions which approximates the space where the exact solution lies and to choose the approximate solution such that it minimizes the functional over this approximation space of functions. For the finite element method, these functions are normally chosen as piecewise polynomials (splines) defined on some partition of the domain in the independent variable(s). Basis functions with local support for the piecewise polynomial space (which give rise to the name "finite element") yield sparse matrix equations for the unknown parameters. Moreover, the matrices are symmetric positive definite, which allows an easier application of sparse matrix techniques for their solution.

For ODEs the task of solving the linear system is much easier and therefore not as important as it is for PDEs. Furthermore, special geometrical flexibility is not needed and, as we have seen in the last section, construction of a high-order spline basis is more or less straightforward. Indeed, the collocation method turns out to usually be more efficient than the traditional finite element methods. Collocation also applies to quite general BVPs for which no variational formulation is available. As a result, these traditional finite element methods which are so important for solving PDEs do not play a correspondingly central role in solving ODEs, except in special situations, and we only mention them briefly.

Some may say that for a method to be called a finite element method it should be derived by using a variational formulation. Others take the broader interpretation and call any method which gives a spline solution a finite element method, so technically a (spline) collocation method is one.

5.7.1 The Ritz method

The Ritz method operates directly on the variational formulation of a BVP. We shall just motivate the method through the simple second-order problem

$$Lu(x) \equiv -u''(x) + r(x)u(x) = q(x) \qquad a < x < b \qquad (5.155a)$$

$$u(a) = 0, \qquad u(b) = 0 \qquad (5.155b)$$

Here $r(x)$ is nonnegative and (5.155a, b) has the solution $u(x)$. Defining the functional

$$I(v) := \frac{1}{2} \int_a^b [(v'(x))^2 + r(x)(v(x))^2 - 2v(x)q(x)]\, dx \qquad (5.156)$$

for any $v(x) \in C^{(1)}[a, b]$, a direct substitution of $d(x) = v(x) - u(x)$ into (5.156) gives

$$I(v) = I(u) + \int_a^b [u'(x) d'(x) + r(x)u(x) d(x) - d(x)q(x)]\, dx \qquad (5.157)$$

$$+ \frac{1}{2} \int_a^b [(d'(x))^2 + r(x)(d(x))^2]\, dx$$

Using integration by parts and (5.155a, b) (see Exercise 24) one can show that the first integral in (5.157) is equal to zero if $v(a) = v(b) = 0$. Thus, $I(v) \geq I(u)$, and the variational formulation for (5.155a, b) is

$$I(u) = \min_{v \in C_0^1[a,b]} I(v) \qquad (5.158)$$

where $C_0^1[a, b] := \{v \in C^{(1)}[a, b] : v(a) = v(b) = 0\}$. Given a mesh π and a spline space of piecewise polynomials (of order k and in $C^{(l-1)}[a, b]$) which are also zero at the end points, $\mathbf{P}^0_{k,\pi,l} := \{s \in \mathbf{P}_{k,\pi,l} : s(a) = s(b) = 0\}$, the corresponding *Ritz solution* $u_\pi(x)$ is defined to satisfy

$$I(u_\pi) = \min_{s \in \mathbf{P}^0_{k,\pi,l}} I(s) \qquad (5.159)$$

Motivated by the BVP (5.155), one would choose $k \geq 3$, but the Ritz solution is defined, following (5.158), even if $u_\pi(x)$ is only linear and piecewise continuous (so $k = 2$). The fact that a Ritz solution may have global continuity and order which are too low for the BVP to make sense leads to the necessity of talking about *weak*, or *generalized solutions*, but this is outside the scope of our brief description.

To solve (5.159), we would first choose a basis $\{\phi_j(x)\}_{j=1}^J$ for $\mathbf{P}^0_{k,\pi,l}$ (e. g., B-splines appropriately modified at the ends so as to satisfy the BC, or a local representation), and write, say, $v(x) = \sum_{j=1}^J \alpha_j \phi_j(x)$ for any $v \in \mathbf{P}^0_{k,\pi,l}$. The necessary first-order conditions for $u_\pi(x) = \sum_{j=1}^J \hat\alpha_j \phi_j(x)$ to be the minimizer of (5.159) are

$$\frac{\partial I(v)}{\partial \alpha_j} = 0 \qquad 1 \leq j \leq N+1$$

The solution $\hat{\boldsymbol\alpha} = (\hat\alpha_1, \ldots, \hat\alpha_J)^T$ can be found by solving

$$A\hat{\boldsymbol\alpha} = \hat{\mathbf{q}} \qquad (5.160)$$

where the elements of A and $\hat{\mathbf{q}}$ are

$$a_{ij} = \int_a^b [\phi_i'(x)\phi_j'(x) + r(x)\phi_i(x)\phi_j(x)]\, dx \qquad (5.161a)$$

$$=: E(\phi_i, \phi_j) \qquad 1 \leq i, j \leq J$$

and

$$\hat q_i = \int_a^b q(x)\phi_i(x)\, dx \qquad (5.161b)$$

Notice that determining these coefficients requires evaluating integrals, and the most efficient way to do this is by using a quadrature method (unless the coefficients are exceedingly simple). Observe also that A is symmetric.

Example 5.19

Taking $k = 2$, and $l = 0$, we find that the B-spline basis is simply the "roof functions"

$$\phi_j(x) = \begin{cases} \dfrac{x - x_j}{h_j} & \text{if } x_j \leq x \leq x_{j+1} \\ \dfrac{x_{j+2} - x}{h_{j+1}} & \text{if } x_{j+1} \leq x \leq x_{j+2} \\ 0 & \text{otherwise} \end{cases}$$

and $\hat\alpha_j$ is $u_\pi(x_{j+1})$ ($1 \leq j \leq J := N-1$). From (5.161a, b),

$$a_{ij} = \begin{cases} \dfrac{1}{h_i} + \dfrac{1}{h_{i+1}} + \int\limits_{x_i}^{x_{i+1}} r(x)(\dfrac{x-x_i}{h_i})^2 dx + \int\limits_{x_{i+1}}^{x_{i+2}} r(x)(\dfrac{x_{i+2}-x}{h_{i+1}})^2 dx & \text{if } j = i \\[2ex] -\dfrac{1}{h_i} + \int\limits_{x_i}^{x_{i+1}} r(x)(\dfrac{x_{i+1}-x}{h_i})(\dfrac{x-x_i}{h_i}) dx & \text{if } j = i-1 \\[2ex] 0 & \text{if } j < i-1 \end{cases}$$

$a_{ij} = a_{ji}$ for $j > i$, and

$$\hat{q}_i = \int\limits_{x_i}^{x_{i+1}} q(x)(\dfrac{x-x_i}{h_i}) dx + \int\limits_{x_{i+1}}^{x_{i+2}} q(x)(\dfrac{x_{i+2}-x}{h_{i+1}}) dx$$

The matrix A is tridiagonal and symmetric positive definite. If the trapezoidal rule is used to approximate the integrals in its coefficients then the resulting matrix A is *identical* to the one resulting from using the simple finite difference method (5.4), (5.5). But note that the Ritz method has been defined on an arbitrary mesh and is essentially no more difficult to apply on a nonuniform mesh than on a uniform one. $\square$

Assuming appropriate smoothness on the coefficients of the ODE (5.155a) [and recalling that $r(x) \geq 0$], we can show for instance

Theorem 5.162 For any integer $n \geq 1$, if $l \geq n$ then the Ritz solution $u_\pi(x) \in \mathbf{P}_{2n,\pi,l}^0$ satisfies

$$\| u_\pi - u \|_2 := [\int\limits_a^b (u_\pi(x) - u(x))^2 dx]^{1/2} = O(h^{2n}) \tag{5.162}$$

Proof outline: For arbitrary $\boldsymbol{\alpha} = (\alpha_1, \ldots, \alpha_J)$, $v(x) = \sum\limits_{j=1}^{J} \alpha_j \phi_j(x) \in \mathbf{P}_{2n,\pi,l}^0$ satisfies

$$\boldsymbol{\alpha}^T A \boldsymbol{\alpha} = \int\limits_a^b [(v'(x))^2 + r(x)(v(x))^2] dx =: \| v \|_A \geq 0$$

Since $\| v \|_A = 0$ iff $v \equiv 0$, A is positive definite, so $u_\pi(x)$ is uniquely defined from (5.160). From (5.157),

$$\| u_\pi - u \|_A = \min_{v \in \mathbf{P}_{2n,\pi,l}^0} \| v - u \|_A$$

and since spline approximation theory guarantees the existence of a $v_\pi \in \mathbf{P}_{2n,\pi,l}^0$ satisfying

$$\| v_\pi - u \|_2 = O(h^{2n})$$

we also have

$$\| v_\pi - u \|_A = O(h^{2n-1})$$

and hence

$$\|u_\pi - u\|_A = O(h^{2n-1}) \tag{5.163}$$

To improve the error estimate (5.163), we use the so-called "Nitsche trick": Let $w(x)$ solve the BVP

$$-w''(x) + r(x)w(x) = (u_\pi(x) - u(x))/\|u_\pi - u\|_2$$

$$w(a) = w(b) = 0$$

Multiplying this ODE by $u_\pi(x) - u(x)$ and integrating, we have

$$\|u_\pi - u\|_2 = \int_a^b (-w''(x) + r(x)w(x))(u_\pi(x) - u(x))\,dx$$

and use of integration by parts and (5.157) show that

$$\|u_\pi - u\|_2 = E(w, u_\pi - u) = E(w - v, u_\pi - u)$$

for any $v \in \mathbf{P}^0_{2n,\pi,l}$. [The quadratic form E is defined in (5.161a).] Choosing for v an appropriate piecewise linear interpolate $\hat{w}$ to w, the Cauchy-Schwarz inequality gives

$$\|u_\pi - u\|_2 \leq \|w - \hat{w}\|_A \|u_\pi - u\|_A = O(h^{2n})$$

□

A linear differential operator L has an *adjoint operator* L^* such that

$$\int_a^b (Lv(x))w(x)\,dx = \int_a^b v(x)(L^* w(x))\,dx$$

for any smooth $v(x)$ and $w(x)$, and L is called *self-adjoint* if $L = L^*$. In the latter case the Green's function $G(x,t)$ is symmetric, satisfying $G(x,t) = G(t,x)$ for $a \leq x, t \leq b$ [cf. Theorem 3.40]. A BVP involving a self-adjoint operator and appropriate BC has an equivalent variational formulation. Generalizing the above, the BVP

$$Lu(x) \equiv \sum_{j=0}^{m/2} (-1)^j (c_{2j}(x) u^{(j)}(x))^{(j)} = q(x) \qquad a < x < b \tag{5.164a}$$

$$u^{(j)}(a) = u^{(j)}(b) = 0 \qquad 0 \leq j \leq m/2 - 1 \tag{5.164b}$$

where m is an even integer, $c_{2j}(x) \geq 0$ ($0 \leq j \leq m/2 - 1$), and $c_m(x) > 0$, has an equivalent variational formulation with

$$I(v) := \int_a^b [\sum_{j=0}^{m/2} c_{2j}(x)(v^{(j)}(x))^2 - 2q(x)v(x)]\,dx \tag{5.165}$$

It is straightforward to show that if $u_\pi(x)$ is the Ritz solution for this problem in $\mathbf{P}^0_{k,\pi,l}$, the subspace of $\mathbf{P}_{k,\pi,l}$ consisting of splines satisfying (5.164b) (for appropriate k and l), and if $\{\phi_j(x)\}_{j=1}^J$ is a basis for $\mathbf{P}^0_{k,\pi,l}$, then

$$\int_a^b [Lu_\pi(x) - q(x)]\phi_j(x)\,dx = 0 \qquad 1 \leq j \leq J \tag{5.166}$$

5.7.2 Other finite element methods

For a BVP consisting of a smooth but otherwise arbitrary linear differential operator L and homogeneous BC, the *Galerkin solution* $v_\pi(x) = \sum_{j=1}^{J} \beta_j \phi_j(x)$ in $\mathbf{P}^0_{k,\pi,l}$ is defined by requiring that

$$\int_a^b [Lv_\pi(x) - q(x)]\phi_j(x)\, dx = 0 \qquad 1 \le j \le J \qquad (5.167)$$

Integration by parts may be used, repeatedly if necessary, to lower the highest derivative appearing in (5.167) and thus relax the minimum smoothness requirement on the approximation space. If L is self-adjoint then the Galerkin solution is the same as the Ritz solution, but the Galerkin solution may be defined for a linear BVP even if there is no corresponding variational formulation. The coefficients $\boldsymbol{\beta} = (\beta_1, \ldots, \beta_J)^T$ for $v_\pi(x)$ are determined by solving the linear system of equations resulting from (5.167). Usually the matrix coefficients $\int_a^b (L\phi_i(x))\phi_j(x)\, dx$ cannot be computed exactly, so a quadrature rule is used to approximate them. The solution to the resulting linear system is called a *discrete Galerkin solution*.

Instead of (5.167), we can solve for $w_\pi(x) \in \mathbf{P}^0_{k,\pi,l}$ defined by

$$\int_a^b [Lw_\pi(x) - q(x)]L\phi_j(x)\, dx = 0 \qquad 1 \le j \le J \qquad (5.168)$$

Both (5.167) and (5.168) may be viewed as *orthogonality conditions*. The solution $w_\pi(x)$ is called the *least squares solution* because (5.168) is equivalent to solving

$$\min_{v \in \mathbf{P}^0_{k,\pi,l}} \int_a^b [Lv(x) - q(x)]^2\, dx$$

As with the Galerkin method, in practice one must generally compute a *discrete least squares solution*. Interestingly, the collocation solution $u_\pi(x) \in \mathbf{P}^0_{k,\pi,l}$ is related naturally to the Galerkin and least squares solutions, where instead of (5.167) or (5.168) we require that the residual $Lu_\pi(x) - q(x)$ satisfy

$$\int_a^b [Lu_\pi(x) - q(x)]\psi_j(x)\, dx = 0 \qquad 1 \le j \le J \qquad (5.169)$$

with $\psi_j(x)$ being a δ–function corresponding to a collocation point z_j [so $\int_a^b f(x)\psi_j(x)\,dx = f(z_j)$]. Using other choices of $\{\psi_j(x)\}$ from *different* spline spaces $\hat{\mathbf{P}}_{k,\pi,\hat{l}}$ defines a general class of methods called *Petrov-Galerkin methods*. Note that if integration by parts cannot be (or is not) used in (5.169) or (5.167) then the minimum smoothness and order requirements on the approximate solution space are not different from those for the least squares method (5.168), the collocation method of Section 5.6, or other difference schemes.

EXERCISES

1. Consider the approximation (5.4) to the problem (5.2).
   (a) Show that if $h \leq 2/\|p\|$ and $r \geq 0$ then $\mathbf{A}$ of (5.5) is diagonally dominant.
   (b) Show that if in addition, $r(x) \geq \delta > 0$ for $x \in [0, 1]$, then (5.9) is satisfied with $K = \max\{1, \delta^{-1}\}$
   (c) Show that
   $$\text{cond}(\mathbf{A}) \approx 4(1 + \frac{1}{\delta h^2}) = O(N^2)$$

2. Describe the application of a finite difference scheme like (5.4) to a nonlinear problem
   $$y'' = f(x, y, y')$$
   subject to the BC (5.2b). How would the scheme look for Example 3.1?

3. Consider the BVP (5.13a) subject to separated BC
   $$B_{a1}\mathbf{y}(a) = \boldsymbol{\beta}_1, \qquad B_{b2}\mathbf{y}(b) = \boldsymbol{\beta}_2$$
   with $B_{a1} \in \mathbf{R}^{(n-v) \times n}$ and $B_{b2} \in \mathbf{R}^{v \times n}$. Show that the linear system of equations (5.17) can be rearranged so that $\mathbf{A}$ is banded with bandwidth $3n - 1$. What would the bandwidth be after Gauss elimination with partial row-pivoting is applied to $\mathbf{A}$?

4. Implement the trapezoidal-Newton algorithm described in Section 5.1.3 for nonlinear first-order BVPs. Your data structures will depend on the linear system solver used. For the latter, use a band solver (Exercise 3), or consult Chapter 7, or treat $\mathbf{A}$ as full (as a last resort). Try your code on Example 5.3. Devise initial guesses which would yield approximations to both exact solutions.

5. Show that both the midpoint scheme and the trapezoidal scheme for the BVP (5.20) are consistent of order 2.

6. Consider the ODE (5.2a) under the BC
   $$u(0) = \beta_1, \qquad u'(1) = \beta_2$$
   Derive appropriate difference approximations for the BC so that, when used with (5.4a), the resulting method is consistent of order 2.

7. Prove Theorem 5.56. (*Hint:* Verify the premises of the Newton-Kantorovich Theorem 2.64. Note that α is bounded by stability, while β can be made small for h and ρ_0 small).

8. Show that the trapezoidal scheme (5.21) is an implicit Runge-Kutta method (5.59), with $k = 2$, $\rho_1 = 0$, $\rho_2 = 1$, $\alpha_{11} = \alpha_{12} = 0$, $\beta_1 = \beta_2 = \alpha_{21} = \alpha_{22} = 1/2$ and that the precision is $p = 2$.

9. Show that (5.76) yields (5.77).

10. (a) Show that the collocation scheme at $k = 3$ Lobatto points per subinterval, derived in Example 5.6, can be written in the one-step form

$$\frac{y_{i+1}-y_i}{h_i} = \frac{1}{6}\{f(x_i, y_i)+f(x_{i+1}, y_{i+1}) + 4f(x_{i+1/2}, \frac{1}{2}(y_i+y_{i+1}))$$

$$+\frac{h}{8}(f(x_i, y_i)-f(x_{i+1}y_{i+1})))\}$$

    (b) Show that, applied to the nonlinear problem (5.20), this scheme is convergent of order 4.

11. Consider Radau collocation with $k = 2$ for $y' = f(x, y)$. The canonical collocation points are given by $\rho_1 = 1/3$, $\rho_2 = 1$.

    (a) Derive the method coefficients α_{jl}, β_l.

    (b) Obtain the method in a one-step form, as in the previous exercise.

    This is a third-order, nonsymmetric scheme with good stability properties for stiff IVPs.

12. Consider a collocation method (5.72) for the linear first-order ODEs $y' = A(x)y + q(x)$ with order $p > k + 1$. Show that the error for $x_i < x < x_{i+1}$ can be written as

$$y(x) - y_\pi(x) = \frac{h_i^{k+1}}{k!}y^{(k)}(x_i) \int_0^{\frac{x-x_i}{h_i}} \prod_{l=1}^{k}(t-\rho_l)\, dt + O(h_i^{k+2}) + O(h^p)$$

    This is a rather strong result, explicitly expressing the leading term of the error. [*Hint:* Write $e_\pi(x) = e_i + \int_{x_i}^{x} e_\pi'$ and use (5.67), (5.78).]

15. **(a)** Show that the local truncation error for the trapezoidal scheme can be expanded as in (5.89), (5.90).

    (b) Show that (5.89) holds for the midpoint scheme as well. What are the T_j in this case?

    (c) Derive the corresponding series expansion for the fourth-order Simpson scheme given in Example 5.6.

16. Show that if the assumptions of Theorem 5.93 hold, except that the quasiuniformity assumption is relaxed to

$$h_i/h \geq Kh^{(r-k+1)/k}$$

    then the trapezoidal solution error can be written as

$$y(x_i) - y_i = \sum_{j=1}^{k-1} h_i^{2j} d_j(x_i) + O(h^{2k}) \qquad 1 \leq i \leq N+1$$

17. Consider the simple finite difference method (5.4).

    (a) Derive a series expansion like (5.89) for the local truncation error.

    (b) Describe a deferred correction method for this basic discretization scheme.

    (c) Write a program implementing a method of order 6, using either extrapolation or deferred correction to accelerate the convergence of (5.4) for the BVP (5.2). Try your program on the problem of Example 5.1 and compare results to those listed in Tables 5.6 and 5.7. [Note that the linear systems of equations (5.5) are of a rather simple, tri-diagonal form.]

18. Show that the local truncation error of the difference scheme (5.137) is $O(h_i^k)$.

19. Prove Theorem 5.147.

20. State and prove a theorem corresponding to Theorem 5.147 for a mixed-order system of ODEs with multipoint BC (1.9a, b).

21. Let us denote the functions ψ_j appearing in (5.132) as ψ_{mj}, bringing out explicitly their dependence on the ODE order m. Show that for the choices (5.133) and (5.138a),
$$\psi_{rj} = \psi_{mj}^{(m-r)}, \qquad 1 \le r \le m, \qquad 1 \le j \le k$$
This does not hold for the choice (5.138b).

22. Construct an algorithm for parameter condensation when using collocation with B-splines, as described in Section 5.6.3. Show that the mesh values y_i can be retrieved from the B-spline parameters remaining after condensation.

23. Constructing the finite difference method (5.107) as outlined, show that the coefficients $\{\alpha_{ij}\}_{j=1}^{2k}$ satisfy (5.108).

24. For the BVP (5.155a, b), show that the solution $u(x)$ satisfies (5.158), for the functional $I(v)$ defined in (5.156).

25. Show that for any $v \in \mathbf{P}_{k,\pi,l}^0$,
$$E(v, u_\pi - u) = 0$$
with u_π defined by (5.159) and E by (5.161a).

6

Decoupling

In Section 3.4 we have seen how dichotomy is intertwined with the conditioning of a given linear BVP. Dichotomy has also been central to the understanding of the initial value methods of Chapter 4. For example, recall the difficulties encountered with the intuitively straightforward approach of superposition (Section 4.2), which attempts to integrate IVPs that may have growing fundamental solutions and hence is unstable in general. Since the methods of Sections 4.3–4.5 are all able to moderate this instability somehow, we are naturally led to ask whether some general structure lies behind this success. In fact, we may also wonder at which point we circumvent an initial value instability when solving a BVP by one of the global methods discussed in Chapter 5.

An indication of what happens has already been given in Section 4.4.2 and in Section 4.5. The algorithm of *compactification* (Section 4.3.4), when applied in a straightforward manner to a *standard multiple shooting* method, essentially retains the instability of single shooting, because increasing and decreasing modes (i.e., fundamental solution modes) are not decoupled. But in Section 4.4.2, having observed that the reorthogonalization marching algorithm of Section 4.4.1 does affect a *decoupling* of increasing and decreasing modes, we are able to supply a *stable compactification* algorithm by applying the recursion in stable directions only. The *Riccati transformation* and *invariant imbedding* ideas in Section 4.5 follow a similar line, but the decoupling is done before discretization, at the ODE level.

As it turns out, *many successful methods discussed in this book utilize decoupling at some stage of the algorithm*: Some decouple before discretization, others via discretization, and yet other methods decouple just via the solution of the resulting linear equations. By decoupling we split the differential equations, or the difference equations resulting from discretization, so that we can effectively and stably compute the increasing modes for decreasing time and the decreasing modes for increasing time (where "time" is our independent variable). This, then, is the basic tool for explaining the stability of boundary value algorithms.

In this chapter we give continuous and discrete versions of a basic decoupling theorem [Theorems 6.15 and 6.30 below], and obtain as corollaries some general stability results covering many numerical methods for BVPs. Since we need a linear structure for this, we assume in this chapter that the BVP is linear. Our arguments apply similarly to the systems resulting from linearizing nonlinear problems.

Recall that by "increasing" modes we mean those modes $\phi(x)$ for which $\dfrac{|\phi(x)|}{|\phi(t)|}$ becomes too large for some points $x \geq t$ in the interval $[a,b]$. (Here, a value being "too large" means that it is unacceptably large as an amplification of roundoff errors, cf. Section 2.1.) Similarly, "decreasing" modes are ones for which $\dfrac{|\phi(x)|}{|\phi(t)|}$ becomes too large for some points $x \leq t$. There are, of course, modes for which $\dfrac{|\phi(x)|}{|\phi(t)|}$ is of moderate size for all $a \leq x, t \leq b$. We call the latter *slow modes*, while the former, of both increasing and decreasing type, are called *fast modes*.

Since our primary concern is to avoid integrating fast modes in the "wrong" direction, we can consider the slow modes in the dichotomy as either nondecreasing or nonincreasing. From the point of view of initial value methods at least, it is most natural to group the slow modes with the fast decreasing ones, but it is not always possible to separate slow modes from fast increasing ones in a global way.

In order to understand decoupling, we first give a geometrical interpretation in Section 6.1. As we have also done on earlier occasions, we deal with the continuous case (i.e., the ODE itself) first, because we can rely on the structure given by the original problem without worrying about possible side effects of a numerical discretization. Therefore, we consider decoupling for ODEs in Section 6.2 and show that suitable transformations do indeed give a system where nondecreasing and nonincreasing modes can be computed separately (and stably). As an instance we consider the Riccati method of Section 4.5. It turns out that the analysis for the continuous case carries over to the discrete case, to which Section 6.3 is devoted. In Section 6.4 we discuss the role of the BC in decoupling and in particular why nonseparated BC should be treated with care. Finally, we remark on continuous decoupling algorithms resulting as a limit (i.e., they are in the closure) of discrete decoupling algorithms.

While applications of decoupling have already been used in Chapter 4, no mention of it was made in Chapter 5. This is because the decoupling when using finite difference methods is done in the solution of the discretized, algebraic equations. This will be discussed in the next chapter.

In this chapter it is convenient to use the Euclidean (2–) norm for vectors and its induced norm for matrices. These will, therefore, be our default norms throughout. Also, we will again use superscripts for partitioned matrices and vectors, while subscripts are used for numerical solutions on a mesh.

6.1 DECOMPOSITION OF VECTORS

In this section we show geometrically, by means of an example, how matrix decomposition affects a decoupling. Consider a homogeneous ODE of order n which admits solutions $\mathbf{y}_1(x)$, $\mathbf{y}_2(x)$ such that

$$|\mathbf{y}_1(x)| \sim e^{10x}, \qquad |\mathbf{y}_2(x)| \sim e^{-10x}$$

Further, consider $\mathbf{y}_1(x)$ and $\mathbf{y}_2(x)$ as vectors in $\mathbf{R}^n$ for a fixed x, and suppose that the angle θ between them [defined as in (2.24)] is larger than, say $\frac{\pi}{4}$, so

$$|\mathbf{y}_1^T(x)\,\mathbf{y}_2(x)| < \sqrt{1/2}\,|\mathbf{y}_1(x)|\,|\mathbf{y}_2(x)| \quad \text{for all } x$$

An arbitrary linear combination of $\mathbf{y}_1(x)$ and $\mathbf{y}_2(x)$, say

$$\mathbf{a}(x) := \alpha_1 \mathbf{y}_1(x) + \alpha_2 \mathbf{y}_2(x) \tag{6.1a}$$

will then grow asymptotically like $\mathbf{y}_1(x)$ (i.e., $|\mathbf{a}(x)| \sim e^{+10x}$) unless $\alpha_1 = 0$; moreover, it will asymptotically have the same direction as $\mathbf{y}_1(x)$. If

$$\mathbf{b}(x) := \beta_1 \mathbf{y}_1(x) + \beta_2 \mathbf{y}_2(x) \tag{6.1b}$$

is another such vector with

$$\det \begin{bmatrix} \alpha_1 & \beta_1 \\ \alpha_2 & \beta_2 \end{bmatrix} \neq 0$$

then clearly $\mathbf{a}(x)$ and $\mathbf{b}(x)$ are independent vectors spanning the same manifold as $\mathbf{y}_1(x)$ and $\mathbf{y}_2(x)$; however, as $x \to \infty$ they generally become numerically dependent. Note that x does not have to be very large for such a numerical dependence to develop (see Fig. 6.1).

This is precisely the reason why superposition (Section 4.2) fails for large shooting intervals when increasing modes like $\mathbf{y}_1(x)$ are present. The vectors $\mathbf{a}(x)$ and $\mathbf{b}(x)$ correspond to the fundamental modes computed by that method. Nevertheless, in exact arithmetic we can retrieve information about the magnitudes of $|\mathbf{y}_1(x)|$ and $|\mathbf{y}_2(x)|$ from $|\mathbf{a}(x)|$ and $|\mathbf{b}(x)|$. To see how, consider a situation as in Fig. 6.1, where $|\mathbf{y}_2| \ll |\mathbf{y}_1|$ and $\mathbf{y}_2$ and $\mathbf{y}_1$ form an angle θ which is not too acute. Here we omit writing x for clarity. Assume that $\mathbf{a}$ has a significant component of $\mathbf{y}_1$ and $\mathbf{b}$ has a significant component of $\mathbf{y}_2$, e.g., $\alpha_1, \beta_2 \approx 1$. Obviously, unless β_1 is extremely small, the angle between $\mathbf{a}$ and $\mathbf{b}$ is very small. A numerically more suitable basis for span $(\mathbf{a},\mathbf{b})$ = span $(\mathbf{y}_1,\mathbf{y}_2)$ would consist of an (almost) orthogonal pair of vectors having (almost) unit length. This can be found as follows: Let $\mathbf{t}_1$ be a unit vector in the direction of $\mathbf{a}$; i.e., define

$$\mathbf{t}_1 = \mathbf{a}/\gamma_1 \qquad \gamma_1 = |\mathbf{a}|$$

Then, let $\mathbf{t}_2$ be a vector in span $(\mathbf{a}, \mathbf{b})$ with $|\mathbf{t}_2| \approx 1$, such that the angle ν between $\mathbf{t}_1$ and $\mathbf{t}_2$ is not small, e.g., $\nu = \pi/2$. Then for some γ_2 and γ_3 we have

$$\mathbf{b} = \gamma_2 \mathbf{t}_1 + \gamma_3 \mathbf{t}_2$$

The vectors $\mathbf{t}_1$ and $\mathbf{t}_2$ do provide a desirable basis. Moreover, it turns out that, to an order of magnitude, $\gamma_1 \approx |\mathbf{y}_1|$ and $|\gamma_3| \approx |\mathbf{y}_2|$, as we show in the next paragraph. The decomposition just described is a generalization of a Gram-Schmidt orthogonalization of $\mathbf{y}_2$ and $\mathbf{y}_1$. In Fig. 6.2 we have sketched this process, where $|\mathbf{t}_1| = |\mathbf{t}_2| = 1$ and $\mathbf{t}_1^T \mathbf{t}_2 = 0$.

Since $\mathbf{a}$ has a significant component of $\mathbf{y}_1$, the angle η between $\mathbf{a}$ and $\mathbf{y}_1$ (so also between $\mathbf{t}_1$ and $\mathbf{y}_1$) is small, whence $\gamma_1 = |\mathbf{a}| \approx |\alpha_1 \mathbf{y}_1| \approx |\mathbf{y}_1|$.

(What is needed, more precisely, is that $\dfrac{|\alpha_1 \mathbf{y}_1|}{|\alpha_2 \mathbf{y}_2|}$ be not too small.) Since η is small, it also follows that the angle between $\mathbf{t}_2$ and $\mathbf{y}_1$ is almost equal to ν. Now since θ is not small and $\mathbf{b}$ has a significant component of $\mathbf{y}_2$ (because $\beta_2 \approx 1$), it can be seen from Fig. 6.2 that the projections of $\mathbf{b}$ and $\mathbf{y}_2$ on $\mathbf{t}_2$ have the same order of magnitude. Since $\cos(\pi/2 - \theta)|\mathbf{y}_2| \approx |\mathbf{y}_2|$, we have $|\gamma_3| \approx |\mathbf{y}_2|$. In summary, the decomposition can be written as

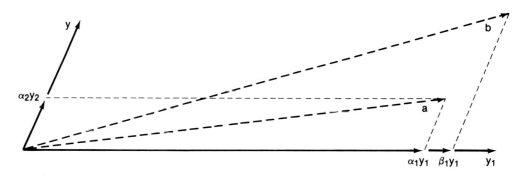

Figure 6.1 Geometric interpretation: $\mathbf{a}$ and $\mathbf{b}$ as linear combinations of $\mathbf{y}_1$ and $\mathbf{y}_2$

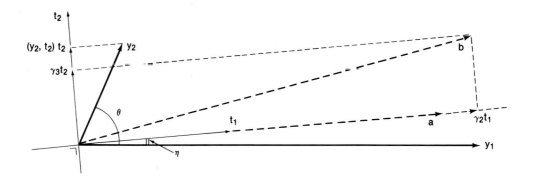

Figure 6.2 Geometric interpretation: Constructing a suitable basis

$$(\mathbf{a}\ \mathbf{b}) = (\mathbf{y}_1\ \mathbf{y}_2)\begin{bmatrix}\alpha_1 & \beta_1\\ \alpha_2 & \beta_2\end{bmatrix} = (\mathbf{t}_1\ \mathbf{t}_2)\begin{bmatrix}\gamma_1 & \gamma_2\\ 0 & \gamma_3\end{bmatrix} \qquad (6.2)$$

The factorization on the right in (6.2) is the QU-decomposition of $(\mathbf{a}\ \mathbf{b})$ if we use the Gram-Schmidt algorithm. But even if not, it contains all the necessary information about $\mathbf{y}_1$ and $\mathbf{y}_2$, namely a stable basis for span $(\mathbf{y}_1, \mathbf{y}_2)$ and estimates for the magnitudes $|\mathbf{y}_1|$ and $|\mathbf{y}_2|$.

What we have done above for a pair of vectors $\mathbf{y}_1$ and $\mathbf{y}_2$ can be extended to a pair of (rectangular) matrices, giving a factorization into a (preferably well-conditioned) matrix and a block upper triangular matrix like $(\mathbf{t}_1\ \mathbf{t}_2)\begin{bmatrix}\gamma_1 & \gamma_2\\ 0 & \gamma_3\end{bmatrix}$ in (6.2). Returning to the ODE case, we may therefore expect a block triangulation of a fairly general fundamental solution to yield diagonal blocks representing the increments of nondecreasing and nonincreasing modes, if the ODE has a dichotomy.

It is interesting to note that the smaller the ratio $\dfrac{|\mathbf{y}_2|}{|\mathbf{y}_1|}$ in (6.2), the *better* we can expect $\left|\dfrac{\gamma_3}{\gamma_1}\right|$ to approximate this ratio. Roughly speaking, this means that the decomposition is more pronounced the worse the "relative" initial value instability is.

6.2 DECOUPLING OF THE ODE

The Riccati method discussed in Section 4.5 employs a transformation which brings the coefficient matrix $A(x)$ in the ODE

$$\mathbf{y}' = A(x)\mathbf{y}, \qquad x > 0 \qquad (6.3)$$

into a block upper triangular form. This is only a special case of a more general method which is now described.

Let $T(x)$ and $U(x)$ be matrix functions chosen so that the following properties hold for all x: (i) $T(x)$ is nonsingular and differentiable; (ii) the lower left $(n-k)\times k$ block of $U(x)$ is zero (for some k independent of x); (iii) T, U, and A in (6.3) satisfy

$$T' = AT - TU \qquad (6.4)$$

[cf. (3.73), (4.72)]. Then the transformed variable

$$\mathbf{w}(x) := T^{-1}(x)\mathbf{y}(x) \qquad (6.5)$$

satisfies the block upper triangular ODE system

$$\mathbf{w}' = U(x)\mathbf{w} \qquad (6.6)$$

Now let $Z(x)$ be a fundamental solution of (6.3) such that $Z(a) = T(a)$. Then $W(x)$, defined by

$$W(x) := T^{-1}(x)Z(x) \qquad (6.7a)$$

is a fundamental solution of (6.6) which satisfies $W(a) = I$. Hence $W(x)$ is block upper triangular for all x. In this way we can find a one-to-one relation between transformed block upper triangular systems like (6.6) and corresponding factorizations of a fundamental solution like (6.7a). We write

$$W = \begin{bmatrix} W^{11} & W^{12} \\ 0 & W^{22} \end{bmatrix} \qquad (6.7b)$$

6.2.1 Consistent fundamental solution

The factorization above is quite general (we have not even specified k yet). It makes sense, however, to choose k such that the information about growing and decaying modes is contained in the appropriate diagonal blocks. For the factorization in (6.2) we had to require that the first column, $\mathbf{a} = \alpha_1 \mathbf{y}_1 + \alpha_2 \mathbf{y}_2$, contain a (significant) nonzero component of $\mathbf{y}_1$, or equivalently, that $\mathbf{a}$ is not purely in span$(\mathbf{y}_2)$, so $\alpha_1 \neq 0$. (In fact, we need $\dfrac{|\alpha_1 \mathbf{y}_1|}{|\alpha_2 \mathbf{y}_2|}$ to be not too small.) We now want to generalize this for the dynamic system (6.3) and the factorization (6.7).

Let us recall the concept of dichotomy from Section 3.4: The ODE (6.3) has a *dichotomy* if for a given fundamental solution $Y(x)$ and for some $0 \leq k \leq n$ there exists a projection P of rank $p = n - k$ and nonnegative constants K, λ and μ, such that

$$\|Y(x) P Y^{-1}(t)\| \leq K\, e^{-\lambda(x-t)}, \qquad x \geq t \qquad (6.8a)$$

$$\|Y(x) (I - P) Y^{-1}(t)\| \leq K\, e^{-\mu(t-x)}, \qquad x \leq t \qquad (6.8b)$$

The dichotomy is *exponential* if λ and μ are positive. We shall further assume that

$$P = \begin{bmatrix} 0 & 0 \\ 0 & I_{n-k} \end{bmatrix} \qquad (6.8c)$$

which is no restriction, as discussed in Section 3.4. Then $Y(x)$ is a *dichotomic fundamental solution*, i.e., upon writing for each x,

$$Y(x) = (Y^1(x) | Y^2(x)), \qquad (6.9)$$

where $Y^1(x) \in \mathbf{R}^{n \times k}$ and $Y^2(x) \in \mathbf{R}^{n \times (n-k)}$, we obtain that the matrix function $Y^1(x)$ represents the nondecaying (nondecreasing) part of the fundamental solution and $Y^2(x)$ the nongrowing part. For the remainder of this chapter, $Y(x)$ refers to this dichotomic fundamental solution.

Example 6.1

If $Y(x) = \begin{bmatrix} e^{10x} & 0 \\ 0 & 1 \end{bmatrix}$, then $P = \begin{bmatrix} 0 & 0 \\ 0 & 1 \end{bmatrix}$

yielding $K = 1$, $\lambda = 0$, and $\mu = 10$. Similar K and μ are obtained for

$$Y(x) = \begin{bmatrix} e^{10x} & 0 \\ 0 & e^{-20x} \end{bmatrix}$$

but now $\lambda = 20$. Note that for initial value problems, these two fundamental solutions have the same kind of instability. □

The dichotomic fundamental solution $Y(x)$ is desirable, but generally unavailable. Given any fundamental solution $Z(x)$ for (6.3), we now want to investigate its suitability to retrieve the growth information of $Y^1(x)$ and $Y^2(x)$ by decoupling. Let us partition $Z(x)$, as in (6.9), as $Z(x) = (Z^1(x) | Z^2(x))$. Corresponding to Section 6.1, where we have required $\alpha_1 \neq 0$ in (6.1a), we make the following

Definition 6.10 Given the dichotomy (6.8), a fundamental solution $Z(x)$ is said to be *consistent with* $Y(x)$ if range $(Z^1(a)) \cap$ range $(Y^2(a)) = \{0\}$. □

To see the connection between this definition and (6.1), let us recall first that, since $Y(x)$ and $Z(x)$ are both fundamental solutions, there is a constant *nonsingular* matrix H such that

$$Z(x) = Y(x)H, \quad \text{all } x \tag{6.11a}$$

Let us further partition the matrix H as

$$H = \begin{pmatrix} H^{11} & H^{12} \\ H^{21} & H^{22} \end{pmatrix} \quad H^{11} \in \mathbf{R}^{k \times k} \tag{6.11b}$$

We then find that H^{11} plays the role of α_1 in (6.1).

Lemma 6.12

(a) $Z(x)$ is consistent with $Y(x)$ iff H^{11} is nonsingular.
(b) $Z(x)$ is consistent with $Y(x)$ iff range $(Z^1(x)) \cap$ range $(Y^2(x)) = \{0\}$ for all $a \leq x \leq b$.

Proof: The proof of (b) follows directly from (6.11a). We now show (a). From (6.11), $Z^1(a) = Y^1(a)H^{11} + Y^2(a)H^{21}$. If H^{11} is nonsingular and for some $\mathbf{v} \in \mathbf{R}^k$, $Z^1(a)\mathbf{v} \in$ range $(Y^2(a))$, then $H^{11}\mathbf{v} = \mathbf{0}$, so $\mathbf{v} = \mathbf{0}$. On the other hand, if H^{11} were singular, then for some nonzero $\mathbf{v} \in \mathbf{R}^k$, $H^{11}\mathbf{v} = \mathbf{0}$. Since H is nonsingular, its first k columns are linearly independent. This gives $H^{21}\mathbf{v} \neq \mathbf{0}$, so $Z^1(a)\mathbf{v} = Y^2(a)H^{21}\mathbf{v} \neq \mathbf{0}$, implying no consistency. □

From Lemma 6.12 we conclude that any vector in $range(Z^1(a))$ must be the initial value vector of a solution which is eventually nondecreasing (for $x - a$ sufficiently large). But this may not be a sufficiently strong property in practice, since the interval $[a,b]$ is generally not infinite. Indeed, the requirement that H^{11} be nonsingular is weak, because a very ill-conditioned matrix cannot be distinguished in numerical computation from a singular one. Just as we required more of α_1 in Section 6.1 than to merely be nonzero, we must make sure here as well that $Y^2(x)H^{21}\mathbf{c}$ does not dominate $Y^1(x)H^{11}\mathbf{c}$ for any vector $\mathbf{c}$. Specifically, the suitability of a particular choice of $Z(x)$ in (6.11) depends upon having

$$\frac{||Y^2(a)H^{21}||}{\text{glb}(Y^1(a)H^{11})} =: L \tag{6.13}$$

where L should be of moderate size [see (2.20) for the definition of glb]. We shall refer to such an L as a *consistency constant* for a given $Z(x)$. In Section 6.4.1 we shall show that in fact $L = 0$ for some common methods.

The proofs of Lemma 6.14 and Theorem 6.15 which follow are somewhat technical. Those readers who wish to skip the details should at least read the statement of the important, pivotal Theorem 6.15. We have

Lemma 6.14 Let $Z(x)$ and $Y(x)$ be fundamental solutions of the ODE (6.3) satisfying (6.11a), and assume that (6.8) holds for $Y(x)$. Further assume that $Z(x)$ is consistent with $Y(x)$. Then

$$\max_{\mathbf{d}\neq 0} \frac{|Z^1(x)\mathbf{d}|}{|Z^1(t)\mathbf{d}|} \leq \tilde{K} e^{\mu(x-t)} \qquad x < t$$

with

$$\tilde{K} \leq K\sqrt{K^2+1}(1+LK^2)$$

where K is the dichotomy constant for $Y(x)$ and L is the consistency constant for $Z(x)$.

Proof: Since $Z^1(x)\mathbf{d} = Y^1(x)H^{11}\mathbf{d} + Y^2(x)H^{21}\mathbf{d}$,

$$|Z^1(x)\mathbf{d}| \leq |Y^1(x)H^{11}\mathbf{d}|(1 + \frac{|Y^2(x)H^{21}\mathbf{d}|}{|Y^1(x)H^{11}\mathbf{d}|})$$

Now, using (3.97) [and assuming without loss of generality that $Y^2(a)H^{21}\mathbf{d}\neq 0$],

$$\frac{|Y^2(x)H^{21}\mathbf{d}|}{|Y^1(x)H^{11}\mathbf{d}|} = \frac{|Y^2(x)H^{21}\mathbf{d}|}{|Y^2(a)H^{21}\mathbf{d}|} \frac{|Y^1(a)H^{11}\mathbf{d}|}{|Y^1(x)H^{11}\mathbf{d}|} \frac{|Y^2(a)H^{21}\mathbf{d}|}{|Y^1(a)H^{11}\mathbf{d}|}$$

$$\leq K^2 e^{(\lambda+\mu)(a-x)} \frac{|Y^2(a)H^{21}\mathbf{d}|}{|Y^1(a)H^{11}\mathbf{d}|} \leq LK^2 e^{(\lambda+\mu)(a-x)}$$

Thus

$$|Z^1(x)\mathbf{d}| \leq |Y^1(x)H^{11}\mathbf{d}|(1 + LK^2 e^{(\lambda+\mu)(a-x)})$$

Further, we write

$$Z^1(t)\mathbf{d} = |Y^1(t)H^{11}\mathbf{d}|(\frac{Y^1(t)H^{11}\mathbf{d}}{|Y^1(t)H^{11}\mathbf{d}|} + \frac{Y^2(t)H^{21}\mathbf{d}}{|Y^1(t)H^{11}\mathbf{d}|})$$

Note that the first vector inside the brackets has length 1 and is in range $(Y^1(t))$, whereas the second vector is in range $(Y^2(t))$. Also, if $\theta(t)$ is the angle between range $(Y^1(t))$ and range $(Y^2(t))$ then by Theorem 3.98, $\cot(\theta(t)) \leq K$. Thus, using Exercises 2 and 3, we obtain

$$|Z^1(t)\mathbf{d}| \geq |Y^1(t)H^{11}\mathbf{d}| / \sqrt{K^2+1}$$

Combining the two bounds, we have

$$\max_{\mathbf{d}\neq 0} \frac{|Z^1(x)\mathbf{d}|}{|Z^1(t)\mathbf{d}|} \leq \max_{\mathbf{c}\neq 0} \frac{|Y^1(x)\mathbf{c}|}{|Y^1(t)\mathbf{c}|} \sqrt{K^2+1}(1+LK^2 e^{(\lambda+\mu)(a-x)})$$

□

We are now able to formulate and prove a *fundamental decoupling theorem*.

Theorem 6.15 Suppose that the ODE (6.3) satisfies the dichotomy (6.8) for the fundamental solution $Y(x)$. Let $Z(x)$ satisfy (6.11) and be consistent with $Y(x)$, and consider the transformed ODE (6) with $T(x)$ and $W(x)$ satisfying (6.4) and (6.7), respectively. Then

$$\|W^{11}(x)[W^{11}(t)]^{-1}\| \leq K_1 \|T(t)\| \, \|T^{-1}(x)\| \, e^{-\mu(t-x)} \quad x \leq t \quad (6.15a)$$

$$\|W^{22}(x)[W^{22}(t)]^{-1}\| \leq K_1 \|T(t)\| \, \|T^{-1}(x)\| \, e^{-\lambda(x-t)} \quad x \geq t \quad (6.15b)$$

for a constant K_1 such that

$$K_1 \leq K(1+LK^2)\sqrt{K^2+1}$$

Proof: For each x, let $T = T(x)$ be partitioned as $T = (T^1 | T^2)$, where T^1 has k columns, and define $\hat{T} := T^{-1} = \begin{bmatrix} \hat{T}^1 \\ \hat{T}^2 \end{bmatrix}$, where $\hat{T}^1$ has k rows. From (6.7),

$$W^{11} = \hat{T}^1 Z^1$$

Also, since W is block upper triangular, W^{11} is nonsingular and

$$T^1 W^{11} = Z^1$$

This gives

$$\|W^{11}(x)[W^{11}(t)]^{-1}\| = \max_{c \neq 0} \frac{|W^{11}(x)[W^{11}(t)]^{-1}\mathbf{c}|}{|\mathbf{c}|} = \max_{d \neq 0} \frac{|W^{11}(x)\mathbf{d}|}{|W^{11}(t)\mathbf{d}|}$$

$$\leq \max_{d \neq 0} \frac{\|\hat{T}^1(x)\| \, |Z^1(x)\mathbf{d}|}{\|T^1(t)\|^{-1} |Z^1(t)\mathbf{d}|}$$

The result (6.15a) now follows from Lemma 6.14.

Next, note that

$$0 = W^{21}(x) = \hat{T}^2(x)Z^1(x) = \hat{T}^2(x)[Y^1(x)H^{11} + Y^2(x)H^{21}]$$

so

$$\hat{T}^2(x)Y^1(x) = -\hat{T}^2(x)Y^2(x)H^{21}[H^{11}]^{-1}$$

and therefore

$$W^{22}(x) = \hat{T}^2(x)[Y^1(x)H^{12} + Y^2(x)H^{22}] = \hat{T}^2(x)Y^2(x)[H^{22} - H^{21}[H^{11}]^{-1}H^{12}]$$

The matrix $H^{22} - H^{21}[H^{11}]^{-1}H^{12}$ is nonsingular (Exercise 1), so

$$\|W^{22}(x)[W^{22}(t)]^{-1}\| = \max_{c \neq 0} \frac{|W^{22}(x)\mathbf{d}|}{|W^{22}(t)\mathbf{d}|} = \max_{d \neq 0} \frac{|\hat{T}^2(x)Y^2(x)\mathbf{d}|}{|\hat{T}^2(t)Y^2(t)\mathbf{d}|}$$

From Exercise 4a we have

$$|\hat{T}^2(t)Y^2(t)\mathbf{d}| \geq \sin \eta(t) \|T(t)\|^{-1} |Y^2(t)\mathbf{d}|$$

where $\eta(t)$ is the angle between range$(Z^1(t))$ and range$(Y^2(t))$. From Exercise 4b we have that $\sin \eta(a) \geq \dfrac{\sin \theta(a)}{\sqrt{L^2+1}}$. By Theorem 3.98 we can bound $\dfrac{1}{\sin \theta(t)}$ by a constant related to the dichotomy constant K as in Lemma 6.14. In a similar way as in the proof

of Lemma 6.14 we then use $\dfrac{|Y^2(x)H^{21}\mathbf{d}|}{|Y^1(x)H^{11}\mathbf{d}|} \leq LK^2$, so $\sin \eta(t) \geq \sin \theta(t) \dfrac{1}{\sqrt{L^2K^4+1}}$
$\geq \dfrac{1}{LK^2+1}$. Finally, $\dfrac{|Y^2(x)\mathbf{d}|}{|Y^2(t)\mathbf{d}|}$ is bounded by (3.97a). Combining all these estimates, (6.15b) is obtained. □

A discrete variant of this theorem will be given in Section 6.3, as Theorem 6.30. These two theorems are utilized in the next two sections.

6.2.2 The basic continuous decoupling algorithm

The bounds on W^{11} and W^{22} in Theorem 6.15 represent, respectively, the growth of the nondecreasing and the nonincreasing modes in appropriate directions. In particular, assuming that the consistency constant L is of moderate size, the lower right block W^{22} does not reflect the growth of the unstable modes as x increases. Therefore, the last $n-k$ ODEs in (6.6) can be stably integrated in the forward direction.

This cannot be generally said about any components of the ODE (6.3). Therefore, a decoupling method does not directly compute a suitable fundamental solution $Z(x)$ or $Y(x)$. Instead, it attempts to first find a block upper triangular system matrix $U(x)$ as in (6.6), with a transformation $T(x)$ satisfying the Lyapunov equation (6.4). Consider the linear two-point BVP

$$\mathbf{y}' = A(x)\mathbf{y} + \mathbf{q}(x), \qquad a < x < b \qquad (6.16)$$
$$B_a \mathbf{y}(a) + B_b \mathbf{y}(b) = \boldsymbol{\beta}$$

Assuming we know $T(x)$ and defining

$$\mathbf{g}(x) = T^{-1}(x)\,\mathbf{q}(x) \qquad (6.17)$$

we are led via (6.5) and (6.6) to the decoupled BVP

$$\dfrac{d}{dx}\mathbf{w}^1 = U^{11}(x)\mathbf{w}^1 + U^{12}(x)\mathbf{w}^2 + \mathbf{g}^1(x) \qquad (6.18a)$$

$$\dfrac{d}{dx}\mathbf{w}^2 = \qquad\qquad U^{22}(x)\mathbf{w}^2 + \mathbf{g}^2(x) \qquad (6.18b)$$

[where U is partitioned like W in (6.7b)], and

$$B_a T(a)\mathbf{w}(a) + B_b T(b)\mathbf{w}(b) = \boldsymbol{\beta} \qquad (6.18c)$$

The superscripts in (6.18a, b) refer, as before, to the first k and last $(n-k)$ components. The decoupling obtained in this way now permits integrating in stable directions only. A superposition method, carried out explicitly or implicitly, can be used to compute the solution $\mathbf{w}(x)$, and the transformation (6.5) gives the desired solution $\mathbf{y}(x)$. There are a number of different ways to achieve this, which can all be described in the following framework:

Outline: Decoupling (continuous case)

Input: A BVP (6.16) and a nonsingular matrix $T(a)$ such that the fundamental solution $Z(x)$, with $Z(a) = T(a)$, is consistent with the dichotomic solution $Y(x)$ satisfying (6.8), and L in (6.13) is of moderate size.
Output: The solution of (6.16).

1. Compute the transformation $T(x)$ and block upper triangular matrix function $U(x)$ that satisfy the Lyapunov equation (6.4).
2. Using forward integration in (6.18b), compute a particular solution $\mathbf{w}^2(x)$ satisfying some suitable initial condition $\mathbf{w}^2(a)$ [possibly found from (6.18c)]. In addition compute, if necessary, a suitable number of fundamental modes satisfying (6.18b) to form $W^{22}(x)$.
3. Using backward integration in (6.18a), compute a particular solution $\mathbf{w}^1(x)$ (corresponding to the particular solution $\mathbf{w}^2(x)$ in step 2). In addition compute, if necessary, the corresponding upper parts of the modes found in step 2 to form $W^{12}(x)$ and the modes satisfying (6.18a) to form $W^{11}(x)$.
4. If the particular solution $\mathbf{w}(x)$ found in steps 2 and 3 cannot be identified with $T^{-1}(x)\mathbf{y}(x)$, then use superposition to find that particular solution $\mathbf{w}(x)$ which satisfies (6.18c).
5. Compute $\mathbf{y}(x) := T(x)\mathbf{w}(x)$. □

Obviously, the above is just a rough outline, omitting many details. In fact, the various methods of implementation may differ quite substantially from one another. Still, Theorem 6.15 guarantees that, for a well-conditioned BVP, *any decoupling method is stable* if it performs step 1 of the outline stably, achieving a smooth transformation T which satisfies

$$||T(t)|| \, ||T^{-1}(x)|| \le C_1, \qquad a \le x, t \le b \tag{6.19a}$$

$$||[T^{-1}(x)T'(x)]^{12}|| \le C_2 ||[T^{-1}(x)A(x)T(x)]^{12}||, \qquad a \le x \le b \tag{6.19b}$$

for some constants C_1, C_2 of moderate size, with the superscripts ^{12} denoting the appropriate upper right blocks.

To see this, note first that the solution of the IVP for (6.18b) is given by

$$\mathbf{w}^2(x) = W^{22}(x)[W^{22}(a)]^{-1}\mathbf{w}^2(a) + \int_a^x W^{22}(x)[W^{22}(t)]^{-1}\mathbf{g}^2(t)\,dt$$

so the bound (6.15b) gives

$$|\mathbf{w}^2(x)| \le K_1[e^{-\lambda(x-a)}|\mathbf{w}^2(a)| + ||T^{-1}(x)||\int_a^x e^{-\lambda(x-t)}||T(t)||\,|\mathbf{g}^2(t)|dt\,] \tag{6.20a}$$

and, upon either integrating the integral in (6.20a) by parts or bounding it directly,

$$|\mathbf{w}^2(x)| \le K_1[e^{-\lambda(x-a)}|\mathbf{w}^2(a)| + C_1||\mathbf{g}^2||\min(\lambda^{-1}, x-a)\,] \tag{6.20b}$$

Then

$$\mathbf{w}^1(x) = W^{11}(x)[W^{11}(b)]^{-1}\mathbf{w}^1(b) - \int_x^b W^{11}(x)[W^{11}(t)]^{-1}[U^{12}(t)\mathbf{w}^2(t)+\mathbf{g}^1(t)]\,dt$$

and the bound (6.15a) gives

$$|\mathbf{w}^1(x)| \le K_1[e^{-\mu(b-x)}|\mathbf{w}^1(b)| + ||T^{-1}(x)||\int_x^b e^{\mu(x-t)}||T(t)||[||U^{12}(t)||\,|\mathbf{w}^2(t)|+|\mathbf{g}^1(t)|]\,dt\,] \quad (6.21a)$$

so

$$|\mathbf{w}^1(x)| \le K_1[e^{-\mu(b-x)}|\mathbf{w}^1(b)| + C_1[||U^{12}||\,||\mathbf{w}^2||+||\mathbf{g}^1||]\,\min(\mu^{-1},b-x)\,] \quad (6.21b)$$

The boundedness of $U^{12}(x)$ follows from (6.4) and (6.19a, b) if $A(x)$ is bounded. [The argument can be extended by using (6.21a) if $A(x)$ has an integrable singularity.]

Let us interpret the stability bounds (6.20b) and (6.21b). As in the previous two chapters, we first consider $||A||(b-a)$ to be a constant of tolerable size, while $e^{||A||(b-a)}$ may be intolerably large. Then $\mathbf{y}(x)$ is agreeably bounded through (6.5), (6.20b) and (6.21b), and the method is stable.

On the other hand, if $||A||(b-a)$ is too large as a roundoff error amplification factor, then the problem is *stiff*, and an additional analysis is needed (cf. Chapter 10). The stiff class includes the case when the problem is defined on an infinite interval; i.e., $b-a$ is "very large." In this case, the bounds (6.20b) and (6.21b) still allow us to conclude stability as before *only if* the ODE has an exponential dichotomy, i.e., λ^{-1} and μ^{-1} are finite. (More precisely, we must require these constants to be of moderate size.) The other stiff case is when the interval length $b-a$ is a constant of moderate size, but $||A||$ is large, say

$$||A|| \approx \varepsilon^{-1}, \quad ||\mathbf{q}|| \approx \varepsilon^{-1}$$

for a parameter $0 < \varepsilon \ll 1$. Then to bound (6.20b), (6.21b) independently of ε we need

$$\lambda^{-1}, \mu^{-1} = O(\varepsilon)$$

Considering a family of stiff BVPs (6.16) depending on a parameter ε in this way, we may speak of *uniform exponential dichotomy* (uniform, that is, in ε).

Example 6.2

As a trivial example, consider the system

$$y_1' = -\varepsilon^{-1}y_1 + \varepsilon^{-1}q_1(x)$$

$$x > 0$$

$$y_2' = -y_2 + q_2(x)$$

$$y_1(0) = y_2(0) = 0$$

with $0 < \varepsilon \ll 1$. For any $\varepsilon > 0$, this is a stable IVP with $\mathbf{y}(x)$ bounded if $\mathbf{q}(x)$ is. Thus, there are no growing components, and a dichotomic fundamental solution is

$$Y(x) = \begin{bmatrix} e^{-x/\varepsilon} & 0 \\ 0 & e^{-x} \end{bmatrix}$$

with $P=I$, $\lambda=0$ and $K=1$ in (6.8). A trivial "continuous decoupling" method is to take $T(x) \equiv I$. Theorem 6.15 then trivially yields a stability constant 1, and hence

$$\|\mathbf{y}\| \leq \max\{\varepsilon^{-1}\|q_1\|, \|q_2\|\}$$

This result is not satisfactory if ε is very small ($\|\mathbf{y}\|$ is not bounded uniformly in ε). That is to say, our framework here admits methods with stability constants of $O(\varepsilon^{-1})$. However, we may consider this to be a *secondary stability effect*.

A closer scrutiny of this simple example reveals that the reason for the insensitivity in the bound on $\mathbf{y}$ is the mixing of a fast solution component and a slow one. Thus we have applied for the fast mode a bound obtained from the slow mode. For if we consider the equation for y_1 alone then the "dichotomy" bounds read

$$\|W^{22}(x)[W^{22}(t)]^{-1}\| \leq e^{(t-x)/\varepsilon} \qquad t < x$$

i.e., $\lambda^{-1} = \varepsilon$, and indeed

$$\left|\int_0^x \varepsilon^{-1} q_1(t) e^{(t-x)/\varepsilon} dt\right| \leq (1 - e^{-x/\varepsilon}) \|q_1\|$$

is certainly bounded if $\|q_1\|$ is. □

Following Example 6.2, it is then tempting to not just decouple nondecreasing and nonincreasing modes, but to separate three ways — into fast increasing, fast decreasing and slow modes. This type of decoupling, however, *cannot be done globally*. Indeed, the IVP for the scalar ODE

$$y' = \lambda(x) y, \qquad a < x < b$$

is stable (well-conditioned) if $\text{Re}(\lambda(x)) < -\delta$ for some positive constant δ, and yet the mode $\exp\{\int_t^x \lambda(\tau)\, d\tau\}$ may be fast (decreasing) or slow in different regions of the interval $[a, b]$, depending on the local size of $|\text{Re}(\lambda(\tau))|$. This is the basic reason that, when such problems are treated in Chapter 10, we divide $[a, b]$ into segments where separate analyses can be carried out. (Another problem is the inadequacy of the dichotomy concept to describe the separation of *three* types of modes.)

Turning to the various methods implementing the decoupling outline, note that the Lyapunov ODE (6.4) generally contains $2n^2$ degrees of freedom, namely n^2 for T and n^2 for U. Since (6.4) can be considered an ODE system of order n^2, it is reasonable to bind the remaining n^2 degrees of freedom by requiring some structure for T and U. If we tie up more of these degrees of freedom with requirements on the structure of T then there are fewer left for U, and vice versa. For instance, if we require T to be orthogonal, then we have $\frac{1}{2} n(n-1)$ degrees of freedom left, just enough to make U upper triangular (see Exercise 8). By restricting T even further to be built up by only k elementary orthogonal transformations, we can make U block upper triangular, with the upper left $k \times k$ block also upper triangular (see Exercise 9).

The Riccati method dealt with in Section 4.5 fits precisely into the framework of this decoupling outline. There we saw that the (simple, but not orthogonal) choice $T(x) = \begin{bmatrix} I & 0 \\ R(x) & I \end{bmatrix}$ transforms the system matrix $A(x)$ kinematically into a block upper triangular form if $R(x)$ satisfies the Riccati differential equation (4.74). Note that the matrix $R(x)$ together with the blocks in $U(x)$ provide for precisely n^2 unknowns, the other n^2 having been fixed by the zero and identity blocks. Also, $U^{12}(x) = A^{12}(x)$.

Moreover, if the BC are separated, then a clever choice for $T(a)$ [cf (4.76a)] directly provides initial values for $\mathbf{w}^2(a)$. (This choice also yields a consistency constant $L = 0$, as discussed in Section 6.4.) The algorithm described in the outline in Section 4.5.2 fits into the basic decoupling outline here; for it we can skip step 4 and the formation of the blocks of $W(x)$ in steps 2 and 3.

We see, however, that the Riccati method also makes sense for more general BC. The problem is deciding how to choose k and $T(a)$, whence $R(a)$. A possible choice is to let the first k columns of $T(a)$ have the direction of the eigenvectors (or principal vectors) belonging to the k absolutely largest eigenvalues of $A(a)$. At least for slowly varying $A(x)$ they should provide good starting values for increasing modes. A practical way to compute them is to use the QR algorithm *for eigenvalues* (cf. Section 2.2.1) which basically computes an orthogonal matrix Q such that $Q^T A(a) Q$ is upper triangular [and hence has the eigenvalues of $A(a)$ on the diagonal]. It is possible to get the eigenvalues on the diagonal of $Q^T A(a) Q$ arranged in order of decreasing real parts. Then a practical determination of k can be made by requiring the $(k+1)$st eigenvalue λ to be the first one such that $e^{\operatorname{Re}\lambda(b-a)}$ is an acceptable magnification of roundoff errors. Then the first k columns of Q can be used to find a matrix $T(a)$. Although this could solve the consistency problem, we may still face the possibility that the upper left $k \times k$ part of Q is singular; this problem is similar to the one mentioned in (4.80a, b). In order to overcome the difficulty it is sometimes advocated to apply the Riccati method to $Q^T A(x) Q$ instead, and choose $T(a) = I$. This corresponds to a *reimbedding*, or a *change of variables* $\mathbf{w} = Q^T \mathbf{y}$.

We must remember, of course, that just making remarks as in the preceding paragraph is far easier than actually producing a general-purpose program. This turns out to be the case here: Whereas the stable decoupling of a well-conditioned BVP with separated BC is practically under control, the situation for nonseparated BC (generally) requires more attention and more computational effort.

6.3 DECOUPLING OF ONE-STEP RECURSIONS

Having discussed dichotomy and decoupling at the ODE level, we now consider the same concepts applied at the discrete level, i.e., after the BVP has been discretized. The entire outline of this section is parallel to that of Section 6.2, thus giving a parallel treatment for the discrete and the continuous cases.

Both in Chapter 4 and in Chapter 5 we have encountered a number of instances where a one-step recursion of the form

$$R_i \mathbf{y}_{i+1} = S_i \mathbf{y}_i + \mathbf{q}_i, \quad 1 \leq i \leq N \quad (6.22a)$$

$$B_a \mathbf{y}_1 + B_b \mathbf{y}_{N+1} = \boldsymbol{\beta} \quad (6.22b)$$

arises when one is discretizing a BVP. Hence, we refer to (6.22a, b) as a *discrete BVP*. Under normal circumstances, both R_i and S_i are nonsingular (and, when properly scaled, well-conditioned) matrices, e.g., as in (4.37) or (5.17)–(5.19) with h small. We shall assume this to be the case in this section. However, other cases may (infrequently) arise as well (see Section 10.4.2). With R_i nonsingular we may in principle assume that it is the identity matrix [and indeed $R_i = I$ in (4.35) and in (5.65a)], although such recursions frequently appear in an implicit form (6.22a) (unlike ODEs).

The discrete BVP may, of course, be written as a system of linear equations

$$A y_\pi = \hat{\beta}$$

[where again A has the familiar form of (4.37) or (5.17)]. Corresponding to the notions of well-conditioning for (continuous) BVPs and stability for one-step difference schemes, we say that the discrete BVP is *well-conditioned* if $||A^{-1}||$ is bounded by a constant of moderate size [when R_i and S_i are scaled for each i such that $q_i \sim q(x_i)$]. In order to be more quantitative, we can also define conditioning constants κ_1 and κ_2 (and a discrete Green's function) in parallel to the definitions in (3.81)–(3.83). We leave this to Exercise 10.

Next we define the notions of fundamental solution and dichotomy for the discrete BVP. The mesh function $\{Y_i\}_{i=1}^{N+1}$ is a *fundamental solution* of (6.22) if Y_1 is nonsingular and

$$R_i Y_{i+1} = S_i Y_i, \qquad 1 \leq i \leq N.$$

Note that since R_i and S_i are nonsingular, so is Y_{i+1}, $1 \leq i \leq N$. Further, we say that the discrete BVP (6.22) has a *dichotomy* if for a fundamental solution $\{Y_i\}_{i=1}^{N+1}$ there exist a projection P (of rank $n-k$, say) and constants K, σ and ρ ($\sigma, \rho \geq 1$), such that

$$||Y_i P Y_{j+1}^{-1}|| \leq K \sigma^{j-i}, \qquad i \geq j+1 \tag{6.23a}$$

$$||Y_i (I-P) Y_j^{-1}|| \leq K \rho^{i-j}, \qquad i \leq j \tag{6.23b}$$

[cf. (6.8a,b)]. Corresponding to the continuous case, it may be shown that a well-conditioned discrete BVP must have a dichotomy (see Exercise 11). Note for instance that if the discrete BVP (6.22) results from using a one-step consistent finite difference method with a uniform step size h for a BVP (6.16) where the dichotomy (6.8a,b) holds, then for h small enough the dichotomy (6.23a,b) holds with the same projection P and with $\sigma = e^{\lambda h}$, $\rho = e^{\mu h}$.

As in the ODE case, we may assume that P is given in the simple form (6.8c), so $\{Y_i\}_{i=1}^{N+1}$ is dichotomic, i.e., writing for each i

$$Y_i = (Y_i^1 \mid Y_i^2)$$

where $Y_i^1 \in \mathbf{R}^{n \times k}$, $Y_i^2 \in \mathbf{R}^{n \times (n-k)}$, the growing (discrete) modes are all contained in $\{Y_i^1\}_{i=1}^{N+1}$ and the rest are all in $\{Y_i^2\}_{i=1}^{N+1}$.

An analogue to the Lyapunov equation (6.4) reads in the discrete case

$$S_i T_i = R_i T_{i+1} U_i, \qquad 1 \leq i \leq N \tag{6.24}$$

where T_i is required to be nonsingular and U_i is required to be block upper triangular (in order to have a decoupled system). Introducing a transformed variable

$$w_i := T_i^{-1} y_i, \qquad 1 \leq i \leq N+1 \tag{6.25}$$

we find that the recursion (6.22) reads after transformation,

$$w_{i+1} = U_i w_i + (R_i T_{i+1})^{-1} q_i \qquad 1 \leq i \leq N \tag{6.26a}$$

$$B_a T_1 w_1 + B_b T_{N+1} w_{N+1} = \beta \tag{6.26b}$$

The explicit form of (6.26a) in the homogeneous case corresponds to (6.6). Note that the marching algorithm in Section 4.4.1 yields relations like (6.26).

6.3.1 Consistent fundamental solutions

As in Section 6.2 we can relate fundamental solutions of (6.22) to those of (6.26) in a simple way. If a fundamental solution $\{Z_i\}_{i=1}^{N+1}$ of (6.22) satisfies $Z_1 = T_1$, then $\{W_i\}_{i=1}^{N+1}$ defined by

$$W_i := T_i^{-1} Z_i \qquad (6.27a)$$

is a fundamental solution of (6.26), with W_i block upper triangular for all i,

$$W_i = \begin{pmatrix} W_i^{11} & W_i^{12} \\ 0 & W_i^{22} \end{pmatrix} \qquad (6.27b)$$

Let H be a fixed, nonsingular matrix such that

$$Z_i := Y_i H \qquad 1 \le i \le N+1 \qquad (6.28)$$

As for $\{Y_i\}$, we partition $Z_i = (Z_i^1 | Z_i^2)$, where Z_i^1 has k columns (and n rows) and use the partitioning notation (6.12) for H. The consistency definition now reads

Definition 6.29. Given the dichotomy (6.23), (6.8c), a fundamental solution $\{Z_i\}$ is said to be *consistent with* $\{Y_i\}$ if range$(Z_1^1) \cap$ range$(Y_1^2) = \{0\}$. Defining

$$\frac{\|Y_1^2 H^{21}\|}{\mathrm{glb}(Y_1^1 H^{11})} =: L$$

then L is called a *consistency constant* for $\{Z_i\}$. □

Discrete analogues of Lemmas 6.12 and 6.14 are readily seen to hold here as well, and the fundamental Theorem 6.15 now reads (cf. Exercise 12)

Theorem 6.30 Let the dichotomy (6.23), (6.8c) be satisfied with the fundamental solution $\{Y_i\}$ and assume that (6.24)–(6.28) hold with $\{Z_i\}$ consistent with $\{Y_i\}$ and with a consistency constant L of moderate size. Then for some constants K_1 and K_2 of moderate size,

$$\|W_i^{11}[W_{j+1}^{11}]^{-1}\| \le K_1 \|T_j\| \, \|T_i^{-1}\| \rho^{i-j}, \qquad i \le j \qquad (6.30a)$$

$$\|W_i^{22}[W_{j+1}^{22}]^{-1}\| \le K_2 \|T_j\| \, \|T_i^{-1}\| \sigma^{j-i}, \qquad i \ge j+1 \qquad (6.30b)$$

□

Remark The major *difference* between the discrete and continuous cases is that the discrete dichotomy assumption does not automatically follow from the well-conditioning of the continuous BVP. In fact, (6.23) can be viewed as a *stability* statement. However, stability of a consistent one-step difference scheme does not necessarily require that R_i and S_i in (6.22) be nonsingular, i.e., it follows from weaker assumptions than those used here. □

6.3.2 The basic discrete decoupling algorithm and additional considerations

Once we have found transformation matrices T_i, we can define a decoupled recursion. Using the same partitioning and superscript notation as before, we write (6.26a) as

$$\mathbf{w}_{i+1}^1 = U_i^{11}\mathbf{w}_i^1 + U_i^{12}\mathbf{w}_i^2 + \mathbf{g}_i^1 \tag{6.31a}$$

$$\mathbf{w}_{i+1}^2 = \phantom{U_i^{11}\mathbf{w}_i^1 + {}} U_i^{22}\mathbf{w}_i^2 + \mathbf{g}_i^2 \tag{6.31b}$$

where we have defined

$$\mathbf{g}_i = T_{i+1}^{-1} R_i^{-1} \mathbf{q}_i \tag{6.31c}$$

The decoupling should be employed to solve the discrete BVP by recurring in stable directions only. Hence we can formulate the following:

Outline: Decoupling (discrete case)

Input: A discrete BVP (6.22a, b) and a nonsingular matrix T_1 such that the fundamental solution $\{Z_i\}$ with $Z_1 = T_1$ is consistent with the dichotomic solution $\{Y_i\}$ satisfying (6.23), (6.8c) (and the consistency constant L is of moderate size).

Output: The solution $\{\mathbf{y}_i\}_1^{N+1}$ of (6.22a, b).

1. Compute the transformation matrices T_i and the block upper triangular matrices U_i, satisfying (6.24).
2. Using forward recursion in (6.31b), compute a particular solution $\{\mathbf{w}_i^2\}$ satisfying some suitable initial condition $\mathbf{w}_1^2$ [possibly found from (6.26b)]. In addition compute, if necessary, a suitable number of homogeneous modes satisfying (6.31b) to form $\{W_i^{22}\}$.
3. Using backward recursion in (6.31a), compute a particular solution $\{\mathbf{w}_i^1\}$ [corresponding to $\{\mathbf{w}_i^2\}$ in step 2)]. In addition compute, if necessary, the corresponding upper parts of the modes found in step 2 to form $\{W_i^{12}\}$, and modes satisfying (6.31a) to form $\{W_i^{11}\}$.
4. If the particular solution $\{\mathbf{w}_i\}$ found in steps 2 and 3 cannot be identified with $\{T_i^{-1}\mathbf{y}_i\}$ then use superposition to find that particular solution $\{\mathbf{w}_i\}$ which satisfies (6.26b).
5. Compute $\mathbf{y}_i = T_i \mathbf{w}_i$, $1 \le i \le N+1$. □

The stability of any discrete decoupling method is guaranteed by Theorem 6.30 (in the sense as in the discussion following the continuous decoupling outline in Section 6.2.2) if step 1 is performed stably, achieving a transformation which satisfies

$$\|T_i\| \, \|T_j^{-1}\| \le C, \quad 1 \le i, j \le N+1$$

The stabilized compactification algorithm in Section 4.4.2 is clearly an instance of the outline above, much in the same way that the Riccati method is an instance of the continuous decoupling outline. Moreover, when the BC are separated (as occurs most frequently in practice), the situation simplifies considerably, as with the Riccati method. For then we are able to find a consistent T_1 with $L = 0$ which yields initial values for $\mathbf{w}_1^2$, and no integration of additional modes is necessary.

In the discrete case there are also many ways to achieve decoupling. In principle we should factorize the matrix $S_i T_i$ [see (6.24)] into a product of two matrices of which the second one is block upper triangular (and hence will be called U_i). We now discuss how this can be done, assuming that T_1 has been properly chosen so that the corresponding fundamental solution has a consistency constant of moderate size. For simplicity of notation, we also take

$$R_i = I$$

in (6.24), for otherwise we can multiply (6.22a) by R_i^{-1} for each i.

One way to proceed is as follows: Since S_i is nonsingular, we may perform an LU-decomposition of $S_i T_i$,

$$S_i T_i = L_i U_i$$

(see Section 2.2.3). Then, to satisfy (6.24) we set

$$T_{i+1} := L_i$$

Note that T_{i+1} is a (possibly permuted) lower triangular, nonsingular matrix. As usual, an advantage of an LU-decomposition is its speed, while a disadvantage is its poor worst-case stability bounds, which makes a sharp analysis difficult.

Another variant is to perform a QU-decomposition (see Section 2.4). Once we have computed, say,

$$S_i T_i = Q_i U_i$$

then T_{i+1} is defined as

$$T_{i+1} := Q_i$$

The main advantage of this (somewhat more expensive) variant is that T_{i+1} is orthogonal and hence ideally conditioned. (In fact, $||T_i|| \, ||T_j^{-1}|| = 1$.)

Implementations may also vary in the form in which an LU- or a QU-decomposition is achieved. (Or a UL- or a UQ-decomposition for that matter.) This could be done with pre- or post-multiplications by elementary transformations, corresponding to elimination by rows or by columns. Different pivoting strategies are also possible, of course. Also note that U_i does not need to be made fully upper triangular as above, only block upper triangular.

We will return to a discussion of decoupling methods in Chapter 7 when we consider block methods for the solution of linear algebraic systems and their stability.

Finally, a word about the numerical stability of the decomposition in (6.24). If we assume that R_i is a well-conditioned matrix, then the computational errors in U_i are of the order of $||S_i|| \, ||T_i|| \, ||T_{i+1}^{-1}|| \varepsilon_M$ (ε_M being the machine constant). We thus see that the U_i are not significantly contaminated by rounding errors made in past computations, despite the recursive nature of the relations we are dealing with. For example, in the variant of multiple shooting we have met in Section 4.3 and in the QU-decomposition variant above, we have $R_i \equiv I$, $||T_i|| = ||T_{i+1}^{-1}|| = 1$.

Remark Armed with our understanding of the decoupling principle, let us look again at the methods of Chapters 4 and 5. Intuitively, the "basic lesson" is that increasing modes should not be computed forward in x, while decreasing modes should not be computed backward. Many of the methods introduced in both chapters yield a discrete BVP of the form (6.22). While stability is a central, nontrivial concern when *obtaining* (6.22) in Chapter 4, it is essentially automatic (though still a central concept) in Chapter 5 — see Theorem 5.38.

Indeed, examining the midpoint or trapezoidal schemes (5.18), (5.19), we realize that when forming (5.17) [i.e., (6.22)], there is no preferred direction of integration! Thus, we are not confronted with the issue of decoupling at this stage. Similarly, for collocation with symmetric points ρ_j satisfying (5.59d), the definition of the method in terms of piecewise polynomial solutions (5.72) is entirely symmetric. However, when we solve the linear system of algebraic equations (6.17), a direction is introduced in the way the matrix **A** is decomposed. Hence, the decoupling principle, which is "dormant" during the *formation* stage of the discrete BVP for symmetric difference schemes, becomes central again in the next chapter, where the solution of the resulting discrete BVP is considered. □

6.4 PRACTICAL ASPECTS OF CONSISTENCY

In the previous two sections we have obtained some rather powerful general results for a variety of decoupling methods, assuming that consistency holds with a constant L of moderate size. To complete this chapter, we consider how this consistency requirement can be fulfilled in practice. It turns out that it is often satisfied under very mild conditions. In particular, it is essentially automatic for separated BC, considered in Section 6.4.1. More delicate situations arise for nonseparated BC, as demonstrated in the sequel.

6.4.1 Consistency for separated BC

We now assume that the BC are separated (and that the BVP is well-conditioned). Then the initial conditions at $x = a$ must control the fast decreasing modes, and the terminal conditions at $x = b$ control the fast increasing modes. Since in our notation the increasing modes appear first in $Y(x)$, we write the BC as

$$B_a = \begin{bmatrix} 0 \\ B_{a1} \end{bmatrix}, \quad B_b = \begin{bmatrix} B_{b2} \\ 0 \end{bmatrix} \qquad (6.32)$$

where B_{a1} and B_{b2} are full rank matrices with m and $n - m$ rows, respectively. (Note the trivial change in writing (6.32) compared to (3.105) - the reason is for notational convenience, as will become apparent later.) Clearly, $n - m \geq k$, i.e., the number of rows in B_{b2} is at least as large as the number of increasing modes it controls. If $n - m > k$ then the additional rows of B_{b2} correspond to slow modes, so we take $m = n - k$ for notational simplicity.

Recall from Section 3.4 that for the case of separated BC, the dichotomy is determined by the BC in an explicit way. Thus, the dichotomic solution $Y(x)$ used in (6.8) can be taken as the fundamental solution of (6.3) which further satisfies

$$I = B_a Y(a) + B_b Y(b) = \begin{bmatrix} B_{b2} Y(b) \\ B_{a1} Y(a) \end{bmatrix}$$

and the projection P of (6.8c) is simply

$$P = B_a Y(a) = \begin{bmatrix} 0 & 0 \\ 0 & I_m \end{bmatrix}$$

Moreover, Theorem 3.107 tells us that no nonzero vector in range $(Y(a)P)$ can be orthogonal to range (B_{a1}^T). Therefore, an initial transformation $T(a)$ (or T_1 for the discrete case) whose first k columns are orthogonal to range (B_{a1}^T) induces a *consistent* fundamental solution $Z(x)$ (or $\{Z_i\}$) with $Z(a) = T(a)$ (or $Z_1 = T_1$).

This, and more, can be seen as follows: If

$$B_{a1} Z^1(a) = 0 \tag{6.33a}$$

then

$$0 = B_a Z^1(a) = B_a Y(a) \begin{bmatrix} H^{11} \\ H^{21} \end{bmatrix} = \begin{bmatrix} 0 \\ H^{21} \end{bmatrix}$$

Therefore

$$H^{21} = 0$$

implying not only that H^{11} is nonsingular (because H is), but also that (6.13) is satisfied with $L = 0$. Formally we have

Theorem 6.33. For a well-conditioned BVP (6.16) with separated BC (6.32), if $Z(x)$ is a fundamental solution satisfying (6.33a) then $Z(x)$ is consistent with the dichotomic solution $Y(x)$, with consistency constant $L = 0$. A direct analogue holds for the discrete BVP case as well. □

A number of methods in Chapter 4 (in particular, reduced superposition and its multiple shooting variants) choose $T_1 = Z_1$ to be an orthogonal matrix. The transformation T_i is ideally conditioned. Therefore, if the first decomposition step of the decoupling outline is carried out stably then such a method is stable. In particular, the stabilized march is stable. (This presupposes that the *computation* of T_i, which may generally involve unstable IVPs, is adequately performed in the first place.)

For the Riccati method of Section 4.5, recall that we choose $R(a) = -D^{-1}C$, where

$$B_{a1} = (C \mid D)$$

with D an $m \times m$ nonsingular matrix. Then

$$T(a) = Z(a) = \begin{bmatrix} I & 0 \\ -D^{-1}C & I \end{bmatrix}$$

and (6.33a) holds. Note the closer $D^{-1}C$ is to 0, the better the conditioning of the transformation near a.

In view of the simplicity of the treatment for separated BC and the natural way in which consistency is fulfilled, it is natural to recall the trick for *separating* the BC introduced in Section 1.1. If we write the BC matrices as

$$B_a = \begin{bmatrix} B_{a1} \\ B_{a2} \end{bmatrix}, \quad B_b = \begin{bmatrix} 0 \\ B_{b2} \end{bmatrix} \tag{6.34}$$

where B_{a1} has m rows and B_{a2}, B_{b2} have $\nu := n - m$ rows, then we can introduce the ν trivial equations

$$\mathbf{z}' = \mathbf{0}$$

and write the BC as

$$\begin{bmatrix} B_{a1} & 0 \\ B_{a2} & -I \end{bmatrix} \begin{bmatrix} \mathbf{y} \\ \mathbf{z} \end{bmatrix}(a) = \begin{bmatrix} \boldsymbol{\beta}_1 \\ \mathbf{0} \end{bmatrix}, \quad \begin{bmatrix} B_{b2} & I \end{bmatrix} \begin{bmatrix} \mathbf{y} \\ \mathbf{z} \end{bmatrix}(b) = \boldsymbol{\beta}_2$$

[cf. (1.9)].

Now we can interpret the effect that this trick has with regard to decoupling. The original (well-conditioned) BVP with (6.34) has k nondecreasing modes ($k \leq \nu$) and $n - k$ nonincreasing ones. The introduction of $\mathbf{z}$ adds ν modes which are neither increasing nor decreasing. The separated BC can be viewed as having allocated these ν modes so that to the left end a there correspond n *nonincreasing* modes while to the right end b there correspond ν *nondecreasing* ones. Thus, the purely increasing and purely decreasing modes are decoupled as far as the BC are concerned, and stable compactification can be applied together with an appropriate discretization algorithm.

6.4.2 Consistency for partially separated BC

It is tempting to try to generalize Theorem 6.33 to the case of a well-conditioned BVP with partially separated BC (6.34). The argument would be that even if we allow $B_{a2} \neq 0$, but retain the 0 block in B_b, then B_{a1} must still control the nonincreasing modes. This argument is false. We shall do two things in this subsection: First we give a counterexample to show that consistency does not always follow from well-conditioning in this case; second, we show that the (generalized) stabilized march is only partially decoupling.

Example 6.3

Consider the simple BVP (cf. Exercise 4.15)

$$\mathbf{y}' = \begin{bmatrix} 10 & 20 & 0 \\ 0 & -10 & 0 \\ 0 & 0 & -20 \end{bmatrix} \mathbf{y} + \begin{bmatrix} -30 \\ 10 \\ 20 \end{bmatrix} \quad 0 < x < b$$

$$\begin{bmatrix} 1 & 0 & 0 \\ 0 & 1 & 0 \\ 0 & 0 & 1 \end{bmatrix} \mathbf{y}(0) + \begin{bmatrix} 0 & 0 & 0 \\ 1 & 0 & 0 \\ 0 & 0 & 1 \end{bmatrix} \mathbf{y}(b) = \begin{bmatrix} 1 \\ 2 \\ 2 \end{bmatrix}$$

The solution is $\mathbf{y}(x) \equiv (1, 1, 1)^T$. It can be readily verified that a fundamental solution is given by

$$Y(x) = \begin{bmatrix} 1 & -1 & 0 \\ 0 & 1 & 0 \\ 0 & 0 & 1 \end{bmatrix} \operatorname{diag}(e^{10x}, e^{-10x}, e^{-20x})$$

and the conditioning constants are of order unity.

Here $k = 1$ and $n - k = 2$, so $Y^2(0)$ has two columns and range $(Y^2(0)) = \operatorname{span}\{\begin{bmatrix}-1\\1\\0\end{bmatrix}, \begin{bmatrix}0\\0\\1\end{bmatrix}\}$. But $B_{a1} = (1, 0, 0)$ has only one row and cannot control the two decaying modes. Thus, we do not have the consistency condition (6.10) for the fundamental solution $Z(x)$ induced by this BC, since range $(Z^1(0)) = \operatorname{span}\{\begin{bmatrix}0\\1\\0\end{bmatrix}, \begin{bmatrix}0\\0\\1\end{bmatrix}\}$, so range $(Z^1(0))$ contains the mode decaying like e^{-20x}. □

Although consistency is not assured in general with the BC (6.34), this does not mean that the stabilized march algorithm of Section 4.4.3 cannot be stable. After all, in Example 3 we generate a partial fundamental solution in which modes growing like e^{10x} are represented (because $k \leq n - m$). Below we show that a careful implementation of the stabilized march can still yield a stable algorithm. Nevertheless, a remaining concern from the previous example is that there may be decreasing modes in the partial fundamental solution which ideally should consist only of increasing ones. Hence, on a smaller scale, lack of consistency will certainly lead to an undesirable situation in the generation of the upper triangular matrices $\overline{\Gamma}_i$ in (4.63). This problem can be solved by permuting columns, as was advocated in Section 6.3.2 for the general algorithm, so that the first k out of $n - m$ columns will correspond to the increasing modes.

Once we have accepted that in the stabilized march algorithm the significant part of the fundamental solution should contain a "consistent" part of the increasing modes supplemented by nonincreasing ones, it is easy to see that a suitable decoupling is assured. Rather than stating it in theorem form, we remark that if the splitting as meant in Theorem 6.15 now corresponds to the one induced by the number of nonzero rows in B_a (rather than to k of the dichotomy), the quantitative result for W^{22} is still correct in general. However, this is not so for W^{11}. Hence the marching is stable, but the resulting reduced order system needs further work. We conclude that the associated linear system (4.65c) *should not* be treated by compactification (as unstable modes may still be present). Instead, the general discrete decoupling algorithm (or any of the linear solvers to be treated in Chapter 7) may be used.

6.4.3 Consistency for general BC

For general BC it is not so obvious how to find a consistent fundamental solution, let alone one such that L is of moderate size. Without precaution one may run into difficulties, as the next example shows.

Example 6.4

Consider the ODE

$$\mathbf{y}' = \begin{bmatrix} 1 - 19\cos 2x & 1 + 19\sin 2x \\ -1 + 19\sin 2x & 1 + 19\cos 2x \end{bmatrix} \mathbf{y} + \mathbf{q}(x), \quad 0 < x < \pi$$

whose fundamental solution $Y(x)$ [with $Y(0) = I$] is given by

$$Y(x) = \begin{bmatrix} \sin x & -\cos x \\ \cos x & \sin x \end{bmatrix} \text{diag}(e^{20x}, e^{-18x})$$

(cf. Example 4.18). Suppose we use $T(0) = I$. This is not a consistent choice, but one might hope that discretization errors would restore consistency, since they cause the discretized problem to be perturbed and since the ODE has an exponential dichotomy. In Table 6.1 we list the actual computed values of the upper triangular matrices U_i and the errors in the solution $\mathbf{y}(x) = e^x(1, 0)^T$ [with suitable inhomogeneity $\mathbf{q}(x)$ and BC $\mathbf{y}(0) + \mathbf{y}(\pi) = \begin{bmatrix} 1 + e^\pi \\ 0 \end{bmatrix}$]. The method used is decoupled marching (see Section 4.4.1), with a tolerance of 10^{-9} and output points $x_i = \frac{(i-1)\pi}{15}$, $i = 1, \ldots, 6$. Note that the U_i are defined only for $i \leq 5$.

TABLE 6.1 Decoupled marching from an inconsistent start

x_i	U_i		$|\text{error}_i|$
0	.15−1	.19−5	.11−3
	0	.11+3	
$\pi/15$	.15−1	.14−1	.23−5
	0	.11+3	
$2\pi/15$	.21−1	.12+3	.39−7
	0	.77+2	
$3\pi/15$	.11+3	.65+2	.70−9
	0	.16−1	
$4\pi/15$	.11+3	.11−5	.95−11
	0	.15−1	
$5\pi/15$			.23−12

We make two observations: First, the appropriately decoupled U_i appears after a few steps (which means that the initially very small component of Z^1 eventually becomes dominant). Second, we have to pay for this bad choice of T_1 by having larger initial errors. This example is expanded upon in Exercise 13.

□

For discrete problems one may check for consistency by trial and error. Indeed, as the previous example shows, an inappropriate choice of T_1 can be monitored by checking for a disorder in the diagonal elements of U_1 (and in some of the succeeding matrices U_i). A way to improve the situation in such a case is to permute the columns of T_1 such that the columns of the permuted matrix are in increasing order.

6.5 CLOSURE AND ITS IMPLICATIONS

We have seen in this chapter a parallel development of continuous and discrete decoupling algorithms. In the discrete case, the discrete BVP (6.22) results from a BVP discretization, say a one-step finite difference scheme or a multiple shooting method. Suppose we consider a decoupling algorithm for a family of discrete BVPs, where the discretization step gets finer and finer as N increases (say like $1/N$). By considering the limit as $N \to \infty$, one can often define a corresponding *continuous* decoupling algorithm. Abusing language somewhat, this algorithm is called the *closure* of the discrete algorithm.

The concept of closure has some interesting implications, and we comment on them here. The consideration of discrete BVPs together with corresponding decoupling algorithms also serves to lead into the next chapter, because a discrete BVP (6.22) is, after all, a system of algebraic equations, and a good discrete decoupling algorithm is a stable factorization method for the solution of the linear system. Conversely, given a discretization of a BVP, a block method for solving the resulting linear system (6.22) can often be interpreted as a discrete decoupling algorithm. We use this to advantage in Sections 7.2 and 7.3.

Thus, we start here with the obvious but important point that the complete description of a method for solving a BVP involves both a basic discretization scheme *and* the numerical method used to solve the discretization equations. The study of *factorization methods* for solving BVPs interprets methods from this viewpoint. We have commented before that if a symmetric difference scheme is used on a fine mesh to discretize a BVP (6.16) obtaining a discrete BVP (6.22), then a decoupling is done via the linear system solver. Looking at the continuous limit, we can interpret the left-to-right (or forward) sweep of the continuous decoupling algorithm as solving IVPs in one direction, and the back sweep as solving IVPs in the other direction. Thus, a finite difference method on a fine mesh can be associated with an initial value method in the sense that the closure of a decoupling factorization for the discretization equations involves IVPs.

This brings up some close connections between initial value approaches and finite difference methods. Recall that some such connections are already pointed out in (5.39), (5.40), where it is noted that a one-step finite difference method approximates a multiple shooting matrix. But there, the method of solution of the linear system [e.g., of (5.17)] is not considered, and the main purpose is bounding the global truncation error in terms of the product of the local truncation error and the conditioning constant of the BVP. Here, on the other hand, the truncation error is zero in the closure limit, and we need to consider the *roundoff error propagation* for the complete description of any resulting method.

Before proceeding we make two observations. Firstly, attention is focused here on the propagation, *not* the accumulation, of roundoff errors. The latter is infinite in the limit (cf. Section 2.8), and indeed all numerical methods considered in this book, even those implementing "continuous decoupling" as in Section 6.2, imply a discretization of the differential equations involved on a finite size mesh. Second, we must note again that it would be a misinterpretation to conclude that finite difference methods can simply be replaced by initial value methods, because a practical step size for a difference scheme may well be rather far from that required for the initial value method, and the difference scheme may be far from its closure limit. Still, it can be fruitful to use the concept of closure in looking for initial value methods involving IVPs for modified ODEs like the Lyapunov equation (6.4), such that these resulting IVPs can be integrated stably (e.g., see Exercise 8).

Sometimes, the closure of a multiple shooting method with a specific matrix factorization can be easily found by analytically accumulating the blocks which are formed in the matrix factorization, i.e., assuming no truncation error. These correspond to a modified part of the fundamental solution, which solves some transformed IVP. The inhomogeneity terms of (6.22), which are modified during the forward recursion, correspond to the particular solution of some IVP, and the back substitution corresponds to some IVP solved in the opposite direction.

As an example of this, note that the closure of a block LU factorization of an appropriately permuted one-step discretization matrix involves the Riccati method (see Exercise 9 in Chap. 7). A simpler example is that the closure of the compactification algorithm of Section 4.3.3 is single shooting [cf. 4.48)]. In view of the many ways to solve the multiple shooting system of equations, it is not surprising that many different initial value methods can be found for solving BVPs, but obviously only stable, efficient ones are sought. Of course, no solution method for a system like (6.22) can improve the condition number of (6.22) itself. Thus, if for instance a standard multiple shooting method is applied to a very stiff BVP, then any factorization method built on it would perform poorly, essentially because the decoupling is begun too late.

We can also try to find the closure of an algorithm consisting of a combined one-step difference discretization and a matrix factorization by taking the multiple shooting method which it approximates [cf. (5.39), (5.40)] and using the above approach. If the closure of such a method preserves the stability properties of the finite difference method, then one may hope that this would lead to a continuous decoupling algorithm with similar properties.

EXERCISES

1. (a) Show that if H is nonsingular and its left upper block H^{11} is nonsingular [cf. (6.11b)], then there exists a block LU - decomposition

$$H = \begin{bmatrix} I_k & 0 \\ L & I_{n-k} \end{bmatrix} \begin{bmatrix} P & Q \\ 0 & R \end{bmatrix}$$

where the matrix R (called the *Schur complement* of H_{11}) is given by

$$R = H^{22} - H^{21}[H^{11}]^{-1}H^{12}$$

(b) Show that R is nonsingular.

2. Let θ be the angle between $Y^1 \in \mathbf{R}^{n \times k}$ and $Y^2 \in \mathbf{R}^{n \times l}$, $k + l \leq n$ (cf. Section 2.2.2). Let $\mathbf{y} = \mathbf{y}^1 + \mathbf{y}^2$ be such that

$$\mathbf{y}^1 \in \text{range}(Y^1), \ |\mathbf{y}^1| = 1, \quad \mathbf{y}^2 \in \text{range}(Y^2)$$

Show that

$$|\mathbf{y}| \geq \sin \theta$$

Hint: Consider the linear subspace $S := \mathbf{y}^1 + \text{range}(Y^2)$. The vector in S with minimal length can be found to be orthogonal to range (Y^2).

3. Assuming a dichotomy (6.8) and writing $Y(x)$ as in (6.9), show that $\sin\theta \geq \dfrac{1}{\sqrt{K^2+1}}$, where θ is the minimal angle (for $a < x < b$) between $Y^1(x)$ and $Y^2(x)$.

4. (a) Let $(Z^1|Y^2)$ be a nonsingular matrix, where Z^1 has k columns, and let η be the angle between Z^1 and Y^2. Let $T = (T^1|T^2)$ be a nonsingular matrix, partitioned in the same way and such that range (T^1) = range (Z^1). Finally, let $\hat{T} = T^{-1}$ be row partitioned as
$$\hat{T} = \begin{bmatrix} \hat{T}^1 \\ \hat{T}^2 \end{bmatrix} \quad (\hat{T}^1 \text{ with } k \text{ rows}). \text{ Show that } |\hat{T}^2 \cdot Y^2 \mathbf{c}| \geq \frac{\sin\eta}{\|T\|}|Y^2\mathbf{c}| \text{ for any } \mathbf{c} \neq 0,$$
$\mathbf{c} \in \mathbf{R}^{n-k}$.

Hint: First use Gram-Schmidt to show that there exist nonsingular upper triangular matrices U^1 and U^2 (of order k and $n-k$ respectively) and matrices with orthogonal rows $\hat{T}^1$ and $\hat{T}^2$ such that $\hat{T}^T = [[\hat{T}^1]^T|[\hat{T}^2]^T]\begin{bmatrix} U^1 & 0 \\ 0 & U^2 \end{bmatrix}$, whence $\text{glb}_2(U^2) \geq \dfrac{1}{\|T\|}$.

Moreover, Y^2 can be orthonormalized, giving a matrix $\bar{Y}^2$ with k orthonormal columns. Then apply Theorem 2.28 for $\hat{T}^2 Z^2$.

(b) Let Z^1 and Y^1 be full rank matrices with k columns and Y^2 a full rank matrix with n−k columns. Denote the angle between range (Z^1) and range (Y^2) by η and between range (Y_1) and range (Y_2) by θ. Let $Z^1 = Y^1 H^{11} + Y^2 H^{22}$ satisfy $\dfrac{\|Y^2 H^{22}\|}{\text{glb}(Y^1 H^{11})} = L\ (<\infty)$. Then $\sin\eta > \dfrac{\sin\theta}{\sqrt{L^2+1}}$.

Hint: Draw a graph for insight! For an arbitrary vector $\mathbf{d}$, prove that the projection of $Z^1\mathbf{d}$ onto $Y^2\mathbf{d}$ has a component whose norm is bounded above by $\sqrt{L^2 + \cos^2\theta}$, whereas the norm of the orthogonal component is bounded below by $\sin\theta$.

5. (a) Indicate why an initial value integration of a decoupled system $\mathbf{y}' = W\mathbf{y}$, W upper triangular and with ordered diagonal elements (with respect to real parts), may work well in a superposition algorithm, assuming we have no overflow problems (and disregarding the practically important question of step size selection!)

(b) Same question for the recursion $\mathbf{y}_{i+1} = W_i \mathbf{y}_i$.

6. Write a program to solve the problem in Example 6.2 by the stabilized march and using
$T_1 = \begin{bmatrix} 0 & 1 \\ 1 & 0 \\ 0 & 0 \end{bmatrix}$. Solve the resulting (multiple shooting) system by

(a) the general decoupling algorithm ($k = 1$) and

(b) an algebraic solver.

Test for various tolerances, both in single and in double precision. What are your conclusions?

7. Consider the second-order scalar ODE
$$u''(x) + p(x)u'(x) + q(x)u(x) = f(x), \qquad 0 < x < 1,$$
where $u(0)$ and $u(1)$ are given. A well-known "trick" to find the general solution of this ODE when one homogeneous solution has already been found is *order reduction*. This is worked out below.

(a) Let $v(x)$ satisfy $v''(x) + p(x)v'(x) + q(x)v(x) = 0$ and assume $v(x) \neq 0$ for $x \in [0, 1]$. Writing $u(x) = v(x)w(x)$, prove that $w'(x)$ satisfies the (first order!) ODE

$$(w'(x))' = -(p(x) + 2\frac{v'(x)}{v(x)})w'(x) + \frac{f(x)}{v(x)}$$

(b) Once we have computed $w'(x)$ we can find $u(x)$ from $u'(x) = u(x)\frac{v'(x)}{v(x)} + v(x)w'(x)$.

(c) If we know $v(0)$ and $w'(0)$, we obviously can compute $v(x)$ and $w'(x)$ by forward integration. We may employ the ODE in (b) then to compute $u(x)$ via backward integration. Now define vectors $\mathbf{y}(x) = \begin{bmatrix} u(x) \\ u'(x) \end{bmatrix}$, $\mathbf{q}(x) = \begin{bmatrix} 0 \\ f(x) \end{bmatrix}$ and matrices

$$A(x) = \begin{bmatrix} 0 & 1 \\ -p(x) & -q(x) \end{bmatrix}, \quad T(x) = \begin{bmatrix} 1 & 0 \\ \frac{v'(x)}{v(x)} & v(x) \end{bmatrix}.$$ Let $v(x)$ be an increasing solution, and assume that the ODE in (a) also has a decreasing solution. Show then that $T(0)$ induces a consistent fundamental solution of the ODE $\mathbf{y}'(x) = A(x)\mathbf{y}(x) + \mathbf{q}(x)$ and that $T(x)$ actually decouples the system.

(d) Show that the method above is equivalent to the Riccati method which, however, involves a transformation $\tilde{T}(x) = T(x)\begin{bmatrix} 1 & 0 \\ 0 & v(x) \end{bmatrix}$.

8. (a) Show that a family of orthogonal matrices $T(t)$ ($T(t) \in \mathbf{R}^{n \times n}$) can be found from $\frac{1}{2}n(n+1)$ equations.
Hint: $T^T T = I$ $[T']^T T + T^T [T'] = 0$, so $[T']^T T$ is skew symmetric.

(b) Let $M \in \mathbf{R}^{n \times n}$ be split as $M = M_L + D + M_U$, where M_L and M_U are the strictly lower and upper triangular parts of M. Define an operator $\phi : \mathbf{R}^{n \times n} \to \mathbf{R}^{n \times n}$ as $\phi(M) = M_L^T + D + M_U$. Show that ϕ is linear. Show also that an orthogonal T gives an upper triangular U in the Lyapunov equation (6.4) iff $U = \phi(T^T A T)$.
Hint: write (6.4) as $U = T^T A T - T^T T'$.

(c) Finally, show that in order to have U upper triangular and T orthogonal in (6.4), we have $\frac{1}{2}n(n-1)$ unknowns for T and $\frac{1}{2}n(n+1)$ unknowns for U.

9. (a) Show that in principle we have $\frac{1}{2}k(2n-k-1)$ unknowns for T in (6.4) if we build it up as k elementary reflections $H^{(1)}, \ldots, H^{(k)}$ like in Section 2.2.4.

(b) Show that we may require the matrix U in (6.4) to be block upper triangular, with the left upper $k \times k$ block upper triangular, if T has the form in (a).

10. (a) Consider the discrete BVP (6.22) with fundamental solution $\{Y_i\}_{i=1}^{N+1}$. Then (6.22) has a unique solution iff $Q := B_a Y_1 + B_b Y_{N+1}$ is nonsingular.

(b) Show that by

$$G_{i,j} = \begin{cases} Y_i Q^{-1} B_a Y_1 Y_{j+1}^{-1} R_j^{-1}, & j \leq i-1 \\ -Y_i Q^{-1} B_b Y_{N+1} Y_{j+1}^{-1} R_j^{-1}, & j \geq i \end{cases}$$

we have defined a Green's function for (6.22), satisfying $B_a G_{1,j} + B_b G_{N+1,j} = 0$.

(c) Show that the solution $\{y_i\}$ of (6.22) is given by

$$\mathbf{y}_i = Y_i Q^{-1}\boldsymbol{\beta} + \sum_{j=1}^{N} G_{i,j}\mathbf{q}_j$$

(d) Show that

$$\max_i |\mathbf{y}_i| \leq \kappa_1 |\boldsymbol{\beta}| + \kappa_2 \max_i |\mathbf{q}_i|$$

where $\kappa_1 := \max_i \|Y_i Q^{-1}\|$, $\quad \kappa_2 := \sum_{j=1}^{N+1} \max_i \|G_{i,j}\|$

11. Show that if the BC of a discrete BVP are separated, they induce a projection P such that

$$\|Y_i P Y_{j+1}^{-1}\| \leq K, \quad j \leq i-1$$

$$\|Y_i (I-P) Y_{j+1}^{-1}\| \leq K, \quad j \geq i,$$

where $K = \max_{i,j} \|G_{i,j}\|$ (see Exercise 10).

12. Using the previous two exercises, prove Theorem 6.30 [the discrete version of Theorem 6.15].

13. Consider the ODE in Example 6.3 with $\mathbf{q}(x) = e^x(-1 + 19(\cos 2x - \sin 2x), 1 - 19(\cos 2x + \sin 2x))^T$ and the BC $\mathbf{y}(0) + \mathbf{y}(\pi) = (1+e^\pi, 1+e^\pi)^T$.

Write or modify a program based on multiple shooting, finite differences, or (condensed) collocation that solves the discrete problem via the basic decoupling algorithm, with $T_0 = \begin{bmatrix} 1 & 0 \\ 0 & 1 \end{bmatrix}$. Take a tolerance *TOL*, smaller than $\sqrt{\varepsilon_M}$. You should get a solution that is less accurate than *TOL* near the boundaries. This can be explained as follows: The choice $T_0 = I$ is an inconsistent one. (Show that.) If no discretization errors are made, the local errors of order ε_M blow up according to the growth of the partitioned parts of the recursion. Indicate by how much. Since we have errors $\approx$ *TOL* (or less), we do not quite have inconsistency, but a large consistency constant $\approx \dfrac{1}{TOL}$. Then show that global errors will in fact be $\approx \dfrac{\varepsilon_M}{TOL}$, implying that if *TOL* gets smaller, the errors get worse!

7

Solving Linear Equations

In previous chapters we have repeatedly encountered the need to solve a large, sparse system of linear equations

$$\mathbf{As} = \mathbf{b} \tag{7.1}$$

resulting from a discretization of a given BVP. Typically, when a multiple shooting method or a finite difference method is applied to the linear BVP (1.4a, b), a system like (7.1) arises with $\mathbf{A}$ of the form [cf. (4.37), (5.17), (6.22)]

$$\mathbf{A} = \begin{bmatrix} S_1 & R_1 & & & \\ & S_2 & R_2 & & \\ & & \ddots & & \\ & & & S_N & R_N \\ B_a & & & & B_b \end{bmatrix} \tag{7.2}$$

where S_i, R_i are $n \times n$ matrices.

In Chapter 2 we discuss various ways to solve a general linear system like (7.1). However, a straightforward version of Gaussian elimination which is suitable for full matrices is not attractive here, especially when N is large. If we keep n fixed and consider the size of $\mathbf{A}$ as a function of N, then clearly the number of nonzero elements in $\mathbf{A}$ is only $O(N)$, whereas the total number of elements in $\mathbf{A}$ is $O(N^2)$. Correspondingly, whereas the *complexity* of a Gaussian elimination algorithm which ignores the sparsity structure of $\mathbf{A}$ is $O(N^3)$ flops, here a complexity of $O(N)$ flops is achievable, as we shall see below. We rely on the fact that, when $\mathbf{A}$ is large and sparse with its nonzero elements concentrated near the main diagonal, one can utilize the structure to reduce the amount of *fill-in* (corresponding to the number of nonzero elements generated during an LU-decomposition of **A** in L or in U at locations where **A** has zero elements). Here is, incidentally, a fundamental difference between the numerical solution of ODEs and of PDEs in more space dimensions. For the latter, immense effort has been invested on the question of just how to solve (7.1), while for the former, simple procedures can be devised to achieve a direct solution method which is asymptotically optimal both in complexity and in storage requirements.

We have already seen the application of Gaussian elimination to a *banded* matrix in Chapter 2. The matrix in (7.2) is also banded, with the exception of the lower left corner. For instance, for $n = 2$ and $N = 3$, **A** is as shown in Fig. 7.1a, where the symbol × stands for a generally nonzero element. Try to picture the form for **A** for the more typical situation where N is much larger, say $N = 100$, but n is still small.

If the BC are separated, as is usually the case in applications, then it is natural to write the BC at the left endpoint a first (and thus rearrange **A** and **b** accordingly). If $n = 2$ and we have one boundary condition at each end, then we obtain the structure shown in Fig. 7.1b. This matrix is banded and is in staircase form, as discussed in Section 7.1 below. Thus, while in previous chapters it has often been convenient to consider general, nonseparated BC, here it is advantageous to first devote our attention to the simpler, separated BC case, returning to nonseparated BC in Section 7.4. Recall also the trick introduced in Chapter 1 (and mentioned again at the end of Chapter 6) for converting nonseparated BC into separated ones (at the cost of increasing the size of the problem).

General banded matrices arise when one is solving a BVP of order n with separated BC (1.5a, d) by a standard multiple shooting or simple finite difference scheme. Using the p left end BC first and the $n-p$ right end BC last, the resulting matrix is

$$\mathbf{A} = \begin{bmatrix} B_{a1} & & & & & \\ S_1 & R_1 & & & & \\ & S_2 & R_2 & & & \\ & & \cdot & \cdot & & \\ & & & \cdot & \cdot & \\ & & & & S_N & R_N \\ & & & & & B_{b2} \end{bmatrix} \qquad (7.3)$$

where $B_{a1} \in \mathbf{R}^{p \times n}$ and $B_{b2} \in \mathbf{R}^{(n-p) \times p}$ (and $rank(B_{a1}) = p$, $rank(B_{b2}) = n - p$). More general block banded, or *staircase,* matrices can occur, as discussed in Section 7.1 below.

For many parts of this chapter, the decoupling concept of Chapter 6 is used. For the reader who is not familiar with that material, it is still possible to read Sections 7.2, 7.4.1, and 7.5. In fact, Sections 7.2.1 and 7.5.1 cover methods for solving (7.1) in forms which include most of the situations which arise in practice. The more traditional elimination methods given in these two sections, though not necessarily the most efficient,

$$\begin{bmatrix} \times & \times & \times & \times & & & & & & \\ \times & \times & \times & \times & & & & & & \\ & & \times & \times & \times & \times & & & & \\ & & \times & \times & \times & \times & & & & \\ & & & & \times & \times & \times & \times & & \\ & & & & \times & \times & \times & \times & & \\ \times & \times & & & & & & & \times & \times \\ \times & \times & & & & & & & \times & \times \end{bmatrix}$$

Figure 7.1a Structure of **A**

$$\begin{bmatrix} \times & \times & & & & & & & \\ \times & \times & \times & \times & & & & & \\ \times & \times & \times & \times & & & & & \\ & & & \times & \times & \times & \times & & \\ & & & \times & \times & \times & \times & & \\ & & & & & \times & \times & \times & \times \\ & & & & & \times & \times & \times & \times \\ & & & & & & & \times & \times \end{bmatrix}$$

Figure 7.1b **A** with separated BC

are reasonably efficient and straightforward, and will suffice for a wide assortment of cases.

This chapter contains the following material:

We begin in Section 7.1 with a discussion of staircase matrices arising in the applications of interest in this book. A number of methods to solve (7.1) for staircase matrices are discussed in Section 7.2. In Section 7.2.1 two methods are presented which amount to the usual Gaussian elimination with (scaled) row pivoting but use efficient data structures. In Section 7.2.2 a method based on alternate row and column elimination is considered, while a block tridiagonal method is presented in Section 7.2.3. In Section 7.2.4 the method of stable compactification is discussed.

The algorithms of Sections 7.2.2, 7.2.3, and 7.2.4 are given for the case when **A** has the form (7.3). Numerical stability for the solution methods of Section 7.2 is discussed in Section 7.3. We take the viewpoint that these are really *discrete BVPs,* and the success (or failure) of the method can often be linked to its ability (or inability) to affect a decoupling. Thus, we invoke the discrete decoupling theory of Chapter 6. The nonseparated BC case is taken up in Section 7.4, where Gaussian elimination with partial pivoting and a stable factorization method which is based on decoupling are considered.

In Section 7.5 we present a matrix decomposition method for the important case of BVPs with parameters and for problems arising from BVPs with multipoint BC. BVPs with nonseparated BC can also be viewed as a special case. It turns out that the decoupling principle becomes much more complicated for these problems (cf. Section 11.9), and it is not discussed at all in this section. We use the straightforward Gaussian elimination with (scaled) row pivoting of Section 7.2.1, but adapted now to the somewhat more complicated form for **A**.

7.1 GENERAL STAIRCASE MATRICES AND CONDENSATION

Matrices like (7.2) and (7.3) are still special instances of more general forms which may be encountered. One generalization of (7.3) occurs when the BVP has multipoint side conditions. Suppose each such condition involves only one point ζ_j, $a \leq \zeta_j \leq b$, as in (1.9). It is natural then to require that ζ_j be part of the mesh underlying the discretized BVP, e.g., $\zeta_j = x_i$, and further to order the BC at ζ_j with respect to its relative location in this mesh. Thus, rows corresponding to the side condition(s) at x_i are added at the beginning of the block S_i, if $i \leq N$.

Example 7.1

Consider the solitary wave example 1.13. Here $n = 4$, one BC is given at $a = -L$, one at $x = 0$ and two at $b = L$, where L is a sufficiently large value. With $N = 3$, for instance, we obtain the structure shown in Fig. 7.2 for **A** when using the midpoint scheme.

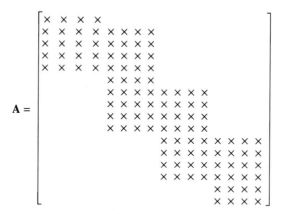

Figure 7.2 Staircase structure

Once again, note that usually in practice $N \gg n$. We have taken N small for illustrative purposes only.

□

Another occasion where a more general matrix form arises (for separated, two-point BC) is when higher-order collocation schemes are used. For the BVP (1.5a, b), using k collocation points in (5.59), we have (before condensation) the form

$$\mathbf{A} = \begin{bmatrix} B_{a1} & & & & & \\ S_1 & M_1 & R_1 & & & \\ & S_2 & M_2 & R_2 & & \\ & & & \ddots & & \\ & & & & S_N & M_N & R_N \\ & & & & & & B_{b2} \end{bmatrix} \quad (7.4)$$

where S_i, $R_i \in \mathbf{R}^{n(k+1) \times n}$, $M_i \in \mathbf{R}^{n(k+1) \times kn}$, $1 \le i \le N$. For the higher-order ODE (5.119), using (5.125) and (5.126) we have the same form as in (7.4) but S_i, $R_i \in \mathbf{R}^{(k+m) \times m}$, $M_i \in \mathbf{R}^{(k+m) \times k}$. Note that a matrix form of (7.4) does not occur with multiple shooting.

In general, matrices which have blocks centered only near the main diagonal are called *staircase matrices*. We shall refer to *block rows* and *block columns* of such matrices in an obvious way.

The matrix $\mathbf{A}$ in the form (7.4) may consume a significant amount of storage. We have already seen *condensation of parameters* as one method to reduce the size of such matrices in (5.64), (5.65), and (5.129)–(5.131). It is based on the observation that the unknowns corresponding to each M_i appear only locally and hence can be locally eliminated. This condensation of parameters brings the matrix into the multiple shooting (or what we have called discrete BVP) form (7.3) with S_i, $R_i \in \mathbf{R}^{n \times n}$. The representation of collocation in terms of B-spline or Hermite-type bases (Section 5.6.2) can also be viewed as a special type of parameter condensation, reducing $\mathbf{A}$ in (7.4) to a matrix of a similar form (not the discrete BVP form) but with smaller blocks: S_i, $R_i \in \mathbf{R}^{k \times m}$, $M_i \in \mathbf{R}^{k \times (k-m)}$, $k \ge m$.

If the matrix $\mathbf{A}$ in (7.3) still consumes too much storage, then in certain circumstances we may go even further and eliminate grid points by assembling subintervals into larger ones. Specifically, if the discrete equations corresponding to (7.3) are

$$S_i \mathbf{y}_i + R_i \mathbf{y}_{i+1} = \boldsymbol{\beta}_i$$

or

$$\mathbf{y}_{i+1} = \Gamma_i \mathbf{y}_i + \mathbf{r}_i, \qquad 1 \le i \le N \quad (7.5)$$

(with $\Gamma_i = -R_i^{-1} S_i$ and $\mathbf{r}_i = R_i^{-1} \boldsymbol{\beta}_i$), then we may form

$$\mathbf{y}_{i+l+1} = (\Gamma_{i+l} \cdots \Gamma_i) \mathbf{y}_i + \sum_{j=0}^{l} (\Gamma_{i+l} \cdots \Gamma_{i+j+1}) \mathbf{r}_{i+j} \quad (7.6)$$

for some $l \ge 1$ and eliminate the corresponding l blocks of rows and columns from (7.3). However, this must be done carefully: Setting $i = 1$ and $l = N$ we recognize (7.6) as being the compactification algorithm of Section 4.3.4, i. e., the discrete analogue of single shooting. Still, by keeping $\| \Gamma_{i+l} \cdots \Gamma_i \|$ below some control parameter (not larger than $\dfrac{TOL}{\varepsilon_M}$, where TOL is the required accuracy and ε_M is the machine accuracy), we can control rounding errors acceptably.

In particular, the stability considerations here are similar to those in Section 4.3.3, because after a partial *subinterval condensation* like (7.6) is performed, we may consider the *resulting* discrete BVP as a multiple shooting scheme. The advantage of this subinterval condensation is its reduction of storage at no increase in the computational complexity. Nonetheless, its utility is limited, not only by the above stability considerations, but also because we are not computing solution values between x_i and x_{i+l+1}, and hence are restricting the advantage of this idea largely to linear problems.

7.2 ALGORITHMS FOR THE SEPARATED BC CASE

We consider here the numerical solution of a linear system representing a stable, discrete BVP. Such a system would typically result from a sufficiently accurate one-step discretization of a well-conditioned BVP (7.1.5a, d). Then, Gaussian elimination with (scaled) row pivoting[1] can be expected to produce a stable method for solving this problem.

7.2.1 Gaussian elimination with partial pivoting

Our first two methods here amount to using special data structures with the basic Gaussian elimination scheme.

I. Banded matrices This is the crudest of all methods in this chapter. The matrix **A** in (7.3) is considered as banded, with lower bandwidth $n+p-1$ and upper bandwidth $2n-p-1$ (see Section 2.2.3). The usual elimination algorithm is applied, thus increasing the upper bandwidth to $2n-1$ (see Fig. 7.3). A convenient way to store **A** is by diagonals, requiring a rectangular array of size $n(N+1)\times(3n+p-1)$. Thus, asymptotically the storage requirements, as well as the computational complexity, are (optimally) *linear in N*.

However, improvements both in storage and in computational effort can still be made. Indeed, the storage required here is almost double that needed to store **A**. This is because when a staircase matrix is considered as a banded matrix, $O(N)$ consistently placed "triangles" of zeros inside the band are ignored, even without pivoting. The next method uses the same elimination scheme, but utilizes better the block structure of **A**.

II. Almost block diagonal form Consider the matrix **A** in (7.3) as a sequence of N *overlapping* blocks, of which the first $N-1$ are of size $(n+p)\times 2n$ and the last is of size $2n \times 2n$. In Fig. 7.4a we show the matrix of Fig. 7.1b and its overlapping blocks.

---

[1] Recall that we are calling *row pivoting* the strategy where a matrix column is searched to find the pivot elements so that rows are interchanged. We assume that row equilibration (explicit or implicit) is used as a matrix scaling in general.

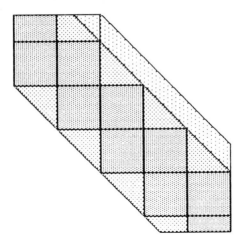

Figure 7.3 A as a banded matrix, with the additional fill-in due to pivoting. Possibly nonzero elements are shaded.

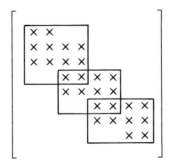

Figure 7.4a A as an almost block diagonal matrix

Now, for each block in turn, we apply n steps of Gaussian elimination with partial row pivoting. The last p rows of the i-th block are then appended to the top of $[S_{i+1} | R_{i+1}]$ to form the next block, and the elimination process is repeated. For the last block we apply $2n - 1$ elimination steps, completing the decomposition.

The results of applying this algorithm to **A** of Fig. 7.1b are shown in Fig. 7.4b, where the symbol + is used for elements arising due to fill-in and a circle is drawn around zeroed elements.

While it gives exactly the same numerical results as the algorithm which treats **A** as a banded matrix, this algorithm improves upon it, especially in terms of storage. The amount of storage required is now seen to be $2n(n+p)N - p(n-p)(N-1) + 2n(n-p)$. This is still about 50 percent more than the minimum of $2n^2N + n^2$ required to store **A** itself. On the other hand, note that the algorithm can be shown to be stable by using the usual backward error analysis (to the extent that the usual algorithm of Gaussian elimination with partial row pivoting is shown stable), and it can be generalized to other staircase matrices (see Exercise 1).

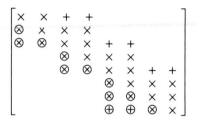

Figure 7.4b Almost block diagonal decomposition

7.2.2 Alternate row and column elimination

When performing Gaussian elimination by rows, we are generating an LU-decomposition,

$$A = LU \qquad (7.7)$$

where **L** is a lower triangular matrix and **U** is an upper triangular matrix. Further, with row pivoting, the decomposition takes the form

$$PA = LU \qquad (7.8)$$

where **P** is a permutation matrix. This is the form of the decomposition obtained in Section 7.2.1. Instead of row pivoting though, we can use *column pivoting* (i.e., search rows for pivot elements and interchange columns) to obtain an LU-decomposition of the form

$$A\hat{P} = LU \qquad (7.9)$$

where $\hat{P}$ is another permutation matrix. Moreover, such a scheme can also be shown to be stable.

While the usual row elimination process corresponds to a *premultiplication* of **A** by elementary transformations [eventually adding up to L^{-1} in (7.8)], we can think alternatively of *column elimination*, which corresponds to a *postmultiplication* of **A** by elementary transformations [and eventually adds up to a matrix U^{-1} in (7.9)]. This leads to the method described here, where we *mix* row and column elimination in order to avoid fill-in for the matrix **A** of (7.3). We begin by applying p column elimination steps (with full column pivoting), transforming B_{a1} into a lower triangular matrix. Since another step of subsequent column pivoting would introduce fill-in, we switch to row elimination with full row pivoting and perform $n-p$ such steps. Then we switch back to perform p column elimination steps, and so on. The process is depicted schematically in Fig. 7.5a, while in Fig 7.5b we have circled the zeroed elements of **A** of Fig. 7.1b when this algorithm is applied to it.

Recall that while row operations on **A** correspond to manipulating the right hand side in (7.1), column operations correspond to transforming the unknowns. Given the matrix of Fig. 7.5b, we can consecutively find the first, third, fifth and seventh transformed unknowns from their corresponding rows, and then consecutively find the eighth, sixth, fourth and second transformed unknowns from their corresponding rows. The original unknowns are subsequently recovered.

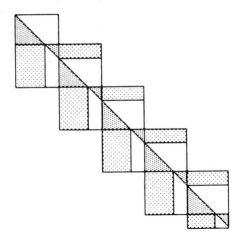

Figure 7.5a Alternating row and column elimination

$$\begin{bmatrix} \times & \otimes & & & & & & & \\ \times & \times & \times & \times & & & & & \\ \times & \otimes & \times & \otimes & & & & & \\ & & \times & \times & \times & \times & & & \\ & & \times & \otimes & \times & \otimes & & & \\ & & & & \times & \times & \times & \times & \\ & & & & \times & \otimes & \times & \otimes & \\ & & & & & & & \times & \times \end{bmatrix}$$

Figure 7.5b Alternating row and column elimination: Example of Fig. 7.1b

Let us consider the algorithm just described from a matrix transformation point of view. Each application of p column steps corresponds to a transformation

$$A \leftarrow A\hat{P}^{(i)}U^{(i)}$$

where $\hat{P}^{(i)}$ is a permutation matrix and $U^{(i)}$ is upper triangular, $0 \leq i \leq N$ [see (7.9)], while each application of $n - p$ row steps corresponds to a transformation

$$A \leftarrow L^{(i)}P^{(i)}A$$

as in (7.8). Thus, the entire process can be described as a decomposition of A in the form

$$PA\hat{P} = L\hat{A}U \tag{7.10a}$$

with P and $\hat{P}$ permutation matrices, L lower triangular and U upper triangular,

$$P := P^{(N)} \ldots P^{(0)}, \qquad L := \tilde{L}^{(0)} \ldots \tilde{L}^{(N)} \tag{7.10b}$$

$$U := \tilde{U}^{(N)} \ldots \tilde{U}^{(0)}, \qquad \hat{P} := \hat{P}^{(0)} \ldots \hat{P}^{(N)} \tag{7.10c}$$

where each $\tilde{L}^{(i)}$ and $\tilde{U}^{(i)}$ has the same nonzero structure as $L^{(i)}$ and $U^{(i)}$. The resulting matrix $\hat{A}$ has the form depicted in Figures 7.5.

In order to solve the system (7.1), given the decomposition (7.10), we solve successively

$$\mathbf{Lt} = \mathbf{Pb} \tag{7.11a}$$

$$\hat{\mathbf{A}}\mathbf{z} = \mathbf{t} \tag{7.11b}$$

$$\mathbf{U}\hat{\mathbf{s}} = \mathbf{z} \tag{7.11c}$$

$$\mathbf{s} = \hat{\mathbf{P}}\hat{\mathbf{s}} \tag{7.11d}$$

Since **L** is lower triangular and **U** is upper triangular, the solutions of (7.11a) and (7.11c) simply correspond to forward and back substitutions, as usual. The solution of (7.11b) is less orthodox, but still straightforward, as discussed below.

Let us write $\hat{\mathbf{A}}$ in *block tridiagonal form*

$$\hat{\mathbf{A}} = \begin{bmatrix} A_1 & D_1 & & & & \\ C_2 & A_2 & D_2 & & & \\ & C_3 & A_3 & D_3 & & \\ & & \cdot & \cdot & \cdot & \\ & & & L_N & A_N & D_N \\ & & & & C_{N+1} & A_{N+1} \end{bmatrix} \tag{7.12}$$

where $A_i, C_i, D_i \in \mathbf{R}^{n \times n}$. This is depicted in Fig. 7.6, where the nonzero elements of $\hat{\mathbf{A}}$ are shaded.

We use the partition notation

$$A_i = \begin{bmatrix} A_i^1 \\ A_i^2 \end{bmatrix} = \begin{bmatrix} A_i^{11} & A_i^{12} \\ A_i^{21} & A_i^{22} \end{bmatrix}, \quad \mathbf{z}_i = \begin{bmatrix} \mathbf{z}_i^1 \\ \mathbf{z}_i^2 \end{bmatrix}, \quad \mathbf{z}^T = (\mathbf{z}_1^T, \ldots, \mathbf{z}_{N+1}^T) \tag{7.13}$$

where $A_i^1 \in \mathbf{R}^{p \times n}$, $A_i^{11} \in \mathbf{R}^{p \times p}$, etc. Since A has the form (7.3), we note that with the similar partitioning for each C_i and D_i, the matrix $\hat{\mathbf{A}}$ satisfies

$$C_i^{12} = C_i^2 = D_i^1 = A_i^{12} = 0$$

A_i^{11} is lower triangular, and A_i^{22} is upper triangular (cf. Fig. 7.6). Thus we can find the $\mathbf{z}_i^1$ by *forward recursion*

$$A_1^{11} \mathbf{z}_1^1 = \mathbf{t}_1^1 \tag{7.14a}$$

$$A_i^{11} \mathbf{z}_i^1 = \mathbf{t}_i^1 - C_i^{11} \mathbf{z}_{i-1}^1 \quad i = 2, \ldots, N+1 \tag{7.14b}$$

and the $\mathbf{z}_i^2$ by *backward recursion*

$$A_{N+1}^{22} \mathbf{z}_{N+1}^2 = \mathbf{t}_{N+1}^2 - A_{N+1}^{21} \mathbf{z}_{N+1}^1 \tag{7.15a}$$

$$A_i^{22} \mathbf{z}_i^2 = \mathbf{t}_i^2 - A_i^{21} \mathbf{z}_i^1 - D_i^2 \mathbf{z}_{i+1}^2 \quad i = N, \ldots, 1 \tag{7.15b}$$

A number of observations are significant here. First, this algorithm preserves storage: The information about the decomposition (7.10) can all be stored in place of the zeroed elements of **A**, thus requiring *no additional storage*. Second, whenever we eliminate, a full partial pivoting is done (i. e., the largest element in the column or row is used as the pivot element). As a result, one can expect the algorithm to have a similar stability property as the usual Gaussian elimination method for almost block diagonal forms in Section 7.2.1 and to be at least as fast. In fact, it turns out that this method has less overhead and vectorizes better than the one for almost block diagonal forms, since no block shifts are required. Third, the decoupled recursion (7.14), (7.15) strongly suggests a connection to decoupling. This will be taken up in the next section. Note that the decomposition information can be stored in place of the zeroed elements of **A**. Finally, recall that in the actual coding it can be important to order operations in a way dependent upon how two dimensional arrays are stored (which depends upon the programming language used) and to avoid indirect array indexing (see Section 2.2.3).

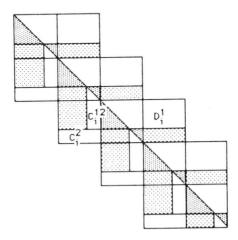

Figure 7.6 $\hat{\mathbf{A}}$ as a block tridiagonal matrix

An intimately related elimination strategy is to select pivot elements precisely as done here (i.e., alternately preforming p searches along rows and then $(n-p)$ searches along columns) but always to do *row elimination*. Instead of (7.10a), this corresponds to an LU factorization of **A** of the form

$$\mathbf{PA\hat{P}} = \mathbf{LU} \tag{7.16}$$

where **P** and $\hat{\mathbf{P}}$ are permutation matrices. We will refer to this variant as *row and column pivoting*. Once again, we expect stability of the algorithm since a full partial pivoting is done.

7.2.3 Block tridiagonal elimination

One should not assume from our previous discussion that the numerical solution of (7.1) is without pitfalls, i.e., that any algorithm which looks reasonable at first sight, necessarily is. To see this, we consider a simpler (and historically earlier) approach than that of the previous subsection, which treats **A** itself in a block tridiagonal form (7.12). This is depicted in Fig. 7.7. Partitioning the blocks as in (7.13), the A_i are full matrices, but we still have

$$C_i^2 = D_i^1 = 0$$

for all i. Note that the matrix of (5.5b) is of a block tridiagonal form with the "blocks" being scalar.

In this setting, it is natural to look for an LU-decomposition in block form. Suppose for the moment that we discard pivoting altogether and consider (7.7) with

$$\mathbf{L} = \begin{bmatrix} I & & & \\ L_2 & I & & \\ & \cdot & \cdot & \\ & & L_{N+1} & I \end{bmatrix}, \quad \mathbf{U} = \begin{bmatrix} U_1 & D_1 & & & \\ & U_2 & D_2 & & \\ & & \cdot & \cdot & \cdot \\ & & & U_N & D_N \\ & & & & U_{N+1} \end{bmatrix} \quad (7.17)$$

where each $L_i, U_i \in \mathbf{R}^{n \times n}$. If a decomposition like this holds (implying that each U_i is nonsingular) then we find directly that

$$U_1 = A_1 \tag{7.18a}$$

$$L_i = C_i U_{i-1}^{-1} \tag{7.18b}$$

$$i = 2, \ldots, N+1$$

$$U_i = A_i - L_i D_{i-1} \tag{7.18c}$$

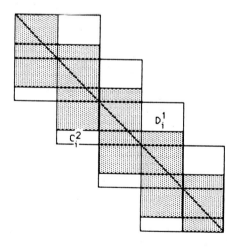

Figure 7.7 A as a block tridiagonal matrix

Of course, the choice of identity blocks on the diagonal for **L** is a special one, inspired by standard Gaussian elimination. Other choices may be used as well. Once we have the decomposition (7.7), (7.17), the solution of (7.1) for a given vector $\mathbf{b}^T = (\mathbf{b}_1^T, \ldots, \mathbf{b}_{N+1}^T)$ proceeds, as expected, by forward and back substitutions in blocks. Thus we set

$$\mathbf{z}_1 = \mathbf{b}_1 \qquad (7.19a)$$

$$\mathbf{z}_i = \mathbf{b}_i - L_i \mathbf{z}_{i-1} \qquad i = 2, \ldots, N+1 \qquad (7.19b)$$

and then

$$\mathbf{s}_{N+1} = U_{N+1}^{-1} \mathbf{z}_{N+1} \qquad (7.20a)$$

$$\mathbf{s}_i = U_i^{-1}(\mathbf{z}_i - D_i \mathbf{s}_{i+1}) \qquad i = N, \ldots, 1 \qquad (7.20b)$$

The method (7.17)–(7.20) is quite intuitive and simple. However, since no pivoting has been done yet, it is not true in general that this decomposition even exists. Some pivoting must be allowed, but here we only wish to consider row pivoting which does not generate any fill-in. Examining the form (7.3), which underlies the tridiagonal form (7.12), we see that row interchanges in $[S_i | R_i]$, i. e., among rows $(i-1)n + p + 1$ through $in + p$ of **A**, do not introduce any fill-in. Also, solving (7.18b), we find that the resulting L_i has the same zero structure as C_i, i. e.,

$$L_i^2 = 0$$

Thus, *no fill-in* is achieved in (7.17) with a decomposition of the form (7.8), so long as the permutation matrix **P** is of the form

$$\mathbf{P} = \mathrm{diag}\,(I_p, P_1, \ldots, P_N, I_{n-p}) \qquad (7.21)$$

where I_l is an identity of size l and each $P_i \in \mathbf{R}^{n \times n}$ is a local permutation matrix. It is in fact not difficult to prove (see Exercise 3) the following:

Theorem 7.22 If **A** is nonsingular then there exist a permutation matrix **P** of the form (7.21) and matrices **L** and **U** of the form (7.17) such that (7.8) holds (i. e., a storage preserving **LU** decomposition with only pivoting inside blocks is possible). □

While Theorem 7.22 asserts the *validity* of the algorithm, it does not say anything about its *stability*. Indeed, consider the following

Example 7.2

Consider the matrix

$$\mathbf{A} = \begin{bmatrix} 1 & 0 & & & & & & & \\ 1 & \varepsilon & -2 & 0 & & & & & \\ 0 & 1 & 0 & -1 & & & & & \\ & & 1 & \varepsilon & -2 & 0 & & & \\ & & 0 & 1 & 0 & -1 & & & \\ & & & & \ddots & & \ddots & & \\ & & & & & \ddots & & \ddots & \\ & & & & & & 1 & \varepsilon & -2 & 0 \\ & & & & & & 0 & 1 & 0 & -1 \\ & & & & & & & & 0 & 1 \end{bmatrix}$$

where $0 < \varepsilon \ll 1$. This matrix can be shown to be well-conditioned (see Exercise 2). However, applying the algorithm (7.17)–(7.18) directly to **A** results in

$$U_1 = \begin{bmatrix} 1 & 0 \\ 1 & \varepsilon \end{bmatrix}, \qquad L_2 = \begin{bmatrix} -1/\varepsilon & 1 \\ 0 & 0 \end{bmatrix}$$

so the algorithm is unstable for ε small, in the sense that $\|L\|\,\|U\| = O(\varepsilon^{-1}) \gg \|A\|$ (cf. Theorem 2.53). □

This example emphasizes that a more careful pivoting, still satisfying (7.21), is needed for stability. One can show (see Exercise 4) that there is a block tridiagonal factorization which introduces no fill-in and is a decoupling algorithm, and hence it is stable under appropriate assumptions about the underlying BVP.

7.2.4 Stable compactification

The linear system (7.1) with $\mathbf{A}$ in the form (7.2) is equivalent to (6.22), so the form (7.3) is in the one-step recursion form (6.22) as well. Hence, (7.1), (7.3) can be viewed as a discrete BVP, and one of the variants of the discrete decoupling algorithm of Section 6.3.2 can be used to solve it. Specifically, consider the discrete BVP (6.22) with R_i nonsingular. For simplicity, we take

$$R_i = I \qquad 1 \leq i \leq N$$

Obviously these equations can be written as (7.1) where

$$\mathbf{A} = \begin{bmatrix} -S_1 & I & & & \\ & -S_2 & I & & \\ & & \ddots & \ddots & \\ & & & -S_N & I \\ B_a & & & & B_b \end{bmatrix} \qquad (7.23)$$

i. e., $\mathbf{A}$ is of the form (7.2). Recall the basic decoupling steps: To obtain the decoupled form (6.26), we redefine the variables by (6.25) and multiply the i-th block row by T_{i+1}^{-1}, $1 \leq i \leq N$. Defining matrices U_{i+1} by

$$S_i T_i = T_i U_i \qquad (7.24)$$

the requirement that the transformation be decoupling is written as

$$U_i = \begin{bmatrix} U_i^{11} & U_i^{12} \\ 0 & U_i^{22} \end{bmatrix} \qquad (7.25)$$

with $U_i^{11} \in \mathbf{R}^{(n-p) \times (n-p)}$, $U_i^{12} \in \mathbf{R}^{p \times (n-p)}$, and $U_i^{22} \in \mathbf{R}^{p \times p}$. The equations (6.26) are then written in the basic linear equation form (7.1), with a matrix

$$\hat{\mathbf{A}} = \begin{bmatrix} -U_1 & I & & & \\ & -U_2 & I & & \\ & & \ddots & & \\ & & & -U_N & I \\ B_a T_1 & & & & B_b T_{N+1} \end{bmatrix} \qquad (7.26a)$$

[the U_i here are generally different than those in (7.17)], where $\hat{\mathbf{A}}$ is related to $\mathbf{A}$ by

$$\hat{\mathbf{A}} = \hat{\mathbf{T}} \mathbf{A} \mathbf{T} \qquad (7.26b)$$

$$\mathbf{T} = \text{diag } \{T_1, T_2, \ldots, T_N, T_{N+1}\} \qquad (7.26c)$$

$$\hat{\mathbf{T}} = \text{diag } \{T_2^{-1}, T_3^{-1}, \ldots, T_{N+1}^{-1}, I\} \qquad (7.26d)$$

Now, since the BC are separated, we move the BC associated with the left end point to the top of the matrix, as in (7.3). Instead of (7.26a) we then have the form

$$\hat{A} = \begin{bmatrix} \hat{B}_{a1} & & & & \\ -U_1 & I & & & \\ & -U_2 & I & & \\ & & \ddots & & \\ & & & -U_N & I \\ & & & & \hat{B}_{b2} \end{bmatrix} \quad (7.27)$$

where we assume without restriction that $\hat{B}_{a1} := B_{a1}T_1 = [OM]$ with nonsingular $M \in \mathbf{R}^{p \times p}$. The zero structure of $\hat{A}$ is depicted in Fig. 7.8.

Recall from Chapter 6 the obvious ways to compute U_i: We use an LU (or block LU) decomposition

$$S_i T_i = L_i U_i \quad (7.28a)$$

or a QU (or block QU) decomposition

$$S_i T_i = Q_i U_i \quad (7.28b)$$

Then $T_{i+1} := L_i$ or $T_{i+1} := Q_i$. The latter is more expensive but gives T_{i+1} which is orthogonal and hence ideally conditioned.

From the structure of $\hat{A}$ it is apparent that we have a partitioned system of the form (6.31), where the unknowns $\mathbf{w}_i^2$, $i = 1, \ldots, N+1$, can be obtained by a forward recursion (corresponding to forward substitution in Gaussian elimination). This can be followed by a backward recursion (back substitution) for the remaining unknowns $\mathbf{w}_i^1$, $i = N+1, \ldots, 1$.

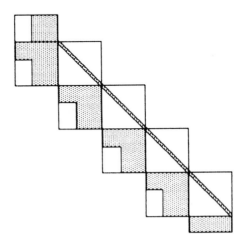

Figure 7.8 The decoupling matrix $\hat{A}$

If we have a well-conditioned BVP with separated BC, then we know from Theorem 6.33 (using its notation) that a consistent fundamental solution with a consistency constant $L = 0$ can be constructed by requiring $B_{a1} T_1 = 0$. (It is easy to construct an orthogonal T_1, for instance, satisfying this condition.) Moreover, recall that for consistency only the span of the first $n-p$ columns of T_1 matters. This implies that the LU (or QU) decompositions used in the decoupling need only reduce the recursion matrices to the *block* upper triangular form. The resulting algorithms for solving (7.1), like those of Sections 7.2.2 and 7.2.3, do not require extra storage.

7.3 STABILITY FOR BLOCK METHODS

When considering the question of stability, we treat the algorithms of Section 7.2.1 by the usual backward error analysis for Gaussian elimination with partial pivoting. For the algorithms in Sections 7.2.2–7.2.4, we view the methods in a block fashion and show them to be instances of discrete decoupling, thereby obtaining stability from the analysis of Chapter 6.

This task is trivial for the stable compactification algorithms considered in Section 7.2.4, because these algorithms are derived directly in Chapter 6. Moreover, we can also consider $\hat{A}$ of (7.27) to be in block tridiagonal form as in (7.12) and obtain an LU-decomposition as in (7.17). To do this and avoid confusion with the U_i in (7.27), let us relabel the U_i in (7.17) as $\hat{U}_i$. Then the block LU-decomposition of $\hat{A}$ is given by

$$\hat{U}_1 = \begin{bmatrix} 0 & M \\ -U_1^{11} & -U_1^{12} \end{bmatrix}, \quad \hat{U}_i = \begin{bmatrix} 0 & I \\ -U_i^{11} & -U_i^{12} \end{bmatrix}, \quad i = 2, \ldots, N, \tag{7.29a}$$

$$\hat{U}_{N+1} = \begin{bmatrix} 0 & I \\ \hat{B}_{b2} & \end{bmatrix}$$

$$L_2 = \begin{bmatrix} -U_1^{22} M^{-1} & 0 \\ 0 & 0 \end{bmatrix}, \quad L_i = \begin{bmatrix} -U_i^{22} & 0 \\ 0 & 0 \end{bmatrix}, \quad i = 3, \ldots, N+1 \tag{7.29b}$$

(verify). It is straightforward to see that the forward and back substitutions of (7.19), (7.20) for this factorization correspond exactly to the forward and backward recursions, respectively, for (6.18) using the decoupling outline. Moreover, the stability of this block LU-decomposition follows immediately from (7.28) if the U_i are obtained in a stable way (e.g., if T_{i+1} is orthogonal, as it is with the QU factorization above, then $\|U_i\|_2 = \|S_i\|_2$). In such a case, this elimination method can be interpreted as a stable version of the block tridiagonal solver in Section 7.2.3 (see Exercise 4).

It is somewhat less immediate, but still straightforward, to interpret the alternate row and column elimination algorithm of Section 7.2.2 as a stable decoupling algorithm. To this end we write the linear system in $\hat{A}$ [see (7.10a), (7.11b)] as the discrete BVP

$$V_i \mathbf{z}_{i+1} = U_i \mathbf{z}_i + \mathbf{t}_i \tag{7.30a}$$

$$\hat{B}_1 z_1 + \hat{B}_2 z_{N+1} = t_{N+1} \tag{7.30b}$$

where in (7.30b) we have formally collected the (separated) BC. From Fig. 7.6 we obtain that the zero-structure of U_i and V_i is as depicted in Fig. 7.9. The lower right $p \times (n-p)$ blocks of both U_i and V_i are 0, and this is consistent with (6.27b), albeit in an altered form.

To obtain (7.30) in the usual decoupling form, we interchange the first p columns with the last $n-p$ columns in both U_i and V_i, and correspondingly interchange the first p elements of z_i with its last $n-p$ elements. The resulting recursion can be written as

$$\tilde{V}_i \tilde{z}_{i+1} = \tilde{U}_i \tilde{z}_i + t_i \tag{7.31}$$

where $\tilde{V}_i$ and $\tilde{U}_i$ are now block upper triangular, like we are used to from Chapters 4 and 6. Multiplying (7.31) by $\tilde{V}_i^{-1}$ directly relates (7.30) to (7.26a). Indeed, the row and column elimination described by (7.10a) can be related in a one-to-one fashion to (7.26b). Consistency follows from Theorem 6.33, because (6.33a) is obtained from the first p column elimination steps of the algorithm. Hence this is a stable decoupling algorithm.

Note that in practice $\tilde{V}_i$ is not generated, let alone inverted. Only the (lower triangular) lower left $p \times p$ block of V_i and the (upper triangular) $(n-p) \times (n-p)$ upper right block of U_i are inverted (i.e., corresponding forward and back substitutions are performed).

Figure 7.9 Block structure of (7.30)

7.4 DECOMPOSITION IN THE NONSEPARATED BC CASE

When the BC are not separated (or partially separated), the matrix **A** of (7.2) can no longer be easily rearranged as a banded matrix. It turns out that this fact is associated with much deeper difficulties, both theoretical and practical, which arise when we attempt to generalize the BC from separated to nonseparated ones. Some such difficulties have already been discussed in Chapters 3, 4 and 6.

Here we consider first how to generalize the straightforward LU-decomposition method to handle nonseparated BC at the cost of extra fill-in. Then we discuss decoupling and a related factorization method (cf. Section 4.4.2) which produces no fill-in.

7.4.1 The LU-decomposition

We consider Gaussian elimination with (scaled) row pivoting for solving the discretization equations (7.2), arising from a BVP with nonseparated BC. (The treatment is almost the same for the case where, except for BC, $\mathbf{A}$ has a more general staircase.) Although not necessary, we first move the BC matrices to the top, so

$$\mathbf{A} = \begin{bmatrix} B_a & & & & B_b \\ S_1 & R_1 & & & \\ & S_2 & R_2 & & \\ & & \ddots & & \\ & & & S_N & R_N \end{bmatrix} \tag{7.32}$$

Then the factorization (7.8) produces fill-in in the last column block, viz.

$$\mathbf{U} = \begin{bmatrix} U_1 & & & & V_1 \\ & U_2 & & & V_2 \\ & & \ddots & & \\ & & & U_N & V_N \\ & & & & U_{N+1} \end{bmatrix} \tag{7.33}$$

The advantages of this factorization are its simplicity and stability; the disadvantage is its fill-in.

One way to avoid the form (7.32) is to transform the BVP to one with separated BC [as done in (1.9)] before discretization. While this results in $\mathbf{A}$ being of the form (7.3), it is a larger system than for the original BVP, and the actual effect is not a reduction in fill-in (see Exercise 5). Despite this, the advantage of this approach is that standard software based upon one of the methods of Section 7.2 can be used.

As an alternative approach, let us consider, mainly for historical reasons, an attempt to generalize the block tridiagonal elimination method (7.17), (7.18) to matrices $\mathbf{A}$ of the form (7.2). It is not difficult to see that some fill-in is unavoidable. However, to save storage we may decide not to use the last n rows for pivoting. This yields the factorization (7.8), where $\mathbf{U}$ is as in (7.17), but $\mathbf{L}$ now reads

$$\mathbf{L} = \begin{bmatrix} I & & & \\ & \ddots & & \\ & & \ddots & \\ & & & \ddots \\ M_1 & \cdots & M_N & I \end{bmatrix} \tag{7.34a}$$

with

$$U_i = A_i \qquad i = 1, \ldots, N \tag{7.34b}$$

$$M_i = (-1)^{i-1} B_a A_1^{-1} D_1 A_2^{-1} \cdots D_{i-1} A_i^{-1} \qquad i = 1, \ldots, N \qquad (7.34c)$$

$$U_{N+1} = B_b + (-1)^N B_a A_1^{-1} D_1 \cdots D_{N-1} A_N^{-1} D_N \qquad (7.34d)$$

It can be proved [recall Theorem 7.22] that the resulting U_i are nonsingular, hence the decomposition is "valid." However the mere form of the expressions in (7.34) suggests that what we have here is yet another discrete form of single shooting, i. e., we are recurring without having decoupled first the increasing modes from the nonincreasing ones. Hence, failure to do full partial pivoting again produces potential instability.

If the BC are only partially separated, e. g.

$$B_a = \begin{bmatrix} B_{a1} \\ B_{a2} \end{bmatrix}, \qquad B_b = \begin{bmatrix} 0 \\ B_{b2} \end{bmatrix} \qquad (7.35)$$

then it is tempting to use the zero rows in B_b to form the first block row of $\mathbf{A}$ and then use some restricted pivoting strategy in order to preserve as many zeroes as possible. It should be clear now that we may not in general expect good results from such a strategy, since the number of rows in B_{a1} may not correspond to the dimension of the nonincreasing solution subspace. For further discussion of this see Exercise 6.

7.4.2 A general decoupling algorithm

We know from Section 3.4.3 that a well-conditioned BVP with nonseparated BC has a dichotomy, so it is natural to look for decoupling algorithms for $\mathbf{A}$ in the form (7.2). But this is exactly what is described in Section 6.2.3, and indeed, in Section 4.4.2. While the decoupling is considered only for a multiple shooting matrix in Chapter 4, with the hindsight of Chapter 5 we see that similar considerations apply if (7.2) arises from one-step difference methods.

Let us recall the basic steps in the decoupling algorithm. For simplicity, assume for now that each $R_i = I$ for $\mathbf{A}$ in the form (7.2). First, the matrix T_1 must be found (cf. Sections 4.4.2 and 6.4.2). Although this is a nontrivial step in the process, assume that a consistent T_1 and the number of increasing modes are known. The next steps proceed as for the case with separated BC (Section 7.2.4): The matrices T_{i+1}, U_{i+1} in (7.24) are formed recursively for $i = 1, \ldots, N$, such that (7.25) holds (e. g., using LU- or QU-factorizations), and the decoupled system (6.31) results.

No fill-in has been necessary in the decomposition, but the solution of the linear system is not simple like in the separated BC case. Now, we must use the stable compactification algorithm of Section 4.4.2; i. e., we must perform (stable) recursions for fundamental modes and not only for a particular solution (which was sufficient for separated BC). The cost of the forward and back-substitutions becomes comparable in order to that of the decomposition itself (cf. Exercise 7).

Not surprisingly, block LU-factorizations which correspond to an obvious decoupling in the separated BC case have a natural analogue in the nonseparated BC case. For instance, consider the alternate row and column elimination algorithm of Section 7.2.2, with $\mathbf{A}$ given by (7.32). Again, suppose that a consistent T_1 and p, the number of *decreasing* modes, are known. (This is for theoretical purposes, as we do not have a theorem like 6.33 handy.) Then $rank(B_a) \geq p$, because the BVP is assumed well-conditioned. Performing p column elimination steps on B_a then, we hope, starts a stable row and column elimination algorithm, as described in Section 7.2.2. The resulting matrix $\hat{\mathbf{A}}$ has the structure depicted in Fig. 7.10, which is to be compared with Fig. 7.5a.

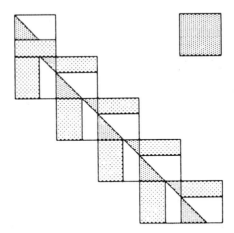

Figure 7.10 Alternating row and column elimination for nonseparated BC

7.5 SOLUTION IN MORE GENERAL CASES

In this section, we discuss the solution of BVPs involving additional parameters or involving multipoint BC. We generalize Gaussian elimination with (scaled) row pivoting for these two cases and see how, although the sparsity structure of **A** can be utilized to varying degrees, fill-in occurs (as in Section 7.4.1). We do not consider decoupling algorithms for these cases because: (i) such algorithms for BVPs involving parameters cannot in general be relied on to be stable (see Exercise 8), and (ii) for multipoint BC, they must be modified to insure stability. The reason is simply that the concept of dichotomy must be generalized to handle these cases, as further discussed in Section 11.9.

7.5.1 BVPs with parameters

We have seen a number of instances where a given BVP depends upon (unknown) parameters, e.g., in Examples 1.4, 1.7, 1.9, 1.11, and 1.20. In the linear BVP case, we have

$$\mathbf{y}' = A(x, \boldsymbol{\lambda})\mathbf{y} + \mathbf{q}(x, \boldsymbol{\lambda}) \qquad a < x < b \qquad (7.36a)$$

$$B_a \mathbf{y}(a; \boldsymbol{\lambda}) + B_b \mathbf{y}(b; \boldsymbol{\lambda}) = \boldsymbol{\beta}(\boldsymbol{\lambda}) \qquad (7.36b)$$

where in (7.36a) there are n ODEs, $\boldsymbol{\lambda} \in \mathbf{R}^m$, and in (7.36b) there are $n+m$ BC. In Chapter 1 our method for converting (7.36a,b) to a standard form was to add the ODEs

$$\boldsymbol{\lambda}' = \mathbf{0} \tag{7.36c}$$

The same trick may of course be used to convert the nonseparated BC to separated BC, as discussed in Section 7.4.1. Since this certainly avoids some of the stability problems we have mentioned, it is a rather useful idea (and carried out, e.g., in Example 1.10). However, there is a price to be paid by an increase in the size of the BVP solved. Indeed, examining a discretization of (7.36), we see that in discretizing (7.36c) a lot of wasted effort might be invested.

Instead, considering a discretization of (7.36) on a mesh π as in (5.1) or (4.31), we may regard the vector of unknowns $\mathbf{s}^T = (\mathbf{y}_1^T, \mathbf{y}_2^T, \ldots, \mathbf{y}_{N+1}^T, \boldsymbol{\lambda}^T)$. This in general yields the matrix form [of order $(N+1)n + m$]

$$\mathbf{A} = \begin{bmatrix} B_a & & & & & B_b & Z_0 \\ S_1 & R_1 & & & & & Z_1 \\ & S_2 & R_2 & & & & Z_2 \\ & & & \ddots & & & \\ & & & & S_N & R_N & Z_N \end{bmatrix} \tag{7.37}$$

At the expense of some fill-in, a simple extension of the (stable) Gaussian elimination with row pivoting algorithm (Section 7.4.1) to (7.37) is straightforward, considering the last $n+m$ columns as full.

7.5.2 Multipoint BC

Recall that the BVP (1.4a, e) with multipoint side conditions, where each condition involves only one point, is discussed in Section 7.1. In that case, $\mathbf{A}$ for most methods would be a staircase matrix, for which Gaussian elimination with (scaled) row pivoting is easy to use (see Section 7.2.1).

When some of the multipoint BC involve more than one point ζ_j, the BVP can be converted to a larger BVP with separated BC using the trick mentioned after (1.9), but its potential disadvantages are even greater here. Instead, we can directly use Gaussian elimination with row pivoting on the discretization equations. For example, if rather than (7.37) we have

$$\mathbf{A} = \begin{bmatrix} B_1 & 0 & \cdots & 0 & B_i & 0 & \cdots & 0 & B_J \\ S_1 & R_1 & & & & & & & \cdot \\ & S_2 & R_2 & & & & & & \cdot \\ & & & \ddots & & & & & \cdot \\ & & & & & & & S_N & R_N \end{bmatrix} \tag{7.38}$$

then nonzero elements would be generated in the column block below each B_i.

EXERCISES

1. **(a)** Illustrate in a diagram the effect of using Gaussian elimination with partial pivoting of Section 7.2.1 to solve (7.1) when A has the form (7.3).

   (b) Then discuss (and illustrate) the effects of instead having A in the form (7.2) (i. e., nonseparated BC).

2. Show that the matrix A in Example 7.2 is well-conditioned for $0 < \varepsilon \ll 1$.

3. Prove Theorem 7.22.

   Hint: Define P_1 using the fact that S_1 has $n-p$ rows independent of those for B_{a1}. Then U_1 and L_2 are easy to define, and this sets the framework for an inductive argument.

4. Show that there is a stable block tridiagonal elimination method for A in (7.3) such that P has the form (7.21).

   Hint: Show that the algorithm described in (7.29) is a decoupling algorithm which can be interpreted as such a method.

5. **(a)** For the matrix A in (7.32), calculate the amount of fill-in which results from using Gaussian elimination with row pivoting.

   (b) Transform the BVP to one with separated BC using (1.9). Find the amount of storage required for the matrix A resulting from the same discretization method (and mesh) for this transformed problem. Compare the results with (a).

6. Let $\mathbf{A}$ be of the form (7.23) with

$$S_i = \frac{1}{4} \begin{bmatrix} 5 & 3 & 0 \\ 3 & 5 & 0 \\ 0 & 0 & 4 \end{bmatrix}$$

   and consider the partially separated BC [cf. (7.35)]

$$B_a = \begin{bmatrix} 0 & 0 & 1 \\ 1 & 0 & 0 \\ 0 & 1 & 0 \end{bmatrix} ; \quad B_b = \begin{bmatrix} 0 & 0 & 0 \\ 1 & 0 & 0 \\ 0 & 1 & 0 \end{bmatrix}$$

   (a) Viewing this as a discrete BVP show that a fundamental solution is given by

$$\Phi_i = \begin{bmatrix} \left(\frac{1}{2}\right)^i & 2^i & 0 \\ -\left(\frac{1}{2}\right)^i & 2^i & 0 \\ 0 & 0 & 1 \end{bmatrix}$$

   (b) Show that the stability constant of this problem (i. e., $\|\mathbf{A}^{-1}\|$) is of order unity (for $N \to \infty$).

   (c) Let us now permute the rows in $\mathbf{A}$ so that the first BC row becomes the first row, while the relative location of the other rows does not change, and call the result $\tilde{\mathbf{A}}$. Further, consider an LU-decomposition of $\mathbf{A}$ where row partial pivoting is performed including all but the last two rows in the pivoting range. This yields an $\mathbf{L}$ of the general form (7.34a). Find the zero-structure of this decomposition.

(d) Show that for the jth block in the last row block of $\mathbf{L}$, call it say M_j, the estimate $\|M_j\| \sim \alpha^j$ holds where $\alpha > 1$. Conclude that the saving of storage results here in a loss of stability.

7. Consider the matrix $\mathbf{A}$ in (7.2).

   (a) Working through the details of the decoupling algorithm of Section 7.4.2 (say, using LU-decompositions) to compute the T_i, determine the number of flops required to solve (7.1).

   (b) Compare this with the number of flops required using Gaussian elimination with partial pivoting.

8. Consider the BVP

$$y' = \frac{1}{\varepsilon} y$$

$$0 < x < 1$$

$$\lambda' = 0$$

$$y(0) = \lambda(0), \quad y(1) = 1$$

This BVP is well-conditioned for $0 < \varepsilon \ll 1$. Using a uniform mesh and the trapezoidal scheme, construct the discretization matrix A of the form (7.39). Show that a simple-minded decoupling algorithm which ignores the last column of A is unstable.

9. Consider the linear system arising from using standard multiple shooting to solve (1.4a, d)

$$\mathbf{A} = \begin{bmatrix} B_{a1} & & & & \\ -Y_1(x_2) & I & & & \\ & & \ddots & & \\ & & -Y_N(x_{N+1}) & I \\ & & & & B_{b2} \end{bmatrix} \begin{bmatrix} y_1 \\ y_2 \\ \cdot \\ \cdot \\ y_{N+1} \end{bmatrix} = \begin{bmatrix} \beta_1 \\ v_1(x_3) \\ \cdot \\ v_N(b) \\ \beta_2 \end{bmatrix}$$

where $B_a = [K_1 K_2] \in \mathbf{R}^{p \times n}$, $B_b = [K_3 K_4] \in \mathbf{R}^{k \times n}$ $(k := n - p)$ with $K_1 \in \mathbf{R}^{p \times p}$.

   (a) Interchange the first p and last k equations corresponding to each block $[-Y_i(x_{i+1}) \quad I]$ for $1 \leq i \leq N$. Then perform an **LU** decomposition of $\mathbf{A}$ as follows: After the first p elimination steps done without pivoting, perform sets of k steps of elimination with row interchanges followed by p steps of elimination with column interchanges. Let the column interchanges be determined from the row interchanges in such a way that it is equivalent to having performed a change of variables for the ODE in that subinterval.

   (b) Show that this elimination process uses pivot elements which are "almost" the largest possible for each step (so that it is "close" to the partial pivoting process with this alternate row and column pivoting).

   (c) Assuming no pivoting, show that the resulting upper blocks of $\mathbf{A}$ and the right-hand side vector during this elimination process are *exactly* the Riccati matrix R and $\mathbf{w}_2$ defined in Section 5.2. Describe how the terms constructed in the elimination process correspond to the solution $\mathbf{y}(x)^T = (\mathbf{y}_1(x)^T, \mathbf{y}_2(x)^T)$ to (1.4a, d), drawing upon the equivalence with the Riccati method of Section 4.5.2. (For the general case where pivoting is done, this corresponds to the Riccati method with reimbedding, mentioned but not discussed in Section 4.5.2.)

Hint: For (a), describe a change of variables for (1.4a) in terms of a permutation matrix involving rows and columns of **A**. For (b), note that the unit elements along the diagonal of the identity matrix, moved during the row interchanges, remain "almost" as large as elements arising in the same block. For (c), use (4.80b).

10. Prove the existence of a UQ-decomposition; i. e., show that for any given square matrix A there exists an orthogonal matrix Q and an upper triangular matrix U such that $A = UQ$. What about a UL-decomposition?

11. Design an analogue of the row and column elimination algorithm, using Householder transformations to replace the Gaussian elimination steps taken within it. What are the advantages and disadvantages of such a variant?

8

Solving Nonlinear Equations

After discretizing a nonlinear BVP in some way, one is faced with the problem of solving a system of nonlinear algebraic equations

$$\mathbf{f}(\mathbf{s}) = \mathbf{0} \tag{8.1}$$

where $\mathbf{f}$ is a vector function, $\mathbf{f} : \mathbf{R}^n \to \mathbf{R}^n$. In Chapters 4 and 5, where a large number of discretization methods are introduced, we always use Newton's method, either directly or in its quasilinearization form, to solve the resulting systems of nonlinear equations. Indeed, in certain circumstances Newton's method has the following desirable properties:

(a) It converges very fast.

(b) It is suitable for problems with a sparse Jacobian matrix

$$J(\mathbf{s}) = \frac{\partial \mathbf{f}}{\partial \mathbf{s}} \tag{8.2}$$

such as those arising from finite difference or multiple shooting discretizations.

(c) It has been found helpful in proofs of existence and convergence of the numerical solution.

Newton's method is therefore our starting point in this chapter. But Newton's method is not without drawbacks. These drawbacks can be roughly divided into two types:

(a) Under the usual assumptions of BVP smoothness, existence of an isolated solution and a sufficiently fine discretization, Newton's method is guaranteed to converge only *locally*. Thus, one needs an initial guess for the solution to the nonlinear system of algebraic equations which is "sufficiently close" to the sought solution. In practice, such an initial guess is frequently unavailable for BVPs (unlike for IVPs, where a guess can be easily constructed locally at any given step). Therefore, we shall seek ways to enlarge the domain of attraction of Newton's method.

(b) Each iteration of the method is rather expensive, requiring an evaluation of the Jacobian matrix $J(s)$ and the solution of a system of linear equations. Therefore, we shall seek cheaper methods.

These two types of drawbacks of Newton's method in some sense "pull" in opposite directions: Intuitively, if we replace Newton's method by a cheaper substitute (such as the secant method — see Section 2.3.3), then we could expect the resulting method to have even poorer global convergence properties. Still, a discussion of cheaper methods is important in our context of BVP solution because, as noted in Chapter 5, after convergence of the nonlinear iteration on a coarse mesh is achieved, we may want to refine the mesh (or to solve on the same mesh for a higher-order method, using a defect correction approach) in order to obtain a more accurate approximation. For the nonlinear iteration on the finer mesh, a good initial guess is available with the (interpolated) coarse mesh solution. Thus, cheaper methods may prove useful at this stage.

Another important observation regarding the type of nonlinear equations encountered in our BVP context is that the Jacobian may be either dense or sparse. Recall that most methods in Chapters 4 and 5 yield large, sparse Jacobians.

The question of finding more robust methods than Newton's method is addressed in Sections 8.1 and 8.3. Finding cheaper alternatives is the topic of Section 8.2. Finally, in Section 8.4 we make further remarks on systems (8.1) arising from decent discretizations of nonlinear BVPs.

This chapter and even more so the next have a different orientation than the previous ones, in that they are more practically oriented and less based on solid theory. One important reason for this is that at present such a theory is not completely developed. Moreover, the need for the techniques of these two chapters often arises when the discretization of a given BVP at the initial stage *does not* model very closely the continuous BVP linearized at the solution. In other words, unlike much of the material covered hitherto, often the situation of interest here is when the standard asymptotic theory is *not* applicable. Consequently, we will mostly discuss *methods* in these chapters, highlighting those which we have found practically useful and illustrating them with computational examples of various applications. The collocation code COLSYS, a version of which is given in the appendix, has been used for the calculations presented in this chapter.

Before proceeding, let us agree to refer to the Jacobian matrix simply as the "Jacobian." Also, note that the notation here is slightly different from that used in Chapters 4 and 5. The system of nonlinear equations there is denoted by $\mathbf{F}(s) = \mathbf{0}$, while here it is $\mathbf{f}(s) = \mathbf{0}$, and the Jacobian there is denoted by $\mathbf{F}'(s)$, while here it is $J(s)$. Also, the number of nonlinear equations n is not necessarily the number of ODEs discretized: It may in fact be much larger.

8.1 IMPROVING THE LOCAL CONVERGENCE OF NEWTON'S METHOD

Recall that Newton's method generates, given an initial guess s^0, a sequence of approximations $s^1, s^2, \ldots, s^m, \ldots,$ to a solution s^* of (8.1) by

$$s^{m+1} := s^m + \xi \tag{8.3}$$

where ξ solves the linear system

$$J(s^m)\, \xi = -\mathbf{f}(s^m) \tag{8.4}$$

and J is the $n \times n$ Jacobian matrix defined in (8.2). We may view this process as follows: The linear system (8.4) defines a direction vector $\xi \in \mathbf{R}^n$, called the *Newton direction*, and then in (8.3) a unit step in this direction is taken in order to move from the current point s^m (in $\mathbf{R}^n$) to the next point s^{m+1}.

If s^0 is not sufficiently close to the isolated solution s^*, then two difficulties may arise. One is that stepping a full unit in the Newton direction may be too much. This is addressed in Section 8.1.1, where the effect of taking possibly smaller steps in the same direction is considered. The resulting *damped Newton* method is seen to have a larger domain of attraction under certain assumptions (which include the nonsingularity of the Jacobian matrix).

The second difficulty is that circumstances may arise when the Newton direction itself is not good; i. e., no step size in that direction leads to a point which is significantly closer to s^* than s^m is. This can be linked to the Jacobian's being effectively singular, as explained in Section 8.1.2. Note that, when attempting to find an isolated solution for a BVP using a reasonably good discretization, we may assume that $J(s^*)$ is nonsingular. However, in some applications Jacobian matrices do appear which are effectively singular, when s^0 is away from s^*. Altering the Newton direction is considered in Section 8.1.2. It turns out that when the matrix J is sparse, this is a tougher task than merely damping the Newton step size.

Before proceeding, we observe that Newton's method is *affine invariant*; i. e., if

$$\hat{\mathbf{f}}(s) = B\,\mathbf{f}(s)$$

where B is a constant, nonsingular matrix, then Newton's method yields precisely the same sequence of iterates for the two functions $\mathbf{f}$ and $\hat{\mathbf{f}}$. (Note that $\mathbf{f}$ and $\hat{\mathbf{f}}$ have the same zeroes.) The Newton-Kantorovich Theorem 2.64 can also be stated in an affine invariant form (see Exercise 1).

8.1.1 Damped Newton

The idea here is to control the step size which is taken in the Newton direction. Thus, with ξ given as the solution of (8.4), we modify (8.3) to read

$$s^{m+1} := s^m + \lambda \xi \tag{8.5}$$

with $0 < \lambda \leq 1$. Of course, both λ and ξ depend on the iteration counter m, but when considering a typical iteration we suppress this dependence for notational simplicity. Thus, we sometimes write (8.4), (8.5) as

$$\xi := -J(s)^{-1} f(s) \tag{8.6a}$$

$$s := s + \lambda \xi \tag{8.6b}$$

The immediate concern is how to choose the *damping factor* λ. Indeed, what is meant by saying that s^{m+1} "improves upon" s^m when we do not know s^* ?

For this purpose we introduce an *objective function*

$$g(s) : \mathbf{R}^n \to \mathbf{R}$$

which relates to some norm of the residual $\mathbf{f}(s)$. The square of the 2-norm is particularly convenient because of its differentiability, so one common choice is

$$g(s) := \frac{1}{2} |\mathbf{f}(s)|_2^2 \equiv \frac{1}{2} \sum_{j=1}^n f_j(s)^2 \tag{8.7}$$

We will use this norm, omitting the subscript 2 for notational convenience, for the rest of this chapter.

An objective function is required to satisfy two conditions:

(a) $g(s) \geq 0$ and $g(s^*) = 0$ iff $\mathbf{f}(s^*) = \mathbf{0}$.

Thus s^* minimizes $g(s)$ if $\mathbf{f}(s^*) = \mathbf{0}$, and we say that one value of λ is *better than* another if its corresponding value of $g(s^m + \lambda \xi)$ is smaller.

(b) Newton's direction is a *descent direction* (or a *downhill* direction) with respect to $g(s)$; i. e.,

$$\xi^T \nabla g < 0$$

where $\xi = \xi(s)$ is defined in (8.6a) and ∇g is the gradient,

$$\nabla g(s) := \left[\frac{\partial g}{\partial s_1}, \frac{\partial g}{\partial s_2}, \ldots, \frac{\partial g}{\partial s_n} \right]^T$$

It is easy to see that these conditions are satisfied by the objective function g defined in (8.7). In particular, since

$$\nabla g = J^T \mathbf{f}$$

the function g satisfies

$$\nabla g^T \xi = -2g < 0 \tag{8.8}$$

(the strict inequality unless $s = s^*$). The significance of having a descent direction is that, for $\lambda > 0$ sufficiently small, Taylor expansion gives

$$g(s^m + \lambda \xi) = g(s^m) + \lambda \xi^T \nabla g(s^m) + O(\lambda^2 |\xi|^2) < g(s^m) \tag{8.9}$$

This means that the next iterate s^{m+1} given by (8.5) can be required to satisfy

$$g(s^{m+1}) < g(s^m) \tag{8.10}$$

i. e., we may require a *monotonic decrease* toward the solution s^*.

We may now consider the problem of minimizing $g(s)$. Proceeding downhill from the point s^m in the Newton direction ξ, the question is to decide which point along the line $s^m + \lambda \xi$ should be chosen as the next iterate. Determining $\min_\lambda g(s^m + \lambda \xi)$, called *exact line search,* is too expensive, indeed often impossible in practice. Instead, we perform a *weak line search,* where we determine an acceptable value of λ, using only a few evaluations of $g(s)$.

An *acceptable* value of λ should basically satisfy (8.10). However, to prevent the method from stalling, we require a slightly more significant reduction in the value of $g(s)$ as follows: Let

$$\phi(\lambda) := \frac{g(s^m + \lambda \xi) - g(s^m)}{\lambda \xi^T \nabla g(s^m)} \tag{8.11}$$

and let σ be a fixed parameter with $0 < \sigma < 1/2$ (e.g., $\sigma = 10^{-2}$). Initially, a test value of λ is accepted if

$$\phi(\lambda) \geq \sigma \tag{8.12a}$$

This requirement ensures that there is a sufficient decrease in g (see Fig. 8.1). If the test (8.12a) fails then we seek a smaller λ satisfying both (8.12a) and

$$\phi(\lambda) \leq 1 - \sigma. \tag{8.12b}$$

This latter requirement insures that λ is not too small (see Fig. 8.1 again); it is also necessary for theoretical purposes. For an objective function satisfying (8.8), e.g., the one given in (8.7), conditions (8.12) thus imply that an acceptable λ should satisfy

$$g(s^{m+1}) \leq (1 - 2\lambda \sigma) g(s^m) \tag{8.13a}$$

$$g(s^{m+1}) \geq (1 - 2\lambda(1 - \sigma)) g(s^m) \tag{8.13b}$$

Standard results from the theory of unconstrained minimization yield

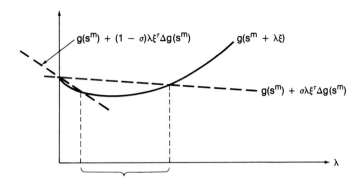

Figure 8.1 An acceptable range for λ

Theorem 8.14 Suppose that for a given starting point s^0, $g(s)$ has continuous second derivatives and is bounded below on the level set $D := \{s \in \mathbf{R}^n : g(s) \leq g(s^0)\}$. Further, assume that there is a constant $\delta > 0$ such that for each $s \in D$ the iterative method produces a direction satisfying

$$\nabla g^T \xi \leq -\delta |\nabla g| \, \|\xi\| \tag{8.14}$$

Then the following hold:

(a) For each $s \in D$ either $\nabla g(s)^T \xi = 0$ or there is an acceptable value of λ.

(b) A sequence of iterates $\{s^m\}$ generated using acceptable values of λ is well-defined, remains in D, and has at least one limit point there. For any such limit point, $\nabla g^T \xi = 0$. □

This theorem is stated for a general objective function and iterative method. To use it for our specialized purposes, note first that any objective function we consider is bounded below (by 0). Also, if (8.8) holds then $\nabla g(s)^T \xi = 0$ implies that $s = s^*$ solves our problem (8.1). Moreover, for the specific objective function given by (8.7), the condition (8.14) holds with $\delta := \max \{1/\text{cond}(J) ; s \in D\}$, because

$$\delta |\nabla g| \, \|\xi\| = \delta |J^T \mathbf{f}| \|J^{-1}\mathbf{f}\| \leq |\mathbf{f}|^2 = -\nabla g^T \xi$$

This gives

Corollary 8.15. Let the objective function g be defined by (8.7). Suppose that $\mathbf{f}$ has bounded second derivatives and that $\text{cond}(J)$ is uniformly bounded in a domain D containing the iterates and one root s^*. Then starting from any point s^0 in D, a damped Newton algorithm which proceeds using acceptable values of λ in (8.5) converges to s^*. □

We note the *global* nature of the convergence statement. Thus, no assumption about the size of $g(s^0)$ is made, in contrast to the assumptions needed for the Newton-Kantorovich Theorem 2.64.

Let us now give a practical procedure for finding, at a point $s = s^m$, an acceptable damping factor $\lambda = \lambda_m$ satisfying (8.12a). We begin with a predicted value for λ_m; e. g., we attempt to use a full Newton step, $\lambda_m = 1$. If (8.12a) is satisfied with this value, then we are done; otherwise, we reduce it. A simple procedure is to repeatedly set $\lambda_m := \rho \lambda_m$ for some $0 < \rho < 1$, until (8.12a) holds. A disadvantage of this procedure is that if ρ is chosen small, then it may yield unnecessarily small values for λ_m and hence slow down convergence, while if ρ is chosen large (too close to 1) then it may require many function evaluations of $g(s)$ to find an acceptable damping value. A more sophisticated approach is to approximate the function $g(s^m + \lambda \xi)$ by a quadratic $\psi(\lambda)$ satisfying

$$\psi(0) = g(s^m), \quad \psi'(0) = \xi^T \nabla g(s^m), \quad \psi(\lambda_m) = g(s^m + \lambda_m \xi) \tag{8.16}$$

[i. e., we interpolate those values of g which are needed anyway to compute $\phi(\lambda)$ in (8.11)]. This quadratic has a minimum at

$$\lambda = \frac{-\lambda_m^2 \, \psi'(0)}{2\,(\psi(\lambda_m) - \psi(0) - \lambda_m \psi'(0))} \leq \frac{\lambda_m}{2(1-\sigma)} \tag{8.17}$$

where the inequality holds because (8.12a) is violated by λ_m (Exercise 2). The value of λ in (8.17) is taken as the new value for λ_m. Then we repeatedly test for satisfaction of (8.12a), reducing λ if needed, using (8.16) and (8.17), until (8.12a) holds. Note that for each *correction* of λ only one evaluation of g is needed, and that the inequality in (8.17) ensures that only a few such corrections are ever needed.

When (8.12a) is satisfied, usually (8.12b) is satisfied as well. If not, then there are techniques to increase λ until both conditions (8.12) are satisfied. But we shall not describe them here, because they will not be necessary in our context.

It can be shown that termination of the resulting algorithm for selecting λ_m subject to (8.12) is guaranteed, so in theory convergence occurs when the conditions of Corollary 8.15 are satisfied. But in practice there are some shortcomings in the outlined method, and certain improvements prove useful.

First, note that if some components of the variable **s** are out of scale compared to others (e. g., because they correspond to poorly scaled dependent variables in the underlying BVP), then the norm of **s** or ξ may virtually ignore the smaller components. Thus, we generally have to consider the variable $D\mathbf{s}$, where D is a diagonal scaling matrix. A simple choice is $D = \operatorname{diag}\{|s_1^0|, \ldots, |s_n^0|\}$. In the sequel we assume that **s** is already reasonably well-scaled and thus avoid cumbersome notation.

Another difficulty with the objective function g of (8.7) is that it is sensitive to rescaling of the *equations* (8.1). Thus, if

$$\hat{\mathbf{f}}(\mathbf{s}) = B\,\mathbf{f}(\mathbf{s})$$

where B is a constant, nonsingular matrix, then the objective function $g(\mathbf{s})$ of (8.7) depends on the matrix B, even though Newton's method itself is affine invariant, as noted earlier.

Difficulty is also seen to occur from condition (8.14) when the Jacobian is ill-conditioned. In such a case δ is very small and the objective function loses its sensitivity to improvements with respect to the problem (8.1) [viz., we see from (8.9) and (8.14) that $g(\mathbf{s})$ is nearly constant near $\mathbf{s}^m$]. We must not forget that, even though the problem of solving (8.1) has been imbedded in a more general minimization problem, it is still the solution of the nonlinear equations (8.1) which we are after. Thus we are led to look for another objective function $g(\mathbf{s})$, one which hopefully overcomes the above difficulties.

It is easy to see that for any fixed, nonsingular matrix M, the function

$$g(\mathbf{s}) := \frac{1}{2}|M\,\mathbf{f}(\mathbf{s})|^2 \tag{8.18}$$

can serve as an objective function according to the two criteria specified earlier [following (8.7)]. In fact, g of (8.18) satisfies (8.8) as well (Exercise 3). A choice for M which leads to an affine invariant objective function is $M = J(\mathbf{s}^0)^{-1}$ [assuming that $J(\mathbf{s}^0)$ is nonsingular]. This gives

$$g(\mathbf{s}) := \frac{1}{2}|J(\mathbf{s}^0)^{-1}\mathbf{f}(\mathbf{s})|^2 \tag{8.19}$$

However, if $\mathbf{s}^0$ is far away from the solution, then the scaling of $\mathbf{f}(\mathbf{s})$ by $J(\mathbf{s}^0)^{-1}$ may not resolve the difficulty arising when the Jacobian is ill-conditioned. A more sensitive objective function may be obtained if we let $g(\mathbf{s})$ vary from iteration to iteration. To distinguish this, we will call such a function a criterion function and denote it by $g(\mathbf{s}^m + \lambda \boldsymbol{\xi})$ at the m th iteration. We define the *natural criterion function* as

$$g(\mathbf{s}^m + \lambda \boldsymbol{\xi}) := \frac{1}{2} |J(\mathbf{s}^m)^{-1} \mathbf{f}(\mathbf{s}^m + \lambda \boldsymbol{\xi})|^2 \qquad (8.20)$$

Note that at $\lambda = 0$,

$$g(\mathbf{s}^m) = \frac{1}{2}|\boldsymbol{\xi}|^2 \quad \text{and} \quad \boldsymbol{\xi}^T \nabla g(\mathbf{s}^m) = \boldsymbol{\xi}^T(-\boldsymbol{\xi}) = -2g(\mathbf{s}^m) \qquad (8.21)$$

Since the decomposition of the Jacobian matrix $J(\mathbf{s}^m)$ is done anyway to obtain the Newton direction $\boldsymbol{\xi}$, the evaluation of g for any given λ is not very expensive, requiring only forward and back substitutions.

The iteration control using this criterion function is indeed affine invariant, just like Newton's method itself. Moreover, by (8.21) Newton's direction is the direction of *steepest descent* with respect to g. Thus, (8.14) is satisfied with $\delta = 1$, even when the Jacobian is ill-conditioned. For all these advantages, the theoretical price to be paid is that Theorem 8.14, and hence Corollary 8.15, are no longer applicable, because the natural criterion function in (8.20) is redefined at each step. Moreover, Example 8.2 shows that the method can *cycle;* i. e., a sequence of iterates $\mathbf{s}^0, \mathbf{s}^1, \ldots, \mathbf{s}^m$ may be generated which are all acceptable using this criterion function, yet $\mathbf{s}^m = \mathbf{s}^0$. Still, it has proven quite useful in BVP applications, and we adopt it for the algorithm which follows.

For a practical application of the damped Newton method with a natural criterion function, we introduce two additional parameters. The first, $\lambda_{\min}$, is a minimum value for the damping factor λ. If the value of λ obtained by (8.17) is below $\lambda_{\min}$, then we have detected an "effective singularity" of the Jacobian (see Section 8.1.2) and, rather than letting the process possibly stall, declare "no convergence" and exit. This effectively eliminates the need to check condition (8.12b), and we may concentrate on (8.12a) alone. The second parameter, τ, limits the amount of change between values of λ and λ_m in (8.17). The reason is that if the change is too large, then we do not trust our quadratic model ψ and hence not its minimum location λ.

Selecting an acceptable damping factor λ can be viewed as a *predictor-corrector* process. The corrector part has been described above. For a predictor, a number of choices can be made. Let us denote this first try for λ_m by $\lambda_m^{(0)}$. The simplest choice is $\lambda_m^{(0)} := 1$. This is normally used for $m = 0$, but for later Newton iterates an attempt can be made to make a better prediction of $\lambda_m^{(0)}$ based on information from previous iterations. For instance, we can choose

$$\lambda_m^{(0)} := \begin{cases} \lambda_{m-1} & \text{if } \lambda_{m-1} < \lambda_{m-2}(1-\sigma) \\ \min(1, 2\lambda_{m-1}) & \text{otherwise} \end{cases} \qquad (8.22a)$$

A more sophisticated choice, considered in Exercise 4, is connected to the natural criterion function: With

$$\mu := \frac{|\xi^{m-1}|}{|\xi^m - J(s^{m-1})^{-1} f(s^m)|} \lambda_{m-1}$$

choose

$$\lambda_m^{(0)} := \max(\lambda_{\min}, \min(\mu, 1)) \tag{8.22b}$$

We are now ready to give a specific damped Newton algorithm.

Algorithm: Damped Newton method

Input: An initial guess for **s**, an iteration limit *limit* and a convergence tolerance *TOL*.

Output: A solution vector **s**, an iteration counter *iter* and a flag indicating "convergence" or "no convergence." If convergence has been detected, then **s** is the correct solution to the extent that the difference between two consecutive Newton iterates is at most *TOL*, and *iter* Newton iterations were required, *iter* ≤ *limit*.

$iter := 0;\ \sigma := 0.01;\ \lambda_{\min} := 0.01;\ \tau := 0.1$
{Nonlinear iteration control}
REPEAT
  {Find Newton direction and predict λ, e.g., using (6.22b)}
  Solve the linear system (8.4) for ξ; calculate $g_0 := g(s) = \frac{1}{2}|\xi|^2$
  Predict λ;
  {Find an acceptable damping factor, or detect no convergence}
  REPEAT
    $\hat{s} := s + \lambda \xi;\quad g_\lambda := g(\hat{s})$
    {Test acceptance; if not, correct damping factor
       using (8.16) and (8.17)}
    IF $g_\lambda \le (1 - 2\lambda\sigma) g_0$ THEN
      accept λ
    ELSE
      $\lambda := \max\left[\tau\lambda,\ \dfrac{\lambda^2 g_0}{(2\lambda - 1) g_0 + g_\lambda}\right]$
      IF $\lambda < \lambda_{\min}$ THEN signal no convergence and EXIT
  UNTIL λ is accepted (or no convergence detected)

  {Update and check convergence}
  $iter := iter + 1$
  $s := \hat{s}$
  IF $|\xi| \le TOL$ THEN signal convergence and EXIT
UNTIL iteration *limit* is reached
{Set flag and exit}
IF iteration limit has been reached THEN signal no convergence   □

Before remarking further about this method, we see how it can perform in a practical situation.

Example 8.1.

Consider the BVP

$$u'' + \frac{2}{x}u' + [u - (1 + \frac{2}{x^2})]u = 0 \qquad 0 < x < \infty$$

$$u \in C[0, \infty), \qquad u(\infty) = 0$$

This problem arises in electromagnetic self-interaction theory. The BVP has the trivial solution $u \equiv 0$, but we are after a nontrivial one.

We have chosen to consider this somewhat nonstandard problem in order to emphasize at the start that preparation of the BVP before applying a numerical method can sometimes save a lot of grief later on. Here, handling the ODE at $x = 0$ is a nontrivial matter, but it simplifies upon using the change of dependent variable $v := \frac{u}{x}$. We obtain the ODE (verify!)

$$v'' + \frac{4}{x}v' + (xv - 1)v = 0 \qquad (8.23a)$$

For this ODE, to control the regular solution behaviour at $x = 0$, the natural condition to impose is

$$v'(0) = 0 \qquad (8.23b)$$

Next, consider the solution behaviour as $x \to \infty$. The solution v is required to decay to 0 faster than $\frac{1}{x}$. Thus, the ODE (8.23a) for x large is approximately $v'' - v = 0$. This has fundamental solution modes e^x and e^{-x}, of which the increasing mode must be effectively eliminated by the right end BC. In practice we replace the infinite interval $[0, \infty)$ by a finite one, $[0, L]$ for a large enough L. To obtain the desired solution decay we may therefore impose the condition

$$v(L) + v'(L) = 0 \qquad (8.23c)$$

which is satisfied by the decaying fundamental mode. (See Section 11.4 for more general comments on how one chooses BC for problems over infinite intervals.)

Having thus prepared the nonlinear BVP (8.23), we are ready to solve it numerically. We apply collocation at four Gaussian points per interval on a uniform mesh of five subintervals. The initial approximation

$$v := \begin{cases} 2, & x \leq 1.5, \\ 2e^{1.5-x}, & x > 1.5 \end{cases}$$

leads, upon convergence, to the desired nontrivial solution. We attempt to solve (8.23) for $L = 10$ and for $L = 20$.

For $L = 10$, no difficulties are encountered. The full Newton method converges in four iterations, and both criterion functions (8.7) and (8.20) decrease monotonically and rapidly.

For $L = 20$, some convergence difficulties are encountered. The damped Newton algorithm described above uses the following relaxation factors: After a step with $\lambda_0 = 1$, $\lambda_1 = 1$ is predicted. But this yields an increase in g of (8.20), so it is successfully corrected to $\lambda_1 = 0.37$. Next $\lambda_2 = 0.11$ is predicted and, failing again to achieve a decrease in g, it is successfully corrected to $\lambda_2 = 0.03$. The next predicted relaxation factor is $\lambda_3 = 0.0023$, and no convergence is declared after four iterations. Then we double the mesh size [i. e., halve the step size, in the hope that the finer mesh would produce a better conditioned problem (8.1), "closer" to the underlying BVP] and continue. A relaxation factor of 0.02 now yields a decrease in g. The next predicted factor is $\lambda = 0.80$. This yields a decrease in g of (8.20) from 0.11 to 0.02, but the criterion function of (8.7) increases from 0.006 to 0.04. Relying on (8.20) and not on (8.7), the step is accepted. After four more full Newton iterations convergence is obtained.

When we attempt to repeat the computation for $L = 20$ using the full Newton iteration without control or damping, the method does not converge: After 40 fruitless iterations we have doubled the number of mesh subintervals as before, from 5 to 10, and continued for 40 additional Newton iterations without convergence. □

Remarks

(a) An important use of the damped Newton iteration, as it turns out, is not necessarily in making the method converge where the full Newton iteration fails (even though we would like that to happen too, of course). A major practical disadvantage of Newton's method is that when it fails, it may take many iterations to discover this. In some cases overflow may unpleasantly terminate the calculation altogether. It is desirable that when a method fails, the failure be detected quickly and elegantly. This is achieved by the damped Newton algorithm (especially when a good prediction for λ is used), and a continuation process may be promptly started (see Section 8.3 below).

(b) Another property of the full Newton iteration is that when there is more than one solution, it may unexpectedly converge to one which is not closest to the initial guess. Thus the user may get a solution, but it is not clear which one, or the obtained solution may not be the desired one. For instance, if the BVP models some physical process, then we would like to obtain a physically meaningful solution, and not every solution to the BVP is necessarily physically meaningful. Another potential difficulty is that the discretized problem (8.1) can have solutions which do not correspond to any solution of the BVP. (These are sometimes called *ghost solutions*.) Yet another instance is Example 8.1 above, where we do not want the trivial solution. The damped Newton iteration, since it is more controlled, gives less room for erratic behaviour. Still, for a closer control of the numerical process in a multiple solution case, we can turn to continuation, as described in Sections 8.3 and 11.5. □

We have mentioned above that the natural criterion function, which has proven useful in practice, is not without flaws. In particular, the damped Newton method with this criterion function may cycle. Thus, an endless sequence of iterates may be generated with a significant reduction in the "objective" value between each two, while no real improvement is being made. Such cases appear to be rare in practice, but we show here that they may indeed occur.

Example 8.2

Consider a function $\mathbf{f}(s)$ of two variables which satisfies

$$\mathbf{f}(\mathbf{0}) = -\frac{1}{10}\begin{bmatrix} 4\sqrt{3} - 3 \\ -4\sqrt{3} - 3 \end{bmatrix} =: -\mathbf{a}, \qquad J(\mathbf{0}) = I$$

$$\mathbf{f}(\mathbf{a}) = \frac{1}{5}\begin{bmatrix} 4 \\ -3 \end{bmatrix}, \qquad J(\mathbf{a}) = \begin{bmatrix} \frac{1}{\sqrt{3}} & \frac{-1}{\sqrt{3}} \\ 1 & 1 \end{bmatrix}$$

Starting from $\mathbf{s}^0 = \mathbf{0}$ we obtain

$$\xi^0 = -\mathbf{f}(\mathbf{s}^0) = \mathbf{a} = \mathbf{s}^1$$

The full Newton step is acceptable because

$$|J(\mathbf{s}^0)^{-1}\mathbf{f}(\mathbf{s}^1)| = |\mathbf{f}(\mathbf{a})| = 1 < |\mathbf{a}| = |\mathbf{f}(\mathbf{s}^0)|$$

The next Newton step yields

$$\xi^1 = -J(\mathbf{s}^1)^{-1}\mathbf{f}(\mathbf{s}^1) = -\mathbf{a}$$

(verify!), so

$$\mathbf{s}^2 = \mathbf{s}^1 + \xi^1 = \mathbf{0} = \mathbf{s}^0 \ .$$

This step is also acceptable with the natural criterion function, because

$$|J(\mathbf{s}^1)^{-1}\mathbf{f}(\mathbf{s}^2)| = |-J(\mathbf{a})^{-1}\mathbf{a}| < 1$$

It may be easily verified that the more restrictive requirements (8.12) are satisfied here as well. □

8.1.2 Altering the Newton direction

The damped Newton scheme just presented works very well in many practical situations, but it breaks down, both theoretically and practically, if at some point $\mathbf{s}^m$ the Jacobian $J(\mathbf{s}^m)$ is effectively singular. Let us explain what we mean here by "effectively singular." If the Jacobian is singular, then the Newton direction is not even well-defined by (8.4). However, it is rare in actual computation to have a precisely singular Jacobian matrix. In general, one says that a matrix is effectively singular if its condition number is "too large." Our context here is more specific than that: We are dealing with the situation where the $O(\lambda^2|\xi|^2)$ term in (8.9) cannot be neglected, compared to the sum of the other two terms in the expression for $g(\mathbf{s}^m + \lambda\xi)$, for any λ, $\lambda_{\min} \le \lambda \le 1$. Thus, the utility of the Newton direction depends upon the Jacobian in relation with the nonlinearity of $\mathbf{f}$, namely the higher (second) derivatives of $\mathbf{f}(\mathbf{s})$. In Theorem 8.14 and Corollary 8.15, which characterize situations where the Newton direction is reliable, we assume that the nonlinearity is bounded, which allows us to concentrate on the Jacobian $J(\mathbf{s})$. If $J(\mathbf{s})$ has a nicely bounded inverse (hence $|\xi|$ is bounded) then we can in principle take λ to be sufficiently small so that the $O(\lambda^2|\xi|^2)$ term is indeed dominated by the other terms in (8.9); thus no effective singularity occurs under these conditions.

In order to realistically quantify effective singularity, it is convenient to use the natural criterion function. First we write [see (5.53c) and Exercise 4] a Taylor expansion for $\mathbf{f}(\mathbf{s} + \lambda \xi)$ with ξ as in (8.6),

$$\mathbf{f}(\mathbf{s} + \lambda \xi) = (1 - \lambda)\mathbf{f}(\mathbf{s}) + \int_0^\lambda [J(\mathbf{s} + t\xi) - J(\mathbf{s})]\xi \, dt \qquad (8.24)$$

Multiplying by $J(\mathbf{s})^{-1}$ and taking 2-norms, we obtain

$$|J(\mathbf{s})^{-1}\mathbf{f}(\mathbf{s} + \lambda \xi)| \leq (1 - \lambda)|\xi| + \frac{\omega}{2}\lambda^2 |\xi|^2 \qquad (8.25)$$

where

$$\omega(s, \lambda) \equiv \omega := \max_{0 < t \leq \lambda} \|J(\mathbf{s})^{-1}[J(\mathbf{s} + t\xi) - J(\mathbf{s})]\| / \left(t|\xi|\right) \qquad (8.26)$$

Note that $\omega \approx \|J^{-1}J'\|$, where J' is the directional derivative of $J(\mathbf{s})$ in the direction ξ. The quantity ω describes an interaction between $J(\mathbf{s})^{-1}$ and the variation of the Jacobian [the nonlinearity in $\mathbf{f}(\mathbf{s})$].

Now, from (8.25) and Exercise 3 the requirement (8.12a) for the criterion function $g^*(\mathbf{s}) := |J(\mathbf{s}^m)^{-1}\mathbf{f}(\mathbf{s})|$ is satisfied for an acceptable value of λ if

$$(1 - \lambda)|\xi| + \frac{\omega}{2}\lambda^2 |\xi|^2 \leq (1 - \lambda\sigma)|\xi|$$

or

$$\frac{\omega}{2}\lambda|\xi| \leq 1 - \sigma$$

Thus, we could claim an *effective singularity* if this condition is violated for $\lambda_{\min} \leq \lambda \leq 1$, i. e., if

$$\omega|\xi| > \frac{2(1 - \sigma)}{\lambda} \qquad \lambda_{\min} \leq \lambda \leq 1 \qquad (8.27)$$

or essentially if $\omega|\xi| > \frac{2}{\lambda_{\min}}$. (A reinterpretation of the damped Newton strategy as it relates to ω is given in Exercise 4.)

Let us assume now that $J(\mathbf{s}^m)$ is effectively singular, perhaps even singular. The Newton direction is not useful, and we ask for ways to alter it. Unfortunately, currently there does not seem to be any method known which satisfactorily deals with this situation when the Jacobian is large and sparse (with the particular sparseness structure arising in Chapters 4 and 5). There are, however, techniques available for the case when J is dense, and we briefly describe some below.

Let us remark at the outset that the situation dealt with here is radically different from the one occurring when $J(\mathbf{s}^*)$ is singular. Here, the effective singularity at $\mathbf{s}^m$ is assumed to happen far away from the solution $\mathbf{s}^*$ (where we still assume that J is well-conditioned), and all we wish to do is move away from the "hole" in an intelligent way. Moreover, there may not even *be* a solution to (8.4). Thus we are naturally led to generalize Newton's method by choosing ξ as a solution of the least squares minimization problem

$$\min |J(\mathbf{s}^m)\,\boldsymbol{\xi} + \mathbf{f}(\mathbf{s}^m)|^2 \qquad (8.28)$$

But the solution of (8.28) is not unique when J is singular, and we must generally be cautious and ensure that the length of the chosen $\boldsymbol{\xi}$ is not too large [also, recall (8.27)].

The simplest approach is to consider minimizing the objective function (8.7) directly, i. e., to imbed the nonlinear equation problem in a nonlinear least squares problem. The *steepest descent* method then suggests an alternative to the generalized Newton direction, namely choosing $\boldsymbol{\xi}$ as the direction of steepest descent

$$\boldsymbol{\xi} := -J^T \mathbf{f} \qquad (8.29)$$

at $\mathbf{s} = \mathbf{s}^m$, and then using a weak line search to determine the next iterate $\mathbf{s}^{m+1}$. This method also utilizes any sparseness structure that the Jacobian J may have. Unfortunately, it is well-known to be rather slowly converging. Our experience has been that where the damped Newton strategy fails, the steepest descent method is not of much help.

A more flexible strategy is then to take a combination of the Newton and the steepest descent directions. The Newton direction is first expressed in terms of a so-called *Gauss-Newton* step for the least squares problem minimizing (8.7). That is, if J has full rank then the normal equations for (8.28) give

$$\boldsymbol{\xi} = -(J^T J)^{-1} J^T \mathbf{f}$$

Then, for a rank-deficient J a combination of this and (8.29) is taken, yielding

$$\boldsymbol{\xi} := -(J^T J + \mu I)^{-1} J^T \mathbf{f} \qquad (8.30)$$

at $\mathbf{s} = \mathbf{s}^m$, with μ a positive control parameter. This is known as the *Marquardt* method. Note that a positive semidefinite matrix $J^T J$ becomes positive definite upon adding μI.

The big question regarding Marquardt's method is how to control the parameter μ. One effective approach is to use a *model-trust region,* where the solution of (8.28) is chosen subject to the requirement that its norm be not too large. Thus we require

$$|\boldsymbol{\xi}| \le \delta \qquad (8.31)$$

where δ is a control parameter representing the region in which the quadratic model, whose minimizer is given by solving (8.28), adequately models the objective function. Practical algorithms to control this latter parameter do exist. This approach also provides a natural way to decide when to alter the Newton direction, namely when (8.31) is violated by the Newton direction.

But if our strategy is to stick with the damped Newton method as much as possible, switching to altering the Newton direction only when $\lambda < \lambda_{\min}$ in the damped Newton algorithm, then we can proceed in a different way. Rather than requiring (8.31), we may attempt to *minimize* the length (norm) of $\boldsymbol{\xi}$ which solves (8.28). This gives

$$\boldsymbol{\xi} := -J^+ \mathbf{f} \qquad (8.32)$$

at $\mathbf{s} = \mathbf{s}^m$, where J^+ is the *pseudo inverse* of J [see (2.15), (2.44)]. This approach also provides an obvious extension to the natural criterion function, replacing J^{-1} by J^+ in (8.20). It has proven useful for BVPs when using single shooting or compactification (Section 4.3.4), although currently there is little theory to support the strategy.

For those discretizations of BVPs which yield large, sparse Jacobians, none of the above modifications of the Newton direction has yet proven to give significant improvement. The additional expense due to the loss of sparseness makes other approaches seem more useful. These include restarting the solution process with a better discretization mesh and, often more importantly, improving the initial guess as discussed in Section 8.3.

8.2 REDUCING THE COST OF THE NEWTON ITERATION

There are a number of ways in which Newton's iteration can be replaced by a cheaper substitute, given that there are no convergence difficulties as discussed in the previous section. We will consider two such iteration modifications, both mentioned in Section 2.3.3 for dense Jacobians but applied here to a large, sparse Jacobian of the form (4.37) or (5.17). The idea of keeping the Jacobian fixed for a few iterations generalizes directly and is considered in Section 8.2.1. The other idea, that of generalizing Broyden's method, is more involved and is discussed in Section 8.2.2.

One conceivable objection to Newton's iteration is that the required Jacobian, even if well-behaved, may be difficult or even impossible to evaluate. A case in point is Example 1.21, where the task of calculating the derivatives is unpleasant and error-prone (though certainly not impossible). This however is generally not a serious obstacle, because we can always use finite differences to approximate the Jacobian. Thus we write

$$\frac{\partial \mathbf{f}}{\partial s_j} \approx \frac{\mathbf{f}(\mathbf{s} + h\mathbf{e}_j) - \mathbf{f}(\mathbf{s})}{h} \tag{8.33}$$

where $\mathbf{e}_j$ is the jth unit vector and h is a small parameter, though not too small compared to the machine precision; e. g., $h \sim \sqrt{\varepsilon_M}$. The more careful choice of h is discussed in the references (see also Exercise 6).

It turns out that, unlike the modifications below, which slow down the convergence of Newton's method, usually the finite difference approximation to the Jacobian is essentially as good as the exact one, when done with sufficient care. The cost of one evaluation of J should not exceed $n + 1$ evaluations of $\mathbf{f}$ in any case, where n is the order of the underlying BVP.

8.2.1 Modified Newton

The idea of this method is very simple: Having evaluated and decomposed the Jacobian at some point, say $\mathbf{s}^0$, use it to generate more than one new iterate. Simplistically, this gives the iteration

$$\mathbf{s}^{m+1} := \mathbf{s}^m - J(\mathbf{s}^0)^{-1}\mathbf{f}(\mathbf{s}^m), \qquad m = 0, 1, 2, \ldots \tag{8.34}$$

Note that $\mathbf{s}^1$ coincides with the original Newton iterate and that the next direction generated at $m = 1$ coincides with the vector used to control the iteration in the Damped Newton Algorithm with the natural criterion function.

The scheme (8.34) is a stationary fixed point iteration; i. e., the approximation to the Jacobian does not get updated. The convergence properties are therefore given according to the Contraction Mapping Theorem 2.63. In particular, if there is convergence at all, then its rate is linear. If $\mathbf{s}^0$ is far away from $\mathbf{s}^*$ then there is no *a priori* reason in general to believe that the mapping will be contracting, i. e., that the iteration will converge. On the other hand, if $\mathbf{s}^0$ is close to $\mathbf{s}^*$ then the convergence, although linear, may be fast, being close to that of Newton's method itself. Because of the small computational effort per iteration [J is not reevaluated and the factorization of the matrix in (8.4) is reused], it is generally preferable to iterate, say, four times with (8.34) than to iterate once more with Newton's method.

Thus we see that the modified scheme (8.34) does not serve as an independent, robust iterative method. Rather, we use it in a given situation by proceeding only as long as convergence seems fast enough. For instance, we may require that

$$g(\mathbf{s}^{m+l+1}) \leq \rho g(\mathbf{s}^{m+l}), \qquad l = 0, 1, \ldots$$

where $\mathbf{s}^m$ is the most recent point where J has been evaluated and decomposed, and ρ is a constant ($0 < \rho < 1$) reflecting the linear rate of convergence of the fixed Jacobian iteration. For the criterion function g we may take (8.19) or (8.7). If this condition is not satisfied, then we switch to the (possibly damped) full Newton method. The choice of value for ρ depends upon the relative expense of the full and modified iterations, hence upon the cost of evaluating the Jacobian J. A reasonable all-purpose value is $\rho = \frac{1}{2}$, or $\rho = \frac{1}{4}$.

Consider now how a finite difference code might utilize these ideas. The nonlinear BVP is first discretized on a relatively coarse mesh, where full damped Newton iterations are not overly expensive. Then, in order to achieve a prescribed overall error tolerance, the discretization has to be made on finer and finer meshes. On each of these finer meshes, however, there is a good initial guess available by interpolating the solution obtained on the previous mesh. This is where the modified Newton iteration is used, allowing solution on each but the coarsest mesh with only a little more effort than needed for one Newton iteration.

Example 8.3

As a concrete example of the above, consider again the BVP (8.23) of Example 8.1. Suppose we wish to determine $v(0)$ to 6 digits. If a uniform mesh is used (and for $L = 10$ this is a reasonable choice), then about 40 mesh subintervals are needed to achieve this accuracy. But rather than iterating four times with Newton's method on a fixed mesh with $N = 40$, we first iterate four times with $N = 5$, then use (8.34) for $N = 10, 20$ and 40, using intermediate results to estimate and control (global) error. As a result, only one full Newton iteration on each of the denser meshes is necessary. Since the cost of an iteration is proportional to N, we have replaced a cost constant of $4 \times 40 = 160$ by one of $4 \times 5 + 10 + 20 + 40 = 90$ plus some overhead.

In certain tougher applications, some of which are listed in Section 1.2, a much greater factor of efficiency often results when this approach is used. □

8.2.2 Rank-1 updates

Let us recall (Section 2.3.3) that for a dense Jacobian, rank-1 updates (Broyden's method) can be applied in order to make both the Jacobian evaluation and the solution of the linear system (8.4) significantly cheaper. But for a large, sparse Jacobian of the form (4.37) or (5.17), no reliable quasi-Newton methods (like Broyden's) are known which retain the sparseness structure. In particular, the solution of the linear system corresponding to (8.4) cannot be done by simple updating of the inverse. Below we make a few comments on these updates. A more extensive treatment can be found in the references.

Rank-1 update techniques can be used, however, to construct an approximate Jacobian of the form (4.37), (5.17), by applying the scheme described in Section 2.3.3 to the individual blocks (see Exercise 7). While this makes the *construction* of the Jacobian cheaper, the techniques described in Chapter 7 still have to be used for the solution of the resulting linear system for ξ. Consequently, the utility of this modification is harder to assess than that in Section 8.2.1. Note that with the rank-1 update, the number of iterations needed for convergence increases (recall that the rate of convergence drops from quadratic to just superlinear). Therefore, the possible advantage of this modification over Newton's method depends on the application: It might be worthwhile if the Jacobian is very complicated, so that Jacobian evaluations form a major part of the total computational cost. This is more the case, for example, with those multiple shooting techniques which attempt to minimize the number of shooting points.

Rank-1 updates should generally be attempted only when the nonlinear problem is not too difficult. In a general-purpose code, some mechanism is needed to detect this and allow use of the updates. The sensitivity measures which are needed anyway for the efficient implementation of a damped Newton scheme can be used for this purpose.

8.3 FINDING A GOOD INITIAL GUESS

There are occasions in practice where the techniques of Section 8.1 are not sufficient to yield a solution of (8.1) from a given starting point s^0. That is to say, the initial guess for the solution of a given discretized BVP may not be sufficiently good to ensure convergence of these iterative methods, even when a solution s^* exists and $J(s^*)$ is well-conditioned.

An intuitive approach for improving the starting point s^0 of Newton's method (say) for (8.1) is to solve first a simpler "nearby" problem. The solution of this simpler problem is then used to produce an improved starting point for the original one. This idea readily generalizes to solving a *chain* of related problems, starting with a simple one and gradually increasing the difficulty, getting closer to s^* with each problem in the chain for which the iteration converges.

We have already seen an example of a chain of related problems in Section 8.2.1, in the form of discretizations of a given BVP on different meshes. Indeed, on all but the coarsest mesh, the starting point s^0 for solving the corresponding problem (8.1) was improved by interpolating the solution on the previous mesh. The efficiency gained was demonstrated in Example 8.3. Here, however, we are concerned not just with efficiency, but with the question of convergence itself for the iterative method on the initial mesh.

The general idea is to *imbed* the given problem (8.1) in a family of related problems

$$\mathbf{h}(\mathbf{s}; \tau) = \mathbf{0}, \qquad 0 \leq \tau \leq 1 \qquad (8.35)$$

where the problem

$$\mathbf{h}(\mathbf{s}; 0) = \mathbf{0} \qquad (8.36)$$

is easy to solve and where

$$\mathbf{h}(\mathbf{s}; 1) \equiv \mathbf{f}(\mathbf{s}) \qquad (8.37)$$

Assuming that there is a *smooth solution path* (or a *homotopy path*) $\bar{\mathbf{s}}(\tau)$, $\bar{\mathbf{s}} : [0, 1] \to \mathbf{R}^n$, defined by the identity

$$\mathbf{h}(\bar{\mathbf{s}}(\tau); \tau) \equiv \mathbf{0}, \qquad 0 \leq \tau \leq 1 \qquad (8.38)$$

we would like to move along this path, with the largest steps possible, from the known $\bar{\mathbf{s}}(0)$ solving (8.36) to the desired $\bar{\mathbf{s}}(1) = \mathbf{s}^*$ solving (8.1).

This is the process of *continuation*, discussed in more detail in Section 8.3.1. In Section 8.3.2 we consider a special continuation process arising naturally for certain applications from physical considerations.

8.3.1 Continuation

A number of basic questions arise from the general description above:

(a) How is the imbedding $\mathbf{h}$ in (8.35) to be chosen, given $\mathbf{f}$ in (8.1)?
(b) Given the solutions $\mathbf{s}^0, \ldots, \mathbf{s}^j$, of (8.35) for parameter values $0 = \tau_1 < \tau_2 < \cdots < \tau_j$, how does one determine
  (i) the next value τ_{j+1} where (8.35) is to be solved, and
  (ii) an initial guess for the iterative method for

$$\mathbf{h}(\mathbf{s}; \tau_{j+1}) = \mathbf{0} \qquad (8.39)$$

based on the available information?

In answering (a), note first that there are many ways to choose $\mathbf{h}$ for a given $\mathbf{f}$. One general, simple choice is to define, given any initial point $\mathbf{s}^0$ for (8.1),

$$\mathbf{h}(\mathbf{s}; \tau) := \mathbf{f}(\mathbf{s}) + (\tau - 1)\mathbf{f}(\mathbf{s}^0), \qquad 0 \leq \tau \leq 1 \qquad (8.40)$$

Clearly then (8.37) holds and $\mathbf{s}^0$ is the solution of (8.36). But not every eligible choice of $\mathbf{h}$ is a practically useful one; indeed, a good choice of $\mathbf{h}$ is *crucial* to the performance of the resulting algorithm. The choice (8.40) is not always very useful for situations where the methods of Section 8.1 fail, because $\frac{\partial \mathbf{h}}{\partial \mathbf{s}} = \frac{\partial \mathbf{f}}{\partial \mathbf{s}}$, so any relative ill-conditioning of the Jacobian still presents difficulties. Fortunately, in practice there often is a natural imbedding offered by the BVP itself.

Example 8.4

Consider Example 1.4, used also for Example 5.17. As noted before, the solution profile of (1.10c)–(1.10b) gets steeper as the Reynolds number R gets larger (see Fig. 5.4). For $R = 25$ the solution profile looks smooth, and this BVP is easy to solve. Thus we may consider the class of BVPs (1.10c)–(1.10b) using the Reynolds number as the imbedding parameter. [Some transformation of the range of R to τ in $[0, 1]$, with $R = 25$ mapped to $\tau = 0$, is done if we wish to conform to (8.35).] If it is required to solve this BVP for $R = 10^5$, say, then we can for instance use the solution for $R = 25$ as an initial guess for the BVP with $R = 10^3$, take the latter solution as the initial guess for the BVP with $R = 10^4$, and so on, until we reach $R = 10^5$.

If we use a general-purpose finite difference code for solving the BVPs, then the mesh too gets gradually updated to reflect the steepening solution profile. It turns out that when we use COLSYS for this particular example, the solution for $R = 10^4$ can be obtained from a fairly crude start, and then one continuation step is needed to reach $R = 10^5$. But using an intermediate continuation step to reach $R = 10^4$ may not be a bad idea anyway. □

Note that the continuation process described in Example 8.4 involves an interplay between nonlinear equation solving and mesh selection, while the description of continuation before assumes that the imbedding (8.35) of the nonlinear equations (8.1) involves fixed unknowns **s**, arising from a fixed discretization of a BVP on a given mesh. In many situations, it is desirable to modify the mesh (and hence the unknowns **s**) along with the continuation parameter τ. If the number of mesh points is required to remain fixed, then this can still be accommodated in the notation (8.35), but such a restriction is artificial. Thus, we are really following a different homotopy path each time the mesh is changed. It is not always obvious how to choose an initial approximation after a mesh is altered. If finite differences are used, interpolation of the previously obtained solution is natural, but the order of interpolation needed is less clear.

Note also that rather crude answers have been provided to the questions in (b) above: The determination of τ_{j+1} is described as a trial-and-error process, while the initial guess for (8.39) is simply $\bar{\mathbf{s}}^j$. In the sequel we give some more sophisticated answers to (b). Still, this relatively primitive continuation process can yield an astonishingly powerful tool, when attached to a good general-purpose BVP solver.

Example 8.5

Consider Example 1.19, which describes a swirling flow over an infinite disk. We solve this BVP for $n = 0.2$ and $s = 0.05$, replacing ∞ by a sufficiently large finite endpoint L. Starting from the initial guess $f = -x^2 e^{-x}$ and $g = 1 - e^{-x}$ [which satisfy the BC (1.40c)], obtaining convergence is extremely difficult. Thus we try continuation in s: For larger values of s the problem is easier to solve numerically. Moreover, the value of L sufficient to represent ∞ for a given accuracy tolerance grows as s decreases. Therefore, we rescale the independent variable for a given L by $x := \dfrac{x}{L}$ and perform continuation on the pair of parameters s and L.

Beginning with the above guess and using COLSYS with four collocation points and an initial mesh of 10 uniform subintervals, we solve the problem first for $s = 0.2$, $L = 60$. The obtained solution (and mesh) are used to start the code for $s = 0.1$, $L = 120$. This is repeated in order to solve for $s = 0.05$ and $L = 200$, obtaining the desired solution to 5 digits accuracy. The obtained solution is graphed in Fig. 8.2. □

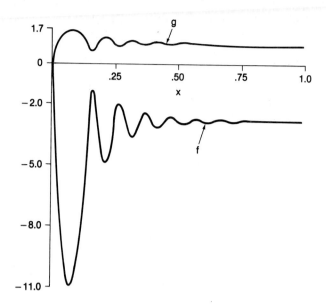

Figure 8.2   Solution profiles for Example 8.5

In Example 8.4 we have a natural parameter for continuation. In Example 8.5 the continuation parameter is a less obvious combination of two parameters appearing in the BVP. In the following example we demonstrate that sometimes the choice of a "natural" parameter for the imbedding may not be efficient at all, and a deeper analysis of the choice of imbedding parameter is rewarding.

Example 8.6

Consider Example 1.24 describing shallow cap dimpling, which we would like to solve for parameter values $\kappa = 1$, $\varepsilon = 10^{-4}$. First, we need to get an idea of the solution behaviour. Setting $\varepsilon = 0$ in (1.48) yields two possibilities, namely either $\phi(x) = 0$ or $\phi(x) = 2\phi_0(x)$, with $\psi(x)$ determined accordingly from (1.48b). It turns out, for reasons beyond the scope of this presentation, that an initial solution profile of interest is obtained by taking $\phi(x) = 2\phi_0(x)$ for $x < x_t$ and $\phi(x) = 0$ otherwise, where x_t is determined by $P(x_t) = 0$. The function $\psi(x)$ determined so as to satisfy the reduced (1.48b) (i. e., with $\varepsilon = 0$) is continuous. We call the result the "limiting case solution."

It turns out that for the full BVP with $0 < \varepsilon \ll 1$, a solution profile with an interior layer roughly at $x = x_t$ may be expected. Thus, it seems reasonable to do continuation in ε, starting with the limiting case solution profile. Since the solution profile is steep around x_T [the point identified by $\phi(x_T) = \phi_0(x_T)$ and presumably occurring near x_t] and gets steeper the smaller ε is, we continue with decreasing values of ε. However, very small continuation steps are needed. The reason is that x_t is not a very good approximation to x_T when $\kappa = 1$. When one is using a code like COLSYS, the mesh refinement process adapts the mesh to be dense where the solution is steep, so difficulties arise because this region keeps changing (see Chapter 9). In contrast, when $\kappa = 1/\varepsilon$, x_t approximates x_T well and the continuation process proceeds with acceptable step sizes.

A deeper analysis of the BVP yields that $x_T \approx x_t(1 + \varepsilon\rho)$, where ρ is proportional to $\frac{\varepsilon}{\kappa}$. Thus, the interior layer does not move much (even though it may get steeper) if we continue with both ε and κ while keeping $\frac{\varepsilon}{\kappa}$ fixed! Performing continuation in this way using COLSYS, we are able to solve the BVP for small values of ε and $\kappa = 1$ with only a few continuation steps. The solution profiles are plotted in Fig. 8.3. □

Let us return now to the question of constructing a sequence of solutions along the homotopy path $\bar{s}(\tau)$ leading from the known $\bar{s}(0) = s^0$ to the sought $\bar{s}(1) = s^*$ [see (8.35)–(8.38)]. We wish to predict from $\bar{s}^j$ and τ_j a value $\hat{s}^{j+1}$ at τ_{j+1}, $0 \leq \tau_j < \tau_{j+1} \leq 1$. Then we turn to an iterative method like Newton's to obtain $\bar{s}^{j+1}$ which solves (8.39), using $\hat{s}^{j+1}$ as an initial guess. This is repeated for $j = 0, 1, \ldots$, until $\tau_{j+1} = 1$.

The *classical continuation method*, which has been used in Examples 8.4–8.6 above, is to take

$$\hat{s}^{j+1} := \bar{s}^j \qquad (8.41a)$$

(see Fig. 8.4a). A more involved scheme is the *continuation with incremental load* (see Fig. 8.4b).

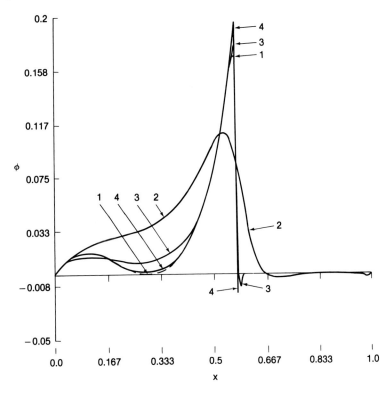

Figure 8.3 Solution profiles for Example 8.6

$$\hat{\mathbf{s}}^{j+1} := \overline{\mathbf{s}}^j - \Delta\tau_j H^{-1}(\overline{\mathbf{s}}^j;\tau_j)\mathbf{h}_\tau(\overline{\mathbf{s}}^j;\tau_j), \qquad \Delta\tau_j := \tau_{j+1} - \tau_j, \qquad (8.41\text{b})$$

with

$$H(\mathbf{s},\tau) := \frac{\partial \mathbf{h}}{\partial \mathbf{s}}(\mathbf{s};\tau) \qquad (8.42\text{a})$$

and

$$\mathbf{h}_\tau(\mathbf{s},\tau) := \frac{\partial \mathbf{h}}{\partial \tau}(\mathbf{s};\tau) \qquad (8.42\text{b})$$

Experience indicates that explicit, more sophisticated approximations than (8.41b) do not seem to be competitive in practice.

It is interesting to note the close connection between these continuation methods. Expanding

$$0 = \mathbf{h}(\overline{\mathbf{s}}^{j+1};\tau_{j+1}) \approx \mathbf{h}(\overline{\mathbf{s}}^j;\tau_{j+1}) + H(\overline{\mathbf{s}}^j;\tau_{j+1})(\overline{\mathbf{s}}^{j+1} - \overline{\mathbf{s}}^j)$$

yields

$$\overline{\mathbf{s}}^{j+1} \approx \overline{\mathbf{s}}^j - H(\overline{\mathbf{s}}^j;\tau_{j+1})^{-1}\mathbf{h}(\overline{\mathbf{s}}^j;\tau_{j+1}) \qquad (8.43)$$

and the right-hand side of (8.43) gives one Newton iteration based on the simple continuation (8.41a). Writing

$$\mathbf{h}(\overline{\mathbf{s}}^j;\tau_{j+1}) \approx \mathbf{h}(\overline{\mathbf{s}}^j;\tau_j) + \Delta\tau_j \mathbf{h}_\tau(\overline{\mathbf{s}}^j;\tau_j) = \Delta\tau_j \mathbf{h}_\tau(\overline{\mathbf{s}}^j;\tau_j)$$

we obtain from (8.43) a variant of continuation with incremental load

$$\hat{\mathbf{s}}^{j+1} := \overline{\mathbf{s}}^j - \Delta\tau_j H^{-1}(\overline{\mathbf{s}}^j;\tau_{j+1})\mathbf{h}_\tau(\overline{\mathbf{s}}^j;\tau_j) \qquad (8.41\text{c})$$

Finally, replacing $H(\overline{\mathbf{s}}^j;\tau_{j+1})$ by $H(\overline{\mathbf{s}}^j;\tau_j)$ we obtain (8.41b). The three variants differ from one another only by second-order terms in the corrections.

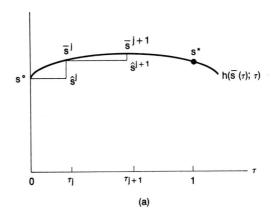

Figure 8.4a Classical continuation

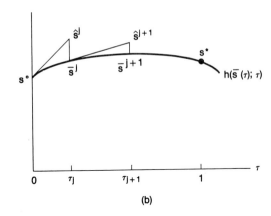

Figure 8.4b Continuation with incremental load

This also sheds light on the question of how to choose an imbedding $\mathbf{h}(\mathbf{s}; \tau)$: We want the Jacobian $H(\bar{\mathbf{s}}(\tau); \tau)$ not to be effectively singular even when $J(\mathbf{s}^0)$ is [e. g., we may want the corresponding $\omega|\xi|$ to be much smaller, cf. (8.26), (8.27)].

Consider now the problem of choosing the continuation step size $\Delta\tau_j$. One wants to choose it so as to ensure convergence of the local iterative method, be it the damped or the pure Newton iteration for (8.39), starting from $\hat{\mathbf{s}}^{j+1}$. A simple approach is to optimistically try $\Delta\tau_j := 1 - \tau_j$ (i. e., $\tau_{j+1} = 1$). If no convergence is detected, then $\Delta\tau_j$ is reduced by half, repeatedly if necessary, until the local iterative method converges. Note that we are assuming here that the imbedded problem (8.38) is parametrized well by τ, i. e., that the difficulty when moving from $\tau = 0$ to $\tau = 1$ increases gradually. Also, if we are using the pure Newton iteration locally, then we must have a device to detect nonconvergence. Recall that when using the damped method of Section 8.1.1 locally, the simple-minded criterion used for local nonconvergence was $\lambda < \lambda_{\min}$.

Another approach, with Newton's iteration as the local scheme, is to attempt to choose τ_{j+1} so as to insure that the (sufficient) conditions of the Newton Kantorovich Theorem 2.64 are satisfied, thus guaranteeing convergence to $\bar{\mathbf{s}}^{j+1}$. The bounds appearing in (2.64) can be rather pessimistic and difficult to estimate, though. Still, a number of efforts in this direction have met with some success.

Remark The concept of continuation as introduced above deals only with finding a solution to a difficult nonlinear system. But continuation may also be a useful tool when considered in a wider context. Often in practice we are not only given a natural imbedding of a BVP, but are also required to construct the homotopy path. For instance, in the semiconductor Example 1.15 we may be required to produce a curve relating the current $J = J_n + J_p$ as an output of the BVP for a given range of the voltage U. Clearly, in such a case we are not just interested in hopping from $\tau = 0$ to $\tau = 1$ with very large steps.

Continuation also arises naturally when considering multiple solutions for a given BVP and in questions of bifurcation. These applications are discussed in Section 11.5. □

8.3.2 Imbedding in a time-dependent problem

One particular way of imbedding the given nonlinear equations (8.1) in a family of related problems is to consider the vector of unknowns $\mathbf{s}$ as a function of a possibly artificial time variable t on the semi-infinite interval $0 \le t < \infty$, and derive an ODE in t for the homotopy path. The idea is natural when the BVP which gives rise to the problem (8.1) is the *steady state* limit of a partial differential equation which does indeed depend on time, i. e., when t is not artificial. We will consider this class of BVPs below, but first let us derive the ODE in t for the general case.

Given an imbedding (8.35) with $H(\mathbf{s}, \tau)$ of (8.42a) nonsingular in a relevant domain, we first imbed in a different parameter range. Thus we choose the transformation $\tau = 1 - e^{-t}$ and consider the imbedding

$$\mathbf{h}(\mathbf{s}; t) = \mathbf{0} \qquad 0 \le t < \infty \tag{8.44a}$$

[We use the same notation as before, but for $\mathbf{h}$, the Jacobian H, and the homotopy path $\bar{\mathbf{s}}(t)$ now functions of t.] Differentiating $\mathbf{h}(\bar{\mathbf{s}}(t); t)$ along the homotopy path (8.38) yields

$$\frac{d\mathbf{s}}{dt} = -H(\mathbf{s}, t)^{-1} \mathbf{h}_t(\mathbf{s}, t) \tag{8.44b}$$

These are the so-called *Davidenko equations*. Starting from $\mathbf{s}(0) = \mathbf{s}^0$, the IVP, it is to be hoped, reaches an asymptotically stable steady state, with $\mathbf{s}(\infty) = \mathbf{s}^*$.

To reach specific conclusions about the usefulness of general imbeddings is extremely difficult. We therefore restrict the class of imbeddings considered so that the ODE (8.44b) is autonomous and the IVP can be written as

$$\frac{d\mathbf{s}}{dt} = M(\mathbf{s})\mathbf{f}(\mathbf{s}) \qquad t \ge 0 \tag{8.45a}$$

$$\mathbf{s}(0) = \mathbf{s}^0 \tag{8.45b}$$

A desired function $M(\mathbf{s}) \in \mathbf{R}^{n \times n}$ should satisfy the following:

(a) $M(\mathbf{s})$ is nonsingular for relevant points $\mathbf{s}$ [so any equilibrium point of (8.45a) solves (8.1)].

(b) The root of interest $\mathbf{s}^*$ is asymptotically stable and is an attractor of the solution of (8.45).

(c) This steady state solution of (8.45) can be efficiently found numerically. □

Finding such a suitable $M(\mathbf{s})$ in general remains a difficult task. Still, for some classes of problems there is a natural way to define one. In particular, there is a choice, considered next, which works when the damped Newton method works.

If the Jacobian $J(\mathbf{s})$ of (8.2) is not effectively singular then we may choose

$$M(\mathbf{s}) := -J(\mathbf{s})^{-1} \qquad (8.46)$$

Then an application of Euler's method (2.2) for solving the IVP (8.45) with stepsize λ_m is easily seen to be the damped Newton method,

$$\mathbf{s}^{m+1} = \mathbf{s}^m - \lambda_m J(\mathbf{s}^m)^{-1} \mathbf{f}(\mathbf{s}^m)$$

If we choose the step size $\lambda_m = 1$ in the ODE integration, then the original Newton's method is retrieved.

This provides an interesting new angle from which to view the damped Newton method. Note that at $\mathbf{s} = \mathbf{s}^*$

$$\frac{\partial}{\partial \mathbf{s}}[J^{-1}\mathbf{f}]_{\mathbf{s}=\mathbf{s}^*} = J(\mathbf{s}^*)^{-1} J(\mathbf{s}^*) = I$$

so the variational problem for the ODE $\dfrac{d\mathbf{s}}{dt} = -J(\mathbf{s})^{-1}\mathbf{f}(\mathbf{s})$ is simply

$$\frac{d\mathbf{s}}{dt} = -(\mathbf{s} - \mathbf{s}^*)$$

and a discretization of the latter using Euler's scheme with step size 1 yields $\mathbf{s}^*$ in one step. This corresponds to the rapid convergence rate of the full Newton's method when $|\mathbf{s}^m - \mathbf{s}^*|$ is small. (With $\lambda_m \neq 1$, Euler's method for the variational problem does not yield $\mathbf{s}^*$ in one step, corresponding to a drop in the convergence rate for the damped Newton method from quadratic to linear.)

But if $\mathbf{s}^m$ is not very close to $\mathbf{s}^*$, in the sense that $\dfrac{\partial}{\partial \mathbf{s}}[J^{-1}\mathbf{f}]$ is far from the identity matrix at $\mathbf{s} = \mathbf{s}^m$, then Euler's method may have both stability and accuracy difficulties for a large step size. The damped Newton strategy in Section 8.1.1 may therefore be interpreted as a clever way to march through regions of difficulty for this IVP. It is not clear how practically useful this interpretation is, though.

Remark The IVP (8.45) with the function M given by (8.46) can be obtained from the Davidenko equations for the imbedding (8.40): In terms of the parameter t this imbedding reads

$$\mathbf{h}(\mathbf{s}, t) = \mathbf{f}(\mathbf{s}) - e^{-t}\mathbf{f}(\mathbf{s}^0)$$

and substitution in (8.44b) verifies our claim. Note that along the homotopy path of this imbedding,

$$\mathbf{f}(\bar{\mathbf{s}}(t)) = e^{-t}\mathbf{f}(\mathbf{s}^0)$$

so the solutions $\bar{\mathbf{s}}^j$ satisfy (with $\Delta t_j := t_{j+1} - t_j$)

$$\mathbf{f}(\bar{\mathbf{s}}^{j+1}) = e^{-\Delta t_j}\mathbf{f}(\bar{\mathbf{s}}^j) = e^{-t_{j+1}}\mathbf{f}(\mathbf{s}^0)$$

A continuation step with incremental load (8.41b) to solve (8.44a) gives

$$\hat{\mathbf{s}}^{j+1} = \bar{\mathbf{s}}^j - \Delta t_j J(\bar{\mathbf{s}}^j)^{-1}\mathbf{f}(\bar{\mathbf{s}}^j)$$

and this is also an explicit Euler step for solving the Davidenko equation. Recall that for our original problem (8.1) this is just a damped Newton step with a damping factor

$\lambda = \Delta t_j$, and we again question whether the imbedding (8.40) is likely to help when the damped Newton strategy fails. □

Let us turn now to the case where (8.46) yields an ill-defined or unstable IVP in (8.45). The choice of the so-called *preconditioner* $M(\mathbf{s})$ becomes trickier, particularly if the requirement of numerical efficiency is to be respected. In order to keep using the forward Euler scheme (2.2), the eigenvalues of $\frac{\partial}{\partial \mathbf{s}}[M\mathbf{f}]$ should be negative and approximately equal to one another. A natural extension of (8.46) is to replace the inverse of the Jacobian by the pseudo-inverse $J^+(\mathbf{s})$, yielding (8.32) for Euler's scheme, but the sparseness of $J(\mathbf{s})$ is somewhat destroyed. Other choices are possible here, but we proceed to discuss only a subclass of BVPs for which the choice $M \equiv I$ is useful when Newton's method has difficulties.

In many applications there arise BVPs which describe the steady state of some physical process. That is, the full problem is indeed time-dependent, and the BVP describes that time when the transient component of the solution has died out. This occurs if the time-dependent problem is parabolic (or rather, its linearization is). In this case the steady state part of the operator is elliptic. Since the eigenvalues of an elliptic operator are (by definition) negative, the Jacobian of $\mathbf{f}(\mathbf{s})$ has eigenvalues with a negative real part if the discretization is fine enough.

Example 8.7

Consider the time-dependent partial differential equation(s)
$$u_t = u_{xx} + f(u, u_x), \qquad a < x < b, \qquad t > 0$$
where subscripts denote differentiation with respect to the indicated variable. This is supplemented by initial and boundary conditions. For simplicity, assume that $u(x,t)$ is given at the initial line $t = 0$ for all $a \le x \le b$, and that at the boundaries, $u(a,t) = 0$ and $u(b,t) = 0$ for all positive t. Suppose we now apply a discretization in x. Thus, a mesh $\{x_i\}_{i=0}^{n+1}$ partitioning $[a,b]$ into $n+1$ subintervals is constructed, and derivatives with respect to x in the differential equation are replaced by difference quotients, obtaining a relationship involving $\{u(x_i,t)\}_{i=0}^{n+1}$. For instance, the simple scheme described in Section 5.1 can be used on a uniform mesh. Incorporating the BC $u(x_0,t) = u(x_{n+1},t) = 0$, this system can be written as

$$\frac{d\mathbf{s}}{dt} = \mathbf{f}(\mathbf{s}), \qquad t \ge 0 \qquad (8.47)$$

where $s_i(t) \approx u(x_i, t)$, $i = 1, \ldots, n$. The ODE (8.47) is supplemented by initial values derived in an obvious way from the known $u(x, 0)$.

The ODE (8.47) is a special case of (8.45a) with $M \equiv I$. Note that if the discretization in space is done by a scheme with unknowns other than just u-values at mesh points (e. g., the collocation schemes of Section 5.6), then the variables $\mathbf{s}(t)$ assume a correspondingly different meaning. It is easy to verify that this IVP is indeed asymptotically stable for a sufficiently fine mesh. It is also stiff [i. e., there is a large spread in size among the eigenvalues of $J(\mathbf{s})$]. We leave the verification of these details to Exercise 8.

The approach of solving the initial-boundary value problem by first discretizing in the space variable x and then solving the IVP in time is called a *method of lines* approach. □

Consider now the stiff IVP for (8.47). The forward Euler method is not effective for stiff problems, and an implicit method must be used to discretize this IVP. Since we are more interested here in stability and simplicity than in accuracy, a backward Euler discretization suggests itself. This yields

$$\mathbf{s}^{m+1} = \mathbf{s}^m + \mu_m \mathbf{f}(\mathbf{s}^{m+1})$$

where μ_m is the step size. To solve this latter system approximately for $\mathbf{s}^{m+1}$, we use one damped Newton iteration with $\mathbf{s}^m$ as a first guess. (This is a standard procedure when solving stiff IVPs.) We obtain (check)

$$\mathbf{s}^{m+1} = \mathbf{s}^m + \lambda_m \mu_m [I - \mu_m J(\mathbf{s}^m)]^{-1} \mathbf{f}(\mathbf{s}^m) \qquad (8.48)$$

with λ_m and μ_m two positive control parameters. This iterative method can be interpreted as a special instance of a Marquardt iteration (8.30), but simpler: Since the eigenvalues of $-J$ are known to have nonnegative real parts, the effect of forming $I - \mu_m J(\mathbf{s}^m)$ is to "push" the matrix to be inverted away from singularity, and we do not need to form $J^T J + \mu I$.

The advantage of (8.48) is that the matrix $I - \mu_m J(\mathbf{s}^m)$ has the same sparseness structure as J itself. So, for the class of BVPs considered, we have succeeded in altering the Newton direction without increasing the cost of the iteration (c.f. Section 8.1.2). The disadvantage is that the convergence rate is only linear. [To hope for a higher rate we must keep M dependent on $\mathbf{s}$ in (8.45a).] A natural hybrid method is to use (8.48) first to obtain a coarse approximation of $\mathbf{s}^*$ and then to switch to Newton's method for refinements. These ideas enjoy considerable use in practice.

8.4 FURTHER REMARKS ON DISCRETE NONLINEAR BVPs

The nonlinear problem (8.1) with which we are concerned in this chapter arises from a discretization of a nonlinear BVP. One would therefore expect some characteristics of the BVP to be inherited by (8.1), at least when the discretization is sufficiently fine and stable. Such characteristics may play an important role in determining how successful the processes discussed in this chapter are. Thus, we remark here on several aspects of discretized nonlinear BVPs, even though this requires a substantial departure from the spirit of the rest of the chapter. We aim at a certain theoretical understanding and do not in general consider the relative practical importance of the points made.

We begin by recalling that, while in this chapter we have treated systems of nonlinear algebraic equations, the basic Newton's method can also be stated in the quasi-linearization version, discussed in Section 2.3.4, where linearization of the BVP precedes discretization. It is not difficult, though, to interpret the various techniques presented here in the quasilinearization context.

Generally, the problem (8.1) may be difficult to solve for various reasons. These include some ill-conditioning of the underlying BVP, and/or instability of the underlying discretization, and/or a discretization which is too coarse to do the BVP justice. But in this section we restrict the discussion to discretizations which are sufficiently fine and stable in the following sense: As for the continuous BVP, we define (locally) the stability (conditioning) constant of a nonlinear iteration as the stability (conditioning) constant of the corresponding linearized problem. Our assumption on the discretizations in this section is that the stability constants encountered are close to the conditioning constant of the underlying BVP itself [see, for instance, Theorem 5.38 or Chapter 6]. We distinguish this by calling (8.1) a *discrete nonlinear BVP*. This discrete BVP may still be difficult to solve, but we may use the connection to the BVP to shed some light on possible sources of difficulty.

In previous sections the effective singularity of the Jacobian has played an important role in the discussion of various approaches and methods. Now, recall from Section 2.8.2 that the matrix condition number χ^h of the Jacobian becomes unbounded as the discretization gets refined to the limit. It may therefore seem at first sight that any discrete nonlinear BVP would have effectively singular Jacobians everywhere. But a closer look reveals that, as before, the matrix condition number of $J(\mathbf{s})$ is a reasonably sharp bound only on roundoff error propagation. For the nonlinear convergence error, as for truncation errors, the smoothing effect of $J(\mathbf{s})^{-1}$ should be taken into account.

Thus consider a linearized BVP (2.73) and its corresponding discrete BVP, and assume for the moment that exact arithmetic is used (i. e., no roundoff error). The basic Newton correction $\boldsymbol{\xi}$ in (8.6) corresponds to $\mathbf{w}$ in (2.73) and solves (a fine discretization of) a linear BVP. Therefore we consider for (8.6), as in (2.166), the condition number

$$\chi(\mathbf{s}) = \|J(\mathbf{s})\|_{U \to V} \|J(\mathbf{s})^{-1}\|_{V \to U} \tag{8.49}$$

with $\boldsymbol{\xi} \in U$, $\mathbf{f}(\mathbf{s}) \in V$, and U is a subspace of V containing only appropriately smooth functions (see Section 2.8). Then $\chi(\mathbf{s})$ is bounded by a constant of moderate size if the linearized BVP is well-conditioned.

A number of interesting observations follow. First, we obtain a new interpretation for the natural criterion function: While the "usual" objective function (8.7) measures progress (or improvement) in V, the natural criterion function (8.20) [and also the objective function (8.19)] measure progress in the smoother space U, where the solution $\mathbf{s}^*$ lies as well.

Another observation is the so-called *mesh independence principle:* Consider a family of meshes with maximum step size $h \to 0$, on which a consistent one-step (say) finite difference scheme is applied. Since the stability constants $\|J(\mathbf{s})^{-1}\|$ and the condition numbers corresponding to (8.49) approach the corresponding values for the BVP as $h \to 0$, one can obtain that the convergence characteristics of the nonlinear iteration (the number of iterations needed) becomes eventually independent of the mesh. (This is true so long as the roundoff error can indeed be neglected.) We hasten to note that this result generally holds only when the meshes considered are already "sufficiently fine"; a counterexample for a rough mesh is given, e. g., by Example 9.9.

The conditioning of a linearized BVP depends, of course, on the function about which the linearization is performed (c.f. Section 3.3.4). We can expect $\|J(\mathbf{s})^{-1}\|$ to be very large if the (local) conditioning constant is very large. For instance, we should expect such difficulties if there is no "appropriate" dichotomy. A particular example where such a case occurs is described next.

Example 8.8

Consider the following BVP, which is a steady state form of Burgers' equation arising in fluid dynamics,

$$u'' + (u^2)' = 0, \quad 0 < x < \infty$$

$$u(0) = 1, \quad u(\infty) = 0$$

Converting to a first-order system, we obtain for $\mathbf{y} := (u, v)^T$,

$$\mathbf{y}' = \begin{bmatrix} -v \\ -2uv \end{bmatrix}$$

The exact solution for this simplified BVP is

$$u = \frac{1}{x+1}, \quad v = u^2$$

For quasilinearization at some point $\hat{\mathbf{y}} = (\hat{u}, \hat{v})^T$, we find

$$\mathbf{y}' = \begin{bmatrix} -\hat{v} \\ -2\hat{u}\hat{v} \end{bmatrix} + J(\hat{\mathbf{y}})(\mathbf{y} - \hat{\mathbf{y}})$$

where

$$J = \begin{bmatrix} 0 & -1 \\ -2\hat{v} & -2\hat{u} \end{bmatrix}$$

Now, if $\hat{\mathbf{y}}$ is chosen as the exact solution given above, then the ODE

$$\mathbf{z}' = J(\hat{\mathbf{y}})\mathbf{z}$$

can be easily seen to have a dichotomic solution space with one increasing and one decreasing mode. This corresponds to the exact solution being isolated, with the variational problem well-conditioned. But this is not necessarily the case for the linearization about any $\hat{\mathbf{y}}$.

Consider the initial approximation $\hat{\mathbf{y}}(x) = (1/(x+1), -1)^T$ (which is exact in the first component, but a poor guess in the second). It can be verified that both eigenvalues of $J(\hat{\mathbf{y}})$ have negative real part. Hence, we lack an initial condition at $x = 0$, and the corresponding linearized problem is ill-conditioned. □

EXERCISES

1. In the affine invariant form of the Newton-Kantorovich Theorem 2.64, assumptions (2.64b, c) are replaced by

$$\|J(\mathbf{s}^0)^{-1}(J(\mathbf{x}) - J(\mathbf{y}))\| \leq \omega |\mathbf{x} - \mathbf{y}| \quad \mathbf{x}, \mathbf{y} \in D$$

for a constant ω which subsequently replaces the product $\beta\gamma$ in (2.64d). Everything else remains the same in (2.64). Prove this theorem.

2. Show that if (8.12a) is violated for an objective function $g(\mathbf{s})$, then (8.17) holds for the quadratic interpolatory function for $g(\mathbf{s})$ defined by (8.16).

3. (a) Show that the objective function (8.18) satisfies

$$\boldsymbol{\xi}^T \nabla g(\mathbf{s}^m) = -2g(\mathbf{s}^m)$$

for any fixed, nonsingular matrix M.

(b) Moreover, the objective function
$$g(\mathbf{s}; M) := |M\mathbf{f}(\mathbf{s})|_2$$
satisfies
$$\boldsymbol{\xi}^T \nabla g(\mathbf{s}^m; M) = -g(\mathbf{s}^m; M)$$
so conditions (8.12a, b) read
$$g(\mathbf{s}^{m+1}; M) \leq (1 - \lambda\sigma)g(\mathbf{s}^m; M)$$
$$g(\mathbf{s}^{m+1}; M) \geq (1 - \lambda(1 - \sigma))g(\mathbf{s}^m; M)$$
[cf. (8.13)].

4. (a) Show that
$$\mathbf{f}(\mathbf{s} + \lambda\boldsymbol{\xi}) = (1 - \lambda)\mathbf{f}(\mathbf{s}) + \int_0^\lambda [J(\mathbf{s} + t\boldsymbol{\xi}) - J(\mathbf{s})]\boldsymbol{\xi}\, dt$$
where $\boldsymbol{\xi}$ is the Newton direction at $\mathbf{s}$. The integral here is an expression for the nonlinearity in $\mathbf{f}$.

(b) Next, for the natural criterion function $g^*(\mathbf{s}) := |J(\mathbf{s}^m)^{-1}\mathbf{f}(\mathbf{s})|_2$, obtain the following quadratic bound in λ:
$$g^*(\mathbf{s}^m + \lambda\boldsymbol{\xi}) \leq (1 - \lambda)g^*(\mathbf{s}^m) + \frac{1}{2}\lambda^2 g^*(\mathbf{s}^m)^2 \omega, \qquad 0 \leq \lambda \leq 1$$
where ω is given in (8.26).

(c) An expression for the relaxation factor λ may be obtained by assuming equality in (b) and minimizing the resulting quadratic. Show that this gives
$$\lambda = \frac{1}{\omega|\boldsymbol{\xi}|}$$
But an estimate of ω is needed:

(d) In a *prediction* step, information from the previous iteration $m - 1$ is used: Show that
$$|J(\mathbf{s}^{m-1})^{-1}[J(\mathbf{s}^m) - J(\mathbf{s}^{m-1})]\boldsymbol{\xi}^m| \leq \omega\lambda_{m-1}|\boldsymbol{\xi}^{m-1}||\boldsymbol{\xi}^m|$$
and that (8.22b) follows. In a *correction* step, we already have $g^*(\mathbf{s}^m + \lambda\boldsymbol{\xi})$ for some λ. Thus ω may be estimated from the equation in (b), and then used in (c) to estimate λ. Show that this is identical to using (8.16), (8.17) with $g = g^*$.

5. Consider the BVP (8.23). Using an available code, find $v(0)$ accurate to 6 digits. (Answer: 2.11997)

6. For $\mathbf{f}$ sufficiently smooth, show that
$$\mathbf{f}(\mathbf{s} + h\mathbf{e}_j) = \mathbf{f}(\mathbf{s}) + h\frac{\partial \mathbf{f}}{\partial \mathbf{s}_j} + \mathbf{r}_j(h)$$
where $|\mathbf{r}_j(h)| = O(h^2)$. Assuming that $|\mathbf{f}(\mathbf{s})| \sim 1$, show that for the approximation (8.33), the optimal choice of h (which minimizes both roundoff and truncation error) is given by $h \sim \sqrt{\varepsilon_M}$.

7. Consider the partial differential equation

$$\frac{\partial u}{\partial t} = \frac{\partial^2 u}{\partial x^2} + g(u,x,t)$$

$$u(x,0) = \phi(x), \qquad a \leq x \leq b$$

$$u(a,t) = \psi_1(t), \qquad u(b,t) = \psi_2(t)$$

where $\lim_{t \to \infty} \frac{\partial u}{\partial t} = 0$ and $\frac{\partial g}{\partial u} < 0$. Choosing a uniform mesh on $[a,b]$ and approximating $\frac{\partial^2 u}{\partial x^2}$ by centered differences, we obtain the ODE

$$\frac{d\mathbf{U}}{dt} = \mathbf{f}(\mathbf{U})$$

where $\mathbf{U}(t)$ approximates $u(x,t)$ values at the mesh points. Give arguments (general or detailed) showing that the Jacobian of $\mathbf{f}$ has eigenvalues in the range $\sim -1/h$ to ~ -1, where h is the step size in x. For h "small," if the resulting IVP is solved using the backward Euler method (say), what does this imply about the size of the time step, both initially and as t becomes large?

9

Mesh Selection

In the previous chapters we have discussed a variety of numerical methods whose success generally depends, in the limit, upon some form of a decoupling of fundamental solution components. Still, there are two further considerations which play a practical role in insuring that difficult problems can be solved at all with reasonable efficiency — solution of nonlinear equations and mesh selection. The first, relevant for nonlinear BVPs, was the topic of the previous chapter, and the second is discussed here.

All numerical methods for solving BVPs on a computer involve, at some stage, discretization of the differential equations on a mesh. The purpose of this chapter is to discuss the practical selection of such a mesh, with the objective of achieving a sufficiently accurate solution as inexpensively as possible. Choosing a good mesh is essential if a method is to be efficient for problems with solutions having narrow regions of rapid change (such as occur even for the simple linear BVP in Example 5.2). Indeed, recall also our discussion in Chapter 8 concerning the importance of a good initial guess to start the (discrete) nonlinear iteration for a nonlinear BVP, where a good initial guess means a satisfactory initial solution profile, discretized *on an appropriate mesh*. Since the computational effort for a given basic discretization method increases with the number of mesh points, we want (in principle) the coarsest possible mesh which still yields an approximate solution with an adequate error. This means, in particular, that the mesh must be sufficiently fine that there *exists* a solution to the discretized system, and that the discrete BVP adequately represents the continuous BVP.

The choice of a mesh should generally be based on two considerations — controlling discretization error and preserving stability. We shall loosely refer to the former as "accuracy" considerations. For initial value methods like multiple shooting, these two types of considerations are to some extent separated: The choice of multiple shooting points is primarily aimed at satisfying stability needs, while the fine mesh generated by the IVP solver between shooting points satisfies accuracy requirements. Since these considerations are discussed in Sections 2.7.4 and 4.2, respectively, in this chapter we consider only the finite difference techniques of Chapter 5.

9.1 INTRODUCTION

Consider the nonlinear BVP

$$\mathbf{y}' = \mathbf{f}(x, \mathbf{y}) \qquad a < x < b \tag{9.1a}$$

$$\mathbf{g}(\mathbf{y}(a), \mathbf{y}(b)) = \mathbf{0} \tag{9.1b}$$

where as usual we assume that there is an isolated solution $\mathbf{y}(x)$ that we wish to compute and that $\mathbf{f}$ and $\mathbf{g}$ are sufficiently smooth. Suppose that a basic method is used which, for a mesh π on $[a, b]$, gives a corresponding discrete or continuous approximate solution $\mathbf{y}_\pi(x)$. We address the following

Mesh Selection Problem Given a BVP (9.1) and an error tolerance TOL, find a mesh

$$\pi: a = x_1 < x_2 < \cdots < x_N < x_{N+1} = b \tag{9.2a}$$

with

$$h = \max_{1 \le i \le N} h_i, \qquad h_i = x_{i+1} - x_i \tag{9.2b}$$

such that N is "small" and the error in $\mathbf{y}_\pi(x)$ as an approximation to $\mathbf{y}(x)$ is less than TOL (in an appropriate norm, measuring absolute or relative errors). □

The notion of a "small" N is certainly more qualitative than quantitative. We can attempt to make this precise in at least two ways. One is to look for an *optimal* mesh, i.e., a mesh which yields a minimal mesh size N for a given error tolerance TOL. In practice, however, the exact solution of the resulting optimization problem turns out to be too expensive, if not impossible. Thus we settle for a quest for *good* meshes, i.e., meshes with N not much larger than the optimal one. To make the notion of a good mesh more quantitative, we could consider a sequence of tolerances and meshes and require that, as $TOL \to 0$, the mesh size $N = N(TOL)$ satisfies

$$N(TOL) \le C\ N^*(TOL)$$

with $N^*(TOL)$ the optimal mesh size for this error tolerance and $C \approx 1$ a constant independent of TOL.

A dual formulation for the mesh selection problem is often found useful. This formulation requires, given a mesh size N, to find a mesh which minimizes the error in some norm.

For a given BVP and numerical method, the appropriateness of the discretized system is directly determined by the mesh. Generally speaking, the mesh must be fine in regions where the desired solution changes rapidly but can be relatively coarse elsewhere, provided that the resulting scheme is sufficiently stable. To see this, let us consider an example.

Example 9.1

Suppose that one of the finite difference schemes of Chapter 5 is applied to a well-conditioned BVP. Recall that if K is the stability constant for a given mesh π [cf. (5.31c) or (5.49)] and $\tau_i[\mathbf{y}]$ is the local truncation error on $[x_i, x_{i+1}]$ [cf. (5.31a) or (5.47a)], then the error satisfies in the maximum norm

$$\|\mathbf{y}_\pi - \mathbf{y}\| \leq K \max_{1 \leq i \leq N} |\tau_i[\mathbf{y}]| \tag{9.3}$$

[cf. (5.34d) or (5.50)].

For simplicity, let us suppose that the BVP is linear, with bounded, slowly varying coefficients in the homogeneous part. For instance, consider

$$u'' = u + q(x), \qquad 0 < x < 1$$

$$u(0) = 0.9129, \qquad u(1) = 0.375$$

with $q(x)$ appropriately chosen to yield the curve of Fig. 9.1 for the solution u. Thus, the sharp variation in the solution profile is caused by the inhomogeneity. Then for virtually any mesh, the stability constant K in (9.3) is of moderate size, being close to the conditioning constant of the BVP. The inhomogeneity $q(x)$ is irrelevant to the stability constant here, but it affects the local truncation error, which involves higher-order derivatives of the solution. For definiteness, suppose that the trapezoidal scheme (5.21) is applied to the corresponding first-order system, yielding

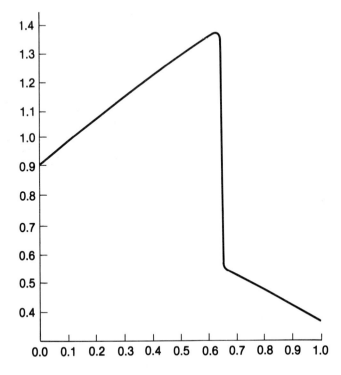

Figure 9.1 A solution profile

$$\tau_i[\mathbf{y}] = -\frac{1}{12}h_i^2 \mathbf{y}'''(x_{i+1/2}) + O(h_i^4)$$

If we wish to ensure that

$$|\tau_i[\mathbf{y}]| \leq \frac{TOL}{K}, \qquad 1 \leq i \leq N$$

then we must choose h_i small where $\mathbf{y}(x)$ varies rapidly (e. g., near $x = 0.63$ in Fig. 9.1) and we may choose h_i larger where $\mathbf{y}(x)$ varies slowly. This is the same conclusion that would have been reached if we were solving a piecewise polynomial approximation problem like interpolation or a quadrature problem for the same solution profile.

But more can happen when one is solving differential equations. Let us assume, more plausibly, that the entire ODE, not just a harmless inhomogeneity, is responsible for the region of rapid variation in the solution profile of Fig. 9.1. For instance, we may consider the BVP (1.35) of the shock wave Example 1.17 (whose solution for $\varepsilon = 10^{-6}$ is indeed what we have plotted in Fig. 9.1). The considerations regarding the control of the size of $\tau_i[\mathbf{y}]$ still hold, but in addition the stability question becomes important. Indeed, our mesh must be chosen such that a stable discretization is obtained, i. e., so that K in (9.3) is still of moderate size. BVPs like Example 1.17 are considered in Chapter 10. □

From this example we see that the choice of a mesh should generally be based on two considerations, namely, controlling the discretization error and preserving stability. The question of stability is rather difficult to handle quantitatively, so in practice it is usually ignored, with the hope that the basic discretization method is sufficiently robust to cover this up in most applications of interest. It remains then to control the discretization error, in principle as for IVPs (cf. Section 2.7.4). But there is one big difference in that for IVPs a *local* control is possible, while here we must consider a *global* strategy.

When attempting to analyze mesh selection, it is easy to see what we want: The role of mesh selection is to obtain a global solution to (9.1) as cheaply as possible for a given accuracy. Unfortunately, the mesh selection analysis is faced with an inevitable dilemma: A strategy is generally more successful if it utilizes the special properties of the particular numerical method being used, namely its error form. However, this error form is normally based upon an asymptotic analysis, and the very fact that mesh selection is important at all is often because one does not yet have a good numerical approximation. In particular, the approximation is *a priori* not yet in the asymptotic range. For a BVP like that in Example 9.1, if our discretization method is a good one in the sense that it allows us to take h large, then the usual theory of Chapter 5 does not hold because h is not "small" with respect to the size of the coefficients of the ODE. (This is also a difficulty for the solution methods for nonlinear equations in Chapter 8, but mesh selection is a newer and less developed field.)

The reader should resist the temptation to indiscriminantly consider mesh selection for the limiting case $h \to 0$, or even $N \to \infty$. For when $N \to \infty$, *all* quasiuniform meshes (i. e., all meshes satisfying

$$h \leq \bar{K} h_{\min}, \quad h_{\min} := \min_{1 \leq i \leq N} h_i \qquad (9.4)$$

for some fixed constant $\bar{K}$) yield an $O(N^{-s})$ error when a difference scheme of order s is used.

Thus, it is not too surprising that there is not much theoretical justification for the different strategies which have been used. Fortunately, some (but not all) strategies based upon asymptotic error formulas perform quite satisfactorily in many practical applications, despite this lack of rigorous theoretical justification. In fact, it was realized early on in the development of mesh selection strategies that the choice of a good mesh is not very sensitive; i. e., often there is a wide range of acceptable meshes of a given size N for a given BVP, even when a uniform mesh of the same size yields poor results.

In general, we distinguish between two types of mesh selection: *a priori* and *adaptive*. The first type is usually relevant in specialized circumstances. For instance, one may have some information (perhaps obtained by a preliminary mathematical analysis) on the shape of the desired solution for a given BVP. Such an initial solution profile can then be used to produce an initial mesh, like an equidistributing mesh as described in Section 9.2. Another idea for *a priori* mesh selection is to design a mesh based on the size and variation of the coefficients of the ODE. We comment on this further in Chapter 10.

In most of this chapter we restrict consideration to adaptive mesh selection, where the mesh and the approximate solution are repeatedly updated until prescribed error criteria are deemed satisfied.

Although our characterization is somewhat artificial, we divide adaptive mesh selection methods into two basic groups: *direct* and *transformation* methods. The latter are considered in several forms and provide a framework with which various methods can be studied. The direct methods have generally been the most useful in applications to date, and in our opinion they are the most important. We describe them in Section 9.2.

In Section 9.3 we discuss a particular direct approach which has been our method of choice for the collocation schemes of Sections 5.4 and 5.6. It has been extensively and successfully used in practice, and we give several numerical examples to show its effectiveness.

In Section 9.4 we discuss transformation methods. While we do not find many good words for some of them, they are not sufficiently well-understood at the present time to make a definitive statement about their potential performance. In Section 9.5 we mention some additional practical mesh selection considerations, including the important question of how the strategy should be coordinated with nonlinear iteration, and briefly describe some other approaches for mesh selection.

9.1.1 Error equidistribution and monitoring

Given a mesh π as in (9.2) and an approximate solution $\mathbf{y}_\pi(x)$, we denote by e_i some measure of the error on the ith subinterval of the mesh, $[x_i, x_{i+1})$. This may be an absolute or a relative error, or a combination of both. For instance, we may set

$$e_i := |\mathbf{y}_i - \mathbf{y}(x_i)|, \qquad 1 \leq i \leq N+1 \qquad (9.5a)$$

when a simple finite difference scheme is used on π, or

$$e_i := \max_{x_i \leq x < x_{i+1}} |\mathbf{y}_\pi(x) - \mathbf{y}(x)| =: \|\mathbf{y}_\pi - \mathbf{y}\|_i, \qquad 1 \leq i \leq N \qquad (9.5b)$$

if the approximate solution is defined continuously. Throughout this chapter we use the max (or sup) function norm.

The error e_i depends on the subinterval $[x_i, x_{i+1})$ and generally increases as h_i increases. It will turn out to be convenient to consider a corresponding error measure d_i which only varies linearly with h_i; i. e., we write

$$d_i = h_i \phi_i \qquad (9.6)$$

with ϕ_i independent of h_i. For example, if the discretization method is a difference scheme of order s then the results in Section 5.5 [cf. (5.91)] tell us (under sometimes unrealistic assumptions) that

$$|\mathbf{y}_i - \mathbf{y}(x_i)| = Ch_i^s |\mathbf{y}^{(s)}(x_i)| + O(h_i^{s+1}) + o(h^s)$$

so an expression like (9.6) can be written where

$$\phi_i \approx C^{1/s} |\mathbf{y}^{(s)}(x_i)|^{1/s}$$

(so $e_i \approx d_i^s$). This expression for ϕ_i does depend on the mesh, but not in a crucial way, and we ignore this dependence when obtaining (9.7) and (9.8) below.

Given N, we can try to choose the mesh π so that $\max_{1 \leq i \leq N} |e_i|$ is minimized. A related, simpler choice is to minimize $\max_{1 \leq i \leq N} |d_i|$ instead. Using this, we obtain a *minimax* problem with only one constraint,

$$\min\{|\mathbf{d}| : \sum_{i=1}^{N} h_i = b - a\} \qquad (9.7)$$

The solution of the trivial optimization problem (9.7) is to make all d_i equal to the same constant λ, which yields

$$h_i = \lambda / \phi_i \quad (1 \leq i \leq N), \qquad \lambda = (b-a) / \sum_{j=1}^{N} \phi_j^{-1} \qquad (9.8)$$

We now generalize the above and consider a smooth function ϕ instead of just a set of discrete values on π.

Definitions 9.9 A function $\phi(x, \mathbf{v})$ is called a *monitor (function)* if it has continuous first partial derivatives on $\{(x, \mathbf{v}) : \mathbf{v} \in S_\rho(\mathbf{y}(x)), a \leq x \leq b\}$, for some sphere S_ρ of radius $\rho > 0$ about $\mathbf{y}(x)$, and if

$$\phi(x, \mathbf{y}(x)) \geq \delta > 0, \qquad a \leq x \leq b \qquad (9.9a)$$

for some $\delta > 0$. A mesh π is *equidistributing* with respect to a monitor function $\phi(x, \mathbf{y}(x))$ (on $[a, b]$) if for some constant λ

$$\int_{x_i}^{x_{i+1}} \phi(x, \mathbf{y}(x)) \, dx \equiv \lambda \qquad 1 \leq i \leq N \qquad (9.9b)$$

It follows that

$$\lambda = \frac{\theta}{N} \qquad (9.9c)$$

where

$$\theta = \int_a^b \phi(x, \mathbf{y}(x))\, dx \qquad (9.9d)$$

Also, the mesh π is equidistributing with respect to a mesh function $\mathbf{d} = (d_1[\mathbf{y}], \ldots, d_N[\mathbf{y}])^T$ if

$$|d_i[\mathbf{y}]| \equiv \text{const} \qquad 1 \le i \le N \qquad (9.9e)$$

A sequence of meshes $\{\pi_N\}_{N=N_0}^\infty$ is *asymptotically equidistributing* (*as.eq.*) with respect to $\phi(x, \mathbf{y}(x))$ if

$$\int_{x_i}^{x_{i+1}} \phi(x, \mathbf{y}(x))\, dx \equiv \lambda(1 + O(h)) \qquad 1 \le i \le N \qquad (9.9f)$$

and asymptotically equidistributing with respect to $\mathbf{d}$ if

$$|d_i[\mathbf{y}]| \equiv \text{const}\,(1 + O(h)) \qquad 1 \le i \le N \qquad (9.9g)$$

□

For convenience, we shall just refer to a mesh π as being *as.eq.*, or to an *as.eq.* mesh, and implicitly assume that N is sufficiently large that an approximate solution in fact exists for an appropriate unspecified sequence of meshes. We are now prepared to investigate various ways to choose monitor functions and select *as.eq.* meshes.

9.2 DIRECT METHODS

A mesh selection method (or rather, a solution method) from the class considered here may be described as follows:

Outline: A direct method   Given a discretization scheme and an initial solution profile

>   REPEAT
>   >   Determine a mesh π from the current solution
>   >   Solve the BVP (9.1) for $\mathbf{y}_\pi(x)$ on the new mesh
>
>   UNTIL error tolerance is satisfied   □

Normally one starts from a given initial (coarse) mesh and first solves (9.1) on it.

The main question is, given an approximate solution $\mathbf{y}_\pi(x)$ on a *current mesh* π, how is a new, *refined mesh* π^* determined? In view of the preceding section, we would know how to proceed if a monitor function ϕ were available: We would determine an *as.eq.* mesh π^* with respect to $\phi(x, \mathbf{y}_\pi(x))$, where N and the constant λ in (9.9c) are determined by using the desired error tolerance. For a monitor function we now consider two choices, one arising from equidistributing the local truncation error and the other, valid for collocation methods, using a localized form of the global error.

9.2.1 Equidistributing local truncation error

Suppose that the discretization method used to solve (9.1) gives an approximate solution $\mathbf{y}_\pi(x)$ which satisfies the error bound (9.3). Under normal circumstances, the local truncation error of a difference scheme of order s satisfies

$$\tau_i[\mathbf{y}] = h_i^s \mathbf{T}(x_i) + O(h_i^p) \qquad (p > s) \tag{9.101}$$

where $\mathbf{T}(x)$ is a continuous function involving high derivatives of $\mathbf{y}(x)$. It is then natural to choose the monitor function

$$\phi(x, \mathbf{y}(x)) := |\mathbf{T}(x)|^{1/s} \tag{9.11a}$$

For example, the trapezoidal scheme yields

$$\phi(x, \mathbf{y}(x)) = \frac{1}{12} |\mathbf{y}'''(x)|^{1/2}$$

The theoretical significance of this is given in

Theorem 9.12 Suppose that an *as.eq.* mesh π with respect to $|\mathbf{T}(x)|^{1/s}$ is determined, i. e.,

$$\int_{x_i}^{x_{i+1}} |\mathbf{T}(x)|^{1/s} \, dx \equiv \lambda(1 + O(h)) \qquad 1 \le i \le N \tag{9.12a}$$

where

$$\lambda := \frac{\theta}{N}, \qquad \theta = \int_a^b |\mathbf{T}(x)|^{1/s} \, dx$$

Then the corresponding discretization method with error satisfying (9.3) and (9.10) gives

$$\|\mathbf{y}_\pi - \mathbf{y}\| \le K \lambda^s (1 + O(h)) + O(h^p) \tag{9.13a}$$

Moreover, if $M_L = \{x : |\mathbf{T}(x)|^{1/s} \ge L\}$, with L chosen such that

$$\int_{M_L} |\mathbf{T}(x)|^{1/s} \, dx = \frac{1}{2} \int_a^b |\mathbf{T}(x)|^{1/s} \, dx = \frac{\theta}{2} \tag{9.12b}$$

then the error satisfies

$$\|\mathbf{y}_\pi - \mathbf{y}\| \le K \left(\frac{2\mu(M_L)}{N} \right)^s \|\mathbf{T}\| (1 + O(h)) + O(h^p) \tag{9.13b}$$

where $\mu(M_L)$ is the measure of M_L.[1]

---

[1] If M_L consists, for example, of a finite number of subintervals of $[a, b]$, then $\mu(M_L)$ is the sum of the lengths of these subintervals.

Proof: From (9.12a),

$$h_i |\mathbf{T}(x_i)|^{1/s} \equiv \frac{\theta}{N}(1+O(h))$$

so

$$h_i^s |\mathbf{T}(x_i)| = (\frac{\theta}{N})^s (1+O(h))$$

and (9.13a) follows immediately from (9.3) and (9.10). Furthermore,

$$\int_{M_L} |\mathbf{T}(x)|^{1/s} \, dx \leq \mu(M_L) \|\mathbf{T}\|^{1/s}$$

so by (9.12b)

$$\theta^s = 2^s [\int_{M_L} |\mathbf{T}(x)|^{1/s} \, dx]^s \leq [2\mu(M_L)]^s \|\mathbf{T}\|$$

and now (9.13b) follows from (9.13a). □

Remarks

(i) For a BVP whose solution $\mathbf{y}(x)$ has a boundary or interior layer (i. e., a small subinterval where the solution varies fast), if (9.10) holds and $\mathbf{T}(x)$ involves some derivative(s) of $\mathbf{y}(x)$ then M_L is often an interval of width proportional to the layer width, say ε. In such a case, the first term in the bound (9.13b) can be independent of the layer width, and the error from an *as.eq.* mesh can be essentially independent of ε. We see this later, in Example 9.6.

(ii) If one wishes to restrict the size of $h = \max_{1 \leq i \leq N} (x_{i+1} - x_i)$ for the meshes chosen, then instead of (9.11a) one can choose

$$\phi(x, \mathbf{y}(x)) := \max\{|\mathbf{T}(x)|^{1/s}, \nu\} \tag{9.11b}$$

for some $\nu > 0$. The theorem still holds, but (9.12) and (9.13) are changed to reflect this new monitor function. Moreover, one can show (see Exercise 1) that the meshes $\{\pi_N\}$ chosen in this way are quasiuniform; i. e., (9.4) holds. This mesh restriction can be desirable in certain contexts such as for the acceleration methods of Section 5.5 (recall the quasiuniform mesh assumption there). Instead of (9.11b), we can also ensure quasiuniformity by using, for instance,

$$\phi(x, \mathbf{y}(x)) := (\alpha^2 + |\mathbf{T}(x)|^2)^{1/2s} \tag{9.11c}$$

for some constant α.

(iii) It is not always obvious how $\mathbf{T}(x)$ can be approximated in practice. If an approximation to $\mathbf{y}(x)$ is available, then its derivatives or divided differences (i. e., derivatives of an interpolant, if it is only a mesh function) can normally be used to approximate the derivatives appearing in $\mathbf{T}(x)$. This is discussed in detail for collocation in the next section.

(iv) Note that θ can be approximated by using a given monitor function. From (9.3), (9.10) and (9.12a) we can roughly estimate the error as $K\lambda^s$, so if (say) a given absolute tolerance *TOL* is desired, then we want $N \geq \theta \cdot (\frac{K}{TOL})^{1/s}$. While K is usually unavailable, recall from Chapter 5 that it is often of the order of the conditioning of the BVP itself and is not sensitive to mesh changes. Thus, we could either ignore it, as many IVP codes do, or estimate it on one mesh by comparing λ^s to an error estimate obtained in another way. The obtained estimate of K can then be used subsequently. Specifically, we can use $\theta(\frac{K}{TOL})^{1/s}$ as a predicted value for N. This in turn gives λ, so a new mesh π^* can be constructed by approximating (9.12a).

(v) It is most important to realize that the results (9.13) of Theorem 9.12 only require that π be *as.eq.*, i. e., that (9.9f) hold. This means that, not only can a somewhat complicated monitor function be selected, but it is *not even desirable* to solve for $\{x_i\}_1^{N+1}$ very accurately, at least if Theorem 9.12 is the theoretical justification for our mesh selection procedure. This agrees with the observed results in practice, too. □

9.3 A MESH STRATEGY FOR COLLOCATION

From the last section we have a general mesh selection approach in which we alternately solve a BVP using a discretization method and adjust the mesh by using a monitor function which itself involves the current approximate solution. Here we choose the discretization method to be spline collocation, which can be viewed as a family of high-order finite difference schemes as described in Section 5.6. The BVP is a higher-order equation, unlike that in the previous section, but the viewpoint taken there easily generalizes, since the determination of the mesh and the solution of the unmodified BVP are done separately. For the monitor function we can, of course, choose the local truncation error as described in Section 9.2.1, but a robust algorithm is obtained by instead using the leading term of the error away from the mesh points. While the resulting strategy is far from being the only reasonable one, it has proven very successful in practice and is the one used in the collocation code of Appendix B. The numerical examples in Section 9.3.2 illustrate the considerable gains one can expect from an adaptive mesh selection strategy.

9.3.1 A practical mesh selection algorithm

We use the spline collocation method presented in Section 5.6 for solving the BVP

$$Lu \equiv u^{(m)} - \sum_{l=1}^{m} c_l(x) u^{(l-1)} = q(x) \qquad a < x < b \qquad (9.14a)$$

$$B_1 \mathbf{y}(a) + B_2 \mathbf{y}(b) = \boldsymbol{\beta} \qquad (9.14b)$$

where $\mathbf{y}(x) := (u(x), u'(x), \ldots, u^{(m-1)}(x))^T$. To recall, the collocation solution $u_\pi(x)$ is determined, for a given mesh π and for k given collocation points on each mesh subinterval, as that piecewise polynomial function of order $k + m$ which has $m - 1$ continuous derivatives {i. e., $u_\pi(x) \in \mathbf{P}_{k+m,\pi} \cap C^{m-1}[a,b]$} and satisfies the differential equation at the collocation points and the BC. Then for $x_i \leq x \leq x_{i+1}$ [cf. (5.142)]

$$u(x) - u_\pi(x) = h_i^{k+m} u^{(k+m)}(x_i) P\left(\frac{x - x_i}{h_i}\right)(1 + O(h_i)) + O(h^p) \quad (9.15a)$$

where

$$P(\xi) = \frac{1}{k!(m-1)!} \int_0^\xi (t - \xi)^{m-1} \prod_{l=1}^k (t - \rho_l)\, dt \quad (9.15b)$$

For the Gauss points $\{\rho_l\}_1^k$, which we shall use here, $p = 2k$. Recall that we have the local representation (9.15a) for the global error, because the scheme is actually superconvergent for certain solution and derivative values.

Thus, corresponding to (9.3) we have

$$\|u_\pi - u\|_i := \max_{x_i \le x \le x_{i+1}} |u_\pi(x) - u(x)| \le C_i h_i^{k+m} \|u^{(k+m)}\|_i + O(h^{2k})$$

where $C_i = \hat{C}(1 + O(h_i))$,

$$\hat{C} := \max_{0 \le \xi \le 1} |P(\xi)|$$

A natural choice for a monitor function is therefore

$$|u^{(k+m)}(x)|^{1/(k+m)} \quad \text{for } x_i \le x < x_{i+1} \quad (9.16)$$

The exact solution $u(x)$ and hence also $u^{(k+m)}(x)$ are unknown, and one cannot directly replace $u(x)$ by $u_\pi(x)$ in (9.16) because $u_\pi^{(k+m)} \equiv 0$. However, suppose that $v(x) \in \mathbf{P}_{2,\pi} \cap C[a,b]$ is the piecewise linear function which interpolates $u_\pi^{(k+m-1)}(x)$ at the subinterval midpoints; i.e., $v(x)$ satisfies $v(x_{i+1/2}) = u_\pi^{(k+m-1)}(x_{i+1/2}), 1 \le i \le N$, and define the monitor function

$$\phi(x, v(x)) := |v'(x)|^{1/(k+m)} \quad (9.17)$$

Since $P^{(k+m-1)}(1/2) = 0$, the estimate (5.142) implies

$$v(x_{i+1/2}) = u_\pi^{(k+m-1)}(x_{i+1/2}) = u^{(k+m-1)}(x_{i+1/2}) + O(h_i^2) \quad 1 \le i \le N$$

and hence $v'(x) = u^{(k+m)}(x)(1 + O(h))$; i.e., $\phi(x, v)$ is an $O(h)$ approximation to the monitor function in (9.16). Moreover, $v'(x)$ is a piecewise constant function, so

$$\theta := \int_a^b |v'(x)|^{1/(k+m)} dx \quad (9.18)$$

is easy to compute. Finding an *as.eq.* mesh (9.2a) such that

$$\int_{x_i}^{x_{i+1}} |v'(x)|^{1/(k+m)} dx \equiv \frac{\theta}{N} \quad 1 \le i \le N$$

is also easy since

$$t(x) := \frac{1}{\theta} \int_a^x |v'(\xi)|^{1/(k+m)} d\xi \quad (9.19a)$$

is piecewise linear. Thus, if we know x_i then finding x_{i+1} such that

$$t(x_{i+1}) = \frac{i}{N} \qquad (9.19b)$$

amounts to inverse interpolation for this simple function. From Theorem 9.12, if π^* is the new mesh determined from (9.19) then the new spline collocation solution $u_{\pi^*}(x)$ satisfies

$$\|u_{\pi^*} - u\|_i \leq \hat{C}(\frac{\theta}{N})^{k+m}(1 + O(h)) + O(h^{2k}) \qquad (9.20)$$

A new mesh size N such that the collocation solution on the corresponding *as.eq.* mesh has uniform error less than tolerance *TOL*, can be predicted from (9.20) by choosing

$$N = \theta(\frac{\hat{C}}{TOL})^{1/(k+m)} \qquad (9.21)$$

Here $\hat{C}$ is known from (9.15). If *TOL* is a relative tolerance, then a trivial modification to (9.21) is made using $\|u_\pi\|$.

We summarize the above in an

Algorithm: Collocation with mesh selection

Input: A BVP (9.14), a collocation method with k, m, $\hat{C}$, an error tolerance *TOL*, and an initial mesh π.

Output: an approximate solution $u_\pi(x)$

1. Set FLAG := false
2. REPEAT
    2.1 Solve the collocation equations for $u_\pi(x)$ (see Section 5.6.1.1).
    2.2 Form the piecewise linear interpolant $v(x)$ to $u_\pi^{(k+m-1)}(x)$ using values at $x_{i+1/2}$ ($1 \leq i \leq N$).
    2.3 Calculate θ from (9.18).
    2.4 IF $\hat{C}(\frac{\theta}{N})^{k+m} \leq TOL$ THEN FLAG := true
    2.5 ELSE
        Find a new N from (9.21).
        Solve for a new mesh π from (9.19).
    UNTIL FLAG = true (or iteration limit exceeded) □

To implement a practical mesh selection strategy, additional bells and whistles are needed to ensure that the strategy does not go awry. We delay a general discussion of special features of mesh selection strategies until Section 9.5, but for concreteness finish this section with an overview of the mesh selection and error estimation strategy specifically used in one particular general-purpose code, COLSYS, for a scalar BVP.

For robustness, errors are estimated for mesh selection and for termination of computation in two different ways. The rough error estimation for mesh selection has already been described. For termination (and to provide error estimates for intermediate meshes), an extrapolation principle is used. Specifically, when two collocation solutions

$u_{\pi_1}(x)$ and $u_{\pi_2}(x)$ are computed on meshes π_1 and π_2, where π_2 is π_1 "doubled" (i. e., each subinterval is halved), the error on π_2 is estimated from (9.15) by

$$u(x) - u_{\pi_2}(x) \approx \frac{1}{1 - 2^{k+m}} (u_{\pi_1}(x) - u_{\pi_2}(x)) \tag{9.22}$$

(see Section 5.6).

Given an approximate solution $u_\pi(x)$, the quantities

$$r_1 = \max_{1 \le i \le N} h_i \left(\frac{\hat{C}\|v'\|_i}{TOL}\right)^{1/(k+m)}, \quad r_2 = \sum_{i=1}^N h_i \left(\frac{\hat{C}\|v'\|_i}{TOL}\right)^{1/(k+m)}, \quad r_3 = \frac{r_2}{N}$$

are computed. The ratio $\frac{r_1}{r_3}$ gives some idea of the gain to be achieved by redistribution. Specifically, if it is large then the maximum subinterval error estimate is significantly larger than the average one, and the mesh is not well-distributed. The code checks to see if

$$\frac{r_1}{r_3} \le 2 \tag{9.23}$$

(a) If (9.23) is satisfied, then the mesh is considered to be sufficiently equidistributed to not warrant construction of an entirely new set of points $\{x_i\}$, and the mesh is "doubled," i. e., $\pi^* = \{x_1, x_{3/2}, x_2, \ldots, x_N, x_{N+1/2}, x_{N+1}\}$. This facilitates error estimation with (9.22).

(b) If (9.23) is not satisfied, then r_2 predicts the number of points needed to satisfy the tolerance [cf. (9.21)]. Normally $N^* = \min\{N, \frac{1}{2}\max(N, r_2)\}$ is used instead of r_2, mainly to prevent jumping to incorrect conclusions early in the mesh selection process. Then the new mesh π^* is chosen using (9.19). However, this strategy is not followed and the mesh is doubled instead if one of the following holds: (i) N is less than it was for the mesh previous to π, (ii) the current N has been used for three consecutive meshes, or (iii) there have been three consecutive equidistributions and doublings (i. e., $N/2$ and N mesh points have been used three consecutive times). These precautions prevent cycling or, again, prematurely jumping to conclusions about what a desirable mesh is, by insuring that N increases gradually. Also, when the choice $N^* = \frac{r_2}{2} < N$ is made, then doubling, it is to be hoped, will provide a solution (and error estimate) satisfying the requested tolerance.

The user also has the option to (i) insert an initial mesh, (ii) insert fixed points (points required to be included in all subsequent meshes), and (iii) turn off the mesh selection so that only mesh doubling is used. These can be particularly useful features if one has special information about the solution, such as the location of boundary layers.

9.3.2 Numerical examples

In this section, we present some examples which are solved by using the code COLSYS. The behaviour of its mesh selection strategy illustrates how the ideas presented in the previous section work in practice and shows how important mesh selection can be in solving BVPs of some difficulty.

Example 9.2

Our first computational example is linear. Consider

$$\varepsilon u'' + xu' = -\varepsilon\pi^2 \cos(\pi x) - (\pi x) \sin(\pi x), \quad -1 < x < 1$$

$$u(-1) = -2, \quad u(1) = 0$$

whose solution is

$$u(x) = \cos(\pi x) + \text{erf}(x/\sqrt{2\varepsilon})/\text{erf}(1/\sqrt{2\varepsilon})$$

For $0 < \varepsilon \ll 1$ this solution has a rapid transition layer at $x = 0$ (see Fig. 9.2).

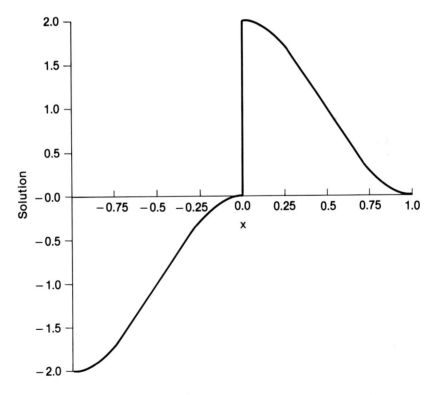

Figure 9.2 Solution to the BVP of Example 9.2 for $\varepsilon = 10^{-6}$

This BVP is solved for $\varepsilon = 10^{-4}$ using COLSYS, with $k = 4$, $TOL = 10^{-6}$ for both u and u', and $N = 5$ for an initial uniform mesh. The mesh sequence is

$$5[0], 10[1], 20[1], 20[1], 20[3], 40[5], 40[19], 30[23], 60[46], 33[21], 66[42], 132[85]$$

where the number of mesh points in the interval [-.05, .05] around the layer is given in brackets. For the last 3 meshes, $h/h_{min} \approx 94.3$ [cf. (9.4)]. The error estimates for u and u', which are available after each mesh doubling, are given in Table 9.1. The actual error in u for the last mesh is .14 – 8. If the problem is run using *no* mesh selection (and $N = 5$ initially), comparable accuracy is not achieved until $N = 1280$.

TABLE 9.1 Error estimates after mesh doubling

N	est (u)	est (u')
10	.272 +0	.101 +2
20	.971 –2	.782 +0
40	.728 –3	.252 +0
60	.116 –6	.235 –3
66	.315 –7	.160 –4
132	.976 –9	.492 –6

□

Example 9.3

Consider Example 1.24 with $\varepsilon = 10^{-2}$ and $\kappa = 10^2$. It is solved by using $TOL = 10^{-5}$ (for all of the components ϕ, ϕ', ψ, and ψ'), $k = 4$, and a uniform initial mesh with $N = 10$. The two initial guesses described in Example 8.6 are used. For the solution with no interior layer, the mesh sequence is

$$10(4), 10(2), 10(2), 20(2), 20(2), 19, 38, 19, 38, 19, 38$$

where the number of modified Newton iterations is listed in parentheses (or omitted, if one iteration is sufficient). For the final mesh, 23 of the mesh points are in (.999, 1]. For the more difficult case involving the solution with an interior layer, the mesh sequence is

$$10(3), 20(4), 20(3), 20(2), 40(2), 40, 40, 80, 40, 80, 40, 80, 160$$

The final mesh has 86 mesh points in (.41, .45) and 37 in (.999, 1]. □

Example 9.4

Here we consider Example 1.6 with $b = 10$, $\beta_0 = 0$, $\beta_1 = -1$, for 3 values of L: $L = 200, 300$ and 400. COLSYS is used with $k = 4$, $TOL = 10^{-6}$ for only the first component P, and an initial uniform mesh with $N = 10$. Because of the complicated form of the Jacobian matrix, derivatives appearing in it are approximated by finite differences. Also, in the course of computation we must keep P nonpositive, so we set it to 0 otherwise. (This introduces sensitivity in the nonlinear iteration.)

The case $L = 200$ is not too difficult, and the mesh sequence is

$$10(9), 20, 10, 20.$$

The number of iterations is in parentheses (omitted if one iteration is sufficient). The final mesh is not extremely far from uniform, with $h/h_{min} = 21.5$.

However, if L is increased to $L = 300$, then the problem becomes much more difficult. Denoting by NC occasions where the damped Newton's method fails to converge, we obtain the mesh sequence

$$10(NC), 20(NC), 40(NC), 80(13), 40(2), 80$$

For the final mesh, $h/h_{min} = .409 + 3$ and the estimated errors for P and P' are $.14 - 8$ and $.11 - 5$, respectively. Other plausible starts may not yield convergence at all. In particular, a simple continuation in L (cf. Section 8.3.1) does not prove fruitful, because the region of high solution activity shifts significantly (from ≈ 2.58 for $L = 200$, to ≈ 1.58 for $L = 300$, to ≈ 1.23 for $L = 400$). This suggests that a more sophisticated continuation scheme is needed, but we will not pursue it further here (see Exercise 6). We just report a mesh sequence obtained for $L = 400$:

$$10(NC), 20(NC), 40(NC), 80(NC), 160(NC), 320(6), 160, 320$$

For the final mesh, $h/h_{min} = .608 + 4$, and the error estimates for the solution and its derivative are $.23 - 10$ and $.27 - 8$, respectively. □

9.4 TRANSFORMATION METHODS

With transformation methods, a change of variables is applied to the given BVP (9.1) to produce a transformed problem which is, we hope, more amenable to effective numerical computation. Thus, a coordinate transformation $x(t)$ of the independent variable is sought such that as a function of t the solution $\mathbf{y}(x(t))$ of (9.1) looks smoother, i. e., varies slowly. We shall see how the direct methods of the previous sections can be viewed as particular ways of achieving such a coordinate transformation [in particular see (9.19)]. Indeed, the monitor function of Definition 9.9 is used to provide a description of suitable coordinate transformations.

Definition 9.24 A coordinate transformation $t(x)$, $t : [a,b] \to [0,1]$, is called *admissible* if it satisfies

$$\frac{dt}{dx} = \frac{1}{\theta}\phi(x, \mathbf{y}(x)), \qquad a < x < b \qquad (9.24a)$$

$$t(a) = 0, \qquad t(b) = 1 \qquad (9.24b)$$

where ϕ is a monitor function [so (9.9a) holds] and θ is given by (9.9d). □

Below we consider various approaches for finding an appropriate transformation for a given BVP.

9.4.1 Explicit method

If some knowledge of the solution $\mathbf{y}(x)$ is available, then an explicit coordinate transformation $x(t)$ can often be chosen to appropriately conform to this solution's behaviour and to furthermore satisfy

$$\frac{dx}{dt} > 0 \qquad 0 < t < 1$$

$$x(0) = a, \qquad x(1) = b$$

The BVP (9.1) can then be rewritten as

$$\hat{\mathbf{y}}'(t) = \hat{\mathbf{f}}(t, \hat{\mathbf{y}}(t)) \cdot x'(t) \qquad 0 < t < 1 \qquad (9.25a)$$

$$\mathbf{g}(\hat{\mathbf{y}}(0), \hat{\mathbf{y}}(1)) = \mathbf{0} \qquad (9.25b)$$

where $\hat{\mathbf{y}}(t) := \mathbf{y}(x(t))$ and $\hat{\mathbf{f}}(t, \hat{\mathbf{y}}(t)) := \mathbf{f}(x, \mathbf{y}(x))$. We want to choose $x(t)$ in such a way that $\hat{\mathbf{y}}(t)$ will be smooth and well-behaved, so that the transformed BVP will be "easy" to solve. Since this frequently involves smoothing out regions of rapid variation that $\mathbf{y}(x)$ may have, $x(t)$ or its inverse $t(x) = x^{-1}(t)$ is often called a *stretched coordinate*. Ideally, the coordinate stretching should be sufficiently successful that the transformed BVP (9.25) could be efficiently solved on a uniform mesh (using our basic numerical method).

Example 9.5

Suppose that the first component of $\mathbf{y}(x)$ on $[0, 1]$ behaves near $x = 1$ like $e^{(x-1)/\varepsilon}$ for some ε, $0 < \varepsilon \ll 1$ (i.e., it has a "boundary layer"), and that the other components are all smooth throughout $[0, 1]$. Apart from satisfying $x(0) = 0$, $x(1) = 1$, a desirable property of $x(t)$ would be that it have at $x = 1$ a similar rapid variation. Hence one might choose $t(x) := t_1(x) \equiv (x + e^{x/\varepsilon} - 1) e^{-1/\varepsilon}$. To prevent $t'(x)$ from being too small, a better choice might be $t(x) := \frac{1}{2}(x + t_1(x))$. Graphically, the result might be as in Fig. 9.3. □

A variety of specially chosen coordinate transformations have been used in various application areas.

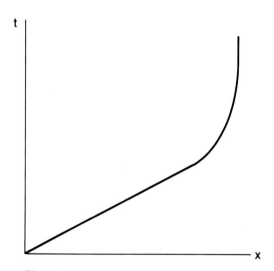

Figure 9.3 Explicit coordinate transformation

Explicit coordinate transformations are considered in a different way in Chapter 10. There, using known behaviour of the solution $\mathbf{y}(x)$, an explicit stretching transformation is defined *locally*, and by analyzing the transformed ODE (9.25a), an appropriate mesh is found for use in solving the original (untransformed) BVP.

9.4.2 Implicit method

Obvious disadvantages of explicit coordinate transformations are that *a priori* knowledge of the behaviour of $\mathbf{y}(x)$ is needed and that choosing an appropriate $x(t)$ is at best very complicated if components of $\mathbf{y}(x)$ are badly behaved in several regions of $[a,b]$. An alternative is to build an implicit mesh selection strategy into the problem statement. There are many approaches to choosing an admissible coordinate transformation $x(t)$, or $t(x)$, for which $\hat{\mathbf{y}}(t)$ is well-behaved. By "well-behaved" we mean here that an equally spaced mesh in the variable t is satisfactory for solving the transformed problem for $\mathbf{y}(t)$ efficiently.

Suppose we have decided upon the choice of the monitor function $\phi(x, \mathbf{y}(x))$. We can use the conditions (9.24a,b) for an admissible coordinate transformation to augment the original BVP (9.1). The requirement of an equidistributing mesh of size N, i. e., a mesh satisfying (9.9b), can be simply expressed in terms of $t(x)$ using (9.19b), by

$$t(x_{i+1}) - t(x_i) = \frac{1}{N} \quad (=\lambda/\theta) \qquad (1 \leq i \leq N-1) \tag{9.26}$$

The mesh selection strategy can thus be accomplished if we solve the BVP (9.1), (9.24) subject to (9.26) with θ an undetermined constant (as, for instance, in Example 1.4). This, however, can be written in an equivalent form as follows: Write (9.24) in terms of t as the independent variable, add a trivial ODE for the constant θ, and add (9.25) to obtain the system

$$\hat{\mathbf{y}}'(t) = \hat{\mathbf{f}}(t, \hat{\mathbf{y}}(t)) x'(t) \tag{9.27a}$$

$$x'(t) = \theta \hat{\phi}(t, \hat{\mathbf{y}}(t))^{-1} \tag{9.27b}$$

$$\theta'(t) = 0 \qquad 0 < t < 1 \tag{9.27c}$$

$$\mathbf{g}(\hat{\mathbf{y}}(0), \hat{\mathbf{y}}(1)) = \mathbf{0} \tag{9.27d}$$

$$x(0) = a, \qquad x(1) = b \tag{9.27e}$$

where $\hat{\phi}(t, \hat{\mathbf{y}}(t)) := \phi(x, \mathbf{y}(x))$. Note that the BVP is in standard form when (9.27b) is substituted into (9.27a). It is also nonlinear even if the original BVP is linear. A nice feature of the BVP (9.27) is that when it is solved, the constraints (9.26) are simply satisfied by letting

$$x_i = x(\frac{i-1}{N}), \qquad 1 \leq i \leq N+1. \tag{9.28}$$

In practice, of course, the BVP (9.27) must be solved approximately. It is to be hoped, though, that the solution is smooth and can thus be computed on the uniform mesh $\{0, 1/N, \ldots, 1\}$, whereby approximate values of $\mathbf{y}(x_i) = \hat{\mathbf{y}}((i-1)/N)$ with x_i defined in (9.28) are obtained.

In general, a monitor function $\phi(x, y(x))$ is chosen as some measure of the change in the behaviour of $y(x)$ which is also reflected in the error behaviour for the numerical solution. One common choice is the *arc length monitor function*,

$$\phi(x, y(x)) = \sqrt{1 + |y'(x)|_2^2} = \sqrt{1 + |f(x, y(x))|_2^2}. \tag{9.29}$$

This is represented graphically in Fig. 9.4 for $N = 7$. The advantages of the arc length monitor in this context are that it does not involve high derivatives of $y(x)$ and that it is a smooth function of its arguments. This is much more important here than in the previous sections, because the monitor function is one of the BVP unknowns to be actually computed.

Example 9.6

We consider the BVP

$$\varepsilon u' = w$$
$$\varepsilon w' = -2uw \qquad 0 < x < 1$$

$$u(0) = 1, \qquad u(1) = 2$$

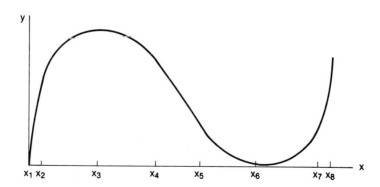

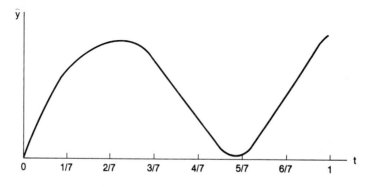

Figure 9.4 Arc length equidistribution

which has a known solution. For $0 < \varepsilon \ll 1$, this solution is given to a very good accuracy by

$$u(x) \approx 2\frac{1 - 1/3 e^{-4x/\varepsilon}}{1 + 1/3 e^{-4x/\varepsilon}}$$

This BVP is modified to the form (9.27) with the arc length monitor (9.29) and then is solved by using the trapezoidal rule and equal spacing. The results for $N = 10$ and for various values of ε are given in Table 9.2. The arc length monitor has a similar form to (9.11c), but with $\phi(x, y(x)) \approx |y'(x)|$ instead of $\frac{1}{12} |y''(x)|^{1/2}$. From remark (i) following Theorem 9.12 and from Exercise 2, one can argue that the error is essentially independent of ε.

TABLE 9.2 Maximum error with $N = 10$.

ε	err(u)	err(w)
1.	.236 − 2	.142 − 1
1. − 2	.106 − 1	.393 − 1
1. − 4	.110 − 1	.402 − 1
1. − 6	.110 − 1	.402 − 1
1. − 8	.110 − 1	.402 − 1

© 1979 by the Society for Industrial and Applied Mathematics. Reprinted with permission from "On Selection of Equidistributing Meshes for Two-Point Boundary Value Problems" by A.B. White Jr. in the SIAM Journal of Numerical Analysis, Vol. 16, no. 3, 1979.

□

Another possible choice of a monitor function would be the one given in the previous sections, namely a localized error term (such as truncation error) for the particular numerical method being used to solve the transformed BVP. But (9.27a, b) becomes extremely complicated if $\phi(t, \hat{y})$ involves any high derivatives of $\hat{y}(t)$, and the actual implementation of such a method is an onerous task.

The continuity conditions used in the definition of an admissible coordinate transformation allow one to prove that the BVP (9.27) has a unique solution which can be approximated numerically under general conditions. In particular, we have the following result, which is stated without proof.

Theorem 9.30 If the original BVP (9.1) has sufficiently smooth coefficients and an isolated solution $y(x)$, then for admissible $t(x)$, the transformed BVP (9.27) has a corresponding isolated solution $\hat{y}(t), x(t), \theta(t)$. If a one-step difference scheme of order s is used to solve (9.27) on a uniform mesh in t, then for N sufficiently large and an initial guess sufficiently close, Newton's method converges quadratically to an $O(N^{-s})$ approximation to $\hat{y}(t), x(t), \theta(t)$. Furthermore, the mesh $\{x_i\}_1^{N+1}$ in (9.28) satisfies

$$t(x_{i+1}) - t(x_i) = \frac{1}{N} + O(h^s) \qquad 1 \le i \le N \qquad (9.30)$$

□

Note that the mesh $\{x_i\}_1^{N+1}$ asymptotically equidistributes $t(x)$, according to (9.30). Theorem 9.30 appears then to be a most satisfying theoretical result. But there turn out to be serious questions about its utility in practice. In particular, it is possible for the transformed BVP (9.27) to be extremely sensitive, as we see from Example 9.7.

Example 9.7

Consider the BVP

$$u' = -w^2$$
$$\varepsilon w' = uw \qquad 0 < x < 1$$

$$u(0) = 2, \qquad u(1) = 1$$

For $0 < \varepsilon \ll 1$, the solution is given to a very good accuracy by

$$u(x) \approx 2\frac{3 - e^{4(x-1)/\varepsilon}}{3 + e^{4(x-1)/\varepsilon}}$$

$$w(x) \approx \pm\sqrt{\frac{4 - u(x)^2}{\varepsilon}}$$

so in fact we have a pair of solutions here. Note that away from $x = 1$, $u(x) \approx 2$ and $w(x) \approx 0$, but near $x = 1$ there is a sharp boundary layer where $u(x)$ drops to 1 and $w(x)$ has an $O(\varepsilon^{-1})$ change over a narrow interval of length $O(\varepsilon)$ in x. Note, in particular, that $w(1) \approx \sqrt{3/\varepsilon}$.

If we apply the arc length transformation directly for this BVP, then difficulty arises because θ becomes unbounded as $\varepsilon \to 0$, so we consider the scaled ODE formulation

$$\varepsilon u' = -v^2$$
$$\varepsilon v' = uv \qquad 0 < x < 1$$

where $v(x) = \sqrt{\varepsilon}\, w(x)$.

First we solve this BVP using COLSYS. Using $k = 2$, $TOL = 10^{-4}$, a linear initial approximation for u satisfying the BC, and a uniform initial mesh with $N = 10$, we apply a simple continuation process (see Section 8.3 and Appendix B). Using ε as the continuation parameter, we start with $\varepsilon = 1$ and reduce it by a factor $1/10$ at each continuation step. The following sequence of mesh sizes is produced for the scaled problem:

$$\varepsilon = 10^0: 10, 20$$

$$= 10^{-1}: 10, 10, 10, 20, 40$$

$$= 10^{-2}: 20, 20, 20, 40, 80$$

$$= 10^{-3}: 40, 27, 54$$

$$= 10^{-4}: 27, 27, 27, 54, 54, 54, 108, 80, 160$$

$$= 10^{-5}: 80, 40, 80$$

$$= 10^{-6}: 40, 40, 40NC, 80NC, \ldots, \text{failure}$$

and there is no convergence for $\varepsilon = 10^{-6}$. Here, as in other examples, NC denotes no convergence of the nonlinear iteration, and failure signals that after a number of mesh doublings the program execution terminates because storage limitations are reached. In the last meshes encountered for each $\varepsilon = 10^{-j}$, $j = 1, \ldots, 5$, all mesh points but one or two are concentrated in the narrow layer region near $x = 1$, and extremely large mesh ratios are generated. (Not only is $\overline{K} = O(\varepsilon^{-1})$ in (9.4), but the *local* mesh variation is very large.) Incidentally, the code COLSYS can also be applied as successfully to the unscaled u, w–formulation.

Now the scaled u, v–problem is converted to the transformed BVP (9.27) with the choice (9.29) for monitor function. Using COLSYS in the same way, we obtain the following mesh sequence:

$$\varepsilon = 10^0: 10, 20$$

$$= 10^{-1}: 10\text{NC}, 20\text{NC}, 40\text{NC}, 80\text{NC}, 160, 80, 160$$

$$= 10^{-2}: 80, 40, 80$$

$$= 10^{-3}: 40, 20, 40$$

$$= 10^{-4}: 20\text{NC}, 40\text{NC}, \ldots, \text{failure},$$

so there is no convergence for $\varepsilon = 10^{-4}$. Thus, not only is this BVP difficult to solve numerically on a uniform mesh, but it is more difficult to solve than either the original BVP, or the modified, scaled one! This is partly indicated by the fact that boundary layer behaviour for u and v is converted to corner layer behaviour for $\hat{u}$, $\hat{v}$, and x (interior layers for $\hat{u}'$, $\hat{v}'$, and x'), whose solution is shown for $\varepsilon = 10^{-3}$ in Fig. 9.5. □

This example illustrates that the direct use of the transformed BVP (9.27) can be risky and is probably not to be recommended in general. Note, however, that in Example 9.7 the solution layer occurs in a higher derivative (a boundary layer in u becomes an interior layer in $\hat{u}'$). There are other disadvantages to solving (9.27) as in this example. One is that the equal significance allotted to finding the solution and to

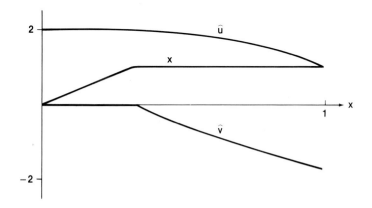

Figure 9.5 Solution and derivative for the transformed BVP for Example 9.7

finding a mesh goes against the fact observed earlier that good meshes are relatively insensitive and need not be computed accurately. Generally, one would want to compute the solution $\mathbf{y}(x)$ much more accurately than the mesh transformation $x(t)$. Also, an inconvenience of the transformed problem is that the only computed solution values in (0, 1) are at the *a priori unknown* points (9.28).

The main advantage of the transformation method in our opinion is that we can examine general mesh selection strategies using this framework. Thus, it can provide the necessary background for studying our other approaches, even though they are implemented in an entirely different way (e. g., see Exercise 4). It also suggests a new *two-phase* variant: Alternately solve for $\hat{\mathbf{y}}(t)$ from the transformed BVP (9.27a, d) and for $x(t)$, $\theta(t)$ from (9.27b, c, e). This would not necessarily require the computation of an accurate coordinate transformation $x(t)$. The variations of this approach have not been extensively tested. Nevertheless, there is still reason for caution. As was pointed out earlier, the choice of $\phi(x, \mathbf{y}(x))$ in (9.27a) is strongly influenced by the need for a simple, differentiable function, not necessarily one which best relates to error equidistribution. But perhaps the worst disadvantage of this transformation method lies again in the need to solve the transformed ODE (9.27a). The experience of Example 9.7, where a very tough nonlinear problem arises, seems to be the rule rather than the exception. The method essentially ignores questions of conditioning and stability that may arise from the transformation [but note that the eigenvalues of $\frac{\partial \mathbf{f}}{\partial \mathbf{y}}$ do not change sign, because $x'(t) > 0$ in (9.27b)]. Moreover, the requirement of a smooth coordinate transformation is questionable: A *local*, rather than global, transformation in the independent variable probably makes more sense.

9.5 GENERAL CONSIDERATIONS

9.5.1 Some practical considerations

One of the most important things to bear in mind when one is producing a nontrivial implementation of mesh selection is this: Before adding sophisticated precautions, it is important that one first have a good understanding of the key elements of the basic strategy. For, while the need to add such precautions can be essential, premature modification of a code may obscure such understanding (e. g., of inherent limitations).

Obviously, the success of a direct strategy depends on the success of the monitor at estimating the errors. Thus, while a crude monitor function may perform satisfactorily at the outset for a crude initial mesh, one which more accurately reflects the actual local error of the numerical method will usually be much more efficient if moderate or high accuracy is desired. One potential pitfall with the latter, however, is that difficulty may occur if the mesh is chosen and the error is estimated using the same monitor function.

One method for reliably estimating errors is related to extrapolation and is described in Section 9.3.1. A possible alternative is to estimate the error in terms of the residual, e. g., (9.15a) can be rewritten as

$$u(x) - u_\pi(x) = h_i^m r(x_i) Q(\frac{x - x_i}{h_i})(1 + O(h_i)) + O(h^p) \qquad (9.31)$$

where $Q(t)$ is a known polynomial and the residual $r(x) = L(u(x) - u_\pi(x)) = f(x) - L_\pi(x)$ is computable. Thus, errors can be *as.eq.* by choosing $s \leq m$ and $T(x) = r(x)$ [cf. (9.10)]. Residuals have been commonly used to estimate local errors for mesh selection. Even for traditional finite difference methods, there are various ways that an approximate residual can be found and used for this purpose.

When one has successfully located regions where a mesh is too coarse by having the large local error estimates there, a variety of different solution procedures may be useful. One is to only solve local problems on refined meshes in such regions — a process sometimes called *chopping* or *rezoning*. For example, if the BVP has been successfully solved on $[a, c]$, where $a < c < b$, then the reduced BVP on $[c, b]$ can be solved by using the needed portion of the computed solution to specify appropriate BC at $x = c$ (e. g., see Exercise 3). After successfully solving on locally refined subintervals whose union gives $[a, b]$, as a precaution one could then solve the problem over $[a, b]$, using the union of the previous local meshes as the final mesh.

Another possibility is to locally adapt a mesh *and* the order of the method. This is similar to what is done in some adaptive quadrature routines. With such a doubly adaptive approach, an added technical problem is that one must check the local smoothness assumptions on the solution. Furthermore, as the order of the method changes, the mesh selection strategy itself usually does too, e. g., if equidistribution of local discretization errors is attempted. This difficulty of constantly changing goals is one that a mesh selection strategy for an acceleration method faces.

While the transformation method of Section 9.4 seems risky, at least until it is better understood, it *is* a method which uses uniform meshes. This is perhaps useful if acceleration methods or finite differences for high-order equations are to be used. (Still, note that uniform meshes with chopping provide another option.)

In the beginning of this chapter we mention that results for mesh selection are mostly heuristic and that solid theory is lacking. Indeed, most examples used here are of the type dealt with in Chapter 10, and the general recipes which have been applied for their mesh selection ignore their very stiff nature. Some special mesh selection strategies for singular perturbation problems and for BVPs over infinite intervals are discussed in later chapters. For the former, one may desire a less stringent error criterion in regions of rapid solution variation than elsewhere. For the mesh strategies recommended in this chapter, it has been assumed that the solution is to be computed with uniform accuracy throughout the interval. Indeed, a side benefit of mesh selection is that a plot of the computed solution on the final mesh is frequently an excellent visual representation of the graph of the solution.

9.5.2 Coordinating mesh selection and nonlinear iteration

Given a mesh of size N, the computational effort to solve the corresponding discrete BVP is proportional to N times the number of Newton iterations required on that mesh. Since the meshes normally become successively finer, it is obviously desirable that any large number of Newton iterations be done on the initial coarse meshes and that few be done on the finer ones. Fortunately, this is usually the case because a good initial approximation is available on the finer meshes. It is common with COLSYS, for example, that most of the work is done on the coarse meshes and only one iteration is required on the finer ones.

Example 9.8

Consider the nonlinear BVP

$$\varepsilon y'' + yy' - y = 0 \qquad 0 < x < 1 \tag{9.32a}$$

$$y(0) = \beta_1, \qquad y(1) = \beta_2 \tag{9.32b}$$

A variety of solution shapes are possible, depending on the boundary data (see Exercise 5). We use COLSYS to solve the problem for $\varepsilon = 2^{-10}$, $\beta_1 = \beta_2 = 1/2$, tolerances 10^{-8} on y and 10^{-4} on y', $k = 4$, and a linear approximation to $y(x)$ satisfying the BC. The solution $y(x)$ is shown in Fig. 9.6. When the problem is solved with continuation for $\varepsilon = 2^{-2}, 2^{-4}, 2^{-6}, 2^{-8}$, and 2^{-10}, starting with $N = 1$, the mesh sequence and number of iterations are as follows:

$$\varepsilon = 2^{-2}: 1(3), 2(1), 4(1), 8(1), 16(1)$$

$$= 2^{-4}: 8(3), 16(1), 9(1), 18(1)$$

$$= 2^{-6}: 9(4), 18(1), 36(1), 18(1), 36(1)$$

$$= 2^{-8}: 18(4), 36(1), 21(1), 42(1)$$

$$= 2^{-10}: 21(4), 21(1), 42(1), 84(1)$$

For the final mesh of 84 points, 34 are in [0, .04), 14 are in (.4, .6), and $h/h_{\min} \approx 126$. (No attempt is made to choose the continuation steps for optimal efficiency.) When the problem

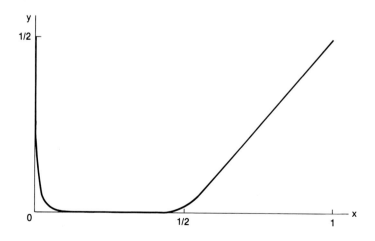

Figure 9.6 Solution for the BVP (9.32) with $\beta_1 = \beta_2 = 1/2$

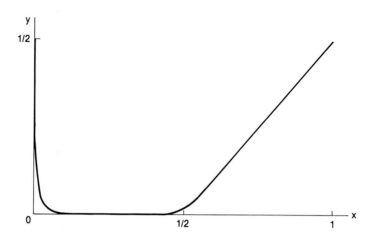

is instead solved directly for $\varepsilon = 2^{-10}$ with $N = 1$ initially, the mesh and iteration sequences are as follows:

$$1(12), 2(6), 4(6), 8(5), 16(\text{NC}), 32(6), 25(2), 50(1), 27(1), 54(1)$$

For this last mesh, 22 points are in $[0, .04)$, 10 are in $(.4, .6)$, and $h/h_{min} \approx 164$. For this example the direct solution for $\varepsilon = 2^{-10}$ requires less storage, but usually continuation parameters can be chosen such that this is not the case (for difficult problems for which a good initial guess is not known *a priori*).

The problem behaves quite differently for $\varepsilon = 2^{-18}$. Then, carrying on the above process with ε decreasing by 0.25 each step, the problem is eventually solved for a final mesh with $N = 72$ and $h/h_{min} \approx .11 + 6$. Attempting to solve directly for $\varepsilon = 2^{-18}$ fails, as there is no convergence on the initial $N = 5$ mesh, convergence on the next $N = 10$ mesh, and no convergence on any subsequent mesh. □

Generally, one is better off choosing a very coarse mesh and letting a code perform most of its iterations on this mesh. Naturally, there is some trade-off point, because a crude mesh can fail to give convergence and/or fail to make use of an accurate initial approximation. All this makes general comparisons between using continuation and using a direct solution difficult.

The study of ways to closely coordinate mesh selection and nonlinear iteration is currently in its preliminary stages, but presumably it will be very important in the future. One way to combine them is to not iterate too much on any given mesh. For example, if we somehow determine that the discretization error for the approximate solution on a mesh π is bounded by $\delta > 0$, then the iteration on the corresponding discrete system could be terminated when consecutive iterates differ by less than δ. Another possibility arises when one is using continuation: If the continuation can be formulated in such a way that the meshes do not change much as the continuation parameter does (so the coordinate transformation is not rapidly changing), then potential savings are considerable. This is done in Example 8.5, where an interior layer is "anchored" for the appropriate choice of continuation parameter.

We mention in the introduction of this chapter that one cannot expect to solve a nonlinear BVP unless the initial mesh provides a sufficiently accurate discretization to the BVP. (But note here that "sufficiently accurate" usually refers to a rather crude discretization and is not related to the required error tolerance.) Interpreting our framework for mesh selection in terms of two-phase methods, we see that there are in a sense two nonlinear BVPs to solve — one for the coordinate transformation and one for the original BVP. A given mesh can be interpreted as being a discrete approximation to the solution of the first one. In other words, an approximate solution to a nonlinear BVP consists in a real sense of *both* a mesh and the approximate solution on that mesh. Note that when we use COLSYS in the continuation examples of this chapter and of Chapter 8, *both the solution and mesh* are gradually refined.

9.5.3 Other approaches

A major difficulty with adaptive mesh selection in general is that regions of fast solution variation can be hard to locate without using an initial mesh that is dense everywhere. The resulting discrete problem can be computationally demanding, especially with regard to storage requirements. The practical remedy for this is usually a continuation process where the mesh is refined gradually as the problem gets tougher, but this does not always work efficiently (and on rare occasions not at all). As we discuss in Section 9.2, a possible solution to this difficulty is to find an appropriate *a priori* mesh. One approach is to locate such regions ahead of time and then build this information into the initial mesh. There are various ways to do this in special circumstances, and several of these are discussed in the next chapter. For instance, recall from Section 9.4.2 the local use of an explicit stretching transformation for selecting an initial mesh. Here we briefly mention two other possibilities.

We consider a linear BVP, having separated BC,

$$\mathbf{y}' = A(x)\mathbf{y} + \mathbf{q}(x) \qquad a < x < b \qquad (9.33a)$$

$$B_{11}\mathbf{y}(a) = \boldsymbol{\beta}_1, \qquad B_{22}\mathbf{y}(b) = \boldsymbol{\beta}_2 \qquad (9.33b)$$

The Riccati transformation method described in Section 4.5.2 can be applied to solve this BVP. As it turns out, for many problems the Riccati matrix $R(x)$ is well-behaved when $\mathbf{y}(x)$ is smooth and has rapid transition regions when $\mathbf{y}(x)$ has. In such cases, if $R(x)$ is computed by using an IVP solver with an automatic mesh selection, then the resulting mesh can be used as an initial mesh for a *global* method (i. e., a method of Chapter 5). The possible advantage is relatively small storage requirements, even for large systems (9.33a). This is achieved when $R(x)$ is computed, if only the mesh points (and not the transformation matrix) are stored. The usefulness of this idea is under investigation at the time of this writing.

A different strategy, perhaps with much the same effect, involves computing the eigenvalues of $A(x)$ throughout $[a, b]$, and using their magnitudes and rates of change to predict possible transition regions for $\mathbf{y}(x)$.

To some extent, however, a major problem remains how to insure that important regions are not missed. For the Riccati method there is the danger that during the transformation computation, large steps are taken and regions of interest are missed (cf. Section 10.4.2). A similar danger exists for the latter approach. Another major obstacle is that the initial mesh generated in this way may be too fine, relating to final accuracy rather than to basic solvability requirements. Recall that for nonlinear problems most Newton iterations are expected to be performed on this initial mesh. For some nonlinear problems this entire initial mesh generation process may in addition depend strongly on the initial solution profile about which we linearize to obtain (9.33). Nevertheless, the basic appeal of these ideas remains: Through some preliminary analysis, an initial mesh adapted to the particular BVP is chosen, eliminating the need to solve large linear systems with a global method before transition regions can be located.

EXERCISES

1. Show that the monitor function (9.11b) will give quasiuniform meshes. *Hint:* If $M = \max_{a \leq x \leq b} |\mathbf{T}(x)|$, show that $\varepsilon = \dfrac{M}{K^p}$, where *as.eq.* meshes satisfy

$$\frac{h}{\min\limits_{1 \le i \le N} h_i} \le K$$

2. Suppose a BVP with solution and derivatives behaving like $y^{(j)}(x) \approx (1/\varepsilon)^j e^{-x/\varepsilon}$ is to be solved with an sth order collocation method. Give an argument why the method might act essentially like a jth order method if the *as.eq.* meshes used the monitor function $|y^{(j)}(x)|^{1/j}$ for $j < s$. *Hint:* Consider the first term in the error bound, assuming that $\tau_i(y) = h_i^3 |y^{(j)}(x_i)|^{1/j}(1 + O(h))$.

3. Write a simple finite difference code based upon (5.4a), and use it to solve the BVP

$$u''(x) + 2\gamma x u'(x) + 2\gamma u(x) = 0, \qquad 0 < x < 1$$

$$u(0) = 1, \qquad u(1) = e^{-\gamma}, \qquad \gamma = 50$$

Now modify the code so that it uses chopping as outlined in Section 9.5.1, using a fairly crude method of estimating the error. Try to solve the BVP again using values of ε for which the unmodified code fails.

4. The mesh selection strategy for collocation in Section 9.3.1 can be interpreted as a two-phase strategy as follows: One alternately solves (i) the unmodified BVP (9.14) by collocation and (ii) the BVP (9.14a, b), $\theta'(x) = 0$ for $t(x)$ and θ using an appropriate monitor function and finite difference scheme. Describe in detail what is involved in (ii) here.

5. Consider the nonlinear BVP of Example 9.8. The solution behaviour is quite different for various choices of β_1 and β_2 in the BC. Interesting cases are (i) $\beta_1 = 1/2$, $\beta_2 = 2$, (ii) $\beta_1 = 1$, $\beta_2 = 3/2$, (iii) $\beta_1 = -3/10$, $\beta_2 = 3/10$, (iv) $\beta_1 = -1$, $\beta_2 = 1$, and (v) $\beta_1 = -2$, $\beta_2 = 1/2$. Solve these problems for $\varepsilon = 2^{-10}$ directly and for continuation from $\varepsilon = 2^{-2}$ with a reduction factor of 2^{-2} for 5 steps. Experimenting with various initial guesses and tolerances, compare the efficiency of these two approaches.

6. Consider application 1.6. The treatment in Example 9.4 suggests that for numerically solving this difficult problem, significant efficiency may be gained by a good choice of initial solution profile and mesh. This may perhaps be achieved by a smart continuation scheme for larger values of L. Try to devise such a scheme. Experiment for smaller (i. e., larger in magnitude) values of β_1 as well.

10

Singular Perturbations

The class of BVPs which we shall consider in this chapter has for many years been a source of difficulty — and entertainment — to applied mathematicians and numerical analysts alike. Mathematically, we may consider a system of ODEs where one or more of the highest derivatives appearing is multiplied by a small parameter ε. If we let ε approach 0, the order of the system reduces, and one cannot in general expect all boundary conditions imposed on the original ODE system to be satisfiable. The perturbation is then "singular." when ε is not 0 but is small, the solution is expected, under certain conditions, to exhibit narrow regions of very fast variation (so-called boundary or interior layers) which connect wider regions where it varies more slowly. A number of the applications listed in Section 1.2 are of this type — e.g., see Examples 1.4, 1.8, 1.14, 1.15–1.17, 1.20, 1.23, 1.24, 8.3, 8.5, 9.2, 9.6, 9.7, and 9.8. If an analytic expression for the solution is desired in the form of an asymptotic expansion, then different asymptotic expansions are needed in different parts of the domain of the independent variable, with no one expansion being uniformly valid everywhere.

Numerically, the situation is different here than it is in Chapters 4 and 5. There we had just one small parameter, the discretization parameter h. For instance, it was assumed for the standard ODE

$$\mathbf{y}' = A(x)\mathbf{y} + \mathbf{q}(x) \tag{10.1}$$

that $h \|A\| \ll 1$, where h is the maximum step size of the discretization. Here there are two small parameters, namely a small parameter ε occurring in the BVP [with $\|A\| = O(\varepsilon^{-1})$, say] and the small discretization parameter h. For efficiency reasons we want, if possible, to consider the case when h is not smaller than ε, or even when $\varepsilon \ll h$ ($\ll 1$). This occurs when some kinematic eigenvalues of $A(x)$ are much larger in magnitude than the inverse of the interval size, $(b-a)^{-1}$ [cf. (2.152)]. The BVP will be called *stiff* in such a case.

Obviously the concept of stiff BVP in numerical analysis relates to the concept of singularly perturbed BVP in applied mathematics, as the following simple example demonstrates.

Example 10.1

Consider the initial value problem for a single ODE

$$\varepsilon y' = -y + q(x) \qquad x > 0$$

$$y(0) = \alpha$$

where ε is positive and small, and the inhomogeneity $q(x)$ is a smooth function bounded by a constant of order 1. This IVP is certainly stable (so as a special case of a BVP we would say that it is well-conditioned).

A well-scaled fundamental solution for this example is $\exp(-x/\varepsilon)$, and the solution y can be written as a sum $y \equiv y_T + y_S$ of two components: A transient component

$$y_T(x) = (\alpha - q(0))\exp(-x/\varepsilon)$$

which dominates the solution behaviour near $x = 0$ (unless $\alpha = q(0)$, a case we shall ignore for now) and a component which can be written as

$$y_S(x) = q(x) + O(\varepsilon)$$

(see Figure 10.1).

We distinguish between two domains in the independent variable x. For

$$0 \leq x \leq K\varepsilon|\ln \varepsilon|,$$

with K a constant of order 1, the transient solution $y_T(x)$ dominates and has derivatives which blow up as $\varepsilon \to 0$. To solve the problem numerically, it seems appropriate because of the high activity to use a fine mesh, so that $h \ll \varepsilon$ there. In fact, applying a *stretching transformation*

$$\xi := x/\varepsilon$$

and writing the ODE in terms of the independent variable ξ, we find

$$\frac{dy}{d\xi} = -y + q(\varepsilon\xi)$$

the step size $\eta := h/\varepsilon$ corresponding to ξ satisfies $\eta \ll 1$. Therefore, we may say that in the stretched variable ξ we have a "regular" numerical discretization of a regular ODE. The usual theory for the numerical error and its locality can be extended for many methods, and this is considered in Section 10.2.4. [Note that the domain in ξ stretches to the entire nonnegative half interval as $\varepsilon \to 0$, so the usual theory has to be extended, strictly speaking, because the mesh size, i. e., the number of mesh points needed, is much larger than η^{-1}. Moreover, in practice we could encounter in this domain step sizes which satisfy only $h = O(\varepsilon)$, i. e., $\eta = O(1)$. Still, the essentials of the usual theory change much less than for the second domain.]

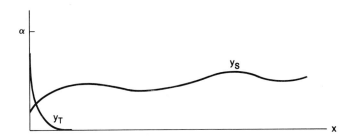

Figure 10.1 Smooth and transient components of the solution.

The second, longer domain is for $x \geq K\varepsilon|\ln\varepsilon|$. We get $\exp(-x/\varepsilon) = O(\varepsilon)$, so for small ε the transient solution component is no longer significant. In this domain the solution is essentially given by $y_S(x)$ and varies slowly, so we expect to be able to use a coarse mesh to approximate it, i. e., $\varepsilon \ll h$ there. However, in terms of the standard linear form (10.1), our ODE in this example satisfies $A = -1/\varepsilon$ for any x, i. e., $h\|A\| \gg 1$, so many of the results of Chapters 4 and 5 do not necessarily apply and an additional, different analysis is needed. For instance, for the midpoint scheme (5.19)

$$\frac{\varepsilon}{h_i}(y_{i+1}-y_i) = -\frac{1}{2}(y_{i+1}+y_i) + q_{i+1/2},$$

the solution in the second domain is (approximately) given in terms of the first term y_l in this domain and a sum of inhomogeneities of the form $2\sum_{j=l}^{i}(-1)^j q_{j+1/2}$. Consequently, the global error is a sum of the form $\sum_{j}(-1)^j \tau_j$ with the local truncation error

$$\tau_j = \frac{1}{8}h_j^2 y''(x_{j+1/2}) + O(h_j^3).$$

This global error is not in the usual form [see Theorem 5.79 and Section 5.5.1]. □

This exemplifies the general approach to be used in the sequel: a stretching transformation is used where the solution has layers, and a separate analysis for h "large" is done in the other region(s).

But the application of numerical methods is not the only matter which requires careful consideration for singularly perturbed BVPs. The conditioning results of Chapter 3 also need to be more delicately treated, as discussed briefly in Section 2.8 and demonstrated in Example 3.9 and in the following one.

Example 10.2

Consider the IVP

$$\varepsilon y' = -(x+\varepsilon)y + q(x), \qquad x > 0$$

$$y(0) = \alpha$$

Unlike Example 10.1, where the solution for any piecewise continuous bounded inhomogeneity $q(x)$ is bounded as $\varepsilon \to 0$, here for boundedness we must require that $q(x) = O(\varepsilon)$ for x near 0. If, for instance, $q(x) \equiv \sqrt{\varepsilon}$ near 0 and $\alpha = 0$, then it can be shown that $\|y\|_\infty = O(\varepsilon^{-1/2})$.

The stability theory of Section 3.3 still holds, of course, but it does not distinguish between Example 10.1 and this one. Indeed, the fundamental solution $Y(x) = \exp(-x/\varepsilon)$ for Example 10.1 and the fundamental solution $Y(x) = \exp(\frac{-x^2}{2\varepsilon} - \frac{x}{\varepsilon})$ here both satisfy $\|Y\|_\infty = 1$. Writing both ODEs in the form (10.1), the solutions are correspondingly bounded by the inhomogeneity $\varepsilon^{-1}\|q\|_\infty$. The *difference* is that the bound in the previous example can be improved to $O(\|q\|_\infty)$, while here it cannot. This can be done by using the L_1 norm, rather than L_∞, to measure the size of $Y(x)$. In the L_1 norm we have for Example 10.1

$$\|Y\|_1 = \int_0^\infty e^{-x/\varepsilon}\, dx = \varepsilon$$

while here $\|Y\|_1 \sim 1$ [the integral of $e^{-x^2/2\varepsilon}$ is proportional to the so-called *error function* $\operatorname{erf}(x/\sqrt{2\varepsilon})$].

The reason for this basic difference is that in Example 10.1 we actually have a trivial case of *uniform exponential dichotomy,* i.e.,

$$|Y(x)| \leq K e^{-cx/\varepsilon}$$

for some positive constants K and c independent of ε, while for this example no such uniform exponential dichotomy holds. The use of the L_1 norm in calculating the conditioning constant κ_2 in (3.89b) is discussed in Section 3.4; see again, in particular, Example 3.9. □

10.1 ANALYTICAL APPROACHES

We begin with a disclaimer: It is not our intention to duplicate, or replace, a text on singular perturbation problems. Rather, we wish to give a taste for the types of problems and solution behaviours which may arise and to prepare the background for our numerical purposes. Our treatment is particularly incomplete with respect to scalar, second-order Dirichlet BVPs, which have received an enormous amount of attention in the literature. Techniques like power series solutions and matched asymptotic expansions will not be dealt with in depth here.

Throughout this chapter, we restrict ourselves to BVPs which are *well-conditioned uniformly in* ε, i.e., whose conditioning constant (when the BVP is suitably scaled) remains bounded as $\varepsilon \to 0$. This excludes many pathological examples from consideration but still encompasses almost all problems of practical interest.

Finally, a word about notation: Our primary aim here is to distinguish between quantities which remain bounded as $\varepsilon \to 0$ and those that do not. So, when we refer to constants or to bounded quantities, *it is to be understood that they are of moderate size.* We will also use K, c, K_j, and const as generic constants, which are meant to be independent of the small parameter ε, as in Examples 10.1 and 10.2. As well, we will not make a distinction between various vector norms, unless specifically mentioned.

10.1.1 A linear second-order ODE

In this subsection we analyze the scalar Dirichlet problem (1.2c,b) where the coefficients become large in magnitude. To express this explicitly and to scale the ODE properly, we rewrite the BVP as

$$\varepsilon u'' - p(x)u' - q(x)u = g(x), \qquad a < x < b \qquad (10.2a)$$

$$u(a) = \beta_1, \qquad u(b) = \beta_2 \qquad (10.2b)$$

where $p(x), q(x)$, and $g(x)$ are smooth, bounded functions, and $0 < \varepsilon \ll 1$. Generally, the inhomogeneities g and β_1, β_2 may depend on ε as well. In the following discussion, we seek results which hold uniformly in ε for ε small enough.

Example 10.3

If we choose $p(x) = -(2 + \cos \pi x)$, $q(x) = 1$, $u(0) = 0$, $u(1) = -1$, and

$$g(x) = -(1 + \varepsilon \pi^2)\cos \pi x - \pi(2 + \cos \pi x)\sin \pi x + (1 + \frac{3}{2\varepsilon}\pi^2 x^2)e^{-3x/\varepsilon}$$

then the solution to (10.2) is

$$u(x) = \cos \pi x - e^{-3x/\varepsilon} + O(\varepsilon^2)$$

As before we see that the solution has a rapidly varying component which is dominant in a narrow layer near $x = 0$ and a slowly varying component which is dominant away from the layer. The *reduced solution* for this problem is obtained by setting $\varepsilon = 0$ in (10.2a) and integrating the resulting first-order ODE with the BC $u(1) = -1$ (where there is no layer). The smooth solution away from the boundary layer at $x=0$, which is sometimes referred to as an *outer solution,* is seen to be approximately equal [up to $O(\varepsilon)$] to the reduced solution. Inside the layer, the situation is different. Here the solution is referred to as an *inner solution* and the appropriate variable to describe it is the stretched one

$$\xi := x/\varepsilon$$

as in Example 10.1. □

Let us now see what can be said in general about the solution of (10.2). We distinguish among a number of different cases, depending upon the properties of $p(x)$.

$$y(0) = \alpha$$

First consider the special case $q(x) \equiv p'(x)$. Then (10.2a) is self-adjoint and can be written as

$$\varepsilon u'' = (p(x)u)' + g(x)$$

Integrating, we obtain

$$\varepsilon u' = p(x)u + v(x) \qquad (10.3)$$

where $v(x) = \int_a^x g(t)\,dt + \alpha$, with α a free parameter. Hence, $v(x)$ is a smooth inhomogeneity. The first-order ODE (10.3) has a simple, familiar form which has been investigated in Chapter 3 (see also Example 10.1). Further, its stability properties relate directly to the conditioning of the BVP (10.2). We obtain the following cases:

(a) $p(x) < 0, \quad a \le x \le b$
Then the ODE (10.3) is stable with the initial value $u(a) = \beta_1$. As in Example 10.1, there is generally a boundary layer of width $O(\varepsilon)$ near $x = a$. The other boundary condition in (10.2b) is at $x = b$ and is satisfied by pinning $v(x)$ down; i.e., it determines α. There is no possibility for a boundary layer near $x = b$ in this case (see Exercise 1).

(b) $p(x) > 0, \quad a \le x \le b$
This is the "mirror image" of case (a): The ODE (10.3) is stable with the terminal value $u(b) = \beta_2$, and there is generally a boundary layer near that end of the interval. By choosing the parameter α for $v(x)$ appropriately, the boundary condition at $x = a$ is satisfied without $u(x)$ having a layer there.

(c) $p(x)$ changes sign in $[a, b]$
In this case the ODE (10.3) is generally unstable under any choice of boundary conditions, because there is no dichotomy. For example, the BVP

$$\varepsilon u'' + (xu)' = 0$$

$$u(-1) = u(1) = 1$$

has the "exploding" solution $u(x) = e^{(1-x^2)/2\varepsilon}$. Note how easy it is to obtain an extremely ill-conditioned problem! We will have more to say about this case below. In this presentation we assume that the (smooth) function $p(x)$ changes sign at any zero it may have.

Next we remove the restriction that $q \equiv p'$. Then (10.2a) reads

$$\varepsilon u'' = (p(x)u)' + (q(x) - p'(x))u + g(x)$$

Integrating, we obtain (10.3) again, where now $v(x)$ satisfies

$$v' = (q(x) - p'(x))u + g(x) \qquad (10.4)$$

Thus, (10.3), (10.4) is a first-order system which is equivalent to the original ODE (10.2a). This system can be written in the form (10.1) for $\mathbf{y}(x) = (u(x), v(x))^T$ with the resulting coefficients

$$A(x) = \begin{bmatrix} p(x)/\varepsilon & 1/\varepsilon \\ q(x)-p'(x) & 0 \end{bmatrix}, \quad \mathbf{q}(x) = \begin{bmatrix} 0 \\ g(x) \end{bmatrix} \quad (10.5)$$

Note that the derivative of $v(x)$ is not multiplied by ε in (10.4), but it depends on $u(x)$ whose derivative is multiplied by ε in (10.3).

To get rid of this ε-dependence, consider the transformation

$$w(x) \equiv v(x) + \varepsilon r(x)u(x) \quad (10.6)$$

in (10.4), yielding (verify!)

$$w' = [q - p' + \varepsilon r' + rp - \varepsilon r^2]u + rw + g$$

This gives

$$w' = r(x)w + g(x) \quad (10.7)$$

if $r(x)$ satisfies

$$\varepsilon r' = -pr + p' - q + \varepsilon r^2 \quad (10.8)$$

To guarantee that a smooth $r(x)$ satisfying (10.8) exists, we must assume that $p(x)$ is bounded away from 0 on the interval $[a,b]$. [In particular, $p(x)$ cannot change sign.] Then we may treat $\varepsilon r' - \varepsilon r^2$ as a small perturbation, obtaining

$$r(x) = \frac{p'(x) - q(x)}{p(x)} + O(\varepsilon) \quad (10.9)$$

(see Exercise 2).

Writing (10.3) but now with $w(x)$ replacing $v(x)$, we have

$$\varepsilon u' = (p(x) - \varepsilon r(x))u + w \quad (10.10)$$

and (10.10), (10.7), (10.2b) is a BVP equivalent to (10.2), where the *slow solution component* $w(x)$ is separated from the *fast solution component* $u(x)$. This slow solution is smoothly integrated, depending on a parameter, so our treatment for the case $p' \equiv q$ is readily extended. We conclude that *cases (a) and (b) above also extend to when $p'(x)$ is not identically equal to $q(x)$*. Example 10.3 is covered by case (a). Note that from (10.6), $v(x) = w(x) + O(\varepsilon)$, so $v(x)$ is also a slow solution component; i.e., it does not vary rapidly anywhere. A BVP like this which has fast and slow solution components is sometimes said to have *two time scales*.

Remark The ODE (10.8) which $r(x)$ satisfies may remind our reader of the Riccati equation. It is. Indeed, the purpose of the transformation is to separate solution components: In Section 4.5 we separate growing and decaying solution components, while here we separate fast and slow components. Here we do not *find* the Riccati transformation $r(x)$; we only capitalize on its existence when it exists in order to explain the solution behaviour for (10.2) (see also Section 10.4.2). □

Consider now the case where $p(x_0) = 0$ at some point x_0, but $q(x) \not\equiv p'(x)$. The construction (10.9), hence the separation of fast and slow solution components, fails to hold. We shall see that the reason is simply because there *are* no such components to separate near x_0. Let us first investigate the simple case where $p(x) \equiv 0$ on the interval $[a,b]$. A suitable conversion of (10.2a) to a first-order system is obtained for the

variables $u(x)$ and
$$v(x) := \mu u'(x), \qquad \mu := \sqrt{\varepsilon}$$
which give
$$\begin{bmatrix} u \\ v \end{bmatrix}' = \mu^{-1} \begin{bmatrix} 0 & 1 \\ q(x) & 0 \end{bmatrix} \begin{bmatrix} u \\ v \end{bmatrix} + \mu^{-1} \begin{bmatrix} 0 \\ g(x) \end{bmatrix} \qquad (10.11)$$

The eigenvalues of the coefficient matrix in (10.11) are $\pm \mu^{-1} \sqrt{q(x)}$. The assumption that $q(x)$ varies smoothly as $\varepsilon \to 0$ implies that the eigenvalues also vary smoothly, and it can be expected that these eigenvalues are indicative of the kinematic eigenvalues (corresponding to a smooth kinematic similarity transformation). Hence the different cases depend upon the sign of $q(x)$.

(d) $q(x) > 0$, $p(x) = 0$, $a \leq x \leq b$
There are two real eigenvalues, one large and positive and one large and negative. The Dirichlet problem (10.2) is therefore well-conditioned, both u and v are fast solution components, and there are possible layers of width $O(\sqrt{\varepsilon})$ at both boundaries.

(e) $q(x) < 0$, $p(x) = 0$, $a \leq x \leq b$
Here the eigenvalues are both imaginary and the solution oscillates rapidly (with period $O(\sqrt{\varepsilon})$) throughout the interval. We will not treat such problems in this chapter (but see Section 11.6).

(f) $q(x)$ changes sign in $[a,b]$, $p(x) \equiv 0$
This is sometimes referred to in the literature as a *classical turning point*. The solution nature changes from rapidly oscillatory to highly evanescent. Again, we will not treat such problems here.

An application where an ODE of type (f) occurs is the seismic problem of Example 1.8.

Let us return now to case (c) above, viz., where $p(x)$ has a zero, say at $x_0 \in (a,b)$, and also changes sign. This case of a so-called *turning point* can lead to a variety of solution behaviours, including a *transition layer* near x_0, where the solution or one of its derivatives varies rapidly. We will not attempt to survey all possibilities, but instead consider some typical cases. To this end, we assume that $a < 0 = x_0 < b$ [so $p(0) = 0$], that 0 is the only turning point [so $p(x) \neq 0$ for $x \neq 0$], and that $p'(0) \neq 0$. Much depends on the sign of $p'(0)$:

(g) $p'(0) < 0$ ($p(0) = 0$; $p(x) \neq 0$ for $x \neq 0$)
Consider the subintervals $[a, -\delta]$ and $[\delta, b]$ for $0 < \delta \ll \min(a,b)$. The ODE on these subintervals is treated by cases (b) and (a), respectively. Thus, the stable direction of IVP integration is $-x$ on $[a, -\delta]$ and $+x$ on $[\delta, b]$. Therefore, there are no layers near the boundaries, and if the BVP is well-conditioned then there is an interior layer of width $O(\sqrt{\varepsilon})$ (a region of rapid variation in u or its derivative) near $x = 0$.

(h) $p'(0) > 0$ ($p(0) = 0$; $p(x) \neq 0$ for $x \neq 0$)
Now the stable IVP integration directions are *towards* $x = 0$. Boundary layers

near $x = a$ and $x = b$ are expected in general, but near $x = 0$ the solution varies smoothly. In fact, it can be shown that if the BVP (10.2) is well-conditioned then the solution for the homogeneous ODE [i.e., with $g \equiv 0$ in (10.2a)] satisfies

$$y(x) \underset{\varepsilon \to 0}{\to} 0 \qquad a + \delta \leq x \leq b - \delta$$

These cases are summarized in Table 10.1 below. They do not cover all possibilities, but are sufficient to give a brief look at the various situations which may arise.

TABLE 10.1 General Solution Behaviour for BVP (10.2)

$p(x) \neq 0, a \leq x \leq b$	(a) $p(x) < 0$ (b) $p(x) > 0$	boundary layer at $x = a$ boundary layer at $x = b$
$p \equiv 0$	(d) $q(x) > 0$ (e) $q(x) < 0$ (f) $q(x)$ changes sign	boundary layers at $x = a$ and $x = b$ rapidly oscillatory solution classical turning point
$p' \not\equiv q$, $p(0) = 0$, $p(x) \neq 0$ for $x \neq 0$	(g) $p'(0) < 0$ (h) $p'(0) > 0$	no boundary layers, interior layer at $x = 0$ boundary layers at $x = a$ and $x = b$, no interior layer at $x = 0$

Remark Consider case (g). For $|x| \gg 0$, one of the eigenvalues of A in (10.5) is approximately equal to $p(x)/\varepsilon$ and the other is ~ 1. Recall that our interest is only in well-conditioned problems, so from Section 3.4 we must have a dichotomy, or basically one nonincreasing and one nondecreasing fundamental solution *throughout* the interval $[a, b]$. The reader may well wonder how this is possible, since $p(x)$ changes sign. The answer is that a rapidly increasing fast solution component for $x < 0$ changes through the turning point layer to become slow, while the slow solution component for $x < 0$ becomes the rapidly decreasing fast solution component for $x > 0$. This is shown by the example below. □

Example 10.4

The ODE

$$\varepsilon u'' + 2xu' = 0 \qquad -1 < x < 1$$

is treated by case (g) since $p'(0) = -2$. The solution for the Dirichlet BC $u(-1) = -1$, $u(1) = 1$ is $u(x) = \text{erf}(x/\sqrt{\varepsilon})$, which has a transition layer at 0, but no boundary layers, as depicted in Figure 10.2. [When such a transition layer approaches a jump discontinuity in $u(x)$ as $\varepsilon \to 0$, the layer is often called a *shock layer*.]

For this simple ODE, we can explicitly calculate the stable modes and watch them rotate and switch roles as fast and slow solution components through the turning point. To do this, write $v(x) := \mu u'(x), \mu = \sqrt{\pi \varepsilon}$, so that

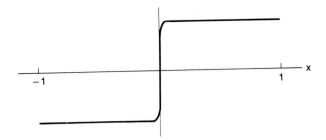

Figure 10.2 Solution profile for Example 10.4

$$\begin{pmatrix} u \\ v \end{pmatrix}' = \begin{pmatrix} 0 & \frac{1}{\sqrt{\pi\varepsilon}} \\ 0 & \frac{-2x}{\varepsilon} \end{pmatrix} \begin{pmatrix} u \\ v \end{pmatrix} =: A \begin{pmatrix} u \\ v \end{pmatrix}$$

A fundamental solution Y is given by (check!)

$$Y(x) := \begin{bmatrix} 1-I(x) & I(x) \\ -E(x) & E(x) \end{bmatrix}$$

where

$$E(x) = e^{-x^2/\varepsilon}, \qquad I(x) = \frac{1}{\sqrt{\pi\varepsilon}} \int_{-\infty}^{x} E(t)\, dt$$

Now consider the transformation

$$T := \begin{pmatrix} \dfrac{1-I}{\sqrt{(1-I)^2+E^2}} & \dfrac{I}{\sqrt{I^2+E^2}} \\ \dfrac{-E}{\sqrt{(1-I)^2+E^2}} & \dfrac{E}{\sqrt{I^2+E^2}} \end{pmatrix}$$

It can be verified that $A(x)$ has kinematic eigenvalues $\sqrt{(1-I(x))^2+E(x)^2}$ and $\sqrt{I^2(x)+E^2(x)}$ corresponding to this transformation, and that $T(x)$ is a well-conditioned matrix uniformly in x (see Exercise 3). □

In the above, we assume that $p(x)$ is independent of ε. Generally we may allow $p = p(x;\varepsilon)$ and/or $q = q(x;\varepsilon)$, provided that the dependence on ε is smooth enough. We see this in the following.

Example 10.5

Consider the BVP

$$\varepsilon u'' = -\mu u' + u$$

$$u(0) = 1, \qquad u(1) = \frac{1}{2}$$

where ε and μ are both small parameters. Various types of behaviour are possible. For example, the asymptotic solution is different if ε/μ^2 tends to 0, 1 or ∞ as $\varepsilon, \mu \to 0$. In all of these cases boundary layers at both ends occur, connecting the boundary values to the zero reduced solution. When $\varepsilon = \mu^2$, the ODE can be written as a first-order system like (10.11), except that now the coefficient matrix is

$$\mu^{-1} \begin{bmatrix} 0 & 1 \\ 1 & -1 \end{bmatrix}$$

This matrix has the eigenvalues $\lambda_+ := -\frac{1}{2\mu}(1-\sqrt{5}) > 0$ and $\lambda_- := -\frac{1}{2\mu}(1+\sqrt{5}) < 0$. The dominant behaviour of the solution near $x = 0$ is therefore $\exp(\lambda_- x/\sqrt{\varepsilon})$, and near $x = 1$ it is $\frac{1}{2}\exp(\lambda_+(1-x)/\sqrt{\varepsilon})$.

□

10.1.2 A nonlinear second-order ODE

As might be expected, the combination of singular perturbations and nonlinearity can yield bizarre examples. To appreciate the new difficulties arising, note that the linearization of the ODE

$$\varepsilon u'' = f(x, u, u') \tag{10.12}$$

about an exact solution yields an ODE like (10.2a), but the coefficients $p(x)$ and $q(x)$ cannot be assumed smooth in layer regions anymore, because they depend on the rapidly varying solution. For instance, sharper transition (turning point) layers may occur here than for the linear problem.

A major source of potential computational difficulty is the fact that the *location* of turning points now depends on the solution and is therefore generally not known in advance. Moreover, unlike for linear problems, the solution profile (and the BVP conditioning) may change drastically with the boundary *values*.

Example 10.6

The ODE

$$\varepsilon u'' + u u' - u = 0 \qquad 0 < x < 1$$

subject to (10.2b) gives rise to 10 characteristically different solution profiles involving boundary layers at 0 and/or at 1 and interior corner layers or shock layers, depending on the values of β_1 and β_2 (see Example 9.8, Exercise 4, and Exercise 9.5 and the references).

□

The dependence of the solution on boundary values in such a dramatic way should not be too surprising. Consider the special case of the ODE (10.12),

$$\varepsilon u'' - p(u)u' - q(x, u) = g(x) \tag{10.13}$$

The profile for $u(x)$ certainly depends on the boundary values β_1 and β_2, so zeros of $p(u)$ may also depend on them. Thus, the number and location of turning points may depend on the boundary values. In general we may expect the possibility of a number of characteristically different solution profiles for the BVP (10.13), (10.2b).

Still, under certain restrictions, a much better characterization of the solution behaviour is possible. It can be shown, for instance, that if

$$\frac{\partial q}{\partial u}(x,u) \geq \delta > 0 \tag{10.14}$$

then the BVP (10.13), (10.2b) has a unique solution for $\varepsilon > 0$ small enough. A turning point here corresponds physically to a stationary shock. For this class of BVPs, the time imbedding (8.43) is possible and sometimes useful.

Most of the applications recalled at the beginning of this chapter involve nonlinear problems, and some of them — e.g., Example 1.17 — are even scalar and second order like (10.12).

10.1.3 Linear first-order systems

Compared to scalar second-order problems, relatively little theoretical analysis has been done for the general case of singularly perturbed first-order BVPs. But we expect similar types of solution behaviour to occur here as well. Thus, consider the ODE (10.1), where $A(x)$ and $\mathbf{q}(x)$ depend on a small parameter ε and may become unbounded as $\varepsilon \to 0$, subject to the well-scaled BC [cf. (3.88)]

$$B_a \mathbf{y}(a) + B_b \mathbf{y}(b) = \boldsymbol{\beta} \tag{10.15}$$

We generally expect a solution profile which has boundary layers and/or interior layers connecting longer subintervals where the solution varies smoothly, as e.g., in Fig. 10.3.

To qualitatively describe a solution like that in Fig. 10.3 we envision having a partition, or *segmentation,*

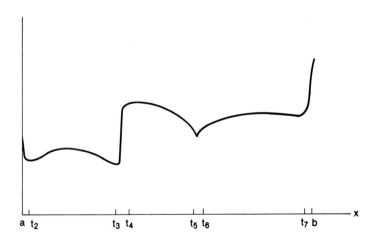

Figure 10.3 A hypothetical solution profile

$$a = t_1 < t_2 < \cdots < t_M < t_{M+1} = b \tag{10.16}$$

such that M is fixed (independent of ε), and on each subinterval $[t_j, t_{j+1}]$, precisely one of the following occurs:

(i) The solution has a boundary layer. Then $j = 1$ (for a layer near a) or $j = M$ (for a layer near b) and $t_{j+1} - t_j \to 0$ as $\varepsilon \to 0$.

(ii) The solution has an interior layer. Here $1 < j < M$ and $t_{j+1} - t_j \to 0$ as $\varepsilon \to 0$.

(iii) The solution is smooth on the subinterval, i.e., for some positive integer p

$$\|\mathbf{y}^{(\nu)}\|_{[t_j, t_{j+1}]} \leq \text{const} \qquad \nu = 0, 1, \ldots, p \tag{10.17}$$

(where const is independent of ε).
Here and in the sequel we use the notation

$$\|\mathbf{z}\|_{[c,d]} := \max_{c \leq x \leq d} |\mathbf{z}(x)| \tag{10.18}$$

Now, the analytic approach we shall use to handle each of the short subintervals of types (i) and (ii) is to apply a stretching transformation to the independent variable, as in Example 10.1 (but possibly more complicated). The choice of such a transformation is not always simple, and we will leave this out of our treatment. The general effect of this stretching is to regularize the ODE on the subinterval, even though the subinterval length in the new variable becomes infinite as $\varepsilon \to 0$. The previous subsections also indicate that for a layer to occur away from the boundaries, some coefficients of the ODE (in $A(x)$ or in $\mathbf{q}(x)$) are not merely large in magnitude, but they also *vary* rapidly. This is what happens to $p(x)/\varepsilon$ near a turning point of (10.2) [where $p(x)$ changes sign], in the sense that $\left|\dfrac{p'(x)}{p(x)}\right|$ is large.

Remark Looking ahead, we note that subintervals of types (i) and (ii) will often be discretized with a fine mesh, which corresponds to applying the stretching transformation. On subintervals of type (iii), however, a coarse mesh is used (e.g., recall Example 9.3), so it is on these subintervals that a nonstandard analysis is most often needed, as already mentioned in Example 10.1. □

Our next task is to see how to identify subintervals of type (iii), assuming that rapid variation in the ODE coefficients does not occur. We begin by examining one stable equation

$$y' = \lambda(x)y + q(x) \qquad c \leq x \leq d \tag{10.19}$$

where we allow the possibility that $\lambda(x)$ and $q(x)$ may become unbounded. The interval $[c,d]$ is a subinterval of $[a,b]$, and we want to specify conditions guaranteeing that the solution is smooth on $[c,d]$. The coefficient $\lambda(x)$ represents an eigenvalue, so it is in general complex.

We make the following assumptions: For simplicity, assume that

$$\max_{c \leq x \leq d} \{\lambda_R(x)\} < 0, \qquad \lambda_R := \text{Re}(\lambda) \tag{10.20a}$$

[cf. Theorem 3.71]. The highly oscillatory case is excluded by further assuming that

$$|\text{Im}(\lambda(x))| \leq \rho |\lambda_R(x)|, \qquad c \leq x \leq d \tag{10.20b}$$

for some constant ρ. We also exclude (relative) rapid coefficient variation by assuming

$$\left\|\frac{\lambda^{(v)}}{|\lambda|+1}\right\|_{[c,d]} \leq K_1, \quad \left\|\frac{q^{(v)}}{|q|+1}\right\|_{[c,d]} \leq K_1, \quad v = 1, \ldots, p \quad (10.21)$$

Finally, to ensure that the solution $y(x)$ is bounded independently of ε, we require that

$$\left|\frac{q(x)}{|\lambda(x)|+1}\right| \leq K_2, \quad c \leq x \leq d \quad (10.22)$$

[The extra term 1 in the denominators of (10.21) and (10.22) is added for technical reasons only, to prevent them from becoming unduly small.]

To see that (10.22) does indeed guarantee a bounded solution (10.19) if $y(c)$ is bounded, note first that (10.22) and (10.20b) yield

$$\left|\frac{q(x)}{|\lambda_R(x)|+1}\right| \leq K_3, \quad c \leq x \leq d \quad (10.23)$$

for an appropriate constant K_3. The solution of (10.19) is

$$y(x) = \exp\{\int_c^x \lambda(\tau) d\tau\} y(c) + \int_c^x \exp\{\int_t^x \lambda(\tau) d\tau\} q(t) dt$$

Assuming without loss of generality that $\lambda_R(\tau) \leq -1$, $c \leq \tau \leq x$, we write

$$|y(x)| \leq \exp\{\int_c^x \lambda_R(\tau) d\tau\}|y(c)| + \int_c^x |\lambda_R(t)| \exp\{\int_t^x \lambda(\tau) d\tau\}| \left|\frac{q(t)}{\lambda_R(t)}\right| dt$$

Noting that

$$\int_c^x |-\lambda_R(t) \exp\{\int_t^x \lambda(\tau) d\tau\}| dt = \int_c^x -\lambda_R(t) \exp\{\int_t^x \lambda_R(\tau) d\tau\} dt$$

$$\int_c^x \frac{d}{dt}[\exp\{\int_t^x \lambda_R(\tau) d\tau\}] dt = 1 - \exp\{\int_c^x \lambda_R(\tau) d\tau\} \leq 1$$

and using (10.23), the desired bound

$$|y(x)| \leq \exp\{\int_c^x \lambda_R(\tau) d\tau\}|y(c)| + 2K_3, \quad c \leq x \leq d \quad (10.24)$$

is obtained (cf. Example 10.2.). This bound shows not only that $y(x)$ is bounded if $y(c)$ is, but also that the effect of the initial value at c rapidly decreases with increasing x, if $\lambda_R \ll -1$.

The assumptions (10.21) are not needed to obtain the bound (10.24); rather, they are used to obtain similar bounds on the derivatives of $y(x)$. To see this, we distinguish between two cases:

(a) If

$$|\lambda_R(x)| \leq K_4, \quad c \leq x \leq d \quad (10.25a)$$

then (10.20b) and (10.22) imply boundedness of the ODE coefficients on this interval. Bounding $y'(x)$ in terms of $y(x)$ through the ODE (10.19) and using

(10.24), we obtain a similar bound for $y'(x)$. Repeated differentiation of the ODE and use of (10.21) yield a similar result for higher-order derivatives. Thus we obtain

$$|y^{(\nu)}(x)| \leq K_5 |y(c)|, \qquad c \leq x \leq d \tag{10.26a}$$

(b) In (10.25a) we have assumed no stiffness on $[c,d]$. Assume now that there is stiffness, or at any rate that λ_R is bounded away from 0, *everywhere* on the interval. It is not too restrictive to assume

$$\lambda_R(x) \leq -1, \qquad c \leq x \leq d \tag{10.25b}$$

Differentiating the ODE (10.19), we find that for $\hat{y} := y'$,

$$\hat{y}' = \lambda(x)\hat{y} + \lambda'(x)y(x) + q'(x)$$

This is an ODE of the form (10.19) for $\hat{y}(x)$, with the inhomogeneity $\lambda'(x)y(x) + q'(x)$. Applying similar arguments as those used to obtain (10.24), we have

$$|\hat{y}(x)| \leq |\hat{y}(c)| + \max_{c \leq t \leq x} \left| \frac{\lambda'(t)y(t) + q'(t)}{\lambda_R(t)} \right|$$

and assumption (10.21) (for $p \geq 1$) implies boundedness of $\hat{y}(x)$ in terms of its initial value. A repeated differentiation of the ODE (10.19) yields

$$\|y^{(\nu)}\|_{[c,d]} \leq K_6 [1 + \sum_{j=0}^{\nu} |y^{(j)}(c)|], \qquad \nu = 0, 1, \ldots, p \tag{10.26b}$$

This latest result tells us that a subinterval of type (iii) is indeed possible provided that the initial data have been "suitably prepared" in the sense that the first $p+1$ derivatives of $y(x)$ can be assumed bounded at c. Moreover, assumption (10.25b) can be relaxed. Thus, given only that (10.20–22) hold, we subdivide the interval $[c,d]$ into as few subintervals as possible such that either (10.25a) or (10.25b) holds on each of them. If the appropriate condition holds at $x = c$ then the corresponding result (10.26a) or (10.26b) applies for the first subinterval. The resulting boundedness of y and its derivatives at the right end of this subinterval allows use of 10.26b for the second subinterval, and so on. We obtain

Theorem 10.27 Assume that the ODE (10.19) satisfies (10.20–22), and that at the left endpoint, $|y(c)|$ is bounded and either $|\lambda_R(c)| \leq K$ or $|y^{(\nu)}(c)| \leq K$, $\nu = 0, 1, \ldots, p$. Then

$$\|y^{(\nu)}\|_{[c,d]} \leq \text{const} \qquad \nu = 0, 1, \ldots, p$$

i.e., $[c,d]$ is a subinterval of type (iii). □

Let us explain how we shall use this theorem to analyze singular perturbation problems. As in Example 10.1 we expect a stable IVP for (10.19) to generally have an initial layer before the solution becomes smooth. If there is such a layer, then the "initial point" c in (10.19) can be chosen as a point already beyond that layer (i.e., the transient solution component has already died away), whence the derivatives will be assumed bounded there. Alternatively, we can analyze the solution in an initial layer by applying an exponential stretching transformation in the layer region. The effect of this transformation is to produce an ODE in the stretched variable with $\lambda_R(c)$ bounded, so the solution and its derivatives are smooth in the stretched variable.

We now wish to extend Theorem 10.27 to the general linear case, where we have n differential equations in (10.1). The discussion in Section 3.4 and in Chapter 6 leads us to consider kinematic eigenvalues; however, the situation here is different in two ways. One is that the decoupling theory distinguishes only between nonincreasing and nondecreasing solution modes, whereas here we want a distinction between three: rapidly increasing, rapidly decreasing, and slow solution modes. We expect such a distinction to be valid only on subintervals of $[a,b]$. The other difference is that here we are stricter in requiring that the solution (and some derivatives) of (10.1), (10.15) be bounded as $\|A(x)\|$ grows unboundedly. An ordinary dichotomy requirement is not sufficient to guarantee this, as Example 10.2 (involving a stable IVP) indicates.

The good news is that when the solution $\mathbf{y}(x)$ is smooth because of no rapid coefficient variation [cf. (10.21)], then the eigenvalues of $A(x)$ are close to the elusive kinematic eigenvalues. To see this, suppose that $T(x)$ is a transformation which brings $A(x)$ to a block upper triangular form as follows: For each $x \in [c,d]$,

$$T^{-1}AT =: \Lambda = \begin{bmatrix} \Lambda^+ & & 0 \\ & \Lambda^- & \\ 0 & & \Lambda^0 \end{bmatrix} \tag{10.28}$$

where Λ^+ is an $n_+ \times n_+$ upper triangular matrix with diagonal elements having positive (unbounded) real part, Λ^- is an $n_- \times n_-$ upper triangular matrix with diagonal elements having negative (unbounded) real part, and Λ^0 is an $n_0 \times n_0$ upper triangular matrix with bounded entries ($n = n_+ + n_- + n_0$). The matrix A is said to have a *hyperbolic splitting* if $n_0 = 0$. As in (3.68), (3.69) we obtain for the transformed unknowns

$$\mathbf{w}(x) := T^{-1}(x)\mathbf{y}(x) \tag{10.29}$$

the transformed ODE

$$\mathbf{w}' = [\Lambda(x) - T^{-1}(x)T'(x)]\mathbf{w} + \mathbf{g}(x) \tag{10.30a}$$

with

$$\mathbf{g}(x) = T^{-1}(x)\mathbf{q}(x) \tag{10.30b}$$

In view of the foregoing treatment, we make the following

Assumptions 10.31

(a) $\|T(x)\|\|T^{-1}(t)\| \leq \tilde{K}$, $c \leq x, t \leq d$, for some moderate constant $\tilde{K}$. [Note that there is a scaling freedom in $T(x)$ which satisfies (10.28).]

(b) The first p derivatives of $T^{-1}(x)T'(x)$ are bounded. [This follows, e.g., if the first $p+1$ derivatives of $T(x)$ are bounded.]

(c) The off-diagonal elements of $\Lambda^+(x)$ are bounded in terms of the diagonal ones, and similarly for $\Lambda^-(x)$. Also, for each of these n_r diagonal elements, the analogue of (10.20b) holds.

(d) $\|\Lambda^0\| \leq K$.

□

These assumptions guarantee that the unboundedness of $A(x)$ is reflected in its eigenvalues, which form the main diagonals of Λ^+ and Λ^- (and represent the rapidly growing and rapidly decaying solution modes), because the matrix $T^{-1}T'$ is a bounded and hence unimportant perturbation to these eigenvalues.

In order to get an idea of what is needed to give a general extension of Theorem 10.27 for systems, we consider a number of special cases first, in increasing complexity. For these cases, the unboundedness of the ODE coefficients is explicitly expressed, and no rapid coefficient variation occurs.

The first case considered is where A is given by

$$A(x) = \varepsilon^{-1}\hat{A}, \qquad a \leq x \leq b \tag{10.32}$$

with $\hat{A}$ a constant, bounded, nonsingular matrix, independent of ε. Corresponding to (10.20b), we require that $\hat{A}$ have no purely imaginary eigenvalues. We can rescale the ODE (10.1) to read

$$\varepsilon \mathbf{y}' = \hat{A}\mathbf{y} + \hat{\mathbf{q}}(x) \tag{10.33}$$

where $\hat{\mathbf{q}} = \varepsilon \mathbf{q}$ is assumed to be bounded and in $C^{(p)}[a,b]$. Examples 10.5 and 10.1 are instances of this special case; in Example 10.5, $n_+ = n_- = 1$ and $n_0 = 0$.

For simplicity of presentation only, let $\hat{A}$ also be nondefective. Then T in (10.28) can be chosen as the eigenvector matrix of $\hat{A}$, yielding a diagonal matrix Λ with $n_0 = 0$. The coefficient matrix in (10.30) is diagonal, so the differential equations for $\mathbf{w}(x)$ decouple. There are no slow solution modes here, only n_+ rapidly increasing and n_- rapidly decreasing modes *throughout* the interval $[a,b]$. The situation is not different from what we have seen in Section 3.4, Chapter 6, and Example 10.1. The boundary conditions control the size of these fast modes. If the BC are separated, then the initial n_- BC control the decreasing modes and the n_+ terminal BC control the increasing modes. Corresponding to the partition (10.16), there are possibly two boundary layers here, connected by a smooth solution curve, so $M = 3$. To find t_2 (qualitatively), note that for an eigenvalue λ with real part $\lambda_R < 0$, the fundamental solution component $w(x) = e^{\lambda x}$ satisfies $|w^{(\nu)}(t_a)| \leq K$, $\nu = 0, 1, \ldots, p$, at

$$t_a = \left| \frac{\ln K - p \ln |\lambda|}{\lambda_R} \right| \tag{10.34}$$

For ε small we see that K does not matter much, and that $t_a = O(\varepsilon |\ln \varepsilon|)$ with the number of derivatives p entering as a linear factor. Obtaining expressions like (10.34) for each eigenvalue in Λ^-, the point t_2 may be taken as their maximum. (Incidentally, note that the stretching transformation $\xi = (x-a)/\varepsilon$ does indeed map the layer interval $[a,t_2]$ to an interval whose size grows unboundedly like $|\ln \varepsilon|$ as $\varepsilon \to 0$.) Similarly, the point t_3 close to the right end point b can be determined by considering the modes emanating from there, corresponding to Λ^+. Theorem 10.27 may now be applied for each of the components of $\mathbf{w}(x)$ to conclude that the solution $\mathbf{y}(x)$ is smooth in the middle subinterval $[t_2, t_3]$. Because of this smoothness (with $p \geq 1$), the solution is not far from the *reduced solution,* that is, the solution of (10.33) with $\varepsilon = 0$. We obtain

$$\mathbf{y}(x) = -A^{-1}\mathbf{q}(x) + O(\varepsilon), \qquad t_2 \leq x \leq t_3$$

Our next task is to extend this simple constant coefficient case. If we let $\hat{A}$ vary smoothly with x (and ε) such that all the other assumptions made above on $\hat{A}$ still hold, then the situation does not change much. (The matrix $T^{-1}T'$ is still a small and unimportant perturbation of Λ and can be dealt with by using a simple contraction argument.) A different extension is to add slow solution components, either explicitly or, more generally, implicitly. If done explicitly, we have an interaction of a fast solution $\mathbf{y}(x)$ and a slow solution $\mathbf{z}(x)$ through the ODE system

$$\varepsilon \mathbf{y}' = \hat{A}^{11}(x)\mathbf{y} + \hat{A}^{12}(x)\mathbf{z} + \hat{\mathbf{q}}^1(x) \tag{10.35}$$
$$\mathbf{z}' = \hat{A}^{21}(x)\mathbf{y} + \hat{A}^{22}(x)\mathbf{z} + \hat{\mathbf{q}}^2(x)$$

where $\hat{A}^{11}$ satisfies the requirements of $\hat{A}$ above, and $\hat{A}^{ij}$ are bounded, $1 \leq i, j \leq 2$. An example of this is the pair (10.3), (10.4) away from turning points.

If $\hat{A} = \hat{A}(x, \varepsilon)$ is rank deficient when $\varepsilon = 0$ then fast and slow solution components arise in the ODE (10.33) implicitly. We will assume here that $\hat{A}(x, \varepsilon)$ has a constant rank independent of x and ε and that it varies smoothly with ε, and proceed to investigate this so-called *singular singularly perturbed* case. Thus, consider (10.28) with assumptions (10.31) holding throughout the interval $[a, b]$, where now $n_0 > 0$. Denote $n_r := n_+ + n_- = n - n_0 = \text{rank}(\hat{A}(x, 0))$, and define $\Lambda^r \in \mathbf{R}^{n_r \times n_r}$ as

$$\Lambda^r := \begin{bmatrix} \Lambda^+ & 0 \\ 0 & \Lambda^- \end{bmatrix} \tag{10.36}$$

To make the notation simpler, assume that $\hat{A} = \hat{A}(x)$ is independent of ε and is nondefective, so that $\Lambda^0 = 0$. Defining $\hat{\Lambda}^r := \varepsilon \Lambda^r$, we have in (10.30)

$$\varepsilon \mathbf{w}' = \begin{bmatrix} \hat{\Lambda}^r(x) & 0 \\ 0 & 0 \end{bmatrix} \mathbf{w} + \varepsilon S(x)\mathbf{w} + T^{-1}(x)\hat{\mathbf{q}}(x, \varepsilon) \tag{10.37}$$

where the eigenvector matrix $T(x)$ in (10.28) is independent of ε, and $S := -T^{-1}T'$.

Now write

$$S = \begin{bmatrix} S^{11} & S^{12} \\ S^{21} & S^{22} \end{bmatrix}, \quad T^{-1}\hat{\mathbf{q}} =: \hat{\mathbf{g}} = \begin{bmatrix} \mathbf{g}^1 \\ \mathbf{g}^2 \end{bmatrix}$$

with $S^{11}(x) \in \mathbf{R}^{n_r \times n_r}$, $S^{22}(x) \in \mathbf{R}^{n_0 \times n_0}$, $\mathbf{g}^1(x, \varepsilon) \in \mathbf{R}^{n_r}$. Suppressing dependence on ε, we write the solution components as

$$\mathbf{w}(x) =: \begin{bmatrix} \mathbf{u}(x) \\ \mathbf{v}(x) \end{bmatrix} =: \begin{bmatrix} \mathbf{u}^+(x) \\ \mathbf{u}^-(x) \\ \mathbf{v}(x) \end{bmatrix} \tag{10.38}$$

with $\mathbf{u}^+(x) \in \mathbf{R}^{n_+}$, $\mathbf{u}^-(x) \in \mathbf{R}^{n_-}$ and $\mathbf{v}(x) \in \mathbf{R}^{n_0}$ for each $x \in [a, b]$. The ODE (10.37) can then be written as

$$\varepsilon \mathbf{u}' = \hat{\Lambda}^r(x)\mathbf{u} + \varepsilon S^{11}(x)\mathbf{u} + \varepsilon S^{12}(x)\mathbf{v} + \mathbf{g}^1(x, \varepsilon) \tag{10.39a}$$

$$\mathbf{v}' = S^{21}(x)\mathbf{u} + S^{22}(x)\mathbf{v} + \varepsilon^{-1}\mathbf{g}^2(x, \varepsilon) \tag{10.39b}$$

which is in the same form as (10.35).

As in Section 10.1.1, our next step is to eliminate the fast components $\mathbf{u}(x)$ from (10.39b). We do this with a Riccati transformation $R(x) \in \mathbf{R}^{n_0 \times n_r}$: Consider

$$\tilde{\mathbf{w}}(x) := \begin{bmatrix} \mathbf{u}(x) \\ \mathbf{v}(x) + \varepsilon R(x)\mathbf{u}(x) \end{bmatrix} \tag{10.40a}$$

[cf. (10.6–10)], where the matrix function $R(x)$ is chosen so that $S^{21}(x)$ becomes zero in the ODEs for $\tilde{\mathbf{w}}$ corresponding to (10.39). As in (10.9), we find

$$R(x) = -S^{21}(x)\hat{\Lambda}^r(x)^{-1} + O(\varepsilon) \tag{10.40b}$$

(see Exercise 2). Thus, since $R(x)$ is well-defined (and bounded), we may assume without loss of generality that $S^{21} \equiv 0$ in (10.39b) and consider $\mathbf{v}(x)$ as the slow solution components.

It is now clear that, while we may allow $\mathbf{g}^1(x,\varepsilon) = O(1)$, the terms in (10.39b) would not balance unless we have the *consistency requirement*

$$\mathbf{g}^2(x,\varepsilon) = O(\varepsilon) \tag{10.41}$$

Thus, not every bounded smooth $\hat{\mathbf{q}}(x)$ in (10.33) is permitted! The requirement (10.41) for the slow components corresponds, in fact, to the requirement (10.22).

Having excluded rapid coefficient variation, we expect the solution of (10.39) to have the following structure: Dominant solution behaviour is potentially determined by fast components in regions of width $O(\varepsilon|\ln\varepsilon|)$ near a and b, and the solution is smooth (slowly varying) in between. To specifically find $\mathbf{w}(x)$, we may use an asymptotic analysis, wherein we refer to the solution in the layer as the *inner solution* and to the solution in the smooth region as the *outer solution*. For the sake of completeness, we give the following

Definition 10.42 If $\mathbf{f}(\varepsilon)$ and $\{\Phi_j(\varepsilon)\}_{j\geq 0}$ are functions such that as $\varepsilon \to 0$, for every N

$$\left|\mathbf{f}(\varepsilon) - \sum_{j=0}^{N} \Phi_j(\varepsilon)\right| = o(|\Phi_N(\varepsilon)|) \tag{10.42a}$$

then $\sum_{j\geq 0} \Phi_j(\varepsilon)$ is called an *asymptotic expansion* of $\mathbf{f}(\varepsilon)$. Often one writes

$$\mathbf{f}(\varepsilon) \sim \sum_{j\geq 0} \Phi_j(\varepsilon) \tag{10.42b}$$

$\square$

Usually one encounters expansions of the form

$$\mathbf{f}(\varepsilon) \sim \sum_{j\geq 0} \varepsilon^j \mathbf{f}_j$$

where $\{\mathbf{f}_j\}_{j\geq 0}$ is some bounded sequence.

Under weak assumptions, it is possible to find an expansion for both the inner solutions and the outer solutions. For the solution $\mathbf{w}$ of (10.37) this can even be done using one formal expression. In particular, we write

$$\mathbf{w}(x) \approx \sum_{j=0}^{p} [\overline{\mathbf{w}}_j(x) + \tilde{\mathbf{w}}_j(\xi) + \hat{\mathbf{w}}_j(\eta)] \varepsilon^j \qquad (10.43a)$$

with a corresponding form (and notation) for $\mathbf{u}$, $\mathbf{v}$, and $\mathbf{y}$, where

$$\xi = (x-a)/\varepsilon, \qquad \eta = (x-b)/\varepsilon \qquad (10.43b)$$

and require

$$\tilde{\mathbf{w}}_j(\xi) \underset{\xi \to \infty}{\to} 0, \qquad \hat{\mathbf{w}}_j(\eta) \underset{\eta \to -\infty}{\to} 0, \qquad 0 \le j \le p \qquad (10.44)$$

Using the *method of matched asymptotic expansions*, we now substitute this expansion into (10.39) and compare equal powers of ε in the three variables separately.

For $j = 0$ in the variable x we obtain

$$0 = \hat{\Lambda}^r(x) \overline{\mathbf{u}}_0(x) + \mathbf{g}^1(x, 0)$$

which determines $\overline{\mathbf{u}}_0(x)$, and with $\mathbf{g}^2(x, 0) = 0$ from (10.41),

$$\overline{\mathbf{v}}_0' = S^{22}(x) \overline{\mathbf{v}}_0 + \frac{\partial}{\partial \varepsilon} \mathbf{g}^2(x, 0).$$

Thus we need n_0 boundary conditions for the nonstiff ODEs for $\overline{\mathbf{v}}_0(x)$ to completely specify the reduced solution $\overline{\mathbf{w}}_0(x)$ [and hence $\overline{\mathbf{y}}_0(x)$]. The inner solution equations are

$$\frac{d}{d\xi} \tilde{\mathbf{u}}_0 = \hat{\Lambda}^r(a) \tilde{\mathbf{u}}_0, \qquad \frac{d}{d\eta} \hat{\mathbf{u}}_0 = \hat{\Lambda}^r(b) \hat{\mathbf{u}}_0$$

and for the slow components, $\tilde{\mathbf{v}}_0 \equiv \hat{\mathbf{v}}_0 \equiv \mathbf{0}$. The layer boundary conditions (10.44) imply that

$$\tilde{\mathbf{u}}_0^+ \equiv \mathbf{0}, \qquad \hat{\mathbf{u}}_0^- \equiv \mathbf{0}$$

This leaves $\tilde{\mathbf{u}}_0$ (hence $\tilde{\mathbf{y}}_0$) represented in terms of n_- parameters, and $\hat{\mathbf{u}}_0$ (hence $\hat{\mathbf{y}}_0$) represented in terms of n_+ parameters. Substitution into the BC (10.15) (which, for notational simplicity only, are assumed to be independent of ε), gives

$$B_a(\overline{\mathbf{y}}_0(a) + \tilde{\mathbf{y}}_0(a)) + B_b(\overline{\mathbf{y}}_0(b) + \hat{\mathbf{y}}_0(b)) = \boldsymbol{\beta}$$

at $\varepsilon = 0$, and elimination of $n_+ + n_- = n_r$ parameters gives n_0 BC for $\overline{\mathbf{v}}_0(x)$.

Higher-order terms in ε in the series expansion (10.43) are determined in a similar way. Note that the outer solution $\sum_{j \ge 0} \varepsilon^j \overline{\mathbf{w}}_j(x)$ is well-defined and represents the smooth solution on the entire interval $[a, b]$.

In short, the general behaviour of $\mathbf{u}(x)$ is very similar to the one described in detail for the case where $\hat{A}$ is constant and nonsingular. The slow variables $\mathbf{v}(x)$ have no boundary layers when $S^{21} \equiv 0$ in (10.39b). But if $S^{21} \not\equiv 0$ then $\mathbf{v}(x)$ is just one derivative smoother than $\mathbf{u}(x)$ [i.e., $\mathbf{v}'(x)$ is as smooth as $\mathbf{u}(x)$]. This does not matter much, though, when one is transforming back to the original solution variables $\mathbf{y}(x)$.

This discussion of singular-singularly perturbed BVPs indicates a way to generalize Theorem 10.27 for ODE systems. We wish to provide conditions under which the solution of (10.1) is smooth on a segment $[c, d] \subseteq [a, b]$. First, let us look at a system with only (rapidly) decaying modes. We consider complex valued vectors and matrices.

Lemma 10.45 Suppose that $A(x)$ can be split as

$$A(x) = \Lambda(x) + B(x)$$

where $\Lambda(x) = \text{diag}(A(x))$, such that the following hold:

(I) $\Lambda_R(x) := \text{Re}(\Lambda(x))$ is a diagonal, real matrix, such that its largest element $\lambda_R(x)$ (smallest in magnitude) is negative.

(ii) There exist positive constants δ and K of moderate size such that

$$\|\Lambda_R^{-1}(x)B(x)\| \leq 1 - \delta \qquad (10.45a)$$

$$x \in [c,d]$$

$$|\Lambda_R^{-1}(x)\mathbf{q}(x)| \leq K \qquad (10.45b)$$

Then a stability bound for (10.1) is given by

$$\|Y(x)Y^{-1}(t)\| \leq 2\delta^{-1} \exp\{\delta/2 \int_t^x \lambda_R(\tau)d\tau\}, \qquad x \geq t \qquad (10.45c)$$

and the solution for the corresponding IVP is bounded by

$$|\mathbf{y}(x)| \leq \delta^{-1}(2|\mathbf{y}(c)|\exp\{\delta/2 \int_c^x \lambda_R(\tau)d\tau\} + K) \qquad (10.45d)$$

$\square$

Remarks

(a) The coefficients and solution of (10.1) considered here are allowed to be complex, because this lemma is used later for a transformed BVP [in Theorem 10.47].

(b) The nonsingularity of $\Lambda_R(x)$ follows from assumption (i), while that of $A(x)$ follows from the diagonal dominance (10.45a).

(c) From (10.45a, b) we can easily conclude boundedness of $|A^{-1}(x)\mathbf{q}(x)|$ on $[c,d]$, so we have in fact a bounded solution for the reduced equation. Also, in (10.45a, b) we are requiring that the real parts of something like the eigenvalues dominate the off-diagonal part in $A(x)$ as well as $\mathbf{q}(x)$.

(d) The bound (10.45c) corresponds to an exponential dichotomy and indicates the way in which the influence of the boundary values decreases away from the boundary.

Proof: We begin by showing (10.45d) for a solution $\mathbf{v}(x)$ of (10.1) satisfying $\mathbf{v}(c) = \mathbf{0}$. First, for the ODE

$$\mathbf{v}' - \Lambda(x)\mathbf{v} = \mathbf{f}(x)$$

we have

$$\mathbf{v}(x) = \int_c^x \exp\{\int_t^x \Lambda(\tau)d\tau\}\mathbf{f}(t)\,dt$$

$$= -\int_c^x \frac{d}{dt}[\exp\{\int_t^x \Lambda_R(\tau)d\tau\}] \exp\{\int_t^x (\Lambda - \Lambda_R)(\tau)d\tau\}\Lambda_R^{-1}(t)\mathbf{f}(t)\,dt$$

and hence, recalling the argument used to obtain (10.24),

$$|\mathbf{v}(x)| \leq \max_{c \leq t \leq x} |\Lambda_R^{-1}(t)\mathbf{f}(t)| \qquad (10.46a)$$

Writing (10.1) for $\mathbf{v}(x)$ as

$$\mathbf{v}'(x) - \Lambda(x)\mathbf{v}(x) = B(x)\mathbf{v}(x) + \mathbf{q}(x) \equiv \mathbf{f}(x)$$

we obtain, using (10.45a, b) and (10.46a),

$$|\mathbf{v}(x)| \leq (1-\delta) \max_{c \leq t \leq x} |\mathbf{v}(t)| + K$$

whence

$$\max_{c \leq x \leq d} |\mathbf{v}(x)| \leq \delta^{-1} K$$

Our next step is to show (10.45c). Let $\mathbf{y}(x)$ be a solution of (1) with $\mathbf{q} \equiv \mathbf{0}$. We write for a fixed t

$$\mathbf{y}(x) = \mathbf{v}(x) + \exp\{\int_t^x \Lambda(\tau)\,d\tau\}\mathbf{y}(t), \qquad x \geq t$$

Then $\mathbf{v}(x)$ satisfies

$$\mathbf{v}' = A(x)\mathbf{v} + B(x)\exp\{\int_t^x \Lambda(\tau)\,d\tau\}\mathbf{y}(t)$$

$$\mathbf{v}(t) = \mathbf{0}$$

so we may apply (10.45d) with t replacing c and zero initial values to obtain

$$|\mathbf{v}(x)| \leq \delta^{-1}(1-\delta)|\mathbf{y}(t)|$$

whence

$$|\mathbf{y}(x)| \leq (\delta^{-1}(1-\delta) + 1)|\mathbf{y}(t)| \leq \delta^{-1}|\mathbf{y}(t)| \qquad (10.46b)$$

Let $p(x) := \delta/2 \int_t^x \lambda_R(\tau)\,d\tau$. Then $\mathbf{z}(x) := e^{-p(x)}\mathbf{y}(x)$ satisfies

$$\mathbf{z}' = (A(x) - \delta/2\lambda_R(x))\mathbf{z}$$

Using (10.46b), we obtain

$$|\mathbf{z}(x)| \leq 2\delta^{-1}|\mathbf{y}(t)|$$

hence

$$|\mathbf{y}(x)| \leq e^{p(x)}|\mathbf{z}(x)| \leq 2\delta^{-1}e^{p(x)}|\mathbf{y}(t)|$$

We then obtain (10.45c) by writing for any fundamental solution matrix $Y(x)$,

$$\|Y(x)Y^{-1}(t)\| = \max_{\mathbf{c} \neq 0} \frac{|Y(x)Y^{-1}(t)\mathbf{c}|}{|\mathbf{c}|} = \max_{\mathbf{d} \neq 0} \frac{|Y(x)\mathbf{d}|}{|Y(t)\mathbf{d}|} \leq 2\delta^{-1} \exp\{\delta/2 \int_t^x \lambda_R(\tau) d\tau\}$$

Finally, to show (10.45d) in general, we write the solution of (10.1) as

$$\mathbf{y}(x) = Y(x)Y^{-1}(c)\mathbf{y}(c) + \mathbf{v}(x)$$

with $\mathbf{v}(c) = \mathbf{0}$, and apply the already obtained results to the two terms on the right hand side. This completes the proof. □

Now we are ready for the generalization of Theorem 10.27.

Theorem 10.47 Consider the ODE (10.1) on the segment $[c, d]$. Assume that there exists a transformation $T(x)$ which brings $A(x)$ to the form (10.28) with Λ^+ and Λ^- upper triangular, such that the following hold for all $c \leq x \leq d$: There are constants K_j of moderate size such that

(a)

$$\|T(x)\| \, \|T^{-1}(t)\| \leq K_1, \qquad c \leq x, t \leq d \qquad (10.47a)$$

$$\|\frac{d^\nu}{dx^\nu}(T^{-1}(x)T'(x))\| \leq K_1, \qquad \nu = 0, \ldots, p \qquad (10.47b)$$

[cf. (10.30)].

(b) Denoting

$$\Lambda_R^+(x) := \text{diag}(\text{Re}(\Lambda^+(x))), \qquad \lambda_R^+(x) := \min_{1 \leq j \leq n_+} \text{Re}(\lambda_j(x)) \qquad (10.47c)$$

$$\Lambda_R^-(x) := \text{diag}(\text{Re}(\Lambda^-(x))), \qquad \lambda_R^-(x) := \max_{n_+ + 1 \leq j \leq n_r} \text{Re}(\lambda_j(x)) \qquad (10.47d)$$

$$\mathbf{g}(x) = \begin{bmatrix} \mathbf{g}^+(x) \\ \mathbf{g}^-(x) \\ \mathbf{g}^0(x) \end{bmatrix} \qquad (10.47e)$$

assume

$$\lambda_R^-(x) \ll -1, \qquad \lambda_R^+(x) \gg 1 \qquad (10.47f)$$

$$\|[\Lambda_R^+(x)]^{-1} \frac{d^\nu}{dx^\nu} \Lambda^+(x)\|, \quad \|[\Lambda_R^-(x)]^{-1} \frac{d^\nu}{dx^\nu} \Lambda^-(x)\| \leq K_2, \qquad \nu = 0, 1, \ldots, p \qquad (10.47g)$$

$$|[\Lambda_R^+(x)]^{-1} \frac{d^\nu}{dx^\nu} \mathbf{g}^+(x)|, \quad |[\Lambda_R^-(x)]^{-1} \frac{d^\nu}{dx^\nu} \mathbf{g}^-(x)| \leq K_3, \qquad \nu = 0, 1, \ldots, p \qquad (10.47h)$$

$$\|\frac{d^\nu}{dx^\nu} \Lambda^0(x)\| \leq K_2, \qquad |\frac{d^\nu}{dx^\nu} \mathbf{g}^0(x)| \leq K_3, \qquad \nu = 0, 1, \ldots, p \qquad (10.47i)$$

Then with the notation (10.29), (10.38), we have the bounds

$$\|\mathbf{y}^{(\nu)}\|_{[c,d]} \leq C[\sum_{j=0}^\nu (|\frac{d^j}{dx^j}\mathbf{u}^-(c)| + |\frac{d^j}{dx^j}\mathbf{u}^+(d)| + |\frac{d^j}{dx^j}\mathbf{v}(c)|) + K_3], \qquad 0 \leq \nu \leq p \qquad (10.47j)$$

Moreover, there is a constant δ, $0 < \delta < 1$, such that

$$|\mathbf{y}(x)| \le C\,[e^{\delta\int_c^x \lambda_R^-(\tau)d\tau} |\mathbf{u}^-(c)| + e^{-\delta\int_x^d \lambda_R^+(\tau)d\tau} |\mathbf{u}^+(d)| + |\mathbf{v}(c)| + K_3], \qquad c \le x \le d \tag{10.47k}$$

Proof: The smoothness of $T(x)$ allows us to look at $\mathbf{w}(x)$ satisfying (10.29), (10.30). Without loss of generality we may assume that the conditions of Lemma 10.45 are satisfied for each of the blocks Λ^+ and Λ^- in place of A there: The requirement (10.45b) follows from (10.47h), while (10.45a) can be achieved by using a further transformation, e.g., of the form $\mathrm{diag}(1, \alpha, \alpha^2, \ldots, \alpha^{n\cdot})$ with $0 < \alpha < 1$, to reduce the size of the off-diagonal elements of the block, since (10.47g) holds.

Furthermore, we may assume that the lower left $n_0 \times n_r$ block of $T^{-1}(x) T'(x)$ is zero, for otherwise we may invoke a Riccati transformation as in (10.40). From the latter simplification, the slow solution components $\mathbf{v}(x)$ are decoupled from the fast ones in (10.30), and the bounds (10.47b) and (10.47i) guarantee that

$$\left\| \frac{d^\nu \mathbf{v}}{dx^\nu} \right\|_{[c,d]} \le K\,[K_3 + |\mathbf{v}(c)|]$$

Now we may consider the terms involving $\mathbf{v}(x)$ in the ODE for $\mathbf{u}(x)$ as smooth, bounded inhomogeneous terms. Considering $T^{-1}(x)T'(x)\mathbf{w}(x)$ as part of the inhomogeneity in (10.30a), we may apply Lemma 10.45 in the forward direction for $\mathbf{u}^-(x)$ and in the backward direction for $\mathbf{u}^+(x)$. Then, applying a contraction argument for $\mathbf{u}(x)$ [essentially as done for proving Theorem 3.71] using this lemma, we obtain (10.47k), and in particular the boundedness of $\mathbf{w}(x)$. Repeated differentiation of the ODE (10.30a) leads, as in Theorem 10.27, to the conclusion (10.47j). □

Theorem 10.47 not only gives conditions under which the solution is smooth in a segment $[c, d]$ of the interval $[a, b]$, but also indicates how to define layers when conditions (10.47a–i) hold. Consider for simplicity an IVP as in Lemma 10.45, but with $|\mathbf{y}^{(\nu)}(c)|$ not necessarily small for $0 \le \nu \le p$. The decay of the influence of the initial values in (10.45d) helps in defining another point $\hat{c}$ where $|\mathbf{y}^{(\nu)}(\hat{c})|$ is of moderate size. The interval $[c, \hat{c}]$ is then a layer region, while a smooth solution interval starts at $x = \hat{c}$.

To see specifically how this is done, write $\mathbf{y}(x)$ as

$$\mathbf{y}(x) = \mathbf{y}_1(x) + \mathbf{y}_2(x)$$

where $\mathbf{y}_1(x)$ is a smooth solution for the inhomogeneous IVP, and $\mathbf{y}_2(x)$ solves the homogeneous problem. The existence of a smooth $\mathbf{y}_1(x)$ is guaranteed by Theorem 10.47 if we can furnish bounded initial values $|\mathbf{y}_1^{(\nu)}(c)|$. But this can be done as follows: Choose the small parameter $\varepsilon := \lambda_R^{-1}(c)$, and define a modified differential equation as in (10.32), (10.33). Then application of a matched asymptotic expansion technique establishes such a set of initial values (see Exercise 5).

We are left with the homogeneous solution $\mathbf{y}_2(x)$, for which repeated differentiation of the ODE yields

$$|\mathbf{y}_2^{(\nu)}(x)| \le C_1 \|\Lambda(x)\|^\nu e^{\int_c^x \lambda_R(\tau)d\tau} |\mathbf{y}_2(c)|$$

Requiring, say, $|\mathbf{y}_2^{(\nu)}(\hat{c})| \le C_2$, $\nu = 0, 1, \ldots, p$, then $\hat{c}$ must satisfy

$$\log C_3 - p \log \|\Lambda(c)\| \approx \int_c^{\hat{c}} \lambda_R(\tau) d\tau \approx (\hat{c} - c) \lambda_R(c)$$

with $C_3 \approx C_2/C_1$ (all C_j being constants of moderate size). So we obtain an estimate for $\hat{c}$,

$$\hat{c} \approx c + \frac{\log C_3 - p \log \|\Lambda(c)\|}{\lambda_R(c)} \tag{10.48}$$

which defines the layer width. This procedure is acceptable if the slow and fast time scales are sufficiently separated.

Remarks

(a) Instead of the block upper triangular form (10.28), one could consider an *essentially diagonally dominant* form for the transformed coefficient matrix $A(x)$: The matrix

$$\Lambda(x) = \begin{bmatrix} \Lambda^{11}(x) & \Lambda^{12}(x) \\ \Lambda^{21}(x) & \Lambda^{22}(x) \end{bmatrix}$$

is essentially diagonally dominant if it satisfies

$$\|D^{-1}\Lambda^{12}\| \leq K, \qquad \|\Lambda^{2j}\| \leq K, \qquad j = 1, 2 \tag{10.49a}$$

where $D(x) := \text{diag}(\Lambda^{11}(x))$, and $\Lambda^{11}(x)$ is *diagonally dominated*. The matrix $\Lambda^{11}(x)$ is diagonally dominated if its diagonal elements $\lambda_i := \Lambda^{11}_{ii}$ each satisfy

$$|\text{Im}(\lambda_i)| \leq \rho |\text{Re}(\lambda_i)|, \qquad c \leq x \leq d \tag{10.49b}$$

[cf. (10.20b)] and there is a constant $\delta > 0$ independent of x such that

$$\sum_{j \neq i} |\Lambda^{11}_{ij}| \leq (1-\delta) |\text{Re}(\lambda_i)|, \qquad i = 1, \ldots, n_r \tag{10.49c}$$

Theorem 10.47 can be appropriately reformulated, and similar results can be obtained, because these conditions, like those of assumptions 10.31, also guarantee that the diagonal elements of the upper left block of $\Lambda(x)$ describe the decoupling of fast modes.

(b) In the sequel we will encounter situations where the discussion will be restricted to BVPs which satisfy conditions (10.47a–i) for $p = 0$ throughout the interval $[a,b]$. The special cases considered before Lemma 10.45 satisfy this, but problems with interior layers do not. Indeed, all we can expect under such a restriction is a smooth solution segment possibly connected to the boundaries by two relatively simple boundary layers. For convenience we shall refer to such a BVP as the *3-segment problem*.

(c) In Theorem 10.47 we have assumed that the slow components are specified at the left end of the segment $[c,d]$, but this is not important at all (as becomes apparent from the proof). In fact, we could specify some components of $\mathbf{v}(x)$ at c and some at d. In particular, we can specify the same number of left end conditions for all segments, by making the total number of conditions specified at the left end of each segment equal to a fixed number of nonincreasing modes, according to the dichotomy which the ODE (10.1) is assumed to have.

□

10.1.4 Nonlinear first-order systems

For nonlinear problems the situation is, as usual, even more complicated than for the corresponding linear ones. The remarks made in Section 10.1.2 apply, of course, but the situation here is significantly more complex in general. The idea of quasilinearization can be found helpful, both for theoretical and for practical purposes. Thus, one may consider the variational problem (cf. Section 3.1.2) or, more generally, a linear ODE (10.1) arising from a linearization about some appropriate function. One possibility is to consider a linearization about a solution of the BVP, but as noted in Section 10.1.2, we can no longer assume that the coefficients of the resulting ODE (10.1) are smooth (vary slowly) in layer regions. Another possibility is to linearize about a "reduced solution," obtaining smooth coefficients in the resulting ODE (10.1). Both linearizations are essentially the same away from layers, and both yield satisfactory results there. But in layer regions the situation is more complicated, as compared to the linear case.

We shall restrict ourselves in this section to a few special cases. Consider first the following relatively simple two time scale BVP

$$\varepsilon \mathbf{y}' = \mathbf{f}(\mathbf{y}, \mathbf{z}, x, \varepsilon) \qquad (10.50a)$$

$$a < x < b$$

$$\mathbf{z}' = \mathbf{g}(\mathbf{y}, \mathbf{z}, x, \varepsilon) \qquad (10.50b)$$

$$\mathbf{h}(\mathbf{y}(a), \mathbf{z}(a), \mathbf{y}(b), \mathbf{z}(b), \varepsilon) = \mathbf{0} \qquad (10.50c)$$

Here $\mathbf{f} \in \mathbf{R}^{n_r}$, $\mathbf{g} \in \mathbf{R}^{n_0}$, and $\mathbf{h} \in \mathbf{R}^n$ ($n = n_r + n_0$) are assumed to be sufficiently smooth functions of their arguments. We assume that $\mathbf{y}$ consists of the fast solution components and $\mathbf{z}$ of the slow ones. This is the case if the Jacobian matrix

$$\Lambda^r := \frac{\partial \mathbf{f}}{\partial \mathbf{y}}(\mathbf{y}, \mathbf{z}, x, 0) \qquad (10.51)$$

has a hyperbolic splitting (i.e., the real parts of all its eigenvalues are bounded away from 0 throughout the interval $[a, b]$). The latter assumption ties this nonlinear problem to its linear counterpart (10.35).

We may make the Ansatz

$$\mathbf{y}(x, \varepsilon) = \overline{\mathbf{y}}(x) + \widetilde{\mathbf{y}}(\xi) + \widehat{\mathbf{y}}(\eta) + O(\varepsilon) \qquad (10.52a)$$

$$a \leq x \leq b$$

$$\mathbf{z}(x, \varepsilon) = \overline{\mathbf{z}}(x) + O(\varepsilon) \qquad (10.52b)$$

where ξ and η are defined in (10.43b). The reduced solution $\overline{\mathbf{y}}(x), \overline{\mathbf{z}}(x)$ solves the reduced equations

$$\mathbf{0} = \mathbf{f}(\overline{\mathbf{y}}, \overline{\mathbf{z}}, x, 0) \qquad (10.53a)$$

$$\bar{z}' = g(\bar{y}, \bar{z}, x, 0) \tag{10.53b}$$

subject to n_0 appropriate BC, as specified below. The assumption that Λ^r is nonsingular allows use of the implicit function theorem to express $\bar{y}(x)$ in terms of $\bar{z}(x)$, and substitution into (10.53b) yields in principle a (nonstiff) ODE system of order n_0 for $\bar{z}(x)$. The left and right end layer correction functions $\tilde{y}(\xi)$ and $\hat{y}(\eta)$ satisfy the ODEs

$$\frac{d}{d\xi}\tilde{y} = f(\bar{y}(a) + \tilde{y}, \bar{z}(a), a, 0), \qquad 0 < \xi < \infty \tag{10.53c}$$

$$\frac{d}{d\eta}\hat{y} = f(\bar{y}(b) + \hat{y}, \bar{z}(b), b, 0), \qquad -\infty < \eta < 0 \tag{10.53d}$$

and decay exponentially to 0 as $\xi \to \infty$, $\eta \to -\infty$, respectively. Equations (10.53) arise from using the representation (10.52) in (10.50a, b), equating terms of comparable magnitude in ε.

To construct the solution as in (10.52), we substitute into the BC to obtain

$$h(\bar{y}(a) + \tilde{y}(0), \bar{z}(a), \bar{y}(b) + \hat{y}(0), \bar{z}(b), 0) = 0 \tag{10.54}$$

The requirement that $\tilde{y}$ and $\hat{y}$ decay exponentially implies that $\tilde{y}(0)$ and $\hat{y}(0)$ must be on the stable manifolds of their corresponding equations, and we write these conditions as

$$\phi_-(\bar{y}(a), \bar{z}(a), \tilde{y}(0)) = 0 \tag{10.55a}$$

$$\phi_+(\bar{y}(b), \bar{z}(b), \hat{y}(0)) = 0 \tag{10.55b}$$

Equation (10.55a) corresponds to the n_- eigenvalues of $\frac{\partial f}{\partial y}$ with negative real part, and (10.55b) to those eigenvalues with a positive real part. Together, there are n_r equations in (10.55) and n in (10.54). Eliminating $\tilde{y}(0)$ and $\hat{y}(0)$ from (10.54) and (10.55) leaves n_0 conditions, and these are the BC for the reduced equations (10.53a, b).

This construction can be carried out in principle also for higher-order terms of a matched asymptotic expansion, as, e. g., in (10.43). But we see from the above description that the actual solution construction following this recipe for a given BVP is far from automatic.

Example 10.7

Consider the nonlinear elastic beam problem of Example 1.14, which is written here with a somewhat different notation than in (1.31),

$$\varepsilon y_1' = -y_2 \tag{10.56a}$$

$$\varepsilon y_2' = \phi(z_1) \cos z_2 - [\sec z_2 + \varepsilon y_2 \tan z_2] y_1 \tag{10.56b}$$

$$z_1' = \sin z_2 \tag{10.56c}$$

$$z_2' = y_1 \tag{10.56d}$$

The function $\phi(z_1)$ is given, and it is smooth. Obviously, $n_r = 2$. Assuming that the beam is simply supported we have the BC

$$y_1(0) = y_1(1) = 0 \tag{10.57a}$$

$$z_1(0) = z_1(1) = 0 \qquad (10.57b)$$

Setting $\varepsilon = 0$ in (10.56), we obtain the reduced equations

$$\bar{z}_1' = \sin \bar{z}_2, \qquad \bar{y}_1 = \phi(\bar{z}_1)\cos^2 \bar{z}_2$$

$$\bar{z}_2' = \bar{y}_1, \qquad \bar{y}_2 = 0$$

which describe a so-called hanging cable. This is a system of two first-order ODEs for the slow components z_1 and z_2. Also,

$$\Lambda^r = \begin{bmatrix} 0 & -1 \\ -\sec z_2 & 0 \end{bmatrix}$$

so the eigenvalues are

$$\pm \sqrt{\sec z_2}$$

The expression under the square root sign is nonnegative, z_2 being an angle which is bounded between $-\pi/2$ and $\pi/2$. Thus we have a hyperbolic splitting. In this case it is also obvious what the BC are for the reduced system: Since y_i is fast and z_i is slow, we simply drop (10.57a) and require only (10.57b). Integrating the hanging cable system (numerically), one can proceed to find the layer correction terms as well.

□

Many problems encountered in practice are (or can be written as) nonlinear first-order stiff BVPs. We refer again to the list of examples in the opening paragraph of this chapter. But of these, few qualify under the assumptions needed for the analysis outlined above. This makes the necessity for good, robust numerical methods all the more important. Mathematical analysis of a given problem usually cannot replace the need for its numerical integration, but it may often yield some further insight. The choice of a way to go about analyzing a given difficult problem can often be a mixture of art with trial and error.

Example 10.8

Consider the semiconductor problem of Example 1.15, which we rewrite here,

$$\lambda^2 \psi'' = n - p - D(x) \qquad (10.58a)$$

$$(n' - n\psi')' = R(n,p) \qquad (10.58b)$$

$$(p' + p\psi')' = R(n,p) \qquad (10.58c)$$

$$\psi(-1) = \ln \frac{\gamma}{p(-1)} + U, \qquad \psi(1) = \ln \frac{n(1)}{\gamma} \qquad (10.58d)$$

$$n(\pm 1)p(\pm 1) = \gamma^2 \qquad (10.58e)$$

$$n(\pm 1) - p(\pm 1) = D(\pm 1) \qquad (10.58f)$$

For simplicity we take $R \equiv 0$. Note that in (10.58) there are two small parameters, λ^2 and γ. But letting $\gamma \to 0$ causes ψ to blow up logarithmically in (10.58d), so we keep it fixed in an analysis and consider λ as the only small parameter. Letting $\lambda \to 0$, we observe that $\psi(x)$ disappears from (10.58a); i.e., what corresponds to Λ^r of (10.51) is 0 here. (We should convert the BVP into first-order form before making such a claim, but this turns out not to matter.) The problem is, in fact, singular singularly perturbed, and ψ, n, and p are all fast solution components.

The transformations (1.35) or (1.36) actually have the effect of "regularizing" the problem, in the sense that the introduced variables ϕ_n and ϕ_p or u and v, replacing n and p, are slow solution components. However, what corresponds to the transformation matrix $\underset{\sim}{T}(x)$ of (10.28) [cf. (10.37), (10.39)] cannot be said to satisfy assumptions 10.31, because K becomes large.

In short, this problem is not standard. Still, some valuable information can be obtained. In particular, a unique solution near thermal equilibrium (i.e., when $|U|$ is small enough) can be shown to exist, and some solution structure can be derived: By (10.58f) the reduced solution satisfies the BC, so there are no layers near ± 1. However, the scaled doping profile $D(x)$ (which is an inhomogeneity) has jump discontinuities at pn- junctions inside the interval $(-1, 1)$, and consequently layers in ψ, n and p are expected to form there. The width of the pn- layer [i.e., the region about the junction in which layer solution components like $\tilde{y}$ of (10.52a) are significantly larger than $O(\lambda)$] can be shown to be $O(\lambda(|\ln \lambda| + \sqrt{[\psi]}))$, where $[\psi]$ is the potential drop (the variation in ψ) at the junction.

□

10.2 NUMERICAL APPROACHES

The previous section of this chapter raises a number of fundamental issues which must be addressed when one is considering the numerical solution of stiff BVPs.

One issue is how to treat the radically different solution behaviour that may occur in different regions of the interval on which the BVP is defined. An example is the solution profile in Fig. 10.3. It stands to reason that different numerical methods and/or analyses should be applied on regions with very different solution characteristics. Thus we consider in Section 10.2.1 a partition of the problem interval into segments, where on each segment a BVP and a numerical method are defined.

Another issue which needs to be addressed is the extent to which fast modes are approximated in smooth regions. Since we know that the solution in smooth regions is composed mainly of contributions from slow modes and from a smooth inhomogeneity, we expect to be able to require only a good pointwise approximation of the latter. This, however, raises questions about the extent to which the stability of the numerical method follows from the well-conditioning of the BVP, and about decoupling. Since decoupling of nonincreasing and nondecreasing solution components is particularly important here, we should note how various numerical methods achieve it. In this respect we can divide the methods in this chapter into two classes, dealt with in Section 10.2.2 and in Section 10.2.3, respectively. One class contains those methods which decouple explicitly, by directly considering a transformed BVP (5.30) and integrating stable IVPs (initial or terminal value problems) in their proper direction. When done right, these methods inherit their stability properties directly from the BVP. The methods in Section 10.3.1 and Section 10.4.2 belong to this class. The other class contains methods where no such explicit decoupling is done (and dichotomy is preserved more weakly, in the sense that exponential dichotomy may be approximated by an ordinary one, as explained later). The schemes of Section 10.3.2 and 10.4.1 belong to this latter class.

A third issue arises with regard to short segments where the solution varies rapidly. A good solution approximation on such segments is sometimes desired, but on other occasions it is not worth its price. (This depends on the application.) Our usual assumption will be that a uniformly accurate approximate solution is desired everywhere, implying use of a dense mesh on these short segments. As observed before, this dense mesh can be viewed as a *stretching transformation*, eliminating unbounded solution derivatives in the stretched variable. This is discussed in Section 10.2.4. In the sequel we also observe the potential of certain methods to produce accurate approximations in regions of solution smoothness even without an accurate approximation of the solution at layers. In particular, see Theorem 10.75 and Example 10.9 in Section 10.2.2.

The treatment of nonlinearity in stiff BVPs, as it turns out, is often not different in essence from that for nonstiff ones. In particular, the nonlinear techniques of Chapter 8 are applicable to stiff problems (see Examples 8.3–8.5). By this we do not mean to imply that such problems are simple, but it partially justifies our concentrating here, unless otherwise noted, on linear BVPs.

In this section, as in Chapter 6, we use the 2–norm for vectors and matrices (but not for functions) as our default norm.

10.2.1 Theoretical multiple shooting

A basic tool for showing stability of our numerical methods shall be a theoretical multiple shooting framework. To describe this, suppose that we have the partition (5.16), where on each segment $[t_j, t_{j+1}]$ the solution of the singularly perturbed BVP (5.1), (5.15) is of one type, be it a boundary layer, an interior layer, or a region of smooth variation. For each segment we want to consider a well-posed BVP and its discretization. Thus we write for $1 \leq j \leq M$

$$\mathbf{y}' = A(x)\mathbf{y} + \mathbf{q}(x), \qquad t_j < x < t_{j+1} \tag{10.59a}$$

$$\mathbf{B}_j \mathbf{y} \equiv B_{1j}\mathbf{y}(t_j^+) + B_{2j}\mathbf{y}(t_{j+1}^-) = \mathbf{s}_j \tag{10.59b}$$

where the matrices B_{1j} and $B_{2j} \in \mathbf{R}^{n \times n}$ and the vectors $\mathbf{s}_j \in \mathbf{R}^n$ are yet to be specified. For notational convenience, we assume that

$$\max_j (\|B_{1j}\|, \|B_{2j}\|) \leq 1 \tag{10.59c}$$

which is commensurate with the assumption that

$$\max (\|B_a\|, \|B_b\|) \leq 1$$

Given B_{1j} and B_{2j}, let $\Phi_j(x)$ be the fundamental solution of (10.59a) satisfying

$$\mathbf{B}_j \Phi_j = I \tag{10.60a}$$

and let $\mathbf{v}_j(x)$ be the particular solution of (10.59a) satisfying

$$\mathbf{B}_j \mathbf{v}_j = \mathbf{0} \tag{10.60b}$$

The solution $\mathbf{y}(x)$ of (10.59) can thus be written for the jth segment as

$$\mathbf{y}(x) = \Phi_j(x)\mathbf{s}_j + \mathbf{v}_j(x), \qquad t_j \leq x \leq t_{j+1}, \quad 1 \leq j \leq M \tag{10.61}$$

The solution of (10.1), (10.15) is then patched together from the pieces in (10.61) by requiring that (10.15) be satisfied and that

$$\mathbf{y}(t_j^-) = \mathbf{y}(t_j^+), \qquad 2 \le j \le M \tag{10.62}$$

obtaining a system of nM linear equations for $\mathbf{s}^T = (\mathbf{s}_1^T, \ldots, \mathbf{s}_M^T)$.

In detail, conditions (10.62) are written as

$$\Phi_j(t_{j+1})\mathbf{s}_j - \Phi_{j+1}(t_{j+1})\mathbf{s}_{j+1} = \boldsymbol{\beta}_j := \mathbf{v}_{j+1}(t_{j+1}) - \mathbf{v}_j(t_{j+1}), \quad 1 \le j \le M-1 \tag{10.63a}$$

and the BC (10.15) give

$$B_a \Phi_1(t_1)\mathbf{s}_1 + B_b \Phi_M(t_{M+1})\mathbf{s}_M = \boldsymbol{\beta}_M := \boldsymbol{\beta} - B_a \mathbf{v}_1(t_1) - B_b \mathbf{v}_M(t_{M+1}) \tag{10.63b}$$

Together we write these equations as

$$\mathbf{As} = \mathbf{b}, \qquad \mathbf{b}^T = (\boldsymbol{\beta}_1^T, \ldots, \boldsymbol{\beta}_M^T) \tag{10.63c}$$

This resembles very much the multiple shooting method described in Section 4.3. But there is a big difference in that here we *do not* require the BC (10.59b) to be initial conditions. Thus, we *do not* necessarily integrate (10.59) numerically using initial value methods. Note that the number of segments M from (10.16) is typically small and independent of $\|A\|$.

Two basic questions arise with regard to the BVPs (10.59):

(i) Given that the BVP (10.1), (10.15) is well-conditioned, what is needed to ensure that the BVPs (10.59) be well-conditioned?

(ii) Given that we have an approximation for each of the quantities appearing on the right-hand side of (10.61) and we use these approximations to define a corresponding numerical solution, what can be said about this numerical solution as an approximation to that of (10.1), (10.15)?

To be able to answer question (ii) intelligently, the basic requirement is that the "theoretical multiple shooting" method described by (10.60)–(10.63) (with Φ_j and $\mathbf{v}_j$ exact), be stable. This is one reason why we require that the newly generated BVPs (10.59) be well-conditioned. The well-conditioning of (10.59) is the key concern in determining the BC matrices of (10.59b) and generally excludes taking (10.59b) simply as initial conditions. (Why?)

To see the connection between the conditioning of these BVPs and that of the original one (10.1), (10.15), let $\Phi(x)$ be the fundamental solution of (10.1) satisfying

$$B_a \Phi(a) + B_b \Phi(b) = I$$

and let $G(x,t)$ be the Green's function for this BVP. It is easy to see that

$$\Phi_j(x) = \Phi(x)(B_j \Phi)^{-1} \qquad 1 \le j \le M \tag{10.64a}$$

From the theoretical multiple shooting framework we can then derive an explicit expression for the unknowns $\mathbf{s}_j$, viz.

$$\mathbf{s}_j = (B_j \Phi)\{\sum_{k=1}^{M-1} \Phi^{-1}(t_j) G(t_j, t_k)\boldsymbol{\beta}_k + \boldsymbol{\beta}_M\} = B_j \{\sum_{k=1}^{M-1} G(\cdot, t_k)\boldsymbol{\beta}_k + \Phi(\cdot)\boldsymbol{\beta}_M\} \tag{10.65}$$

(see Exercise 6a). Moreover, it is simple to see that

$$\mathbf{v}_j(x) = \int_{t_j}^{t_{j+1}} G_j(x,s)\mathbf{q}(s)\,ds \tag{10.66}$$

where for each j, $1 \le j \le M$, $G_j(x,s)$ is the Green's function of the corresponding BVP (10.59). It can also be verified (Exercise 6b) that

$$G_j(x,s) = G(x,s) - \Phi_j(x)\,\mathbf{B}_j G(\cdot,s) \tag{10.64b}$$

As for any BVP we can associate conditioning constants with (10.59) by using maximum norm estimates for Φ_j and G_j. In particular, let us define

$$\kappa_{1j} := \|\Phi_j\|_{[t_j, t_{j+1}]} \tag{10.67a}$$

$$\kappa_{2j} := \|G_j\|_{[t_j, t_{j+1}]} \tag{10.67b}$$

We obtain

Lemma 10.68 Let the BVP (10.1), (10.15) have a conditioning constant κ. Then

$$\kappa_{2j} \le \kappa(1 + 2\kappa_{1j}), \qquad 1 \le j \le M \qquad \square$$

The proof of this lemma follows immediately from (10.64b) and (10.59c).

Lemma 10.68 implies that if the original BVP (10.1), (10.15) is well-conditioned and if the BC (10.59b) are properly chosen for each j so that κ_{1j} is of moderate size [with Φ_j satisfying (10.60a)], then the BVP (10.59) is well-conditioned. Thus we obtain a "local" dichotomy with a moderate dichotomy constant on the segment (t_j, t_{j+1}). Assuming that κ_{1j} in (10.67a) are bounded, we can provide an answer to the basic questions posed earlier.

Theorem 10.69 Suppose that there is a moderate constant K_1 such that for each j, $1 \le j \le M$,

$$\kappa_{1j} \le K_1. \tag{10.69a}$$

Further suppose that the BVP (10.1), (10.15) is well-conditioned (uniformly in ε, i.e., there is a constant K of moderate size and some $\varepsilon_0 > 0$ such that K bounds the conditioning constants of the BVP for all $0 < \varepsilon \le \varepsilon_0$). Then the following hold (also uniformly in ε):

(i) The BVPs (10.59) are well-conditioned, with conditioning constants bounded by
$$\kappa_{2j} \le \kappa(1 + 2K_1).$$

(ii) The theoretical multiple shooting method (10.60)–(10.63) is stable, i.e., there is a moderate constant $K_2 = 2\kappa K_1$ such that

$$\text{cond}(A) \le K_2 M \tag{10.69b}$$

(iii) The vector $\mathbf{s}$ is bounded in terms of the original data by

$$|\mathbf{s}| \le \kappa(|\boldsymbol{\beta}| + 2\sum_{j=1}^{M} \kappa_{2j}\,\|\mathbf{q}\|_{1,[t_j,t_{j+1}]}) \le \kappa(|\boldsymbol{\beta}| + 2\kappa(1+2K_1)\,\|\mathbf{q}\|_1) \tag{10.69c}$$

Proof: Conclusion (i) follows directly from Lemma 10.68 and (10.69a). To see (ii), note that the rightmost expression in (10.65) is in fact a detailed expression of $\mathbf{A}^{-1}\mathbf{b}$ [cf. (10.63)]. Observing (10.59c) we have

$$\|\mathbf{A}^{-1}\| \leq M\kappa$$

and (10.69b) follows.

For conclusion (iii) we again use (10.65), this time to obtain

$$|\mathbf{s}_j| \leq \kappa \sum_{k=1}^{M} |\boldsymbol{\beta}_k|$$

The advertised bound then results directly from (10.63), (10.66) and (i). □

The utility of Theorem 10.69 still depends on finding suitable local BC for (10.59) so that (10.69a) holds with a constant K_1 of moderate size. In order to construct such local BC we use the dichotomic structure of the problem. Since the BVP (10.1), (10.15) is assumed to be well-conditioned (uniformly in ε), (10.1) has such a dichotomy. Thus, there are p nonincreasing and $n-p$ nondecreasing modes. So, just like for the stable IVP integration after decoupling in Chapter 6, we specify on each segment $[t_j, t_{j+1}]$ values for the p nonincreasing modes at t_j and values for the $n-p$ nondecreasing modes at t_{j+1}. For the transformed form (10.28–30) in particular, this corresponds to specifying $\mathbf{u}^-(t_j)$ and $\mathbf{u}^+(t_{j+1})$. We now work out the details in general.

Because there is a dichotomy, there exists a constant K of moderate size and a projection P of rank p such that

$$\|\Phi(x)P\Phi^{-1}(s)\| \leq K \qquad x \geq s \qquad (10.70a)$$

$$\|\Phi(x)(I-P)\Phi^{-1}(s)\| \leq K \qquad x < s \qquad (10.70b)$$

We have shown in Section 3.4 that without loss of generality we may take

$$P = \begin{bmatrix} 0 & 0 \\ 0 & I_p \end{bmatrix} \qquad (10.70c)$$

Writing

$$\Phi(x) = (\Phi^1(x) | \Phi^2(x))$$

where $\Phi^1(x) \in \mathbf{R}^{n \times (n-p)}$, $\Phi^2(x) \in \mathbf{R}^{n \times p}$, let $Q_{1j} \in \mathbf{R}^{n \times p}$ and $Q_{2j} \in \mathbf{R}^{n \times (n-p)}$ be two matrices with orthonormal columns such that

$$Q_{1j}^T \Phi^1(t_j) = 0, \qquad Q_{2j}^T \Phi^2(t_{j+1}) = 0 \qquad (10.71a)$$

We then define the local BC

$$B_{1j} := \begin{bmatrix} 0 \\ Q_{1j}^T \end{bmatrix}, \qquad B_{2j} := \begin{bmatrix} Q_{2j}^T \\ 0 \end{bmatrix} \qquad (10.71b)$$

It follows that $\mathbf{B}_j \Phi$ is a block diagonal matrix with full rank blocks, hence $\mathbf{B}_j \Phi$ is nonsingular and by (10.64a) $\Phi_j(x)$ is well-defined. Moreover, we have

Lemma 10.72 If (10.70) holds and the local BC are chosen as in (10.71) then

$$\|\Phi_j(x)\| \leq 2K, \qquad t_j \leq x \leq t_{j+1} \tag{10.72a}$$

$$\|G_j(x,s)\| \leq K \qquad t_j \leq x, s \leq t_{j+1} \tag{10.72b}$$

Proof: The result (10.72b) follows directly from (10.70), upon writing $G_j(x,s)$ in terms of $\Phi(x)$, using (10.64a). Now, for (10.72a) note that the matrix $[B_{1j}|B_{2j}]$ clearly has orthonormal rows, hence $\|[B_{1j}|B_{2j}]\| = 1$. It follows that

$$\|\Phi_j(x)\| = \|\Phi_j(x)[B_{1j}|B_{2j}]\| \leq \|\Phi_j(x)B_{1j}\| + \|\Phi_j(x)B_{2j}\| = \|G_j(x,t_j)\| + \|G_j(x,t_{j+1})\| \leq 2K$$

□

Now we can substitute the bounds of Lemma 10.72 in Theorem 10.69 to obtain

Corollary 10.73 With the dichotomy (10.70) holding, and choosing the local BC as in (10.71), Theorem 10.69 holds with $K_1 = 2K$, $K_2 = 4\kappa K$, and

$$|s| \leq \kappa(|\boldsymbol{\beta}| + 2K\|\mathbf{q}\|_1) \tag{10.73}$$

□

Now we can consider the process of patching together a numerical solution for (10.1), (10.15) from segment approximations. (This process is considered here for *theoretical* purposes.) Suppose that we have approximations $\Phi_j^h(x)$ to $\Phi_j(x)$ and $\mathbf{v}_j^h(x)$ to $\mathbf{v}_j(x)$, where $\Phi_j^h(x)$ satisfy

$$\mathbf{B}_j \Phi_j^h = I, \qquad 1 \leq j \leq M$$

These approximations are assumed to be defined continuously on $[t_j, t_{j+1}]$; for a mesh function we require that the points t_j be part of the mesh and postulate (for notational simplicity only) some interpolation in between mesh points. Corresponding to (10.63) we define a vector $\mathbf{s}^h$ which solves (10.63c) with a matrix $\mathbf{A}^h$ and vector $\mathbf{b}^h$ formed by replacing the quantities in (10.63a,b) by their approximations. Then we construct the numerical solution $\mathbf{y}^h(x)$ by

$$\mathbf{y}^h(x) = \Phi_j^h(x)\mathbf{s}_j^h + \mathbf{v}_j^h(x), \qquad t_j \leq x \leq t_{j+1}, \quad 1 \leq j \leq M \tag{10.74}$$

[cf. (10.61)]. Let us call this process an "approximate theoretical multiple shooting method." Depending on the numerical method used (and the choice of points t_j), we may or may not have $\mathbf{A}^h$ closely approximating $\mathbf{A}$ element-wise. In the following two subsections we discuss these two cases separately.

10.2.2 Stability and global error analysis, I

If the approximate theoretical multiple shooting matrix $\mathbf{A}^h$ accurately approximates the theoretical multiple shooting matrix $\mathbf{A}$ element by element, then the stability of the numerical method is inherited from the conditioning of the approximated BVP, much as in Theorem 5.38. We also obtain directly a localized pointwise error estimate.

Theorem 10.75 Suppose that the assumptions of Theorem 10.69 hold. In addition suppose that there are constants δ_1, δ_2 such that

$$\|\Phi_j^h(t_j) - \Phi_j(t_j)\|, \qquad \|\Phi_j^h(t_{j+1}) - \Phi_j(t_{j+1})\| \le \delta_1 \qquad (10.75a)$$

$$2\kappa M \delta_1 =: \gamma < 1 \qquad (10.75b)$$

$$|\mathbf{v}_j^h(t_j) - \mathbf{v}_j(t_j)|, \qquad |\mathbf{v}_j^h(t_{j+1}) - \mathbf{v}_j(t_{j+1})| \le \delta_2 \qquad (10.75c)$$

Then the approximate theoretical multiple shooting method is well defined and stable. Furthermore,

$$\|[\mathbf{A}^h]^{-1}\| \le \frac{\kappa M}{1-\gamma} \qquad (10.75d)$$

$$|\mathbf{s}^h - \mathbf{s}| \le \frac{1}{1-\gamma}(\hat{\gamma} + \gamma|\mathbf{s}|) \qquad (10.75e)$$

where $\hat{\gamma} := 2\kappa M \delta_2$, and

$$|\mathbf{y}^h(x) - \mathbf{y}(x)| \le \frac{K_1}{1-\gamma}(\hat{\gamma} + \gamma|\mathbf{s}|) + |(\Phi_j^h(x) - \Phi_j(x))\mathbf{s}_j^h| + |\mathbf{v}_j^h(x) - \mathbf{v}_j(x)| \quad (10.75f)$$

$$t_j \le x \le t_{j+1}, \qquad 1 \le j \le M$$

Proof: From (10.75a),

$$\|\mathbf{A}^h - \mathbf{A}\| \le 2\delta_1$$

Then (10.75b) guarantees the existence and boundedness of $(\mathbf{A}^h)^{-1}$ as in (10.75d). The result (10.75e) follows from simple algebraic manipulations (reminiscent to those in Section 2.2.7). Finally, to obtain (10.75f) we write

$$\mathbf{y}^h(x) - \mathbf{y}(x) = \Phi_j^h(x)\mathbf{s}_j^h - \Phi_j(x)\mathbf{s}_j + \mathbf{v}_j^h(x) - \mathbf{v}_j(x)$$

$$= \Phi_j(x)(\mathbf{s}_j^h - \mathbf{s}_j) + (\Phi_j^h(x) - \Phi_j(x))\mathbf{s}_j^h + \mathbf{v}_j^h(x) - \mathbf{v}_j(x)$$

and take norms. □

The importance of Theorem 10.75 is that it allows us to concentrate on well-conditioned BVPs each defined on a segment of one type only, for instance with the solution smooth throughout the segment, if the local BC are chosen as in (10.71). These BC yield well-conditioned BVPs with (10.69a) holding. Note that $\mathbf{s}$ (and $\mathbf{s}^h$) appearing in (10.75f) can be bounded using (10.73). Examining (10.75a) and (10.75f) we see that slow modes and smooth inhomogeneities (but not necessarily fast modes) are required to be approximated well. Moreover, according to (10.75f), approximation errors made in a segment affect the errors in neighboring segments *only through their values at the segment ends*.

Example 10.9

Consider applying the backward Euler scheme (2.155) with a uniform step size h to the IVP of Example 10.1

$$\varepsilon y' = -y + q(x) \qquad 0 < x < 1$$

$$y(0) = q(0) + 1$$

with $q(x)$ a smooth function, $\|q\| = 1$, and $0 < \varepsilon \ll h \ll 1$. This scheme yields

$$y_{i+1} = \frac{\varepsilon h^{-1}}{1+\varepsilon h^{-1}} y_i + \frac{1}{1+\varepsilon h^{-1}} q(x_{i+1}), \qquad i = 1, \ldots, h^{-1}$$

The exact solution has, to recall, a boundary layer of the form $e^{-x/\varepsilon}$ at $x = 0$ and is smooth away from 0. Using $h \gg \varepsilon$, this layer is skipped over by the mesh (see Fig. 10.4).

Consider the segments defined by $t_1 = 0$, $t_2 = h$, $t_3 = 1$, i.e., $M = 2$, and define approximate quantities in between mesh points using linear interpolation. We make the stable choice $B_{11} = B_{12} = 1$, $B_{21} = B_{22} = 0$. Clearly, in the segment $[t_1, t_2]$ we do not have a pointwise accurate approximation to $\Phi_1(x) = e^{-x/\varepsilon}$, but at $x = t_2$ we do have

$$\Phi_I^h(t_2) = \frac{\varepsilon h^{-1}}{1+\varepsilon h^{-1}} \approx 0 \approx e^{-h/\varepsilon} = \Phi_1(t_2)$$

and similarly $v_I^h(t_2) \approx v_1(t_2)$. Therefore δ_1 and δ_2 in (10.75) are very small (in fact they shrink to 0 as $\varepsilon \to 0$). According to (10.75f), then, the approximation of the smooth solution on $[h, 1]$ depends only on accuracy considerations for this segment, and is essentially not affected by the poor pointwise approximation in the layer segment $[0, h)$.

□

Theorem 10.75 also allows for use of different discretization schemes in different segments of the interval $[a, b]$. We have obtained a rather powerful framework for analyzing numerical methods. With BC as specified above, we may consider the numerical discretization separately in regions of smooth solution variation and in layer regions.

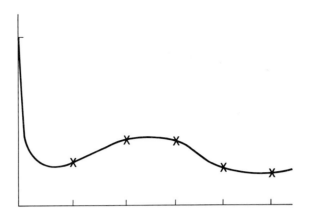

Figure 10.4 Coarse mesh skipping layer

Suppose that we use a finite difference or marching discretization method for a stiff BVP (10.1), (10.15) which gives rise to a one-step recursion of the form

$$R_i \mathbf{y}_{i+1} = S_i \mathbf{y}_i + \mathbf{q}_i, \quad 1 \leq i \leq N \tag{10.76a}$$

$$B_a \mathbf{y}_1 + B_b \mathbf{y}_{N+1} = \boldsymbol{\beta} \tag{10.76b}$$

[cf. (6.22)]. Here $\mathbf{y}_i \sim \mathbf{y}(x_i)$ for a mesh

$$\pi : a = x_1 < x_2 < \cdots < x_{N+1} = b \tag{10.77}$$

$$h_i := x_{i+1} - x_i, \quad h := \max_{1 \leq i \leq N} h_i$$

To fit this into the framework presented here, we can take a subset of the points of π and conceive an "approximate theoretical multiple shooting method" in a variety of ways. If the conditions of Theorem 10.75 are fulfilled, then a general error estimate is given in (10.75f). This applies to a number of methods discussed in Sections 10.3.1, 10.3.3, and 10.4.2, where δ_1 and δ_2 are very small in magnitude for very stiff problems. The error is then localized. Moreover, on a smooth solution segment $[t_j, t_{j+1}]$, $\|\Phi_j^h(x) - \Phi_j(x)\|$ need not be small for a small error to be obtained in $\mathbf{y}(x)$; only $|(\Phi_j^h(x) - \Phi_j(x))\mathbf{s}_j^h|$ and $|\mathbf{v}_j^h(x) - \mathbf{v}_j(x)|$ matter. These correspond to the smooth components of the solution only, according to the representations (10.61) and (10.74).

If we want a uniformly accurate approximate solution on $[a,b]$ then we must have a fine mesh in layer regions, where the solution varies rapidly, e.g. $h_i = O(\varepsilon)$, as is further discussed in Section 10.2.4. But in segments where the solution varies slowly, we want h_i only commensurate with the accurate approximation of the smooth solution, i.e., $h_i \gg \varepsilon$ there. The mesh is therefore extremely nonuniform when ε is very small, and indeed the extreme examples in Chapter 9 are of this type. This is also true of the methods covered by the analysis in the following subsection.

10.2.3 Stability and global error analysis, II

When considering a method like (10.76) for a stiff problem, we must discuss basic stability questions. In Chapter 5 (and in Section 2.7.2) stability of one-step schemes follows automatically from consistency, with the stability constant being equal up to $O(h)$ to the BVP conditioning constant. This is essentially because the well-scaled fundamental matrix $\Phi(x)$ of (10.1) is well approximated everywhere. Here, $\Phi(x)$ is still approximated well in layer regions, but its part corresponding to fast modes is not necessarily approximated well in the smooth regions, where $h_i \gg \varepsilon$.

For stiff IVPs it is still possible to approximate the fundamental solution well at mesh points (but not everywhere in between) when $h_i \gg \varepsilon$ by applying a scheme which has stiff decay (cf. Section 2.7.3). Such a scheme must be one-sided, and its application in a BVP context must generally be preceded by an explicit decoupling of fast increasing and decreasing modes, as demonstrated in Example 10.10 and discussed in Chapter 6 and in Section 10.3.1. Symmetric schemes like those discussed in Chapter 5 will not yield an accurate approximation of $\Phi(x)$ where the mesh is coarse; but then, in regions where the solution is smooth, the necessity of such an accurate approximation may be questioned, because the contribution of the fast modes must be very small.

As before, we wish to be able to analyze a given scheme (or a combination of schemes) on one segment at a time [with the joints $t_2, \ldots, t_M$ included in the mesh π of (10.77)], where on each segment the solution has one behaviour type. Theorem 10.75 provides us with a stability result (10.75d), much as Theorem 10.38 does for the nonstiff case, if our scheme essentially produces accurate approximations for $\Phi(x)$ at the joints of the segments. For, if (10.75a,b) hold in addition to the assumptions of Theorem 10.69, then the approximate multiple shooting matrix is close to the exact one [as in the proof of Theorem 10.75], so the well-conditioning of the BVP implies the boundedness of its inverse, with a stability constant of moderate size.

But if $\Phi_j(x)$ is not approximated well at mesh points of a segment $[t_j, t_{j+1}]$ with a smooth solution, then in general (10.75a,b) does not hold. To be able to say more in this case, we assume that a three-way splitting of modes into rapidly increasing, rapidly decreasing, and slow ones is possible. Thus, for a family of ODEs (10.1) depending on a singular perturbation parameter ε, $0 < \varepsilon \leq \varepsilon_0$, there are positive constants $\lambda_+, \lambda_-,$ and K, projections

$$P_+ = \begin{bmatrix} I_{n_+} & 0 \\ 0 & 0 \end{bmatrix} \quad P_- = \begin{bmatrix} 0 & 0 \\ I_{n_-} & 0 \\ 0 & 0 \end{bmatrix} \quad P_0 = \begin{bmatrix} 0 & 0 \\ 0 & I_{n_0} \end{bmatrix}$$

(with $P_+ + P_- + P_0 = I_n$) and a fundamental matrix $Z(x)$ such that for all $0 < \varepsilon \leq \varepsilon_0$,

$$\|Z(x)P_+Z^{-1}(s)\| \leq K e^{\lambda_+(x-s)/\varepsilon} \qquad x < s \qquad (10.78\text{a})$$

$$\|Z(x)P_-Z^{-1}(s)\| \leq K e^{\lambda_-(s-x)/\varepsilon} \qquad x > s \qquad (10.78\text{b})$$

$$\|Z(x)P_0Z^{-1}(s)\| \leq K \qquad \text{all } x, s \qquad (10.78\text{c})$$

The three-way splitting (10.78) holds for a *3-segment problem*, as described in Remark (b) at the end of Section 10.1.3.

This restriction allows us to conclude that the solution of the homogeneous problem is smooth except possibly near the boundaries. The widths of the boundary layers are given as in (10.48). Denoting the left end point of the smooth segment by c and the right end point by d, $a < c < d < b$, the solution $y(x)$ is thus smooth on $[c,d]$ and potentially has layers on $[a,c]$ and $[d,b]$. Correspondingly, we assume that the mesh π is dense on $[a,c]$ and on $[d,b]$, and coarse on $[c,d]$, as in Fig. 10.5. A basic question is, to what extent does a given scheme on the coarse mesh segment $[c,d]$ preserve the splitting (10.78)? (The meshes on the short segments $[a,c]$ and $[d,b]$ are assumed to be sufficiently fine to essentially reproduce a well-scaled fundamental solution there.)

Figure 10.5 Mesh for the 3-segment problem

Example 10.10

Consider the scalar ODE

$$\varepsilon y' = y - 1 \qquad 0 < x < 1$$

subject to two sets of BC: (i) $y(1) = 2$; (ii) $y(0) + y(1) = 1$. Both BVPs are easily seen to be well-conditioned. Consider the backward Euler scheme (which works well for stiff IVPs) on a uniform mesh:

$$\frac{\varepsilon}{h}(y_{i+1} - y_i) = y_{i+1} - 1 \qquad 1 \leq i \leq N. \tag{10.79}$$

Case (i). To (10.79) we add

$$y_{N+1} = 2$$

It is easy to see that the resulting system of equations for $y_1, \ldots, y_{N+1}$ is almost singular for $\varepsilon \ll h$. In fact, as $\varepsilon \to 0$ it is required that both $y_{N+1} = 2$ and $y_{N+1} = 1$, and $y_1 \sim (h/\varepsilon)^N$. But note that even if we had

$$y_{N+1} = 1$$

specified (as is the effect of placing a fine mesh with $h_i \ll \varepsilon$ near $x = 1$), the system of equations is still nearly singular, i.e., unstable.

Indeed, a transformation $t := 1 - x$ which reverses the direction of integration shows that we are really trying to apply the *forward* Euler method to a stiff IVP with a coarse step size, and this is not absolutely stable.

The trouble encountered can be described as follows: The backward Euler scheme applied to find the fundamental solution $e^{x/\varepsilon}$, which increases rapidly, gives

$$\frac{\varepsilon h^{-1}}{1 - \varepsilon h^{-1}} \approx 0$$

Thus, the damping property which is applauded in Example 10.9 is undesirable here, because it distorts (10.78), approximating a rapidly increasing mode by a rapidly decreasing one.

Case (ii). To (10.79) we add

$$y_1 + y_{N+1} = 1$$

The discrete system is now well-conditioned, but while the exact solution is

$$y(x) \approx 1 - e^{(x-1)/\varepsilon}$$

the approximate one is

$$y_1 \approx 0, \qquad y_2 \approx y_3 \approx \cdots \approx y_{N+1} = 1$$

which suggests that a layer occurs near $x = 0$ and not near $x = 1$. This may be regarded in a sense as being worse than the "exploding" case (i), because of the deceptive look of the approximate solution.

□

This example serves to illustrate that rapidly decreasing modes must be approximated by nonincreasing ones and rapidly increasing modes must be approximated by nondecreasing ones; i.e., the essence of the continuous decoupling must be preserved by the discrete BVP (cf. Chapter 6). This is achieved by all methods discussed in the remainder of this chapter, but in various ways and to varying degrees. Thus, the symmetric (centered) schemes discussed in Section 10.3.2 approximate fast modes, both increasing and decreasing, by slow ones, thereby achieving only a "weak" decoupling. Since Theorem 10.75 may not be used in this case, it is not *a priori* clear that the resulting scheme is stable, even though we would not expect in any event to encounter with symmetric schemes such extreme (exponential) instability as in Example 10.10.

Unlike for Theorem 10.75, we do not know that $\|\mathbf{A}^h - \mathbf{A}\|$ is small. But, assuming that Φ_j^h exist, $1 \leq j \leq M$, consider the matrices

$$\mathbf{QA} \text{ and } \mathbf{QA}^h$$

where $\mathbf{Q}$ is a block diagonal matrix

$$\mathbf{Q} = \text{diag}(Q_1, Q_2, \ldots, Q_{M-1}, I) \tag{10.80a}$$

with

$$Q_j := \begin{bmatrix} Q_{2j}^T \\ Q_{1,j+1}^T \end{bmatrix} \tag{10.80b}$$

[see (10.71)]. The jth row blocks of $\mathbf{A}^h$ and $\mathbf{A}$ are

$$\Psi_j^h := (\Phi_j^h(t_{j+1}) \mid -\Phi_{j+1}^h(t_{j+1})) \tag{10.81a}$$

and

$$\Psi_j := (\Phi_j(t_{j+1}) \mid -\Phi_{j+1}(t_{j+1})) \tag{10.81b}$$

respectively. By definition we have the structure

$$Q_j \Psi_j^h = \begin{bmatrix} I & 0 & \mid & U_{j2}^h & V_{j2}^h \\ V_{j1}^h & U_{j1}^h & \mid & 0 & -I \end{bmatrix} \tag{10.82a}$$

where U_{j1}^h, U_{j2}^h are nonsingular blocks. A similar structure holds for Ψ_j as well, but for the latter more can be specified as discussed below: Examining the blocks

$$Q_{1,j+1}^T \Phi_j^1(t_{j+1}), \qquad Q_{2j}^T \Phi_{j+1}^2(t_{j+1}) \tag{10.83a}$$

we recall first that

$$Q_{1,j+1}^T \Phi_{j+1}^1(t_{j+1}) = 0, \qquad Q_{2j}^T \Phi_j^2(t_{j+1}) = 0 \tag{10.83b}$$

But by (10.64a)

$$\Phi_j(x) = \Phi_{j+1}(x)[(\mathbf{B}_{j+1}\Phi)(\mathbf{B}_j\Phi)^{-1}] \tag{10.84}$$

Since $\mathbf{B}_j\Phi$ is block diagonal with the same block structure for each j, the matrix $(\mathbf{B}_{j+1}\Phi)(\mathbf{B}_j\Phi)^{-1}$ also has this block structure, namely

$$(\mathbf{B}_{j+1}\Phi)(\mathbf{B}_j\Phi)^{-1} = \begin{bmatrix} R_j^1 & 0 \\ 0 & R_j^2 \end{bmatrix}$$

where $R_j^1 \in \mathbf{R}^{(n-p) \times (n-p)}$, $R_j^2 \in \mathbf{R}^{p \times p}$ are nonsingular. Hence

$$\Phi_j^1(x) = \Phi_{j+1}^1(x) R_j^1, \qquad \Phi_j^2(x) = \Phi_{j+1}^2(x) R_j^2$$

Substituting this into (10.83a, b) for $x = t_{j+1}$ we obtain

$$Q_j \Psi_j = \begin{bmatrix} I & 0 & | & U_{j2} & 0 \\ 0 & U_{j1} & | & 0 & -I \end{bmatrix} \qquad (10.82b)$$

for some nonsingular blocks U_{j1}, U_{j2}.

Suppose now that the BVP (10.1), (10.15) is well-conditioned and consider a one-step method (10.76) for the three-segment problem, where the mesh on the layer regions is sufficiently fine that the fundamental solutions $\Phi_1(x), \Phi_3(x)$ of (10.59), (10.60) are accurately recovered. The BC $\mathbf{B}_j, 1 \le j \le 3$, are chosen as described in (10.71). We will show that if the method is stable for the BVP (10.59) on the middle segment $[c,d]$, then it is stable. For this we should show that the theoretical multiple shooting matrix

$$\mathbf{A}^h = \begin{bmatrix} \Phi_1(c) & -\Phi_2^h(c) & 0 \\ 0 & \Phi_2^h(d) & -\Phi_3(d) \\ B_a \Phi_1(a) & 0 & B_b \Phi_3(b) \end{bmatrix}$$

is boundedly invertible, where $\{\Phi_2^h(x_j)\}$ is the approximation of $\Phi_2(x)$ satisfying (10.60a), which exists and is bounded by assumption. Since (10.1), (10.15) is well-conditioned we may consider only $B_a = B_{11}, B_b = B_{32}$. Then

$$B_a \Phi_1(a) = \begin{bmatrix} 0 & 0 \\ 0 & I_p \end{bmatrix}, \quad B_b \Phi_3(b) = \begin{bmatrix} I_{n-p} & 0 \\ 0 & 0 \end{bmatrix}$$

and (10.82) give the block structure

$$\mathbf{Q}\mathbf{A}^h = \begin{bmatrix} I & 0 & U_{12}^h & V_{12}^h & & \\ 0 & U_{11} & 0 & -I & & \\ & & I & 0 & U_{22} & 0 \\ & & V_{21}^h & U_{21}^h & 0 & -I \\ 0 & 0 & & & I & 0 \\ 0 & I & & & 0 & 0 \end{bmatrix}$$

It is not difficult to see that $\mathbf{QA}^h$ is nonsingular. In fact, upon permuting the rows so that the identity blocks end up on the main diagonal, and subsequently performing a Gauss-Jordan row elimination (i.e., reducing the permuted matrix to an identity — not merely an upper triangular one as in Gauss elimination), we can obtain an explicit expression for $[\mathbf{QA}^h]^{-1}$. It follows that there is a moderate constant c such that

$$\|[\mathbf{A}^h]^{-1}\| \le c \max \{\kappa, \|\Phi_2^h(c)\|, \|\Phi_2^h(d)\|\}. \qquad (10.85)$$

This proves our claim that the method is stable if (and only if) it is stable on the middle segment $[c,d]$ with the BC (10.71).

It is important to stress that the stability of the scheme on the coarse mesh segment, subject to the BC $\mathbf{B}_2$, *does not follow* automatically from the well-conditioning of (10.59), if the exponential dichotomy part of (10.78) is not preserved by the discretization. In particular we give in Section 10.3.2 an example where such instability arises for symmetric schemes. Fortunately, this type of instability is quite rare in practice, as discussed later.

Remark As we have observed in Section 10.1.4, for BVPs (10.1), (10.15) arising as linearizations of nonlinear ones we generally do not expect the conditions of Theorem 10.47 to hold on the entire interval $[a,b]$, because at layer regions the coefficients in $A(x)$ involve the rapidly varying solution [so $T(x)$ is not smooth]. Still, under reasonable restrictions we may obtain (10.78) on $[a,b]$, assuming the conditions of Theorem 10.47 hold only on $[a+\varepsilon\alpha_1, b-\varepsilon\alpha_2]$ for α_1, α_2 appropriate positive constants. □

The result (10.85) can be generalized for theoretical multiple shooting formulations with more than 3 segments (i.e., $M > 3$), given that (10.78) holds.

Theorem 10.86 Suppose that the BVP (10.1), (10.15) is well-conditioned and that (10.78) holds on $[a,b]$. Consider a one-step method (10.76) in the theoretical multiple shooting framework (10.59), (10.63), (10.71), (10.74), with M odd ($M \leq N$), such that the solution on each segment $[t_{2j}, t_{2j+1}]$ is smooth, $j = 1, \ldots, \hat{M}$, $\hat{M} := \frac{1}{2}(M-1)$. [By this latter requirement we specifically mean that all fast modes are $O(\varepsilon)$ on these segments.] Assume further that the mesh on each segment $[t_{2j-1}, t_{2j}]$, $j = 1, 2, \ldots, \hat{M}+1$, is fine so that the fundamental solutions $\Phi_{2j-1}(x)$ of (10.59), (10.60) are accurately recovered. Finally, assume that Φ_{2j}^h exist, $j = 1, 2, \ldots, \hat{M}$.

Then there is a constant c of moderate size such that the following stability estimate holds:

$$\|[\mathbf{A}^h]^{-1}\| \leq c \; \max \{\kappa, \max_{1 \leq j \leq \hat{M}} \{\|\Phi_{2j}^h(t_{2j})\|, \|\Phi_{2j}^h(t_{2j+1})\|\}\} \quad (10.86)$$

Proof: As before, we consider $\mathbf{QA}^h$, with $\mathbf{Q}$ given by (10.80), and assume without loss of generality that $p = n_-$ (i.e., the slow modes are grouped with the rapidly increasing ones), that $B_a = B_{11}$, $B_b = B_{M2}$, and that the modes are already separated so that $U_{2j-1,1} = O(\varepsilon)$, $j = 1, 2, \ldots, \hat{M}$. The latter estimate is possible because $U_{2j-1,1}$ involves rapidly decaying modes evaluated at the right end of their segment [cf. (10.81), (10.82)]. Permuting row blocks of $\mathbf{QA}^h$ so that the identity blocks form the main diagonal, and considering again Gauss-Jordan elimination (with identity blocks as pivots), we find that attention is reduced to super-blocks of the form

$$\begin{bmatrix} I & U_{2j+1,1} & 0 & 0 \\ 0 & I & 0 & V_{2j+1,2}^h \\ V_{2j,1}^h & 0 & -I & 0 \\ 0 & 0 & U_{2j+1,1} & -I \end{bmatrix}$$

[everything else being the same as when we derive (10.85)]. But the above is an $O(\varepsilon)$ perturbation of the super-block

$$\begin{bmatrix} I & U_{2j,2} & 0 & 0 \\ 0 & I & 0 & V_{2j+1,2}^h \\ V_{2j,1}^h & 0 & -I & 0 \\ 0 & 0 & 0 & -I \end{bmatrix}$$

for which a Gauss-Jordan elimination can obviously proceed without difficulty, starting from the bottom. A contraction argument for ε small enough concludes the proof. □

Having shown stability, we now turn to error estimation. Given that (10.78) holds, e. g., for the 3-segment problem, the above considerations apply in principle for ε small enough to any method (10.76) which preserves the splitting in a discretized form [i. e., if (10.78) holds at all mesh points for some positive constants λ_+^h, λ_-^h replacing λ_+, λ_-, with the approximate fundamental solution $Z^h(x)$ replacing $Z(x)$].

Furthermore, if the stability conditions of Theorem 10.86 hold, with the right-hand side of (10.86) being of moderate size, then we may substitute (local truncation and boundary) errors in place of solutions in stability bounds, in the usual way. The stability and error analysis may be done on each segment separately. The so-obtained segment error estimates are then magnified in the global error estimate by at most the bound in (10.86). Again we note the crucial point that on the smooth solution segments like $[c,d]$ in the 3-segment problem, where the mesh is coarse, the error is driven by the local truncation error, which depends on the smooth components of the solution alone (and not on $\|\Phi_{2j}^h(x) - \Phi_{2j}(x)\|$).

Observe that there is one difference between this segment analysis and the usual analysis of Chapter 5 for one-step schemes. There we automatically assume that the BC are satisfied exactly by the approximate solution, while here we must admit boundary errors at segment ends. For instance, in the approximation error for the smooth segment $[c,d]$ of the 3-segment problem we have an error at c and d, which depends on the approximate solution values emanating from the layers at these points. The precise effect of these boundary errors on the error in (c,d) depends on the kind of stability bound we have (e. g., the layer error is well damped in Example 10.9); in general we do know that it is bounded.

10.2.4 The discretization mesh as a stretching transformation

In a good portion of Chapter 9 a discretization mesh is interpreted in terms of a transformation of the independent variable. This is also done, in the singular perturbation context, in the simple Example 10.1. We now want to generalize these considerations and give an analysis of one-step difference schemes for a simple IVP with a layer segment. This is used as a building block for analyses in later sections.

Consider the scalar ODE (10.19), where now the segment $[c,d]$ is a (short) layer region. We assume that (10.20) holds (so the IVP is stable) and that for some (possibly small) $\mu > 0$,

$$\|\lambda^{(\nu)}\|_{[c,d]}, \quad \|q^{(\nu)}\|_{[c,d]} \leq K\mu^{-\nu-1}, \quad \nu = 0, 1, \ldots, p \qquad (10.87)$$

For instance, in Examples 10.1 and 10.3, $\mu = \varepsilon$ in the boundary layer region near 0, while in Examples 10.2 [with an additional assumption on $q(x)$] and 4, $\mu = \sqrt{\varepsilon}$ near 0. In the variable

$$\xi := x/\mu \qquad (10.88)$$

the ODE reads

$$\frac{dy}{d\xi} = \mu\lambda(\mu\xi)y + \mu q(\mu\xi), \qquad c/\mu \leq \xi \leq d/\mu \qquad (10.89)$$

and from (10.87), obviously (10.21), (10.22), and (10.25a) hold in ξ. This allows us to conclude (10.26a); i.e., the solution and its derivatives are bounded in the stretched variable. The IVP for (10.89) is then not singularly perturbed, but the interval length may become large; i.e., $\mu^{-1}(d-c) \to \infty$ as $\mu \to 0$.

For a numerical discretization of (10.89), we may use any stiff IVP method as described in Sections 2.7.3 and 10.3 and below, provided that the step size is small enough. If the step size in ξ is η, then in the variable x we obtain the corresponding step size

$$h = \mu\eta$$

i.e., the step size in the original variable is very small, or the mesh on $[c,d]$ is very dense (fine). Conversely, if a fine mesh is used in a layer interval in x (say, the mesh is allocated by an *a priori* or adaptive mesh selection algorithm), then this corresponds to a stretching transformation which smooths out the solution (as discussed in Chapter 9).

At this point one may wonder if a nonstiff IVP scheme, say an explicit Runge-Kutta scheme, may be used to solve (10.89) [i.e., (10.19)]. Indeed, we are using a dense mesh to solve an ODE which in the stretched variable does not look stiff. But the usual theory for numerical IVPs needs to be supplemented when an "infinite" (or large) interval length $\mu^{-1}(d-c)$ is considered. This is because in the usual error analysis, the number of discretization steps is inversely proportional to the step size, but the constant of proportionality grows with the interval length. What can save the numerical schemes, though, is the fact that the solution of the stable IVP, hence also the solution for a fine discretization, decays, and so the accumulation of local errors is more favourable than merely linear.

The numerical discretization of the IVP for (10.89) should be considered carefully for efficiency reasons. A large portion of the interval in ξ involves an exponentially decaying tail which may well be the terminating end of the transient component of the solution of (10.19) where the smooth, stationary component is already dominant. Thus we may be able to choose relatively large steps η in the variable ξ, which are still quite small steps h in the variable x. To do so, we must ensure that the scheme used does not develop stability difficulties.

To be more specific, consider a one-step scheme

$$y_{i+1} = \gamma(\zeta_i)y_i + h_i r_i, \qquad 1 \leq i \leq N \qquad (10.90a)$$

on a mesh $c = x_1 < \cdots < x_{N+1} = d$, for the ODE (10.19) which is rewritten here,

$$y' = \lambda(x)y + q(x), \qquad c \leq x \leq d$$

The argument ζ of γ relates to $\lambda(x)h$, with $\zeta_i = \lambda(x_i)h_i$ if $\lambda(x)$ is constant on $[x_i, x_{i+1}]$, and

$$|r_i| \leq \overline{K}\, \|q\|_{[x_i, x_{i+1}]} \qquad (10.90b)$$

for a constant $\bar{K} \sim 1$. We require

Assumption 10.91 There are two nonnegative constants α_0 and β_0 such that for any set S of the form

$$S = S(\alpha_1, \alpha_2, \beta) = \{\zeta \colon 0 < |\zeta| \leq \beta, \frac{\pi}{2} + \alpha_1 \leq \arg \zeta \leq \frac{3\pi}{2} - \alpha_2\} \qquad (10.91a)$$

with $\alpha_1, \alpha_2 \geq \alpha_0$, $\alpha_1 + \alpha_2 \leq \pi$, $\beta \leq \beta_0$, there exists a positive constant $s = s(\alpha_1, \alpha_2, \beta)$ such that

$$|\gamma(\zeta)| \leq e^{s \, \text{Re}(\zeta)}, \qquad \zeta \in S \qquad (10.91b)$$

□

The requirement (10.91) is related to, but is not the same as, the concept of $A(\alpha_0)$-stability mentioned in Section 2.7.3 (see Fig. 10.6).

Note that α_1 and α_2 are determined by the ratio of the imaginary and the real part of $\lambda(x)$, and this bounds α_0. Given β_0, for the argument of (10.90a) to satisfy the assumption we must have

$$h_i |\lambda(x_i)| \leq \beta \leq \beta_0 \qquad (10.92)$$

i.e., $h_i = O(\mu)$. We wish to use a discretization such that $s = s(\alpha_1, \alpha_2, \beta)$ is bounded away from 0 with β sufficiently large that (10.92) will not restrict the choice of h_i more than accuracy requirements do.

Theorem 10.93 Let the ODE (10.19) satisfy (10.87) with $p = 0$, and assume that $\lambda(x)$ satisfies (10.20b) and

$$\text{Re}(\lambda(x)) \leq -\bar{\lambda} < 0, \qquad c \leq x \leq d \qquad (10.93a)$$

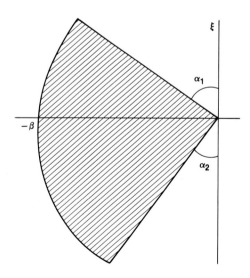

Figure 10.6 Region of decay for (10.91)

Suppose that a one-step scheme (10.90) satisfying Assumption 10.91 with $\alpha_0 < \tan^{-1}\rho$ is applied (with $y_1 := y(c)$) to discretize this ODE. Let $\alpha_1, \alpha_2 > 0$ be defined by

$$\tan \alpha_1 = \min \frac{\operatorname{Im}(\lambda(x))}{\operatorname{Re}(\lambda(x))}, \qquad \tan \alpha_2 = \max \frac{\operatorname{Im}(\lambda(x))}{\operatorname{Re}(\lambda(x))} \qquad (10.93b)$$

and assume further that the mesh is chosen such that

$$\max_i |\zeta_i| =: \beta \leq \beta_0 \qquad (10.93c)$$

with the corresponding $s = s(\alpha_1, \alpha_2, \beta)$ satisfying (10.91b). Then the following stability result holds:

$$\max_{1 \leq i \leq N} |y_i| \leq |y_1| + \frac{\overline{K}}{s\overline{\lambda}} \|q\|_{[c,\hat{a}]}$$

Proof: First note that by (10.20b) and (10.93b), $\alpha_1, \alpha_2 > \alpha_0$, so from (10.93c), (10.91b) holds.

Solving the recurrence relation (10.90a), we have

$$y_{i+1} = [\prod_{l=1}^{i} \gamma(\zeta_l)] y_1 + \sum_{j=1}^{i} [\prod_{l=j+1}^{i} \gamma(\zeta_l)] h_j r_j$$

Using (10.91b) and (10.93a),

$$|\prod_{l=j+1}^{i} \gamma(\zeta_l)| \leq \prod_{l=j+1}^{i} e^{-s\overline{\lambda} h_i} = e^{-s\overline{\lambda}(x_{i+1} - x_{j+1})} \qquad (\leq 1)$$

and this with (10.90b) yields

$$|\sum_{j=1}^{i} [\prod_{l=j+1}^{i} \gamma(\zeta_l)] h_j r_j| \leq \sum_{j=1}^{i} e^{-s\overline{\lambda}(x_{i+1} - x_{j+1})} h_j |r_j|$$

$$\leq \max_j |r_j| |\int_{x_1}^{x_{i+1}} e^{-s\overline{\lambda}(x_{i+1} - t)} dt \leq \frac{\overline{K}}{s\overline{\lambda}} \|q\|_{[c,\hat{a}]} \qquad (10.93d)$$

$\square$

This result closely resembles the corresponding analytical one (if $\dfrac{\overline{K}}{s}$ is a constant of moderate size).

For a given IVP (10.19) satisfying the assumptions of Theorem 10.93, the angles α_1 and α_2 are determined, and the question becomes: which schemes satisfy Assumption 10.91 for β sufficiently large? Certainly the absolute stability region of the scheme (cf. Section 2.7) contains $S(\alpha_1, \alpha_2, \beta)$. Thus it can be seen that explicit Runge-Kutta schemes allow only small values of β; e.g. $\beta < 1$ for (the forward) Euler's scheme when $\operatorname{Im}(\lambda) = \operatorname{Re}(\lambda)$. In certain cases a rough indication of the desired size of β is given by $\beta_0 \approx |\ln \varepsilon|$ (cf. Section 10.3.2). For the usual explicit Runge-Kutta schemes we therefore have an undesirable restriction in (10.92). On the other hand, any of the implicit Runge-Kutta schemes mentioned in Section 2.7 or in Chapter 5 (and in Section 10.3) is a suitable scheme here, being A-stable and satisfying Assumption 10.91 for any ρ of moderate size in (10.20b) and a much larger β.

Extensions of Theorem 10.93 to layer regions of a BVP (10.1), (10.15) are discussed in Section 10.3.

Remarks

(a) It is important to understand that the contribution of Theorem 10.93 to the usual stability theory in the stretched variable is that it deals with "long" intervals (infinite as $\mu \to 0$). Thus we may use a constant number of steps with a given scheme inside the layer region, and then apply the theorem to analyze the error on the remaining portion. This may give a more practically relevant value for $\bar{\lambda}$ in (10.93a). For instance we get $\bar{\lambda} = O(\varepsilon^{-1/2})$ in Examples 10.2 and 10.4. Another instance where a more relevant $\bar{\lambda}$ is obtained only away from c is when (10.19) is a linearization of a nonlinear problem, in which case $\lambda(x)$ involves the solution in the layer and hence varies rapidly. This corresponds to the coefficient in (10.89) not being close to constant. But even for nonlinear problems, we expect the solution tail for ξ large to approach a constant at an exponential rate under reasonable assumptions (cf. Section 11.4.2), and the relevant $\bar{\lambda}$ can be derived from the values of $\lambda(x)$ at the tail range.

(b) While the above analysis is only done for one-step schemes, the principle behind it extends also to multistep schemes. For instance, BDF schemes of order >2, which are not absolutely stable everywhere in the negative half-plane of λh, are $A(\alpha)$-stable and so would work well provided ρ in (10.20b) is small enough that $|\tan \alpha| < \rho$.

(c) In general, a singularly perturbed problem may have more than two scales, and this may cause two types of layer near the same location in the interval. For instance, we may have $O(\varepsilon)$ and $O(\sqrt{\varepsilon})$ layers near $x = 0$, say, where ε is a small parameter appearing in a given BVP. For this situation we would have a number of stretched variables in succession, beginning with the one that resolves the steepest layer. In (10.16) we would then have a number of short segments in succession. □

10.3 DIFFERENCE METHODS

Most efforts to solve practical stiff BVPs reported to date have used finite difference methods. Many methods exist which are supported by a fairly thorough analysis, or by a good practical experience, or by both. In this section we discuss some of these methods, as usual highlighting what we think is important from the viewpoint of generality and practicality.

10.3.1 One-sided schemes

We have mentioned before how useful even the simple backward Euler scheme (2.155) can be for a very stiff IVP. A fast mode like $e^{\lambda x}$, $\text{Re}(\lambda) \ll -1$, is approximated very well by using this scheme with a large step size, viz. $|\lambda| h \gg 1$ (cf. Example 10.9). Higher-order extensions exist: Multistep extensions are the BDF schemes, and Runge-Kutta extensions are collocation at Radau points and various other possibilities. These extensions are more complicated as well, and it is worthwhile to devote attention to the backward Euler scheme for a scalar problem before proceeding to more complex schemes and problems.

(i) Backward Euler for a scalar ODE. Consider the scalar ODE (10.19) with (10.20–22) holding. Like any one-step scheme, the backward Euler approximation can be written as (10.90a), i.e.,

$$y_{i+1} = \gamma(\zeta_i) y_i + h_i r_i \qquad 1 \le i \le N$$

where

$$\gamma(\zeta) = \frac{1}{1-\zeta}, \quad \zeta_i = h_i \lambda(x_{i+1}), \qquad r_i = \gamma(\zeta_i) q(x_{i+1}) \tag{10.94}$$

with the solution

$$y_{i+1} = [\prod_{l=1}^{i} \gamma(\zeta_l)] y_1 + \sum_{j=1}^{i} [\prod_{l=j+1}^{i} \gamma(\zeta_l)] h_j r_j \tag{10.95}$$

This solution form is good for any one-step scheme (10.90a) for a scalar ODE (10.19), and yields a stability bound

$$|y_{i+1}| \le |y_1| + \bar{K} \|q\|_\infty (b-a) \qquad 1 \le i \le N \tag{10.96a}$$

if

$$|\gamma(\zeta_i)| \le 1, \qquad i = 1, \ldots, N \tag{10.96b}$$

However, for the backward Euler scheme, not only does (10.96b) hold, but $|\gamma(\zeta)| \approx \frac{1}{|\zeta|}$ for $|\zeta|$ large, so a large step size h such that $|\lambda h| \gg 1$ yields a small $|\gamma(\zeta)|$ and improves the stability bound. This, in turn, yields improved errors: Inserting as usual the error $e_i := y_i - y(x_i)$ in place of the approximate solution y_i in (10.90a), the inhomogeneity $q(x_{i+1})$ is replaced by an $O(h_i)$ local truncation error term (if y has two bounded derivatives at x_{i+1}), and r_i is then replaced by $\gamma(\zeta_i) O(h_i)$. This gives in place of (10.95)

$$e_{i+1} = [\prod_{l=1}^{i} \gamma(\zeta_l)] e_1 + \sum_{j=1}^{i} [\prod_{l=j}^{i} \gamma(\zeta_l)] O(h_j^2) \tag{10.97}$$

So we see that if, for instance, $\gamma(\zeta_l) = O(h)$ (which happens if $|h_l \lambda(x_{l+1})| > h^{-1}$), where h is the maximum step size, then $e_{i+1} = O(h^3), 2 \le i \le N$. Note that e_1 does not need to be small — it may be a large (~1) error emanating into the smooth solution interval $[c,d]$ from a layer region where the solution is not approximated well at all. From (10.97) we also see that if $|h_i \lambda(x_{i+1})| \to \infty$ then $e_{i+1} \to 0, 1 \le i \le N$. In singular perturbation terminology, the scheme reproduces the reduced solution at mesh points.

If $|h_l \lambda(x_{l+1})| \ll 1, 1 \le l \le i$, then (10.97) shows that the scheme is only of order 1 [if $e_1 = O(h)$], and this indeed is the order if the method is applied to a stiff IVP involving both fast and slow components. But the properties just described, which the backward Euler scheme exhibits for fast modes, can be extremely useful. For comparison note that the second-order symmetric midpoint scheme, which is of the form (10.90a) with

$$\gamma(\zeta) = \frac{1 + \zeta/2}{1 - \zeta/2}, \quad \zeta_i = h_i \lambda(x_{i+1/2}), \quad r_i = \frac{q(x_{i+1/2})}{1 - \zeta_i/2} \tag{10.98}$$

and whose solution therefore satisfies (10.95) as well, does not improve as $|\lambda h|$ grows. Since $\gamma \to -1$ as $\text{Re}(\zeta) \to -\infty$, the effect of a layer error e_1 in an expression corresponding to (10.97) does not die away either. Similar behaviour is exhibited by the (also symmetric second-order) trapezoidal scheme.

(ii) A scalar, second-order BVP. Our next step is to consider a BVP (10.2) of the type discussed in Section 10.1.1. Special 3-point schemes for such problems have been popular in the literature of numerical analysis of stiff BVPs. Researchers in computational fluid dynamics observed that when applying the usual 3-point scheme (5.6) with a uniform step size $h \gg \varepsilon$, oscillations occur in the solution throughout the interval $[a, b]$. (Note that the condition for stability in Section 5.1.1 is also strongly violated.) But upon replacing the symmetric approximation to $p(x)u'$ in (5.6) by a one-sided one,

$$[pu']_{x_i} := \begin{cases} p(x_i)(u_{i+1} - u_i)/h & \text{if } p(x_i) \leq 0 \\ p(x_i)(u_i - u_{i-1})/h & \text{if } p(x_i) > 0 \end{cases} \tag{10.99a}$$

keeping everything else the same, the spurious oscillations in the computed solution disappear. The difference equations then read

$$\frac{\varepsilon}{h^2}(u_{i+1} - 2u_i + u_{i-1}) - [pu']_{x_i} - q(x_i)u_i = g(x_i), \quad 2 \leq i \leq N \tag{10.99b}$$

$$u_1 = \beta_1, \quad u_{N+1} = \beta_2 \tag{10.99c}$$

The resulting system of linear equations also gains some nice properties (described later). The discretization (10.99), which does have the one disadvantage of yielding only a first-order method, is defined also in the presence of turning points, and yields a one-sided scheme in the appropriate direction. This was (and is) called *upwind*, or *upstream differencing*.

To understand this description in our previous context, we proceed along the route taken in Section 10.1.1. Considering (10.2) as a system of two first-order ODEs (10.3), (10.4), we recall from Section 5.6.1 that a scheme similar to (5.6) is obtained by applying symmetric discretizations to (10.3), (10.4) and then substituting one into the other. The spurious solution oscillations mentioned above can therefore be explained by the fact that a symmetric scheme for a stiff ODE like (10.3) does not damp layer errors. Rather, $\gamma \to -1$, and this produces the oscillations. Layers are not represented at all by the coarse, uniform mesh, as in Fig. 10.4.

Rather than using a symmetric scheme, the previous discussion suggests using a one-sided scheme. Then the effect from layers being ignored will not be felt away from them. The ODE (10.3) is a special case of (10.19-22) (with $v(x)$ treated as a smooth inhomogeneity) if $p(x) < 0$. If $p(x) > 0$ instead then (10.20a) does not hold and the stable direction of integration is in reverse (from right to left). This yields the scheme

$$\frac{\varepsilon}{h}(u_{i+1} - u_i) = \begin{cases} p(x_{i+1})u_{i+1} + v_{i+1} & \text{if } p(x_{i+1/2}) < 0 \\ p(x_i)u_i + v_i & \text{if } p(x_{i+1/2}) > 0 \end{cases} \tag{10.100a}$$

where turning points are assumed to occur at mesh points. For (10.4) we may use a symmetric scheme because $v(x)$ is a slow solution component. For instance, the midpoint scheme gives

$$\frac{v_{i+1}-v_i}{h} = \frac{1}{2}\phi(x_{i+1/2})(u_i+u_{i+1}) + g(x_{i+1/2}) \qquad (10.100\text{b})$$

where

$$\phi(x) := q(x) - p'(x) \qquad (10.101)$$

We can now use (10.100a) to express $\{v_i\}$ in terms of $\{u_i\}$. For simplicity, assume that $p(x)$ does not change sign on $[a,b]$. Then we obtain if $p < 0$

$$\frac{\varepsilon}{h^2}(u_{i+1}-2u_i+u_{i-1}) = \frac{1}{h}(p(x_{i+1})u_{i+1}-p(x_i)u_i) + \frac{1}{2}\phi(x_{i+1/2})(u_i+u_{i+1}) + g(x_{i+1/2}) \qquad (10.102\text{a})$$

and if $p > 0$,

$$\frac{\varepsilon}{h^2}(u_{i+1}-2u_i+u_{i-1}) = \frac{1}{h}(p(x_i)u_i - p(x_{i-1})u_{i-1}) + \frac{1}{2}\phi(x_{i-1/2})(u_i+u_{i-1}) + g(x_{i-1/2}) \qquad (10.102\text{b})$$

An equally good, and slightly simpler scheme reads for $p < 0$

$$\frac{\varepsilon}{h^2}(u_{i+1}-2u_i+u_{i-1}) = p(x_{i+1/2})\frac{u_{i+1}-u_i}{h} + \frac{1}{2}q(x_{i+1/2})(u_i+u_{i+1}) + g(x_{i+1/2}) \qquad (10.103\text{a})$$

and for $p > 0$

$$\frac{\varepsilon}{h^2}(u_{i+1}-2u_i+u_{i-1}) = p(x_{i-1/2})\frac{u_i-u_{i-1}}{h} + \frac{1}{2}q(x_{i-1/2})(u_i+u_{i-1}) + g(x_{i-1/2}) \qquad (10.103\text{b})$$

The schemes (10.102) and (10.103) turn out to have some similar properties to those of the scheme (10.99). These include convergence bounds which are uniform in ε; i.e., for some ε_0 and all $0 < \varepsilon \leq \varepsilon_0$ the same convergence bounds hold for a fixed h. The advantage of (10.99) is that only coefficient values at mesh points are used. The advantage of (10.102) and (10.103) over (10.99) is that they are more accurate as $\varepsilon \to 0$, because the reduced equation is integrated by the second-order midpoint scheme, while in (10.99) it is integrated by a first-order backward Euler scheme.

Example 10.11

We compare the schemes (5.6), (10.99) and (10.103) for the BVP of Example 10.3 with $\varepsilon = 10^{-6}$, using uniform meshes with $h = 0.1, 0.05$, and 0.025. Using the symmetric scheme (5.6), we find that the maximum error is 4.7 on all three meshes. The error oscillates widely. Using the upwinded scheme (10.99) instead, we find that the maximum error for $h = 0.1$ is 0.12, and the oscillation disappears from the error (which is mainly the contribution of the truncation error for the reduced equation). For $h = 0.05$ the error is 0.062, and for $h = 0.025$ it is 0.031. The linear order of convergence is apparent. Using the upwinded scheme (10.103), we find that the maximum error for $h = 0.1$ is 0.54−2, that for $h = 0.05$ is 0.14−2 and that for $h = 0.025$ is 0.36−3. The second-order convergence of this method is apparent, as is its superiority for this example. □

One way to interpret the one-sided scheme (10.99) is to write it as

$$\frac{\varepsilon+\sigma_i h}{h^2}(u_{i+1}-2u_i+u_{i-1}) = p(x_i)\frac{u_{i+1}-u_{i-1}}{2h} + q(x_i)u_i + g(x_i) \quad (10.104)$$

with $\sigma_i = \frac{1}{2}|p(x_i)|$ (verify!) This is a symmetric scheme for a modified problem where the layer has been "diffused." Thus, the scheme (10.104) [or (10.99)] is a first-order approximation to (10.2) and a second-order approximation to the ODE

$$(\varepsilon+\sigma(x)h)u'' = p(x)u' + q(x)u + g(x)$$

with $\sigma(x)$ defined smoothly so that $\sigma(x_i) = \sigma_i$, $2 \le i \le N$. The additional term $h\sigma(x)u''$ is an "artificial diffusion" or "artificial viscosity" term.

A useful property for theoretical purposes is expressed in the following.

Definition 10.105 A difference operator L_h of the form

$$L_h = a_i u_{i-1} + b_i u_i + c_i u_{i+1}$$

is of a *positive type* if

$$a_i + b_i + c_i \le 0, \qquad a_i > 0, \qquad c_i > 0$$

□

The symmetric difference operator is not of a positive type, but those of (10.103) and (10.99) are whenever $q \ge 0$. For such operators a discrete maximum principle holds and the existence of a unique solution to the discrete system (consisting of the difference equations plus the BC (10.99c)) can be shown for each $\varepsilon > 0, h > 0$.

Corresponding schemes have been developed for some nonlinear scalar ODEs under fairly restricted conditions. We do not discuss these here. Our approach is to consider one-sided schemes for (10.2) in such a way as to capture that aspect of them which is generalizable to first-order systems, and which is in this sense the essential one. This is perhaps an ODE-oriented approach. Many of the other considerations just mentioned, including the insistence on an inadequate uniform mesh, are motivated by intentions to extend methods and results directly to problems with more space variables.

(iii) Simple schemes for first-order systems. For the ODE

$$\varepsilon(y^1)' = -y^1 + q^1(x) \qquad a < x < b \quad (10.106a)$$

$(0 < \varepsilon \ll 1)$, the backward Euler scheme has been shown to be stable and accurate, the error being $O(\varepsilon h)$. For the ODE

$$\varepsilon(y^2)' = y^2 + q^2(x) \qquad a < x < b \quad (10.106b)$$

the stable direction of integration is backwards, from b to a, so corresponding to the above one would use a *forward* Euler scheme, again obtaining an $O(\varepsilon h)$ error. Much in the same spirit as the upwind differencing (10.99), the combined scheme reads

$$\frac{\varepsilon}{h}(y^1_{i+1} - y^1_i) = -y^1_{i+1} + q^1(x_{i+1})$$

$$2 \le i \le N$$

$$\frac{\varepsilon}{h}(y^2_{i+1} - y^2_i) = y^2_i + q^2(x_i)$$

on a mesh π of (10.77).

This can be generalized in various directions. In place of (10.106a) and (10.106b) we can have blocks, where the modes are all of one (fast) type. Also, slow solution components can be included. The key to such generalizations is the remarkable way in which transformations at the ODE level are preserved at the discrete level by using the backward Euler scheme (and also by its one-sided extensions mentioned earlier). To see this, consider an ODE

$$\varepsilon \mathbf{y}' = A(x)\mathbf{y} + \mathbf{q}(x) \tag{10.107a}$$

where $\mathbf{y}(x) \in \mathbf{R}^n$, and its backward Euler discretization

$$\frac{\varepsilon}{h_i}(\mathbf{y}_{i+1} - \mathbf{y}_i) = A(x_{i+1})\mathbf{y}_{i+1} + \mathbf{q}(x_{i+1}), \qquad 1 \le i \le N \tag{10.107b}$$

In (10.107) $A(x)$ and $\mathbf{q}(x)$ are assumed smooth and bounded, and $0 < \varepsilon \ll h \ll 1$. Let $T(x)$ be a transformation such that

$$\|T^{-1}T^{(\nu)}\| \le K \qquad \nu = 0, 1, 2$$

and denote $\Lambda := T^{-1}AT$, $\mathbf{w} := T^{-1}\mathbf{y}$, $\mathbf{g} := T^{-1}\mathbf{q}$. Then (10.107a) transforms into

$$\varepsilon \mathbf{w}' = (\Lambda(x) - \varepsilon T^{-1}(x)T'(x))\mathbf{w} + \mathbf{g}(x) \tag{10.108a}$$

[cf. (10.30)]. Now, defining $\mathbf{w}_i := T^{-1}(x_i)\mathbf{y}_i$, multiplying (10.107b) by $T^{-1}(x_{i+1})$ and observing that

$$T^{-1}(x_{i+1})\mathbf{y}_i = T^{-1}(x_{i+1})(T(x_{i+1}) - h_i T'(x_{i+1}) + O(h_i^2))T^{-1}(x_i)\mathbf{y}_i$$
$$= (I - h_i T^{-1}(x_{i+1})T'(x_{i+1}) + O(h_i^2))\mathbf{w}_i$$

we have

$$\frac{\varepsilon}{h_i}(\mathbf{w}_{i+1} - \mathbf{w}_i) = \Lambda(x_{i+1})\mathbf{w}_{i+1} - (\varepsilon T^{-1}(x_{i+1})T'(x_{i+1}) + O(\varepsilon h_i))\mathbf{w}_i + \mathbf{g}(x_{i+1}) \tag{10.108b}$$

and this is, up to $O(\varepsilon)$ terms, the same as the backward Euler scheme applied to the transformed ODE (10.108a).

Now, suppose that $A(x)$ has eigenvalues $\lambda_1(x), \ldots, \lambda_n(x)$ satisfying $\text{Re}(\lambda_i(x)) \le \lambda_R$, $a \le x \le b$, $1 \le i \le N$, for some negative constant λ_R. If $T(x)$ is (for simplicity) an eigenvector matrix of $A(x)$ then in (10.108b) we can consider each row separately, ignoring the small coupling terms first. These are added later on by a perturbation (contraction) argument. The theoretical multiple shooting framework allows us to concentrate on one segment at a time, and we consider a long segment with a smooth solution first. Taking $h_i \gg \varepsilon$, the results at the beginning of this section show that (10.107b) yields an $O(\varepsilon h)$ approximation to the solution of (10.107a).

The analysis yielding (10.108b) also shows that if (10.107a) has both fast (decreasing) and slow modes and $T(x)$ is a transformation which decouples (separates) modes of the two types, then the backward Euler scheme applied to (10.107a) implicitly separates these modes as well, up to $O(\varepsilon)$ terms. In Exercise 7 the reader is invited to show that this is not true for the trapezoidal scheme. However, for slow modes, as for layer regions if they are to be resolved, the backward Euler scheme has only an $O(h)$ error.

What the backward Euler scheme cannot achieve implicitly is the decoupling of rapidly increasing and rapidly decreasing modes (and their stable integration). Indeed, the upwind differencing demonstrated for (10.106) can be applied only *after* explicit decoupling of different fast modes has been accomplished. This is the one major disadvantage of one-sided schemes! There are applications where it is not needed, or where it can be done ahead of time by a patient user, but if a general-purpose approach is contemplated, then it must include some mechanism to achieve the decoupling numerically.

Two alternatives exist, each with its advantages and disadvantages. One is to just decouple nonincreasing and nondecreasing modes, as discussed at length in Chapter 6. Some properties of the Riccati method in the context of this chapter are discussed in Section 10.4.2. However, if this approach is taken then the backward Euler scheme must be replaced by a higher-order alternative (e.g., a higher-order BDF scheme), because the retained mix of fast and slow modes makes it inaccurate [being only $O(h)$]. The other approach is to split modes into more classes, separating slow and fast ones as well. This gives a 3-way splitting, as in (10.28), when there are two time scales. To the fast modes we may then apply an Euler scheme in the appropriate direction, and this is accurate in case that the solution is smooth and the step size is large. To the slow modes it is safe, and second-order accurate, to apply a simple symmetric scheme, and the trapezoidal scheme is a natural choice (since it utilizes values at mesh points only).

The transformation required in the second approach is more detailed and, worse, possibly changes drastically when moving from segment to segment of the interval $[a,b]$. For instance, consider Example 10.4, which is defined on $[-1,1]$ with a turning point at $x = 0$. For the first approach we have one nonincreasing and one nondecreasing mode throughout the interval $[-1,1]$, but for the second approach the number of increasing, decreasing, and slow modes changes from $1,0,1$ for $x<0$ to $0,1,1$ for $x>0$ (away from the layer). Still, this is possible to do, though not simple.

For the solution in layer regions, if it is to be resolved and not ignored, the backward Euler scheme does not offer much. Obviously, small step sizes have to be taken, and the first-order nature of the scheme again shows up. Higher-order extensions may be used, but they do not possess any advantage over symmetric schemes. Recall from Section 10.2.4 that the fine mesh used to resolve the solution in a layer region affects a stretching transformation, and that a symmetric scheme works well in such a case. More details on this will be delayed until Section 10.3.2; here we contemplate the use of a hybrid scheme employing one-sided upwinded Euler schemes for fast solutions in smooth solution regions and the trapezoidal scheme in layers as well as for slow modes anywhere. Such a scheme is (at least) second-order everywhere if the stiffness is large compared to the step size whenever the one-sided scheme is used [e.g., $\varepsilon = O(h)$ for (10.107a)].

A concern when one is using one-sided schemes is abrupt switches from one direction to another. This is particularly important when the three-way mode splitting is used, and to this extent the use of a "neutral" symmetric scheme in interior layer regions is beneficial.

The use of an explicit transformation for the dependent variables yields another choice to make: After a one-sided scheme is applied, the resulting system of equations can either be solved and the obtained values then transformed back to the original solution variables, or a back transformation to the original unknowns can be made first and only then the system is solved. We prefer the second way, which avoids unnecessary trouble with poorly scaled quantities in some extreme cases.

The stability of the above method in smooth segments has been shown. Stability in layer regions (if they are resolved) is proved in Section 10.3.2 (in fact, essentially already in Section 10.2.4). To assemble the pieces we next invoke the multiple shooting framework. The method falls under the conditions in Section 10.2.2 if we pick the joints t_j as points from the mesh π. These include the boundary points and the mesh points in smooth regions which are closest to a layer or a boundary (see Fig. 10.7). The coarse mesh is chosen fine enough to approximate slow solution components well. Theorem 10.75 can then be invoked, and it yields

Theorem 10.109 The hybrid method composed of one-sided Euler (upwinded) differences to approximate fast modes in smooth regions and second-order symmetric scheme to accurately approximate slow modes everywhere, with layer regions being either well-approximated or not at all, is stable with a stability constant directly related to the BVP conditioning constant. Moreover, it yields an overall $O(h^2)$ error. □

Most experience gathered to date with general-purpose one-sided schemes is for linear problems. For nonlinear problems, one proceeds with a quasilinearization approach. The location of interior layers for a linearized problem obtained by this process depends upon the previous solution approximation (about which the linearization is made). It is not clear what the effect is if layers are placed "wrongly" in this way. In some instances nonlinear convergence is easier if a one-sided (upwinded) scheme is used with a coarse mesh, *ignoring* layer regions. This is not always the case, though, and the solution obtained this way may not be as detailed as desired. Generally, experience suggests that, regardless of the method used, continuation (in ε, or a combination of critical paramaters) is often needed. Thus, when the problem gets very stiff, we already have a fairly good idea of solution profile *and mesh* to be used.

(iv) Higher-order schemes for first-order systems. The second order achieved in the scheme just described is not all that high. To achieve higher-order schemes, one can use Runge-Kutta techniques. Use of collocation at appropriately upwinded Radau points suggests itself at first, but a second look reveals that these points are not the same in both directions; i.e., the canonical points ρ_j of (5.59) give a completely different sequence than the points $1 - \rho_j$, so interpolation of missing values must be used.

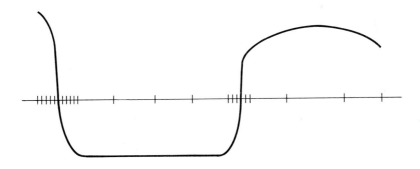

Figure 10.7 Segment ends for theoretical multiple shooting

Instead, we use a set of symmetric collocation points which include the endpoints in a nonsymmetric way. Thus we have points

$$0 = \rho_1 < \rho_2 < \cdots < \rho_k = 1$$

satisfying $\rho_{k+1-j} = 1 - \rho_j$, which are then mapped into each mesh subinterval $[x_i, x_{i+1}]$ to form collocation points x_{ij}, $j = 1, \ldots, k$, such that $x_{i1} = x_i$, $x_{ik} = x_{i+1}$. Then, for slow modes we use collocation on all k points x_{ij}, for fast increasing modes we collocate only at x_{ij}, $j=1, \ldots, k-1$ (excluding x_{i+1}), and for fast decreasing modes we use collocation only at x_{ij}, $j = 2, \ldots, k$ (excluding x_i), as in Fig. 10.8.

The actual choice for $\{\rho_j\}_1^k$ can be Lobatto points or even simply equidistant points. As it turns out, the properties of the one-sided schemes constructed this way are similar in essence to those of the backward Euler scheme in that they have stiff decay (but the precise statements are far from obvious). The order of the obtained method (which reduces to the one considered before for $k = 2$) is at least $O(h^k)$ for k collocation points.

10.3.2 Symmetric schemes

The big attraction of symmetric schemes is that explicit decoupling of increasing and decreasing modes before application of the discretization scheme is not necessary. Indeed, such schemes read the same when used to integrate forward or backward, and they look the same whether used to integrate a stiff or a nonstiff problem. For a k-stage Runge-Kutta scheme (5.59) (see Fig. 5.2) the symmetry requirement is expressed as

$$\rho_l = 1 - \rho_{k+1-l} \qquad 1 \leq l \leq k \qquad (10.110a)$$

$$\alpha_{k+1-j,k+1-l} + \alpha_{jl} = \beta_l = \beta_{k+1-l} \qquad 1 \leq j, l \leq k \qquad (10.110b)$$

As in Sections 5.3 and 5.4, we restrict attention to some particular families of symmetric Runge-Kutta schemes. One such family consists of collocation schemes at Gaussian points. The simplest member of this family (obtained for $k = 1$) is the midpoint scheme. Another family of symmetric schemes which will receive less attention here is that of collocation at Lobatto points. The simplest member of this family (obtained for $k = 2$) is the trapezoidal scheme.

Their robustness makes symmetric schemes a favoured choice when implementing a general purpose BVP code, despite a number of possible difficulties to be discussed below. The examples in Chapters 8 and 9, which are mentioned at the beginning of this

Figure 10.8 One-sided collocation

chapter and are solved using the code described in Appendix B (which is based on collocation at Gaussian points), provide a significant number of applications of symmetric schemes for difficult, non-artificial stiff BVPs.

The analysis of Section 10.2.3, but not of Section 10.2.2, applies here. Thus, we consider individual solution segments, and concentrate on the 3-segment problem for most of the discussion.

(i) Transformations; stability difficulties. As in (10.107), (10.108), it is important to see how various schemes "preserve" (mode separating) transformations. The midpoint scheme for (10.107a)

$$\frac{\varepsilon}{h_i}(\mathbf{y}_{i+1} - \mathbf{y}_i) = \frac{1}{2} A(x_{i+1/2})(\mathbf{y}_i + \mathbf{y}_{i+1}) + \mathbf{q}(x_{i+1/2}), \qquad 1 \le i \le N \qquad (10.111a)$$

can be multiplied by $T^{-1}(x_{i+1/2})$ to yield (Exercise 7)

$$\frac{\varepsilon}{h_i}(\mathbf{w}_{i+1} - \mathbf{w}_i) = \frac{1}{2}(\Lambda(x_{i+1/2}) - \varepsilon T^{-1}(x_{i+1/2})T'(x_{i+1/2}))(\mathbf{w}_{i+1} + \mathbf{w}_i) \qquad (10.111b)$$

$$+ \mathbf{g}(x_{i+1/2}) + \Lambda(x_{i+1/2})h_i \, \boldsymbol{\phi}_M(\mathbf{w}_i, \mathbf{w}_{i+1}) + \varepsilon h_i \, \boldsymbol{\psi}_M(\mathbf{w}_i, \mathbf{w}_{i+1})$$

where $\boldsymbol{\phi}_M$ and $\boldsymbol{\psi}_M$ are bounded operators. The two rightmost terms in (10.111b) are artifacts of the transformation, and without them the expression in (10.111b) is the midpoint discretization of the transformed system (10.108a). The extra terms are small compared to others on the right hand side of (10.111b), and they can be dealt with (carefully) as perturbations. In particular, separation of fast and slow modes is preserved by the midpoint scheme. On the other hand, the trapezoidal scheme

$$(10.112a)$$
$$\frac{\varepsilon}{h_i}(\mathbf{y}_{i+1} - \mathbf{y}_i) = \frac{1}{2}(A(x_i)\mathbf{y}_i + A(x_{i+1})\mathbf{y}_{i+1}) + \frac{1}{2}(\mathbf{q}(x_i) + \mathbf{q}(x_{i+1})), \qquad 1 \le i \le N$$

is not so easily dealt with. We can still write $A(x_i)\mathbf{y}_i = A(x_i)T(x_i)\mathbf{w}_i = T(x_i)\Lambda(x_i)\mathbf{w}_i$ and similarly for $A(x_{i+1})\mathbf{y}_{i+1}$. Multiplying by $T^{-1}(x_{i+1/2})$, this gives (Exercise 7)

$$\frac{\varepsilon}{h_i}(\mathbf{w}_{i+1} - \mathbf{w}_i) = \frac{1}{2}(\Lambda(x_i)\mathbf{w}_i + \Lambda(x_{i+1})\mathbf{w}_{i+1} - \varepsilon T^{-1}(x_i)T'(x_i)\mathbf{w}_i \qquad (10.112b)$$

$$- \varepsilon T^{-1}(x_{i+1})T'(x_{i+1})\mathbf{w}_{i+1}) + \frac{1}{2}(\mathbf{g}(x_i) + \mathbf{g}(x_{i+1}))$$

$$+ h_i \, \boldsymbol{\phi}_T(\Lambda(x_i)\mathbf{w}_i, \Lambda(x_{i+1})\mathbf{w}_{i+1}, \mathbf{g}(x_i), \mathbf{g}(x_{i+1})) + \varepsilon h_i \, \boldsymbol{\psi}_T(\mathbf{w}_i, \mathbf{w}_{i+1})$$

where $\boldsymbol{\phi}_T$ and $\boldsymbol{\psi}_T$ are some bounded operators. Now, if h_i is small [e.g., $h_i = O(\varepsilon)$] then (10.112b) is fine, but if $0 < \varepsilon \ll h_i$ then the term $h_i \boldsymbol{\phi}_T$ may destroy the separation of fast and slow modes. [Recall that for a singular singularly-perturbed problem we essentially divide the last n_0 rows of (10.108a) by ε, when the corresponding rows of Λ are $O(\varepsilon)$, to obtain the separated equations (10.39). This can also be boundedly done in (10.108b) and (10.111b), but not in (10.112b).]

At first glance it may seem that the preservation of mode separation properties of the backward Euler scheme and of the midpoint scheme are unique to collocation schemes with only one collocation point [because the effect of the transformation $T^{-1}(x)$ can be concentrated at one point per subinterval]. However, a closer scrutiny shows that all collocation schemes based on Gauss points have similar properties in this respect to those of the midpoint scheme, while all collocation schemes based on Lobatto points have similar properties to those of the trapezoidal scheme (Exercise 8). The property of correctly separating fast and slow modes as $\varepsilon \to 0$ is related to the concept of *D-stability*, which we do not discuss further.

With symmetric schemes, rapidly increasing and rapidly decreasing modes are not approximated well when the mesh is coarse, and the stability of the scheme is in a sense only marginal. This causes possible nonlocal effects to appear in the error, and results in the need for a more careful analysis. In particular, as previously mentioned and as demonstrated in Example 10.11, neglecting to cover layer regions with a fine mesh results in layer errors possibly polluting the solution everywhere. With collocation at Gauss points, any error from the boundary is kept essentially at the same magnitude on neighboring segments with a coarse mesh [see Theorem 10.129 below]. But with collocation at Lobatto points, layer errors might even be magnified by $O(\varepsilon^{-1} h^2)$, because of the incomplete separation of fast and slow modes.

Another stability property which distinguishes collocation schemes at Gauss points is AN-stability and algebraic stability (cf. Section 2.7.3). Considering the homogeneous ODE (10.19)–(10.20) with $|\lambda h| \gg 1$, we have for the midpoint scheme

$$|y_{i+1}| \leq \left| \frac{1 + 1/2 h_i \lambda(x_{i+1/2})}{1 - 1/2 h_i \lambda(x_{i+1/2})} \right| \, |y_i| \leq |y_i|$$

but for the trapezoidal scheme

$$y_{i+1} \approx -\frac{\lambda(x_i)}{\lambda(x_{i+1})} y_i \approx \frac{\lambda(x_{i-1})}{\lambda(x_{i+1})} y_{i-1} \approx \cdots \approx (-1)^i \frac{\lambda(x_1)}{\lambda(x_{i+1})} y_1$$

The approximate solution in the latter case may therefore grow while the exact solution decays.

Example 10.12

The BVP

$$\varepsilon u'' = xu' + \frac{1}{2} u, \quad -1 < x < 1$$

$$u(-1) = 1, \quad u(1) = 2$$

has the solution

$$u(x) = e^{-(x+1)/\varepsilon} + 2 e^{(x-1)/\varepsilon} + O(\varepsilon)$$

Thus there are boundary layers at the ends, but $u(x)$ is smooth near the "turning point" $x = 0$. The numerical error propagates from the boundaries to the middle of the interval $[-1, 1]$. Note that when $|x|$ is bounded away from 0, there is only one fast mode. We rewrite this ODE as a first-order system as in rewriting (10.2) as (10.3), (10.4), and apply collocation at k Gauss or Lobatto points per mesh subinterval. The meshes are determined as follows: For a given error tolerance δ, dense meshes near -1 and near 1 are constructed according to Algorithm 10.115 described later, such that the error emanating from the boundary layers is $O(\delta)$. This is overlaid by a uniform mesh with $h = 0.2$. The resulting meshes are the same for a k-stage Gauss scheme and for a $k + 1$-stage Lobatto scheme, and usually the errors are too. But here the maximum errors on the mesh, listed in Table 10.2 under "e", are quite different.

TABLE 10.2 Numerical Results for Example 10.12; $\varepsilon = 10^{-6}$

k	Scheme	δ	N	e
1	Gauss (midpoint)	.1−1	24	.14−1
2	Lobatto (trapez.)	.1−1	24	.10+2
3	Gauss	.1−5	28	.19−5
4	Lobatto	.1−5	28	.59−1

Clearly, the boundary layer errors are magnified by the Lobatto schemes, but not by the Gauss schemes. □

The good stability properties of the collocation schemes at Gauss points lead us to consider these schemes alone in the remainder of this section.

The nonlocal behaviour of errors causes difficulties to a general-purpose code using adaptive mesh selection, when one attempts to start "cold," i.e., from essentially no information about solution or mesh; see, e.g., Example 9.2. The problem is that in the absence of localization of the error on a poor mesh there is little information on how to refine it. We will discuss an improved mesh selection strategy below, but note that for many nonlinear problems one needs to use continuation, and in such a situation the "cold" start assumption is inappropriate.

(ii) Layer regions. In a layer region we use a fine mesh, so that a stretching transformation is affected. Under such circumstances we know no better general alternative than a symmetric scheme. We may consider (in principle) the effect of a decoupling transformation (6.4–7) on a layer segment $[t_j, t_{j+1}]$. The block of nonincreasing modes is controlled by Q_{1j}^T in the BC (10.71) and the nondecreasing ones by Q_{2j}^T. The ODEs for the block of nonincreasing modes may be further separated into individual ODEs for the fast modes and for the slow modes. Ignoring connecting terms first, we may apply Theorem 10.93 to the individual ODEs for the fast modes, yielding stability for the entire block. Since the numerical scheme essentially reproduces the continuous transformation when h_i is so small, we obtain

Theorem 10.113 Suppose that a k-stage collocation scheme at Gauss points is applied to the ODE (10.1) with BC (10.71) on a layer segment $[t_j, t_{j+1}]$. Let the segment be covered by a mesh

$$t_j = x_1 < x_2 < \cdots < x_N < x_{N+1} = t_{j+1}$$

and $h_L := \max_{1 \leq i \leq N} h_i$. If h_L is sufficiently small that in the stretched variable $\xi := x/h_L$ the ODE (10.1) has bounded coefficients, then the difference scheme is stable, satisfying

$$|\mathbf{y}_i| \leq K(|\mathbf{B}_j \mathbf{y}| + \|\mathbf{q}\|_{[t_j, t_{j+1}]}), \quad 1 \leq i \leq N$$

for a constant K of moderate size. □

This stability bound may be further utilized to obtain error bounds. There are actually two errors of interest here, and they should not be confused with one another. The first is the error in the layer region itself, and the second is the error emanating from the layer. The latter has more to do with the size of the fast modes which decrease in the direction of the layer's edge at the end of the fine portion of the mesh than with how fine the layer mesh is [cf. (10.48)]. The mesh in a layer region may be determined by the fast decreasing modes (in the appropriate direction) which cause the layer to form.

To see how this layer mesh selection can be done, consider first a scalar, constant coefficient equation

$$\varepsilon y' = \lambda y, \qquad y(0) = 1$$

with $\lambda_R := \operatorname{Re}(\lambda) < 0$. The difference scheme (10.90a) for k Gaussian points gives

$$\gamma(\frac{\lambda h}{\varepsilon}) = \exp\{\frac{\lambda h}{\varepsilon}\}[1 + c_\gamma(\frac{\lambda h}{\varepsilon})^{2k+1} + O((|\lambda|\frac{h}{\varepsilon})^{2k+2})]$$

(with $|\lambda| h \leq K \varepsilon$) where c_γ is a known constant. (Incidentally, precisely the same expression holds for $k+1$ Lobatto points.) By (10.95) we have for the absolute error

$$|e_{i+1}| \approx |\exp\{\frac{\lambda x_{i+1}}{\varepsilon}\} c_\gamma \sum_{j=1}^{i}(\frac{\lambda h_j}{\varepsilon})^{2k+1}| \leq \exp\{\frac{\lambda_R x_{i+1}}{\varepsilon}\}|c_\gamma| \sum_{j=1}^{i}(\frac{|\lambda| h_j}{\varepsilon})^{2k+1}$$

Suppose that an (absolute) error tolerance δ is to be satisfied. We may determine the layer region to be $[0, T\varepsilon]$, where $T = \lambda_R^{-1} \ln \delta$, because $|y(T\varepsilon)| = \delta$.

Lemma 10.114 The mesh defined by

$$h_1 := \frac{\varepsilon}{|\lambda|}[\frac{-\lambda_R}{|\lambda c_\gamma|}]^{1/2k} \delta^{1/2k} \tag{10.114a}$$

$$h_i := h_{i-1} \exp\{\frac{1}{2k}\frac{-\lambda_R}{\varepsilon}h_{i-1}\} = h_1 \exp\{\frac{1}{2k}\frac{-\lambda_R}{\varepsilon}x_i\} \qquad i = 2, \ldots, N \tag{10.114b}$$

(where $x_{N+1} \geq T\varepsilon$) yields

$$|e_i| \leq \delta$$

for the scalar, constant coefficient IVP. □

The proof of this lemma is left to the reader as Exercise 9.

The N mesh points (and N itself) for the layer region are determined automatically, as defined in the lemma, and it is easily seen that $N = O(\delta^{-1/2k})$, independent of ε. This simple strategy gives a limited, but sometimes useful, *a priori mesh selection algorithm*:

Algorithm 10.115: *a priori* boundary layer mesh selection

Input: A 3-segment ODE (10.1), a tolerance δ, a number of stages k, and a coarse mesh away from boundaries.

Output: A mesh with fine portions at layer regions and a coarse portion in between.

1. Find the eigenvalues $\lambda_1(a), \ldots, \lambda_n(a)$ of $A(a)$, ordered according to increasing real parts (the most negative first).

2. Find n_- as the first nonnegative integer such that $-\operatorname{Re}(\lambda_{n_-+1})$ is not too large. IF $n_- > 0$ THEN set

$$\mu := \max\{|\lambda_j(a)|, j = 1, \ldots, n_-\}$$

$$\nu := \min\{-\text{Re}(\lambda_j(a)), j = 1, \ldots, n_-\} > 0$$

and DO the following two steps.

3. Set

$$h_1 := \varepsilon\mu^{-1}[\frac{\nu}{\mu|c_\gamma|}]^{1/2k} \delta^{1/2k}$$

$$x_1 := a, \qquad T := \varepsilon\nu^{-1}|\ln\delta|$$

4. FOR $i = 1 \ldots$, UNTIL $x_{i+1} \geq T$ DO
    Set

$$x_{i+1} := x_i + h_i$$

$$h_{i+1} := h_i \exp\{\frac{1}{2k}\frac{\nu}{\varepsilon}h_i\}$$

5. Combine the coarse mesh portion.
6. Repeat steps 1–4 for the right end, replacing a with b and concentrating on the last n_+ eigenvalues of $A(b)$, where n_+ is the first nonnegative integer such that $\text{Re}(\lambda_{n-n_+})$ is not too large. Combine the resulting fine mesh portion. □

This algorithm has already been used in Example 10.12. It can also be applied for some simple nonlinear problems, using for $A(x)$ a linearization about a reasonable guess.

Example 10.13

Algorithm 10.115 for *a priori* mesh selection works very well for Example 10.7 (nonlinear beam with simple supports) with the initial guess

$$y_1 = x(1-x), \qquad y_2 = 0, \qquad z_1 = \sin\pi x, \qquad z_2 = \frac{1}{2}x^2 - \frac{1}{3}x^3$$

Only three Newton iterations were needed for convergence for any value of ε tried, and very accurate results are obtained with mesh sizes roughly as in Table 10.2. No continuation is necessary here (somewhat uncharacteristically, perhaps). Representative values obtained for $\varepsilon > 0$ very small are

$$y_2(0) = .863899, \qquad z_2(0) = .434520, \qquad y_1(0.5) = -.891686, \qquad z_1(0.5) = .108314.$$

□

Unfortunately, the incorporation of this algorithm into a general-purpose code having *a posteriori* error estimation and mesh selection is not easy, and it is certainly not usable on its own for most difficult applications.

While our treatment for layer regions has been in terms of collocation at Gaussian points, the treatment using Algorithm 10.115 for a $k+1$-stage Lobatto scheme is identical to that of a k-stage Gauss scheme (see N in Table 10.2).

(iii) Instabilities on smooth solution segments. We are considering an ODE (10.1), written more conveniently as (10.107a), under the special BC (10.71), on a segment $[c,d] = [t_j, t_{j+1}]$ where the solution is smooth and bounded, and we use collocation at Gaussian points with a large step size. Once again, recall that when the step size is large, symmetric schemes do not approximate fast modes well. Indeed, for the scalar ODE (10.19), a k-stage Gaussian scheme can be brought into the form (10.90a) with

$$\gamma(\zeta) = (-1)^k(1 + 2k(k+1)\zeta^{-1} + O(\zeta^{-2})) \underset{\zeta \to -\infty}{\to} (-1)^k \qquad (10.116)$$

This, while the exact mode decays exponentially. In Section 10.2.3 we show that the entire scheme is stable if it is stable on segments with a coarse mesh, but there is no guarantee for the latter to occur. To see this, consider again the test equation

$$\varepsilon y' = \lambda y$$

and the difference scheme

$$y_{i+1} = \gamma(\lambda h_i/\varepsilon) y_i$$

Letting the step size h be constant and defining

$$\sigma := \varepsilon h^{-2} \qquad (10.117)$$

which we assume to be bounded, (10.116) gives the difference equation

$$y_{i+1} = (-1)^k(1 + 2k(k+1)\sigma\lambda^{-1}h + O(h^2))y_i$$

Defining $\hat{y}_i := (-1)^{ik} y_i$, we write this difference equation in terms of $\hat{y}_i$. If we let $h \to 0$, keeping σ fixed (i.e., we let ε and h both approach zero, keeping the relation (10.117) between them fixed), then the difference scheme for $\hat{y}_i$ is seen to approximate the ODE

$$\hat{y}' = 2k(k+1)\sigma\lambda^{-1}\hat{y}$$

This in itself is no disaster. But a similar result can be obtained for an ODE system (10.107a) where $A(x)$ has a hyperbolic splitting. Thus, the stability of the k-stage Gauss collocation scheme depends on the well-conditioning of the *auxiliary* BVP

$$\hat{\mathbf{y}}' = 2k(k+1)\sigma A^{-1}\hat{\mathbf{y}} \qquad (10.118a)$$

$$\mathbf{B}_j \hat{\mathbf{y}} = 0 \qquad (10.118b)$$

While the BVP (10.118) is usually well-conditioned, this does not follow from the well-conditioning of (10.107a), (10.71).

Example 10.14

In (10.107a) let

$$A(x) = \begin{bmatrix} -\theta \sin 2x & -1 - \theta \cos 2x \\ 1 - \theta \cos 2x & \theta \sin 2x \end{bmatrix}, \qquad \mathbf{q}(x) = \varepsilon(A^{-1}\mathbf{g})' - \mathbf{g}$$

and consider the BC

$$B_a = \begin{bmatrix} v & 0 \\ 0 & 1 \end{bmatrix}, \quad B_b = \begin{bmatrix} 0 & 0 \\ v\cos 1 - \sin 1 & -v\sin 1 - \cos 1 \end{bmatrix}, \quad \beta = 0$$

for $a = 0$, $b = 1$, where

$$\theta = \mu(\mu^2-1)^{-1/2}, \quad v = \mu - \sqrt{\mu^2-1}, \quad \mu = \frac{\pi}{2}, \quad g(x) = (\sin \pi x, \sin \pi x)^T$$

The exact solution is

$$y(x) = A^{-1}(x)g(x)$$

(i.e., there are no boundary layers because of the particular choice of the boundary values), the eigenvalues of $A(x)$ are $\pm(\mu^2-1)^{-1/2}$, the BVP is well-conditioned and the auxiliary problem (10.118) is singular when $\sigma = \dfrac{1}{2k(k+1)}$ (check!). Therefore, we expect stability troubles for these values of ε/h^2. In Table 10.3 we list some numerical results. The meshes were constructed with a uniform step size h, superimposed on layer meshes constructed by using Algorithm 10.115 with $\delta = .01$. The maximum error in the first solution component over the uniform part of the mesh is denoted e. Note that when we specify h and σ, $\varepsilon = \sigma h^2$ is specified as well.

TABLE 10.3 Numerical Results for Example 10.14

σ	h	e ($k=1$)	e ($k=2$)	e ($k=3$)
1/4	.1	.79	.30−1	.13−3
1/4	.05	1.62	.73−2	.74−5
1/12	.1	.25	.46+1	.24−3
1/12	.05	.67−1	.25+1	.19−4
1/24	.1	.24	.1	.27−1
1/24	.05	.60−1	.25−1	.21−2

From Table 10.3 it is apparent that the errors deteriorate when σ attains a critical value.

If we remove the layer meshes and apply the midpoint scheme ($k = 1$) with $h = .1$, $\sigma = 1/4$, we obtain $e = .89-1 (< .79)$. So, the error when we use a coarse overall mesh is better (smaller) on the common part of the two meshes! This is because the mesh structure allowing for the peculiar singularity demonstrated in Table 10.3 has been violated (and, at the same time, y is smooth everywhere). □

The instability demonstrated in Example 10.14 is more a theoretical curiosity than a practical problem. Indeed, in the highly unlikely event that it occurs at all for a given BVP, if a code with an adaptive mesh selection algorithm is used, then it would usually respond to obtaining the poorer approximation by changing the mesh, thus inadvertently eliminating the instability, because this instability depends on a particular ratio of ε and h. Of more interest is the case when we let $\varepsilon \to 0$ keeping h fixed (or just looking at $\varepsilon \ll h^2$). This corresponds to the BVP (10.118) having the eigenvalue $\sigma = 0$. It is easy to see that the well-conditioning of the BVP (10.118) with $\sigma = 0$ translates to the algebraic condition that the matrix

$$\begin{pmatrix} Q_{1j}^T \\ Q_{2j}^T \end{pmatrix}$$

[cf. (10.71)] have a bounded inverse.

Example 10.15

In (10.107a) let

$$A(x) = T(x;\nu) \begin{bmatrix} -1 & 0 \\ 0 & 2 \end{bmatrix} T^{-1}(x;\nu), \quad \mathbf{q}(x) = T(x;\nu)\,[\varepsilon \mathbf{w}_p{}'(x) - \hat{A}\,\mathbf{w}_p(x)]$$

where T is the reflection matrix

$$T(x;\nu) = T^{-1}(x;\nu) = \begin{bmatrix} \sin \nu x & \cos \nu x \\ \cos \nu x & -\sin \nu x \end{bmatrix}$$

$$\hat{A} = \begin{bmatrix} -1 & \nu\varepsilon \\ -\nu\varepsilon & 2 \end{bmatrix}, \quad \mathbf{w}_p(x) := \begin{pmatrix} \sin 5x + \cos 10x \\ \cos 7x \end{pmatrix}$$

and consider the BC for $a = 0$, $b = 1$,

$$B_a = \begin{bmatrix} 0 & 0 \\ 0 & 1 \end{bmatrix} \quad B_b = \begin{bmatrix} \cos \nu & -\sin \nu \\ 0 & 0 \end{bmatrix}, \quad \beta = 0$$

These BC correspond in fact to (10.71). Using the transformation (10.29) we have

$$\varepsilon \mathbf{w}' = \hat{A}\,\mathbf{w} + T^{-1}(x;\nu)\mathbf{q}(x)$$

for which $\mathbf{w}_p(x)$ is a particular solution. The fundamental solution for the ODE in $\mathbf{w}$ can be easily found, so $\mathbf{y}(x)$ can be found as well, and the BVP can be verified to be well-conditioned for $0 < \varepsilon \ll \nu$ (see Exercise 10). However, the matrix $B_a + B_b$ is singular for $\nu = \pi(l + 1/2)$, $l = 0, 1, \ldots$, so we expect stability problems with symmetric collocation schemes for these values of ν.

Below we list numerical results using collocation at k Gaussian points per subinterval for two values of ν: In Table 10.4 the results are for $\nu = \pi/4$, where no stability trouble occurs, whereas in Table 10.5 the results are for the potentially troublesome case $\nu = \pi/2$. This is a 3-segment problem with only fast modes and two boundary layers. We have used Algorithm 10.115 with tolerance δ to choose the mesh in the layers, superimposed on a coarse uniform mesh of N_U subintervals to cover the smooth solution segment. The resulting number of mesh points is denoted by N, the maximum errors in the two components of $\mathbf{y}(x)$ away from the layers are denoted by e_1 and e_2, and their calculated rates of convergence by r_1 and r_2, respectively. The numerical results in Table 10.4 show that for this value of ν there are no serious numerical difficulties when one is using symmetric schemes for a very stiff problem. The number of mesh points needed to obtain a given accuracy is independent of ε for ε small enough (although the mesh itself does depend on ε). The errors on the smooth solution segment appear to behave like $O(h^{k+\eta})$, where $\eta = 0$ if k is even and $\eta = 1$ if k is odd. This observation is made precise in Theorem 10.129. Using $\delta = 10^{-6}$, we find that the convergence rates deteriorate for $k = 5$, $N_U = 20$ (and also for $k = 4$, $N_U = 40$), because the dominant error term is δ, which emanates from the boundary layers and does not change much when h is halved. Imposing the stricter tolerance $\delta = 10^{-10}$ for $k = 5$ improves the errors. (A corresponding reduction in δ for $k = 1, 2$ or 3 does not alter the errors away from the layers for the given uniform coarse meshes.)

TABLE 10.4 Numerical Results for Example 10.15, $\nu = \pi/4$

ε	δ	k	N_U	N	e_1	e_2	r_1	r_2
10^{-6}	10^{-4}	1	5	125	.13+1	.18+1		
			10	130	.28	.23	2.2	3.0
			20	140	.66−1	.73−1	2.1	1.7
			40	160	.16−1	.18−1	2.0	2.0
10^{-10}	10^{-4}	1	5	125	.13+1	.18+1		
			10	130	.28	.23	2.2	3.0
			20	140	.66−1	.73−1	2.1	1.7
			40	160	.16−1	.18−1	2.0	2.0
10^{-6}	10^{-6}	2	5	57	.25	.19		
			10	62	.58−1	.64−1	2.1	1.6
			20	72	.14−1	.16−1	2.0	2.0
			40	92	.35−2	.40−2	2.0	2.0
10^{-6}	10^{-6}	3	5	23	.57−1	.64−1		
			10	28	.20−2	.19−2	4.8	5.1
			20	38	.11−3	.14−3	4.2	3.8
			40	58	.51−5	.75−5	4.4	4.2
10^{-10}	10^{-6}	3	5	23	.57−1	.64−1		
			10	28	.20−2	.19−2	4.8	5.1
			20	38	.11−3	.14−3	4.2	3.8
			40	58	.51−5	.75−5	4.4	4.2
10^{-6}	10^{-6}	4	5	17	.36−2	.34−2		
			10	22	.21−3	.25−3	4.1	3.8
			20	32	.10−4	.14−4	4.3	4.1
			40	52	.19−5	.10−5	2.6	3.8
10^{-6}	10^{-6}	5	5	13	.73−3	.73−3		
			10	18	.71−5	.63−5	6.7	6.9
			20	28	.21−5	.11−5	1.8	2.5
			40	48	.19−5	.10−5	0.1	0.1
10^{-6}	10^{-10}	5	5	27	.73−3	.73−3		
			10	32	.51−5	.53−5	7.2	7.1
			20	42	.68−7	.97−7	6.2	5.7
			40	62	.12−8	.15−8	5.8	6.0

The results in Table 10.5 are much less satisfactory than those in Table 10.4, as predicted by theory. The large errors that the instability generates appear to concentrate, for some reason, in the first component of $\mathbf{y}(x)$ alone, where they are often intolerably large. The calculated rates of convergence are less meaningful here, because the error e_1 improves as h decreases not only because of reduced truncation error but also because the instability weakens as $\varepsilon^{-1}h^2$ decreases.

Let us again emphasize that instabilities like the one demonstrated in Table 10.5 are, in our experience, rare in practice. On the other hand, note that this peculiar difficulty is a feature of the discretization scheme used: If an upwinded one-sided scheme of Section 10.3.1 (or a Riccati method of Section 10.4.2) is used instead, then the case for $\nu = \pi/2$ is not more difficult than the case for $\nu = \pi/4$.

□

Note that the instability demonstrated in Examples 10.14 and 10.15 depends on the coarse mesh being coarse throughout the smooth solution segment. A simple way to get rid of this instability in both of these examples is to add a few mesh points $O(\varepsilon)$ apart in the middle of the interval. This creates a "layer" region there which separates the original long segment into two segments with coarse meshes. We may then consider the ODE on each of these segments, with BC as in (10.71). With the changed segment lengths, σ is no longer an eigenvalue for any of the corresponding BVPs (10.118). Theorem 10.86 may now be invoked to prove that the resulting scheme is stable.

The discussion above is for ODEs (10.107a) with only fast modes on a smooth solution segment. If slow modes are present as well, then the situation does not get simpler, of course. The assumption which we must generally make—that a symmetric scheme used with a coarse mesh on a smooth segment with the BC (10.71) is stable—can be characterized for a more general singular-singularly perturbed BVP, even though the characterization is not as simple as (10.118) and does not reduce to an algebraic condition for $\sigma = 0$. We obtain (see Exercise 11).

Theorem 10.119 Consider a k-stage Gauss collocation scheme applied for a singular-singularly perturbed ODE (10.1) on a smooth solution segment $[c,d]$ with a coarse, uniform mesh. Using the notation of (10.28–30), (10.36–39), and (10.117), the scheme is stable on this segment for positive ε and h small enough, if the auxiliary BVP

TABLE 10.5 Numerical Results for Example 10.15, $\nu = \pi/2$

ε	δ	k	N_U	N	e_1	e_2	r_1	r_2
10^{-6}	10^{-4}	1	5	125	.15+5	.12+1		
			10	130	.77+3	.19	4.3	2.6
			20	140	.49+2	.59−1	4.0	1.7
			40	160	.31+1	.16−1	4.0	1.9
10^{-10}	10^{-4}	1	5	125	.15+9	.11+1		
			10	130	.78+7	.19	4.3	2.5
			20	140	.49+6	.60−1	4.0	1.7
			40	160	.31+5	.16−1	4.0	1.9
10^{-6}	10^{-6}	2	5	57	.83+3	.16		
			10	62	.56+2	.42−1	3.9	1.9
			20	72	.36+1	.11−1	4.0	1.9
			40	92	.23	.27−2	4.0	2.0
10^{-6}	10^{-6}	3	5	23	.16+3	.41−1		
			10	28	.10+1	.22−2	7.3	4.2
			20	38	.14−1	.14−3	6.1	4.0
			40	58	.14−3	.82−5	6.7	4.1
10^{-10}	10^{-6}	3	5	23	.16+7	.41−1		
			10	28	.10+5	.22−2	7.3	4.2
			20	38	.14+3	.14−3	6.1	4.0
			40	58	.14+1	.82−5	6.7	4.1

$$\hat{u}' = (S^{11}(x) + 2k(k+1)\sigma[\Lambda^r(x)]^{-1})\hat{u} \qquad (10.119a)$$

$$\hat{u}^-(c) = 0, \qquad \hat{u}^+(d) = 0 \qquad (10.119b)$$

is well-conditioned [and in particular (10.119) has only the trivial solution]. □

(iv) Convergence for collocation at Gaussian points. As in Section 10.1.3, we consider a scalar ODE first, to get a feeling of what we may expect more generally. We therefore consider (10.19), assuming (10.20) and (10.22), and restrict λ to be constant as well. Since the solution is smooth, it is close to the reduced solution

$$y(x) \approx -\lambda^{-1} q(x), \qquad c \le x \le d \qquad (10.120)$$

Considering first the midpoint scheme (collocation at one Gauss point per subinterval)

$$h_i^{-1}(y_{i+1} - y_i) = \frac{1}{2}\lambda(y_i + y_{i+1}) + q_{i+1/2}, \qquad 1 \le i \le N \qquad (10.121a)$$

we obtain, when $|\lambda h_i| \gg 1$,

$$y_{i+1} \approx -y_i + \frac{2}{\lambda} q_{i+1/2} \approx (-1)^i y_1 + \frac{2}{\lambda} \sum_{j=1}^{i} (-1)^{i-j} q_{j+1/2} \qquad (10.121b)$$

so the reduced solution (10.120) is not approximated well at mesh points. Note, however, that for

$$y_{i1} := \frac{1}{2}(y_i + y_{i+1})$$

we do obtain [using the marginal stability of (10.121b) in (10.121a)]

$$y_{i1} = -\lambda^{-1} q_{i+1/2} + O\left(\frac{1}{|\lambda h_i|}\right) \qquad (10.122)$$

so *the reduced solution is approximated very well at the collocation point when $|\lambda h_i|$ is large*.

Still, we are interested in the mesh point errors $e_i := y_i - y(x_i)$. From (10.121) we obtain (recall Example 10.1)

$$|e_{i+1}| \lesssim |e_1| + \left|\frac{2}{\lambda}\sum_{j=1}^{i}(-1)^{i-j}\tau_j\right| \qquad (10.123a)$$

The error $|e_1|$ is what emanates from the boundary layer. We see that although a layer error δ does not get damped, it does not grow throughout the interval $[c,d]$ either. To further assess the error $|e_{i+1}|$, note that $\tau_j = O(h_j^2)$ and that τ_j varies smoothly with j.

Definition 10.124 The mesh $c = x_1 < x_2 < \cdots < x_N < x_{N+1} = d$ is *locally almost uniform* if

$$h_{i+1} = h_i(1 + O(h_i)) \qquad \text{for all } i \text{ odd } or \text{ for all } i \text{ even.}$$

□

Note that any mesh generated by halving each subinterval of another, arbitrary mesh, is locally almost uniform. We now have from (10.123a) that the error at mesh points when using the midpoint scheme on a coarse mesh is

$$|e_i| \leq \delta + O(h^{1+\eta}), \qquad 1 \leq i \leq N+1 \qquad (10.123b)$$

where $\delta = |e_1|$ is the error from the boundary layer and $\eta = 1$ if the mesh is locally almost uniform, $\eta = 0$ otherwise.

In (10.123b) we see some possible order reduction, as compared to the usual, nonstiff case, on a general coarse mesh. More significant order reduction occurs at mesh points for higher-order Gauss collocation schemes: The superconvergence phenomenon (5.79a) essentially disappears, because it is obtained from orthogonality when integrating $y(x)$, whereas in (10.120) there is no integration. To see this more precisely, consider a k-stage scheme written as

$$h_i^{-1}(y_{ij} - y_i) = \sum_{l=1}^{k} \lambda \alpha_{jl} y_{il} + g_{il}, \qquad 1 \leq j \leq k \qquad (10.125a)$$

$$h_i^{-1}(y_{i+1} - y_i) = \sum_{l=1}^{k} \lambda \beta_l y_{il} + g_{i+1} \qquad (10.125b)$$

$(g_{ij} = \sum_{l=1}^{k} \alpha_{jl} q(x_{il}), g_{i+1} = \sum_{l=1}^{k} \beta_l q(x_{il}))$. Eliminating $y_{i1}, \ldots, y_{ik}$ from (10.125a) and substituting in (10.125b), we obtain (10.90a) with

$$\gamma(\zeta) = 1 + \boldsymbol{\beta}^T (\zeta^{-1}I - \alpha)^{-1} \mathbf{1}, \qquad \zeta_i = \lambda h_i \qquad (10.126a)$$

where

$$\boldsymbol{\beta} = (\beta_1, \ldots, \beta_k)^T, \qquad \mathbf{1} = (1, \ldots, 1)^T \qquad (10.126b)$$

and $\alpha \in \mathbf{R}^{k \times k}$ is the coefficient matrix of (5.59). The inhomogeneity r_i in (10.90a) is

$$r_i = \boldsymbol{\beta}^T (\zeta_i^{-1} I - \alpha)^{-1} \mathbf{g}_i + g_{i+1}, \qquad \mathbf{g}_i = (g_{i1}, \ldots, g_{ik})^T \qquad (10.126c)$$

The solution of (10.90a) is given in (10.95). Substituting the error e_i in place of y_i, we find that the local truncation error in (10.126c) is $r_i = O(h_i^k)$ (even though $g_{i+1} = O(h^{2k})$). Using (10.116), we obtain the generalization of (10.123b)

$$|e_i| \leq \delta + O(h^{k+\eta})$$

where

$$\eta := \begin{cases} 1 & \text{if } k \text{ is odd and the mesh is locally almost uniform} \\ 0 & \text{otherwise} \end{cases} \qquad (10.127)$$

This is a clear reduction in order compared to the superconvergence order (5.79a).

These observations generalize to 3-segment problems. The proofs are rather technical and involved, though, so we merely state the results.

Lemma 10.128 Let the assumptions of Theorem 10.47 hold on a smooth solution segment $[c,d]$ with $p = k + \eta$, and consider a k-stage Gauss collocation scheme on a mesh satisfying (in the notation of Theorem 10.47) for $1 \leq i \leq N$

$$h_i \|\Lambda^0\| \leq K_2 h_i \ll 1 \qquad (10.128a)$$

$$\|h_i \lambda_R^+\|^{-1}, \quad \|h_i \lambda_R^-\|^{-1} \leq K_4 h_i \ll 1 \qquad (10.128b)$$

Then the collocation method with $\mathbf{y}_1$ specified has a unique solution, provided that h, K_2, and K_4 are sufficiently small. This solution satisfies

$$\max_i |\mathbf{y}_i| \leq \text{const} \{|\mathbf{y}_1| + \|\mathbf{q}\|_{[c,d]} + \max_i |\sum_{j=1}^{i} (-1)^{jk} \mathbf{g}^r(x_j)|\} \qquad (10.128c)$$

where $\mathbf{g}^r(x) := (\mathbf{g}^+(x), \mathbf{g}^-(x))$. □

The stability bound (10.128c) has the same annoying term as in (10.123a). Now, *assuming* that the scheme is stable on the coarse mesh part and that the layer regions are resolved well, we can invoke Theorem 10.86 to obtain

Theorem 10.129 Suppose that the BVP (10.1), (10.15) is well-conditioned and that the assumptions of Theorem 10.47 hold throughout the interval $[a,b]$ for $p = k + \eta$ [η as defined in (10.127)]. Consider a k-stage Gauss collocation scheme on a 3-segment mesh (see Fig. 10.5), where the fine meshes in the boundary layer regions are constructed such that the errors there are $O(\delta)$ for a given tolerance δ [this can be done, e.g., by using Algorithm 10.115, and the coarse mesh in between satisfies (10.128a,b)]. Assume further that Φ_2^h [see (10.74) and Section 10.2.3] exists and is bounded. Then for h, K_4, K_2 sufficiently small, there is a unique collocation solution which satisfies

$$\max_i |\mathbf{y}_i - \mathbf{y}(x_i)| \leq \text{const}(\delta + h^{k+\eta}) \qquad (10.129)$$

□

The numerical results in Tables 10.4 and 10.2 indicate that the error bound in Theorem 10.129 is sharp. While the error bound (10.129) is not as nice as the ones available in Theorem 5.79 for the nonstiff case, it does indicate the convergence of symmetric schemes for stiff BVPs, provided that a fine mesh is used in layer regions. Furthermore, analysis reveals that the quantity which is being equidistributed in the adaptive mesh selection algorithm of Section 9.3 is still a relevant one here [cf. (10.123) and Example 10.1], and this is one reason for the demonstrated success of that algorithm for many practical, stiff BVPs.

(v) Utilizing the solution at collocation points. In (10.122) we have already seen an indication that the solution of a Gauss collocation scheme is better localized at collocation points than elsewhere (including at mesh points). In fact, as the stiffness is increased while the mesh is held fixed, the approximation error at collocation points for a BVP with only fast modes is easily seen to improve. When slow modes are also present, the error for very stiff problems depends mainly on that of the slow modes (and does not improve as the stiffness increases), but the latter is essentially a "nonstiff error" and as such is better localized. Perhaps the more disturbing feature of the error bound (10.129) is not the order reduction, but rather the global error term δ which emanates from layers. At collocation points, however, it can be seen that the contribution from this term is bounded by $\delta K_4 h$ [see (10.128b)].

This is particularly important when designing an adaptive, general-purpose code (see Sections 9.2, 9.3). Using a poor initial mesh for a stiff BVP may introduce a layer error δ which is not very small, and then it is desirable to be able to evaluate the solution and derivatives in smooth solution regions in such a way that this smoothness is indicated to the code, enabling it to concentrate mesh points in layers. Preliminary studies indicate that rather impressive improvements to COLSYS (Appendix B) can be made in this way.

10.3.3 Exponential fitting

Of the many special schemes available in the literature for a scalar, second-order singularly perturbed ODE, we have briefly mentioned a few in Section 10.3.1. We now introduce yet another such scheme, which works fairly well when it is applicable, and has a somewhat different flavor than the ones discussed earlier. We demonstrate it on an important application in Example 10.16 below, but we do not know of other difficult applications where this scheme works (or can be applied at all). Our development of the scheme will parallel that in Section 5.6.1, but for simplicity we will do it on a uniform mesh with step size h, even though an extension to arbitrary meshes is possible.

Consider the singularly perturbed ODE

$$\varepsilon u'' = (p(x)u)' + r(x)u + q(x) \tag{10.130}$$

where $0 < \varepsilon \ll 1$ and p, r and q are bounded and smooth. We assume that $p(x)$ is bounded away from 0 (no turning points). The dependence on ε may be more involved in general, and is made simple and explicit here for transparency. For this ODE we seek a 3-point discretization on a mesh

$$\pi : a = x_1 < x_2 < \cdots < x_N < x_{N+1} = b$$

$$h =: h_i = x_{i+1} - x_i, \quad 1 \leq i \leq N$$

As in Sections 10.1.1 and Section 5.6.1, we write (10.130) as a first-order system

$$\varepsilon u' = p(x)u + v \tag{10.131a}$$

$$v' = r(x)u + q(x) \tag{10.131b}$$

[cf. (10.2)–(10.4)] and note that while $u(x)$ is a fast variable, $v(x)$ is slow. Thus we may use a usual symmetric discretization for (10.131b), but for (10.131a) we replace the symmetric scheme of Section 5.6.1 by doing an exponential fitting: Since $p(x)$ is smooth and bounded away from 0, and since $v(x)$ varies slowly, it is reasonable to approximate them to first order by piecewise constant functions,

$$p(x) = p_{i+1/2}, \quad v(x) = v_{i+1/2} \quad x_i \leq x \leq x_{i+1}$$

Then the transformation

$$u = \exp\{\varepsilon^{-1} p_{i+1/2}(x - x_i)\} y$$

yields

$$\varepsilon y' = \exp\{-\varepsilon^{-1} p_{i+1/2}(x - x_i)\} v_{i+1/2} \quad x_i \leq x \leq x_{i+1}$$

which *we can integrate exactly*, obtaining

$$\varepsilon(y(x_{i+1}) - y(x_i)) = \int_{x_i}^{x_{i+1}} \exp\{-\varepsilon^{-1} p_{i+1/2}(x - x_i)\} v_{i+1/2} \, dx$$

$$= -\varepsilon p_{i+1/2}^{-1} v_{i+1/2} [\exp\{-\varepsilon^{-1} p_{i+1/2}(x-x_i)\}]_{x_i}^{x_{i+1}}$$

$$= -\varepsilon p_{i+1/2}^{-1} v_{i+1/2} (\exp\{-\varepsilon^{-1} h p_{i+1/2}\} - 1).$$

Thus, transforming back from y to u we have

$$v_{i+1/2} = \frac{p_{i+1/2}(\exp\{-\varepsilon^{-1} p_{i+1/2} h\} u_{i+1} - u_i)}{-\exp\{-\varepsilon^{-1} p_{i+1/2} h\} + 1}$$

or

$$v_{i+1/2} = \frac{1}{\varepsilon^{-1} h}[B(\varepsilon^{-1} h p_{i+1/2}) u_{i+1} - B(-\varepsilon^{-1} h p_{i+1/2}) u_i)] \qquad (10.132\text{a})$$

where

$$B(x) := \frac{x}{e^x - 1} \qquad (10.133)$$

is the *Bernoulli function*. Note that

$$B(\varepsilon^{-1} h p_{i+1/2}) \to \begin{cases} 1 - \frac{1}{2}\varepsilon^{-1} h p_{i+1/2} & h \ll \varepsilon/|p_{i+1/2}| \\ 0 & \varepsilon \ll h\, p_{i+1/2},\, p > 0 \\ -\varepsilon^{-1} h p_{i+1/2} & \varepsilon \ll h\, |p_{i+1/2}|,\, p < 0 \end{cases}$$

So, if $\varepsilon \ll h$ and $p_{i+1/2} < 0$, then

$$v_{i+1/2} \approx -p_{i+1/2} u_{i+1}$$

while if $\varepsilon \ll h$ and $p_{i+1/2} > 0$ then

$$v_{i+1/2} \approx -p_{i+1/2} u_i$$

If $h \ll \varepsilon$ we essentially have a centered, second-order scheme. We see, therefore, that this scheme does automatic upwinding!

The next step is to approximate (10.131b) by the second-order scheme

$$h^{-1}(v_{i+1/2} - v_{i-1/2}) = r(x_i) u_i + q(x_i) \qquad (10.132\text{b})$$

Substituting (10.132a), we obtain the 3-point formula

$$a_i u_{i-1} + b_i u_i + c_i u_{i+1} = q(x_i) \qquad 2 \le i \le N \qquad (10.134\text{a})$$

$$a_i = \frac{\varepsilon}{h^2} B(-\varepsilon^{-1} h p_{i-1/2}), \qquad c_i = \frac{\varepsilon}{h^2} B(\varepsilon^{-1} h p_{i+1/2}) \qquad (10.134\text{b})$$

$$b_i = -\frac{\varepsilon}{h^2}(B(\varepsilon^{-1} h p_{i-1/2}) + B(-\varepsilon^{-1} h p_{i+1/2})) - r(x_i) \qquad (10.134\text{c})$$

This scheme is *exact* if $v(x)$ and $p(x)$ are constant. But in general its accuracy is only $O(h)$ uniformly in ε (i.e., for any $0 < \varepsilon \le \varepsilon_0$). If $h \ll \varepsilon$ then the scheme is second-order accurate.

Example 10.11 (continued)

For the BVP of Example 10.3 with $\varepsilon = 10^{-6}$, using uniform meshes with $h = .1, 0.05$, and 0.025 (i.e., $\varepsilon \ll h$), the scheme (10.134) with (10.99c) has no advantage over the simpler method (10.99): The maximum errors on the three meshes are 0.2, 0.1, and 0.051, respectively. However, when we use uniform meshes with $h = 0.025$ and 0.0125 for $\varepsilon = 10^{-2}$ (i.e., ε is comparable in size to h), the exponential fitting method yields maximum errors 0.033 and 0.011, while (10.99) and (10.103) both yield errors 0.12 and 0.19 for the two meshes, respectively. The scheme (10.134) is therefore seen to better adapt to different ratios of ε and h. For $h \ll \varepsilon$, its convergence rate is even higher. □

The advantage of this method is its ability to adapt to various rates of change of modes, giving the "structurally correct" solution independent of the mesh (in particular, regardless of whether layers are resolved or not).

Example 10.16

Consider the steady state semiconductor equations (10.58) of Example 10.8. We recall that ψ, n, and p are all fast solution components. This BVP is not precisely in the form (10.130), yet we note that (i) the usual 3-point formula for (10.58a) does reproduce the reduced solution at mesh points in the limit $\lambda \to 0$; (ii) if we consider a piecewise linear approximation for ψ in each of (10.58b) and (10.58c), then these ODEs do look like (10.130) with what corresponds to $p(x)$ in (10.130) being piecewise constant; and (iii) the "regularizing" transformation (1.36) is in essence none other than the exponential transformation from u to y used to derive (10.132a). Thus we apply exponential fitting to each of (10.58b) and (10.58c), obtaining the difference scheme for $2 \leq i \leq N$

$$\frac{\lambda^2}{h^2}(\psi_{i+1} - 2\psi_i + \psi_{i-1}) = n_i - p_i - D(x_i) \qquad (10.135a)$$

$$\frac{1}{h^2}(B(\psi_{i+1} - \psi_i)n_{i+1} - B(\psi_i - \psi_{i+1})n_i - B(\psi_i - \psi_{i-1})n_i + B(\psi_{i-1} - \psi_i)n_{i-1}) = R(n_i, p_i) \qquad (10.135b)$$

$$\frac{1}{h^2}(B(\psi_i - \psi_{i+1})p_{i+1} - B(\psi_{i+1} - \psi_i)p_i - B(\psi_{i-1} - \psi_i)p_i + B(\psi_i - \psi_{i-1})p_{i-1}) = R(n_i, p_i) \qquad (10.135c)$$

The latter two equations are obtained from (10.134) upon noting that if ψ is linear on $[x_i, x_{i+1}]$ then what corresponds to $\varepsilon^{-1} h p_{i+1/2}$ in (10.134) is

$$h_i \frac{\psi_{i+1} - \psi_i}{h_i} = \psi_{i+1} - \psi_i$$

The scheme (10.135) has been used successfully in many practical calculations. If $R \equiv 0$ then its error is $O(h^2)$, because the current densities J_n and J_p in (1.33), which correspond to v in (10.131), are constant. □

10.4 INITIAL VALUE METHODS

Most research efforts for solving (very) stiff BVPs have concentrated on difference methods, like the ones described in the previous section, and not on extending the initial value approaches of Chapter 4. A quick glance at Examples 5.2 and 4.3 shows why: While basic symmetric difference methods perform well when the stiffness (λ in these examples) is increased (provided that appropriate meshes are selected), the methods of Sections 4.3 and 4.4 do not. For instance, $O(\varepsilon^{-1})$ shooting points are required with these methods to solve the BVP in Example 10.5 with $\varepsilon = \mu^2$. Moreover, of the general purpose BVP software available to date, codes based on initial value techniques perform by and large poorly for stiff problems, while those based on (symmetric) difference schemes do much better, even if the theory on which they are based does not strictly hold for stiff BVPs.

Nonetheless, initial value methods for solving singularly perturbed BVPs with reasonable efficiency *can* be constructed if certain precautions are taken. In this section we consider two such methods.

The basic difficulty with the various variants of the multiple shooting method of Chapter 4 is that they attempt to integrate rapidly increasing modes when computing part of the fundamental solution. These methods may achieve decoupling on a discrete level, but stability requirements restricting the growth of modes impose a very fine mesh of shooting points everywhere, regardless of the smoothness of the solution. One way to circumvent this is to give up the "exact" integration of fast modes in segments where the solution is smooth, since in such segments fast modes do not contribute much to the solution anyway. In Section 10.4.1 we describe such a marching algorithm. An integrator using implicit symmetric schemes gives a method which in some theoretical aspects is reminiscent of the ones in Section 10.3.2. Choice of a coarse mesh in smooth solution segments is then sometimes possible, and we describe an algorithm for this purpose based on a so-called *pathfinder* technique.

Another possible approach here is to perform the decoupling on the continuous (ODE) level. In Section 10.4.2 we reconsider the Riccati method of Sections 4.5.2 and 6.2.2, where the IVPs to be integrated are generally stable. Such a method does not need to be changed when applied to stiff BVPs, but the theory to support it needs to be refined, and nontrivial implementation questions do arise (for, after all, the continuous decoupling is also done by discretization in practice). We consider only linear BVPs in this section, partly to emphasize the basic ideas of the methods involved, but also because these methods are still undergoing investigation, and computational experience is currently limited.

10.4.1 Sequential shooting

The methods described in Sections 4.3 and 4.4 all require the number of integration steps $\tilde{N}$ in (4.38) to increase as the stiffness of the BVP increases, e. g., $\tilde{N} \sim 1/\varepsilon$. This is costly for ε small. The problem can be circumvented if we abandon the notion of closely integrating fast increasing modes on segments where the solution is smooth, because the fast modes contribute little to the exact solution there. But we do require that the difference scheme used to integrate the fast modes on a coarse mesh preserve dichotomy. Corresponding to the concepts of absolute stability and A-stability (cf.

Section 2.7) we have *dichotomic* stability, which for simplicity we define for one-step schemes only:

Definition 10.136   The one-step scheme

$$y_{i+1} = y_i + h\,\Phi(x_{i+1}, x_i; y_{i+1}, y_i; h)$$

is called *di-stable* on a region $\mathbf{D} \subset \mathbf{C}$ if for the model equation $y' = \lambda y$ the resulting recursion $y_{i+1} = \gamma(\lambda h) y_i$ has the property that for all $\zeta \in \mathbf{D}$ satisfying $|\gamma(\zeta)| < \infty$,

$$|\gamma(\zeta)| \begin{cases} \le 1 & \text{Re}(\zeta) < 0 \\ \ge 1 & \text{Re}(\zeta) > 0 \end{cases} \qquad (10.136)$$

The scheme is referred to simply as *di-stable* if it is di-stable on $\mathbf{C}$. □

This definition is of a particular importance for $|\lambda h|$ large. [For $|\lambda h|$ small we actually expect the inequalities in (10.136) to be sharp, because of accuracy requirements.] Since symmetric schemes, like the Gauss or Lobatto collocation (implicit Runge-Kutta) schemes discussed in Section 10.3.2, are di-stable, we are naturally led to use them for the purpose of marching through a smooth solution segment with a coarse mesh. The stability of such an integration scheme is described by Lemma 10.128. The implementation of these schemes is much less straightforward than for the standard IVP context. In particular, we must use a special predictor that mimics well the asymptotic growth behaviour (as $|\lambda h| \to \infty$) of discrete counterparts of the fast modes.

Having decided to march through a smooth solution segment $[c, d]$ ($a < c < d < b$) with a coarse mesh, the next question is how to find such a mesh. The idea of a *pathfinder* is to find initial conditions at $x = c$ corresponding to a smooth solution $\bar{y}(x)$ which satisfies the stiff ODE (10.1) [not necessarily the one corresponding to the given BC (10.15)], and to find a suitable mesh while integrating $\bar{y}(x)$ on $[c, d]$. The underlying idea behind the procedure for finding such initial values at c is to exploit (as in Section 4.4 and Section 11.4.2) the fact that when one is integrating a fundamental solution which involves fast modes, they can be discovered and decoupled from the rest. The faster they grow, the faster they can be decoupled.

Let us assume that we have a 3-segment problem [i. e., (10.78) holds for the entire interval $[a, b]$]. We begin by marching through the left end layer from a to c, starting with some initial values for a fundamental solution and a particular solution. A marching preocedure with the aforementioned distable integrator often works reasonably well for this purpose, provided that appropriate measures are taken to make the step size choice efficient. The step size control can be done by computing the eigenvalues of $A(x)$ at some shooting points; they provide information about the layers [cf. (10.48), Section 10.3.2] and about the possibility for increasing the minimum step size, once the layer has been passed.

The point c is now chosen such that the influence of the fast decreasing modes has died away (below a given tolerance *TOL*) there. The next step is to find suitable initial values at $x = c$ for the pathfinder $\bar{y}(x)$. This is done by using an *explicit* scheme, with an *a priori* chosen step size which is found using eigenvalue information at c as well as the behaviour of the forcing term; we omit details. Given a trial value $\mathbf{s}$ for $\mathbf{y}(c)$, we integrate over a *few* steps only, say till $x = e$ ($<d$). Since we have used an explicit scheme, *all* fast modes will have grown in magnitude more than TOL^{-1} if e is large

enough, so that separating them from the slow ones is possible. We then perform a QU-decomposition of the computed fundamental solution $\bar{Y}(e)$,

$$\bar{Y}(e) = QU$$

Taking for simplicity $\bar{Y}(c) = I$, we can write

$$\bar{y}(x) = \bar{Y}(x)\mathbf{s} + \mathbf{v}(x)$$

(with $\bar{y}(c) = \mathbf{s}$), so we have

$$Q^T \bar{y}(e) = U\mathbf{s} + Q^T \mathbf{v}(e) \tag{10.137}$$

Partitioning U, and correspondingly vectors like $\mathbf{s}$, according to a splitting between the diagonal elements of U whose real parts are much larger than 1 and the rest,

$$U = \begin{bmatrix} U^{11} & U^{12} \\ 0 & U^{22} \end{bmatrix}, \quad \mathbf{s} = \begin{bmatrix} \mathbf{s}^1 \\ \mathbf{s}^2 \end{bmatrix}$$

we can annihilate the fast increasing modes by requiring

$$[Q^T \bar{y}(e)]^1 = \mathbf{0}$$

and using this with (10.137) to express $\mathbf{s}^1$ in terms of $\mathbf{s}^2$. For $\mathbf{s}^2 = [\bar{y}(c)]^2$, we use the values obtained from the march through the left boundary layer.

Having thus prepared the data at $x = c$, we can integrate the pathfinder $\bar{y}(x)$ on $[c, d]$, using a di-stable integrator (based on a symmetric scheme like collocation at Gauss points, e.g. the midpoint scheme) with an adaptive step size control. The mesh may be found in the course of this march, even though there are more difficulties here than with the usual BDF error control techniques (cf. Sections 10.3.2 and 2.7.4).

The integration of the pathfinder makes sense only on a smooth solution segment, so we should not proceed with it beyond the point d and into a layer. For the 3-segment problem we can let d coincide with the midpoint of the interval $\frac{a+b}{2}$, and repeat the above procedure integrating now from b backward to $\frac{a+b}{2}$. Assembling the pieces, we obtain a multiple shooting system, which we solve by stable compactification (cf. Section 4.4.2).

Before solving the linear system as above, note that since we obtain a decoupled recursion [cf. (4.57)] we can find suitable approximations of the modes. Since the fast decreasing modes are important only in the left end layer, and the fast increasing ones are important only in the right end layer, in the discrete system we can explicitly set the fast modes to 0 in regions where they are not of importance. Only then the compactification is performed.

When the BVP of Examples 4.3 and 5.2 is solved by using a sequential shooting algorithm, the number of shooting points remains essentially constant as the stiffness parameter λ is increased!

10.4.2 The Riccati method

We have seen several instances where the Riccati transformation is a useful tool for decoupling. For singular perturbation problems, it is used both as an analytical tool [cf. (10.4a), (10.40b)] and as a numerical tool (in Section 10.3.1) to decouple fast modes from slow ones. For the Riccati method, which has been described in some detail in Sections 4.5.2 and 6.2.2, it is used as a continuous decoupling technique to separate nonincreasing from nondecreasing modes for BVPs with separated BC. We show here that this latter use of the Riccati transformation has certain desirable features in the context of singular perturbation problems.

We consider the Riccati method for solving a BVP with separated BC, specifically

$$\mathbf{y}' = A\,\mathbf{y} + \mathbf{q} \qquad (10.138a)$$

$$B_{a1}\mathbf{y}(a) = \boldsymbol{\beta}_1, \qquad B_{b2}\mathbf{y}(b) = \boldsymbol{\beta}_2 \qquad (10.138b)$$

where $B_{a1} = (C \mid D) \in \mathbf{R}^{(n-k)\times n}$ with $D \in \mathbf{R}^{(n-k)\times(n-k)}$ nonsingular, and $B_{b2} = (E \mid F) \in \mathbf{R}^{k\times n}$. Briefly recalling the method from Section 4.5.2 (and using the same partitioning), we solve the IVPs

$$R' = A^{21} + A^{22}R - RA^{11} - RA^{12}R \qquad (10.139a)$$

$$R(a) = -D^{-1}C \qquad (10.139b)$$

$$\mathbf{w}^{2\prime} = (A^{22} - RA^{12})\mathbf{w}^2 - R\mathbf{q}^1 + \mathbf{q}^2 \qquad (10.140a)$$

$$\mathbf{w}^2(a) = D^{-1}\boldsymbol{\beta}_1 \qquad (10.140b)$$

from a to b, the IVP

$$\mathbf{w}^{1\prime} = (A^{11} + A^{12}R)\mathbf{w}^1 + A^{12}\mathbf{w}^2 + \mathbf{q}^1 \qquad (10.141a)$$

$$\mathbf{w}^1(b) = [E + FR(b)]^{-1}(\boldsymbol{\beta}_2 - F\mathbf{w}^2(b)) \qquad (10.141b)$$

from b to a, and then let

$$\begin{bmatrix} \mathbf{y}^1(x) \\ \mathbf{y}^2(x) \end{bmatrix} := \begin{bmatrix} \mathbf{w}^1(x) \\ R(x)\mathbf{w}^1(x) + \mathbf{w}^2(x) \end{bmatrix}$$

or

$$\mathbf{y}(x) = T(x)\mathbf{w}(x)$$

where

$$T(x) = \begin{bmatrix} I_k & 0 \\ R(x) & I_{n-k} \end{bmatrix} \qquad (10.142)$$

Like the other methods of Chapter 4, the Riccati method is conceptually exact, and its stability as a continuous decoupling algorithm is described in Section 6.2.2. It is recommended that the reader review (6.3–21), upon which much of the following material relies. If we assume a well-conditioned BVP (10.138), the separated BC impose a particular decoupling of the k nondecreasing modes $\tilde{Y}^1(x) = \begin{bmatrix} \tilde{Y}^{11}(x) \\ \tilde{Y}^{21}(x) \end{bmatrix}$ with $B_{a1}\tilde{Y}^1(a) = 0$

from the $n-k$ nonincreasing modes $\tilde{Y}^2(x) = \begin{bmatrix} \tilde{Y}^{12}(x) \\ \tilde{Y}^{22}(x) \end{bmatrix}$ with $B_{b2}\tilde{Y}^2(b) = 0$, and

$$R(x) = \tilde{Y}^{21}(x)\tilde{Y}^{11}(x)^{-1} \tag{10.143}$$

[We deviate slightly from the notation of Lemma 6.14.] When the dichotomy (10.6.8) holds, $\tilde{Y} := [\tilde{Y}^1 \mid \tilde{Y}^2]$ is a consistent fundamental solution with consistency constant $L = 0$ such that

$$\|\tilde{Y}(x)\, P\, \tilde{Y}(t)^{-1}\| \leq \tilde{K}\, e^{-\lambda(x-t)} \qquad x \geq t \tag{10.144a}$$

$$\|\tilde{Y}(x)\, (I-P)\, \tilde{Y}(t)^{-1}\| \leq \tilde{K}\, e^{-\mu(t-x)} \qquad x < t \tag{10.144b}$$

$$P = \begin{bmatrix} 0 & 0 \\ 0 & I_{n-k} \end{bmatrix} \tag{10.144c}$$

with $\tilde{K} = K\sqrt{K^2+1}$.

We use the theoretical multiple-shooting framework to show stability of the Riccati method. For this we show the stability of the IVPs (10.139–141). If $R(x)$ exists and is bounded on $[a,b]$ then the transformation $T(x)$ of (10.142) is also bounded and satisfies (6.19a). Subsequently, the bounds (6.20), (6.21) are obtained with $U^{12} = A^{12}$ in (6.21). These bounds are desirable, even though not sufficient in themselves to show stability of the IVPs (10.140), (10.141) uniformly in (the stiffness parameter) ε. We therefore concentrate first on obtaining a bounded $R(x)$ and then consider further the stability of the entire method.

(i) Stability of the Riccati method. The Riccati transformation $R(x)$ is obtained by integrating the nonlinear stiff IVP (10.139). By (10.143), its boundedness depends on that of $\tilde{Y}^{11}(x)^{-1}$. Assuming boundedness of the latter we consider first the linearization of (10.139), namely

$$(\delta R)' = -\delta R \cdot U^{11} + U^{22} \cdot \delta R \tag{10.145}$$

where

$$U^{11} = A^{11} + A^{12}R, \qquad U^{22} = A^{22} - RA^{12}$$

[cf. (6.6), (6.18), (10.140a), (10.141a)]. We show that the IVP for (10.145) is stable. Then we discuss ways to keep $\tilde{Y}^{11}(x)^{-1}$ bounded.

Theorem 10.146 If (10.144) holds and $\tilde{Y}^{11}(x)^{-1}$ is bounded on $[a,b]$, then the linearization (10.145) of the IVP (10.139) is stable, with

$$\|\delta R(x)\| \leq (1+\|R\|)\tilde{K}^2 \, \|\delta R(a)\| \, e^{-(\lambda+\mu)(x-a)} \tag{10.146}$$

Proof: Note first that by (10.143), $R(x)$ is bounded. To show boundedness of $\delta R(x)$ in terms of $\delta R(a)$, we apply another Riccati transformation to bring U into a block diagonal form: Defining the transformation

$$\hat{T}(x) = \begin{bmatrix} I_k & \hat{R}(x) \\ 0 & I_{n-k} \end{bmatrix}$$

the transformed variables

$$\begin{bmatrix} \hat{\mathbf{w}}^1(x) \\ \hat{\mathbf{w}}^2(x) \end{bmatrix} \equiv \hat{\mathbf{w}}(x) := \hat{T}(x)^{-1}\mathbf{w}(x) = \begin{bmatrix} \mathbf{y}^1(x) - \hat{R}(x)\mathbf{w}^2(x) \\ \mathbf{w}^2(x) \end{bmatrix}$$

satisfy

$$\hat{\mathbf{w}}'(x) = \begin{bmatrix} U^{11}(R(x)) & 0 \\ 0 & U^{22}(R(x)) \end{bmatrix} \hat{\mathbf{w}} + \hat{T}(x)^{-1}\mathbf{g} \quad (10.147a)$$

$$D\,\hat{\mathbf{w}}^2(a) = \boldsymbol{\beta}_1, \qquad [E + FR(b)]\hat{\mathbf{w}}^1(b) = \boldsymbol{\beta}_2 \quad (10.147b)$$

if $\hat{R}(x)$ satisfies the Lyapunov equation

$$\hat{R}' = U^{11}\hat{R} - \hat{R}U^{22} + A^{12} \quad (10.148a)$$

$$\hat{R}(b) = -[E + FR(b)]^{-1}F \quad (10.148b)$$

Since $\mathbf{y}(x) = T\hat{T}\hat{\mathbf{w}}(x)$ where $T\hat{T} = \begin{bmatrix} I_k & \hat{R} \\ R & I_{n-k} + R\hat{R} \end{bmatrix}$ and $(T\hat{T})^{-1} = \begin{bmatrix} I_k + \hat{R}R & -\hat{R} \\ -R & I_{n-k} \end{bmatrix}$,
the ODEs (10.138a) and (10.147a) are kinematically similar if $\|\hat{R}\|$ is bounded. It can be shown that

$$\hat{R}(x) = \tilde{Y}^{21}(x)[\tilde{Y}^{22}(x) - R(x)\tilde{Y}^{12}(x)]^{-1} \quad (10.149)$$

(Exercise 13), so this similarity follows since $\|R\|$ is bounded. Now let

$$Z(x) = \begin{bmatrix} Z^{11} & 0 \\ 0 & Z^{22} \end{bmatrix} := (T\hat{T})^{-1}\tilde{Y} = \begin{bmatrix} \tilde{Y}^{11} & 0 \\ 0 & \tilde{Y}^{22} - R\tilde{Y}^{12} \end{bmatrix}$$

so that $Z(t)$ is a fundamental solution of (10.147a) with $Z(a) = I$. Then

$$\delta R(x) = Z^{22}(x)Z^{22}(a)^{-1}\,\delta R(a)\,Z^{11}(a)Z^{11}(x)^{-1} = [\tilde{Y}^{22}(x) - R(x)\tilde{Y}^{12}(x)]\,\delta R(a)\,\tilde{Y}^{11}(x)^{-1}$$

and, after some manipulation, (10.146) can be shown to follow from (10.144). □

Of course, as discussed in Section 4.5.2, there is no guarantee in general that $\tilde{Y}^{11}(x)^{-1}$, whence $R(x)$, exists and remains bounded throughout $[a,b]$ (or even through a segment where the solution is known to be smooth). This can happen because $\tilde{Y}^{11}(x)$ becomes singular. In practice, it is necessary to perform a reimbedding (here, a reordering of the variables) to make $\tilde{Y}^{11}(x)$ nonsingular for the new variables, and then the integration is continued. We outline one possible reimbedding process below.

Let $P_1 := I_n$, and suppose that $R_1 = R$ and $\mathbf{w}_1^2 = \mathbf{w}^2$ are the computed solutions of the IVPs (10.139), (10.140) starting at $a =: s_1$. Further, suppose that it is determined that a reimbedding is necessary at a point s_2 because $\|R_1(x)\|$ is becoming large. Using row pivoting, we form the LU factorization

$$\begin{bmatrix} I_k \\ R_1(s_2) \end{bmatrix} = P_2^T \begin{bmatrix} L^{11} \\ L^{21} \end{bmatrix} U$$

If

$$P_2 P_1^{-1} =: \begin{bmatrix} P_2^{11} & P_2^{21} \\ P_2^{12} & P_2^{22} \end{bmatrix}$$

$$A_2 := P_2 A P_2^T$$

and

$$\mathbf{q}_2 := P_2 \mathbf{q}$$

then $R_2(x)$ and $\mathbf{w}_2^2(x)$, the solutions to the IVPs corresponding to

$$\mathbf{y}' = A_2 \mathbf{y} + \mathbf{q}_2$$

are computed by using initial values $R_2(s_2) = L^{21}(L^{11})^{-1}$ and $\mathbf{w}_2^2(s_2) = [P_2^{22} - R_2(s_2) P_2^{12}] \mathbf{w}_1^2(s_2)$. Continuing this process, suppose that $\{R_i, \mathbf{w}_i^2\}_1^J$ are computed before $x = b$ is finally reached. The return integration for $\{\mathbf{y}_i^1\}_1^J$ is then performed with

$$\mathbf{y}_J^1(b) = ([E \mid F]P_J^T \begin{bmatrix} I_k \\ R_J(b) \end{bmatrix})^{-1} (\boldsymbol{\beta}_2 - [C \mid D]P_J^T \begin{bmatrix} \mathbf{w}_J^2(b) \\ 0 \end{bmatrix})$$

$$\mathbf{y}_j^1(s_{j+1}) = [P_{j+1}^{11} + P_{j+1}^{12} R_j(s_{j+1})]^{-1} (\mathbf{y}_{j+1}^1(s_{j+1}) - P_{j+1}^{12} \mathbf{w}_j^2(s_{j+1})) \qquad j = J-1, \ldots, 1$$

Finally, $\mathbf{y}_j^2 := R_j \mathbf{y}_j^1 + \mathbf{w}_j^2$ and $\begin{bmatrix} \mathbf{y}_j^1 \\ \mathbf{y}_j^2 \end{bmatrix} \equiv P_j \mathbf{y}$ $(1 \leq j \leq J)$. The basic stability property of the IVPs is unaffected by this reimbedding process because the fundamental solution $\tilde{Y}_j(x)$ for the transformed ODE satisfies

$$\tilde{Y}_j(x) = P_j \tilde{Y}(x)$$

for the *permutation* matrix P_j and hence satisfies the dichotomy relation (10.144), too.

It is useful to interpret this multiple imbedding procedure in terms of the theoretical multiple shooting framework of Section 10.2. We assume a segmentation (10.16) which contains the reimbedding segmentation and has in addition the structure specified following (10.16). For simplicity, let the reimbedding segmentation coincide for now with the segmentation (10.16). The above procedure can be shown to be theoretically equivalent to solving on each segment $[t_j, t_{j+1}]$ $(1 \leq j \leq M)$ the BVP

$$\mathbf{y}_j' = A_j \mathbf{y}_j + \mathbf{q}_j \tag{10.150a}$$

$$B_{1j} \mathbf{y}_j(t_j) + B_{2j} \mathbf{y}_j(t_{j+1}) = \boldsymbol{\beta}_j , \tag{10.150b}$$

where

$$B_{1j} = \begin{bmatrix} 0 & 0 \\ (P_j^{12})^T - R_j(t_j) P_j^{11} & (P_j^{22} - R_{j-1}(t_j) P_j^{12})^{-1} \end{bmatrix} \tag{10.150c}$$

$$B_{2j} = \begin{bmatrix} P_{j+1}^{11} & P_{j+1}^{12} \\ 0 & 0 \end{bmatrix} \tag{10.150d}$$

[except for $B_{11} = \begin{bmatrix} 0 & 0 \\ C & D \end{bmatrix}$, $B_{2M} = \begin{bmatrix} E & F \\ 0 & 0 \end{bmatrix} P_M^{-1} \begin{bmatrix} I_k & 0 \\ R_M(b) & I_{n-k} \end{bmatrix}$]. The argument that

$$T_j(x) = \begin{bmatrix} I_k & 0 \\ R_j(x) & I_{n-k} \end{bmatrix}$$

decouples on $[t_j, t_{j+1}]$ is as in the non-reimbedding case. To interpret the subproblems (10.150) in terms of Theorem 10.69, we assume, for notational convenience, that each $P_j = I$. Then from (10.150c, d)

$$B_j \tilde{Y}_j = \begin{bmatrix} \tilde{Y}_j^{11}(t_{j+1}) & \tilde{Y}_j^{12}(t_{j+1}) \\ 0 & -R_j(t_j)\tilde{Y}_j^{12}(t_j) + \tilde{Y}_j^{22}(t_j) \end{bmatrix}$$

so Φ_j in (10.60a) is

$$\Phi_j =: [\Phi_j^+ \mid \Phi_j^-] = \tilde{Y}_j (B_j \tilde{Y}_j)^{-1}$$

where $\Phi_j^+(x)$ and $\Phi_j^-(x)$ are the nondecreasing and nonincreasing modes, respectively, on $[t_j, t_{j+1}]$. Moreover, $\Phi_j^+(t_{j+1}) = \begin{bmatrix} I \\ R_j(t_{j+1}) \end{bmatrix}$ and $\Phi_j^-(t_j)$ can be nicely bounded by using dichotomy, so $\Phi_j(x)$ can be bounded as well. Since

$$B_j \mathbf{y} = B_j \begin{bmatrix} \mathbf{y}_j^1 \\ R_j \mathbf{y}_j^1 + \mathbf{w}_j^2 \end{bmatrix} = \begin{bmatrix} \mathbf{w}_j^2(t_j) \\ \mathbf{y}_j^1(t_{j+1}) \end{bmatrix}$$

we have

$$\mathbf{v}_j(x) = \mathbf{y}(x) - \Phi_j(x)(B_j \mathbf{y})$$

in (10.60b). As a result, Theorem 10.69 is applicable here. There is a slight degeneration in the bounds in Corollary 10.73 [where the ordering of the blocks Φ^1, Φ^2 has now been reversed using the BC matrices (10.150c, d) partitioned as in Section 6.4.1 instead of as in (10.71b)], because (10.150c, d) do not have orthogonal columns [even assuming that the original BC (10.138b) are well-scaled as in (3.89)], and because an extra term involving $\|R\|$ appears in the bound.

Given the segmentation (10.16), let us assume further that on long segments the conditions of Theorem 10.47 hold in the variable x. [The transformation T of (10.28–30), (10.47), will be called S here, since it is in general different from the decoupling transformation T of (10.142) and Chapter 6.] These conditions limit the number of reimbeddings that may be needed to maintain boundedness of $R(x)$, because the rate at which the modes may rotate is asssumed bounded independent of the stiffness. For instance, the parameter ω in the unfriendly Example 4.13 [where $O(\omega)$ reimbeddings are needed] is bounded.

Having obtained the decoupling transformation $T(x)$ of (10.142) stably, we find that the IVPs (10.140) and (10.141) are stable in the sense that (6.20), (6.21) hold. This assures that no exponential magnification of errors occurs but may leave the possibility of stability constants which deteriorate, say linearly in the stiffness parameter ε^{-1}. The reason for this is that while nondecreasing and nonincreasing modes have been decoupled by $T(x)$, fast and slow modes have not [recall the discussion following (6.21b), where this is called a secondary stability effect]. Thus we may have $\min(\lambda^{-1}, x-a) \sim 1$ and/or $\min(\mu^{-1}, b-x) \sim 1$ in (6.20), (6.21), when considering the solution of a stiff BVP on a "long" segment (not a layer) and when λ and μ are not very large because of the presence of slow modes.

Fortunately, the hypotheses needed for the segmented approach of Sections 10.1.3 and 10.2. guarantee that this is not the case. For let us assume that the conditions of Theorem 10.47 hold on a given smooth solution segment $[c,d]$. This means in particular that there is a transformation $S(x)$ [substitute S for T and

$$\mathbf{x}(x) := S^{-1}(x)\mathbf{y}(x)$$

for $\mathbf{w}$ in (10.28–30), (10.47)], which splits the modes to fast increasing, fast decreasing, and slow ones. Comparing $\mathbf{x}$ and $\mathbf{w}$, we easily see that $S^{-1}(x)T(x)$ induces a bounded decoupling transformation of fast and slow modes in each of the blocks of nondecreasing and nonincreasing modes. If, for instance, the slow modes are all in $\mathbf{w}^2$ and we write

$$\mathbf{x} = \begin{bmatrix} \mathbf{x}^+ \\ \mathbf{x}^- \\ \mathbf{x}^0 \end{bmatrix}$$

with an obvious notation, then the stiff IVP for $\begin{bmatrix} \mathbf{x}^- \\ \mathbf{x}^0 \end{bmatrix}$ is stable (uniformly in ε, because the fast and slow modes have been decoupled), and therefore so is the IVP for $\mathbf{w}^2$. The terminal value problem for $\mathbf{w}^1$ is also stable, because there are only fast increasing modes. This can be readily generalized for the case where slow modes are present both in $\mathbf{w}^1$ and in $\mathbf{w}^2$.

(ii) Computational aspects. It is important to realize that while the decoupling transformation $T(x)$ is computed, the 3-way splitting $S(x)$ is only considered for theoretical purposes. In order to use the ODE-stability result successfully, we must of course solve the IVPs (10.139–141) sufficiently accurately, choosing an IVP integration scheme which implicitly decouples fast and slow modes as well. We use an IVP integrator based on a BDF scheme which achieves this (cf. Section 10.3.1), assuming (reasonably) also that the discretization that it selects appropriately matches the segmentation (10.16). Then, adding the reimbedding points to the appropriate subset of the other discretization points, we can use Theorem 10.75. The solutions R, $\mathbf{w}^2$, and $\mathbf{y}^1$ naturally must reflect any solution behaviour such as boundary or interior layers, so an appropriate segmentation is found if they are computed accurately. (Note also that a reimbedding can introduce a layer in the Riccati transformation.) But such an integration *does not necessarily* proceed without incident, despite the stability of the IVPs. We have chosen two examples which demonstrate this below.

In the numerical results which follow we use an IVP code with a BDF scheme and a mixed absolute/relative error tolerance at each step (viz., $|\text{local error}| \leq (1+|z|)TOL$ for each unknown component z). The Riccati ODE is converted to a vector ODE, and a maximum value of $\rho = (n-k)k\,2^{k-1}$ is used for $\|R_j\|$ along with the reimbedding strategy outlined above. For interpolation of the components of R_j and $\mathbf{w}_j^2$, a cubic Hermite representation is applied, filtering out points used by the stiff integrator which are unnecessary for preserving the accuracy TOL. The required points are called "output points."

Example 10.17

Consider first the simple BVP

$$\varepsilon u'' + u' = 0 \qquad 0 < x < 1$$
$$u(0) = 0, \qquad u(1) = 1$$

with exact solution $u(x) = \dfrac{1-e^{-x/\varepsilon}}{1-e^{-1/\varepsilon}}$. We convert to a first-order system with $\mathbf{y} = (u', u)^T$ [recall that D in (10.138b) must be nonsingular]. Since the fast mode is decreasing, it must be detected by the forward integration. In particular, (10.139) reads $R' = -\varepsilon + \varepsilon e^{x/\varepsilon}$. This becomes unbounded, and a reimbedding (the only possible one, since $n = 2$) gives $\tilde{R}' = \dfrac{1}{\varepsilon}\tilde{R} - \tilde{R}^2$ with stable trajectory $\tilde{R} = 0$. We solve the problem easily as long as TOL is small enough. But if, for example, $\varepsilon = 10^{-7}$ and $TOL = 10^{-5}$, then $R = -\varepsilon$ is (in a sense, correctly) detected as the Riccati solution throughout $[0, 1]$. The return integration blows up, signalling trouble.

A better choice of variables (and one for which the hypotheses of Theorem 10.47 hold) is $\mathbf{y} = (\varepsilon u' + u, u)^T$, for now $R(x) = 1 - e^{-x/\varepsilon}$ and the solution is easily found for $10^{-1} \leq \varepsilon \leq 10^{-9}$, regardless of what TOL is. $\square$

Example 10.18

Consider the BVP

$$\varepsilon u'' - \frac{x}{2}u' + \frac{x}{2}v' + v = \varepsilon\pi^2\cos(\pi x) + \frac{1}{2}(\pi x)\sin(\pi x) \qquad -1 < x < 1$$
$$\varepsilon v'' - v = 0$$
$$u(-1) = -1, \qquad v(-1) = 1, \qquad u(1) = v(1) = e^{-2/\sqrt{\varepsilon}}$$

with exact solution $u(x) = \mathrm{erf}(x/\sqrt{2\varepsilon})/\mathrm{erf}(1/\sqrt{2\varepsilon}) + v(x) + \cos\pi x$, $v(x) = e^{-(x+1)/\sqrt{\varepsilon}}$. Both u and v have a boundary layer of width $\sqrt{\varepsilon}$ at $x = -1$, and u has an interior layer of width $\sqrt{\varepsilon}$ at $x = 0$. Using $\mathbf{y} = (\varepsilon v', \varepsilon u' + \dfrac{x}{2}u, v, u)^T$, the solution is straightforward to compute for $10^{-1} \leq \varepsilon \leq 10^{-6}$, although accuracy degenerates slightly near $x = 0$. Integrating (10.139), a reimbedding is necessary almost immediately, and for the new system with

$$A(x) = \begin{pmatrix} -x/2\varepsilon & 0 & 1/\varepsilon & 0 \\ 0 & 0 & 0 & 1/\varepsilon \\ 1/2 & 1 & 0 & x/2\varepsilon \\ 0 & 1 & 0 & 0 \end{pmatrix}$$

$$R(x) = \begin{pmatrix} -\dfrac{\varepsilon}{x} & \dfrac{2\sqrt{\varepsilon}+x}{2(1-x\sqrt{\varepsilon})} \\ 0 & \sqrt{\varepsilon} \end{pmatrix}$$

For a negative x in an $O(\sqrt{\varepsilon})$ neighborhood of $x = 0$, a reimbedding occurs and the stable Riccati trajectory is

$$R(x) = \begin{bmatrix} x/\varepsilon & \sqrt{\varepsilon} \\ 0 & \sqrt{\varepsilon} \end{bmatrix}$$

For $\varepsilon = 10^{-4}$ and $TOL = 10^{-5}$, 300 output points for R and $\mathbf{w}^2$ are needed to preserve the accuracy TOL.

$\square$

However, it is again important to keep TOL small enough. Otherwise, the integration of (10.139) may miss interior reimbeddings. What can happen is that a region where a stable Riccati trajectory becomes unstable (and an unstable trajectory becomes stable) is missed, because a large step size may jump over the blowup region, so that the current trajectory [i.e., the IVP for (10.145)] is very unstable. However, a BDF initial value solver interprets this now wrong trajectory as still being stable! (For the model equation $y' = \lambda y$, the approximate solution using a BDF scheme is damped for *any* λ when $|\lambda h|$ is large.) This difficulty can be met by keeping TOL less than ε or by requiring small integration steps (or restarting) at the points where this phenomenon can occur, assuming that their location is known. For instance, for a scalar second order linear problem, this can happen only at potential turning points.

A very desirable feature of the method is that, if successful, the (final) mesh is generated in one shot, through the integration of stable IVPs. As seen from the examples, though, the integrator's error tolerance is not independent of the size of small problem parameters. This is perhaps not surprising, in that it seems reasonable to expect that, without any *a priori* information about the solution, any general method must be told how often to sample the problem data in (a, b) and that the sampling points must increase as $\varepsilon \to 0$ (say).

The Riccati method has proven quite successful for solving many singular perturbation problems of low dimension ($n \leq 4$), but it remains to be tested on larger systems with a greater variety of fast and slow modes and on nonlinear problems. Recall from Chapter 6 that there do exist other continuous decoupling methods, although the Riccati method has been the most popular.

EXERCISES

1. Show that if $p(x) < 0$, $a \leq x \leq b$, then the BVP (10.2) (with smooth coefficients) may have a boundary layer at $x = a$, but not at $x = b$.

2. Show that the Riccati transformation separating fast and slow solution components in (10.2a) can be written as in (10.9), and that the same transformation for the system (10.39) can be written as in (10.40b).

3. Verify the claims in Example 10.4.

4. In Exercise 9.5 we list 5 pairs of boundary values for Example 10.6, which result in five rather different solution profiles with different layer structures. Find the asymptotic structure of these solutions for $\varepsilon > 0$ small. Compare to the numerical results of Exercise 5 in Chapter 9.

5. Apply a matched asymptotic expansion technique to obtain bounded initial values $|\mathbf{y}_1^{(v)}(c)|$, thus establishing the existence of a smooth solution $\mathbf{y}_1(x)$, as described following Theorem 10.47.

6. (a) Verify that (10.65) holds.

   (b) Verify that (10.64b) holds.

7. (a) Show that (10.111b) and (10.112b) hold.

   (b) Furthermore, show that for a singular-singularly perturbed BVP an $O(\varepsilon^{-1} h_i)$ term appears in the transformed trapezoidal scheme (10.112b).

8. Show that collocation schemes at Gauss points are D-stable, while those at Lobatto points are not.

9. Prove Lemma 10.114.

10. With reference to Example 10.15, show that a well-scaled fundamental solution for **w** is given by

$$\Psi(x) = \begin{pmatrix} \varepsilon \exp\{(x-1)(2+\varepsilon\alpha)/\varepsilon\} & -\dfrac{v\varepsilon}{\alpha}\exp\{-x(1+\varepsilon\alpha)/\varepsilon\} \\ -\dfrac{v\varepsilon}{\alpha}\exp\{(x-1)(2+\varepsilon\alpha)/\varepsilon\} & \varepsilon\exp\{-x(1+\varepsilon\alpha)/\varepsilon\} \end{pmatrix}$$

where

$$\alpha = \frac{1}{2\varepsilon}[-3+\sqrt{9-4\varepsilon^2 v^2}] = -\frac{1}{3}\varepsilon v^2(1+\frac{1}{9}\varepsilon^2 v^2) + O(\varepsilon^5)$$

Further, use this to construct the general solution of the ODE for **y**, and an approximation to the exact solution **y**(x) of the given BVP. Show that the BVP is well-conditioned as well.

11. Prove Theorem 10.119.

12. Derive an exponential fitting method like (10.134) for a general, nonuniform mesh, using the methodology of Section 5.6.1. Then extend (10.135) in a similar way.

13. Let $\widetilde{W}(x) = \begin{pmatrix} \widetilde{W}^{11}(x) & 0 \\ \widetilde{W}^{12}(x) & \widetilde{W}^{22}(x) \end{pmatrix}$ be the fundamental solution to (6.18) satisfying $\widetilde{W}(b) = \begin{pmatrix} I_k & 0 \\ -(E+FR(b))^{-1}F & I_{n-k} \end{pmatrix}$. Show that $\hat{R}(x)$ defined by (10.148) satisfies $\hat{R}(x) = \widetilde{W}^{12}(x)\widetilde{W}^{22}(x)^{-1}$. Now writing $\widetilde{W}(x) = W(x)C$ for a constant nonsingular matrix C, show using (10.148b) that (10.149) holds. Note that boundedness of $\|T\hat{T}\|\,\|(T\hat{T})^{-1}\|$ now follows from (10.144).

11

Special Topics

In this chapter we have accumulated a number of topics which are related to the main theme of the book but which have not received full attention in the previous chapters. Some of these topics are specialized in one way or another, and some are wide enough to span a book in their own right. Our intention here is merely to describe them briefly, in an attempt to achieve an overview of BVPs and their relation with other subjects. We will often describe, or highlight, the issues involved by means of an example, rather than applying a full, rigorous analysis. The sections in this chapter are each concerned with a separate topic, so the relationships between sections are looser than in the other chapters.

11.1 REFORMULATION OF PROBLEMS IN "STANDARD" FORM

Many problems arise in practice in a form which at first sight does not look "standard"; i.e., available software may not be directly applied to it. In this section we list "tricks" to convert some such problems into the standard form

$$\mathbf{y}' = \mathbf{f}(x, \mathbf{y}), \qquad a < x < b \tag{11.1a}$$

$$\mathbf{g}_1(\mathbf{y}(a)) = \mathbf{0}, \qquad \mathbf{g}_2(\mathbf{y}(b)) = \mathbf{0} \tag{11.1b}$$

Thus, our standard form is a first-order system subject to separated two-point BC. Some of the tricks are rather straightforward, or have been mentioned before, so we will only briefly refer to them here. Other less obvious ones will receive more attention. Still others will be delayed and discussed in the later sections of this chapter, within the context of the general class of problems to which they fall.

The approach of problem reformulation into standard form has obvious advantages. Not only can we use available software but, more generally, this modular approach enables us to capitalize on existing *knowledge* of how to handle certain problems numerically. It is important to note, though, that the approach is not without disadvantages. First, a transformation of a well-conditioned problem into standard form is not always possible (see, e.g., Sections 11.2 and 11.5). Less transparent is the argument that even when possible, such a transformation may not always be desirable: The algebraic structure of the converted problem could become more complicated, and the "degree of nonlinearity" may increase. Indeed, in some applications direct transformation of a nonlinear problem may be difficult, and delaying essentially the same transformation technique until after linearization is easy, e.g., involving merely a matrix inversion. (The latter does put the transformation inside the numerical algorithm, though.) A rule of thumb is to try to convert and use a general-purpose code first, and seek alternatives only if this does not work.

11.1.1 Higher-order equations, parameters, nonseparated BC, multipoint BC

In this section we briefly mention four fairly straightforward conversions, all of which have already been encountered in Chapter 1.

Recall that converting higher-order systems to first-order systems is straightforward. A high-order ODE $u^{(m)} = f(x, u, \ldots, u^{(m-1)})$ can be simply converted to a first-order system (11.1a) by (for example) defining $y_1 := u$, $y_2 := u'$, ..., $y_m := u^{(m-1)}$. The extension for a system of higher-order ODEs (and BC) is easy (cf. Section 1.1.2). The ability to convert such systems to the equivalent form (11.1a,b) is in fact the major justification for our defining key concepts such as dichotomy *only* for first-order systems.

We have seen several examples where a BVP with nonseparated BC is converted to one with separated BC (e.g., Example 1.10). This conversion can usually be done at the cost of adding one trivial ODE for each nonseparated boundary condition. (In any case, at most n additional trivial ODEs are needed for a system of order n.) In Section 1.1.3 we show how this is done for the general linear BC case [and obtain (1.9a,b)]. Often this same basic device can be used to also handle unknown parameters in the BVP. A parameter appearing in the ODE is eliminated this way in Examples 1.4 and 1.9.

A BVP with separated, multipoint BC at $a = s_0 < s_1 < \cdots < s_{J+1} = b$ can be easily converted to the form (11.1a,b) by using linear changes of variables mapping each subinterval $[s_j, s_{j+1}]$ to $[0, 1]$. To avoid introducing an unnecessary nonseparation in the BC, this is done by mapping s_j to 0 for j odd and s_j to 1 for j even. This specific transformation in a similar but more complicated context is given in the next section. The cost of this conversion is that the resulting BVP on $[0, 1]$ involves a system with Jn solution components. If the multipoint BC are not separated, then obviously some of the conversion techniques here can be combined, each having the effect of further increasing the size of the final ODE system.

11.1.2 Conditions at special points

BVPs can satisfy special conditions at unknown points. For instance, a BVP may involve a BC at an unknown point (see Examples 1.7 and 1.21). The trick of adding a trivial ODE, used in the last section to deal with unknown parameters, works with a slight modification here as well, as we demonstrate below.

Example 11.1.

A BVP arising in lubrication theory is

$$(h(x) - \frac{h(x)^3}{12}u'(x))' = -Ku(x) \qquad -\infty < x < c$$

$$u(-\infty) = 0, \qquad u(c) = u'(c) = 0$$

where $h(x)$ and K are known but the boundary point c is unknown. This problem can be converted to standard form (but still on a semi-infinite interval) by letting $t := x/c$ and writing, with an obvious redefinition of notation,

$$\frac{1}{c}[h(tc) - \frac{h(tc)^3}{12c}u'(t)]' = -Ku(t) \qquad -\infty < t < 1$$

$$c' = 0$$

$$u(-\infty) = 0, u(1) = u'(1) = 0$$

(We could use here the translation transformation $t = x - c$ instead, because the interval in x is semi-infinite.) □

BVPs commonly arise for optimal control applications where at unknown points the ODE and corresponding solution change discontinuously or some additional condition is satisfied. An approach like the one above, whereby these so-called *switching points* are mapped to known points, can be used. This is done for one switching point in Example 1.22. When the number of points is known, the idea extends naturally.

For instance, consider a BVP with J switching points

$$a = s_0 < s_1 < \cdots < s_{J+1} = b,$$

$$\mathbf{y}' = \mathbf{f}_j(x, \mathbf{y}) \qquad s_{j-1} < x < s_j \qquad j = 1, \ldots, J+1$$

$$h_j(s_j, \mathbf{y}(s_j)) = 0 \qquad j = 1, \ldots, J \qquad (11.2)$$

$$\mathbf{g}(\mathbf{y}(a), \mathbf{y}(b)) = \mathbf{0}$$

and $\mathbf{y} \in C[a, b]$. Define the mapping

$$t = \begin{cases} \dfrac{x - s_{j-1}}{s_j - s_{j-1}} & s_{j-1} \le x \le s_j, \ j \text{ odd} \\ \dfrac{s_j - x}{s_j - s_{j-1}} & s_{j-1} \le x \le s_j, \ j \text{ even} \end{cases} \qquad (11.3)$$

With $\mathbf{y}_j(t) := \mathbf{y}(x)$ for $s_{j-1} \le x \le s_j$ we obtain a system of order $n(J+1) + J$,

$$\mathbf{y}_j'(t) = (s_j - s_{j-1})\mathbf{f}_j(s_{j-1} + (s_j - s_{j-1})t, \mathbf{y}_j(t)) \qquad j \text{ odd} \qquad 1 \le j \le J+1$$

$$\mathbf{y}_j'(t) = (s_{j-1} - s_j)\mathbf{f}_j(s_j + (s_{j-1} - s_j)t, \mathbf{y}_j(t)) \qquad j \text{ even} \qquad 1 \le j \le J+1$$

$$s_j'(t) = 0 \qquad 1 \le j \le J$$

$$h_j(s_j, \mathbf{y}_j(1)) = 0, \quad \mathbf{y}_j(1) = \mathbf{y}_{j+1}(1) \qquad j \text{ odd}, \qquad 1 \le j \le J$$

$$h_j(s_j, \mathbf{y}_j(0)) = 0, \quad \mathbf{y}_j(0) = \mathbf{y}_{j+1}(0) \qquad j \text{ even}, \qquad 1 \le j \le J$$

$$\mathbf{g}(\mathbf{y}_1(0), \mathbf{y}_{J+1}(p)) = \mathbf{0}$$

where $p = 0$ if J is odd, $p = 1$ if J is even.

If we are using BVP software which handles multipoint BC, then instead of (11.3) a transformation like

$$t = j - 1 + \frac{x - s_{j-1}}{s_j - s_{j-1}}, \qquad s_{j-1} \le x \le s_j, \qquad j+1, \ldots, J+1 \qquad (11.4)$$

may be used to map each switching point s_j to the known point j, $0 \le j \le J+1$. The size of the resulting system is then only $n+J$, so this transformation would usually be more efficient. Still, for both transformations [especially (11.3)] a price is paid in terms of lost problem structure (and in storage).

Sometimes there arise BVPs having interface conditions at *known* points, and various transformations are again needed. For instance, consider the BVP

$$u'' = f(x, u, u') \qquad 0 < x < 1$$

$$u(0) = \alpha, \qquad u(1) = \beta, \qquad u(\tfrac{1}{2}^-) = u(\tfrac{1}{2}^+), \qquad u'(\tfrac{1}{2}^-) = u'(\tfrac{1}{2}^+) + 1$$

Transforming $[0, \tfrac{1}{2}] \to [0, 1]$ and $[\tfrac{1}{2}, 1] \to [1, 0]$ gives an enlarged BVP in the form (11.1a,b). Another possibility is to consider the multipoint BVP

$$u_1'' = \begin{cases} f(x, u_1, u_1') & x < \tfrac{1}{2} \\ 0 & x > \tfrac{1}{2} \end{cases}$$

$$u_2'' = \begin{cases} 0 & x < \tfrac{1}{2} \\ f(x, u_2, u_2') & x > \tfrac{1}{2} \end{cases}$$

$$u_1(0) = \alpha, \quad u_2(1) = \beta, \quad u_1(\tfrac{1}{2}) = u_2(\tfrac{1}{2}), \qquad u_1'(\tfrac{1}{2}) = u_2'(\tfrac{1}{2}) + 1$$

for which

$$u(x) = \begin{cases} u_1(x) & x \le \tfrac{1}{2} \\ u_2(x) & x > \tfrac{1}{2} \end{cases}$$

This second modification has the advantage that if any other ODEs appear in the original problem, they need not necessarily be changed.

Remark When a BVP involves an interface condition or a side condition at a point c, $a < c < b$, the solution may generally have fewer continuous derivatives at c than elsewhere. Hence c should be a mesh point if a finite difference scheme, or a multiple shooting method, is employed [e.g., see Theorem 5.134 and Remark (d) following it]. □

11.1.3 Integral relations

A BVP for (11.1a) can have a term

$$\gamma = \int_a^b G(t, \mathbf{y}(t))\, dt \tag{11.5}$$

occurring either (i) as a quantity to be determined after the BVP is solved or (ii) as an integral constraint to be satisfied instead of a BC. In such a case, defining $w(x) = \int_a^x G(t, \mathbf{y}(t))\, dt$, we can add equations

$$w'(x) = G(x, \mathbf{y}(x)) \tag{11.6a}$$

$$w(a) = 0 \tag{11.6b}$$

to the BVP, and then either (i) evaluate $\gamma = w(b)$ after solving this extended BVP or (ii) use $w(b) = \gamma$ as a BC replacing the integral constraint. An important special case of this is when the norm of the solution is desired, or specified. Setting

$$G(x, \mathbf{y}(x)) := \mathbf{y}^T(x)\, \mathbf{y}(x) \tag{11.7}$$

yields $w(b) = \|\mathbf{y}\|_{2,[a,b]}^2$. (Also, Example 1.23 involves a minor variation on the above.)

A similar trick can be applied for a nonlinear Fredholm integral equation with a separable kernel, i.e.,

$$y(x) = \int_a^b K(x)\, G(t)\, F(y(t))\, dt + f(x)$$

where K, G, F, and f are all known functions. Letting

$$w(x) := \int_a^x G(t)\, F(y(t))\, dt$$

consider the BVP

$$w' = G(x)\, F(cK(x) + f(x))$$

$$c' = 0$$

$$w(a) = 0, \qquad w(b) = c(b)$$

Since $w(x) \equiv \int_a^x G(t)\, F(cK(t) + f(t))\, dt$, we have $y(x) = cK(x) + f(x)$. This is not the only possible such conversion, but while such ideas can be useful, they are also of restricted applicability, and we should keep in mind that numerical methods specifically designed for integral equations have of course been developed.

Conversions in other related contexts have been used. For example, similar tricks have been applied to convert integro-differential equations to BVPs, and problems in optimal control are commonly converted to BVPs using standard techniques. In Sections 11.3, 11.5, and 11.7 below we see further such conversions to the standard form (11.1).

11.2 GENERALIZED ODES AND DIFFERENTIAL ALGEBRAIC EQUATIONS

Often in applications, the differential problem is naturally cast in the form

$$\phi(x, y, y') = 0 \quad a < x < b \tag{11.8}$$

i.e., y' appears implicitly, rather than explicitly as in (11.1a). Since standard software usually requires the explicit form (11.1a), there is a strong incentive to convert (11.8) to the more standard form. This is frequently, but *not always,* possible and desirable. Here we distinguish some cases where (11.8) is truly more general than (11.1a) and mention some numerical approaches for such cases.

When we consider classes of problems of the form (11.8), the matrix

$$E(x) := \frac{\partial \phi(x, y(x), y'(x))}{\partial y'} \tag{11.9}$$

plays a crucial role.

Case I: $E(x)$ is nonsingular, $a < x < b$

The variational form of (11.8) involves the linear operator

$$L y \equiv E(x) y' + J(x) y \tag{11.10}$$

where

$$J(x) = \frac{\partial \phi(x, y(x), y'(x))}{\partial y} \tag{11.11}$$

Upon multiplication by $E^{-1}(x)$, we obtain the explicit linear form (11.1a), so this case can in theory be treated by the usual analysis and methods given throughout this book. In practice, however, a conversion of the nonlinear ODE (11.8) to the explicit form (11.1a) may not always be simple or even possible. Still, while standard software may be inapplicable, we know at least how to choose and implement suitable methods.

Case II: $E(x)$ is singular, and the problem has *index 1;* i.e., the matrix $U^{11}(x)$ defined below is nonsingular.

This defines a manageable class of *differential algebraic equations* (DAEs). The simplest instance is when the differential and the algebraic equations are separated, so we have

$$0 = f(x, u, v) \tag{11.12a}$$

$$v' = g(x, u, v) \tag{11.12b}$$

where $u(x) \in \mathbf{R}^{n_r}$, $v(x) \in \mathbf{R}^{n_0}$, $n = n_0 + n_r$ and we assume that $U^{11}(x) \equiv \frac{\partial f(x, u(x), v(x))}{\partial u}$ is nonsingular. Hence $E(x) = \begin{bmatrix} 0 & 0 \\ 0 & I_{n_0} \end{bmatrix}$. The system (11.12a, b) is subject to n_0 BC involving $\mathbf{u}$ and $\mathbf{v}$. Such a system arises, for example, when we attempt to find the reduced solution of a singularly perturbed system with separated (decoupled) fast and slow components, e. g., when we set $\varepsilon = 0$ in (10.50) or (10.35).

But the numerical solution of (11.12) can be much easier than that of (10.50) or (10.35). First, it is often possible and desirable to use (11.12a) to eliminate $\mathbf{u}$ in terms of $\mathbf{v}$ and substitute into (11.12b) to obtain a system of order n_0 in standard form. If such an explicit elimination is not possible then an implicit one is, upon noticing that in (10.12) $\mathbf{v}(x)$ is generally one derivative smoother than $\mathbf{u}(x)$. It is natural to require that this be reflected in the numerical approximation, as is suggested when collocation is considered: In Section 5.6 we use an approximation from $\mathbf{P}_{k+m,\pi} \cap C^{(m-1)}[a,b]$ for an ODE involving derivatives of order m. In (11.12a) the "order" is 0, so we require the collocation approximation $\mathbf{u}_\pi(x)$ of $\mathbf{u}(x)$ to be in the piecewise *discontinuous* space $\mathbf{P}_{k,\pi}$. Following the recipe in Section 5.6.1, we can locally eliminate the k coefficients of the representation of each component of $\mathbf{u}_\pi(x)$ in each subinterval of the mesh, after linearization. The nonsingularity of $U^{11}(x)$ guarantees that this is possible to do.

It is important to realize that while (11.12) can be viewed as a limit of a stiff BVP, it is not necessarily "stiff" in itself, provided that the lower smoothness of $\mathbf{u}(x)$ [hence of its approximant $\mathbf{u}_\pi(x)$] is recognized.

Note that (11.12) is subject to n_0 (not n) BC. These correspond to the integration constants in (11.12b), so for an IVP, $\mathbf{v}(a)$ alone [and not $\mathbf{u}(a)$] should be prescribed. For the linear BVP

$$\mathbf{0} = U^{11}(x)\mathbf{u} + U^{12}(x)\mathbf{v} + \mathbf{g}^1(x) \tag{11.13a}$$

$$\mathbf{v}' = U^{21}(x)\mathbf{u} + U^{22}(x)\mathbf{v} + \mathbf{g}^2(x) \tag{11.13b}$$

$$B_a \begin{bmatrix} \mathbf{u}(a) \\ \mathbf{v}(a) \end{bmatrix} + B_b \begin{bmatrix} \mathbf{u}(b) \\ \mathbf{v}(b) \end{bmatrix} = \boldsymbol{\beta}, \quad B_a, B_b \in \mathbf{R}^{n_0 \times n} \tag{11.13c}$$

with $U^{11}(x)$ nonsingular, we have from (11.13a)

$$\mathbf{u}(x) = -[U^{11}(x)]^{-1}[U^{12}(x)\mathbf{v}(x) + \mathbf{g}^1(x)] \tag{11.13d}$$

and this can be substituted into (11.13b) and (11.13c) to obtain a regular BVP of order n_0 for $\mathbf{v}(x)$. If this latter BVP has a unique solution then the BC (11.13c) are said to be *consistent* for the full problem (11.13a, b, c).

The situation for DAEs gets much more complicated when the differential and algebraic solution components are mixed together. Thus, consider a linear BVP

$$E(x)\mathbf{y}' = A(x)\mathbf{y} + \mathbf{q}(x) \tag{11.14a}$$

$$B_a \mathbf{y}(a) + B_b \mathbf{y}(b) = \boldsymbol{\beta} \tag{11.14b}$$

where $E(x)$ has a smooth nontrivial null space, which happens if there are smooth, nonsingular matrix functions $S(x), T(x)$ such that

$$E(x) = S(x) \begin{bmatrix} 0 & 0 \\ 0 & I \end{bmatrix} T^{-1}(x) \tag{11.15}$$

It is not difficult to see that with the notation

$$U(x) := S^{-1}(x)A(x)T(x) \tag{11.16a}$$

$$\begin{pmatrix} \mathbf{u}(x) \\ \mathbf{v}(x) \end{pmatrix} \equiv \mathbf{w}(x) := T^{-1}(x)\mathbf{y}(x) \tag{11.16b}$$

the condition of solvability is again that the upper left $n_r \times n_r$ block of U, $U^{11}(x)$, be nonsingular. If this condition holds, then $\mathbf{u}(x)$, the "algebraic part" of the solution, can in principle be eliminated, and a standard ODE for the "differential part" results. Consistent initial and boundary conditions can be derived along similar lines. The theory for DAEs with index 1 is in a satisfactory condition, but we avoid going into further detail.

For the numerical solution of such problems, we have two basic approaches. One is to decouple the differential and the algebraic parts of the solution explicitly. Then a piecewise discontinuous approximation space for $\mathbf{u}(x)$ may be used, and the problem becomes simple. However, the decoupling (11.16) may be expensive.

More popular in the literature hitherto has been the approach to not attempt an explicit decoupling, but to proceed with a method for stiff ODEs. Such a method necessarily considers $\mathbf{y}(x)$ to be piecewise continuous, hence approximates $\mathbf{u}(x)$ [say by $\mathbf{u}_\pi(x)$] in a "wrong" space, in the sense that $\mathbf{u}_\pi(x)$ is as smooth as the approximation to $\mathbf{v}(x)$. But if the method damps out the local error contributions arising from this excess continuity in $\mathbf{u}_\pi(x)$, then it will still work, because an implicit, approximate decoupling of algebraic and differential components results. Most popular for such IVPs is, not surprisingly, the BDF family of schemes (cf. Section 2.7). Using, for instance, the backward Euler scheme for

$$y' = \lambda y$$

with a step size h such that $|\lambda|h \to \infty$, we have the damping factor $|\frac{1}{1-h\lambda}| \to 0$ *regardless* of the sign of $\text{Re}(\lambda)$. Care should be taken in designing a local error control, though, because only controlling the error in $\mathbf{v}(x)$ makes sense.

For BVPs, the BDF schemes can be used to advantage as well in a single shooting—or multiple shooting setting—provided that there is no excess stiffness (in both directions) in the ODEs for $\mathbf{v}(x)$. The procedure is not entirely straightforward, because only consistent initial and boundary conditions play a part, strictly speaking, in the shooting matching. The additional details appear in the literature, and will not be discussed here.

The possibility that the ODE part in a DAE would have a BVP stiffness (thus preventing an efficient use of a typical IVP stiff solver) has received little attention in the literature. If a symmetric scheme is used to solve (11.14), then it should be able to implicitly decouple (cf. Section 10.3), so for instance the trapezoidal scheme is inappropriate, but the midpoint scheme still is appropriate. Yet, since the growth factor of any symmetric scheme has modulus 1 (and not <1) in the limit, any local error (e.g., BC inconsistency) will not be damped out. An explicit decoupling of the differential and algebraic components may then become the favored option.

Case III: $E(x)$ is singular, and the problem has a higher index; viz., $U^{11}(x)$ is singular.

This class of problems is generally ill-conditioned and not well understood theoretically. To illustrate the potential trouble, we consider a simple example.

Example 11.2

Consider the ODE of order ν

$$E\mathbf{y}' = \mathbf{y} - \mathbf{q}(x)$$

where

$$E = \begin{pmatrix} 0 & 1 & & & 1 \\ & & \cdot & & \\ & & & \cdot & \\ & & & & 1 \\ 0 & & & & 0 \end{pmatrix}$$

i.e., E is a $\nu \times \nu$ Jordan block for the eigenvalue 0. Componentwise, the system is

$$y_2' = y_1 - q_1(x)$$
$$y_3' = y_2 - q_2(x)$$
$$\vdots$$
$$y_\nu' = y_{\nu-1} - q_{\nu-1}(x)$$
$$0 = y_\nu - q_\nu(x)$$

Letting for simplicity $q_1(x) = \cdots = q_{\nu-1}(x) = 0$, $q_\nu(x) \equiv q(x)$, the solution is seen to be

$$y_\nu(x) = q(x), \; y_{\nu-1}(x) = q'(x), \; \ldots, \; y_1(x) = q^{(\nu-1)}(x)$$

and no BC are needed. Note that the problem is solvable only if $q(x)$ has $\nu - 1$ derivatives. The *index* of this DAE is ν.

□

This is a perturbing example, despite its simplicity. Recall our discussion from Section 5.6: If L is a scalar differential operator of order m with appropriate BC and $Lu = q$, then $u(x)$ generally has m more derivatives than $q(x)$. The well-conditioning of the BVP is expressed in terms of a corresponding reasonable bound on L^{-1}, and this gives a reasonable stability constant for a decent numerical discretization, and subsequently a reasonable bound on the truncation error. But in Example 11.2 we see essentially a "differential operator" of order $-\nu + 1$, which is negative when $\nu > 1$. Thus, $q(x)$ has $\nu - 1$ more derivatives than $y_1(x)$. So, just as the differential operator L itself is unbounded in Section 5.6 (in the sup norm), here the conditioning constant of the problem is unbounded! The problem is ill-posed, and a direct approach of simply replacing derivatives by difference quotients may yield an unstable process in general. (Sometimes, however, meaningful approximate solutions can nonetheless be salvaged.

Consider next the system (11.12) with $\dfrac{\partial \mathbf{f}}{\partial \mathbf{u}}$ singular. We can usually differentiate (11.12a) with respect to x, obtaining

$$\mathbf{0} = \frac{\partial \mathbf{f}}{\partial x} + \left(\frac{\partial \mathbf{f}}{\partial \mathbf{u}}\right)\mathbf{u}' + \left(\frac{\partial \mathbf{f}}{\partial \mathbf{v}}\right)\mathbf{v}'$$

and using (11.12b),

$$\left(\frac{\partial \mathbf{f}}{\partial \mathbf{u}}\right)\mathbf{u}' = -\frac{\partial \mathbf{f}}{\partial x} - \left(\frac{\partial \mathbf{f}}{\partial \mathbf{v}}\right)\mathbf{g}(x,\mathbf{u},\mathbf{v})$$

If this latter system has index 1, the original is said to have *global index* 2; otherwise, the differentiation can be repeated. The number $\nu - 1$ of differentiations needed to obtain a problem with index 1 determines a global index ν.

Practically, if the process just described to determine the global index can be explicitly carried out with no major pain, then generally *it should be carried out:* This amounts to replacing an ill-posed problem by a well-posed one with the same solution. The resulting index 1 problem can then be solved numerically as described before. But in some cases, explicit differentiation and substitution are not possible. In such an event one may utilize some regularization techniques to solve the ill-posed problem approximately, by perturbing it to form a stiff BVP.

11.3 EIGENVALUE PROBLEMS

In this section we discuss eigenvalue problems for BVPs. In the linear case, we are given a homogeneous ODE and homogeneous BC, one or both of which depends upon a parameter, and values of this parameter are desired such that the BVP has nontrivial solutions. Such parameter values and corresponding solutions are called *eigenvalues* and *eigenfunctions*, respectively. In Section 11.3.1 we describe an important special class of second-order problems called Sturm-Liouville problems. Then eigenvalue problems for first-order systems are considered. The discussion here is very brief, even though many aspects of eigenvalue problems have been extensively studied elsewhere. Several important topics, such as global methods for eigenvalue problems for higher-order ODEs, are essentially omitted altogether (but see the references at the end of the chapter).

11.3.1 Sturm-Liouville problems

The underlying problem in the study of many physical phenomena, such as the vibrations of strings, the interaction of atomic particles, or the Earth's free oscillations, yields a *Sturm-Liouville (SL) eigenvalue problem.* A general form for this self-adjoint problem is

$$-(pu')' + qu = \lambda r u \qquad a < x < b \qquad (11.17a)$$

$$c_1 u(a) + p(a)u'(a) = 0, \; c_2 u(b) - p(b)u'(b) = 0 \qquad (11.17b)$$

where $-\infty < a < b < \infty$, $c_1, c_2 > 0$, and where p', q, r are continuous functions with $p > 0$, $q > 0$, and $r \geq 0$ on $[a,b]$. (Various extensions are of course possible, e.g., in the so-called *singular* case, one allows (a,b) to be an infinite interval, p to be 0 at an endpoint, or q to become unbounded at an endpoint — cf. Section 11.4.) Solutions to (11.17) consisting of values of λ and corresponding nontrivial solutions $u(x)$ are the eigenvalues and eigenfunctions, respectively. Note that if u is an eigenfunction then so is αu, $\alpha \in \mathbf{R}$, so we do not have uniqueness. There is a well-developed theory for SL problems. It can be shown that there are an infinite number of eigenvalues $\{\lambda_k\}_{k=1}^{\infty}$ for (11.1), they are real, and they can be ordered such that $0 < \lambda_1 < \lambda_2 < \cdots < \lambda_k \to \infty$.

For each λ_k, the corresponding eigenfunction oscillates and has $k-1$ zeros in (a,b). Thus, for large k, an eigenfunction is highly oscillatory. The eigenfunctions can be chosen uniquely to within signs by requiring that

$$\int_a^b r u_k^2 \, dx = 1, \qquad \int_a^b r u_k u_j \, dx = 0 \qquad \text{for } j \neq k \tag{11.18}$$

Bounds on the size of λ_k and $\|u_k\|_2$ can be given.

One way to solve (11.17), much in the spirit of the way we treat unknown parameters in Section 11.1, is to convert the problem to a "standard" BVP. This can be done by adding

$$\lambda' = 0 \tag{11.19a}$$

$$w' = ru^2 \tag{11.19b}$$

$$w(a) = 0, \qquad w(1) = 1 \tag{11.19c}$$

Now, any of our standard methods can be used to solve the *nonlinear* BVP (11.17), (11.19) because its solutions, which are eigenvalue-eigenfunction pairs, are isolated.

This approach can be extremely convenient in the common case where only the first one or two eigenvalues and eigenfunctions are desired, and it is unclear how such an approach then compares with the methods designed specifically for eigenvalue problems. Its usefulness is more questionable in the cases where one seeks a number of eigenvalues or some large eigenvalues, and (possibly) the corresponding eigenfunctions.

Most of the basic methods for solving BVPs can be modified to solve (11.17) directly. Perhaps the simplest way is to use single shooting (cf. Section 4.1.2). For a fixed guess for λ, we solve the IVP

$$v_1' = v_2/p$$

$$\qquad \qquad \qquad \qquad \qquad \qquad x > a \tag{11.20a}$$

$$v_2' = qv_1 - \lambda r v_1$$

$$c_1 v_1(a) + v_2(a) = 0, \qquad d_1 v_1(a) - v_2(a) = 1 \tag{11.20b}$$

where $d_1 \neq c_1$, and denote the resulting solution by $\mathbf{v}(x;\lambda)$. Then the nonlinear equation

$$\phi(\lambda) := c_2 v_1(b;\lambda) - v_2(b;\lambda) = 0 \tag{11.21}$$

is satisfied iff λ and $v_1(x;\lambda)$ are an eigenvalue and corresponding eigenfunction for (11.17). The only real difference between the treatment in Sections 4.2.1 and 4.6.1, and here is that, instead of changing an initial condition and keeping the ODE fixed each step, we keep the initial conditions fixed and change the ODE (by adjusting λ). In addition to its standard potential disadvantages, the shooting problem now involves solving a nonlinear problem in λ.

Finite difference methods can also be applied directly to solve SL problems. Consider the problem

$$Lu := -(pu')' + qu = \lambda r u \qquad a < x < b \tag{11.22a}$$

$$u(a) = u(b) = 0 \tag{11.22b}$$

where $r > 0$. Following basically the approach of Section 5.1.1, we define a step size $h := \frac{b-a}{N}$, the uniform mesh $x_i = a + (i-1)h$, $i = 1, 2, \ldots, N+1$, and approximate $Lu(x_i) = \lambda r(x_i)u(x_i)$ by

$$L_\pi u(x_i) := -\frac{1}{h}\left[p(x_{i+1/2})\frac{u(x_{i+1})-u(x_i)}{h} - p(x_{i-1/2})\frac{u(x_i)-u(x_{i-1})}{h}\right] + q(x_i)u(x_i)$$

[cf. (5.6)]. We seek a value λ and mesh function $\mathbf{u}_\pi := (u_1, \ldots, u_{N+1})^T$ satisfying

$$L_\pi u_i = \lambda r(x_i)u_i \quad i = 2, \ldots, N \tag{11.23a}$$

$$u_1 = u_{N+1} = 0 \tag{11.23b}$$

or in matrix form

$$A\,\mathbf{u}_\pi = \lambda B\,\mathbf{u}_\pi \tag{11.24}$$

Here by assumption on p, q, and r, A is tridiagonal and strictly diagonally dominant, and B is diagonal and nonnegative definite. In general (11.24) is a so-called *generalized eigenvalue problem*, and it can be solved using standard linear algebra routines (e.g., if $B = I$ then we can use the QR algorithm — cf. Section 2.2.6). Note that this gives us $N-1$ eigenvalues and eigenvectors. An analogous analysis to that in Section 5.1.1 shows that the eigenvalues $\{\hat\lambda_\pi^k\}_{k=1}^N$ and corresponding appropriately normalized eigenvectors $\{\mathbf{u}_\pi^{(k)}\}_{k=1}^N$ for (11.24) converge to the eigenvalues and eigenfunctions for (11.22) and for k fixed the rate of convergence is $O(h^2)$. Various generalizations are possible. For example, if A and B are symmetric positive definite, then the error in the positive values $\{\hat\lambda_\pi^k\}_{k=1}^N$ is proportional to the truncation error associated with the particular finite difference approximation used.

There are other well-studied methods for calculating a set of eigenvalue approximations. The Rayleigh-Ritz method, similar in spirit to the Ritz method of Section 5.7.1, utilizes a variational characterization of the eigenvalues and, after selection of an approximation space for the eigenfunctions (e.g., piecewise polynomials) involves solving a resulting matrix eigenvalue problem just as in (11.24). Positive definiteness of A and B is again guaranteed under reasonable assumptions. A different approach is to approximate the coefficients p, q, and r by splines, and the eigenvalue problem for the resulting BVP is then solved exactly.

All of these methods can satisfactorily solve a large class of SL problems if only a few eigenvalues are sought, which is often the case. For example, for an organ pipe, the first few eigenvalues (related to the energy of the resonant states) are sufficient because only the first few harmonics are observed in practice. The corresponding eigenfunctions are slowly varying.

As previously mentioned, a large number of eigenvalues are sometimes sought. A famous example of this is the radial Schrödinger equation

$$-u'' + V(x)u = \lambda u$$

$$u(-\infty) = u(\infty) = 0$$

with $V(x)$ a known potential function, which arises in quantum mechanics in the study of atom-ion interaction. Here, over 20 (say) eigenvalues are sought, and the corresponding eigenfunctions become increasingly oscillatory. Even if the approximate eigenvalues converge at a rate like $O(h^p)$ in theory, the error constant grows rapidly with k. For the spline or finite difference methods, it seems clear that N must be fairly large before a rapidly oscillatory eigenfunction can be adequately approximated (cf. also Section 11.6). For such situations, it is often advantageous to first employ the Liouville transformation $t = \int_a^x (\frac{r}{p})^{1/2} dx$, $z = u/w$, $w := (pr)^{-1/4}$, which transforms (11.17) to the *Liouville normal form*

$$-z'' + sz = \lambda z \qquad 0 < t < T \qquad (11.25a)$$

$$c_1^* z(0) + z'(0) = 0, \ c_2^* z(T) - z'(T) = 0 \qquad (11.25b)$$

where

$$T = \int_a^b (\frac{r}{p})^{1/2}, \quad s = \frac{p}{r} - (\frac{w'}{w}) + (\frac{w'}{w})^2, \quad c_1^* = c_1 w^2(0) - (\frac{w'}{w})(0),$$

and $c_2^* = c_2 w^2(T) - (\frac{w'}{w})(T)$. This device has been used with considerable success, presumably because oscillations in the eigenfunctions for (11.25) are more regular than for (11.17).

One popular way to solve (11.17) is to use the *Prüfer transformation*. We suppose for convenience that the SL problem has been translated to $(-1,1)$, so that $a = -1$, $b = 1$. The ODE (11.17a) is transformed into two first-order ODEs by using a polar representation for the eigenfunction $u(x)$, viz.,

$$u(x) = \rho(x) \sin \theta(x), \qquad p(x) u'(x) = \rho(x) \cos \theta(x) \qquad (11.26)$$

where $\rho(x)$ and $\theta(x)$ are the amplitude and phase of $u(x)$. Substituting into (11.17), we obtain

$$\theta'(x) = \frac{1}{p(x)} \cos^2 \theta(x) + [-q(x) + \lambda r(x)] \sin^2 \theta(x) \qquad (11.27a)$$

$$\rho'(x) = [\frac{1}{p(x)} + q(x) - \lambda r(x)] \sin \theta(x) \cos \theta(x) \rho(x) \qquad (11.27b)$$

$$\theta(-1) = \beta_1 \qquad (11.27c)$$

$$\theta(1) = \beta_2 + (k-1)\pi \qquad (11.27d)$$

where $\beta_1 := \arctan(-\frac{1}{c_1})$, $0 \leq \beta_1 < \pi$ and $\beta_2 := \arctan(-c_2)$, $0 < \beta_2 \leq \pi$, and $k \geq 1$ corresponds to the solution pair $\lambda_k, u_k(x)$. Now single shooting can be applied by solving the nonlinear IVP (11.27a,c) and adjusting the parameter λ until (11.27d) is satisfied for $\lambda = \lambda_k$. Alternatively, we can add the trivial ODE $\lambda' = 0$ to (11.27a,c,d) and solve this standard BVP with any other stable method. If the eigenfunction is desired, then (11.27b) can be solved with an initial condition which plays the role of pinning down the arbitrary constant factor for $u_k(x)$. Certain modifications of the

Prüfer transformation (11.26) are sometimes used. This method seems to be successful for the typical SL problems arising in practice, as evidenced by the fact that there are two extensively tested implementations of this method.

11.3.2 Eigenvalues of first-order systems

We consider the general eigenvalue problems for a first-order system of order n of the form

$$\mathbf{y}' = [\lambda A(x) + E(x)]\,\mathbf{y}(x) \qquad a < x < b \qquad (11.28a)$$

$$B_{a1}\,\mathbf{y}(a) = \mathbf{0} \qquad (11.28b)$$

$$B_{b2}\,\mathbf{y}(b) = \mathbf{0} \qquad (11.28c)$$

where $B_{a1} \in \mathbf{R}^{p \times n}$ and $B_{b2} \in \mathbf{R}^{q \times n}$, $q = n - p$. Under appropriate assumptions on the $n \times n$ matrices $A(x)$, $E(x)$ and the BC, certain general properties of the eigenvalues and eigenfunctions can be shown to hold.

A similar trick to adding (11.19) to (11.17) can be used to construct a standard BVP for finding simple eigenvalues. By adding to (11.28) the equations

$$\lambda' = 0 \qquad (11.29a)$$

$$w' = \mathbf{y}^T \mathbf{y} \qquad (11.29b)$$

$$w(a) = 0,\ w(b) = 1 \qquad (11.29c)$$

we construct a nonlinear BVP to which our standard methods can be applied, provided that the sought eigenvalue has algebraic multiplicity 1.

In principle, there is no reason why a finite difference or spline collocation method as in Chapter 5 cannot be applied directly to (11.28) to give a matrix eigenvalue problem, although this has not been widely done. The methods of Chapter 4 can also be fairly directly adapted for solving (11.28).

Consider, for instance, the Riccati method first described in Section 4.5.2. We partition $\mathbf{y}$ and $B_{a1} = (C \mid D)$ as in that section and let $R(x;\lambda) \in \mathbf{R}^{q \times (n-q)}$ be the Riccati matrix satisfying the equivalent here of (11.4.74) with λ fixed. Assume for convenience that $R(x,\lambda)$ exists on $[a,b]$, i.e., that no reimbedding is necessary. Since the ODE is homogeneous, $\mathbf{y}^2(x;\lambda) = R(x;\lambda)\,\mathbf{y}^1(x;\lambda)$, and it is easy to see that the initial condition for $\mathbf{w}^1 = \mathbf{y}^1$ at b is

$$[C + D\ R(b;\lambda)]\,\mathbf{y}^1(b) = \mathbf{0}$$

Thus, a necessary and sufficient condition for a nontrivial solution to exist is that

$$\phi(\lambda) := \det[C + D\ R(b;\lambda)] = 0 \qquad (11.30)$$

and the problem of determining eigenvalues is again reduced to finding the zeros of a function. [The choice of the function ϕ is far from ideal, since all eigenvalues of $C + DR(b;\lambda)$ change as λ changes and so their product, being equal to $\phi(\lambda)$, may behave very erratically.] The method can be modified to incorporate reimbeddings [as in Section 10.4.2)], and we illustrate it with an example.

Example 11.3

The transverse vibrations of a beam which is clamped at both ends are governed by the BVP

$$u^{(iv)} - \lambda^4 u = 0 \qquad 0 < x < 1 \tag{11.31a}$$

$$u(0) = u'(0) = 0, \qquad u(1) = u'(1) = 0 \tag{11.31b}$$

The problem can be solved analytically, with the eigenvalue satisfying $\cosh \lambda \cos \lambda = 1$. We compute the solution for the Riccati method, using the secant method with initial guesses of 4 and 5 to solve (11.30). The IVPs are solved by using a fourth-order Runge-Kutta method with $TOL = 10^{-7}$ on a computer with roughly 16 decimal digits, and the stopping criterion is $|\phi(\lambda)| \leq 10^{-10}$. The result is that six secant iterations are required (with three reimbeddings for the last iteration) to compute the eigenvalue approximation 4.730083. This problem is also straightforward when using the trick of adding (11.29) and solving with the collocation code of Appendix B. An advantage of the latter is that the accuracy of the approximate eigenvalue is easier to ensure (and in fact $\lambda \approx 4.730041$).

□

One popular method to approximate many eigenvalues is the spectral method. It is generally used for a higher-order ODE [like (11.31)], and the solution $u(x)$ is approximated by a finite sum of so-called Chebyshev polynomials. Using the orthogonality properties of these polynomials, a matrix eigenvalue problem is constructed. The advantage of the spectral method [and indeed of the other global methods applied directly to (11.28)] is that a set of eigenvalue approximations are produced. On the other hand, while using (11.28), (11.29) or applying initial value methods directly to (11.28) can be used to refine eigenvalue approximations, they would generally be unsatisfactory if a large number of eigenvalues are sought and initial approximations are unavailable.

We conclude this section by mentioning the general nonlinear eigenvalue problem

$$\mathbf{y}' = \mathbf{f}(x, \mathbf{y}, \boldsymbol{\lambda}) \qquad a < x < b \tag{11.32a}$$

$$\mathbf{g}(\mathbf{y}(a), \mathbf{y}(b), \boldsymbol{\lambda}) = \mathbf{0} \tag{11.32b}$$

where $\mathbf{y} \in \mathbf{R}^n$, $\boldsymbol{\lambda} \in \mathbf{R}^m$, and $\mathbf{g} \in \mathbf{R}^{n+m}$. Values of the eigenvalue vector $\boldsymbol{\lambda}$ are to be determined such that (11.15) has a solution (possibly trivial in the nonlinear case). Now it is not necessarily true that an eigenfunction is only unique to within a multiplicative constant, but the methods for the linear eigenvalue problem can often be extended to this case in a natural way.

11.4 BVPS WITH SINGULARITIES; INFINITE INTERVALS

Up to now we have assumed "sufficient regularity" of the differential operator when developing theory for various numerical methods. This may not always be a reasonable assumption, though: For a number of practical problems singularities may arise in a natural way, say at a boundary (or an interface) point, where the differential operator becomes undefined. Thus consider the following ODE,

$$\mathbf{N}\mathbf{y}(x) := x^\alpha \mathbf{y}' - \mathbf{f}(x, \mathbf{y}) = \mathbf{0}, \qquad 0 < x < 1 \tag{11.33}$$

where **f** is continuous in its arguments and **y** is required to be in $C[0,1] \cap C^1(0,1]$. Cases of interest are when the scalar α is positive. If $\alpha = 1$ then **N** is said to have a *singularity of the first kind*, whereas if $\alpha > 1$ then **N** is said to have a *singularity of the second kind*.

Throughout the book we have seen examples where singularities of both types occur. Problems with a singularity of the first kind arise naturally when a PDE is formulated in polar (or cylindrical) coordinates and use of some symmetry allows reduction of the differential equation to an ODE; for instances, see Examples 1.5, 1.23, and 1.24. We briefly discuss such singularities in Section 11.4.1. A singularity of the second kind arises when a change of independent variable is defined, where an infinite interval is mapped onto the finite interval $(0,1]$. This is further discussed in Section 11.4.2.

11.4.1 Singularities of the first kind

We assume here that $\alpha = 1$ in (11.33). The assumptions on the regularity of a solution $\mathbf{y}(x)$ of (11.33) imply that $\lim \mathbf{y}(x)$ exists, as x decreases to 0. This is needed in order to make the BVP for (11.33) meaningful, and is reasonable in most applications. These assumptions further yield that

$$\mathbf{f}(0, \mathbf{y}(0)) = \mathbf{0} \tag{11.34}$$

which must be compatible with the prescribed BC. In fact, the requirement (11.34) is often used to *determine* part of the BC, as e. g., in Example 8.1.

To be somewhat more specific, let us consider now the linear BVP

$$x\mathbf{y}' = A(x)\mathbf{y} + \mathbf{q}(x), \qquad 0 < x < 1 \tag{11.35a}$$

$$B_0 \mathbf{y}(0) + B_1 \mathbf{y}(1) = \boldsymbol{\beta} \tag{11.35b}$$

where A and $\mathbf{q}$ are assumed analytic around $x = 0$. From (11.34) we have

$$A(0)\mathbf{y}(0) = -\mathbf{q}(0) \tag{11.36}$$

Given a Jordan form for $A(0)$, which we identify for notational simplicity with $A(0)$ itself, we may assume that it is ordered and partitioned as follows: The eigenvalues with positive real parts appear in one block exclusively, the nondefective zero eigenvalues appear next, followed by the defective zero eigenvalues and then the purely imaginary ones, and the eigenvalues with negative real parts appear in the last block. This gives the structure

$$A(0) = \begin{bmatrix} A(0)^+ & & & \\ & 0 & A(0)^0 & \\ & & A(0)^d & \\ & & & A(0)^- \end{bmatrix} \tag{11.37a}$$

with column-block sizes n_+, n_0, n_d, and n_- respectively, $n = n_+ + n_0 + n_d + n_-$. (The geometric multiplicity of the zero eigenvalue is n_0.) If we now partition $\mathbf{q}(x)$ correspondingly as

$$\mathbf{q}(0) = \begin{bmatrix} \mathbf{q}(0)^+ \\ \mathbf{q}(0)^0 \\ \mathbf{q}(0)^d \\ \mathbf{q}(0)^- \end{bmatrix} \qquad (11.37b)$$

then by (11.36) it must hold that $\mathbf{q}(0)^0 \in \text{range}\,(A(0)^0)$ and $\mathbf{q}(0)^d \in \text{range}\,(A(0)^d)$. It can also be shown that if S is the subspace of basis solutions of (11.35a) whose limit at $x = 0$ exists, then $\dim(S) = n_+ + n_0$.

Before turning to numerical methods, it is important to note that there is often a marked difference between the smoothness of the solution allowed by the restrictions above and the one observed in most practical applications. While in Section 3.1 p continuous derivatives of the coefficients of the ODE imply $p+1$ continuous derivatives of the solution, here smoothness of $\mathbf{f}(x, \mathbf{y})$ in (11.33) [or of $A(x)$ and $\mathbf{q}(x)$ in (11.35)] does not imply corresponding smoothness of $\mathbf{y}(x)$ near $x = 0$. For example, the IVP

$$xy' = \frac{1}{2}y, \qquad y(0) = 0$$

has the solution $y(x) = \sqrt{x}$, which has an unbounded first derivative at $x = 0$. However, examples like this are not common in practice, where often the solution $\mathbf{y}(x)$ is nonetheless smooth near the singularity. In particular, when the singular ODE arises as a result of exploitation of a polar or a cylindrical symmetry in a PDE, the point of singularity $x = 0$ corresponds to the origin of the original domain, where usually no particularly high solution activity is expected. Therefore it is legitimate and useful to contemplate the performance of numerical methods for problems with singularities of the first kind where the solution is smooth at the singularities.

In the latter situation, not much practical difficulty is expected with the finite difference (and collocation) methods of Chapter 5. This expectation is borne out by the ease with which Examples 5.1 and 8.1 are solved. Methods like those using the midpoint scheme, and more generally collocation at Gaussian points, can be directly applied as usual, without any modification. For other schemes like the trapezoidal scheme, which attempt to discretize the ODE at the boundary, some simple adjustment is needed, as described for Example 5.1. The basic convergence result (from stability + consistency) which claims an $O(h^k)$ error for collocation at k points per subinterval [cf. Theorem 5.75] holds here as well. An $O(h^{k+1})$ error can also be obtained in case the collocation points satisfy some minimal orthogonality requirement, as in the regular case. This covers the usual convergence of the midpoint and the trapezoidal schemes, but for higher-order superconvergence [cf. Theorem 5.79] the situation is more delicate, and stronger assumptions may be needed, or else weaker estimates than in the regular case are obtained. We will not go into more detail on this here.

The situation is much less straightforward for some of the initial value approaches of Chapter 4. This is because not all fundamental solution components of (11.35a) may be expected to be as smooth near the singularity as the solution $\mathbf{y}(x)$ is. For example, the IVP

$$y' = -\frac{2}{x}y + \frac{2}{x}$$

$$y(0) = 1$$

has the solution $y(x) \equiv 1$ and a fundamental solution $Y(x) = \dfrac{1}{x^2}$. Therefore a special treatment near $x = 0$ is often required before a code based on an initial value approach can be used. Such a special treatment may consist of an analytic power expansion of a fundamental solution in the vicinity of $x = 0$, followed by use of an initial value code when we are sufficiently far away from the singularity. Once a fundamental solution $Y(x), 0 < x < \delta$, has been found in this way, an appropriate particular solution can be found as well, and BC

$$B_\delta \mathbf{y}(\delta) + B_1 \mathbf{y}(1) = \tilde{\boldsymbol{\beta}} \qquad (11.38)$$

can be constructed to replace (11.35b). The location of the joint $\delta > 0$ has to be small enough so that the power series expansion for $Y(x)$ on $[0, \delta]$ can be easily and efficiently constructed, and at the same time large enough so that the BVP (11.35a), (11.38) on $[\delta, 1]$ can be solved by a standard initial value method without difficulty.

11.4.2 Infinite interval problems

In previous chapters we have frequently seen problems on (semi-) infinite intervals; e. g., see Examples 1.8, 1.18, and 1.19. Suppose that we have the ODE

$$\mathbf{y}' = \mathbf{f}(x, \mathbf{y}), \qquad a < x < \infty \qquad (11.39)$$

Then a transformation

$$t := a/x \qquad (11.40a)$$

reformulates (11.39) as an ODE defined on the interval $(0, 1]$, namely

$$t^2 \frac{d\mathbf{y}}{dt} = -a\, \mathbf{f}(a/t, \mathbf{y}) \qquad (11.40b)$$

in which we recognize an ODE with a singularity of the second kind. [In (11.40a) we have tacitly assumed that $a > 0$. If $a \leq 0$ then the transformation $t := 1/(x + 1 - a)$ may be used.] Since the formulation (11.39) is more natural, and since it turns out to be usually preferable for numerical discretization as well, we shall consider this form only.

Often in applications, part of the BC for (11.39) is given in the form

$$B_a \mathbf{y}(a) = \boldsymbol{\beta}_a \qquad (11.41)$$

with $\text{rank}(B_a) < n$. To complement these, one normally has to require some regular behaviour of the solution $\mathbf{y}(x)$ as $x \to \infty$, like $\lim_{x \to \infty} \mathbf{y}(x) = \mathbf{0}$ or, more generally, that $\mathbf{y}(x)$ have a bounded limit. Therefore, one is interested only in nonincreasing solutions with a limit. Once such appropriate complementary BC at ∞ have been chosen — given that such a choice is possible, and in particular that B_a controls the *stable manifold* — the resulting BVP should be well-conditioned. In other words, the efforts to compute solutions of (11.39), (11.41) are in finding ways to approximate this stable manifold, and part of this is done by prescribing BC which put the solution on that manifold.

We now consider the question of finding appropriate terminal BC for the linear ODE

$$\mathbf{y}' = A(x)\mathbf{y} + \mathbf{q}(x), \qquad a < x < \infty \qquad (11.42)$$

If we assume for the moment that $\lim_{x \to \infty} A(x)$ exists then we can consider the transformed problem for $\mathbf{w}(x) := T^{-1}\mathbf{y}(x)$, where T specifies the similarity transformation which brings $A(\infty)$ into a Jordan canonical form. As before, we identify $\mathbf{y}$ with $\mathbf{w}$ to simplify notation, and assuming for now that A is constant, have the form (11.37) for it. Then the fundamental solution $Y(x)$ satisfying $Y(a) = I$ can be partitioned accordingly, say

$$Y = [\, Y^+ \mid Y^0 \mid Y^d \mid Y^- \,] \qquad (11.43)$$

for each x, where $\|Y^+(x)\| \to \infty$ as $x \to \infty$, $\|Y^d(x)\|$ has no limit or grows polynomially, $\|Y^-(x)\| \to 0$, and only $Y^0(x)$ may approach a (nonzero) constant. We conclude that the stable manifold of the homogeneous part of (11.42) whose elements approach a limit as $x \to \infty$ is characterized by $Y^0 \oplus Y^-$. Hence, natural terminal conditions are specified by

$$B_{\infty 1} \mathbf{y}(\infty) = \mathbf{0} \qquad (11.44a)$$

with a boundary matrix

$$B_{\infty 1} = [\, \times \mid 0 \mid \times \mid 0 \,] \qquad (11.44b)$$

having $n_+ + n_d$ rows, with the partition structure similar to that in (11.43) and (11.37). There may be additional terminal conditions

$$B_{\infty 2} \mathbf{y}(\infty) = \boldsymbol{\beta}_{\infty 2} \qquad (11.44c)$$

with a boundary matrix

$$B_{\infty 2} = [\, 0 \mid \times \mid 0 \mid 0 \,] \qquad (11.44d)$$

having at most n_0 rows.

The form (11.44a, b) effectively requires that the *terminal conditions annihilate the unstable and the "limitless" mode directions*. The total number of terminal conditions in (11.44) is complementary to the maximal number of initial conditions which are required by B_a, that is $n_- + n_0$. Conditions in which Y^0 is involved may appear at both interval ends. This BC construction which has been considered for the constant coefficient case may be extended to the variable coefficient case, with $\lim_{x \to \infty} A(x)$ replacing $A(0)$ in (11.37). By expansion arguments it can be shown that the stable manifold is still characterized by a partitioned solution like (11.43), and therefore BC can be found in a similar way.

Of course, when it comes to numerical discretization the infinite interval $[a, \infty)$ has to be replaced by a finite one, say $[a, b]$ where b is "large". An instance where terminal BC are formulated in this way is Example 8.1, which contains singularities of both the first and the second kinds, and which is easily solved numerically after the preliminary analysis is performed. Another example is the following.

Example 11.4

For the solitary wave problem of Example 1.13 we have in (1.28a, b, e) a fourth-order ODE system, and in (1.28c, d, f) 5 linearly dependent BC. This is a 3-point BVP (cf. Section 11.9), where the BC (1.28f) is introduced to specify a unique solution. We now wish to analyze the remaining BC in order to extract an appropriate linearly independent subset, so that the resulting BVP becomes well-conditioned. Introducing the variable $z(x) \equiv v'(x)$, we write (1.28a, b) as

$$\mathbf{y}' \equiv \begin{bmatrix} v \\ z \\ r \end{bmatrix}' = \begin{bmatrix} 0 & 1 & 0 \\ V_R^2-1 & c & 1 \\ \phi/c & 0 & -\phi b/c \end{bmatrix} \begin{bmatrix} v \\ z \\ r \end{bmatrix} + \begin{bmatrix} 0 \\ 1/3 v^3 + V_R v^2 \\ 0 \end{bmatrix} =: A\,\mathbf{y} + \mathbf{g}(x,\mathbf{y})$$

Note that this problem does not quite fit our description in a number of ways: Not only are the BC prescribed at three points (and not just two), also the interval is infinite (and not just semi-infinite), and the problem is nonlinear. However, it is typical for nontrivial problems to be somehow non-standard, and we may still apply the principles described above to advantage, as follows. Firstly, note that by (1.28c) we can neglect $\mathbf{g}(x,\mathbf{y})$ for $|x|$ large. This leaves us with the ODE

$$\mathbf{y}' = A\,\mathbf{y}$$

but A still depends on the unknown constant c. There turns out to exist a transformation $T = T(c)$ consisting of the eigenvectors of A such that for each $c > 0$ in a reasonable range,

$$T^{-1}AT = \Lambda = \mathrm{diag}\,(\lambda_1, \lambda_2, \lambda_3)$$

This transformation can be found either analytically as a function of c, or numerically for a given c. One obtains that

$$\mathrm{Re}\,(\lambda_1) < 0, \ \mathrm{Re}\,(\lambda_2) < 0, \ \mathrm{Re}\,(\lambda_3) > 0$$

Defining the transformed variables

$$\mathbf{w}(x) := T^{-1}\mathbf{y}(x)$$

it is then clear that for boundedness at $-\infty$ we must require

$$w_1(-\infty) = 0, \qquad w_2(-\infty) = 0$$

and for boundedness at ∞ we must require

$$w_3(\infty) = 0$$

Expressing the w_i as linear combinations of v, v', and r, we obtain so-called *radiative BC* for the original unknowns. Before attempting approximation, we must replace the infinite interval by a finite one, say $[-L, L]$. The radiative BC may then be imposed at the interval ends, provided that we know (a reasonable approximation of) c. But in fact there is no need to insist on the radiative BC. Since we now know that the signs of the real parts of the eigenvalues λ_i are as above, we can simply prescribe the following subset of (1.28c, d)

$$v(-L) = r(-L) = v(L) = 0$$

The BVP consisting of (1.28a, b, c, e, f) can subsequently be shown to be well-conditioned. Once we solve this BVP we obtain c and can then derive radiative BC more precisely. [For this purpose the solution of (1.28a, b, c, e, f) does not have to be very accurate; in particular L does not need to be very large.] These BVPs can be numerically solved satisfactorily using the code described in Appendix B, obtaining $c \approx 0.8117656$. They are not easy, though.

□

In many other applications, the explicit determination of the transformation T which brings $A(\infty)$ into Jordan form is not really necessary. We only need that the BC specified at b would yield a well-conditioned BVP for any b large enough. The determination of b can be done by a simple comparison of solutions for different values of b until they do not differ by much. This primitive approach is surprisingly effective in many applications, but a more sophisticated one is described in the next paragraph. Once an appropriate b and appropriate BC have been determined, the numerical approximation of the resulting BVP can proceed by using a symmetric difference scheme, as discussed in Sections 10.2.4 and 10.3.2. The algorithm for *a priori* mesh selection in Section 10.3.2 can be used here as well (for a stretched variable). It often allows for gross overestimation of the value of b needed for a particular accuracy, with little penalty in computational cost.

Another approach is to employ a marching procedure jointly with decoupling. Suppose we employ the stabilized march of Section 4.4.3 or stable compactification of Section 4.4.1 up to an initially guessed terminal point $\hat{b}$. Due to the decoupling of the incremental matrix we can estimate the growth of the unstable modes. In particular, let Γ_i be the upper triangular matrix as in (4.51a). Having arrived at $\hat{b} = x_N$ safely and cheaply (a somewhat nontrivial assumption in itself), we can obtain a reasonable idea of the growth of the various modes by computing the diagonal of $\Gamma_N \cdots \Gamma_1$. If the dimension of the increasing mode space is denoted by n_+ and Γ_i^{11} denotes the upper left $n_+ \times n_+$ block of Γ_i, then $\text{glb}(\Gamma_N^{11} \cdots \Gamma_1^{11})$ is fairly well estimated by the smallest element on the diagonal of $\Gamma_N^{11} \cdots \Gamma_1^{11}$ – call it τ. Assuming an exponential character for these growing modes we see that they roughly have grown by (at least) a factor $e^{\lambda \hat{b}}$, where

$$\lambda := \frac{\ln \tau}{\hat{b}} \tag{11.45a}$$

Now suppose we want to compute the solution $\mathbf{y}(x)$ on the interval $[a, b]$. Then we determine a point c such that the increasing modes grow at least by a factor TOL^{-1} (TOL being the required tolerance) when going from b to c. Identifying $\hat{b}$ with b, we find the point c from (11.45a) as

$$c = b - \frac{\ln TOL}{\lambda} \tag{11.45b}$$

We now continue the marching procedure until we arrive at $x = c =: x_M$, for some $M > N$. If $A(\infty)$ has no defective zero eigenvalues and no purely imaginary ones (i.e., $n_d = 0$) then we may impose a straightforward (because of the decoupling) terminal condition. Using the notation of (4.56) we then require

$$\mathbf{s}_M^1 = \mathbf{0}$$

[cf. (11.44), truncated at $x = c < \infty$]. Note that this means that we do not exclude increasing components from our desired solution; in fact, they may have a magnitude of $O(1)$ at c. However, backward recursion via (4.57a) ensures that these components have damped out by a factor $\approx TOL$ by the time we arrive at $x = b$.

This marching algorithm is fairly generally applicable; in particular, it does not need the matrix $A(x)$ to be asymptotically constant. It would also work if $A(\infty)$ has defective zero eigenvalues, but the formulas (11.45) have to be adjusted to accommodate polynomially growing modes, and c may then have to be much larger than b, with a corresponding reduced efficiency. The treatment of purely imaginary eigenvalues in $A(\infty)$ is more delicate (cf. Example 1.8), and will not be discussed further here.

11.5 PATH FOLLOWING, SINGULAR POINTS, AND BIFURCATION

In Section 8.3.1 we discuss the method of continuation as a possible means for finding solutions of a difficult nonlinear problem

$$\mathbf{f(s)} = \mathbf{0} \tag{11.46}$$

The basic idea there is to imbed this problem in a family of problems depending on a *continuation parameter* λ,

$$\mathbf{h}(\mathbf{s}, \lambda) = \mathbf{0}, \qquad \lambda_\alpha \leq \lambda \leq \lambda_\beta \tag{11.47}$$

such that the solution of $\mathbf{h}(\mathbf{s}, \lambda_\alpha) = \mathbf{0}$ is known (or is easy to obtain), and the original problem (11.46) coincides with $\mathbf{h}(\mathbf{s}, \lambda_\beta) = \mathbf{0}$. Then we may attempt to find a solution to (11.46) by *following the path* $\bar{\mathbf{s}}(\lambda)$ along which

$$\mathbf{h}(\bar{\mathbf{s}}(\lambda), \lambda) \equiv \mathbf{0} \tag{11.48}$$

from $\lambda = \lambda_\alpha$ to $\lambda = \lambda_\beta$.

But an investigation of families of nonlinear equations like (11.47) (with perhaps even a number of parameters) may be desired, or required, for other reasons as well. Before giving instances, we note that many applications yield BVPs with parameters, which can be naturally posed (after discretization) as (11.47).

Let us denote

$$H(\mathbf{s}, \lambda) := \frac{\partial \mathbf{h}}{\partial \mathbf{s}}(\mathbf{s}, \lambda) \tag{11.49}$$

Recall from Section 8.3.1 that occasionally it is required to reproduce (an approximation of) the path $\bar{\mathbf{s}}(\lambda)$, or some value $u(\lambda)$ related to it, for some range $\lambda_\alpha \leq \lambda \leq \lambda_\beta$, where $H(\bar{\mathbf{s}}(\lambda), \lambda)$ is nonsingular along the path, and the problem coefficients are smooth. Thus, unlike the case in Section 8.3, where we try to reach from λ_α to λ_β with essentially as few λ values in between as we can get away with, here we need to solve (11.47) for a sequence of values $\lambda_\alpha = \lambda_1 < \lambda_2 < \cdots < \lambda_{J+1} = \lambda_\beta$ placed sufficiently densely that, e.g., an interpolation of the resulting approximations to $\bar{\mathbf{s}}(\lambda_1), \bar{\mathbf{s}}(\lambda_2), \ldots, \bar{\mathbf{s}}(\lambda_{J+1})$ [or $u(\lambda_1), \ldots, u(\lambda_{J+1})$] yields an acceptable approximation to the path $\bar{\mathbf{s}}(\lambda)$ ($u(\lambda)$). An instance of this occurs in Example 1.15, where the semiconductor device equations may be solved in order to find the performance curve of the current versus voltage.

Another major reason for following the homotopy path of an equation like (11.47) is to find *multiple solutions* of (11.46). The techniques of Section 8.1 (like Newton's method alone) are useful for finding *a* solution, but not for switching intelligently from one solution to another (when there are more than one). It may occur that several initial guesses for (a damped) Newton's method yield convergence to the same solution, and initial guesses which would yield convergence to another are hard to figure out. Indeed, frequently it is not clear in practice whether or not there *is* another solution in addition to one already found. It often turns out that by following the homotopy path of an imbedding (11.47), naturally suggested by the application, it is possible to switch solution branches of (11.47), and thereby obtain multiple solutions to (11.46).

Example 11.5

A classical example of continuation is given by the BVP

$$u'' + \lambda e^u = 0 \qquad (\lambda > 0) \tag{11.50a}$$

$$u(0) = u(1) = 0 \tag{11.50b}$$

(see Example 3.2, where we give the solutions $u(x;\lambda)$ of (11.50) explicitly). Graphing $\|u(\cdot;\lambda)\| = u(1/2;\lambda)$ as a function of λ, we obtain the curve in Fig. 11.1. Thus we see not only that, as observed in Example 3.2, there are two solutions to this BVP for each λ, $0 < \lambda < \lambda_c$, but also that these solutions (or rather, solution branches) are connected by one homotopy curve. Continuing with λ along the lower branch [e.g., starting with the known solution $u(\cdot;0) \equiv 0$], we might hope to switch to the other solution branch near $\lambda = \lambda_c$. However, since the number of solutions to the BVP changes in the passage in λ through λ_c, it follows (from the implicit function theorem) that the BVP (11.50) must be singular (not well-posed) at $\lambda = \lambda_c$ (and thus ill-conditioned in a neighborhood of $\lambda = \lambda_c$). The point $\lambda = \lambda_c$ is a *singularity point* along the homotopy path: In terms of the notation (11.47–49) we have that $H(\bar{\mathbf{s}}(\lambda_c); \lambda_c)$ is singular.

This type of singularity point (sometimes called a *turning point*, which is a *normal limit point*) does not present much difficulty, once we realize that we cannot just continue past it with λ. A general treatment will be considered later, but here the following simple procedure can be applied: Observe that the curve in Fig. 11.1 is single-valued in the "variable" $\|u\|$. Hence we can augment the original BVP (11.49) to read (cf. Section 11.1)

$$u'' + \lambda e^u = 0 \tag{11.51a}$$

$$\lambda' = 0 \tag{11.51b}$$

$$w' = u^2 \tag{11.51c}$$

$$u(0) = u(1) = 0, \qquad w(0) = 0, \qquad w(1) = \gamma \, (= \|u\|^2) \tag{11.51d}$$

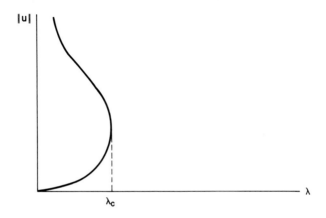

Figure 11.1 A turning point for Example 11.5

and do continuation in γ. The augmented BVP (which is in standard form) is well-conditioned in the passage through λ_c.

□

Turning points are not the only kind of singularity which may occur on a homotopy curve. Bifurcations may occur as well, as demonstrated in the following example.

Example 11.6

Consider a thin rod of length 1 which is fixed at the left end and is free to move at the right end in the axial direction. If we compress the rod it will initially remain undeformed (i.e., straight), but eventually it will buckle (or break, but we assume such a case does not happen). This can be modelled by the BVP

$$\phi'' + \lambda \sin \phi = 0, \quad 0 < x < 1 \quad (11.52a)$$

$$\phi'(0) = \phi'(1) = 0 \quad (11.52b)$$

where ϕ is the angle between the tangent to the buckled rod and its undeformed direction, x is the arclength measured from the left, and λ is a parameter.

Equation (11.52a) is a well-known ODE describing the motion of a simple pendulum. If the angle ϕ remains small then it is meaningful to linearize (11.52a) by replacing $\sin \phi$ by ϕ. This yields the linear eigenvalue problem

$$\phi'' + \lambda \phi = 0 \quad (11.53)$$

subject to the homogeneous BC (11.52b).

The BVP (11.53), (11.52b) has eigenvalues

$$\lambda = k^2 \pi^2, \quad k = 0, 1, 2, \ldots$$

with corresponding eigenfunctions

$$\phi(x; \lambda) = c \, \cos \sqrt{\lambda} x, \quad c \in \mathbf{R}$$

The first eigenvalue corresponds to the unbuckled state. For $k \geq 1$ we may obtain buckling if $c \neq 0$, but the unbuckled state is also possible. Apparently we have a multiple solution at such points, unlike the previous example.

From the viewpoint of continuation we may consider continuing with λ along the solution branch $u(\cdot; \lambda) \equiv 0$. At each eigenvalue $\lambda = k^2 \pi^2$, we have a singular point and another, nontrivial solution branch emanates, as depicted in Fig. 11.2 (solid lines). Using

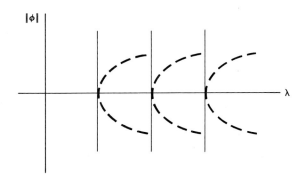

Figure 11.2 Bifurcation diagram for Example 11.6

the notation (11.47–49) [i.e., after discretizing the BVP (11.53), (11.52b)] we have the usual linear eigenvalue problem

$$\mathbf{h}(\mathbf{s}, \lambda) = \mathbf{A}\mathbf{s} - \lambda \mathbf{s}$$

$$H(\mathbf{s}, \lambda) = \mathbf{A} - \lambda I$$

Observe that the "linear eigenvalue problem" is linear in **s** but not in λ, hence not in the variable vector

$$\mathbf{z} := \begin{bmatrix} \mathbf{s} \\ \lambda \end{bmatrix}$$

and that the matrix $H(\mathbf{s}, \lambda)$ is singular precisely at the eigenvalues.

Returning to the buckling problem, it is not always reasonable to assume that ϕ remains small, so we must consider the nonlinear problem (11.52) as such. It turns out that this BVP can be solved in terms of elliptic functions, but this is not of interest here. We just note that since the eigenfunctions of the linear eigenvalue problem depart from the trivial solution at small ϕ values, the eigenvalues "generate" new branches of solutions for the BVP (11.52) as well. Since $-\pi < \phi < \pi$ we obtain the dotted curves in Fig. 11.2. The points $\lambda = k^2\pi^2$ are called *bifurcation points*. Such a point on a solution branch of (11.47) is defined by the property that every neighborhood of it contains a solution **z** not lying on this branch. Obviously, a bifurcation point must be a singular point. □

Continuation, path following, singular points, and bifurcation form a subject which at the time of this writing is receiving a tremendous amount of attention; in fact, it is the subject of many specialized books. Necessarily our presentation here is brief and sketchy, even more so than in other sections of this chapter. In the next subsection we give some analytical preliminaries, and this is followed by a discussion of numerical techniques in Section 11.5.2.

11.5.1 Branching and stability

As in Chapter 8 we restrict most of the discussion to finite dimensional problems,

$$\mathbf{h} : \mathbf{R}^{n+1} \to \mathbf{R}^n$$

At a singular point $\mathbf{z}_0 = (\mathbf{s}_0, \lambda_0)$ along a homotopy path we have rank $(H(\mathbf{z}_0)) < n$, and we further assume that

$$\text{rank}(H(\mathbf{z}_0)) = n - 1 \qquad (11.54)$$

This assumption excludes, for instance, multiple eigenvalues for a linear eigenvalue problem, but leaves in simple bifurcation points and turning points (as in Examples 11.5 and 11.6). The basic approach is to reimbed (11.46) [or (11.47)] in a problem having a nonsingular $(n+1) \times (n+1)$ Jacobian matrix at $\mathbf{z}_0$, and this is possible to do because of (11.54). The following lemma is easy to show:

Lemma 11.55 Let $A \in \mathbf{R}^{n \times n}$, $\mathbf{c}, \mathbf{b} \in \mathbf{R}^n$, $d \in \mathbf{R}$, and

$$\bar{A} = \begin{bmatrix} A & \mathbf{b} \\ \mathbf{c}^T & d \end{bmatrix}$$

(i) If A is nonsingular then $\bar{A}$ is nonsingular iff
$$d - \mathbf{c}^T A^{-1}\mathbf{b} \neq 0$$

(ii) If rank $(A) = n-1$ then $\bar{A}$ is nonsingular iff
$$\mathbf{b} \notin \text{range}(A), \qquad \mathbf{c} \notin \text{range}(A^T)$$

□

Consider the case of a turning point. Here $\mathbf{h}_\lambda(\mathbf{z}_0) \notin \text{range}(H(\mathbf{z}_0))$ and so rank $(H \mid \mathbf{h}_\lambda) = n$. We can augment (11.47) by one more scalar relation, and if the gradient of this additional relation with respect to $\mathbf{s}$ is not in range $(H^T(\mathbf{z}_0))$ then by Lemma 11.55 the singularity is removed. The BVP (11.51) corresponds to using this idea in Example 11.5. In more general situations an imbedding like (11.51) may not work; however, parametrization in the *arclength* of the homotopy path (i.e., using arclength continuation) will. We consider this continuation further in Section 11.5.2.

Geometrically, it is apparent that arclength continuation removes the singularity in the case of a turning point but does not remove it in the case of a bifurcation point. There are a variety of bifurcation types, and the one demonstrated in Example 11.6, called *primary bifurcation,* is the simplest. Here a primary solution branch (the trivial solution) is known at every point λ. More generally we may have *secondary bifurcation* points, where none of the solution branches is known, and their computation near the bifurcation point is therefore more difficult.

When the mathematical problem under consideration arises from an application, an obvious question is which solution branches are physical and which are not. For the buckling rod of Example 11.6, it seems reasonable to expect the nontrivial solution branch originating at $\lambda = \pi^2$ to be the physical one. A mathematical way to investigate this question is to perform a stability analysis on the ODE

$$\frac{d\mathbf{s}}{dt} = \mathbf{h}(\mathbf{s}, \lambda) \qquad (11.56)$$

viewing a point $(\bar{\mathbf{s}}(\lambda), \lambda)$ on a branch as an equilibrium solution of (11.56). Such an equilibrium solution can then be examined using standard stability techniques (cf. Section 3.3.4). This is particularly natural if (11.46) or (11.47) arises from (a discretization of) an elliptic differential equation, whereby (11.56) is a parabolic equation reflecting the dissipation in the system. We touch upon this briefly in Section 8.3.2 and again in Section 11.8.

Example 11.7

We consider several simple examples:

$$h(s, \lambda) = \lambda - s^2 \qquad (11.57a)$$

$$h(s, \lambda) = \lambda s - s^2 \qquad (11.57b)$$

$$h(s, \lambda) = \lambda s - s^3 \qquad (11.57c)$$

$$\mathbf{h}(\mathbf{s}, \lambda) = \begin{bmatrix} -s_2 + s_1(\lambda - (s_1^2 + s_2^2)) \\ s_1 + s_2(\lambda - (s_1^2 + s_2^2)) \end{bmatrix} \qquad (11.57d)$$

In (11.57a) we have a turning point at $s = \lambda = 0$, in (11.57b) we have a *transition* bifurcation point at the same place, and in (11.57c) we have there a *pitchfork* bifurcation point. The bifurcation diagrams for these are displayed in Fig. 11.3 (a–c). Performing stability

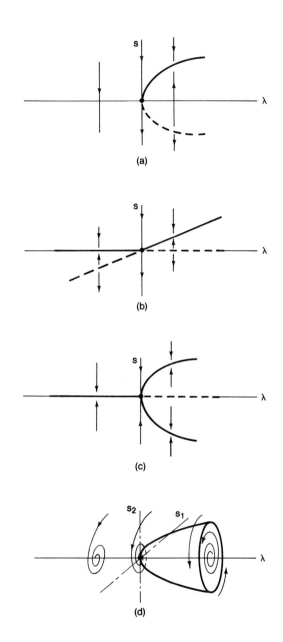

Figure 11.3 Bifurcation diagrams and stability analysis

analysis for these three simple examples yields the picture sketched using arrows in Fig. 11.3. If the arrows point in the direction of a branch then this is a stable branch, and if they point away then it is unstable.

The case for (11.57d) is more complicated. It turns out that appending the $\frac{d\mathbf{s}}{dt}$ term in (11.56) gives periodic solutions on the paraboloid as sketched in Fig. 11.3d. For each positive λ we obtain a stable limit cycle. A point like $(0,0,0)$ in Fig. 11.3d from which purely imaginary eigenvalues emanate (of course pairwise) is called a *Hopf bifurcation point*. Hopf bifurcations are in a class by themselves, and we shall not deal with them further, except to mention that in their stability analysis one has to solve a BVP with periodic BC.

□

11.5.2 Numerical techniques

In this section we briefly consider the following numerical issues raised by the discussion so far:

   (i) Simple path following

   (ii) Continuation past a turning point

   (iii) Determination of singular points

   (iv) Branch switching at a bifurcation point

(i) Simple path following. The problem is to construct an approximation to a homotopy path $\bar{s}(\lambda)$ [or some related path $u(\lambda)$] along which (11.48) holds, where for now we assume no singular points. The numerical issue arising in this case is not that of solving tough nonlinear equations (11.46), but the efficient calculation of $\bar{s}(\lambda)$ ($u(\lambda)$). We are solving a sequence of closely related problems (11.47), for $\lambda = \lambda_1, \lambda_2, \ldots, \lambda_{J+1}$, and so should capitalize on the closeness of these problems to one another. Thus, assuming that our basic iterative method is Newton's method, and using the classical continuation method (8.41a), we may presume that only one Newton iteration per λ_j suffices. Choosing $\lambda_{j+1} - \lambda_j \leq \Delta\lambda$ small enough, the quadratic convergence of Newton's method gives an $O(\Delta\lambda^2)$ overall method. (Note that the closely related Euler scheme applied to the Davidenko equation is only first-order: Here we do not have the error accumulation typical of ODE integration.)

Indeed, the situation is similar to that considered in Section 8.2, and other considerations given there apply here as well. In particular, an analogue to the modified Newton method (8.34) suggests itself; namely, we evaluate and decompose $H(\mathbf{s}, \lambda)$ only once every few continuation steps, using the fixed, decomposed Jacobian from a previous λ for a few successive λ values. This gives a certain type of *chord* iteration in between Jacobian evaluations.

(ii) Continuation past a turning point. In Example 11.5 we have been able to get rid of the turning point singularity by switching the continuation parameter from λ to $\|u\|$. But more generally, it is not difficult to imagine curves in $\mathbf{R}^2$ which are not uniquely parametrizable in any of the obvious variables globally.

Example 11.8

Consider the following thin spherical shallow shell equations,

$$u'' = -144v + uv - \frac{3}{x}u' - 2\lambda \tag{11.58a}$$

$$v'' = 144u - \frac{1}{2}u^2 - \frac{3}{x}v' \tag{11.58b}$$

$$u'(0) = v'(0) = u(1) = v'(1) + \frac{2}{3}v(1) = 0 \tag{11.58c}$$

The load parameter λ has a critical value λ_c beyond which there is no solution. Following (11.51) in Example 11.5, we add to the above equations

$$\lambda' = 0 \tag{11.58d}$$

$$w' = u^2 \tag{11.58e}$$

$$w(0) = 0 \tag{11.58f}$$

$$w(1) = \gamma (= \|u\|^2) \tag{11.58g}$$

and continue in γ. However, unlike Example 11.5, the augmented BVP eventually becomes singular again, having a turning point in $\|u\|$. This is apparent in Fig. 11.4, which depicts the first two solution branches for this problem. (There is no bifurcation in this example; i.e., the solution branches do not cross each other.)

A brute force solution to this new difficulty is to prescribe

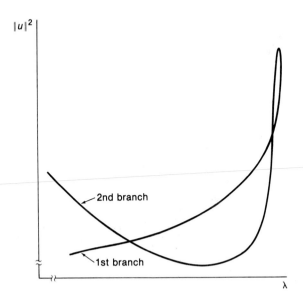

Figure 11.4 Two solution branches for Example 11.8 (no bifurcation)

$$\lambda(1) = \hat{\gamma} \tag{11.58h}$$

and alternate between continuation with (11.58h) and with (11.58g) as the need arises. This can hardly be described as elegant, but it works and has the advantage that the usual BVP software can be utilized (i.e., these BVPs are in standard form). A more elegant approach is described below.

□

The failure to follow a general curve in $\mathbf{R}^2$ in either one of the obvious variables suggests the idea of parametrizing with a new variable "on the curve," i.e., to use arclength continuation, as mentioned earlier. Since the actual arclength is not easily manageable, *pseudo-arclength* is used. For instance, we require in addition to (11.47) satisfaction of

$$\theta_1 \dot{\mathbf{s}}_0^T (\mathbf{s} - \mathbf{s}_0) + \theta_2 \dot{\lambda}_0 (\lambda - \lambda_0) + \sigma - \sigma_0 = 0, \quad (\theta_1 + \theta_2 = 1) \tag{11.59}$$

where $\mathbf{z}_0 = (\mathbf{s}_0, \lambda_0)$ is a known point on the homotopy path [with $H(\mathbf{z}_0)$ either singular or not] from which the next continuation step is taken, $\theta_{1,2}$ are positive weights and σ is the arclength. The derivative (˙) is with respect to the arclength σ. We have

$$H(\mathbf{z}_0)\dot{\mathbf{s}}_0 + \mathbf{h}_\lambda(\mathbf{z}_0)\dot{\lambda}_0 = \mathbf{0} \tag{11.60a}$$

$$\|\dot{\mathbf{z}}_0\|^2 > 0 \tag{11.60b}$$

Since $\mathbf{h}_\lambda \notin \text{range}(H)$ at a turning point, (11.60) and Lemma 11.55 imply that the $(n+1) \times (n+1)$ Jacobian matrix

$$\bar{A} = \begin{pmatrix} H & \mathbf{h}_\lambda \\ \theta_1 \dot{\mathbf{s}}_0^T & \theta_2 \dot{\lambda}_0 \end{pmatrix}$$

of the new system is nonsingular at $\mathbf{z}_0$ and by continuity also in a neighborhood around it. Hence we can base a general continuation scheme on this pseudo-arclength continuation.

It is popular in practice to first predict the new point on the path, given $\mathbf{z}_0$ and $\dot{\mathbf{z}}_0$, by

$$\hat{\mathbf{z}}_1 = \mathbf{z}_0 + (\sigma_1 - \sigma_0)\dot{\mathbf{z}}_0, \tag{11.61}$$

and then apply Newton's iteration to (11.47), (11.59) as a corrector.

(iii) Determination of singular points. While in many applications the main concern with singular points is how to proceed past them when they are encountered, or how to avoid them, it is sometimes desired to find their location more precisely. For instance, we may want to know the actual critical load values for Examples 11.8 or 11.6. Consider the case for Example 11.5. We know the value of λ_c, derived by special considerations in Example 3.2. But it would have been difficult to calculate that value within an accuracy of, say, 5 digits, by simple continuation in λ using trial and error as the BVP gets singular.

A simple general approach suggests itself. Since the BVP is singular at the desired critical point, its variational problem (see Section 3.1) must have a nontrivial solution there. Therefore, we may augment the original BVP with its variational problem, adding conditions like (11.51c,d) or (11.58e,f,g) to insure a nontrivial solution for the variational problem. The obtained system has the advantages of being in standard form and of directly yielding the desired result when it works. For Example 11.8 it does work, and yields the value

$$\lambda_c = 19{,}969.63$$

(try to obtain this value by a simple-minded search!).

But for other problems, such as in Example 11.6, the above approach produces an augmented problem which is still singular at the original singular point. A more general approach for (11.47) is to use the augmented system

$$\mathbf{h}(\mathbf{z}) - \alpha \mathbf{y} = \mathbf{0} \qquad (11.62\text{a})$$

$$H^T(\mathbf{z})\mathbf{y} = \mathbf{0} \qquad (11.62\text{b})$$

$$\mathbf{h}_\lambda^T(\mathbf{z})\mathbf{y} = 0 \qquad (11.62\text{c})$$

$$\mathbf{y}^T \mathbf{y} = 1 \qquad (11.62\text{d})$$

where $\mathbf{y} \in \mathbf{R}^n$ and $\alpha \in \mathbf{R}$ are new unknowns. The Jacobian of this system is guaranteed to be nonsingular at a simple bifurcation or turning point, and the solution of (11.62) should yield $\alpha = 0$ at such a point.

(iv) Branch switching at a bifurcation point. Whether we use a simple-minded continuation in λ or a more sophisticated scheme like arclength continuation, the system becomes singular as we approach a bifurcation point. Unlike a turning point, however, the solution branch along which we are continuing has an extension past the singular point. If we wish to stay on the same solution branch, then we may attempt to choose a larger step size than the one which brings us to the singular point, hoping that this still gives a reasonable guess for a point on the curve on the other side of the singular point (i.e., such that it is in the attraction region of Newton's method or of one of its variants, at that other nonsingular point). This may result in a fairly good approximation for one solution branch. (Alternatively, we may have a primary bifurcation, where we know one solution branch in advance.)

In order to find another solution branch, we should switch branches at a bifurcation point. This is an important, nontrivial task which must be performed, for instance, to find multiple solutions to a problem (11.46) when (11.47) has bifurcation points. A variety of algorithms have been proposed, but the background needed for the detailed description of most of them is beyond the scope of this presentation. The basic idea is usually to perform an educated search, perturbing the problem in an intelligent way so that the solution of the perturbed problem puts us on another branch near but away from the singular point. Since we have a reasonably good idea of one solution branch through the bifurcation point, we also know the tangent $\dot{\mathbf{z}}_0 = (\dot{\mathbf{s}}(\sigma_0), \dot{\lambda}(\sigma_0))$, so we may seek solutions on some subset "parallel" to the tangent but displaced from the bifurcation point in some direction "normal" to the tangent.

Thus we may replace (11.61) by

$$\hat{\mathbf{z}}_1 = \mathbf{z}_0 + \varepsilon + \Delta\sigma \hat{\mathbf{z}}_0,$$

where ε is some "reasonably small" perturbation vector, and $\hat{\mathbf{z}}_0^T \dot{\mathbf{z}}_0 = 0$. A combination of luck and good will does the rest.

We conclude this section by remarking that there are various other ways to form an augmented system which has a nonsingular Jacobian matrix away from bifurcation points (not necessarily turning points). One natural possibility in a BVP context is to replace some of the BC by suitable new ones. The information about which new BC to choose (and which to delete) may be gleaned from the previous point on the solution branch.

11.6 HIGHLY OSCILLATORY PROBLEMS

The problems which we consider here, unlike most others considered in this chapter, are not really *extensions* of BVPs in standard form. However, they are treated separately because the difficulties associated with them are not the usual ones. They arise in a singular perturbation context when some of the eigenvalues of large magnitude are (almost) purely imaginary. When all large eigenvalues are almost purely imaginary, such BVPs do not have any particular stability difficulties compared with IVPs, and the main difficulty is in efficiently obtaining *accurate* solutions, as the following example demonstrates.

Example 11.9

Consider the scalar singularly perturbed BVP of Section 10.1.1 with $p(x) \equiv 0$,

$$\varepsilon u'' - q(x)u = g(x), \qquad a < x < b$$

$$u(a) = \beta_1, \qquad u(b) = \beta_2$$

and assume further that $|q(x)| \equiv |q| \sim 1$. Here $0 < \varepsilon \ll 1$ and $g(x)$ is smooth. The two modes of the ODE are

$$e^{\pm \sqrt{q/\varepsilon}}$$

The case considered in Section 10.1.1 is with $q > 0$, where one of the modes is fast increasing and the other is fast decreasing. But now consider

$$q < 0$$

Then $|e^{\pm \sqrt{q/\varepsilon}}| = |e^{\pm i \sqrt{-q/\varepsilon}}| = 1$, so on one hand there are no fast increasing modes and the stability characteristics of the BVP are similar to those of a corresponding IVP, but on the other hand the general solution

$$u(x) = \alpha_1 e^{i\sqrt{-q/\varepsilon}} + \alpha_2 e^{-i\sqrt{-q/\varepsilon}} - g(x)/q + O(\varepsilon)$$

is *not smooth anywhere* on $[a,b]$, unless $\alpha_1 = \alpha_2 = 0$. In other words, unlike the case treated in Chapter 10, the rapid variation of the solution is not restricted to narrow boundary layers, but occurs throughout the interval.

For a numerical discretization of the ODE, potential approximation difficulties exist, because all of our methods are based somehow on quadrature, which necessitates $O(\varepsilon^{-1/2})$ mesh points for a good approximation of these modes.

Consider now what happens if $q(x)$ varies smoothly from negative to positive, e.g.,

$$q(x) = \frac{2x - (a+b)}{b - a}$$

The point $x = \dfrac{a+b}{2}$ is called a *classical turning point*. Generally we expect the solution to be highly oscillatory for $x < \frac{a+b}{2}$, and to be smoothly varying for $x > \frac{a+b}{2}$ sufficiently far from $\frac{a+b}{2}$ and from b.

□

We consider now a linear singularly perturbed BVP without turning points, involving an ODE of the form

$$\varepsilon \mathbf{y}' = A(x)\mathbf{y} + \mathbf{q}(x) \tag{11.63}$$

and possibly having a highly oscillatory solution. Given no more *a priori* information, if an accurate, general solution is desired everywhere on the interval of definition, then there is little we can do but produce it at a large expense. There are more specialized applications, though, where something better can be done. For instance, the problem may have (almost) constant coefficients which yield a regular pattern in the rapid oscillation. The rapid oscillation can then be detected at the beginning with relatively small effort, and be filtered out. It may also be that an accurate account of all the wiggles in the solution is not needed.

One such case is when the solution is known to be smooth. The general solution of the linear BVP can then be written as

$$\mathbf{y}(x) = \Phi(x)\mathbf{c} + \phi(x)$$

where under certain mild conditions $\phi(x)$ can be chosen to be smooth [so all the highly oscillatory activity is in $\Phi(x)$]. If the smooth solution is sought, then the BC must be such that the part of $\mathbf{c}$ corresponding to the highly oscillatory part of $\Phi(x)$ is very small in magnitude. Hence we wish to obtain the smooth solution without approximating all of $\Phi(x)$ well everywhere, so the BC must be such that the highly oscillatory modes are suppressed. It is often possible to *prepare the boundary data* such that this is the case.

To illustrate the point, assume that there are only fast oscillatory modes (of one scale) for (11.63). Thus, the eigenvalues of $A(x)$ are all (almost) purely imaginary and bounded away from 0. (If there are also slow solution components in $A(x)$ then an explicit decoupling transformation of fast oscillatory and slow components needs to be done ahead of time, and the slow components present no further exceptional difficulty.) To first order in ε, the smooth solution is the reduced one,

$$\mathbf{y}(x) \approx \bar{\mathbf{y}}_0(x) \equiv -A^{-1}(x)\mathbf{q}(x)$$

and the initial value

$$\mathbf{y}(a) := \bar{\mathbf{y}}_0(a)$$

puts us on the smooth track [up to $O(\varepsilon)$]. This process may be continued repeatedly, so if

$$\left\| \frac{\partial^\nu A^{-1}}{\partial x^\nu} \right\|, \left| \frac{\partial^\nu \mathbf{q}}{\partial x^\nu} \right| \leq C, \qquad 0 \leq \nu \leq p$$

then we may write

$$\mathbf{y}(x) = \sum_{\nu=0}^{p-1} \varepsilon^\nu \bar{\mathbf{y}}_\nu(x) + O(\varepsilon^p)$$

with
$$\mathbf{y}(a) := \sum_{v=0}^{p-1} \varepsilon^v \bar{\mathbf{y}}_v(a)$$

Next we want to use difference schemes with a step size $h \gg \varepsilon$ to approximate the smooth solution curve. For the test equation
$$\varepsilon y' = \lambda(x) y$$
with $|\lambda(x)| \approx 1$ and λ purely imaginary, the analysis yields similar damping factors as in Section 10.3. Thus, for the backward Euler method we have
$$|y_{i+1}| = |1 - \lambda(x_{i+1}) h/\varepsilon|^{-1} |y_i| < |y_i|$$
for the midpoint scheme
$$|y_{i+1}| = \frac{|1 + \lambda(x_{i+1/2}) h/2\varepsilon|}{|1 - \lambda(x_{i+1/2}) h/2\varepsilon|} |y_i| = |y_i|$$
and for the trapezoidal scheme
$$|y_{i+1}| = \frac{|1 + \lambda(x_i) h/2\varepsilon|}{|1 - \lambda(x_{i+1}) h/2\varepsilon|} |y_i| = \cdots = \frac{|1 + \lambda(x_1) h/2\varepsilon|}{|1 - \lambda(x_{i+1}) h/2\varepsilon|} |y_1|$$

While these results are essentially the same as in Section 10.3, their interpretation is different, because the exact solution now satisfies
$$|y(x)| = |e^{\int_a^x \lambda(t)/\varepsilon dt}| \; |y(a)| = |y(a)|$$

The midpoint scheme is therefore preferred. (Note that the backward Euler scheme now introduces artificial dissipation which, while being an accuracy rather than a stability issue, can still produce misleading results, especially in the context of hyperbolic PDEs.)

In other situations, the solution is known to be rapidly oscillatory (and not smooth), but some *analytical* approach may be used to find the oscillatory part. This brings about a variety of analytical-numerical methods, and we demonstrate one by example.

Example 11.10

Consider the SH-problem in theoretical seismograms (Example 1.8) with constant material properties, when the S-wave velocity β is smaller than the phase velocity ω/k. The matrix A in (1.20a) has the eigenvalues
$$\pm (k^2 - \frac{\omega^2}{\beta^2})^{1/2}$$
and this is a pair of imaginary eigenvalues under our assumption here. For high frequencies (i.e., when ω is large), the two modes are rapidly oscillatory [with wavelength $O(\omega^{-1})$]. Thus, the finite difference approximations considered in Chapter 5 (and those underlying Chapter 4 as well) all have difficulties with the highly oscillatory solution, essentially necessitating $O(\omega)$ mesh points.

To write the ODE as in (11.63), define the *slowness*

$$p := k/\omega$$

Then (1.20a) can be written, with an obvious change in meaning of A and $\mathbf{y}$, as

$$\mathbf{y}' = \omega A(x)\mathbf{y} \qquad (11.64)$$

where

$$A = \begin{bmatrix} 0 & \mu^{-1} \\ \mu p^2 - \rho & 0 \end{bmatrix}$$

Let $T(x)$ be the eigenvector matrix for $A(x)$, namely

$$T^{-1}AT = \Lambda$$

with

$$\Lambda(x) = \text{diag}(\nu_\beta, -\nu_\beta), \qquad \nu_\beta = (p^2 - \beta^{-2})^{1/2}$$

Then for

$$\mathbf{w}(x) := T^{-1}(x)\mathbf{y}(x)$$

we have the (almost) diagonal ODE

$$\mathbf{w}' = (\omega \Lambda - T^{-1}T')(x)\mathbf{w} \qquad (11.65)$$

If the problem actually had constant coefficients, then the exact fundamental solution for (11.65) would be

$$W(z;\zeta) = e^{\omega \Lambda(z-\zeta)}$$

and the corresponding fundamental solution for (11.64) would be

$$Y(z;\zeta) = T(z)W(z;\zeta)T^{-1}(\zeta) \qquad (11.66)$$

In practice, the medium (the earth) cannot be considered homogeneous (constant) throughout, but often stratified media approximations are used. In these, the material properties are considered to be piecewise constant. Then for each uniform material layer the exact fundamental solution can be obtained, and an exact multiple shooting scheme connects the pieces by requiring solution continuity and satisfaction of the boundary conditions. Note that the number of multiple shooting intervals does not relate directly to the small parameter ω^{-1}, and that in piecing together the nondecoupled solutions $Y(z;\zeta)$, we are relying on the eigenvalues of $A(x)$ not having a significant real component, so $\|Y(z;\zeta)\|$ is nicely bounded.

But the piecewise constant approximation for the medium may often be unsatisfactory, necessitating many layers (many shooting points) for approximation reasons. Worse still, an inappropriate solution may well look like a "physical" solution and therefore may be accepted as a good approximation by geophysicists, even though it is not.

Thus we consider an alternative approach, assuming that $A(x)$ varies slowly in each material layer, with its eigenvalues well separated and having only small real parts. Then we write

$$W(z;\zeta) = \left(\sum_{l=0}^{\infty} P_l(z)\omega^{-l}\right) \exp\{\omega Q_0(z) + Q_1(z)\} \qquad (11.67)$$

where the P_l are bounded matrices, Q_0 and Q_1 are diagonal matrices and $\text{Re}(Q_0(z))$

$= O(\omega^{-1})$. Substituting into (11.65) and equating coefficients for equal powers of ω, we obtain

$$P_0 = I, \qquad Q_0(z) = \int_\zeta^z \Lambda(t)\,dt$$

and expressions for Q_1, P_1, etc. The practical use of such an expansion depends on how quickly the sum in (11.67) converges. An $O(\omega^{-2})$ approximation to $W(z;\zeta)$ is obtained by

$$\widehat{W}(z;\zeta) = (I + \omega^{-1} P_1)\exp\{\omega Q_0(z) + Q_1(z)\} \tag{11.68}$$

and we can use this if the difference can be neglected. This in turn depends on the gradient of the material properties. As it turns out, for gentle gradients such an approximation can be found much more efficiently than a corresponding finite difference approximation. This is demonstrated in Fig. 11.5, where the method described above (denoted "body wave" method) uses two approximations like (11.68) to obtain a solution with an accuracy comparable to that produced by a collocation method using a mesh of 28 subintervals with 3 Gauss points per subinterval. In this figure the left part of each panel shows the model velocities and the right part shows the meshes and solution methods chosen. Frequency is held constant at 9.0 Hz, so $\omega = 18\pi$, and the wave number k and phase velocity $c = 1/p$ corresponding to each panel are shown on the right.

□

Most of the numerical techniques which have been proposed for highly oscillatory problems deal with IVPs, but many are also useful for BVPs. The situation is reminiscent in this respect to that for differential/algebraic equations, cf. Section 11.2. And just

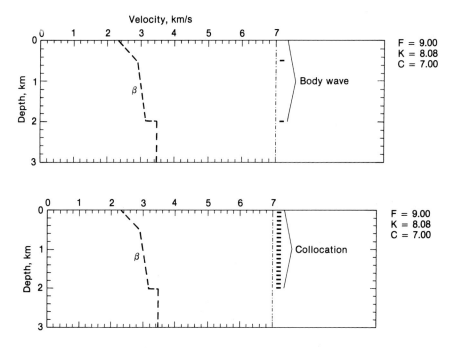

Figure 11.5 Solution schemes and meshes for the SH problem

like DAEs, little has appeared in the literature to accommodate a situation where a BVP is also stiff, adversely affecting an IVP-type method. In this case, an explicit decoupling of modes of different types is necessary, as demonstrated in the following example.

Example 11.11

We consider the theoretical seismograms problem of Example 1.8 again, but this time we concentrate on the more complicated P-SV equations. They can also be brought into the form (11.64), with

$$A = \begin{pmatrix} 0 & (\rho\alpha^2)^{-1} & q_1 & 0 \\ -\rho & 0 & 0 & p \\ -p & 0 & 0 & \mu^{-1} \\ 0 & -q_1 & q_2 & 0 \end{pmatrix}$$

$$q_1 = p(1 - 2\beta^2/\alpha^2)$$
$$q_2 = 4\mu p^2(1 - \beta^2/\alpha^2) - \rho$$

As for the SH problem, there is a smooth eigenvector matrix $T(x)$ bringing $A(x)$ into diagonal form

$$\Lambda = \mathrm{diag}(\nu_\alpha, \nu_\beta, -\nu_\alpha, -\nu_\beta)$$

with ν_β as in Example 11.10 and ν_α similarly defined (with α replacing β), so long as the eigenvalues stay away from 0. Now, if the phase velocity is such that

$$\beta(z) < p^{-1} < \alpha(z)$$

then there is one pair of highly oscillatory modes and one pair of increasing and decreasing modes. The latter must be decoupled (from one another) for stability reasons, but if we agree to decouple explicitly then we might as well decouple the fast oscillatory modes from the other two. Then, in a multiple shooting setting, *an appropriate approximation scheme can be applied for each type of mode.* For instance, the scheme described in Example 11.10 can be applied to the fast oscillatory modes, and a symmetric or upwinded difference scheme can be applied to the others. All this can then be mixed back via $T(x)$ as in (11.66) to yield a special multiple shooting scheme which works well. A treatment in turning point regions along similar lines is also possible, but we refer the interested to the literature.
□

11.7 FUNCTIONAL DIFFERENTIAL EQUATIONS

There is a variety of applications which are more naturally modelled as *functional differential equations* (FDEs) rather than ODEs. In such equations dependent variables are concurrently evaluated at more than one value of the independent variable.

Functional differential equations are more difficult to solve numerically, and correspondingly, a less complete theory is known for them. A generic form for such equations considered here is

$$\mathbf{y}'(x) = \mathbf{f}(x, \mathbf{y}(x), \mathbf{y}(\omega_1(x)), \mathbf{y}(\omega_2(x))), \qquad a < x < b \qquad (11.69)$$

where

$$\omega_1(x) < x < \omega_2(x)$$

When there are $\omega_1(x)$ terms (corresponding to so-called *delays*), but no $\omega_2(x)$ terms (corresponding to so-called *advances*), the the FDE is called a *delay differential equation* (DDE). The FDE here is subject to the linear BC

$$P_1 \mathbf{y}(x) = \phi(x), \qquad \underline{a} \le x \le a \qquad (11.70\text{a})$$

$$P_2 \mathbf{y}(x) = \psi(x), \qquad b \le x \le \overline{b} \qquad (11.70\text{b})$$

$$B_a \mathbf{y}(a) + B_a^1 \mathbf{y}(\omega_1(a)) + B_a^2 \mathbf{y}(\omega_2(a)) + B_b \mathbf{y}(b) + B_b^1 \mathbf{y}(\omega_1(b)) + B_b^2 \mathbf{y}(\omega_2(b)) = \boldsymbol{\beta} \qquad (11.70\text{c})$$

Here $\mathbf{y}$, $\mathbf{f}$ and $\boldsymbol{\beta}$ are vectors in $\mathbf{R}^n$, P_1, P_2 are projectors from $\mathbf{R}^n$ into $\mathbf{R}^n$ and $\omega_{1,2}(x)$ are continuous scalar functions satisfying

$$\min_{x \in [a,b]} \omega_1(x) = \underline{a}, \qquad \max_{x \in [a,b]} \omega_2(x) = \overline{b}$$

A solution of the BVP (11.69), (11.70) is in general only piecewise smooth on $[\underline{a},\overline{b}]$, and may be discontinuous at $x = a$ and at $x = b$. If we use a numerical discretization of the BVP consisting of, say, some finite difference scheme, and if the points of lower smoothness (e.g. where we have fewer than p continuous derivatives, for some positive integer p) are

$$a = s_1 < s_2 < s_3 < \cdots < s_{J+1} = b \qquad (11.71)$$

then recalling that high solution derivatives appear in the usual local truncation error forms, it becomes clear that the points s_j should be included as part of any discretization mesh. Note that since ω_1 and ω_2 are functions of x only, it is possible to compute the points s_j of (11.71) in advance.

As in other sections of this chapter, there are two basic numerical approaches which we mention: Converting the problem into a BVP in standard form, and applying special methods. For a conversion into standard form to be possible and useful, some restrictive conditions must be met. For instance, assume that $\omega_{1,2}$ are just translations by a constant c and that $b - a$ is an integer multiple of c, namely

$$\omega_1(x) = x - c, \qquad \omega_2(x) = x + c, \qquad b - a = Jc$$

Then we may "fold" the interval into overlapping segments of length c using transformations like

$$x_j(t) = a + (j - 1 + t)c, \qquad 0 < t < 1$$

similarly to what is done in Section 11.1.2. Rather than giving the details again, let us demonstrate by example.

Example 11.12

The BVP

$$u''(x) = -\frac{1}{16} \sin u(x) - (x+1)u(x-1) + x, \qquad 0 < x < 2$$

$$u(x) = x - \frac{1}{2}, \quad -1 \le x \le 0, \quad u(2) = -\frac{1}{2}$$

has been a standard test problem for some methods for DDEs. It is trivially converted into the following BVP in standard form

$$u_1''(t) = -\frac{1}{16} \sin u_1(t) - (t+1)(t - \frac{3}{2}) + t, \quad 0 < t < 1$$

$$u_2''(t) = -\frac{1}{16} \sin u_2(t) - (t+2)u_1(t) + t + 1, \quad 0 < t < 1$$

$$u_1(0) = -\frac{1}{2}, \quad u_2(1) = -\frac{1}{2}$$

$$u_1(1) = u_2(0), \quad u_1'(1) = u_2'(0)$$

The latter BVP does not have a delay any more and, as it turns out, is very easy to solve numerically.

□

The simplicity of this example should not mislead the reader into thinking that all DDEs can be handled that easily. In addition to the special form of (11.69), (11.70) required to apply the conversion trick, note also that if J is large, then a large system of ODEs arises. (Interestingly, if we let c decrease, keeping the interval length $b - a$ fixed, then the complexity of the problem using the conversion trick increases, while the DDE itself "approaches" an ODE in some sense.)

Of the direct methods which have been proposed for solving FDEs, we consider briefly piecewise polynomial collocation (see Section 5.4). The appeal here is that, while the method has the efficiency and generality of a finite difference scheme, it also gives conceptual simplicity because its solution is continuously defined. Since the approximate solution is defined for all $x \in [a, b]$, we can simply write down the discretization equations directly as they are posed in (11.69), (11.70). Denoting the piecewise polynomial solution based on a mesh π by $\mathbf{y}_\pi(x)$, and considering collocation at k points per mesh subinterval as described in (5.59), we obtain

$$\mathbf{y}_\pi'(x_{ij}) = \mathbf{f}(x_{ij}, \mathbf{y}_\pi(x_{ij}), \mathbf{y}_\pi(\omega_1(x_{ij})), \mathbf{y}_\pi(\omega_2(x_{ij}))), \quad 1 \le j \le k, \, 1 \le i \le N$$

The BC (11.70) are discretized in a similar way.

Assuming sufficient smoothness of the problem coefficients and of the exact solution [except at the points (11.71) which form part of the mesh π], it is possible to derive the usual existence and convergence results. In particular, an analogue to Theorems 5.56 and 5.75 is obtained, yielding an $O(h^k)$ convergence (where h is the maximum step size of the mesh π). However, recall that the usual collocation theory yields also a higher order convergence (superconvergence) at mesh points. This estimate at special points is lost in the FDE case, unless the problem and the mesh are chosen such that mesh points get mapped into mesh points (and collocation points into collocation points) by the delays $\omega_{1,2}$. The latter is a fairly special case, though. Still, the basic $O(h^k)$ error estimate is good enough if we take, say, $k = 4$.

We make a final comment concerning the implementation of collocation for solving FDEs as compared to ODEs (which is carried out in Appendix B). While the nonlinear iteration control does not change in a significant way, other major considerations do. Since local parameter elimination as in Section 5.6.1 becomes difficult, use of basis functions as described in Section 5.6.2 becomes preferable. Even more importantly, the linear system solvers discussed in Chapter 7 are not applicable any more, because the sparseness structure of the Jacobian matrix changes. A typical sparseness structure for collocation is depicted in Figure 11.6. Such a system of linear equations may be treated as banded, or by a more sophisticated so-called *frontal* method, on which we do not elaborate. Note that, unlike when we convert to an ODE system, here the smaller $a - \underline{a}$ and $\bar{b} - b$ are, the better the structure of the Jacobian matrix becomes. Finally, note that mesh selection also becomes a more difficult question for FDEs.

11.8 METHOD OF LINES FOR PDES

A common approach for solving partial differential equations is to discretize the problem in terms of one of the independent variables and then consider solving the resulting problem of reduced dimensionality. If the PDE involves two independent variables, then this reduced problem involves an ODE system.

For example, consider the simple parabolic PDE

$$u_t = u_{xx} + f(x,t,u,u_x), \qquad 0 \leq t \leq T \qquad (11.72a)$$

$$u(x,0) = \phi(x) \qquad (11.72b)$$

$$u(a,t) = u(b,t) = 0 \qquad (11.72c)$$

where $\phi(x)$ is the known initial value of $u(x,t)$. If we seek an approximate solution on the lines $t = t_n = (n-1)k$, $n = 1, \ldots, M+1$ ($k := \frac{T}{M}$) and replace the derivative with respect to t at $t = t_{n+1}$ by the backward difference $u_t(x, t_{n+1}) \approx \frac{u(x, t_{n+1}) - u(x, t_n)}{k}$, then (11.72) can be approximated by the BVP system

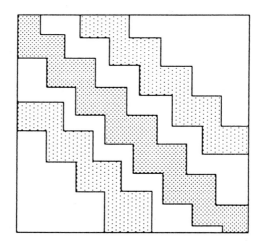

Figure 11.6 Sparseness structure for FDE collocation

$$u''_{n+1}(x) + f_n(x, u_{n+1}, u'_{n+1}) = [u_{n+1}(x) - u_n(x)]/k \qquad a < x < b \qquad (11.73a)$$

$$u_{n+1}(a) = u_{n+1}(b) = 0, \qquad n = 1, 2, \ldots, N \qquad (11.73b)$$

where $u_1(x) = \phi(x)$ and $u_n(x) \approx u(x, t_n)$ $(n = 1, \ldots, N)$. Alternatively, if the derivatives with respect to x are approximated by centered finite differences on the mesh lines $x = x_i = (i-1)h$, $i = 1, \ldots, N+1$ $(h := \frac{b-a}{N})$, then one instead obtains an IVP for the approximate solution $u(x_i, t)$ $(i = 2, \ldots, N)$ (cf. Example 8.7 and Exercise 7 in Chap. 8). These are both examples of the appropriately named *method of lines*. The methods resulting from first discretizing in t (and then solving a BVP) or first discretizing in x (and then solving an IVP) are sometimes called *transverse* and *longitudinal* methods of lines, respectively.

If the BVP (11.73) is solved by using centered differences and if the IVP is solved by using backward Euler, then the resulting discretizations are mathematically equivalent, even though their implementations would generally differ.

The method of lines can also be applied to solve hyperbolic and elliptic PDEs. For instance, for the elliptic PDE

$$u_{xx} + u_{yy} = f(x, y) \qquad a \le x \le b, \qquad c \le y \le d$$

$$u(a, y) = \phi_1(y), \qquad u(b, y) = \phi_2(y) \qquad c \le y \le d$$

$$u(x, c) = \psi_1(x), \qquad u(x, d) = \psi_2(x) \qquad a \le x \le b$$

a finite difference discretization in either of the spatial variables results in a BVP system involving the unknown solution along lines in the other spatial variable. While the method of lines can be used to compute the solution along parallel lines when the region is a rectangle, it can also be used to reduce an elliptic PDE in cylindrical coordinates on a subregion of a circle to BVPs for ODEs (along rays emitting from the center).

The early development of the method of lines for parabolic problems treated both transverse and longitudinal methods. However, in computational practice the longitudinal approach has become more popular. This is in part due to a natural physical interpretation (integrating to steady state) and in part due to the ease of implementation, given the wide availability of stiff IVP software, with the BDF formulas being particularly common.

For illustration, consider how (11.72) is solved by using spline collocation for the space discretization (see also the end-of-chapter references). Given $\pi : a = x_1 < \cdots < x_{N+1} = b$ and a basis $\{\psi_{i,k}\}_{i=1}^J$ for $\mathbf{P}_{k,\pi,2}$ (cf. Section 2.5.2), we write

$$u(x, t) \approx \sum_{i=1}^{J} a_i(t) \psi_{i,k}(x) \qquad (11.74)$$

Labelling the (appropriately chosen) collocation points $a < \xi_2 < \cdots < \xi_{J-1} < b$ (see Section 5.6.2), and assuming $\psi_{i,k}(\xi_i) \neq 0$, we require

$$\sum_{i=1}^{J} \psi_{i,k}(\xi_j) \frac{da_i}{dt} = u_{xx}(\xi_j, t) + f(\xi_j, t, u, u_x) \qquad j = 2, \ldots, J-1 \qquad (11.75)$$

By differentiating the BC (11.72c) with respect to t,

$$\frac{du(a,t)}{dt} = \frac{du(b,t)}{dt} = 0$$

and replacing $u(x,t)$ with (11.74), we obtain two more ODEs. This with (11.75) can be written as

$$A \frac{d\mathbf{a}}{dt} = \mathbf{f}(t, \mathbf{a}), \qquad 0 \le t \le T \tag{11.76}$$

and initial conditions for $\mathbf{a} = (a_1, \ldots, a_J)^T$ are easily derived from (11.72b). This method can be generalized in several ways, e.g., for a PDE system

$$\mathbf{u}_t = \mathbf{f}(x, t, \mathbf{u}, \mathbf{u}_x, \mathbf{u}_{xx})$$

with $\mathbf{u}(x,t) \in \mathbf{R}^n$. As it stands, the method has the obvious drawbacks that the spatial mesh π does not change. This means that for problems with transient regions like a moving wave front or a shock layer, a fine mesh would be needed throughout $[a,b]$. For other problems, moving boundaries occur, where a and b *themselves* change with t (see Example 11.13 below). For problems where the solution becomes very smooth, it may even be desirable to *coarsen* the grid. In other words, while the step size in t can be adapted when solving the IVP (11.76), we in some sense have the wrong flexibility if variation in $u(x,t)$ is more in the direction of x than of t.

These situations (and indeed the overall viewpoint developed in this book) lead us naturally to consider a transverse approach instead. For this, consider now the general PDE of the form

$$C(\mathbf{u})\mathbf{u}_t = (D(\mathbf{u})\mathbf{u}_x)_x + \mathbf{f}(\mathbf{u}, \mathbf{u}_x) \qquad a < x < b, \qquad 0 \le t \le T \tag{11.77a}$$

$$\mathbf{g}(\mathbf{u}(a,t), \mathbf{u}(b,t)) = \mathbf{0} \tag{11.77b}$$

$$\mathbf{u}(x,0) = \boldsymbol{\phi}(x) \tag{11.77c}$$

where $\mathbf{u}, \boldsymbol{\phi}, \mathbf{g} \in \mathbf{R}^n$ and where $C, D \in \mathbf{R}^{n \times n}$ are positive definite diagonal matrices (depending on x and t as well). We assume that (11.77) is a meaningful problem in the sense that a unique isolated solution exists. Choosing a temporal mesh and then replacing $\mathbf{u}_t(x,t)$ at a fixed time level $t = t^*$ by a linear combination of values of $\mathbf{u}(x, t^*)$ and $\mathbf{u}(x,t)$ at previous time levels, we arrive at a BVP for $u(x,t)$ ($a < x < b$).

We now illustrate the transverse method of lines for a specific problem.

Example 11.13

Many PDEs involve problems with boundary and interface conditions. A well-known type is the two-phase Stephan problem for heat transfer with change of phase. For example, suppose that a sphere is alternately heated and cooled on its surface. This can be described by the equations

$$\frac{\partial u_1}{\partial t} = \frac{\partial^2 u_1}{\partial x^2} + \frac{2}{x}\frac{\partial u_1}{\partial x} - q_1(x,t) \tag{11.78a}$$

$$0 < x < 1, t \ge 0$$

$$\frac{\partial u_2}{\partial t} = \frac{\partial^2 u_2}{\partial x^2} + \frac{2}{x}\frac{\partial u_2}{\partial x} - q_2(x,t) \qquad (11.78\text{b})$$

$$\frac{\partial u_1}{\partial x}(0,t) = 0, \quad u_2(1,t) = c(t) \qquad t > 0 \qquad (11.78\text{c})$$

$$u_1(x,0) = \phi_1(x), \qquad u_2(x,0) = \phi_2(x) \qquad 0 < x < 1 \qquad (11.78\text{d})$$

with the interface conditions

$$u_1(s(t),t) = 0 = u_2(s(t),t) \qquad (11.78\text{e})$$

$$t > 0$$

$$\frac{ds}{dt} + \frac{\partial u_1}{\partial x}(s(t),t) - \frac{\partial u_2}{\partial x}(s(t),t) = d(s(t),t) \qquad (11.78\text{f})$$

Here u_1 and u_2 denote the temperature above and below melting on $[0, s(t)]$ and $[s(t), 1]$, respectively, where the interface $s(t)$ is an unknown.

If we discretize in time with the backward Euler scheme and step size k, then we can generate a sequence of BVPs as follows: Suppose that at time step $t_{n-1} = (n-1)k$ we have the approximations $u_1^{n-1}(x) \approx u_1(x, t_{n-1})$, $u_2^{n-1}(x) \approx u_2(x, t_{n-1})$, $s^{n-1} \approx s(t_{n-1})$. Then from (11.78a–d),

$$\frac{v_1(s) - u_1^{n-1}(x)}{k} = v_1''(x) + \frac{2}{x}v_1'(x) - q_1(x,t_n) \qquad (11.79\text{a})$$

$$0 < x < 1$$

$$\frac{v_2(s) - u_2^{n-1}(x)}{k} = v_2''(x) + \frac{2}{x}v_2'(x) - q_2(x,t_n) \qquad (11.79\text{b})$$

$$v_1'(0) = 0, \qquad v_2(1) = c(t_n) \qquad (11.79\text{c})$$

and from the interface conditions (11.78e, f)

$$v_1(s) = v_2(s), \quad \frac{s - s^{n-1}}{k} + v_1'(s) - v_2'(s) = d(s, t_n) \qquad (11.79\text{d})$$

where the time derivative in (11.78f) has been replaced by a forward difference. Now (11.79) is a BVP which can be solved in a variety of ways for approximations $u_1^n \approx v_1$, $u_2^n \approx v_2$ and $s^n \approx s$. For instance, (11.79) can be converted to a standard BVP by using the trick (11.2), (11.3) of Section 11.1.2.

□

In some respects, the simplest way to solve a resulting sequence of BVPs when applying the transverse approach is to use a BVP code. This approach has been used with success for certain nontrivial problems (and is related to the approach of integrating to steady state discussed in Section 8.3.2). For a variety of reasons it can be very inefficient, though, mainly because such a code is designed to solve a BVP once only. For instance, it may be quite costly to perform an elaborate mesh selection process at each new time step, and one may be able to instead predict a satisfactory mesh by using information about the meshes from previous time steps. Of course, simple continuation (cf. Section 8.3.1) from one time step to the next is useful here.

While the transverse method can be improved by designing a BVP solver explicitly for this purpose, ultimately it is better to borrow ideas from both transverse *and* longitudinal approaches and develop fully adaptive methods. This then allows automatic selection of meshes in both x and t using the relationships between these derivatives given through the PDE itself. It is then natural to not only change the spatial mesh smoothly with time, but we can even consider the mesh as a function of time. In fact, given a mesh selection strategy (such as error equidistribution) with the number of mesh points fixed, differential equations describing the evolution of these points in time can be derived. Certain numerical experiments and theoretical results in the literature suggest that it is preferable not to couple these differential equations for the mesh with the PDE for the solution variables, but rather to compute them separately. This is consistent with what we might expect from the BVP case discussed in Section 9.4.2.

11.9 MULTIPOINT PROBLEMS

We have encountered problems with multipoint BC before in a number of places. In Section 1.1 we show how such a problem can be converted to a two-point BVP, and an application where multipoint conditions arise is given in Example 1.13 (and Example 11.4). A direct treatment of such conditions in the context of solving linear equations is given in Section 7.5.2. Here we briefly consider the theoretical implications arising when solution values are prescribed at locations other than the two end points.

In particular, the concept of dichotomy needs to be extended, as the following simple example indicates.

Example 11.14

The problem
$$\varepsilon y' = -y, \quad -1 < x < 1$$
$$y(0) = \beta$$

is ill-conditioned for $0 < \varepsilon \ll 1$, because on the segment $[-1, 0]$ we have a terminal value prescribed for a fast decreasing mode.

On the other hand, the problem
$$\varepsilon y' = \begin{cases} y & x < 0 \\ -y & x > 0 \end{cases} \quad -1 < x < 1$$
$$y(0) = \beta$$

is well-conditioned. To see this, let $y_1(x) \equiv y(-x)$ and $y_2(x) \equiv y(x)$ on $[0, 1]$. We obtain the uniformly stable IVP

$$\varepsilon y_1' = -y_1$$
$$\quad\quad\quad\quad\quad\quad 0 < x < 1$$
$$\varepsilon y_2' = -y_2$$
$$y_1(0) = y_2(0) = \beta$$

But the (only) mode for this problem on the original interval $[-1, 1]$ has the shape depicted in Fig. 11.7. It is neither nondecreasing nor nonincreasing everywhere, but rather it decreases rapidly away from $x = 0$.

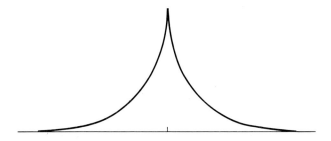

Figure 11.7 A fundamental solution corresponding to a side condition

We see then that a side condition given in the middle of the interval of integration should control a mode which does not increase rapidly in *both* directions away from its location. This therefore indicates the need to extend the theory of dichotomy for well-conditioned multipoint BVPs.

□

The Fitzhugh-Nagumo equations described in Example 1.13 (cf. also Example 11.4) form a typical instance where a PDE is reduced to an ODE by looking for solutions along a characteristic. The extra degree of freedom is then fixed by requiring another side condition to be satisfied, giving rise to the multipoint BVP. Another example occurs when we study the shock structure for Burgers' equation:

Example 11.15

Consider the ODE
$$\varepsilon u'' = uu' - cu$$
Here c is an unknown parameter, the wave speed. Suppose for simplicity that the BC are
$$u(-\infty) = 1, \qquad u(\infty) = -1$$
Then a solution is given by
$$u_s(x) = -1 + \frac{2}{1 + e^{s+x}}, \qquad c = 0$$
where s is arbitrary. As in Example 1.13 we may make a solution like this isolated by choosing the translation parameter s; but without pinning s down the BVP is underdetermined (so a direct approximation will result in an almost singular discrete BVP). Practically, we may obtain a BVP with an isolated solution by prescribing a "middle" value for u. Because of the boundary values we realize that the solution must go through 0 somewhere, so we can specify
$$u(0) = 0$$
(which gives $s = 0$). Note that we do not need to know the solution in order to make this choice of BC. If we linearize the equation around some $u_0(x)$, we obtain
$$\varepsilon u'' = u_0(x) u' + u_0'(x) u$$
It is easy to see that this equation has solutions $z_1(x)$, $z_2(x)$ with the property
$$z_1(x) \sim \exp\{-\frac{1}{\sqrt{\varepsilon}} |x|\}, \qquad |x| \to \infty$$

$$z_2(x) \sim 1, \qquad |x| \to \infty$$

Hence the solution space is not dichotomic, just like in Example 11.14.

□

For such problems we have the following generalization of the dichotomy notion.

Definition 11.80 Suppose $Y(x)$ is a fundamental solution of the ODE

$$\mathbf{y}' = A(x)\mathbf{y}$$

where $A(x) \in C[a,b]$. The ODE has a *polychotomy* if there exist points $\zeta_1, \ldots, \zeta_J$ and projection matrices $P_1, \ldots, P_J \in \mathbf{R}^{n \times n}$ of ranks $\rho_1, \ldots, \rho_J$, respectively, such that $\sum_{j=1}^{J} P_j = I$, $\sum_{j=1}^{J} \rho_j = n$, and for a positive constant K of moderate size,

$$\left\| Y(x) \sum_{j=1}^{l} P_j Y^{-1}(t) \right\| \leq K, \qquad \zeta_l \leq t \leq \zeta_{l+1}, \qquad x \geq t \tag{11.80a}$$

$$\left\| Y(x) \sum_{j=l+1}^{J} P_j Y^{-1}(t) \right\| \leq K, \qquad \zeta_l \leq t \leq \zeta_{l+1}, \qquad x \leq t \tag{11.80b}$$

□

For Example 11.15 we note that $J = 3$, $\zeta_1 = -\infty$, $\zeta_2 = 0$, $\zeta_3 = \infty$, where the choice for ζ_2 makes K optimally small. The theory of Section 3.4 can be generalized for multipoint problems in the sense that well-conditioning implies polychotomy and, if it is assumed that the BC are chosen correctly, that polychotomy implies well-conditioning.

With regards to numerical methods for dealing with multipoint problems, it is important to realize that the "standard" decoupling described in Chapter 6 is no longer meaningful (cf. Section 7.5 and Exercise 8 in Chap. 7). In particular, the number of nondecreasing and nonincreasing modes may change in the passage through a point ζ_j. This is easy to see for the separated (linear) multipoint BC case, for which the BC are

$$\sum_{j=1}^{J} B_j \mathbf{y}(\zeta_j) = \boldsymbol{\beta}$$

where only ρ_j rows of B_j are nonzero and the corresponding rows in B_l, $l \neq j$, are zero. It is possible to show then that the BC induce the projection matrices of (11.80) by

$$P_j := B_j Y(\zeta_j)$$

where $\sum_{j=1}^{J} B_j Y(\zeta_j) = I$. As a corollary we obtain that the dimension of the nondecreasing mode space on $[\zeta_l, \zeta_{l+1}]$ is given by $\sum_{j=l+1}^{J} \rho_j$. Similar statements hold for nonseparated cases.

This change in the underlying theory has ramifications not only on the theory of Chapter 6 (and much of Chapter 7) but also on some of the actual algorithms proposed in those two chapters which exploit the dichotomic structure explicitly. In particular, the decoupling blocks now vary in size. However, a generalization is possible and fairly straightforward, even though not immediate. If the BC are separated, then the natural placement of corresponding rows in the system of linear equations resulting from a discretization is according to increasing x. This then results in a staircase matrix as discussed in Section 7.1. Generalizations of the almost block diagonal procedure and of the row-and-column elimination procedure are straightforward. The latter procedure performs the more general decoupling.

11.10 ON CODE DESIGN AND COMPARISON

The state of the art of BVP software is at the time of this writing not as advanced as that of software for IVPs, and this relative situation may never change because of the increased complexity and variability of the class of problems involved. Still, some useful general-purpose BVP codes have been written in recent years, and it is worthwhile to say a few words about the design and comparison of such codes.

A major decision when one is designing software is what performance criteria to use. Taking for granted that a code must be portable and free of errors (nontrivial assumptions to implement in themselves), standard measures of performance are computer time and storage requirements. But there are other criteria which can well be equally important. These include the following:

(i) Ease of use — common examples where the use of a code is made less easy include instances when the user is required to provide an extensive list of parameter values or a subroutine defining an analytical Jacobian matrix (both of these being common sources of programming errors). Also, the user may have to provide detailed starting information on the problem before the code is able to solve it.

(ii) Robustness — a code should solve a large set of problems from the class of "solvable" BVPs without an excessive "hand holding" by the user. It should solve all the problems that it is intended to solve and stop gracefully when it cannot solve a problem.

(iii) User feedback — a variety of different features can be made available, such as an ability to provide a continuous form of the approximate solution, or an *a posteriori* error estimate, or a detailed account of how the nonlinear iteration is proceeding, or diagnostic information like conditioning constants and indications of how to proceed when failure has been encountered.

(iv) Flexibility — important aspects include an ability of the user to use special options, e.g., special treatment of problems with parameters or multipoint BC, or an ability of a user to insert an initial approximation and mesh or to use a special termination criterion, or to initiate a continuation process tailored for a specific problem.

Some of these design criteria may overlap, but this causes no harm. More importantly, some may be in conflict with one another. For instance, providing the user with many handles and options may contrast ease of use. A way to resolve this conflict is to have more than one user interface. Thus a user desiring a simple usage may do so, but the route to reach special buttons in the code is not blocked from another user needing a deeper layer of involvement (for increased efficiency or because a tougher problem is to be solved). An example is an optional numerical differentiation routine to generate the Jacobian matrix; another is default values for an input parameter list.

At the heart of a BVP code lies the discretization method used. A preliminary comparison of methods can be made by performing a rough count of arithmetic operations needed for these methods. But much more is involved. A number of distinct and fundamental components such as those discussed in Chapters 4, 7, 8, and 9 (each of which can be dealt with in a variety of ways) has to be chosen and implemented. Thus, one must keep in mind that a code is only one particular implementation of a method.

From the discussion so far, it should be clear that the task of code comparison for BVPs is nontrivial and can easily yield a thankless effort. The first point to note in this respect is that a comparison of two codes is obviously biased by the class of test problems that are given, and one can easily modify the conclusions of a study if the problems or criteria are changed. A second popular mistake is to make such a comparison too "general." In view of the many factors involved, one should avoid the temptation to make sweeping statements about the superiority of one code to another.

Nevertheless, even though complex codes like those written for BVPs do not lend themselves easily to comparison, this is not to say that attempting a comparison is fruitless. One can often specify the types of problems for which a given method is adequate or lacking and give recommendations to users with special needs (as done in Chapter 10 and in Sections 11.2, 11.4, 11.5, and 11.6, for instance). In our opinion it is generally more appropriate to view codes as complementing one another rather than competing.

Appendix A

A Multiple Shooting Code

The two codes described here are adaptations from the package BOUNDPACK, which consists of a number of routines for BVPs all based on multiple shooting in some form. The first one, MUSL, is designed for nonstiff linear problems, and the second one, MUSN, is for nonstiff nonlinear problems. It is written in FORTRAN 77 and is available through NETLIB or from

G.W.M. Staarink
Economisch Instituut
Katholieke Universiteit
Thomas van Aquinostraat 6
6525 GD Nijmegen, The Netherlands

MUSL solves the linear two-point BVP of order n (1.4a, b) as rewritten here:

$$\mathbf{y}' = A(x)\mathbf{y}(x) + \mathbf{q}(x), \quad a < x < b$$

$$B_a \mathbf{y}(a) + B_b \mathbf{y}(b) = \boldsymbol{\beta}$$

The code employs the explicit initial value integrator RKF45 (cf. Table 2.1) and performs QU factorizations to obtain a stably decoupled upper triangular multiple shooting recursion, along the lines of Section 4.4.2 (but with more fine-tuning, cf. [1, 2]). An important feature of the code is that it has a number of error indicators. The output parameters also contain important diagnostic information (like conditioning constant estimates). Other features are the low memory requirements and the automatic selection of shooting points (none of which needs to be an output point).

Below we describe the calling sequence of MUSL.

```
      SUBROUTINE MUSL(FLIN,FDIF,N,IHOM,A,B,MA,MB,BCV,AMP,ER,
     1               NRTI,TI,NTI,Y,U,NU,Q,D,KPART,PHI,W,LW,IW,LIW,IERROR)
C     INTEGER N,IHOM,NRTI,NTI,NU,KKK(N),IP1(N),IP2(N),IERROR
C     DOUBLE PRECISION A,B,MA(N,N),MB(N,N),BCV(N),AMP,ER(5),TI(NTI),
C    1               Y(N,NTI),U(NU,NTI),Q(N,N,NTI),D(N,NTI),PHI(NU,NTI),W(LW)
C     EXTERNAL FLIN,FDIF
```

FLIN   SUBROUTINE, supplied by the user with specification:

         SUBROUTINE FLIN(X,Y,F)
         DOUBLE PRECISION X,Y(N),F(N)

where N is the order of the system. FLIN must evaluate the homogeneous part of the differential equation, $A(x)\mathbf{y}(x)$, for $x = X$ and $\mathbf{y}(x) = Y$, and place the result in $F(1), F(2), \ldots, F(N)$.

FDIF   SUBROUTINE, supplied by the user, with specification:

         SUBROUTINE FDIF(X,Y,F)
         DOUBLE PRECISION X,Y(N),F(N)

where N is the order of the system. FDIF must evaluate the righthand side of the inhomogeneous differential equation, $A(x)\mathbf{y}(x)+\mathbf{q}(x)$, for $x = X$ and $\mathbf{y}(x) = Y$, and place the result in $F(1), F(2), \ldots, F(N)$.
In the case that the system is homogeneous, FDIF is the same as FLIN.

N   The order of the system.

IHOM   Indicates whether the system is homogeneous (= 0) or inhomogeneous (= 1).

A, B   The two boundary points ($A = a$, $B = b$).

MA,MB   BC arrays ($MA = B_a$, $MB = B_b$).

BCV   array of boundary values ($BCV = \beta$).

The boundary conditions are defined by
MA Y(A) + MB Y(B) = BCV .

AMP   On entry AMP must contain the allowed incremental factor of the homogeneous solutions between two successive output points. If

	AMP ≤ 1 the defaults are: NRTI = 0: max(ER(1), ER(2)/ER(3)); NRTI ≠ 0 : infinity.
ER	On entry ER(1) and ER(2) must contain a relative and an absolute tolerance, respectively, for solving the differential equation. On entry ER(3) must contain the machine constant. On exit ER(4) contains an estimation of the condition number of BVP. On exit ER(5) contains an estimated error amplification factor.
NRTI	On entry: NRTI = 0, in this case the subroutine determines automatically the output-points using AMP. NRTI = 1, in this case the output-points are supplied by the user in the array TI. NRTI > 1, in this case the subroutine computes the output-points TI(k) by TI(k) = A + (k − 1) * (B − A)/NRTI, so TI(1) = A and TI(NRTI + 1) = B. On exit NRTI contains the total number of output-points.
TI	On entry: if NRTI = 1, TI must contain the output-points in strictly increasing order: A=TI(1) < TI(2) < ⋯ < TI(N) = B. On exit: TI(i), i = 1, 2, ..., NRTI, contains the output-points.
NTI	The dimension of TI and one of the dimensions of the arrays Y, U, Q, D, PHI. NTI must be greater than the total number of output-points + 3.
Y	On exit Y(i, k), i = 1, 2, ..., N contains the solution of the BVP at the output-point TI(k), k = 1, ..., NRTI.
U	On exit U(i, k) i = 1, 2, ..., NU contains the relevant elements of the upper triangular matrix U_k, k = 2, ..., NRTI.
NU	One of the dimensions of U and PHI. NU must be at least equal to N*(N + 1)/2.
Q	On exit Q(i, j, k) i = 1, 2, ..., N, j = 1, 2, ..., N contains the N columns of the orthogonal matrix Q_k, k = 1, ..., NRTI.
D	If IHOM = 1 then on exit D(i, k) i = 1, 2, ..., N contains the inhomogeneous term d_k, k = 1, 2, ..., NRTI, of the multiple shooting recursion.
KPART	On exit KPART contains the order (size) of the left upper blocks of the matrices U_k, corresponding to the globally increasing modes.

PHI   On exit PHI contains an upper triangular fundamental solution of the multiple shooting recursion.

W,IW   Used as work space.

LW,LIW   Array dimensions; LW $\geq$ 8 * N + 2 * N * N, LIW $\geq$ 3 * N.

IERROR   Error indicator; if IERROR = 0 then there are no errors detected.

Example: MUSL

Consider the linear BVP given by

$$A(x) = \begin{bmatrix} 1-2\cos 2x & 0 & 1+2\sin 2x \\ 0 & 2 & 0 \\ -1+2\sin 2x & 0 & 1+2\cos 2x \end{bmatrix}, \quad q(x) = \begin{bmatrix} (-1+2\cos 2x - 2\sin 2x)e^x \\ -e^x \\ (1-2\cos 2x - 2\sin 2x)e^x \end{bmatrix}, \quad \beta = \begin{bmatrix} 1+e^6 \\ 1+e^6 \\ 1+e^6 \end{bmatrix}$$

for $0 \leq x \leq 6$, and $B_a = B_b = I$. The solution for this problem is

$$y(x) = (1, 1, 1)^T e^x$$

The following driver is used:

$\square$

```
      DOUBLE PRECISION A,B,MA(3,3),MB(3,3),BCV(3),AMP,ER(5),TI(15),
     1 Y(3,15),U(6,15),Q(3,3,15),D(3,15),PHI(6,15),W(42),
      INTEGER IW(9)
      EXTERNAL FLIN,FDIF
C
      N = 3
      IHOM = 1
      ER(1) = 1.D-11
      ER(2) = 1.D-6
      ER(3) = 1.D-16
      AMP =0.D0
      NRTI = 10
      NTI = 15
      NU = 6
      LW = 42
      LIW = 9
      A = 0.D0
      B = 6.D0
C
      DO 1100 I = 1,N
        DO 1000 J = 1,N
          MA(I,J) = 0.D0
          MB(I,J) = 0.D0
1000    CONTINUE
        MA(I,I) = 1.D0
        MB(I,I) = 1.D0
```

```
1100        CONTINUE
            BCV(1) = 1.D0 + DEXP(6.D0)
            BCV(2) = BCV(1)
            BCV(3) = BCV(1)
C
            CALL MUSL(FLIN,FDIF,N,IHOM,A,B,MA,MB,BCV,AMP,ER,NRTI,TI,NTI,Y,
   1             U,NU,Q,D,KPART,PHI,W,LW,IW,LIW,IERROR)
            ⋮
            ⋮
            END
              SUBROUTINE FLIN(X,Y,F)
            DOUBLE PRECISION X,Y(3),F(3)
            DOUBLE PRECISION TI,SI,CO
            TI = 2.D0 * X
            SI = 2.D0 * DSIN(TI)
            CO = 2.D0 * DCOS(TI)
            F(1) = (1.D0 - CO) * Y(1) + (1.D0 + SI) * Y(3)
            F(2) = 2.D0 * Y(2)
            F(3) = (-1.D0 + SI) * Y(1) + (1.D0 + CO) * Y(3)
            RETURN
            END
              SUBROUTINE FDIF(X,Y,F)
            DOUBLE PRECISION X,Y(3),F(3)
            DOUBLE PRECISION TI,SI,CO
            CALL FLIN(X,Y,F)
            TI = 2.D0 * X
            SI = 2.D0 *DSIN(TI)
            CO = 2.D0 * DCOS(TI)
            TI = DEXP(X)
            F(1) = F(1) + (-1.D0 + CO - SI) * TI
            F(2) = F(2) - TI
            F(3) = F(3) + (1.D0 - CO - SI) * TI
            RETURN
            END
```

The second code, MUSN, is based on ideas outlined in Sections 4.6.3 and 8.1. It solves nonlinear two-point BVPs of the form (1.3a, b), viz.,

$$\mathbf{y}' = \mathbf{f}(x, \mathbf{y}), \qquad a < x < b$$

$$\mathbf{g}(\mathbf{y}(a), \mathbf{y}(b)) = \mathbf{0}$$

MUSN computes discrete approximations of the Jacobians (when necessary) based on decoupled discrete BVPs, for which the upper triangular matrices are found directly, as in MUSL. To some extent it also utilizes techniques reminiscent of codes based on global methods (of Chapter 5). These include initially using a low tolerance, and refinement only when convergence on the thus obtained coarse mesh has been achieved. Also, the mesh is kept "fixed" during the iteration (only a subset of the mesh is stored, though); cf. [3]. The calling sequence of MUSN is similar to that of MUSL, and below we comment only on the parameters where there are differences.

```
          SUBROUTINE MUSN(FDIF,YOT,G,N,A,B,ER,TI,NTI,NRTI,AMP,ITLIM,
     1         Y,Q,U,NU,D,PHI,KPART,W,LW,IW,LIW,WG,LWG,IERROR)
C
C         DOUBLE PRECISION A,B,ER(5),TI(NTI),AMP,Y(N,NTI),Q(N,N,NTI),
C    1         U(NU,NTI),D(N,NTI),PHI(NU,NTI),W(LW,WG(LWG)
C         INTEGER N,NTI,NRTI,ITLIM,NU,KPART,IW(LIW),LIW,IERROR
C         EXTERNAL FDIF,YOT,G
```

FDIF   SUBROUTINE, supplied by the user with specification:

SUBROUTINE FDIF(X,Y,F)
DOUBLE PRECISION X,Y(N),F(N)

where N is the order of the system. FDIF must evaluate the right hand side of the differential equation, $\mathbf{f}(x,\mathbf{y})$ for $x = X$ and $\mathbf{y} = Y$ and place the result in F(1), ..., F(N).

YOT   SUBROUTINE, supplied by the user with specification:

SUBROUTINE YOT(X,Y)
DOUBLE PRECISION X,Y(N)

YOT must evaluate the initial approximation $\mathbf{y}^0(x)$ of the solution, for any value $x = X$, and place the result in Y(1), ..., Y(N).

G   SUBROUTINE, supplied by the user with specification:

SUBROUTINE G(N,YA,YB,FG,DGA,DGB)
DOUBLE PRECISION YA(N),YB(N),FG(N),DGA(N,N),DGB (N,N)

G must evaluate $\mathbf{g}(\mathbf{y}(a),\mathbf{y}(b))$ for $\mathbf{y}(a) = YA$ and $\mathbf{y}(b) = YB$ and place the result in FG(1), ..., FG(N). In addition, it must evaluate the Jacobians $\frac{\partial \mathbf{g}(\mathbf{u},\mathbf{v})}{\partial \mathbf{u}}$ for $\mathbf{u} = YA$ and $\frac{\partial \mathbf{g}(\mathbf{u},\mathbf{v})}{\partial \mathbf{v}}$ for $\mathbf{v} = YB$, and place the result in the arrays DGA and DGB, respectively.

ER   On entry ER(1) must contain the required tolerance for solving the differential equation
On entry ER(2) must contain the initial tolerance with which a first approximate solution will be computed. This approximation is then used as an initial approximation for the computation of a solution

	with a tolerance ER(2) * ER(2), and so on, until the required tolerance is reached. As an initial tolerance, max (ER(1), min (ER(2), 1.d − 2)) is used.
AMP	On entry AMP must contain the allowed incremental factor of the solution and the fundamental solution of the first variational problem between two successive shooting points. Unless 1 < AMP < .25 * SQRT (ER(1)/ER(3)) the default value .25 * SQRT (ER(1)/ER(3)) is used.
ITLIM	Maximum allowed number of iterations.
LW,LIW	array dimension with LW ≥ 7 * N + 3 * N * NTI + 4 * N * N and LIW ≥ 3 * N + NTI
WG	Used as work space for storing the fixed mesh.
IERROR	On entry: if ERROR = 1, diagnostics during computation are given. On exit: error indicator, IERROR = 0, no errors detected.

Example: MUSN

Consider the differential equations

$$u' = .5u(w-u)/v$$
$$v' = -0.5(w-u)$$
$$w' = (0.9 - 1000(w-y) - 0.5w(w-u))/z$$
$$z' = 0.5(w-u)$$
$$y' = -100(y-w)$$

subject to the boundary conditions

$$u(0) = v(0) = w(0) = 1, z(0) = -10, w(1) = y(1).$$

As an initial guess for the solution we take

$$u(x) = 1; \quad v(x) = 1; \quad z(x) = -10;$$
$$w(x) = -4.5x^2 + 8.91x + 1; \quad y(x) = -4.5x^2 + 9x + 0.91.$$

The following driver is used:

```
      IMPLICIT REAL*8 (A-H,O-Z)
      DIMENSION ER(5),TI(12),X(5,12),Q(5,5,12),U(15,12),D(5,12),
     1               PHIREC(15,12),W(315),WGR(20)
      INTEGER IW(27)
      EXTERNAL FDIF,YOT,G
C
      N = 5
      NU = 15
      NTI = 12
```

Appendix A

```
              LW = 315
              LIW = 27
              ER(3) = 1.1D – 15
              LWG = 20
              ER(1) = 1.D – 6
              ER(2) = 1.D – 2
              JFC = 0
              A = 0.D0
              B = 1.D0
              NRTI = 10
              AMP = 100
              ITLIM = 20
       C
              CALL MUSN(FDIF,YOT,G,N,A,B,ER,TI,NTI,NRTI,AMP,ITLIM,X,Q,U,NU,D,
          1              PHIREC,KPART,W,LW,IW,LIW,WGR,LWG,IERROR)
              :
              :
              END
                  SUBROUTINE FDIF(X,Y,F)
              IMPLICIT REAL*8(A-H,O-Z)
              DIMENSION Y(5),F(5)
              Y3MY1 = Y(3) – Y(1)
              Y3MY5 = Y(3) – Y(5)
              F(1) = 0.5D0 * Y(1) * Y3MY1/Y(2)
              F(2) = – 0.5D0 * Y3MY1
              F(3) = (0.9D0 – 1.D3 * Y3MY5 – 0.5D0 * Y(3) * Y3MY1)/Y(4)
              F(4) = 0.5D0 * Y3MY1
              F(5) = 1.D2 * Y3MY5
              RETURN
              END
                  SUBROUTINE YOT(X,Y)
              IMPLICIT REAL*8 (A-H,O-Z)
              DIMENSION Y(5)
              Y(1) = 1.D0
              Y(2) = 1.D0
              Y(4) = – 10.D0
              Y(3) = – 4.5D0*X*X + 8.91D0 * X + 1.D0
              Y(5) = – 4.5D0*X*X + 9.D0 * X + 0.91D0
              RETURN
              END
                  SUBROUTINE G(N,XA,XB,FG,DGA,DGB)
              IMPLICIT REAL*8 (A-H,O-Z)
              DIMENSION XA(N),XB(N),FG(N),DGA(N,N),DGB(N,N)
              DO 1100 I = 1,N
              DO 1100 J = 1,N
                 DGA(I,J) = 0.D0
                 DGB(I,J) = 0.D0
       1100   CONTINUE
              DGA(1,1) = 1.D0
              DGA(2,2) = 1.D0
```

```
DGA(3,3) = 1.D0
DGA(4,4) = 1.D0
DGB(5,3) = 1.D0
DGB(5,5) = -1.D0
FG(1) = XA(1) - 1.D0
FG(2) = XA(2) - 1.D0
FG(3) = XA(3) - 1.D0
FG(4) = XA(4) + 10.D0
FG(5) = XB(3) - XB(5)
RETURN
END
```

□

Appendix B

A Collocation Code

The package briefly described here is a modification of the package COLSYS by Ascher, Christiansen, and Russell [1]. Its implementation is described in Bader and Ascher [2]. It is written in FORTRAN IV (compatible with FORTRAN 77) and is available through NETLIB (under the name COLNEW) or from

Dr. G. Bader
Institut f. Angewandte Mathematik
University of Heidelberg
Im Neuenheimer Feld 294
D-6900 Heidelberg 1, Germany.

The package solves a mixed-order system of ODEs subject to separated, multipoint boundary conditions, given by (1.8a,b) as rewritten here:

$$u_i^{(m_i)} = f_i(x, u_1, \ldots, u_1^{(m_1-1)}, u_2, \ldots, u_d^{(m_d-1)})$$
$$= f_i(x, \mathbf{z}(\mathbf{u})) \quad 1 \le i \le d \quad a < x < b$$
$$g_j(\mathbf{z}(\mathbf{y}(\zeta_j))) = 0 \quad 1 \le j \le m^*$$

where $\mathbf{u}(x) = (u_1(x), \ldots, u_d(x))^T$, $m^* = \sum_{j=1}^{d} m_i$, $a = \zeta_1 \le \zeta_2 \le \cdots \le \zeta_{m^*} = b$, and
$\mathbf{z}(\mathbf{u}(x)) := (u_1(x), u_1'(x), \ldots, u_1^{(m_1-1)}(x), u_2(x), \ldots, u_2^{(m_2-1)}(x), \ldots, u_d^{(m_d-1)}(x))^T$.
The orders m_i of the differential equations satisfy $1 \le m_i \le 4$, $i = 1, \ldots, d$. The functions f_i and g_j are generally nonlinear.

The method used to approximate the solution is collocation at Gaussian points, requiring $u_i(x) \in C^{(m_i-1)}[a,b]$, $i=1, \ldots, d$. It is described in detail in Sections 5.6.2–5.6.4. The number of collocation points (stages) per subinterval k is chosen such that $\max_{1 \leq i \leq d} m_i \leq k \leq 7$.

COLSYS was written in 1980, and has enjoyed considerable usage. The package described here, which may be referred to as either COLNEW or COLSYS, incorporates a new basis representation replacing B-splines and improvements for the linear and nonlinear algebraic equation solvers.

When one considers a general purpose collocation code, the following aspects arise:

1. Solution representation
2. Linear system solution
3. Error estimation and mesh selection
4. Nonlinear problem solution

The techniques used in this code have been discussed in various chapters of the text. The method used for solution representation is described in Section 5.6.2. For the solution of linear systems of equations, which are cast in an almost block diagonal form (see Section 7.2.1), we have modified the code SOLVEBLOK of de Boor and Weiss [3] (see [2, Section 6]). The method implemented for error estimation and mesh selection is the same as in [1] and is described and demonstrated in Section 9.3. Finally, for nonlinear problems a damped Newton strategy is used, essentially as described in Section 8.1.1. Care has to be taken when one is scaling the dependent variables for the latter purpose, as described in [2, Section 6].

Below we describe the calling sequence of COLSYS (COLNEW).

```
      SUBROUTINE COLSYS (NCOMP, M, ALEFT, ARIGHT, ZETA, IPAR, LTOL,
1                       TOL, FIXPNT, ISPACE, FSPACE, IFLAG,
2                       FSUB, DFSUB, GSUB, DGSUB, GUESS)
```

*   variables

*   NCOMP = d – number of differential equations (≤ 20)

*   M(j) – order of the jth differential equation

*   ALEFT = a, ARIGHT = b – interval ends.

*   ZETA(j) = ζ_j, $1 \leq j \leq \sum_{l=1}^{NCOMP} M(l)$. Must be a mesh point in all meshes used, see description of IPAR(11) and FIXPNT below.

*   IPAR – an integer array dimensioned at least 11. A list of the parameters in IPAR and their meaning follows:

*   IPAR(1) = 0 if the problem is linear
                       = 1 if the problem is nonlinear

*   IPAR(2) = number of collocation points per subinterval ($=k$).

*   IPAR(3) = number of subintervals in the initial mesh ($=N$).

*   IPAR(4) = number of solution and derivative tolerances, $0 <$ IPAR(4) $\le m^*$.

*   IPAR(5) = dimension of FSPACE.

*   IPAR(6) = dimension of ISPACE.

*   IPAR(7) – output control
                       = –1 for full diagnostic printout
                       = 0 for selected printout
                       = 1 for no printout

*   IPAR(8) = 0 causes COLSYS to generate a uniform initial mesh.
                          = 1 if the initial mesh is provided by the user. It is defined in FSPACE as follows: The mesh $a = x_1 < x_2 < \cdots < x_N < x_{N+1} = b$ will occupy FSPACE(1), ..., FSPACE(N+1).
                          = 2 if the initial mesh is supplied by the user as with IPAR(8) = 1, and in addition no adaptive mesh selection is to be done.

*   IPAR(9) = 0 if no initial guess for the solution is provided.
                          = 1 if an initial guess is provided by the user in subroutine GUESS
                          = 2 if an initial mesh and approximate solution coefficients are provided by the user in FSPACE. (The former and new mesh are the same.)
                          = 3 if a former mesh and approximate solution coefficients are provided by the user in FSPACE, and the new mesh is to be taken twice as coarse, i.e., every second point from the former mesh.
                          = 4 if in addition to a former initial mesh and approximate solution coefficients, a new mesh is provided in FSPACE as well. (See description of output for further details on IPAR(9) =2, 3, 4.)

*   IPAR(10)= 0 if the problem is regular
                          = 1 if the first relax factor is small, and the nonlinear iteration does not rely on past covergence (use for an extra sensitive nonlinear problem only).
                          = 2 if we are to return immediately upon (a) two successive nonconvergences, or (b) after obtaining an error estimate for the first time.

*   IPAR(11)= number of fixed points in the mesh other than ALEFT and ARIGHT.

*   LTOL   – an INTEGER array of dimension IPAR(4). LTOL(j) = 1 specifies that the jth tolerance in TOL controls the error in the lth component of $\mathbf{z}(\mathbf{u})$.

*   TOL   – a REAL array of dimension IPAR(4). TOL(j) is the error tolerance on the LTOL(j)th component of $\mathbf{z}(\mathbf{u})$. Thus, the code attempts to satisfy for $j = 1, \ldots,$ IPAR(4) on each subinterval $|(\mathbf{z}(v) - \mathbf{z}(\mathbf{u}))_{LTOL(j)}| \leq TOL(j)\,(|\mathbf{z}(\mathbf{u})_{LTOL(j)}| + 1)$ if $\mathbf{v}(x)$ is the approximate solution vector.

*   FIXPNT – an array of dimension IPAR(11). It contains the points, other than ALEFT and ARIGHT, which are to be included in every mesh.

*   ISPACE – an INTEGER work array of dimension IPAR(6). Its size provides a constraint on the maximum mesh size.

*   FSPACE – a REAL work array of dimension IPAR(5). Its size provides a constraint on the maximum mesh size.

*   IFLAG   – the mode of return from COLSYS.
                          = 1   for normal return
                          = 0   if the collocation matrix is singular.
                          = −1   if the expected number of subintervals exceeds storage specifications.
                          = −2   if the nonlinear iteration has not converged.
                          = −3   if there is an input data error.

******************   user supplied subroutines   *****************
*
*
*   The following subroutines must be declared external in the main program which calls COLNEW (COLSYS).

*   FSUB   – name of subroutine for evaluating $f_i(x, \mathbf{z}(\mathbf{u}(x)))$. It should have the heading

                      SUBROUTINE FSUB (X , Z , F)

             where X = x, Z = $\mathbf{z}$ and F is the vector containing the values of f_i, as defined above.

*   DFSUB – name of subroutine for evaluating the Jacobian of F at a point X. it should have the heading

Appendix B

SUBROUTINE DFSUB (X , Z , DF)

where $Z = \mathbf{z}(\mathbf{u}(x))$ is defined as for FSUB and the $d \times m^*$ array DF should be filled by the partial derivatives of F, namely, for a particular call one calculates

$$DF(i,j) = \frac{\partial f_i}{\partial z_j}, \; i=1, \ldots, d, \; j = 1, \ldots, m^*$$

* GSUB – name of subroutine for evaluating the jth side condition g_j at a point $x = \text{ZETA}(j)$, $1 \leq j \leq m^*$. It should have the heading

SUBROUTINE GSUB (J , Z , G)

where Z is as for FSUB, and j and the scalar G are as above.

* DGSUB – name of subroutine for evaluating the jth row of the Jacobian of g_j. It should have the heading

SUBROUTINE DGSUB (J , Z , DG)

where Z is as for FSUB, J as for GSUB and the m^*-vector DG gives $DG(l) = \frac{\partial g_j}{\partial z_l}, \; j = 1, \ldots, m^*$.

* GUESS – name of subroutine to evaluate the initial approximation for $Z = \mathbf{z}(\mathbf{u}(x))$ and for DMVAL = vector of m_j^{th} derivatives of $\mathbf{u}(x)$. It should have the heading

SUBROUTINE GUESS (X , Z , DMVAL)

Note that this subroutine is needed only for nonlinear problems if using IPAR(9) = 1.

*
****************** use of output from COLSYS ************
*
******************** solution evaluation ***************
*

On return from COLSYS, the arrays FSPACE and ISPACE contain information specifying the approximate solution. The user can produce the solution $\mathbf{z}(\mathbf{u}(x))$ at any point $x \in [a,b]$ by the statement,

CALL APPSLN (X, Z, FSPACE, ISPACE)

******************** simple continuation ***************
*

A formerly obtained solution can easily be used as the first approximation for the non-linear iteration for a new problem by setting IPAR(9) = 2, 3, or 4.
*
****

Example

Consider the linear BVP

$$\varepsilon u_1' = -u_1 + u_2 + q_1(x)$$

$$0 < x < 1$$

$$u_2^{(iv)} = u_1 + u_2 + q_2(x)$$

$$u_1(0) = 2, \quad u_2(0) = u_2(1) = u_2''(0) = u_2''(1) = 0$$

where we choose the inhomogeneities $q_1(x)$ and $q_2(x)$ such that the solution is

$$u_1(x) = e^{-x/\varepsilon} + \cos \pi x, \quad u_2(x) = \sin \pi x$$

The parameter ε is small and positive; we take $\varepsilon = 10^{-4}$.

In this example we have a 4th-order ODE for a slow solution $u_2(x)$, but the mesh must be very fine near $x = 0$ in order to resolve the boundary layer in $u_1(x)$. We attempt to solve this problem using our collocation code on a machine with 14 hexadecimal digit mantissa, imposing an error tolerance of 10^{-7} on u_1, u_2, and u_2'', and starting from a uniform initial mesh with 10 subintervals. The following driver is used:

```
      REAL*8 ZETA(5), FSPACE(52000), TOL(3), FIXPNT, X, Z(6)
      REAL*8 ALEFT, ARIGHT, EPS, PI
      INTEGER M(2), ISPACE(3000), LTOL(3), IPAR(11)
      COMMON  EPS , PI
      EXTERNAL FSUB, DFSUB, GSUB, DGSUB, GUESS
      EPS = 1.D-4
      PI = 3.1415926535898D0
C
      NCOMP = 2
      M(1) = 1
      M(2) = 4
      ALEFT = 0.D0
      ARIGHT = 1.D0
C
      ZETA(1) = 0.D0
      ZETA(2) = 0.D0
      ZETA(3) = 0.D0
      ZETA(4) = 1.D0
      ZETA(5) = 1.D0
C
      IPAR(1) = 0
      IPAR(2) = 5
      IPAR(3) = 10
      IPAR(4) = 3
      IPAR(5) = 52000
```

```fortran
            IPAR(6) = 3000
            IPAR(7) = -1
            IPAR(8) = 0
            IPAR(9) = 0
            IPAR(10) = 0
            IPAR(11) = 0
C
            LTOL(1) = 1
            LTOL(2) = 2
            LTOL(3) = 4
            TOL(1) = 1.D-7
            TOL(2) = 1.D-7
            TOL(3) = 1.D-7
C
            CALL COLSYS (NCOMP, M, ALEFT, ARIGHT, ZETA, IPAR, LTOL,
           +        TOL, FIXPNT, ISPACE, FSPACE, IFLAG,
           +        FSUB, DFSUB, GSUB, DGSUB, GUESS)
C

                ⋮

            END
                SUBROUTINE GUESS (X, Z, DMVAL)
            END
                SUBROUTINE FSUB (X, Z, F)
            REAL*8 Z(6), F(2) , EPS, PI, PIX, X
            COMMON  EPS, PI
            PIX = PI * X
            F(1) = (-Z(1) + Z(2) + DCOS(PIX) - (1.D0 + EPS *PI) * DSIN (PIX)) / EPS
            F(2) = Z(1) + Z(2) + (PI**4 - 1.D0)* DSIN(PIX) - DCOS(PIX) - DEXP(-X/EPS)
            RETURN
            END
                SUBROUTINE DFSUB (X, Z, DF)
            REAL*8 Z(5), DF(2,5), EPS, PI, X
            COMMON  EPS, PI
            DO 10 I=1,2
               DO 10 J=1,5
         10      DF(I,J) = 0.D0
            DF(1,1) = -1.D0 / EPS
            DF(1,2) = 1.D0 / EPS
            DF(2,1) = 1.D0
            DF(2,2) = 1.D0
            RETURN
            END
                SUBROUTINE GSUB (I, Z, G)
            REAL*8 Z(1), G
            GO TO (1, 2, 3, 2, 3), I
          1 G = Z(1) - 2.D0
            RETURN
          2 G = Z(2)
            RETURN
```

```
      3 G = Z(4)
        RETURN
        END
                SUBROUTINE DGSUB (I, Z, DG)
        REAL*8 Z(1), DG(1)
        DO 10 J=1,5
     10   DG(J) = 0.D0
        GO TO (1, 2, 3, 2, 3), I
      1 DG(1) = 1.D0
        RETURN
      2 DG(2) = 1.D0
        RETURN
      3 DG(4) = 1.D0
        RETURN
        END
```

The code performs very well (even though the original code of [1] does not): It automatically adapts the mesh to the boundary layer and terminates with a mesh of 80 subintervals of which more than half are in $[0, 10^{-3}]$, and a maximum error of .79−14 for $u_2(x)$.

□

The robustness of the code for nonlinear BVPs has been demonstrated through a number of examples in Chapters 8 and 9. On the other hand, it does not always exit in the most graceful way. For instance, when solving Example 4.11 with $T = 2.5$ there is no difficulty, but when solving the same (but now rather ill-conditioned) problem for $T = 3$, the code is unable to obtain accurate solutions with the same machine precision as before, and no indication of the reason for the difficulty is given.

References

CHAPTER 1

The physical details of Example 1.4 are described in

1. C.L. Huang, "Application of quasilinearization technique to the vertical channel flow and heat convection," *Int. J. Non-Linear Mech.* 13 (1978), 55–60.

Some plots of the solution for large Reynolds numbers are given in:

2. U. Ascher, "Solving boundary-value problems with a spline-collocation code," *J. Comp. Phys.* 34 (1980), 401–413.

The numerical method used in [2] is described in Chapter 5. Example 1.5 appears in many variations in the chemical engineering literature; see

3. B.A. Finlayson, *The Method of Weighted Residuals and Variational Principles.* New York: Academic Press, 1972.

4. G.F. Carey and B.A. Finlayson, "Orthogonal collocation on finite elements," *Chem. Eng. Sci.* 30 (1975), 587–596.

For Example 1.6 see:

5. B.A. Finlayson, "Water movement in dessicated soils," in *Finite Elements in Water Resources,* ed. W. Gray, G. Pinder, C. Brebbia, London: Pentech Press, 1977.

The ray tracing problem presented in Example 1.7 is derived and discussed in:

6. W.K.H. Lee and S. Stewart, *Principles and Applications of Microearthquake Networks,* New York: Academic Press, 1981.

The formulation (1.17a) is a variant, due to R. Comer, of the formulation in:

7. V. Pereyra, W.K.H. Lee, and H.B. Keller, "Solving seismic ray tracing problems in a heterogeneous medium," *Bull. Seism. Soc. Amer.* 70 (1980), 79–99.

The derivation of the ODEs of Example 1.8 is given in:

8. H. Takeuchi and M. Saito, "Seismic surface waves," in *Methods of Computational Physics,* 11. New York: Academic Press, 1972.

There are many papers dealing with the numerical solution of this problem by a variety of methods. We mention only:

9. B.L.N. Kennett, *Seismic Wave Propagation in Stratified Media.* Cambridge: Cambridge University Press, 1983.

10. P. Spudich and U. Ascher, "Calculation of complete theoretical seismograms in vertically varying media using collocation methods," *Geoph. J. R. Astr. Soc.* 75 (1983), 101–124.

Our description of Example 1.9 is taken from:

11. P. Concus, "Static menisci in a vertical right circular cylinder," *J. Fluid Mech.* 34 (1968), 481–495.

Some of the early work on the numerical solution of IVPs by Runge and by Bashford and Adams was done in an attempt to solve this problem by using the shooting technique described in Chapter 4.

The measles problem is described in:

12. I..B. Schwartz, "Estimating regions of existence of unstable periodic orbits using computer-based techniques," *SIAM J. Numer. Anal.* 20 (1983), 106–120.

As the title implies, the ambitions (and achievements) of this paper run much higher than the mere solution of Example 1.10.

The kidney problem of Example 1.11 was described by R. Mejia. The model is in:

13. J. L. Stephenson, R. P. Tewarson and R. Mejia, "Quantitative analysis of mass and energy balance in non-ideal models of the renal counterflow system," *Proc. Nat. Acad. Sci. USA* 71 (1974), 1618–1622,

and numerical solutions appear in

14. R. Mejia, J. L. Stephenson and R. J. Le Veque, "A test problem for kidney model," manuscript,

15. G. Bader and U. Ascher, "A new basis implementation for a mixed order boundary value ODE solver," *SIAM J. Sci. Stat Com.* 8 (1987), 483-500.

The magnetic monopoles example is described in:

16. K. O. Olynyk, "Spherically-symmetric monopole solutions in SU(2) and SU(3) gauge theories," Physics MSc, Simon Fraser Univ., 1978.

The description of Example 1.13 is taken from:

17. R. Miura, "Accurate computation of the stable solitary wave for the FitzHugh-Nagumo equations," *J. Math. Biology* 13 (1982), 247–269.

See also Example 7.1 of:

18. U. Ascher and R. D. Russell, "Reformulation of boundary value problems into 'standard form'," *SIAM Review* 23 (1981), 238–254.

The nonlinear beam problem was described by J. Flaherty; see:

19. J. F. Flaherty and R. E. O'Malley, "Singularly perturbed boundary value problems for nonlinear systems, including a challenging problem for a nonlinear beam," in *Springer Lecture Notes in Math* 942 (1981).

The numerical solution of this problem for very small ε is discussed in:

20. U. Ascher and R. Weiss, "Collocation for singular perturbation problems III: Nonlinear problems without turning points," *SIAM J. Sci. Stat. Comp.* 5 (1984), 811–829.

21. U. Ascher, "On some difference schemes for singular singularly-perturbed boundary value problems," *Numer. Math.* 46 (1985), 1–30.

Another singular perturbation problem with no turning points is Example 1.15. In our presentation we have followed:

22. P. A. Markowich and C. A. Ringhofer, "A singularly perturbed boundary value problem modelling a semiconductor device," *SIAM J. Appl. Math.* 44 (1984), 231–256.

23. P. A. Markowich, *The Stationary Semiconductor Device Equations*. New York: Springer, 1986.

An earlier mathematical analysis for a more restricted case was given by Vasil'eva and Stel'makh; it, and much more, appears in the manuscript,

24. A. B. Vasil'eva and V. F. Butuzov, "Singularly perturbed equations in the critical case," MRC Tech. Rep. 2039 (1980), Madison, Wisconsin.

A number of special techniques were devised to solve this problem numerically; see [23] and [21]. Example 1.16 is presented in:

25. B. Sigfridsson and J. L. Lindstrom, "Electrical properties of electron-irradiated n-type silicon," *J. Appl. Phys.* 47 (1976), 4611–4620,

and a numerical discussion is given in:

26. L. Abrahamsson, "Numerical solution of a boundary value problem in semiconductor physics," Uppsala Computer Science Rep. 81, (1979).

See also:

27. M. R. Maier, "An algorithm for singularly perturbed boundary value problems," manuscript (1983),

where a variant of Example 1.15 is discussed as well. A reference for the problem of shock wave in one-dimensional nozzle flow is:

28. C. E. Pearson, "On nonlinear ordinary differential equations of boundary layer type," *J. Math. and Phys.* 47 (1968), 351–358,

where the problem is also solved numerically.

A legendary volume of literature exists on the swirling flow problems. It has proved to be challenging, both analytically and numerically. The basic similarity transformation of Example 1.18 was done in:

29. T. Von Karman, "Über laminare und turbulente Reibung," *ZAMM* 1 (1921), 232–252.

Von Karman actually did the derivation only for $\gamma = 0$, but the extension is obvious. For Example 1.18, asymptotic analysis and computational analysis are given, respectively, in:

30. P. A. Markowich, "Asymptotic analysis of von Karman flows," *SIAM J. Appl. Math.* 42 (1982), 549–557.

31. M. Lentini and H. B. Keller, "The von Karman swirling flows," *SIAM J. Appl. Math.* 38 (1980), 52–64.

Further references are also given in these papers. Example 1.19, known in some circles as "Holt's problem," has been frequently used as a tough test problem for BVP codes. Relevant references include:

32. K. Stewartson and B. A. Troesch, "On a pair of equations occurring in swirling viscous flow with an applied magnetic field," *ZAMP* 29 (1977), 951–963.

33. J. F. Holt, "Numerical solution of nonlinear two-point boundary value problems by finite difference methods," *Comm. ACM* 7 (1964), 366–373.

and papers by B. A. Troesch and by U. Ascher, J. Christiansen and R. D. Russell in:

34. B. Childs, et al., eds., *Codes for Boundary Value Problems*, Springer Lecture notes in Computer Science 76 (1979).

Example 1.20 has possibly the richest history of controversy of all BVP applications. Following von Karman's approach [29], G. K. Batchelor and K. Stewartson considered the problem in the early 1950's, reaching different conjectures regarding the nature of the flow away from the disk (Stewartson's turned out to be correct). Many asymptotic analyses, as well as numerical studies, have been made since. One published solution, found numerically by a dubious method, was proved analytically impossible by J. B. McLeod and S. V. Parter, causing some people to realize that applied analysis was not a total waste of time, even in the computer era. A summary and references are provided in:

35. S. V. Parter, "Swirling flow between rotating coaxial disks," *Lecture Notes in Math.* 942, Berlin: Springer, 1981.

The reentry control problem of Example 1.21 is described, with the numerical solution carefully explained, in section 7.3.7 of:

36. J. Stoer and R. Bulirsch, *Introduction to Numerical Analysis*. Berlin: Springer Verlag, 1980.

The optimal harvesting problem was formulated in:

37. D. Ludwig, "Optimal harvesting of a randomly fluctuating resource I: Application of perturbation methods," *SIAM J. Appl. Math.* 37 (1979), 166–184.

See also:

38. D. Ludwig and J. Varah, "Optimal harvesting of a randomly fluctuating resource II: Numerical methods and results," *SIAM J. Appl. Math.* 37 (1979), 185–205.

Examples 1.23 and 1.24 come from the theory of shells of revolution. For Example 1.23 see:

39. F. Y. M. Wan, "Polar dimpling of complete spherical shells," in *Theory of Shells, Proc. IUTAM Symp.*, W. T. Koiter and G. K. Mikhailov, eds., North Holland, 1979.

and the numerical treatment in [2]. For Example 1.24 see:

40. F. Y. M. Wan, "The dimpling of spherical caps," *Mech. Today*, 5 (1980), 495–508.

41. F. Y. M. Wan and U. Ascher, "Horizontal and flat points in shallow cap dimpling," Tech. rep. 80–5 (1980), IAMS, Univ. of B. C., Vancouver.

A brief description of a problem like the flame Example 1.25, techniques for its numerical solution, and further references are given in:

42. Y. Reuven, M. D. Smooke, and H. Rabitz, "Sensitivity analysis of boundary value problems: application to nonlinear diffusion systems," *J. Comp. Phys.* 64 (1986), 27–55.

Here is a list of references where more applications, not included in our chosen 25, are given. Needless to say, there are many more.

The conference proceedings [34] contain a large number of papers with applications. In particular, we mention the references of two of us (Ascher and Russell) there,

43. M. Kubiček, V. Hlaváček, and M. Holodniok, "Test examples for comparison of codes for nonlinear boundary value problems in ordinary differential equations," in [34], 325–346,

and

44. I. Gladwell, "The development of the boundary-value codes in the ordinary differential equations chapter of the NAG library," in [34], 122–143.

In addition, [2], [18], and

45. M. Kubiček and V. Hlaváček, *Numerical Solution of Nonlinear Boundary Value Problems with Applications*. Englewood Cliffs, NJ: Prentice-Hall, 1983.

contain other examples of BVPs. Another application is in:

46. U. Ascher and F. Y. M. Wan, "Numerical solutions for maximum sustainable consumption growth with a multi-grade exhaustible resource," *SIAM J. Sci. Stat. Comput.* 1 (1980), 160–172.

A collection of test problems is contained in:

47. H.-J. Diekoff, P. Lory, H. J. Oberle, H.-J. Pesch, P. Rentrop, and R. Seydel, "Comparing routines for the numerical solution of initial value problems of ordinary differential equations in multiple shooting," *Numer. Math.* 27 (1977), 449–469.

Of course, the various collections of BVPs in these references are not necessarily mutually disjoint. Other papers considering the same types of problems include:

48. P. Deuflhard, "A stepsize control for continuation methods with special application to multiple shooting techniques," Tech. rep. TUM 7627 (1976), Technische Universitat München.

49. R. Rentrop, "Numerical solution of the singular Ginzburg-Landau equations by multiple shooting," *Computing* 16 (1976), 61–67.

50. P. Deuflhard, H.-J. Pesch, and R. Rentrop, "A modified continuation method for the numerical solution of nonlinear two-point boundary value problems by shooting techniques," *Numer. Math.* 26 (1976), 327–343.

A number of interesting BVPs appear in:

51. F. Milinazzo and P. G. Saffman, "Turbulence model predictions for the inhomogeneous mixing layer," *Studies Appl. Math.* 55 (1976), 45–63.

CHAPTER 2

Much of the material reviewed in this chapter appears in introductory numerical analysis texts, such as:

1. R. C. Burden and J. D. Faires, *Numerical Analysis*. Prindle, Weber, and Schmidt, 1985.

2. S. D. Conte and C. de Boor, *Elementary Numerical Analysis*, 2nd ed. New York: McGraw-Hill, 1980.

3. G. Dahlquist and A. Björck (translated by N. Anderson), *Numerical Methods*. Englewood Cliffs, N.J.: Prentice-Hall, 1974.

A good text on numerical linear algebra is:

4. G. H. Golub and C. F. van Loan, *Matrix Computations*. Baltimore: Johns Hopkins Univ. Press, 1983.

For nonlinear equation solution, see:

5. J. E. Dennis, Jr., and J. J. Moré, "Quasi-Newton Methods, Motivation and Theory," *SIAM Review* 19 (1977), 48–89.

6. J. M. Ortega and W. C. Rheinboldt, *Iterative Solution of Nonlinear Equations in Several Variables*. New York: Academic Press, 1970.

For polynomial and spline approximation, see:

7. C. de Boor, *A Practical Guide to Splines*. Berlin: Springer-Verlag, 1978.

A thorough survey of quadrature techniques is found in:

8. P. J. Davis and P. Rabinowitz, *Methods of Numerical Integration,* 2nd ed. Orlando: Academic Press, 1985.

Some texts on numerical solution of IVPs are:

9. P. Henrici, *Discrete Variable Methods in Ordinary Differential Equations.* New York: John Wiley & Sons, 1962.

10. C. W. Gear, *Numerical Initial-Value Problems in Ordinary Differential Equations.* Englewood Cliffs, N.J.: Prentice-Hall, 1971.

11. J. D. Lambert, *Computational Methods in Ordinary Differential Equations. New York:* John Wiley & Sons, 1973.

12. L. F. Shampine and M. K. Gordon, *Computer Solution of Ordinary Differential Equations: The Initial Value* Problem. San Francisco: W. H. Freeman, 1975.

CHAPTER 3

The material in Section 3.1.1 can be found in many textbooks, e. g.,

1. E. A. Coddington and N. Levinson, *The Theory of Ordinary Differential Equations.* New York: McGraw Hill, 1955.
2. P. Hartman, *Ordinary Differential Equations.* New York: Wiley, 1964.

An introductory treatment of BVPs is much more difficult to find. An early treatment is:

3. P. Bailey, L. F. Shampine, and P. Waltman, *Nonlinear Two Point Boundary Value Problems.* New York: Academic Press, 1968.

Much of the basic theory was developed and presented in:

4. H. B. Keller, *Numerical Methods for Two-Point Boundary Value Problems.* London: Blaisdell, 1968.

A fairly complete treatment of Green's functions can be found in:

5. I. Stakgold, *Green's Functions and Boundary Value Problems.* New York: Wiley, 1979,

in addition to [1] and [2]. Most of the stability theory in Section 3.3 is in:

6. L. Cesari, *Asymptotic Behaviour and Stability Problems in Ordinary Differential Equations.* Berlin: Springer, 1971.

Also, for a discussion on uniform stability, see:

7. W. A. Coppel, *Dichotomies in Stability Theory*. Lecture Notes in Math. 629. Berlin: Springer, 1978.

An example of an ODE which possesses stability but not uniform stability can be found in:

8. M. Rao, *Ordinary Differential Equations Theory and Applications*. London: Edward Arnold, 1980.

A more extensive discussion of the material in section 3.3.3 can be found in:

9. C. J. Harris and J. F. Miles, *Stability of Linear Systems: Some Aspects of Kinematic Similarity*. New York: Academic Press, 1980.

10. G. Söderlind and R. M. M. Mattheij, "Stability and asymptotic estimates in nonautonomous linear differential systems," *SIAM J. Math. Anal.* 16 (1985), 69–92.

Theorem 3.65 is new.
A more comprehensive treatment of the nonlinear case than section 3.3.4 is available from a wide variety of sources, e. g. [1], [6], and

11. D. W. Jordan and P. Smith, *Nonlinear Ordinary Differential Equations*. Oxford: Clarendon Press, 1977.

The material in Section 3.4 has not been brought together before, and some of the results are new. A presentation of dichotomy is given in [7] and

12. J. L. Massera and J. J. Schäeffer, "Linear differential equations and functional analysis, I.," *Ann. of Math.* 67 (1958), 517–573.

Most of the material in Section 3.4.3 is drawn from

13. R. M. M. Mattheij, "The conditioning of linear boundary value problems," *SIAM J. Numer. Anal.* 19 (1982), 963–978.

14. F. de Hoog and R. M. M. Mattheij, "On dichotomy and well-conditioning in BVP," SIAM J. Numer. Anal. 24 (1987), 89-105.

15. F. de Hoog and R. M. M. Mattheij, "The role of conditioning in shooting techniques," in *Numerical Boundary Value ODEs*; eds. U. Ascher and R. D. Russell, New York: Birkhäuser, 1986.

CHAPTER 4

Since the development of sophisticated numerical methods for solving IVPs (Section 2.7) predates that for BVPs, it is not surprising to find that early efforts of solving BVPs centered around the shooting approach. Two of the first textbooks on numerical ODEs,

1. L. Fox, *The Numerical Solution of Two-Point Boundary Problems in Ordinary Differential Equations*. Oxford: Oxford University Press, 1957.

and

2. L. Collatz, *The Numerical Treatment of Differential Equations*. Berlin: Springer, 3rd ed., 1960.

discuss shooting methods. One of the first numerical papers discussing general shooting methods is:

3. T. R. Goodman and G. N. Lance, "The numerical integration of two-point boundary value problems," *MTAC* 10 (1956), 82–86,

although this was preceded by many applications of single shooting for particular problems. An excellent early discussion of initial value methods is in the book:

4. N. Bakhvalov, *Numerical Methods*. Moscow: MIR, 1977 (Russian edition, 1975).

The basic convergence result for shooting for nonlinear problems can be found in:

5. R. Weiss, "The convergence of shooting methods," *BIT* 13 (1973), 470–475.

Development of general-purpose codes using single shooting has been done for various mathematical libraries, especially NAG; see I. Gladwell's article in:

6. B. Childs, et al., *Codes for Boundary Value Problems in Ordinary Differential Equations*. Lecture Notes in Computer Science 76. Berlin: Springer-Verlag, 1979.

Multiple shooting appeared in early forms in:

7. L. Fox, "Some numerical experiments with eigenvalue problems in ordinary differential equations," in *Boundary Problems in Differential Equations*, ed. R. E. Langer. Madison: Univ. Wisconsin Press, 1960.

8. D. D. Morrison, J. D. Riley and J. F. Zancanaro, "Multiple shooting method for two-point boundary value problems," *Comm. ACM* 5 (1962), 613–614.

9. A. Kalnins, "Analysis of shells of revolution subjected to symmetrical and nonsymmetrical loads," *Trans. ASME*, ser. E, *J. Appl. Mech.* 31 (1964), 467–476.

A general analysis was done independently by:

10. H. B. Keller, *Numerical Methods for Two-Point Boundary Value Problems*. Waltham: Blaisdell, 1968.

and

11. M. R. Osborne, "On shooting methods for boundary value problems," *J. Math. Anal. Appl.* 27 (1969), 417–433.

The stability results collected in Theorems 4.42 and 4.46 are discussed to various extents in:

12. M. R. Osborne, "Aspects of the numerical solution of boundary value problems with separated boundary conditions," Computing Research Group Report, Australian National Univ., 1978.

13. R. M. M. Mattheij, "Estimates for the errors in the solutions of linear boundary value problems, due to perturbations," *Computing* 27 (1981), 299–318.

14. M. Lentini, M. R. Osborne and R. D. Russell, "The close relationships between methods for solving two-point boundary value problems," *SIAM J. Numer. Anal.* 21 (1985), 280-309.

A number of general-purpose codes based on multiple shooting have been written. We mention efforts by England and Reid, and by Enright, Keener, and Krogh. Perhaps the most famous multiple shooting code is the one by Bulirsch, Stoer, and Deuflhard (see Deuflhard's article in [6]). This code has a sophisticated nonlinear equation solver and was successfully used for a number of interesting applications — see refs. [34, 44–47] of Chapter 1. This, despite the fact that it uses the potentially unstable compactification algorithm (recall section 4.3.4), as do the IMSL code of Sewell and the bifurcation code AUTO of Doedel. It is interesting to note that the possible practical instability of this algorithm was noted as early as 1953 in,

15. N. A. Haskell, "The dispersion of surface waves in multilayered media," *Bull. Seism. Soc. Am.* 43 (1953), 17–34,

a paper which considers our Example 1.8. A multiple shooting code which does not use compactification is:

16. M. Hermann and H. Berndt, "RWPM, a multiple shooting code for nonlinear two-point boundary value problems." Preprint 67, 68, 69, Friedrich Schiller Universität, Jena , 1982.

The multiple shooting codes described in :

17. R. M. M. Mattheij and G. W. Staarink, "An efficient algorithm for solving general linear two point BVP ," *SIAM J. Sci. Stat. Comput.* 5 (1984), 745–763.

and

18. R. M. M. Mattheij and G. W. Staarink, "A multiple shooting code for nonlinear problems," manuscript, 1987.

use compactification, but with decoupling to stabilize the process. The codes MUSL and MUSN in Appendix A are based on the latter two papers.

The stabilized march algorithm of section 4.4.3 was introduced in:

19. S. K. Godunov, "Numerical solution of boundary value problems for systems of linear ordinary differential equations," *Usp. Mat. Nauk* 16 (1961), 171–174.

20. S. D. Conte, "The numerical solution of linear boundary value problems," *SIAM Review* 8 (1966), 309–321.

Stability was shown for the linear case in:

21. M. R. Osborne, "The stabilized march is stable," *SIAM J. Numer. Anal.* 16 (1979), 923–933.

and

22. R. M. M. Mattheij, "On conditioning of linear boundary value problems," *SIAM J. Numer. Anal.* 19 (1982), 963–978,

and for the nonlinear case in some unpublished notes of Osborne. An implementation of this algorithm is described in [6] and

23. M. R. Scott and H. A. Watts, "Computational solution of linear two point boundary value problems via orthonormalization," *SIAM J. Numer. Anal.* 14 (1977), 40–70.

There is an extensive literature on invariant imbedding; e.g. see the references in:

24. I. Babuška, M. Prager, and E. Vitasek, *Numerical Processes in Differential Equations*. New York: Wiley, 1966,

25. M. R. Scott, *Invariant Imbedding and its Applications to Ordinary Differential Equations*. Reading, Mass.: Addison-Wesley, 1973.

26. G. H. Meyer, *Initial-Value Methods for Boundary Value Problems*. New York: Academic Press, 1973.

The relationship between the Riccati matrix and the fundamental solution (e.g. as in Exercise 7.9) is investigated for various forms of invariant imbedding in Denman's article in [6]. For papers showing interrelationships among the multiple shooting, stabilized march, and invariant imbedding methods, along with a discussion of possible variants of the methods and related stability questions, see Chapter 6, [14] and

27. R. M. M. Mattheij, "Decoupling and stability of BVP algorithms," *SIAM Review* 27 (1985), 1–44.

Finally, the theoretical development of section 4.6.3 can be found in [5].

CHAPTER 5

Many aspects of finite difference methods for two-point BVPs are introduced in:

1. L. Fox, *The Numerical Solution of Two-Point Boundary Value Problems in Ordinary Differential Equations*. London: Oxford University Press, 1957

2. L. Collatz, *The Numerical Treatment of Differential Equations*. Berlin: Springer, 1960.

Most basic books on numerical analysis which discuss difference methods for two-point BVPs at all, treat only the scheme of section 5.1.1, to varying detail. A thorough treatment of this scheme can be found in:

3. H. B. Keller, *Numerical Methods for Two-Point Boundary Value Problems*. Waltham: Blaisdell, 1968.

Treatment of the more general case of first-order systems or even the scheme of section 5.6.1 is less common in texts. Much of the theory of section 5.2 for one-step schemes was developed by H. B. Keller, and is presented in his monograph,

4. H. B. Keller, *Numerical Solution of Two Point Boundary Value Problems*. CBMS Regional Conference Series in Applied Mathematics, 24. Philadelphia: SIAM, 1976.

The theory of section 5.2 differs from that in [4] only in not assuming that the mesh is quasiuniform and, more importantly, in connecting stability to the conditioning of the approximated BVP. For this aspect, see :

5. C. de Boor, F. de Hoog, and H. B. Keller, "The stability of one-step schemes for first-order two-point boundary value problems," *SIAM J. Numer. Anal.* 20 (1983), 1139–1146.

Our presentation also uses:

6. M. Lentini, M. R. Osborne, and R. D. Russell, "The close relationships between methods for solving two-point boundary value problems," *SIAM J. Numer. Anal.* 22 (1985), 280–309.

The Runge-Kutta schemes of section 5.3 have already been introduced in section 2.7, for the treatment of IVPs. Most early efforts on implicit Runge-Kutta schemes were by J. Butcher. The connection between the Runge-Kutta schemes of section 5.3.2 and the collocation methods of section 5.4, which follows naturally from the approach taken here, has been observed in a number of papers. We mention only:

7. R. Weiss, "The application of implicit Runge-Kutta and collocation methods to boundary-value problems," *Math. Comp.* 28 (1974), 449–464.

Implementation of Runge-Kutta schemes which are not equivalent to collocation has been considered in:

8. S. Gupta, "An adaptive boundary value Runge-Kutta solver for first-order boundary value problems," *SIAM J. Numer. Anal.* 22 (1985), 114–126.

9. P. Muir, Implicit Runge-Kutta methods for Two-point boundary value problems, Ph.D. thesis, Dept. of Computer Science, Univ. of Toronto, 1984.

10. J. R. Cash, "On the numerical integration of nonlinear two-point boundary value problems using iterated deferred corrections, Part I," *Comput. Maths. Appl.*, 1986.

11. J. R. Cash, "On the numerical integration of nonlinear two-point boundary value problems using iterated deferred corrections, Part II," 1986.

The collocation theory for first-order systems was developed, using different points of view, in [7] and in:

12. R. D. Russell, "Collocation for systems of boundary value problems," *Numer. Math.* 23 (1974), 119–133.

This work was preceded by the related work [19] below.

Many of the ideas presented in section 5.5 may again be found in [4]. In particular, most of the material of sections 5.5.1–5.5.3 is covered there. A thorough treatment of asymptotic expansions and discretization methods can be found in:

13. H. Stetter, *Analysis of Discretization Methods for Ordinary Differential Equations*. Berlin: Springer-Verlag, 1973.

The adaptation of the well-known extrapolation principle to BVPs was done by Keller. First works on deferred correction were in [1]. A deep theoretical development of the deferred correction method was carried out by V. Pereyra, and a general-purpose implementation is described in:

14. M. Lentini and V. Pereyra, "An adaptive finite difference solver for nonlinear two-point boundary problems with mild boundary layers," *SIAM J. Numer. Anal.* 14 (1977), 91–111.

This paper also contains references to the theoretical development. The code is available in the NAG library. The alternative theoretical viewpoint outlined in section 5.5.4 which avoids asymptotic expansions is found in:

15. J. Christiansen and R. D. Russell, "Deferred corrections using uncentered differences," *Numer. Math.* 35 (1980), 21–33,

and the general framework with (5.114) to more completely describe convergence is in:

16. R. D. Skeel, "The order of accuracy for deferred corrections using uncentered end formulas," *SIAM J. Numer. Anal.* 23 (1986), 393–402.

Implementation aspects of various implicit Runge-Kutta methods with deferred corrections are considered in [10, 11]. A related treatment to that of section 5.6.1, and generalizations in references therein, is:

17. T. A. Manteuffel and A. B. White, "On the efficient numerical solution of systems of second-order boundary value problems," *SIAM J. Numer. Anal.* 23 (1986), 996–1007.

The general theory of polynomial collocation for a higher-order ODE is given in:

18. G. M. Vainniko, "On convergence of the collocation method for nonlinear differential equations," *USSR Comp. Math. & Math. Phys.* 6 (1966), 35–42,

and the spline collocation case with superconvergence is treated in:

19. C. de Boor and B. Swartz, "Collocation at Gaussian points," *SIAM J. Numer. Anal.* 10 (1973), 582–606.

Our presentation of spline collocation follows an idea originally due to M. Osborne, as developed in:

20. U. Ascher, "Collocation for two-point boundary value problems revisited," *SIAM J. Numer. Anal.* 23 (1986), 596–609.

The various implementations of sections 5.6.2 and 5.6.3 are discussed in:

21. U. Ascher, S. Pruess, and R. D. Russell, "On spline basis selection for solving differential equations," *SIAM J. Numer. Anal.* 20 (1983), 121–142,

and

22. J. Paine and R. D. Russell, "Conditioning of collocation matrices and discrete Green's functions," *SIAM J. Numer. Anal.* 23 (1986), 376–392.

An implementation of collocation at Gaussian points for mixed-order systems of BVPs is discussed in:

23. U. Ascher, J. Christiansen, and R. D. Russell, "Collocation software for boundary value ODEs," *Trans. Math. Software* 7 (1981), 209–222.

and references therein. The code can be found in the ACM collection (see [23]). It uses B-splines. The implementation described in Appendix B, which uses the local representation of section 5.6.2 instead, is in:

24. G. Bader and U. Ascher, "A new basis implementation for a mixed-order boundary value ODE solver," *SIAM J. Sci. Stat. Comp.* 8 (1987), 483-500.

The discussion of conditioning in section 5.6.4 is from:

25. U. Ascher and G. Bader, "A note on conditioning, stability and collocation matrices," Tech. Rep. CMA, Australian National University, Canberra (1986).

A number of codes and applications, including the two mentioned above in [14], [23], are discussed in:

26. B. Childs, et al., *Codes for Boundary Value Problems in Ordinary Differential Equations.* Lecture Notes in Computer Science 76. Berlin: Springer-Verlag, 1979.

There is an abundance of literature dealing with the finite element method. The ODE case is usually treated as being the natural lead-in to the PDE case. For a discussion of the topic, see

27. G. Strang and G. J. Fix, *An Analysis of the Finite Element Method.* Englewood Cliffs, N. J.: Prentice-Hall, 1973.

CHAPTER 6

Decoupling as an important tool in the analysis of general ODEs has been employed for a long time, although mainly in the context of constant coefficients and Jordan forms. This notion is generalized as dichotomy in:

1. I. L. Massera and I. I. Schäffer, *Linear Differential Equations and Function Spaces.* New York: Academic Press, 1966.

2. W. A. Coppel, *Dichotomies in Stability Theory.* Berlin: Lecture Notes in Mathematics, Vol. 629. Berlin: Springer-Verlag, 1978.

Related material can be found in:

3. H. Bart, I. Gohberg, and M. A. Kaashoek, *Minimal Factorization of Matrix and Operator Functions.* Basel: Birkhaüser Verlag, 1979.

4. R. J. Sacker and G. R. Sell, "Existence of dichotomies and invariant splittings for linear differential systems I, II, III," *J. Diff. Equa.* 15 (1974), 429–458, and 22 (1976), 478–522.

A related factorization approach can also be found in:

5. I. Babuška, M. Prager, and E. Vitasek, *Numerical Processes in Differential Equations.* London: Wiley, 1966.

A theory for decoupling of recursions appears in:

6. R. M. M. Mattheij, "Characterizations of dominant and dominated solutions of linear recursions," *Numer. Math.* 35 (1980), 421–442,

and relations between various approaches in:

7. R. M. M. Mattheij, "On decoupling of linear recursions," *Bull. Amer. Math. Soc.* 27 (1983), 347–360.

Related topics on splitting can be found in:

8. G. W. Reddien, "Projection methods for two point BVP," *SIAM Review* 22 (1980), 156–171.

9. M. Lentini, M. R. Osborne and R. D. Russell, "The close relationships between methods for solving two-point boundary value problems," *SIAM J. Numer. Anal.* 22 (1985), 280–309.

A fairly general overview on decoupling and BVPs appears in:

10. R. M. M. Mattheij, "Decoupling and stability of algorithms for boundary value problems," *SIAM Review* 27 (1985), 1–44.

Our presentation here is closest to this paper, but improves upon it in certain aspects.
An extensive literature exists on decoupling of problems with more time scales, a subject treated in Chapter 10. For example, see:

11. R. E. O'Malley Jr. and L. R. Anderson, "Time-scale decoupling systems," *Optimal Control Appl. and Methods* (1982), 133–153.

The concept of closure is an old one, representing a viewpoint often expressed in the Soviet literature, where the algorithms are frequently interpreted as factorization or sweep methods. For instance, see:

12. I. Babuška and S. L. Sobolev, "The optimization of numerical processes," *Apl. Mat.* 10 (1965), 96–130,

13. N. Bakhvalov, *Numerical Methods*. Moscow: MIR, 1977 (Russian edition 1975),

and references in:

14. L. Dieci and R. D. Russell, "Riccati and other methods for singularly perturbed BVP," Proceedings of BAIL IV, ed. J. Miller. Dublin: Boole Press, 1986.

Closure is also used in [9]. For a very general presentation of factorization methods, see:

15. I. Babuška and V. Majer, "The factorization method for the numerical solution of TPBVP for linear ODE's," *SIAM J. Numer. Math.* (1987).

CHAPTER 7

Probably no subject has received as much attention in numerical analysis as linear algebra, and in particular the solution of linear systems. Correspondingly, there is a choice of excellent textbooks on this matter, e. g.,

1. G. W. Stewart, *Introduction to Matrix Computations*. New York: Academic Press, 1973.

2. G. H. Golub, C. F. van Loan, *Matrix Computations*. Baltimore, Md.: John Hopkins University Press, 1983.

A good survey paper on elimination methods for staircase matrices is:

3. R. Fourer, "Staircase Matrices and Systems," *SIAM Review* 26, (1984) 1–70.

The general almost block diagonal algorithm of section 7.2.1 is presented, along with a code, in:

4. C. de Boor and R. Weiss, "SOLVEBLOK: A package for solving almost block diagonal linear systems," *ACM TOMS* 6 (1980), 80–87.

This code is used in the BVP solver COLSYS,

5. U. Ascher, J. Christiansen, and R. D. Russell, "Collocation software for boundary value ODE's," *ACM TOMS* 7 (1981), 209–220.

The alternate row and column elimination method of section 7.2.2 is originally due to D. Lam and is discussed in:

6. J. M. Varah, "Alternate row and column elimination for solving linear systems," *SIAM J. Numer. Anal.* 13 (1976), 71–75.

A package based on this method is given in:

7. J. C. Diaz, G. Fairweather, and P. Keast, "FORTRAN packages for solving certain almost block diagonal linear systems by modified alternate row and column elimination," *ACM TOMS* 9 (1983), 358–375.

Adaptations of this code and of SOLVEBLOK [4] to vector machines, and performance comparisons (in which the alternate row and column elimination method is faster) are made in:

8. R. I. Hay and I. Gladwell, "Solving almost block diagonal linear equations on the CDC Cyber 205," NA Report 98, Dept. of Math., University of Manchester, 1985.

The block tridiagonal method of section 7.2.3 is discussed in:

9. J. M. Varah, "On the solution of block tridiagonal systems arising from certain finite difference equations," *Math. Comp.* 26 (1972), 859–868.

10. H. B. Keller, *Numerical Solution of Two Point Boundary Value Problems. CBMS Regional Conference Series in Applied Math*, 24. Philadelphia: SIAM, 1976.

This method is used in the general-purpose code PASVA3,

11. M. Lentini and V. Pereyra, "An adaptive finite difference solver for nonlinear two-point boundary value problems with mild boundary layers", *SIAM J. Numer. Anal.* 14 (1977), 91–111.

Regarding stability of the methods discussed in this chapter, one may also consult:

12. M. van Veldhuizen, "A note on partial pivoting and Gaussian elimination," *Numer. Math.* 29 (1977), 1–10.

13. R. M. M. Mattheij, "The stability of LU-decomposition of block tridiagonal matrices," *Bull. Austr. Math. Soc.* 29 (1984), 177–205.

14. R. M. M. Mattheij, "Stability of block LU-decompositions of matrices arising from BVP," *SIAM J. Disc. Math.* 5 (1984), 314–331.

15. R. M. M. Mattheij, "Decoupling and stability of algorithms for boundary value problems," *SIAM Review* 15 (1985), 1-44.

16. M. Lentini, M. R. Osborne and R. D. Russell, "The close relationships between methods for solving two-point boundary value problems," *SIAM J. Numer. Anal.* 22 (1985), 280-309.

The latter two papers also discuss the relationships between various methods, as is the case with:

17. H. B. Keller and M. Lentini, "Invariant imbedding, the box scheme and an equivalence between them," *SIAM J. Numer. Anal.* 19 (1982), 942–962.

It is interesting to note that two-point BVPs and methods to solve them, even multiple shooting, have inspired authors to employ related techniques for solving more general types of linear equations; see, e. g.,

18. R. E. Bank and D. J. Rose, "Marching algorithms for elliptic boundary value problems I, the constant coefficient case," *SIAM J. Numer. Anal.* 14 (1977), 792–829.

19. R. E. Bank, "Marching algorithms for elliptic boundary value problems II, the variable case," *SIAM J. Numer. Anal.* 14 (1977), 950–970.

CHAPTER 8

Good, comprehensive treatments of the numerical solution of nonlinear equations are contained in the books

1. J. E. Dennis and R. B. Schnabel, *Numerical Methods for Unconstrained Optimization and Nonlinear Equations*. Englewood Cliffs, N.J.: Prentice-Hall, 1983.
2. R. Fletcher, *Practical Methods of Optimization. Vol. 1.* Chichester: Wiley, 1980.

Chapter 6 of [1] is particularly relevant for our section 8.1. In particular, Theorem 8.15 here is a special adaptation of the material in their section 6.3.1. The choice of h in (8.33) is discussed in [1] as well.

In developing the procedure to satisfy (8.12) we have also followed:

3. M. R. Osborne, "An efficient weak line search with guaranteed termination," *Tech. Rep. 1870* (1978). Madison: MRC.

These references do not contain, however, the natural criterion function. The latter has been developed by P. Deuflhard and can be found, together with the justification for (8.22b) (cf. Exercise 4), in:

4. P. Deuflhard, "A relaxation strategy for the modified Newton method," Lecture Notes in Math. 447. Berlin: Springer, eds. R. Bulirsch, W. Oettli, and J. Stoer, 1975.

Variations of the damping strategy proposed in [4] have been successfully used in the multiple shooting code BOUNDS (see references of Chapter 4) and in the collocation code COLSYS (see [7]). The unfriendly Example 8.2 is given in:

5. U. Ascher and M. R. Osborne, "A note on solving nonlinear equations and the 'natural' criterion function," to appear in *JOTA*.

A discussion on affine invariance of Newton's method and related issues, including the solution (by reference) of exercise 1 can be found in:

6. P. Deuflhard and G. Heindl, "Affine invariant convergence theorems for Newton's method and extensions to related methods," *SIAM J. Numer. Anal.* 16 (1979), 1–10.

Most issues reviewed in section 8.1.2 can be found in [1], [2], and [4].

The modified Newton scheme of section 8.2.1 is routinely used in many IVP as well as BVP codes. In particular, see:

7. U. Ascher, J. Christiansen and R. D. Russell, "COLSYS — A collocation code for boundary-value problems," *Lecture Notes in Computer Science* 76. Childs et al., eds., Berlin: Springer, 1979.

The BVP used for Examples 8.1 and 8.3 is treated as described in:

8. U. Ascher and R. D. Russell, "Reformulation of boundary value problems into 'standard' form," *SIAM Review* 23 (1981), 238–254.

The updates discussed in section 8.2.2 are used in the code BOUNDS. A more extensive discussion can be found in:

9. A. Griewank, "The solution of boundary value problems by Broyden-based secant methods," Centre for Math. Anal. Research Rep. CMA-R22–85, Australian National Univ, 1985.

which follows an earlier work of:

10. W. E. Hart and S. O. W. Soul, "Quasi-Newton methods for discretized non-linear boundary problems," *J. Inst. Maths. Applics.* 11 (1973), 351–359.

There is a large amount of literature available on continuation methods. A paper most relevant to our treatment here and containing other references is:

11. P. Deuflhard, "A stepsize control for continuation methods and its special application to multiple shooting techniques," *Numer. Math.* 33 (1979), 115–146.

The time imbedding problem of section 8.3.2 has been studied in many places, e. g.,

12. J. P. Abbott and R. P. Brent, "Fast local convergence with single and multistep methods for nonlinear equations," *J. Austral. Math. Soc.* 19 (Series B), (1975), 173–199.

Examples 8.4–8.6 have all been listed in section 1.2, where references are given as well. Material related to some of the comments in section 8.4 can be found in:

13. U. Ascher and G. Bader, "A note on conditioning, stability and collocation matrices," Research Rep CMA-R16–86, Australian National University, Canberra, 1986.

For the mesh independence principle, see:

14. E. Allgower, K. Böhmer, F. A. Potra, and W. C Rheinboldt, "A mesh-independence principle for operator equations and their discretizations," *SIAM J. Numer. Anal.* 23 (1986), 160–169,

and references therein.

CHAPTER 9

One of the early key papers on mesh selection for BVPs is:

1. V. Pereyra and G. Sewell, "Mesh selection for discrete solution of boundary value problems in ordinary differential equations," *Numer. Math. 23* (1975) 261–268,

in which Theorem 9.12 (in a slightly modified form) can be found as well. The general-purpose code PASVAR, described in:

2. M. Lentini and V. Pereyra, "An adaptive finite difference solver for nonlinear two-point boundary value problems with mild boundary layers", *SIAM J. Numer. Anal.* 14 (1977), 91–111,

has a mesh selection strategy which produces quasiuniform meshes and involves also varying the order of the method adaptively (based on the techniques of section 5.5.3). Mesh points are never taken out as the meshes are refined, and this may reduce efficiency for tough problems.
The earliest use of adaptive mesh selection for spline collocation is probably due to:

3. C. de Boor, "Good approximation by splines with variable knots. II," in *Lecture Notes in Mathematics* 353. New York: Springer-Verlag (1973).

A broader and more specific investigation is reported in:

4. R. D. Russell and J. Christiansen, "Adaptive mesh selection strategies for solving boundary value problems," *SIAM J Numer Anal* 15 (1978) 59–80,

where numerical examples for some of the mesh selection strategies mentioned in section 9.5.1 are given. A strategy which produces *locally* quasiuniform meshes appears in:

5. J. Kautsky and N. K. Nichols, "Equidistributing meshes with constraints," *SIAM J. Sci. Stat. Comput.* 1 (1980), 499–511.

An early survey of the literature for most of the methods appearing in sections 9.2 and 9.4 is:

6. R. D. Russell, "Mesh selection methods," in *Codes for Boundary Value Problems,* Lecture Notes in Computer Science 74 Childs et al., eds., Berlin: Springer, 1979.

The papers [3, 4] lay the foundations for the mesh selection strategy in the general-purpose code COLSYS, which is described in:

7. U. Ascher, J. Christiansen, and R. D. Russell, "A collocation solver for mixed order systems of boundary value problems," *Math. Comp.* 33 (1979), 659–679.

The newer version of this code given in Appendix B is not significantly different from the original regarding the mesh selection strategy.
The field of mesh selection is currently one of intense activity, e. g., as it relates to solving PDEs with finite element methods. Frequently this involves first analyzing the ODE case. A discussion of the mesh insensitivity from a somewhat different perspective than in this chapter is given in:

8. I. Babuška and W. C. Rheinbolt, "Analysis of optimal finite-element meshes in $\mathbf{R}^1$," *Math. Comp.* 33 (1979), 435–463.

The role of mesh equidistribution schemes for ODEs as they relate to solving PDEs in one space dimension is discussed in:

9. J. M. Hyman, "Moving mesh methods for initial BVPs," Los Alamos National Labs Report LA-UR-84-61, 1984.

and stability properties of such schemes in this context are described in:

10. J. M. Coyle, J. E. Flaherty and R. Ludwig, "On the stability of mesh equidistribution on strategies for time dependent PDEs," *J. Comp. Phys.* 62 (1986), 26–39.

The implicit method of section 9.4.2 is presented in the paper,

11. A. White, Jr., "On selection of equidistributing meshes for two-point boundary-value problems," *SIAM J. Numer. Anal.* 16 (1979), 472–502,

which also gives numerical examples (including Examples 9.6 and 9.7) using the implicit method, mostly for the arc-length monitor function. Table 9.2 is from [11].
The BVP in Example 9.8 is analyzed in detail in:

12. J. Kevorkian and J. Cole, *Perturbation Methods in Applied Mathematics*. New York: Springer-Verlag, 1981.

CHAPTER 10

There are a number of books available on the analytical treatment of singularly perturbed problems. They vary in emphasis, some being more oriented toward basic theory, others being more specialized and more theoretically advanced, and yet other books exist which emphasize "engineering" aspects of just constructing reasonable solution profiles. One of the earliest basic books is:

1. R. E. O'Malley, *Introduction to Singular Perturbations*. New York: Academic Press, 1974.

Another is:

2. J. Kevorkian and J. D. Cole, *Perturbation Methods in Applied Mathematics*. New York: Springer-Verlag, 1981,

where Example 10.6 can be found (cf. Exercise 4). A more recent book is:

3. D. R Smith, *Singular Perturbation Theory*. Cambridge: Cambridge University Press, 1985.

The monograph,

4. K. W. Chang and F. A. Howes, *Nonlinear Singular Perturbation Phenomena: Theory and Application*. Springer Applied Math. Series 56, 1984,

covers in depth problems of the form of those in section 10.1.2. For the material in section 10.1.3 and section 10.1.4 we have relied mainly on papers. Much of the basic theory in section 10.1.3 is our adaptation of material from:

5. H.-O. Kreiss, N. K. Nichols and D. L. Brown, "Numerical methods for stiff two-point boundary value problems," *SIAM J. Numer. Anal.* 23 (1986), 325–368.

An important reference for singular singularly perturbed problems is:

6. A. B. Vasileeva and V. F. Butuzov, "Singularly perturbed equations in the critical case," Nauka 1973, Translated from Russian as MRC Tech. Report 2039, Madison, Wis., 1980.

The main effort involved in Example 10.7 is apparently in posing the problem as such. This, together with an asymptotic analysis, was done in:

7. J. E. Flaherty and R. E. O'Malley, "Singularly perturbed boundary value problems for nonlinear systems, including a challenging problem for a nonlinear beam," in *Lecture Notes in Math* 942. Berlin: Springer, 1982.

The singular perturbation analysis of the semiconductor device problem of Example 10.8 was done by:

8. P. A. Markowich and C. A. Ringhofer, "A singularly perturbed boundary value problem modelling a semiconductor device," *SIAM J. Appl. Math.* 44 (1984), 231–256.

Analysis of the conditioning of this problem can be found in:

9. U. Ascher, P. A. Markowich, C. Schmeiser, H. Steinrück and R. Weiss, "Conditioning of the steady state semiconductor device problem," Tech. Rep. 86–18, *Computer Science*, Univ. of B. C., Vancouver, 1986.

Many of the results in section 10.2 are new and appear in:

10. U. Ascher and R. M. M. Mattheij, "General framework, stability and error analysis, for numerical stiff boundary value methods," Tech. Rep. 87-28, Univ. of B.C., Vancouver, 1987.

The material in section 10.2.4 is adapted from:

11. U. Ascher and R. Weiss, "Collocation for singular perturbation problems II: Linear first order systems without turning points," *Math. Comp.* 43 (1984), 157–187.

The thesis,

12. P. W. Hemker, *A Numerical Study of Stiff Two-Point Boundary Value Problems.* Amsterdam: Math. Centrum, 1977,

provides a good description of methods and many examples of problems of the form (10.2).
The hybrid Euler method for first-order linear systems is from [5], where a very complicated *a priori* mesh selection strategy in a general-purpose setting is described as well. The idea of upwinded collocation at common points has been analyzed in:

13. C. Ringhofer, "On collocation schemes for quasilinear singularly perturbed boundary value problems," *SIAM J. Numer. Anal.* 21 (1984), 864–882.

An extension of the work of [5] in this direction has been carried out by:

14. D. Brown and J. Lorenz, "A higher-order method for stiff boundary-value problems with turning points," *SIAM J. Sci. Stat. Comp.* 8 (1987), 790-805.

A paper which describes many of the aspects of our treatment of symmetric schemes in section 10.3.2 is:

15. U. Ascher, "Two families of symmetric difference schemes for singular perturbation problems," in *Numerical Boundary Value ODEs,* eds. U. Ascher and R. D. Russell, Boston: Birkhäuser, 1985.

Examples 10.12 and 10.14, and in particular Tables 10.2 and 10.3, are taken from [15], and further references are given there as well. See also:

16. U. Ascher and G. Bader, "Stability of collocation at Gaussian points," *SIAM J. Numer. Anal.* 23 (1986), 412–422.

D-stability is defined and discussed in:

17. M. van Veldhuizen, "D-stability," *SIAM J. Numer. Anal.* 18 (1981), 45–64.

Algorithm 10.115 for *a priori* boundary layer mesh selection appears in:

18. U. Ascher and R. Weiss, "Collocation for singular perturbation problems I: first order systems with constant coefficients," *SIAM J. Numer. Anal.* 20 (1983), 537–557.

The potential instability of symmetric schemes was noticed first by H.-O. Kreiss in 1972; see discussion and references in [15]. Weiss gave a detailed analysis for the midpoint scheme applied to singularly perturbed problems of the form (10.35) in:

19. R. Weiss, "An analysis of the box and trapezoidal schemes for linear singularly perturbed boundary value problems," *Math. Comp.* 42 (1984), 41–68.

Example 10.15 is an adaptation of the one in [19]. The use of solution values at collocation points for adaptive mesh selection is described in:

20. U. Ascher and S. Jacobs, "On collocation implementation for singularly perturbed two-point problems," Tech. Rep. 86–19, Dept. Computer Science, Univ. of BC, Vancouver, 1986.

The reference [12] is a good source of information about exponential fitting methods, even though the approach taken there is different from ours. The related scheme (10.135) for the semiconductor problem was derived by:

21. D. L. Scharfetter and H. K Gummel, "Large signal analysis of a silicon read diode oscillator," *IEEE Trans. Electron Devices*, ED-16 (10.1969), 64–77,

and is known as the *Scharfetter-Gummel scheme*.
Most of the material in section 10.4.1 is from:

22. R. England and R. M. M. Mattheij, "A sequential multiple shooting strategy for stiff boundary value problems," in preparation.
   Most of the material in section 10.4.2 is from:

23. L. Dieci, M. R. Osborne and R. D. Russell, "A Riccati transformation method for solving boundary value problems, I: theoretical aspects" and "II: computational aspects," to appear in SIAM J. Numer. Anal.,

where a general analysis of the Riccati method is given which uses a theoretical multiple shooting approach but which is in certain respects different from the approach in this chapter.

Recall that other continuous decoupling methods are referenced in Chapter 6. A related strategy to the one in section 10.4.2 which involves solving Riccati equations in both directions is in:

24. I. Babuška and V. Majer, "The factorization method for the numerical solution of two-point boundary value problems for linear ordinary differential equations," *SIAM J. Numer. Anal.* (1987).

CHAPTER 11

The material in section 11.1 is largely based on the survey paper,

1. U. Ascher and R. D. Russell, "Reformulation of boundary value problems into 'standard form'," *SIAM Review* 23 (1981), 238–254.

Differential algebraic equations of section 11.2 have captured the attention of many researchers in recent years, especially for IVPs. An early such paper is:

2. C. W. Gear, "Simultaneous numerical solution of differential/algebraic equations," *IEEE Trans. Circuit Theory*, CT-18 (1971), 89–95.

Later important work appears, e. g., in:

3. C. W. Gear and L. Petzold, "ODE methods for the solution of differential/algebraic systems," *SIAM J. Numer. Anal.* 21 (1984), 716–728.
4. S. Campbell, "Index two linear time-varying singular systems of differential equations," *SIAM J. Alg. & Discrete Methods* 4 (1983), 237–243.

There are many other papers. A good general-purpose IVP code based on BDF methods is reported in:

5. L. Petzold, "A description of DASSL: A differential/algebraic system solver," SAND82–8637, Sandia Labs, Livermore, 1982.

R. März has made major contributions for understanding DAEs for BVPs; see, e. g.,

6. R. März, "On boundary value problems in differential-algebraic equations," Preprint 124, Humboldt-Universität zu Berlin, 1986,

and the most comprehensive treatment we know for both initial and boundary value DAEs,

7. E. Griepentrog and R. März, *Differential-Algebraic Equations and their Numerical Treatment*. Texte zur Math., Band 88, Leipzig: Teubner, 1986.

Our presentation was helped by:

8. G. Bader, private communication.

For section 11.3 there are many extensive treatments of eigenvalue problems, especially Sturm-Liouville problems. The trick of converting to a standard BVP is discussed in [1], where numerical results are also given. Most of the other material of section 11.3.1 is in:

9. J. W. Paine, "Numerical approximation of Sturm-Liouville eigenvalues," Ph.D. Thesis, Australian National University, 1979.
10. H. B. Keller, *Numerical Methods for Two-Point Boundary-Value Problems*. Waltham, Mass.: Blaisdell, 1968.

or referenced therein. Two shooting codes which use Prüfer transformations are:

11. P. B. Bailey, M. K. Gordon, and L. F. Shampine, "Automatic solution of the Sturm-Liouville problem," *ACM Trans. on Math. Software* 4 (1978), 193–208.
12. J. D. Pryce, "Software for Sturm-Liouville eigenvalues with error bounds," C. S. Tech. Rep. 79–02, Bristol Univ., 1979.

A general theory of finite difference methods for eigenvalue problems for high order equations appears in:

13. H.-O. Kreiss, "Difference approximations for boundary and eigenvalue problems for ordinary differential equations," *Math. Comp.* 26 (1972), 605–624,

and a corresponding (even more general) theory for spline collocation is in:

14. C. deBoor and B. Swartz, "Collocation approximation to eigenvalues of an ODE: the principle of the thing," *Math. Comp.* 35 (1980), 679–694.

A discussion of the theory of linear eigenvalue problems for first-order systems is given in:

15. F. V. Atkinson, *Discrete and Continuous Boundary Problems*. New York: Academic Press, 1964.

The Riccati approach and example in section 11.3.2 are from:

16. L. Dieci, "Theoretical and computational aspects of the Riccati transformation for solving differential equations," Ph.D. thesis, University of New Mexico, 1986.

For a discussion of other initial value methods and the spectral method, see, for example,

17. J. S. Bramley and S. C. R. Dennis, "The calculation of eigenvalues for the stationary perturbation of Poiseuille flow using initial value methods," *J. Math. Anal. & Appl.* 101 (1984), 30–38.

A description of shooting for the nonlinear eigenvalue problem appears in [10].

Singular problems of section 11.4.1 have been discussed frequently in the literature. We mention only:

18. F. de Hoog and R. Weiss, "Difference methods for boundary value problems with a singularity of the first kind," *SIAM J. Numer. Anal.* 13 (1976), 775–813.

19. D. C. Brabston and H. B. Keller, "A numerical method for singular two-point boundary value problems," *SIAM J. Numer. Anal.* 14 (1977), 779–791,

and a paper of a later vintage,

20. E. Weinmüller, "Collocation for singular boundary value problems of second order," *SIAM J. Numer. Anal.* 23 (1986), 1062–1095.

Problems on infinite intervals have been investigated, among others, by

21. M. Lentini and H. B. Keller, "Boundary value problems on semi-infinite intervals and their numerical solution," *SIAM J. Numer. Anal.* 17 (1980), 577–604.

22. F. de Hoog and R. Weiss, "An approximation theory for boundary value problems on infinite intervals," *Computing* 24 (1980), 227–239.

23. P. A. Markowich, "A theory for the approximation of solutions of boundary value problems on infinite intervals," *SIAM J. Math. Anal.* 13 (1982), 484–513.

Example 11.4 is taken from [1]. Some numerical methods are described in [21] and in:

24. A. Bayliss, "A double shooting scheme for certain unstable and singular boundary value problems," *Math. Comp.* 32 (1978), 61–71.

25. T. N. Robertson, "The linear two-point boundary value problem on an infinite interval," *Math Comp.* 25 (1971), 475–481.

Aspects of using symmetric collocation schemes for problems on an infinite intervals (plus the *a priori* mesh selection algorithm mentioned here) have been considered by:

26. P. A Markowich and C. Ringhofer, "Collocation methods for boundary value problems on 'long' intervals," *Math. Comp.* 40 (1983), 123–150.

The marching idea described in section 11.4.2 and a strategy for more general problems are considered in

27. R. M. M. Mattheij, "On the computation of BVP on infinite intervals," *Math. Comp.* 48 (1987), 533-549.

A good introduction to analytical aspects of the material in section 11.5 is :

28. I. Stakgold, "Branching of solutions of nonlinear equations," *SIAM Review* 13 (1971), 289–332.

Further up-to-date theory and applications can be found in:

29. J. Guckenheimer and P. Holmes, *Nonlinear Oscillation, Dynamical Systems and Bifurcation of Vector Fields.* Springer, 1983.
30. P. H. Rabinowitz, ed. *Applications of Bifurcation Theory.* New York: Academic Press, 1977.

One of the first, and best, numerical survey papers is:

31. H. B. Keller, "Numerical solution of bifurcation and nonlinear eigenvalue problems," in [19].

There is a regular series of proceedings on the subject; for one of them see:

32. T. Küpper, H. D. Mittelmann and H. Weber, eds. *Numerical Methods for Bifurcation Problems.* ISNM 70, Basel: Birkhäuser, 1984.

For an applied textbook see:

33. M. Kubiček and M. Marek, *Computational Methods in Bifurcation Theory and Dissipative Structures.* New York: Springer-Verlag, 1983.

Finally, we mention the existence of several codes that are capable of producing bifurcation diagrams. Two generally available ones are:

34. E. J. Doedel, "AUTO, a program for the automatic bifurcation analysis of autonomous systems," *Congressus Numerantium* 30 (1981), 165–184.
35. R. Seydel, "BIFPAC—a program for calculating bifurcations," November 1983.

Most of the literature concerning special methods for problems with highly oscillatory solutions is aimed at IVPs, but the ideas are sometimes applicable to BVPs as well. The book,

36. W. Wasow, *Linear Turning Point Theory*. New York: Springer-Verlag, 1985.

discusses the theory behind turning points as in Example 11.9. The judicious choice of initial data so as to obtain a smooth solution is described in:

37. H.-O. Kreiss, "Problems with different time scales for ordinary differential equations," *SIAM J. Numer. Anal.* 16 (1979), 980–998.

Methods for filtering out the rapid oscillations are also considered, by:

38. V. Amdursky and A. Ziv, "On the numerical solution of stiff linear systems of the oscillatory type," *SIAM J. Appl. Math* 33 (1977), 593–606.

and

39. G. Majda, "Filtering techniques for systems of stiff ordinary differential equations II. error estimates," *SIAM J. Numer. Anal.* 22 (1985), 1116–1134.

Analytical-numerical approaches for IVPs (including nonlinear problems) where the presence of rapid oscillations cannot be ignored or suppressed are considered, by:

40. W. L. Miranker, *Numerical Methods for Stiff Equations and Singular Perturbation Problems*. Dordrecht: Reidel, 1981.
41. W. L. Miranker and M. van Veldhuizen, "The method of envelopes," *Math. Comp.* 32 (1978), 453–498.
42. R. E. Scheid, Jr., "The accurate numerical solution of highly oscillatory ordinary differential equations," *Math. Comp.* 41 (1983), 487–509.

Examples 11.10 and 11.11 are documented in significantly more detail, including schemes for turning points and mesh selection for a hybrid method, in:

43. U. Ascher and P. Spudich, "A hybrid collocation method for calculating complete theoretical seismograms in vertically varying media," *Geophys. J. R. Astr. Soc.* 86 (1986), 19–40.

See also:

44. P. Spudich and U. Ascher, "Calculation of complete theoretical seismograms in vertically varying media using collocation methods," *Geophys. J. R. Astr. Soc.* 75 (1983), 101–124.

The expansion in (11.67) is due to:

45. P. G. Richards, "Elastic wave solutions in stratified media," *Geophysics* 36 (1971), 798–809.

In our description of functional differential equations, we have followed mainly:

46. G. Bader, "Solving boundary value problems for functional differential equations by collocation," in *Numerical Boundary Value ODEs,* eds. U. Ascher and R. D. Russell. Birkhauser 1985.

This paper summarizes some results from a Ph.D. thesis which include a collocation code FDECOL for such problems. Numerical results for Example 11.12 are given in [1]. Some theoretical results are given in:

47. A. Bellen and M. Zennaro, "A collocation method for boundary value problems of differential equations with functional arguments," *Computing* 32 (1984), 307–318.

An excellent survey of early (mainly Soviet) work on the method of lines (of section 11.8) is:

48. O. A. Liskovets, "The method of lines (review)," *Differential Equations* 1 (1965), 1308–1323 (translated from the Russian).

A widely used method of lines code which uses B-spline collocation with the longitudinal approach for solving parabolic and hyperbolic PDEs is:

49. N. K. Madsen and R. F. Sincovec, "PDECOL: General collocation software for partial differential equations," *ACM Trans. on Math. Software* 5 (1979), 326–351.

The method of lines for solving elliptic PDEs seems not to have been pursued very much; see:

50. D. J. Jones and J. C. South, Jr., "Application of the method of lines to the solution of elliptic partial differential equations," NRC Canada Aeronautical Report LR-599, Ottawa, 1979.

Example 11.13 is taken from:

51. G. H. Meyer, *Initial Value Methods for Boundary Value Problems: Theory and Applications of Invariant Imbedding.* New York: Academic Press, 1973,

where an invariant imbedding technique is used to solve the problem. Numerical experiments using the transverse approach with BVODE software are carried out in:

52. R. M. Corless, "A feasibility study of the use of COLSYS in the solution of systems of partial differential equations," M. Math., Univ. Waterloo, 1982.

A fully adaptive finite difference code designed to solve combustion problems is described in:

53. M. D. Smooke and M. L. Koszykowski, "Fully adaptive solutions of one-dimensional mixed initial-boundary value problems with applications to unstable problems in combustion," *SIAM J. Sci. Stat. Comp.* 7 (1986), 301-329.

This paper gives results for the calculation of the moving spatial mesh, and an in-depth study of the stability of this problem is in:

54. J. M. Coyle, J. E. Flaherty, and R. Ludwig, "On the stability of mesh equidistribution strategies for time-dependent PDEs," *J. Comp. Phys.* 62 (1986), 26–39.

As relates to section 11.10, there have been a number of papers concerned with comparison of methods (especially finite difference and finite element methods) which compare the relative operation counts. Other papers have attempted limited comparisons of codes, including:

55. V. Pereyra and R. D. Russell, "Difficulties of comparing complex mathematical software: general comments and the BVODE case," *Acta Cient. Venezolana* 33 (1982), 15–22,

which uses the codes COLSYS and PASVA for illustrating various pitfalls.

APPENDIX A

1. R. M. M. Mattheij and G. W. M. Staarink, "On optimal shooting intervals," *Math. Comp.* 42 (1984), 25–40.

2. R. M. M. Mattheij and G. W. M. Staarink, "An efficient algorithm for solving general linear two point BVP," *SIAM J. Sci. Stat. Comp.* 5 (1984), 745–763.

3. R. M. M. Mattheij and G. W. M. Staarink, "A multiple shooting like algorithm for solving nonlinear BVP," in preparation.

APPENDIX B

1. U. Ascher, J. Christiansen and R. D. Russell, "Collocation software for boundary-value odes," *ACM Trans. Math Software* 7 (1981), 209–222.

2. G. Bader and U. Ascher, "A new basis implementation for a mixed order boundary value ode solver," *SIAM J. Sci. Stat. Comp.* 8 (1987) 483-500.

3. C. de Boor and R. Weiss, "SOLVEBLOK: A package for solving almost block diagonal linear systems," *ACM Trans. Math. Software* 6 (1980), 80–87.

Bibliography

ABBOTT, J. P., "An efficient algorithm for the determination of certain bifurcation points," *J. Comp. & Appl. Math.* 4 (1978), 19–27.

ABBOTT, J. P., AND R. P. BRENT, "Fast local convergence with single and multistep methods for nonlinear equations," *J. Austral. Math. Soc.* 19 (Series B), (1975), 173–199.

ABRAHAMSSON, L., "Numerical solution of a boundary value problem in semiconductor physics," Uppsala Computer Science Rep. 81, (1979).

ABRAHAMSSON, L., H. B. KELLER, AND H.-O. KREISS, "Difference approximations for singular perturbations of systems of ordinary differential equations," *Num. Math.* 22 (1974), 367–391.

ABRAMOV, A. A., "On the transfer of boundary conditions for systems of ordinary linear differential equations (a variant of the dispersive method)," *USSR Comp. Math. & Math. Phys.* 1 (1962), 617–622.

ABRAMOV, A. A., "Numerical solution of some algebraic problems arising in the theory of stability," *USSR Comp. Math. & Math. Phys.* 24 (1984), 1–6.

AGARWAL, R. P., *Boundary Value Problems for Higher Order Differential Equations*. Singapore: World Scientific, 1985.

AHLBERG, J. H., AND T. ITO, "A collocation method for two-point boundary value problems," *Math. Comp.* 29 (1975), 761–776.

AKTAS, Z., AND J. H. STETTER, "A classification and survey of numerical methods for boundary value problems in ordinary differential equations," *Int. J. Num. Meth. Engr.* 11 (1977), 771–796.

ALBASINY, E. L., AND W. D. HOSKINS, "Cubic spline solutions to two-point boundary value problems," *Comput. J.* 12 (1969), 151–153.

ALLEN, R. C., JR., AND G. M. WING, "Generalized trigonometric identities and invariant imbedding," *J. Math. Anal. & Appl.* 42 (1973), 397–408.

ALLEN, R. C., JR. AND G. M. WING, "An invariant imbedding algorithm for the solution of inhomogeneous two-point boundary value problems," *J. Comp. Phys.* 14 (1974), 40–58.

ALLGOWER, E. L., AND S. F. MCCORMICK, "Newton's method with mesh refinement for numerical solution of nonlinear two-point boundary value problems," *Num. Math.* 29 (1978), 237–260.

ALLGOWER, E. L., AND S. F. MCCORMICK, "A mesh independence principle for operator equations and their discretizations," *SIAM J. Numer. Anal.* 23 (1986), 160–169.

ALLGOWER, E. L., S. F. MCCORMICK, AND D. V. PRYOR, "A general mesh independent principle for Newton's method applied to second order boundary value problems," *Computing* 23 (1979), 233–246.

AMDURSKY, V., AND A. ZIV, "On the numerical solution of stiff linear systems of the oscillatory type," *SIAM J. Appl. Math.* 33 (1977), 593–606.

ARCHER, D., AND J. C. DIAZ, "A family of modified collocation methods for second order two-point boundary value problems," *SIAM J. Numer. Anal.* 15 (1978), 242–254.

ARNOLD, V. I., *Geometrical Methods in the Theory of Ordinary Differential Equations.* New York: Springer, 1983.

ASCHER, U., "Discrete least squares approximations for ordinary differential equations," *SIAM J. Num. Anal.* 15 (1978), 478–496.

ASCHER, U., "Solving boundary-value problems with a spline-collocation code," *J. Comp. Phys.* 34 (1980), 401–413.

ASCHER, U., "On some difference schemes for singular singularly-perturbed boundary value problems," *Num. Math.* 46 (1985), 1–30.

ASCHER, U., "Collocation for two-point boundary value problems revisited," *SIAM J. Numer. Anal.* 23 (1986), 596–609.

ASCHER, U., "Two families of symmetric difference schemes for singular perturbation problems," in [Ascher-Russell].

ASCHER, U., AND G. Bader, "Stability of collocation at Gaussian points," *SIAM J. Numer. Anal.* 23 (1986), 412–422.

ASCHER, U., AND G. BADER, "A note on conditioning, stability and collocation matrices," *Appl. Math. & Comp.* (1988).

ASCHER, U., J. CHRISTIANSEN, AND R. D. RUSSELL, "A collocation solver for mixed order systems of boundary value problems," *Math. Comp.* 33 (1979), 659–679.

ASCHER, U., J. CHRISTIANSEN, AND R. D. RUSSELL, "COLSYS — A collocation code for boundary-value problems," in [Childs et al.].

ASCHER, U., J. CHRISTIANSEN, AND, R. D. RUSSELL, "Collocation software for boundary value ODEs," *ACM Trans. Math. Software* 7 (1981), 209–222.

ASCHER, U., AND S. JACOBS, "On collocation implementation for singularly perturbed two-point problems," *Computer Science Tech. Rep.* 86–19, Univ. of B.C., 1986.

ASCHER, U., P. A. MARKOWICH, C. SCHMEISER, H. STEINRÜCK, AND R. WEISS, "Conditioning of the steady state semiconductor device problem," *Computer Science Tech. Rep.* 86–13, Univ. of B.C., Vancouver, Canada, 1986.

ASCHER, U., AND R. M. M. MATTHEIJ, "General framework, stability and error analysis, for numerical stiff boundary value methods," *Computer Science Tech. Rep.* 87–28, Univ. of B.C., 1987.

ASCHER, U., AND M. R. OSBORNE, "A note on solving nonlinear equations and the 'natural' criterion function," *J. Opt. Theory & Applic.* (1987).

ASCHER, U., S. PRUESS, AND R. D. RUSSELL, "On spline basis selection for solving differential equations," *SIAM J. Numer. Anal.* 20 (1983), 121–142.

ASCHER, U., AND R. D. RUSSELL, "Reformulation of boundary value problems into 'standard' form," *SIAM Review* 23 (1981), 238–254.

ASCHER, U., AND R. D. RUSSELL, EDS., *Numerical Boundary Value ODEs,* Progress in Scientific Computing Vol. 5. Boston: Birkhauser, 1985.

ASCHER, U., AND P. SPUDICH, "A hybrid collocation method for calculating complete theoretical seismograms in vertically varying media," *Geophys. J. R. Astr. Soc.* 86 (1986), 19–40.

ASCHER, U., AND F. Y. M. WAN, "Numerical solutions for maximum sustainable consumption growth with a multi-grade exhaustible resource," *SIAM J. Sci. Stat. Comp.* 1 (1980), 160–172.

ASCHER, U., AND R. WEISS, "Collocation for singular perturbation problems I: first order systems with constant coefficients," *SIAM J. Numer. Anal.* 20 (1983), 537–557.

ASCHER, U., AND R. WEISS, "Collocation for singular perturbation problems II : linear first order systems without turning points," *Math. Comp.* 43 (1984), 157–187.

ASCHER, U., AND R. WEISS, "Collocation for singular perturbation problems III: Nonlinear problems without turning points," *SIAM J. Scient. Stat. Comp.* 5 (1984), 811–829.

ATKINSON, F. V., *Discrete and Continuous Boundary Problems.* New York: Academic Press, 1964.

AZIZ, A. K., ED. *Numerical Solutions of Boundary Value Problems for Ordinary Differential Equations.* New York: Academic Press, 1975.

BABUŠKA, I., "The connection between the finite difference like methods and the methods based on initial value problems for O.D.E.," in [Aziz].

BABUŠKA, I., AND V. MAJER, "The factorization method for numerical solution of TPBVP for linear ODEs," *SIAM J. Numer. Anal.* (1987), to appear.

BABUŠKA, I., M. PRAGER, AND E. VITASEK, *Numerical Processes in Differential Equations.* New York: Wiley, 1966.

BABUŠKA, I., AND W. C. RHEINBOLT, "Analysis of optimal finite-element meshes in $\mathbf{R}^1$," *Math. of Comp.* 33 (1979), 435–463.

BABUŠKA, I., AND S. L. SOBOLEV, "The optimization of numerical processes," *Apl. Mat.* 10 (1965), 96–130.

BADER, G., "Solving boundary value problems for functional differential equations by collocation," in [Ascher–Russell].

BADER, G., AND U. ASCHER, "A new basis implementation for a mixed order boundary value ODE solver," *SIAM J. Scient. Stat. Comp.* 8 (1987) 483–500.

BAILEY, P. B., M. K. GORDON, AND L. F. SHAMPINE, "Automatic solution of the Sturm-Liouville problem," *ACM Trans. on Math. Software* 4 (1978), 193–208.

BAILEY, P. B., L. F. SHAMPINE, AND P. E. WALTMAN, *Nonlinear Two Point Boundary Value Problems.* New York: Academic Press, 1968.

BAKHVALOV, N., *Numerical Methods.* Moscow: MIR, 1977 (Russian ed. 1975).

BANK, R. E., "Marching algorithms for elliptic boundary value problems II, the variable case," *SIAM J. Numer. Anal.* 14 (1977), 950–970.

BANK, R. E., AND D. J. ROSE, "Marching algorithms for elliptic boundary value problems I, the constant coefficient case," *SIAM J. Numer. Anal.* 14 (1977), 792–829.

BANKS, H. T., AND F. KAPPEL, "Spline approximations for functional differential equations," *J. Diff. Equations* 34 (1979), 496–522.

BART, H., I. GOHBERG, AND M. A. KAASHOEK, *Minimal Factorization of the Matrix and Operator Functions.* Basel: Birkhäuser Verlag, 1979.

BAYLISS, A., "A double shooting scheme for certain unstable and singular boundary value problems," *Math. Comp.* 32 (1978), 61–71.

BELLEN, A., "One step collocation for delay differential equations," *J. Comp. & Appl. Math.* 10 (1984), 275–283.

BELLEN, A., "A Runge-Kutta-Nystrom method for delay differential equations," in [Ascher–Russell].

BELLEN, A., AND M. ZENNARO, "A collocation method for boundary value problems of differential equations with functional arguments," *Computing* 32 (1984), 307–318.

BELLMAN, R. E., *Stability Theory of Differential Equations.* New York: McGraw Hill, 1953.

BELLMAN, R. E., AND K. L. COOK, "On the computational solution of a class of functional differential equations," *J. Math. Anal. & Appl.* 12 (1965), 495–500.

BELLMAN, R. E., AND R. E. KALABA, *Quasilinearization and Nonlinear Boundary-Value Problems.* New York: American Elsevier, 1965.

BELLMAN, R. E., AND G. M. WING, *An Introduction to Invariant Imbedding.* New York: John Wiley and Sons, 1975.

BELTRAMI, E. J., *An Algorithmic Approach to Nonlinear Analysis and Optimization.* New York: Academic Press, 1970.

BERGER, A. E., H. HAN, AND R. B. KELLOGG, "A priori estimates and analysis of a numerical method for a turning point problem," *Math. Comp.* 42 (1984), 465–492.

BEYN, W.-J., "Discrete Green's functions and strong stability properties of the finite difference method," *Appl. Anal.* 14 (1982), 73–98.

BEYN, W.-J., "Zur Stabilität von Differenzenverfahren für Systemen linearer gewohnlicher Rantwertaufgaben," *Num. Math.* 29 (1978), 209–226.

BEYN, W.-J., AND E. BOHL, "Organizing centers for discrete reaction diffusion models," in [Küpper et al.].

BEYN, W.-J., AND E. DOEDEL, "Stability and multiplicity of solutions to discretizations of nonlinear differential equations," *SIAM J. Sci. Stat. Comp.* 2 (1981), 107–120.

BIRKHOFF, G., C. DE BOOR, B. SWARTZ, AND B. WENDROFF, "Rayleigh-Ritz approximation by piecewise cubic polynomials," *SIAM J. Num. Anal.* 3 (1966), 188–203.

BLAIR, J. J., "Error bounds for the solution of nonlinear two-point boundary value problems by Galerkin's method," *Num. Math.* 19 (1972), 99–109.

BLOTTNER, F. G., "Nonuniform grid method for turbulent boundary layers," in *Proc. 4th Internat. Conf. on Num. Methods in Fluid Dynamics*, ed. R. D. Richtmyer. Berlin: Springer-Verlag, 1975.

BOCK, H. G., "Recent advances in parameter identification techniques for O.D.E.," in [Deuflhard-Hairer].

BÖHMER, K., "Discrete Newton methods and iterated defect corrections," *Num. Math.* 37 (1981), 167–192.

BOLAND, W. R., AND P. NELSON, "Critical lengths by numerical integration of the associated Riccati equation to singularity," *Appl. Math. and Comp.* 1 (1975), 67–82.

DE BOOR, C., "The method of projections as applied to the numerical solutions of two-point boundary value problems using cubic splines," Ph.D. thesis, Univ. of Michigan, 1966.

DE BOOR, C., "Good approximation by splines with variable knots. II," in *Springer Lecture Note Series 363*, Berlin: Springer-Verlag, 1973.

DE BOOR, C., ED., *Mathematical Aspects of Finite Elements in Partial Differential Equations*. New York: Academic Press, 1974.

DE BOOR, C., "Package for calculating with B-splines," *SIAM J. Numer. Anal.* 14 (1977), 441–472.

DE BOOR, C., *A Practical Guide to Splines*. New York: Springer-Verlag, 1978.

DE BOOR, C., AND F. DE HOOG, "Stability of finite difference schemes for two-point boundary-value problems," *SIAM J. Numer. Anal.* 23 (1986), 925–935.

DE BOOR, C., F. DE HOOG, AND H. B. KELLER, "The stability of one-step schemes for first-order two-point boundary value problems," *SIAM J. Numer. Anal.* 20 (1983), 1139–1146.

DE BOOR, C., AND B. SWARTZ, "Collocation at Gaussian points," *SIAM J. Numer. Anal.* 10 (1973), 582–606.

DE BOOR, C., AND B. SWARTZ, "Comments on the comparison of global methods for linear two-point boundary value problems," *Math. Comp.* 31 (1977), 916–921.

DE BOOR, C., AND B. SWARTZ, "Collocation approximation to eigenvalues of an ODE: the principle of the thing," *Math. Comp.* 35 (1980), 679–694.

DE BOOR, C., AND R. WEISS, "SOLVEBLOK: A package for solving almost block diagonal linear systems," *ACM TOMS* 6 (1980), 80–87.

BRABSTON, D., AND H. B. KELLER, "A numerical method for singular two point boundary value problems," *SIAM J. Numer. Anal.* 14 (1977), 779–791.

BRAMBLE, J. H., ED., *Numerical Solution of Partial Differential Equations*. New York: Academic Press, 1966.

BRAMLEY, J. S., AND S. C. R. DENNIS, "The calculation of eigenvalues for the stationary perturbation of Poiseuille flow using initial value methods," *J. Math. Anal. & Appl. 101* (1984), 30–38.

BROWN, D. L., "A numerical method for singular perturbation problems with turning points," in [Ascher–Russell].

BROWN D., AND J. LORENTZ, "A high order method for stiff BVPs with turning points," *SIAM J. Sci. Stat. Comp.* 8 (1987), 798-805.

BROWN, R. R., "Numerical solution of boundary value problems using nonuniform grids," *SIAM J. Appl. Math.* 10 (1962), 475–495.

BURCHARD, H. G., "Splines (with optimal knots) are better," *J. Appl. Anal.* 3 (1974), 309–319.

BURDEN, R. C., AND J. D. FAIRES, *Numerical Analysis*. Boston: Prindle, Weber, and Schmidt, 1985.

BURRAGE, K., AND J. C. BUTCHER, "Stability criteria for implicit Runge-Kutta methods," *SIAM J. Numer. Anal.* 18 (1979), 46–57.

BUTCHER, J. C., "Implicit Runge-Kutta processes," *Math. Comp.* 18 (1964), 50–64.

BYKOV, A. A., "A stable numerical method of solving boundary-value problems for systems of linear ordinary differential equations," *Soviet Math. Dokl.* 21 (1980), 559–562.

BYKOV, A. A., "A stable numerical method for constructing a fundamental matrix for a system of ordinary differential equations," *Soviet Math. Dokl.* 29 (1984), 1–4.

BYKOV, A. A., AND A. S. IL'INSKII, "Solution of boundary value problems for linear systems of ordinary differential equations by the method of directed orthogonalization," *USSR Comp. Math. & Math. Phys.* 19 (1979), 74–82.

CAMPBELL, S., "Index two linear time-varying singular systems of differential equations," *SIAM J. Alg. & Discr. Meth.* 4 (1983), 237–243.

CAREY, G. F., "A mesh refinement scheme for finite element calculations," *Comp. Meth. in Appl. Mech. and Engr.* 7 (1976), 93–105.

CAREY, G. F., AND B. A. FINLAYSON, "Orthogonal collocation on finite elements," *Chem. Eng. Sci.* 30 (1975), 587–596.

CASH, J. R., "Diagonally implicit Runge-Kutta formulae for the numerical integration of nonlinear two-point boundary value problems," *Comp. & Maths. with Appls.* 10 (1984), 123–137.

CASH, J. R., "Adaptive Runge-Kutta methods for nonlinear two-point boundary value problems with mild boundary layers," *Comp. & Maths. with Appls.* 11 (1985), 605–619.

CASH, J. R., "On the numerical integration of nonlinear two point boundary value problems using iterated deferred corrections, Part I," *Comput. & Maths. with Appls.* (1986).

CASH, J. R., "On the numerical integration of nonlinear two point boundary value problems using iterated deferred corrections, Part II," *SIAM J. Numer. Anal.* (1988).

CASH, J. R., AND D. R. MOORE, "A high order method for the numerical solution of two-point boundary value problems," *BIT* 20 (1980), 44–52.

CEBECI, T., AND A. M. D. SMITH, "A finite difference method for calculating compressible laminar turbulent flows," *J. of Basic Engr.* 92 (1970), 523–535.

CERUTTI, J. H., Collocation for systems of ordinary differential equations, *Comp. Sci. Tech. Rept.* 230, Univ. of Wisc., 1974.

CHANG, K. W., AND F. A. HOWES, *Nonlinear Singular Perturbation phenomena: Theory and Application*, Springer Applied Math. Series 56, New York: Springer-Verlag, 1984.

CHAWLA, M. M., AND C. D. KATTI, "Finite difference methods and their convergence for a class of singular two point boundary value problems," *Num. Math.* 39 (1982), 341–350.

CHILDS, B., H. H. DOIRON, AND C. C. HOLLOWAY, "Numerical solution of multipoint boundary value problems in linear systems," *Int. J. Systems Science* 2 (1971), 58–66.

CHILDS, B., D. LUCKINBILL, J. BRYAN, AND J. H. BOYD, JR., "Numerical solution of multipoint boundary value problems in linear systems," *Int. J. Systems Sci.* 2 (1971), 49–57.

CHILDS, B., M. SCOTT, J. W. DANIEL, E. DENMAN, AND P. NELSON, EDS., *Codes for Boundary Value Problems in Ordinary Differential Equations*, Lecture Notes in Computer Science 76. Berlin: Springer-Verlag, 1979.

CHIN, R. C. Y., G. W. HEDSTROM AND F. A. HOWES, "A survey of analytical and numerical methods for multiple scale problems," preprint, Lawrence Livermore Lab., 1984.

CHOCHOLATY, P., AND L. SLAHOR, "A method to boundary value problems for delay equations," *Num. Math.* 33 (1979), 69–75.

CHRISTIANSEN, J., AND R. D. RUSSELL, "Adaptive mesh selection strategies for solving boundary value problems," *SIAM J. Num. Anal.* 15 (1978), 59–80.

CHRISTIANSEN, J., AND R. D. RUSSELL, "Error analysis for spline collocation methods with applications to knot selection," *Math. Comp.* 32 (1978), 415–419.

CHRISTIANSEN, J., AND R. D. RUSSELL, "Deferred corrections using uncentered differences," *Num. Math.* 35 (1980), 21–33.

CIARLET, P. G., M. H. SCHULTZ, AND R. S. VARGA, "Numerical methods of high order accuracy for nonlinear boundary value problems. I. one dimensional problem," *Num. Math.* 9 (1967), 394–430.

CIARLET, P. G., M. H. SCHULTZ, AND R. S. VARGA, "Numerical methods of high order accuracy for nonlinear boundary value problems. II. nonlinear boundary conditions," *Num. Math.* 11 (1968), 331–345.

CIARLET, P. G., M. H. SCHULTZ, AND R. S. VARGA, "Numerical methods of high order accuracy for nonlinear boundary value problems. IV. periodic boundary conditions," *Num. Math.* 12 (1968), 266–279.

CIARLET, P. G., M. H. SCHULTZ, AND R. S. VARGA, "Numerical methods of high order accuracy for nonlinear boundary value problems. V. monotone operators," *Num. Math.* 13 (1969), 51–77.

CLENSHAW, C. W., "The solution of van der Pol's equation in Chebyshev series," in [Greenspan].

COLE, J. D. *Perturbation Methods in Applied Mathematics.* Waltham: Blaisdell, 1968.

COLLATZ, L., *The Numerical Treatment of Differential Equations.* Berlin: Springer, 3rd ed., 1960.

CONCUS, P., "Static menisci in a vertical right circular cylinder," *J. Fluid Mech.* 34 (1968), 481–495.

CONTE, S. D., "The numerical solution of linear boundary value problems," *SIAM Review* 8 (1966), 309–321.

CONTE, S. D., AND C. DE BOOR, *Elementary Numerical Analysis,* 2nd ed. New York: McGraw-Hill, 1980.

COPPEL, W. A., *Dichotomies in Stability Theory.* Lecture Notes in Mathematics, Vol. 629. Berlin: Springer-Verlag, 1978.

CORLESS, R. M., "A feasibility study of the use of COLSYS in the solution of systems of partial differential equations," M.Sc Thesis, Univ. Waterloo, 1982.

COX, M. G., "The numerical evaluation of B-splines," *J. Inst. Maths. Applics.* 10 (1972), 134–149.

COYLE, J. M., J. E. FLAHERTY, AND R. LUDWIG, "On the stability of mesh equidistribution on strategies for time dependent PDEs," *J. Comp. Phys.* 62 (1986), 26–39.

CRYER, C. W., "The numerical solution of boundary value problems for second order functional differential equations by finite differences," *Num. Math.* 20 (1973), 288–299.

DAHLQUIST, G., AND A. BJÖRCK (translated by N. Anderson), *Numerical Methods.* Englewood Cliffs, N.J.: Prentice-Hall, 1974.

DANIEL, J. W., "Extrapolation with spline-collocation methods for two-point boundary value problems I: proposals and justifications," *Aeq. Math.* 16 (1977), 107–122.

DANIEL, J. W., AND A. J. MARTIN, "Numerov's method with deferred corrections for two-point boundary value problems," *SIAM J. Numer. Anal.* 14 (1977), 1033–1050.

DANIEL, J. W., AND B. K. SWARTZ, "Extrapolated collocation for two-point boundary value problems using cubic splines," *J. Inst. Maths. Applics.* 16 (1965), 161–174.

DAVEY, A., "A simple numerical method for solving Orr-Sommerfeld problems," *Quart. J. Mech. Appl. Math.* 26 (1973), 401–411.

DAVEY, A., "An automatic orthonormalization method for solving stiff BVPs," *J. Comp. Phys.* 51 (1983), 343–356.

DAVIS, P. J., AND P. RABINOWITZ, *Methods of Numerical Integration,* 2nd ed. New York: Academic Press, 1985.

DENNIS, J. E., JR., AND J. J. MORÉ, "Quasi-Newton Methods, Motivation and Theory," *SIAM Review,* 19 (1977), pp. 48–89.

DENNIS, J. E., AND R. B. SCHNABEL, *Numerical Methods for Unconstrained Optimization and Nonlinear Equations.* Englewood Cliffs, N.J.: Prentice-Hall, 1983.

DENMAN, E. D., "An overview of invariant imbedding algorithms and two-point boundary value problems," in [Childs et al.].

DENNY, V. E., AND R. B. LANDIS, "A new method for solving two-point boundary value problems using optimal node distributions," *J. Comp. Phys.* 9 (1972), 120–137.

DEUFLHARD, P., "A modified Newton method for the solution of ill conditioned systems of nonlinear equations with applications to multiple shooting," *Num. Math.* 22 (1974), 289–315.

DEUFLHARD, P., "A relaxation strategy for the modified Newton method," in Springer Lecture Notes in Math. 447 (1975), R. Bulirsch, W. Oettli, and J. Stoer, eds.

DEUFLHARD, P., "A stepsize control for continuation methods and its special application to multiple shooting techniques," *Num. Math.* 33 (1979), 115–146.

DEUFLHARD, P., "Nonlinear equation solvers in boundary value problem codes," in [Childs et al.].

DEUFLHARD, P., "Recent advances in multiple shooting techniques," in *Computational Techniques for Ordinary Differential Equations*. New York: Academic Press, 1980.

DEUFLHARD, P., AND G. BADER, Multiple-shooting techniques revisited," in [Deuflhard–Hairer].

DEUFLHARD, P., AND E. HAIRER, EDS., *Numerical Treatment of Inverse Problems in Differential and Integral Equations*. Boston: Birkhauser, 1983.

DEUFLHARD, P., AND G. HEINDL, "Affine invariant convergence theorems for Newton's method and extensions to related methods," *SIAM J. Numer. Anal.* 16 (1979), 1–10.

DEUFLHARD, P., H.-J. PESCH, AND R. RENTROP, "A modified continuation method for the numerical solution of nonlinear two-point boundary value problems by shooting techniques," *Num. Math.* 26 (1976), 327–343.

DIAZ, J. C., "A collocation-Galerkin method for two-point boundary value problems using continuous piecewise polynomial spaces," *SIAM J. Numer. Anal.* 14 (1977), 844–858.

DIAZ, J. C., G. FAIRWEATHER, AND P. KEAST, "FORTRAN packages for solving certain almost block diagonal linear systems by modified alternate row and column elimination," *ACM Trans. on Math. Software* 9 (1983), 358–375.

DIAZ, J. C., G. FAIRWEATHER, AND P. KEAST, "COLROW and ARECO: Fortran packages for solving certain almost block diagonal linear systems by modified row and column elimination," *ACM Trans. Math. Software* 9 (1983), 376–380.

DICKMANNS, E. D., AND H. J. PESCH, "Influence of a reradiative heating constraint on lifting entry trajectories for maximum lateral range," in *11th Int. Symp. on Space Tech. and Sci.*, Tokyo, 1975.

DICKMANNS, E. D., AND K. H. WELL, "Approximate solution to optimal control problems using third order Hermite polynomial functions," in *Springer Lecture Notes in Comp. Sci.* 27, Berlin: Springer-Verlag, 1975.

DIECI, L., Theoretical and computational aspects of the Riccati transformation for solving BVP," Ph.D. thesis, Univ. of New Mexico, 1986.

DIECI, L., "Some aspects of using IV software to solve BVP via a Riccati transformation," *Appl. Math. & Comp.* (1988).

DIECI, L., M. R. OSBORNE, AND R. D. RUSSELL, "A Riccati transformation method for solving boundary value problems, I: theoretical aspects," *SIAM J. Numer. Anal.* (1988).

DIECI, L., M. R. OSBORNE, AND R. D. RUSSELL, "A Riccati transformation method for solving BVPs, II: computational aspects, *SIAM J. Numer. Anal.* (1988).

DIECI, L., AND R. D. RUSSELL, "Riccati and other methods for singularly perturbed BVP," in *Proceedings of BAIL IV*, ed. J. Miller, Dublin: Boole Press, 1986.

DIEKOFF, H.-J., P. LORY, H. J. OBERLE, H.-J. PESCH, P. RENTROP, AND R. SEYDEL, "Comparing routines for the numerical solution of initial value problems of ordinary differential equations in multiple shooting," *Num. Math.* 27 (1977), 449–469.

DODSON, D. S., Optimal order approximation by polynomial spline functions, Ph.D. thesis, Purdue Univ., 1972.

DOEDEL, E. J., "The construction of finite difference approximations to ordinary differential equations," *SIAM J. Numer. Anal.* 15 (1978), 450–465.

DOEDEL, E. J., "Finite difference methods for nonlinear two-point boundary value problems," *SIAM J. Numer. Anal.* 16 (1979), 173–185.

DOEDEL, E. J., "Auto: a program for the automatic bifurcation analysis of autonomous systems," *Congressus Numerantium* 30 (1981), 265–284.

DOEDEL, E. J., "The computer-aided bifurcation analysis of predator-prey models," *J. Math. Biology* 20 (1984), 1–14.

DOEDEL, E. J., AND R. F. HEINEMANN, "Numerical computation of periodic solution branches and oscillatory dynamics of the stirred tank reactor with A → B → C reactions," *Chem. Eng. Sci.* 38 (1983), 1493–1499.

DOEDEL, E. J., AND J. P. KERNEVEZ, "Software for continuation problems in ordinary differential equations," *SIAM J. Numer. Anal.*, (1988).

DOOLAN, E. P., J. J. H. MILLER, AND W. H. A. SCHILDERS, *Uniform Numerical Methods for Problems with Initial and Boundary Layers.* Dublin: Boole Press, 1980.

DOUGLAS, J., JR., AND T. DUPONT, "Superconvergence for Galerkin methods for the two-point boundary value problem via local projections," *Num. Math.* 21 (1973), 270–278.

DOUGLAS, J., JR., AND T. DUPONT, "Galerkin approximations for the two-point boundary value problem using continuous, piecewise polynomial spaces," *Num. Math.* 22 (1974), 99–109.

DOUGLAS, J., JR., AND T. DUPONT, *Collocation Methods for Parabolic Equations in a Single Space Variable Based on C^1 Piecewise-Polynomial Spaces.* Berlin: Springer-Verlag, 1974.

DOUGLAS, J., JR., T. DUPONT, AND LARS WAHLBIN, "Optimal L_∞ error estimates for Galerkin approximations to solutions of two-point boundary value problems," *Math. Comp.* 29 (1975), 475–483.

DRURY, L. O'C., "Numerical solution of Orr-Sommerfeld-type equations," *J. Comp. Phys.* 37 (1980), 133–139.

DUNN, R. J., JR., AND M. F. WHEELER, "Some collocation-Galerkin methods for two-point boundary value problems," *SIAM J. Numer. Anal.* 13 (1976), 720–733.

DUPONT, T., "A unified theory of superconvergence for Galerkin methods for two-point boundary value problems," *SIAM J. Numer. Anal.* 13 (1976), 362–368.

EIROLA, T., "Convergence of the back-and-forth shooting method," *J. Opt. Theory & Appl.* 41 (1983), 559–572.

EIROLA, T., "A study of the back-and-forth shooting method," Helsinki U. of Tech. Systems Theory Lab Rep. A 3, 1985.

ENGLAND, R., A program for the solution of boundary value problems for systems of ordinary differential equations, Culham Laboratory Report, PDN 3/73, 1976.

ENGLAND, R., AND R. M. M. MATTHEIJ, "Discretizations with dichotomic stability for two-point boundary value problems," in [Ascher–Russell].

ENGLAND, R., AND R. M. M. MATTHEIJ, "On dichotomic stability," report, Math. Inst. Nijmegen.

ENGLAND, R., AND R. M. M. MATTHEIJ, "A sequential multiple shooting strategy for stiff boundary value problems," in preparation.

ENRIGHT, W. H., "Improving the performance of numerical methods for two point boundary value problems," in [Ascher–Russell].

FAIRWEATHER, G., P. KEAST, AND J. C. DIAZ, "On the H^{-1} Galerkin method for second order linear two point boundary value problems," *SIAM J. Numer. Anal.* 21 (1984), 314–326.

FARAHZAD, P., AND R. P. TEWARSON, "An efficient numerical method for solving the differential equation of renal counterflow systems," *Comp. Bio. Med.* 8 (1978), 57–64.

FARRINGTON, C. C., R. T. GREGORY, AND A. H. TAUB, "On the numerical solution of Sturm-Liouville differential equations," *Math. Comp.* 11 (1957), 131–150.

FERGUSON, N. B., AND B. A. FINLAYSON, "Error bounds for approximate solutions to nonlinear ordinary differential equations," *AIChE J.* 18 (1972), 1053–1059.

FINLAYSON, B. A., *The Method of Weighted Residuals and Variational Principles.* New York: Academic Press, New York, 1972.

FINLAYSON, B. A., "Water movement in dessicated soils," in *Finite Elements in Water Resources,* eds. W. Gray, G. Pinder, and C. Brebbia. London: Pentech Press, 1977.

FLAHERTY, J. E., "A rational function approximation for the integration point in exponentially weighted finite element methods," *Int. J. Num. Meth. Eng.* 19 (1982), 782–791.

FLAHERTY, J. E., AND W. MATHON, "Collocation with polynomial and tension splines for singularly-perturbed boundary value problems," *SIAM J. Sci. Stat. Comp.* 1 (1980), 260–289.

FLAHERTY, J. E., AND R. E. O'MALLEY, JR., "The numerical solution of boundary value problems for stiff differential equations," *Math. Comp.* 31 (1977), 66–93.

FLAHERTY, J. E., AND R. E. O'MALLEY, JR., "Singularly perturbed boundary value problems for nonlinear systems, including a challenging problem for a nonlinear beam," in *Springer Lecture Notes in Math* 942 (1981).

FLAHERTY, J. E., AND R. E. O'MALLEY, JR., "Numerical methods for stiff systems of two-point boundary value problems," *SIAM J. Sci. Stat. Comp.* 5 (1984), 865–886.

FLETCHER, R. *Practical Methods of Optimization*, Vol. I. Chichister: Wiley, 1980.

FOURER, R., "Staircase matrices and systems," *SIAM Review* 26 (1984), 1–70.

FOX, L., *The Numerical Solution of Two-Point Boundary Problems in Ordinary Differential Equations*. Oxford: Oxford University Press, 1957.

FOX, L., "Some numerical experiments with eigenvalue problems in ordinary differential equations," in [Langer].

FRANK, R., "The method of iterated defect correction and its application to two-point boundary value problems, I & II," *Num. Math.* 25 (1976), 409–419, & 27 (1977), 407–420.

GARTLAND, E. C., JR, "Computable pointwise error bounds and the Ritz method in one dimension," *SIAM J. Numer. Anal.* 21 (1984), 84–100.

GARY, J., "Computing eigenvalues of ordinary differential equations by finite differences," *Math. Comp.* 19 (1965), 365–379.

GEAR, C. W., *Numerical Initial-Value Problems in Ordinary Differential Equations*. Englewood Cliffs, N.J.: Prentice-Hall, 1971.

GEAR, C. W., "Simultaneous numerical solution of differential/algebraic equations," *IEEE Trans. Circuit Theory*, CT-18 (1971), 89–95.

GEAR, C. W., AND L. PETZOLD, "ODE methods for the solution of differential/algebraic systems," *SIAM J. Numer. Anal.* 21 (1984), 716–728.

GEORGE, J. H., AND R. W. Gunderson, "Conditioning of linear boundary value problems," *BIT* 12 (1972), 172–181.

GLADWELL, I., "Shooting codes in the NAG library," in [Childs et al.].

GLADWELL, I., AND D. J. MULLINGS, "On the effect of boundary conditions in collocation by polynomial splines for the solution of boundary value problems in ordinary differntial equations," *J. Inst. Maths. Appl.* 16 (1975), 93–107.

GODUNOV, S., "On the numerical solution of boundary-value problems for systems of linear ordinary differential equations," *Uspekhi Mat. Nauk.* 16 (1961), 171–174.

GOLDBERG, M. A., "Some functional relationships for two-point boundary value problems," *J. Math. Anal. & Appl.* 45 (1974), 199–209.

GOL'DIN, V.YA., AND A. V. KOLPAKOV, "Solution of stiff boundary value problems by differential pivotal condensation," *USSR Comp. Math. & Math. Phys.* 22 (1982), 85–92.

GOLUB, G. H., AND C. F. VAN LOAN, *Matrix Computations*. Baltimore: Johns Hopkins Univ. Press, 1983.

GOODMAN, T. R., AND G. N. LANCE, "The numerical integration of two-point boundary value problems," *MTAC* 10 (1956),82–86.

GOUGH, D. O., E. A. SPIEGEL, AND J. TOOMRE, "Highly stretched meshes as functionals of solutions," in *Proc. 4th Internat. Conf. Num. Methods in Fluid Dynamics*, ed. R. D. Richtmyer. Berlin: Springer-Verlag, 1975.

GREBENNIKOV, A. I., "The choice of nodes in the approximation of functions by splines," *USSR Comp. Math. & Math. Phys.* 16 (1976), 208–213.

GREENSPAN, D., ED., *Numerical Solutions of Nonlinear Differential Equations*. New York: John Wiley, 1966.

GRIEPENTROG, E., AND R. MÄRZ, *Differential-Algebraic Equations and their Numerical Treatment*. Teubner - Texte zur Math., Band 88, Leipzig: 1986.

GRIEWANK, A., "The solution of boundary value problems by Broyden based secant methods," *Centre for Math. Anal. Rep.* CMA-R22-85, Austral. Nat. U., 1985.

GRIEWANK, A., AND G. W. REDDIEN, "The calculation of Hopf points by a direct method," *IMA J. Numer. Anal.* 3 (1983), 295–303.

GRIEWANK, A., AND G. W. REDDIEN, "Characterization and computation of generalized turning points," *SIAM J. Numer. Anal.* 21 (1984), 176–185.

GRIEWANK, A., AND G. W. REDDIEN, "The approximation of simple singularities," in [Ascher–Russell].

GRIGORIEFF, R. D., "Convergence of discrete Green's functions for finite difference schemes," *Applic. Anal.* 19 (1985), 233–250.

GRIGORIEFF, R. D., "On the convergence of stability constants," *SIAM J. Numer. Anal.* 23 (1986), 832–836.

DE GROEN, P. N., AND M. HERMANN, "Bidirectional shooting: a strategy to improve the reliability of shooting methods for ODE," *SIAM J. Sci. Stat. Comp.* 5 (1984), 360–369.

GUCKENHEIMER, J., AND P. HOLMES, *Nonlinear Oscillation, Dynamical Systems and Bifurcation of Vector Fields.* New York: Springer, 1983.

GUDERLEY, K. G., "A unified view of some methods for stiff two point boundary value problems," *SIAM Review* 17 (1975), 416–442.

GUPTA, S., "An adaptive boundary value Runge-Kutta solver for first order boundary value problems," *SIAM J. Numer. Anal.* 22 (1985), 114–126.

HALL G., AND J. M. WATT, EDS. *Modern Numerical Methods for Ordinary Differential Equations.* Oxford: Clarendon Press, 1976.

HANKE, M., R. LAMOUR, AND R. WINKLER, "The program system RWA for the solution of two-point boundary-value problems — foundations, algorithms, comparisons," Humboldt U. Math. Rep. 67, 1985.

HART, W. E., AND S. O. W. SOUL, "Quasi-Newton methods for discretized nonlinear boundary problems," *J. Inst. Maths. Appl.* 11 (1973), 351–359.

HARTMAN, P. *Ordinary Differential Equations.* New York: John Wiley, 1964.

HASKELL, N. A., "The dispersion of surface waves in multilayered media," *Bull. Seism. Soc. Am.* 43 (1953), 17–34.

HASSARD, B. D., N. D. KAZARINOFF, AND Y. H. WAN, *Theory and Applications of Hopf Bifurcations.* Cambridge: Cambridge Univ. Press, 1981.

HAY, R. I., AND I. GLADWELL, "Solving almost block diagonal linear equations on the CDC Cyber 205," NA Report 98, Dept. of Math., University of Manchester, 1985.

HEMKER, P. *A Numerical Study of Stiff Two-point Boundary Problems.* Amsterdam: Math. Centrum, 1977.

HEMKER, P. W., AND J. J. H. MILLER, EDS. *Numerical Analysis of Singular Perturbation Problems.* London: Academic Press, 1979.

HENRICI, P., *Discrete Variable Methods in Ordinary Differential Equations.* New York: John Wiley, 1962.

HERBOLD, R. J., M. H. SCHULTZ, AND R. S. VARGA, "The effect of quadrature errors in the numerical solution of boundary value problems by variational techniques," *Aeq. Math.* 3 (1969), 247–270.

HERMANN, M., "Schiesstechnieken fuer Zweipunkt Randwertprobleme: eine Uebersicht," Humboldt Universität zu Berlin, Sektion Mathematik, Seminar- bericht 55, in *Numerical methods for ordinary boundary value problems*, ed., R. März, Berlin (1983).

HERMANN, M., AND H. BERNDT, "RWPM, a multiple shooting code for nonlinear two-point boundary value problems". Preprint 67, 68, 69, FSU Jena ,1982.

HOLODNIOK, M., AND M. KUBIČEK, "New algorithms for evaluation of complex bifurcation points in ordinary differential equations. A comparative study," manuscript, 1981.

HOLODNIOK, M., AND M. KUBIČEK: "Derper — an algorithm for continuation of periodic solutions in ordinary differential equations," *J. Comp. Phys.* 55 (1985), 254–267.

HOLT, J. F., "Numerical solution of nonlinear two-point boundary value problems by finite difference methods," *Comm. ACM* 7 (1964), 366–373.

DE HOOG, F., AND D. JACKETT, "On the rate of convergence of finite difference schemes on nonuniform grids," *J. Austral. Math. Soc.* Series B (1986–1987).

DE HOOG, F., AND R. M. M. MATTHEIJ, "Conditioning, dichotomy, and scaling for two-point BVPs," in [Ascher–Russell].

DE HOOG, F., AND R. M. M. MATTHEIJ, "On dichotomy and well-conditioning in BVP," *SIAM J. Numer. Anal.*, 24 (1987), 89–105.

DE HOOG, F., AND R. M. M. MATTHEIJ, "An algorithm for solving multipoint boundary value problems," *Computing* 38 (1985), 219–234.

DE HOOG, F., AND R. M. M. MATTHEIJ, "On the conditioning of multipoint boundary and integral value problems," (1986), to appear.

DE HOOG, F., AND R. WEISS, "Difference methods for boundary value problems with a singularity of the first kind," *SIAM J. Numer. Anal.* 13 (1976), 775–813.

DE HOOG, F., AND R. WEISS, "An approximation theory for boundary value problems on infinite intervals," *Computing* 24 (1980), 227–239.

DE HOOG, F., AND R. WEISS, "On the boundary value problem for systems of ordinary differential equations with a singularity of the second kind," *SIAM J. Math. Anal.* 11 (1980), 41–60.

HOUSTIS, E. N., "A collocation method for systems of nonlinear ordinary differential equations," *J. Math. Anal. & Appl.* 62 (1978), 24–37.

HOWES, F. A., "Some old and New results on singularly perturbed boundary value problems," in [Meyer-Parter].

HUANG, C. L., "Application of quasilinearization technique to the vertical channel flow and heat convection," *Int. J. Non-Linear Mech.* 13 (1978), 55–60.

HUMPHREY, D., AND G. F. CAREY, "Adaptive mesh refinement algorithm using element residuals," in [Childs et al.].

HYMAN, J. M., "Moving mesh methods for initial boundary value problems," Los Alamos Nat. Labs Rep. LA-UR-84,61, 1984.

ISAACSON, E., AND H. B. KELLER, *Analysis of Numerical Methods.* New York: John Wiley, 1966.

JONES, D. J., AND J. C. SOUTH, JR., "Application of the method of lines to the solution of elliptic partial differential equations," NRC Canada Aeronautical Report LR-599, Ottawa, 1979.

JORDAN, D. W., AND P. SMITH, *Nonlinear Ordinary Differential Equations.* Oxford: Clarendon Press, 1977.

KALNINS, A., "Analysis of shells of revolution subjected to symmetrical and nonsymmetrical loads," *Trans. ASME, ser. E, J. Appl. Mech.* 31 (1964), 467–476.

KANTOROVICH, L. V., AND G. P. AKILOV, *Functional Analysis in Normed Spaces.* New York: Pergamon Press, 1964.

KANTOROVICH, L. V., AND G. P. AKILOV, *Approximate Methods of Higher Analysis.* New York: Interscience, 1964.

VON KARMAN, T., "Über laminare und turbulente Reibung," *ZAMM* 1 (1921), 232–252.

KAUTSKY, J., AND N. K. NICHOLS, "Equidistributing meshes with constraints," *SIAM J. Sci. Stat. Comp.* 1 (1980), 499–511.

KEAST, P., G. FAIRWEATHER, AND J. C. DIAZ, "A computational study of finite element methods for second order linear two-point boundary value problems," *Math. Comp.* 40 (1983), 499–518.

KEDEM, G., "A posteriori error bounds for two-point boundary value problems," *SIAM J. Numer. Anal.* 18 (1981), 431–448.

KELLER, H. B., "On the accuracy of finite difference approximations to the eigenvalues of differential and integral operators," *Num. Math.* 7 (1965), 412–419.

KELLER, H. B., "Existence theory for two point boundary value problems," *Bulletin AMS* 72 (1966), 728–731.

KELLER, H. B., *Numerical Methods for Two-Point Boundary Value Problems.* Waltham: Blaisdell, 1968.

KELLER, H. B., "Accurate difference methods for linear ordinary differential systems subject to linear constraints," *SIAM J. Numer. Anal.* 6 (1969), 8–30.

KELLER, H. B., "Accurate difference methods for nonlinear two point boundary value problems," *SIAM J. Numer. Anal.* 11 (1974), 305–320.

KELLER, H. B., "Numerical solution of boundary value problems for ordinary differential equations: survey and some recent results on difference methods," in [Aziz].

KELLER, H. B., "Approximation methods for nonlinear problems with application to two-point boundary value problems," *Math. Comp.* 29 (1975), 464–474.

KELLER, H. B., *Numerical Solution of Two Point Boundary Value Problems.* CBMS Regional Conference Series in Applied Math, 24, SIAM, Philadelphia, 1976.

KELLER, H. B., "Constructive methods for bifurcation and nonlinear eigenvalue problems," in *Proc. 3rd Int. Symp. on Comp. Methods in Appl. Sci. and Engr.*, Versailles, 1977.

KELLER, H. B., "Numerical solution of bifurcation and nonlinear eigenvalue problems," in [Rabinowitz].

KELLER, H. B., AND M. LENTINI, "Invariant imbedding, the box scheme and an equivalence between them," *SIAM J. Numer. Anal.* 19 (1982), 942–962.

Keller, H. B., and V. Pereyra, "Difference methods and deferred corrections for ordinary boundary value problems," *SIAM J. Numer. Anal.* 16 (1979), 241–259.

Keller, H. B., and A. B. White, Jr., "Difference methods for boundary value problems in ordinary differential equations," *SIAM J. Numer. Anal.* 12 (1975), 791–801.

Kellogg, R. B., and A. Tsan, "Analysis of some difference approximations for a singular perturbation problem without turning points," *Math. Comp.* 32 (1977), 1025–1039.

Kendall, R. P., and M. F. Wheeler, "A Crank-Nicolson H^1 Galerkin procedure for parabolic problems in a single space variable," *SIAM J. Numer. Anal.* 13 (1976), 861–876.

Kennett, B. L. N., *Seismic Wave Propagation in Stratified Media.* Cambridge: Cambridge University Press, 1983.

Kevorkian, J., and J. Cole, *Perturbation Methods in Applied Mathematics.* Berlin: Springer-Verlag, 1981.

Kreiss, B., and H.-O. Kreiss, "Numerical methods for singular perturbation problems," *SIAM J. Numer. Anal.* 18 (1981), 262–276.

Kreiss, H.-O., "Centered difference approximation to singular systems of ODEs," Symposia Mathematica X, Inst. Nazionale di Alta Math., 1972.

Kreiss, H.-O., "Difference approximations for boundary and eigenvalue problems for ordinary differential equations," *Math. Comp.* 26 (1972), 605–624.

Kreiss, H.-O., "Difference methods for stiff ordinary differential equations," *SIAM J. Numer. Anal.* 15 (1978), 21–58.

Kreiss, H.-O., "Problems with different time scales for ordinary differential equations," *SIAM J. Numer. Anal.* 16 (1979), 980–998.

Kreiss, H.-O., T. A. Manteuffel, B. Swartz, B. Wendroff, and A. B. White, "Supra-convergent schemes on irregular grids," *Math. Comp.* (1987).

Kreiss, H.-O., N. K. Nichols, and D. L. Brown, "Numerical methods for stiff two-point boundary value problems," *SIAM J. Numer. Anal.* 23 (1986), 325–368.

Krogh, F. T., J. P. Keener, and W. H. Enright, "Reducing the number of variational equations in the implementation of multiple shooting," in [Ascher–Russell].

Kubíček, M., and V. Hlavaček, "A review of application of shooting methods for nonlinear chemical engineering problems," Scientific Papers of Prague Inst. Chem. Tech. K9 (1974), 5–35.

Kubíček, M., V. Hlavaček, and M. Holodniok, "Test examples for comparison of codes for nonlinear boundary value problems in ordinary differential equations," in [Childs et al.].

Kubíček, M., and M. Marek, *Computational Methods in Bifurcation Theory and Dissipative Structures.* New York: Springer, 1983.

Küpper, T., H. D. Mittelmann, and H. Weber, eds., *Numerical Methods for Bifurcation Problems.* Boston: ISNM 70, Birkhäuser, 1984.

Lambert, J. D., *Computational Methods in Ordinary Differential Equations.* New York: John Wiley & Sons, 1973.

Langer, R. E., ed., *Boundary Problems in Differential Equations.* Madison: Univ. Wisconsin Press, 1960.

Langford, W. F., "A shooting algorithm for the best least squares solution of two-point boundary value problems," *SIAM J. Num. Anal.* 14 (1977), 527–542.

Langford, W. F., "Numerical solution of bifurcation problems for ordinary differential equations," *Numer. Math.* 28 (1977), 171–190.

Lee, W. K. H., and S. Stewart, *Principles and Applications of Microearthquake Networks.* New York: Academic Press, 1981.

Lees, M., "Discrete methods for nonlinear two-point boundary value problems," in [Bramble].

Lentini, M., and H. B. Keller, "Boundary value problems on semi-infinite intervals and their numerical solution," *SIAM J. Numer. Anal.* 17 (1980), 577–604.

Lentini, M., and H. B. Keller, "The von Karman swirling flows," *SIAM J. Appl. Math.* 38 (1980), 52–64.

Lentini, M., M. R. Osborne, and R. D. Russell, "The close relationships between methods for solving two-point boundary value problems," *SIAM J. Numer. Anal.* 22 (1985), 280–309.

LENTINI, M., AND V. PEREYRA, "A variable order finite difference method for nonlinear multipoint boundary value problems," *Math. Comp.* 28 (1974), 981–1004.

LENTINI, M., AND V. PEREYRA, "Boundary problem solvers for first order systems based on deferred corrections," in [Childs et al.].

LENTINI, M., AND V. PEREYRA, "An adaptive finite difference solver for nonlinear two-point boundary value problems with mild boundary layers", *SIAM J. Numer. Anal.* 14 (1977), 91–111.

LISKOVETS, O. A., "The method of lines (review)," *Differential Equations* 1 (1965), 1308–1323 (translated from Russian).

VAN LOON, P., "Riccati transformations: when and how to use?," in [Ascher–Russell].

VAN LOON, P. M., AND R. M. M. MATTHEIJ, "Stable continuous orthonormalization techniques for linear boundary value problems," *J. Austr. Math. Soc.* Ser. B (1987).

LUCAS, T. R., AND G. W. REDDIEN, "Some collocation methods for nonlinear boundary value problems," *Num. Math.* 9 (1972), 341–356.

LUCAS, T. R., AND G. W. REDDIEN, "A high order projection method for nonlinear two-point boundary value problems," *Num. Math.* 20 (1973), 257–270.

LUDWIG, D., "Optimal harvesting of a randomly fluctuating resource I: Application of perturbation methods," *SIAM J. Appl. Math.* 37 (1979), 166–184.

LUDWIG, D., AND J. VARAH, "Optimal harvesting of a randomly fluctuating resource II: Numerical methods and results," *SIAM J. Appl. Math.* 37 (1979), 185–205.

LYNCH, R. E., AND F. R. RICE, "A high order difference method for differential equations," *Math. Comp.* 34 (1980), 333–372.

MADSEN, N. K., AND R. F. SINCOVEC, "PDECOL: General collocation software for partial differential equations," *ACM Trans. on Math. Software* 5 (1979), 326–351.

MAIER, M. R., "An adaptive shooting method for singularly perturbed boundary value problems," *SIAM J. Scient. Stat. Comp.* 7 (1986), 418–440.

MAIER, M. R., "Numerical solution of singular perturbed boundary value problems using a collocation method with tension splines," in [Ascher–Russell].

MAIER, M. R., AND D. R. SMITH, "Numerical solution of a symmetric diode model," *J. Comp. Phys.* 42 (1981), 309–326.

MAJDA, G., "Filtering techniques for systems of stiff ordinary differential equations II. error estimates," *SIAM J. Numer. Anal.* 22 (1985), 1116–1134.

MANTEUFFEL, T. A., AND A. B. WHITE, JR., "On the efficient numerical solution of systems of second order boundary value problems," *SIAM J. Numer. Anal.* 23 (1986), 996–1007.

MARKOWICH, P. A., "A theory for the approximation of solutions of boundary value problems on infinite intervals," *SIAM J. Math. Anal.* 13 (1982), 484–513.

MARKOWICH, P. A., "Asymptotic analysis of von Karman flows," *SIAM J. Appl. Math.* 42 (1982), 549–557.

MARKOWICH, P. A., "A finite difference method for the basic stationary semiconductor device equations," in [Ascher–Russell].

MARKOWICH, P. A., *The Stationary Semiconductor Device Equations.* Wien: Springer, 1986.

MARKOWICH, P. A., AND C. A. RINGHOFER, "Collocation methods for boundary value problems on 'long' intervals," *Math. Comp.* 40 (1983), 123–150.

MARKOWICH, P. A., AND C. A. RINGHOFER, "A singularly perturbed boundary value problem modelling a semiconductor device," *SIAM J. Appl. Math.* 44 (1984), 231–256.

MARKOWICH, P. A., C. A. RINGHOFER, S. SELBERHERR, AND M. LENTINI, "A singular perturbation approach for the analysis of the fundamental semiconductor equations," *IEEE Trans. Electron. Devices*, Vol. ED-30, No. 9 (1983), 1165–1180.

MÄRZ, R., "On boundary value problems in differential-algebraic equations," Preprint 124, Humboldt-Universität zu Berlin, 1986.

MÄRZ, R., "On the numerical treatment of differential-algebraic equations," *ZAMM* 67 (1986).

MASSERA, I. L., AND I. I. SCHÄFFER, *Linear Differential Equations and Function Spaces.* New York: Academic Press, 1966.

MATTHEIJ, R. M. M., "On approximating smooth solutions of linear singularly perturbed ODE," in [Hemker-Miller].

MATTHEIJ, R. M. M., "Characterizations of dominant and dominated solutions of linear recursions," *Num. Math.* 35 (1980), 421–442.

MATTHEIJ, R. M. M., "Estimates for the errors in the solutions of linear boundary value problems, due to perturbations," *Computing* 27 (1981), 299–318.

MATTHEIJ, R. M. M., "Stable computation of solutions of unstable linear initial value recursions," *BIT* 22 (1982), 79–93.

MATTHEIJ, R. M. M., "The conditioning of linear boundary value problems," *SIAM J. Numer. Anal.* 19 (1982), 963–978.

MATTHEIJ, R. M. M., "On decoupling of linear recursions," *Bull. Aust. Math. Soc.* 27 (1983), 347–360.

MATTHEIJ, R. M. M., "The stability of LU-decomposition of block tridiagonal matrices," *Bull. Austr. Math. Soc.* 29 (1984), 177–205.

MATTHEIJ, R. M. M., "Accurate estimates for the fundamental solutions of discrete BVP," *J. Math. Anal. Appl.* 101 (1984), 444–464.

MATTHEIJ, R. M. M., "Stability of block LU-decompositions of matrices arising from BVP," *SIAM J. Disc. Math.* 5 (1984), 314–331.

MATTHEIJ, R. M. M., "Decoupling and stability of algorithms for boundary value problems," *SIAM* Review (1985), 1–44.

MATTHEIJ, R. M. M., "On the computation of BVP on infinite intervals," *Math. Comp.* 48 (1987), 553-579.

MATTHEIJ, R. M. M., AND F. R. DE HOOG, "On non-invertible boundary value problems," in [Ascher–Russell].

MATTHEIJ, R. M. M., AND R. E. O'MALLEY, JR., "Decoupling of boundary value problems for two time scale systems," manuscript, 1982.

MATTHEIJ, R. M. M., AND R. E. O'MALLEY, JR., "On solving boundary value problems for multi-scale systems using asymptotic approximations and multiple shooting," *BIT* (1984).

MATTHEIJ, R. M. M., AND M. D. SMOOKE, "Estimates for the inverse of tridiagonal matrices arising in boundary value problems," *Lin. Alg. Appl.* 73 (1986), 33–57.

MATTHEIJ, R. M. M., AND G. W. STAARINK, "BOUNDPACK user's manual," Report 84–01, Math. Inst. Kath. Univ., Nijmegen, 1984.

MATTHEIJ, R. M. M., AND G. W. M. STAARINK, "On optimal shooting intervals," *Math. Comp.* 42 (1984), 25–40.

MATTHEIJ, R. M. M., AND G. W. STAARINK, "An efficient algorithm for solving general linear two point BVP," *SIAM J. Sci. Stat. Comp.* 5 (1984), 745–763.

MEJIA, R., J. L. STEPHENSON, AND R. J. LEVEQUE, "A test problem for kidney model". Manuscript.

MELHEM, R. G., AND W. C. RHEINBOLT, "A comparison of methods for determining turning points of nonlinear equations," *Computing* 29 (1982), 201–226.

MEYER, G. H., *Initial-Value Methods for Boundary Value Problems.* New York: Academic Press, 1973.

MEYER, G. H., "Continuous orthonormalization for boundary value problems," *J. Comp. Phys.* 62 (1986), 248–262.

MEYER, R. E., AND S. V. PARTER, EDS. *Singular Perturbations and Asymptotics.* New York: Academic Press, 1980.

MILINAZZO, F., AND P. G. SAFFMAN, "Turbulence model predictions for the inhomogeneous mixing layer," *Studies Appl. Math.* 55 (1976), 45–63.

MIRANKER, W. L., *Numerical Methods for Stiff Equations and Singular Perturbation Problems.* Dordrecht: Reidel, 1981.

MIRANKER, W. L., AND M. VAN VELDHUIZEN, "The method of envelopes," *Math. Comp.* 32 (1978), 453–498.

MITTELMANN, H. D., AND H. WEBER, EDS. *Bifurcation Problems and Their Numerical Solution.* ISNM 54, Boston: Birkhäuser, 1980.

MIURA, R., "Accurate computation of the stable solitary wave for the FitzHugh-Nagumo equations," *J. Math. Biology* 13 (1982), 247–269.

MOORE, G., "Defect correction from a Galerkin viewpoint," Brunel U. Math. Rep. TR/21/85, 1985.

MOORE G., AND A. SPENCE, "The calculation of turning points of nonlinear equations," *SIAM J. Numer. Anal.* 17 (1980), 567–576.

MORRISON, D. D., J. D. RILEY, AND J. F. ZANCANARO, "Multiple shooting method for two-point boundary value problems," *Comm. ACM* 5 (1962), 613–614.

MUIR, P., Runge-Kutta methods for two point boundary value problems," Ph.D. thesis, Dept. Computer Science, U. Toronto, 1984.

NATTERER, F., "Uniform convergence of Galerkin's method for splines on highly nonuniform meshes," *Math. Comp.* 31 (1977), 457–468.

NELSON, P., JR., and I. T. ELDER, "Calculation of eigenfunctions in the context of integration-to-blowup," *SIAM J. Numer. Anal.* 14 (1977), 124–136.

DE NEVERS, K., AND K. SCHMITT, "An application of the shooting method to boundary value problems for second order delay equations," *J. Math. Anal. & Appl.* 36 (1971), 588–597.

NG, B. S., AND W. H. REID, "The compound matrix method for ordinary differential equations," *J. Comp. Phys.* 58 (1985), 209–228.

NICHOLS, N. K., AND R. ENGLAND, "Numerical solution of an elastic boundary layer problem using multiple shooting techniques," *J. Comp. Phys.* 46 (1982), 369–389.

OLYNYCK, K. O., "Spherically-symmetric monopole solutions in SU(2) and SU(3) gauge theories," M.Sc. Thesis, Physics, Simon Fraser U., 1978.

O'MALLEY, R. E., JR., *Introduction to Singular Perturbations*. New York: Academic Press, 1974.

O'MALLEY, R. E., JR., "On the simultaneous use of asymptotic and numerical methods to solve two-point problems with boundary and interior layers," in [Ascher–Russell].

O'MALLEY, R. E., JR., AND L. R. ANDERSON, "Time-scale decoupling systems," *Optimal Control Appl. and Methods* (1982), 133–153.

ORAVA, P. J., AND P. A. J. LAUTALA, "Back-and-forth shooting method for solving two-point boundary-value problems," *J. Opt. Theory & Appl.* 18 (1976), 485–498.

ORSZAG, S. A., "Accurate solution of the Orr-Sommerfeld stability equation," *J. Fluid Mech.* 50 (1971), 689–703.

ORTEGA, J. M., AND W. C. RHEINBOLDT, *Iterative Solution of Nonlinear Equations in Several Variables*. New York: Academic Press, 1970.

OSBORNE, M. R., "Minizing truncation error in finite difference approximations to ordinary differential equations," *Math. Comp.* 21 (1967), 133–145.

OSBORNE, M. R., "On shooting methods for boundary value problems," *J. Math. Anal. & Appl.* 27 (1969), 417–433.

OSBORNE, M. R., "Collocation, difference equations, and stitched function representations," in *Topics in Numerical Analysis II*, ed. J. Miller. New York: Academic Press, 1974.

OSBORNE, M. R., "On the numerical solution of boundary value problems for ordinary differential equations," in *Information Processing '74*. Amsterdam: North Holland, 1974.

OSBORNE, M. R., "An efficient weak line search with guaranteed termination," Tech. Rep. 1870, Math. Res. Center, Madison, Wis., 1978.

OSBORNE, M. R., "Aspects of the numerical solution of boundary value problems with separated boundary conditions," Computing Research Group Report, Australian National Univ., 1978

OSBORNE, M. R., "The stabilized march is stable," *SIAM J. Numer. Anal.* 16 (1979), 923–933.

OSBORNE, M. R., AND R. D. RUSSELL, "The Riccati transformation in the solution of BVPs," *SIAM J. Numer. Anal.* 23 (1986), 1023–1033.

OSHER, S., "Nonlinear singular perturbation problems and one-sided difference schemes," *SIAM J. Numer. Anal.* 18 (1981), 129–144.

PAINE, J. W., "Numerical approximation of Sturm-Liouville eigenvalues," Ph.D. thesis, Australian National University, 1979.

PAINE, J. W., AND A. L. ANDREW, "Bounds and higher order estimates for Sturm-Liouville eigenvalues," *J. Math. Anal. & Appl.* 96 (1983), 388–394.

PAINE, J., AND R. D. RUSSELL, "Conditioning of collocation matrices and discrete Green's functions," *SIAM J. Numer. Anal.* 23 (1986), 376-392.

PARTER, S. V., "Swirling flow between rotating coaxial disks," in *Springer Lecture Notes in Math* 942, 1981.

PEARSON, C. E., "A numerical method for ordinary differential equations of boundary-layer type," *J. Math. and Phys.* 47 (1968), 134–154.

PEARSON, C. E., "On nonlinear ordinary differential equations of boundary layer type," *J. Math. and Phys.* 47 (1968), 351–358.

PEREYRA, V., "The difference correction method for nonlinear two point boundary value problems of class M," *Rev. Union Mat.* (Argentina) 22 (1965), 184–201.

PEREYRA, V., "On improving an approximate solution of a functional equation by deferred corrections," *Num. Math.* 8 (1966), 376–391.

PEREYRA, V., "Iterated deferred corrections for nonlinear operator equations," *Num. Math.* 10 (1967), 316–323.

PEREYRA, V., "Iterated deferred corrections for nonlinear boundary value problems," *Num. Math.* 11 (1968), 111–125.

PEREYRA, V., "High order finite difference solution of differential equations," Comp. Sci. Dept. Report 73-348, Stanford Univ., 1973.

PEREYRA, V., W. K. H. LEE, AND H. B. KELLER, "Solving seismic ray tracing problems in a heterogeneous medium," *Bull. Seism. Soc. Amer.* 70 (1980), 79–99.

PEREYRA, V., AND R. D. RUSSELL, "Difficulties of comparing complex mathematical software: general comments and the BVODE case," *Acta Cient. Venezolana* 33 (1982), 15–22.

PEREYRA, V., AND E. G. SEWELL, "Mesh selection for discrete solution of boundary value problems in ordinary differential equations," *Num. Math.* 23 (1975), 261–268.

PERRIN, F. M., H. S. PRICE, AND R. S. VARGA, "On higher-order methods for nonlinear two-point boundary value problems," *Num. Math.* 13 (1969), 180–198.

PETZOLD, L., "A description of DASSL: A differential/algebraic system solver," SAND82–8637, Sandia Labs, Livermore, Cal., 1982.

PÖNISCH, G., AND M. SCHWETLICK, "Computing turning points of curves implicitly defined by nonlinear equations depending on a parameter," *Computing* 26 (1981), 107–121.

PRAGER, W., "A note on the optimal choice of finite element grids," *Comp. Meth. Appl. Mech. Engr.* 6 (1975), 363–366.

PRUESS, S., "Interpolation schemes for collocation solutions of two point boundary value problems," *SIAM J. Sci. Stat. Comp.* 7 (1986), 322–333.

PRYCE, J. D., "Software for Sturm-Liouville eigenvalues with error bounds," C.S. Tech. Rep. 79–02, Bristol Univ., 1979.

RABINOWITZ, P., ED. *Applications of Bifurcation Theory.* New York: Academic Press, 1977.

RACHFORD, H. H., JR., AND M. F. WHEELER, "An H^1 Galerkin procedure for the two-point boundary value problem," in [de Boor].

RAO, M., *Ordinary Differential Equations Theory and Applications.* London: Edward Arnold, 1980.

REDDIEN, G. W., "Approximation methods for two-point boundary value problems with nonlinear boundary conditions," *SIAM J. Numer. Anal.* 13 (1976), 405–411.

REDDIEN, G. W., "Approximation methods and alternative problems," *J. Math. Anal. & Appl.* 60 (1977), 139–149.

REDDIEN, G. W., "Collocation at Gauss points as a discretization in optimal control," *SIAM J. Control & Opt.* 17 (1979), 298–306.

REDDIEN, G. W., "Projection methods for two point BVP," *SIAM Review* 22 (1980), 156–171.

REDDIEN, G. W., AND C. C. TRAVIS, "Approximation methods for boundary value problems of differential equations with functional arguments," *J. Math. Anal. & Appl.* 46 (1974), 62–74.

RENTROP, R., "Numerical solution of the singular Ginzburg-Landau equations by multiple shooting," *Computing* 16 (1976), 61–67.

RENTROP, R., "A Taylor series method for the numerical solution of two-point boundary value problems," *Num. Math.* 31 (1979), 359–375.

REID, W. T , *Riccati Differential Equations*. New York: Academic Press, 1972.

REUVEN, Y., M. D. SMOOKE, AND H. RABITZ, "Sensitivity analysis of boundary value problems: applications to nonlinear diffusion systems," *J. Comp. Phys.* 64 (1986), 27–55.

RICHARDS, P. G., "Elastic wave solutions in stratified media," *Geophysics* 36 (1971), 798–809.

RINGHOFER, C., "On collocation schemes for quasilinear singularly perturbed boundary value problems," *SIAM J. Numer. Anal.* 21 (1984), 864–882.

RINZEL, J., AND R. MILLER, "Numerical calculation of stable and unstable periodic solutions to the Hodgkins-Huxley equations," *Math. Bios.* 49 (1980), 27–60.

DE RIVAS, E. M., "On the use of nonuniform grids in finite-difference equations," *J. Comp. Phys.* 10 (1972), 202–210.

ROBERTS, G. O., "Computational meshes for boundary value problems," in *Proc. 2nd Int. Conf. Num. Meth. Fluid Dyn.*, ed. M. Holt. Berlin: Springer-Verlag, 1970.

ROBERTS, S. M., AND J. S. SHIPMAN, *Two-Point Boundary Value Problems: Shooting Methods*. New York: American Elsevier, 1972.

ROBERTSON, T. N., "The linear two-point boundary value problem on an infinite interval," *Math. Comp.* 25 (1971), 475–481.

ROOSE, D., "An algorithm for the computation of Hopf bifurcation points," manuscript, Leuven, 1984.

ROOSE, D., AND R. PIESSENS, "Numerical Computation of nonsimple turning points and cusps," *Num. Math.* 46 (1985), 189–211.

RUSSELL, R. D., "Collocation for systems of boundary value problems," *Num. Math.* 23 (1974), 119–133.

RUSSELL, R. D., "Efficiencies for B-spline methods for solving differential equations," in *Proc. 5th Conf. on Num. Math.*, Manitoba, 1975.

RUSSELL, R. D., "A comparison of collocation and finite differences for two point boundary value problems," *SIAM J. Numer. Anal.* 14 (1977), 19–39.

RUSSELL, R. D., AND J. CHRISTIANSEN, "Adaptive mesh selection strategies for solving boundary value problems," *SIAM J. Numer. Anal.* 15 (1978), 59–80.

RUSSELL, R. D., AND L. F. SHAMPINE, "A collocation method for boundary value problems," *Num. Math.* 19 (1972), 13–36.

RUSSELL, R. D., AND J. M. VARAH, "A comparison of global methods for two-point boundary value problems," *Math. Comp.* 29 (1975), 1007–1019.

SACKER, R. J., AND G. R. SELL, "Existence of dichotomies and invariant splittings for linear differential systems I, II, III," *J. Diff. Equa.* 15 (1974), 429–458 and 22 (1976), 478–522.

SAKAI, M., "Piecewise cubic interpolation and two-point boundary value problems," Publ. RIMS, Kyoto Univ. 7 (1971/72), 345–362.

SAMMON, P. H., "A discrete least squares method," *Math. Comp.* 31 (1977), 60–65.

SCHARFETTER, D. L., AND H. K. GUMMEL, "Large signal analysis of a silicon read diode oscillator," *IEEE Trans. Electron Devices*, ED-16 (1969), 64–77.

SCHEID, R. E., "The accurate numerical solution of highly oscillatory ordinary differential equations," *Math. Comp.* 41 (1983), 487–509.

SCHMEISER, C., "Uniform dichotomies for singularly perturbed problems," Manuscript, 1985.

SCHMEISER, C., AND R. WEISS, "Asymptotic and numerical methods for singular singularly perturbed boundary value problems in ordinary differential equations," *Proc. BAIL III*, ed. J. J. H. Miller, 1984.

SCHMEISER, C., AND R. WEISS, "Asymptotic analysis of singular singularly perturbed boundary value problems," *SIAM J. Math. Anal.* 17 (1986), 560–579.

SCHREIBER, R., "Finite element methods of high order accuracy for singular two point boundary value problems with non smooth solutions," *SIAM J. Numer. Anal.* 17 (1980), 547–566.

SCHRÖDER, J., "Upper and lower bounds for solutions of generalized two-point boundary value problems," *Num. Math.* 23 (1975), 433–457.

SCHULTZ, M. H., *Spline Analysis*. Englewood Cliffs, N.J.: Prentice-Hall, 1973.

SCHWARTZ, I. B., "Estimating regions of existence of unstable periodic orbits using computer-based techniques," *SIAM J. Numer. Anal.* 20 (1983), 106–120.

SCOTT, M. R., *Invariant Imbedding and its Applications to Ordinary Differential Equations*. Reading, Mass: Addison-Wesley, 1973.

SCOTT, M. R., "A bibliography on invariant imbedding and related topics," SLA-74–0284, Sandia Labs., Albuquerque, 1974.

SCOTT, M. R., "On the conversion of boundary value problems into stable initial value problems via several invariant imbedding algorithms," in [Aziz].

SCOTT, M. R., AND W. H. VANDEVENDER, "A comparison of several invariant imbedding algorithms for the solution of two-point boundary-value problems," *Appl. Math. and Comp.* 1 (1975), 187–218.

SCOTT, M. R., AND H. A. WATTS, "A systematized collection of codes for solving two-point boundary-value problems," in [Aziz].

SCOTT, M. R., AND H. A. WATTS, "Computational solution of nonlinear two-point boundary value problems," in *Proc. 5th Symp. of Comp. in Chem. Engr.*, 1977.

SCOTT, M. R., AND H. A. WATTS, "Computational solution of linear two point boundary value problems via orthonormalization," *SIAM J. Numer. Anal.* 14 (1977), 40–70.

SERBIN, S. M., "Computational investigations of least-squares type methods for the apporximate solution of boundary value problems," *Math. Comp.* 29 (1975), 777–793.

SEYDEL, R., "Numerical computation of branch points in ordinary differential equations," *Num. Math.* 32 (1979), 51–68.

SEYDEL, R., "Numerical computation of periodic orbits that bifurcate from stationary solutions of ordinary differential equations," *Appl. Math. Comp.* 9 (1981), 257–271.

SEYDEL, R., "BIFPAC — a program package for calculating bifurcations," manuscript, 1983.

SHAMPINE, L. F., AND M. K. GORDON, *Computer Solution of Ordinary Differential Equations: The Initial Value Problem*. San Francisco: W. H. Freeman, 1975.

SIGFRIDSSON, B., AND J. L. LINDSTROM, "Electrical properties of electron-irradiated n-type silicon," *J. Appl. Phys.* 47 (1976), 4611–4620.

SINCOVEC, R. F., "On the relative efficiency of higher order collocation methods for solving two-point boundary value problems," *SIAM J. Numer. Anal.* 14 (1977), 112–123.

SKEEL, R. D., "A theoretical framework for proving accuracy results for deferred corrections," *SIAM J. Numer. Anal.* 19 (1982), 171–196.

SKEEL, R. D., "The order of accuracy for deferred corrections using uncentered end formulas," *SIAM J. Numer. Anal.* 23 (1986), 393–402.

SLEPTSOV, A. G., "The spline-collocation and the spline-Galerkin methods for Orr-Sommerfeld problem," in [Ascher–Russell].

SLOANE, D. M., "Eigenfunctions of systems of linear ordinary differential equations with separated boundary conditions," *J. Comp. Phys.* 24 (1977), 320–330.

SMITH, D. R., *Singular Perturbation Theory*. Cambridge: Cambridge University Press, 1985.

SMOOKE, M. D., "Numerical solution of burner stabilized premixed laminar flames by boundary value methods," *J. Comp. Phys.* 48 (1982), 72–105.

SMOOKE, M. D., "An error estimate for the modified Newton method with applications to the solution of nonlinear two point boundary value problems," *J. Opt. Theory and Appl.* 39 (1983), 489–502.

SMOOKE, M. D., AND M. L. KOSZYKOWSKI, "Fully adaptive solutions of one-dimensional mixed initial-boundary value problems with applications to unstable problems in combustion," *SIAM J. Sci. Stat. Comp.* 7 (1986), 301-329.

SMOOKE, M. D., AND R. M. M. MATTHEIJ, "On the solution of nonlinear two point boundary value problems on successively refined grids," *Applied Num. Math.* 1 (1985), 463–487.

SMOOKE, M. D., J. A. MILLER, AND R. J. KEE, "Solution of premixed and counterflow diffustion flame problems by adaptive boundary value methods," in [Ascher–Russell].

SOBOLEV, S. L., *The Closure of Computational Algorithms and Some of its Applications.* Moscow: USSR Academy of Sciences, 1955 (in Russian).

SÖDERLIND, G., AND R. M. M. MATTHEIJ, "Stability and asymptotic estimates in nonautonomous linear differential systems," *SIAM J. Math. Anal.* 16 (1985), 69–92.

SPENCE, A., "Non simple turning points and cusps," *IMA J. Numer. Anal.* 2 (1982), 413–427.

SPUDICH, P., AND U. ASCHER, "Calculation of complete theoretical seismograms in vertically varying media using collocation methods," *Geoph. J. R. Astr. Soc.* 75 (1983), 101–124.

STAKGOLD, I., "Branching of solutions of nonlinear equations," *SIAM Review* 13 (1971), 289–332.

STEPHENSON, J. L., R. P. TEWARSON, AND R. MEJIA, "Quantitative analysis of mass and energy balance in non-ideal models of the renal counterflow system," *Proc. Nat. Acad. Sci. USA* 71 (1974), 1618–1622.

STETTER, H., "Asymptotic expansions for the error of discretization algorithms for non-linear functional equations," *Num. Math.* 7 (1965), 18–31.

STETTER, H., *Analysis of Discretization Methods for Ordinary Differential Equations.* Berlin: Springer-Verlag, 1973.

STEWART, G. W., *Introduction to Matrix Computations.* New York: Academic Press, 1973.

STEWARTSON, K., AND B. A. TROESCH, "On a pair of equations occurring in swirling viscous flow with an applied magnetic field," *ZAMP* 29 (1977), 951–963.

STOER, J., AND R. BULIRSCH, *Introduction to Numerical Analysis,* (English version). New York: Springer Verlag, 1980.

STRANG, G., AND G. J. FIX, *An Analysis of the Finite Element Method.* Englewood Cliffs, N.J.: Prentice-Hall, 1973.

SYLVESTER, S., AND F. MEYER, "Two-point boundary problems by quasilinearization," *J. Appl. Math.* 13 (1965), 586–602.

TAKEUCHI, H., AND M. SAITO, "Seismic surface waves"," in *Methods of Computational Physics,* 11. New York: Academic Press, 1972.

TANG, J. W., AND D. J. TURCKE, "Characteristics of optimal grids," *Comp. Meth. Appl. Mech. Engr.* 11 (1977), 31–37.

TAUFER, J., "On the factorization method," *Apl. Mat.* 11 (1966), 427–450.

TIMOTHY, L. K., AND B. E. BONA, *State Space Analysis, an Introduction.* New York: McGraw-Hill, 1968.

TROESCH, B. A., "Intrinsic difficulties in the numerical solution of a boundary value problem," Space Tech. Labs., Tech. Note NN-142, 1960.

TROESCH, B. A., "A severe test problem for two-point boundary-value routines," in [Childs et al.].

URABE, M., "Galerkin's procedure for nonlinear periodic systems and its extension to multipoint boundary value problems for general nonlinear systems," in [Greenspan].

URABE, M., "Numerical solution of multi-point boundary value problems in Chebychev series. Theory and method," *Num. Math.* 9 (1967), 341–366.

VAINNIKO, G. M., "On the stability and convergence of the collocation method," *Differentsial'nye Uravneniya,* 1 (1965), 244–254.

VAINNIKO, G. M., "On convergence of the collocation method for nonlinear differential equations," *USSR Comp. Math. & Math. Phys.* 6 (1966), 35–42.

VARAH, J. M., "On the solution of block tridiagonal systems arising from certain finite difference equations," *Math. Comp.* 26 (1972), 859–868.

VARAH, J. M., "Alternate row and column elimination for solving linear systems," *SIAM J. Numer. Anal.* 13 (1976), 71–75.

VARGA, R. S., "Hermite interpolation-type Ritz methods for two point boundary value problems," in [Bramble].

VASIL'EVA, A. B., AND V. F. BUTUZOV, *Asymptotic Expansions of Solutions of Singularly Perturbed Equations.* Moscow: Nauka, 1973.

VASIL'EVA, A. B., AND V. F. BUTUZOV, "Singularly perturbed equations in the critical case," MRC Tech. Rep. 2039, Madison, Wisconsin, 1980.

VAN VELDHUIZEN, M., "A note on partial pivoting and Gaussian elimination," *Num. Math.* 29 (1977), 1–10.

VAN VELDHUIZEN, M., "D-stability," *SIAM J. Numer. Anal.* 18 (1981), 45–64.

VAN VELDHUIZEN, M., "Higher order methods for a singularly perturbed problem," *Num. Math.* 30 (1978), 267–279.

VAN VELDHUIZEN, M., "Asymptotic expansions of the global error for the implicit midpoint rule," *Computing* 33 (1984), 185–192.

WALLISCH, W., AND M. HERMANN, *Schiessverfahren zur Loesung von Rand un Eigenwertaufgaben.* Leipzig: Teubner, 1985.

WAN, F. Y. M., "Polar dimpling of complete spherical shells," in *Theory of Shells,* Proc. IUTAM Symp., eds. W. T. Koiter and G. K. Mikhailov. Amsterdam: North Holland, 1979.

WAN, F. Y. M., "The dimpling of spherical caps," *Mech. Today,* 5 (1980), 495–508.

WAN, F. Y. M., AND U. ASCHER, "Horizontal and flat points in shallow cap dimpling," Tech. rep. 80–5, IAMS, Univ. of B. C., Vancouver, Canada, 1980.

WASOW, W., *Linear turning point theory.* New York: Springer, 1985.

WEBER, H., "On the numerical approximation of secondary bifurcation problems," in *Numerical Solution of Nonlinear Equations,* Lecture Notes in Math. 878. Berlin: Springer-Verlag, 1981.

WEINMÜLLER, E., "A difference method for a singular boundary value problem of second order," *Math. Comp.* 42 (1984), 441–464.

WEINMÜLLER, E., "Collocation for singular boundary value problems of second order," *SIAM J. Numer. Anal.* 23 (1986), 1062–1095.

WEISS, R., "The convergence of shooting methods," *BIT* 13 (1973), 470–475.

WEISS, R., "The application of implicit Runge-Kutta and collocation methods to boundary-value problems," *Math. Comp.* 28 (1974), 449–464.

WEISS, R., "Bifurcation in difference approximations to two point boundary value problems," *Math. Comp.* 29 (1974), 746–760.

WEISS, R., "An analysis of the box and trapezoidal schemes for linear singularly perturbed boundary value problems," *Math Comp.* 42 (1984), 41–67.

WERNER, B., AND A. SPENCE, "The computation of symmetry-breaking bifurcation points," *SIAM J. Numer. Anal.* 21 (1984), 388–399.

WHITE, A., JR, "On selection of equidistributing meshes for two point boundary value problems," *SIAM J. Numer. Anal.* 16 (1979), 472–502.

WRIGHT, K., "Some relationships between implicit Runge-Kutta, collocation and Lanczos $tau-$ methods and their stability properties," *BIT* 20 (1970), 217–227.

ZENNARO, M., "One-step collocation: uniform superconvergence, predictor-corrector method, local error estimate," *SIAM J. Numer. Anal.* 22 (1985), 1135–1152.

Index

Adam's scheme, **70**
Adjoint, 270
   -operator, 270
   self-, 97, 270
Admissible transformation (of coordinates), 373
Affine invariant, 329
Almost block diagonal matrix, 308
Alternate row and column:
   elimination, 310
   pivoting, 310
Angles
   -between matrices, 38
   -between subspaces, 38, 119
Arclength, 494, 498
Asymptotically equidistributing, 364
Asymptotic expansion, 404

Backsubstitution, 424
Backward:
   -differentiation formulas, 74
   -error analysis, 46
   -Euler scheme, 74, 139, 215, 433
Banded matrix, 43, 304, 308
Bandwidth, 43
Basis:
   B-spline-, 61
   Hermite type-, 59, 60
BC, (*see* Boundary condition)

BDF, 74
Bifurcation, 490
   Hopf-, 496
   pitchfork-, 495
   -point, 493
   primary-, 494
   secondary-, 494
   transition-, 495
Block:
   -banded matrix, 304
   -diagonal matrix (almost), 308
   -tridiagonal matrix, 312
Boundary condition (BC), 3, 5
   multipoint-, 6, 92, 323, 470, 512
   nonseparated-, 5, 470
   partially separated-, 5
   scaling of, 114
   separated-, 5
   special point-, 471
   two point-, 2
Boundary layer, 8
Boundary value problem (BVP):
   discrete-, 288
   first order formulation, 3
   -on infinite intervals, 483, 486
   multipoint-, 6
   -with oscillatory modes, 500
   -with parameters, 322, 470
   singular- (first kind), 484

Boundary value problem (BVP) (*cont.*)
   singular- (second kind), 478, 480
   singularly perturbed-, 113, 386
   standard form, 469
Box scheme: 191,
Branch, 491
   -switching, 499
Branching, 493
Broyden update, 52
B-spline, 61, 260
Butcher array, 69
BVP, 1

Cauch-Schwartz inequality, 77
Characteristic polynomial, 32
Classical:
   continuation, 347
   turning point, 393
Closure, 297
Code design, 515
Collocation, 210, 218, 222, 247
   conditioning of-, 252
   -for high order DE, 244, 251
   -matrix, 252, 260
   -point, 210, 248
   -solution, 218
   -using splines, 259
Compactification, 153
   -for a decoupled system, 158, 316
   stable-, 316
Companion matrix, 32
Complexity, 31, 35, 303
Composite formulas, 66
Cond (A), 39
Condensation of parameters, 212, 307
Condensing method, 153, 306
Conditioned (of BVP):
   ill-, 79, 94
   well-, 79, 94
Conditioned (of matrices):
   ill-, 39, 114
   well-, 39, 114
Conditioned (of problem):
   ill-, 29
   well-, 29
Conditioning constant, 78, 95, 111
Condition number, 39, 78, 81
Conjugate transpose, 34
Consistency (of difference methods), 71, 188
   -of BVP, 198, 199
   -of IVP, 71
   -order, 71
Consistency (of fundamental solutions), 280, 281, 290

-constant, 281, 290
Consistency requirement (for reduced solution), 404
Consistent DAE, 475
Continuation, 344
   classical-, 347
   -with incremental load, 347
   parameter-, 490
Continuous decoupling, 279, 282
Contraction mapping, 49, 50
Convergence (of difference methods):
   -of BVP, 189, 205, 255, 451
   -of IVP, 71
   super-, 211
Convergence (of iteration):
   global-, 328, 332
   linear-, 49
   local-, 328
   quadratic-, 49
   rate of-, 48
   superlinear-, 52
Convergent of order p, 71, 189, 190, 199
Criterion function, 334

DAE, 474, 504
Damped Newton, 329
Damping factor, 330
Davidenko equation, 350
Decomposition, 319
   almost block-, 308
   LU-, 40, 320
   QU-, 35, 43
Decoupling, 157
   continuous-, 279, 282
   discrete-, 288
Decoupling algorithm:
   continuous case, 284
   discrete case, 291, 321
Defect correction, 234
Defective eigenvalues, 33
Deferred correction, 228, 235, 238, 240
Delay differential equation, 506
Descent direction, 330
Diagonally dominant, 42
   essentially-, 410
   strictly-, 107, 120
Diagonally dominated, 410
Dichotomic fundamental solution, 118, 280
Dichotomy, 115, 280, 289
   discrete case, 289
   exponential-, 116
   ordinary-, 116
Difference:
   finite (method), 185

-operator, 188
-operator (positive), 436
-scheme (explicit), 69
-scheme (implicit), 69
-scheme (one step), 68
Differential:
-algebraic equation, 474, 504
-operator, 93, 188
Discrete:
-BVP, 288, 305, 354
-decoupling, 288
-fundamental solution, 289
-Green's function, 138
Discretization error, 29
Divided differences, 56
Dominant (strictly diagonally), 107, 120
D-stability, 442

Effective singularity, 339
Eigenvalue, 32
   defective-, 33
   generalized (problem), 480
   kinematic-, 105
   -problem (ODE), 478
Eigenvector, 32
Elementary:
   -hermitian matrix, 43
   -lower triangular matrix, 40
   -reflector, 43
Elimination:
   alternate row-column-, 310
   block tridiagonal-, 313
   column-, 310
   Gaussian-, 39
   row-, 310
Equidistributing mesh, 62, 363
Equidistribution, 62, 362
Equilibration, 42
Error:
   -analysis, 28
   -analysis (backward), 46
   -control (IVP), 74
   discretization-, 29
   -equidistribution, 362
   global-, 30
   input-, 30
   local-, 30, 72, 81
   -per unit step control, 75
   rounding-, 30
   truncation-, 29
Euler backward, 424
Euler's method, 29, 68, 424
Exact line search, 331
Expansion:
   asymptotic-, 404
   matched asymptotic-, 405
Exponential:
   dichotomy, 389
   fitting, 454
Extrapolation, 228, 230

Factorization, 33, 40
-methods, 298, 319
-Fehlberg schemes, 69
Finite difference method, 185
Finite element method, 266
Fixed point, 49
   -iteration, 49
Flop, 42
Forward substitution, 42
Fréchet derivative, 53
Function:
   criterion-, 334
   monitor-, 363
   objective-, 330
Functional differential equation, 505
Fundamental solution (matrix), 86
   dichotomic-, 118, 280
   discrete-, 289

Galerkin's method, 271
Gauss:
   -formulas, 64
   -(Legendre) points, 65, 215
Gaussian:
   -elimination, 39
   quadrature, 64
Gauss Newton, 340
Generalized solution, 268
Gershgorin, 33, 268
Glb, 37
Global:
   -discretization error, 29
   -error, 71, 256
   -method, 185
Grade of a vector, 33
Gram-Schmidt, 34
Greatest lower bound, 37
Green's function, 94
   discrete-, 138
   first order system, 94, 95
   high order ODE, 96
Grid staggered-, 246

Hermite type basis, 59, 60, 260
Hessenberg form, 46

Highly oscillatory BVP, 500
High order schemes, 208
Homotopy path, 344, 490
Hopf bifurcation, 496
Hyperbolic splitting, 401

Ill-conditioning, 29, 39, 79, 94, 114
Imbedding:
   invariant-, 164, 167
   parameter-, 165
   -in a time dependent problem, 350
Index of DAE, 474, 478
Infinite interval BVP, 483, 486
Initial layer, 398
Initial value:
   -methods, 68, 457
   -problems, 2, 67, 85
Inner solution, 390, 404
Interpolation, 55
   Lagrange form, 56
   Newton form, 57
   osculatory-, 58
   polynomial-, 55
   spline-, 62
Invariant imbedding, 164
Isolated solution, 91
IVP, 2, 67, 85

Jacobian (matrix), 90, 109
Jordan:
   -block, 33, 102
   -canonical form, 33, 102

Kinematic:
   -eigenvalue, 105
   -similarity, 105

Lagrange form, 56
Layer:
   boundary-, 386, 398
   initial-, 398
   interior, 386
   -region, 409, 443
   shock-, 394, 396
   transition-, 393
Least:
   -squares, 45
   upper bound, 37
Legendre polynomial, 65
Limit point, 49

Line search:
   exact-, 331
   weak-, 331
Lines (method of):
   longitudinal-, 509
   transverse-, 509
Liouville normal form, 481
Lipschitz:
   -condition, 88
   -constant, 85
   -continuous, 85
Lobatto points, 66, 215
Local:
   -error, 72
   -representation, 59
   -support (discretization), 61, 259
   -truncation error, 71, 204
Longitudinal method of line, 509
Lub, 37
LU decomposition, 40, 320
Lyapunov equation, 109, 289

Machine epsilon, 29
Matching technique, 146, 155
March (stabilized), 61, 163
Marquardt algorithm, 340
Matched asymptotic expansion, 405, 409
Matrix:
   collocation-, 252, 260
   companion-, 32
   diagonally dominant-, 42
   elementary lower triangular-, 40
   hermitian-, 43
   Hessenberg-, 46
   -norm, 37
   orthogonal-, 33
   permutation-, 41
   positive definite, 35
   staircase-, 304, 306
   strictly diagonally dominant-, 42, 107, 120
   totally positive-, 62
   upper triangular-, 34
   Vandermonde-, 56
Maximizing vector, 37
Mesh:
   equidistributing-, 363
   -independence principle, 354
   locally almost uniform-, 451
   optimal-, 359
   quasi uniform-, 227, 361
   -selection, 358, 381, 444
   staggered-, 246
Method of lines, 508
Midpoint:

-rule, 64
-scheme, 70, 73, 191, 215
Minimizing vector, 37
Mixed order system, 3
Modes, 104
    fast-, 116, 276, 392
    slow-, 116, 276, 392
Modified Euler, 69
Monitor function, 363, 376
Monomials:
    B-spline representation, 59
    Hermite type representation, 59
Moore-Penrose pseudo inverse, 36
Multiple shooting, 145
    standard-, 146, 147, 175
    theoretical-, 415, 418
Multiple solution, 490
Multiplicity:
    algebraic-, 33
    geometric-, 33
Multipoint BVP, 92, 512
Multistep method, 70

Newton:
    -Cotes formulas, 64
    damped-, 329
    -direction, 329, 338
    Kantorowich, 51
    -method, 48, 176, 195, 327, 335
    modified-, 341
    quasi-, 51, 52
Nonlinear equations, 48
Nonseparated BC, 5, 470
Norm, 36, 37
    1-, 37
    2-, 37
    Euclidean-, 37
    function-, 77
    Hölder-, 35
    induced-, 37
    matrix-, 37
    max-, 37
    Sobolev-, 77
    sup-, 37, 79
    vector-, 36
Normal equation, 45

Objective function, 330
One sided schemes, 116
One step schemes, 68
Operator:
    boundary, 93
    difference-, 188
    differential-, 93, 188
Ordinary dichotomy, 116
Ordinary differential equation (ODE), 3
    delay-, 506
    first order-, 3
    functional-, 505
    generalized-, 474
    high order-, 3, 244, 470
    homogeneous-, 3
    -inhomogeneous, 3
    mixed order-, 4
    nonlinear-, 3
Orthogonal:
    matrix, 33
    projection, 36, 117
Orthogonalization, 33, 43
Orthonormal polynomial, 64
Oscillatory (highly), 500
Osculatory interpolation, 58
Outer solution, 390, 404

Parameter(s):
    BVP with-, 322, 470
    condensation of-, 212, 261, 307
    imbedding-, 165
Partially separated, 5
Partial pivoting, 41, 308
Path:
    -finder, 4, 58
    -following, 490, 496
    homotopy-, 344
PDE, 508
Petrov-Galerkin, 271
Piecewise polynomial, 59, 218
Pitchfork bifurcation, 495
Pivot, 41
Pivoting:
    alternate row and column, 310
    column-, 310
    partial-, 41, 308
    row-, 310
Polychotomy, 514
Polynomial:
    interpolation-, 55
    orthonormal-, 64
    piecewise-, 59
Positive definite symmetric, 35
Precision of quadrature, 64, 211
Predictor-correct (Newton), 334
Preprocessing, 237
Primary bifurcation, 64, 211
Principal:
    -error function, 227
    -vector, 33

Projection (orthogonal), 36, 125
Prüfer transformation, 481
Pseudo-arclength, 498
Pseudo-inverse, 36, 340

QR algorithmn, 45
Quadrature, 63
   composite formulas, 66
   Gaussian-, 64
   Gauss point-, 65
   Lobatto point-, 66
   precision of-, 64, 211
   Radau point-, 55
   -rule, 63
Quasi:
   -Newton algorithm, 51, 52
   -uniform mesh, 227, 361
Quasilinearization, 53, 222
QU decomposition, 35

Rank-1 update, 52, 343
Rate of convergence, 48
Rayleigh quotient, 35
Recursion:
   -backward, 312
   -forward, 312
Reduced:
   -equation, 411, 413
   -solution, 390, 451
   -superposition, 143
Reflection, 33, 43
Reimbedding, 168
Representation:
   global-, 59
   local-, 59
Riccati:
   -differential equation, 165, 166
   -method, 64, 167, 460
   -transformation, 166, 392
Richardson extrapolation, 228
Ritz method, 267
Rotation, 33
Roundoff (error), 29
Rünge-Kutta, 68, 210
   explicit-, 210
   implicit-, 210
   k-stage-, 210
   -representation, 252
   -scheme, 213

Scaling of BC, 114
Scheme:
   higher order-, 208
   one-sided-, 432
   symmetric-, 440
Schur, 35
Secondary bifurcation, 494
Segmentation, 397, 410, 423
Separated BC, 5
Sequential shooting, 457
Shift, 45
Shock layer, 394, 396
Shooting, 133
   bidirectional, 178
   multiple-, 145, 175
   sequential, 457
   simple-, 91
   single-, 91, 137, 170
   standard multiple-, 146, 147
Similar, 32
Similarity:
   kinematic-, 105
   -transformation, 32
Singular:
   -BVP (first kind), 450, 484
   -BVP (second kind), 478, 480
   -perturbation, 386
   -point, 490, 498
   -value, 35
   -value decomposition (SVD), 35
Singularity (essential), 339
Singularly perturbed:
   -BVP, 113, 386
   -IVP, 112, 116
   singular-, 403
Sobolev norm, 77
Solution:
   collocation-, 218
   consistent (fundamental)-, 280, 281, 290
   fast-, 392
   fundamental-, 86
   generalized-, 268
   highly oscillatory-, 500
   inner-, 390, 404
   isolated-, 91
   multiple-, 490
   outer-, 390, 404
   reduced-, 390
   slow-, 392
   weak-, 268
Spectral radius, 32
Spline, 59
   B-, 61
   Hermite-, 59
   smooth-, 59
Stability:
   A-, 73
   A(α)-, 74

algebraic-, 73
AN-, 73
asymptotic-, 100
-of bifurcation, 493
-of block methods, 318
-of collocation, 262
-constant, 81, 265
di(chotomic)-, 458
-of decoupling, 385
-of finite difference methods, 200, 201
-of initial value problems, 99, 100
L-, 74
-of multiple shooting, 149
relative-, 100
-of single shooting, 137
uniform-, 100
Stabilized march, 161, 163
Staggered mesh, 246
Staircase matrix, 304, 306
Standard form, 469
Steepest descent, 334, 340
Stiff, 73, 286, 387
-decay, 74
Stretching transformation, 387, 415, 428
Strictly diagonally dominant, 107, 120
Sturm-Liouville, 480
Substitution:
    back-, 42
    forward-, 42
Superconvergence, 211
    -Order, 255
Superposition, 135
    reduced-, 143
SVD, 35
Symmetric:
    -positive definite, 35
    -schemes, 440

Time scales, 392
Tolerance, 31
Totally positive matrix, 62
Transformation:
    admissible-, 373
    elementary-, 40

Householder-, 43
    -methods, 373
    reflection-, 33
    rotation-, 34
    similarity-, 32
    stretching, 386
Transient, 388
Transition:
    -bifurcation, 495
    -layer, 393, 396
Tranverse method of lines, 509
Trapezoidal scheme, 190, 196, 215
Tridiagonal matrix, 187
Truncation (error), 71, 188, 199, 204
Turning point:
    for singularly perturbed problem, 393, 396, 442
    singular point, 491, 496

Unitary, 34
Update (rank 1-), 52
Upstream, 434
Upwind, 434

Vandermonde matrix, 56
Variational:
    -equation, 109
    -formulation, 267
    -problem, 90
Vector:
    grade of-, 33
    maximizing-, 37
    minimizing-, 37
    principal-, 33

Weak:
    -line search, 331
    -solution, 268
Weight function, 331
Well-conditioning, 39, 79, 94, 114
Wronskian, 98